Advances in Bioremediation and Phytoremediation for Sustainable Soil Management

Junaid Ahmad Malik
Editor

Advances in Bioremediation and Phytoremediation for Sustainable Soil Management

Principles, Monitoring and Remediation

Springer

Editor
Junaid Ahmad Malik
Department of Zoology
Government Degree College
Bijbehara, Jammu and Kashmir, India

ISBN 978-3-030-89983-7 ISBN 978-3-030-89984-4 (eBook)
https://doi.org/10.1007/978-3-030-89984-4

This Springer imprint is published by the registered company Springer Nature Switzerland AG
The registered company address is: Gewerbestrasse 11, 6330 Cham, Switzerland

Preface

Owing to the exponential growth of industrialization, urbanization, advanced agricultural production and power generation, environmental pollution has become a significant global concern. These also led to the indiscriminate extraction of natural resources in order to meet human interests and needs, which have contributed to disrupting the ecological equilibrium on which the quality of our ecosystem depends. Human beings are the product of their ecosystem in the real sense; the interaction between humans and the environment suggests that contamination has a societal basis. Modern technical developments in chemical processes have given rise to new chemicals and new toxins, which are beyond the self-cleaning potential of the environment, at a very abundant level. One of the main challenges the developing world faces today is the pollution of drinking water, surface water, soil and air with dangerous and harmful chemicals. Such chronically polluted substances are also handled using strict chemical procedures or physical procedures, such as encapsulation, for solid waste. Though efficient, such procedures are also costly, in turn environmentally hazardous, and involve polluted material instead of removing it. With regard to the solution to this issue, there is an immediate need to build solutions that are less resource-intensive, less time-consuming, and environmentally sustainable. Consequently, in recent years, biological methods have gained substantial interest. Bioremediation is one of the successful biological methods to combat environmental pollution and has been proven to be effective in treating soil and water pollution at various sites worldwide.

The implementation of a bioremediation strategy involves a comprehension both in the short and in the long term of the relationship of the individual pollutant, or combination, with the ecosystem into which it is released. This is also based on the persistence or bioavailability of the individual compound, its reaction to the physicochemical conditions under which it is contained and the remediation of the association between it and the microbial properties found in the environment.

The living soil is not only the foundation of our food chain, but also of our culture as a whole. We ought to keep it clean, realizing that the soil is a foundation of our food. As the global population and demands for food supply grow, keeping our soil fertile and sustainable is of vital importance. Industrialization has become the biggest contributor to contamination by releasing xenobiotics, which invade our environments and damage the soil.

Fertility is reduced by toxic chemicals, making the soil unfit for natural plant cover or agricultural growth.

To add value to the method, the introduction of molecular techniques into bioremediation protocols would require a variety of existing problems to be overcome. Practically speaking, the effectiveness of these methods for the elimination of toxins in the field demands that degradation levels can be analysed.

This book discusses the different variants and a mixture of different processes that will make the method of bioremediation and phytoremediation more influential. The book is an effort to resolve the problems in a scientific and ecological way, taking into account the effect of pollution on diverse habitats. The book discusses the numerous remediation approaches designed across the globe to tackle land and agricultural depletion. In order to obtain a thorough insight into the current research and technology status in these fields, bio-fertilizers and phytoremediation have been also discussed.

This book includes contributions from researchers in the field of microbiology, agronomy, edaphology, horticulture, agriculture, biotechnology and bioremediation. The authors, with fundamental, applied and industrial science backgrounds, come from numerous institutions, universities, government laboratories and industries. This book should prove to be useful for biotechnology, microbiology, biochemistry, soil and environmental sciences and engineering undergraduate and graduate students. I hope that students, scientists and engineers will find the content, including its basic and practical elements, helpful, be it in academia, business or government.

I would like to acknowledge my family members with love and affection in particular my wife, father and mother for their continuous support and constant encouragement throughout the process. I strongly believe that the successful completion of this manuscript is possible because of the blessings of "Almighty God."

Bijbehara, India Junaid Ahmad Malik

Contents

Editor and Contributors

About the Editor

Dr. Junaid Ahmad Malik received B.Sc. (2008) Science from the University of Kashmir, Srinagar, J&K; M.Sc. (2010) in Zoology from Barkatullah University, Bhopal, Madhya Pradesh; and Ph.D. (2015) in Zoology from the same university. He completed his B.Ed. program in 2017 from the University of Kashmir, Srinagar, J&K. He started his career as Lecturer in School Education Department, Government of J&K for 2 years. Dr. Malik is now working as a Lecturer in Department of Zoology, Government Degree College, Bijbehara, Kashmir (J&K) and actively involved in teaching and research activities. He has more than 8 years of research experience. His areas of interest are ecology, soil macrofauna, wildlife biology, conservation biology etc. Dr. Malik has published more than 20 research papers in various national and international peer-reviewed journals. He has authored four books, 16 book chapters, edited five books, and more than 10 popular editorial articles. He is also serving as editor and reviewer of several journals with a reasonable repute. He has participated in several State, National, and International conferences, seminars, workshops, and symposia and more than 20 conference papers are to his credit. He is the life member of SBBS (Society for Bioinformatics and Biological Sciences) with membership id LMJ-243.

Readers may contact him at: editor@worldbiologica.com

Contributors

Imran Ahmad Department of Geology, University of Malakand, Dir (Lower) Chakddara, Pakistan

Muhammad Arif Department of Biotechnology, Abdul Wali Khan University, Mardan, Pakistan

Avijit Bakshi Ecotoxicology, Fisheries and Aquaculture Extension Laboratory, Department of Zoology, University of Kalyani, Kalyani, India

Manickam Dakshinamoorthi Balakumaran Department of Biotechnology, Dwaraka Doss Goverdhan Doss Vaishnav College, Chennai, Tamil Nadu, India

Velramar Balasubramanian Department of Biotechnology, Amity Institute of Biotechnology (AIB), Amity University, Raipur, Chhattisgarh, India

Subia Bano Elvesys Microfluidics Innovation Centre, Paris, France

Sagar H. Barage Amity Institute of Biotechnology, Amity University, Panvel, Mumbai, Maharashtra, India

H. Baraiya Divyeksha Department of Life Sciences, CHRIST (Deemed To Be University), Bangalore, Karnataka, India

Brij Bhushan Department of Chemistry, Graphic Era University, Dehradun, India

Trinath Biswal Department of Chemistry, VSS University of Technology, Burla, India

Manish Biyani Department of Bioscience and Biotechnology, Japan Advanced Institute of Science and Technology, Ishikawa, Japan

Vijaya Geetha Bose Department of Biotechnology, Rajalakshmi Engineering College (Autonomous), (Affiliated to Anna University), Thandalam, Chennai, Tamilnadu, India

Arghya Chattopadhyay Department of Soil Science and Agricultural Chemistry, Benaras Hindu University, Varanasi, India

Amit Chattree Department of Chemistry, Sam Higginbottom University of Agriculture, Technology and Sciences, Prayagraj (Allahabad), UP, India

Karthikeyan Subbiahanadar Chelladurai Regeneration and Stem Cell Biology Lab, Centre for Molecular and Nanomedical Sciences, International Research Centre, Sathyabama Institute of Science and Technology (Deemed to be University), Chennai, Tamil Nadu, India

Jackson Durairaj Selvan Christyraj Regeneration and Stem Cell Biology Lab, Centre for Molecular and Nanomedical Sciences, International Research Centre, Sathyabama Institute of Science and Technology (Deemed to be University), Chennai, Tamil Nadu, India

Johnson Retnaraj Samuel Selvan Christyraj Regeneration and Stem Cell Biology Lab, Centre for Molecular and Nanomedical Sciences, International Research Centre, Sathyabama Institute of Science and Technology (Deemed to be University), Chennai, Tamil Nadu, India

Ujjwal Ranjan Dahiya CSIR—Institute of Genomics and Integrative Biology, New Delhi, India

Shabnam Dan Department of Chemistry, Sam Higginbottom University of Agriculture, Technology and Sciences, Prayagraj (Allahabad), UP, India

Aurobinda Das Department of Zoology, Centurion University of Technology and Management, Bhubaneswar, Odisha, India

Bappa Das Division of Agricultural Physics, ICAR-Indian Agricultural Research Institute, New Delhi, India

Jhumki Das Department of Pediatrics, PGIMER Chandigarh, Chandigarh, India

Sonia Devi Division of Agricultural Physics, ICAR-Indian Agricultural Research Institute, New Delhi, India

Ramamurthy Dhandapani Fermentation Technology Laboratory, Department of Microbiology, School of Biosciences, Periyar University, Salem, Tamil Nadu, India

Varun Dhiman Department of Environmental Sciences, Central University of Himachal Pradesh, Dharamshala, India

K. R. Don Department of Oral Pathology, Saveetha Dental College, Saveetha Institute of Medical and Technical Sciences, Saveetha University, Chennai, Tamil Nadu, India

Md. Maharub Hossain Fahim Department of Biotechnology and Genetic Engineering, Faculty of Biological Science, Islamic University, Kushtia, Bangladesh

Rashi Garg Department of Biotechnology, Biyani Girls College, Jaipur, Rajasthan, India

Shreya Gawas Amity Institute of Biotechnology, Amity University, Panvel, Mumbai, Maharashtra, India

T. Gopishankar Department of Life Sciences, CHRIST (Deemed To Be University), Bangalore, Karnataka, India

Nandini Gowala Department of Environmental Science, Tezpur University, Napaam, Tezpur, Assam, India

Iram Gul Department of Earth and Environmental Sciences, Hazara University, Mansehra, Pakistan

Zahurul Haque Department of Biotechnology and Genetic Engineering, Faculty of Biological Science, Islamic University, Kushtia, Bangladesh

Nazmul Huda Department of Biotechnology and Genetic Engineering, Faculty of Biological Science, Islamic University, Kushtia, Bangladesh

Md. Amdadul Huq Department of Food and Nutrition, College of Biotechnology and Natural Resource, Chung-Ang University, Anseong-si, Republic of Korea

Mohd. Imran Department of Chemistry, Chandigarh University, Mohali, Punjab, India

Muhammad Iqbal Department of Environmental Sciences and Engineering, Government College University, Faisalabad, Pakistan

Shafeeqa Irfan Institute of Food Science and Nutrition, University of Sargodha, Sargodha, Pakistan

Deepak Kumar Jha P.C. Vigyan Mahavidyala, Chapra, Bihar, India

R. Kalidoss Department of Applied Biology, College of Sciences, University of Sharjah, Sharjah, United Arab Emirates

Nirat Kandwani Department of Biotechnology, Biyani Girls College, Jaipur, Rajasthan, India

A. Karthika Department of Microbiology, The Standard Fireworks Rajarathnam College for Women, Sivakasi, Tamil Nadu, India

Nalinee Kumari Department of Zoology, DPG Degree College, Gurugram, Haryana, India

Tarun Kumar Kumawat Department of Biotechnology, Biyani Girls College, Jaipur, Rajasthan, India

Varsha Kumawat Naturilk Organic and Dairy Foods Pvt. Ltd, Jaipur, Rajasthan, India

Suseela Lanka Department of Bio-Sciences and Biotechnology, Krishna University, Machilipatnam, Andhra Pradesh, India

Junaid Ahmad Malik Department of Zoology, Government Degree College, Bijbehara, Kashmir (J&K), India

Melinda Grace Rossan Mathews Regeneration and Stem Cell Biology Lab, Centre for Molecular and Nanomedical Sciences, International Research Centre, Sathyabama Institute of Science and Technology (Deemed to be University), Chennai, Tamil Nadu, India

M. Mizanur Rahman Department of Biotechnology and Genetic Engineering, Faculty of Biological Science, Islamic University, Kushtia, Bangladesh

Bhabani Prasad Mondal Division of Agricultural Physics, ICAR-Indian Agricultural Research Institute, New Delhi, India

Sowjanya Goud Murari Department of Bio-Sciences and Biotechnology, Krishna University, Machilipatnam, Andhra Pradesh, India

Manjunath K. Naik University of Agricultural and Horticultural Sciences, Shivamogga, Karnataka, India

Anandkumar Naorem Indian Council of Agricultural Research-Central Arid Zone Research Institute, Regional Research Station, Bhuj, Gujarat, India

Arunima Nayak Department of Chemistry, Graphic Era University, Dehradun, India

K. R. Padma Department of Biotechnology, Sri Padmavati Mahila Visva-Vidyalayam (Women's) University, Tirupati, AP, India

Suresh Kumar Panda Department of Ecology and Environmental Sciences, Pondicherry University, Puducherry, India

Ashis Kumar Panigrahi University of Burdwan, Burdwan, India

Deepak Pant Department of Environmental Sciences, Central University of Himachal Pradesh, Dharamshala, India

Priya Paul Division of Plant Physiology, ICAR-Indian Agricultural Research Institute, New Delhi, India

Shivanand Pradeep Organic Farming Research Center, University of Agricultural and Horticultural Sciences, Shivamogga, Karnataka, India

Kumar Pranay Department of Biochemistry, Indira Gandhi Institute of Medical Sciences, PatnaBihar, India

N. Praveen Department of Life Sciences, CHRIST (Deemed To Be University), Bangalore, Karnataka, India

Nandini Ramesh Center for Climate Resilient Agriculture, University of Agricultural and Horticultural Sciences, Shivamogga, Karnataka, India

Muhammad Modassar Ali Nawaz Ranjha Institute of Food Science and Nutrition, University of Sargodha, Sargodha, Pakistan

Mamoona Rauf Department of Botany, Abdul Wali Khan University, Mardan, Pakistan

Sanjoy Sadhukhan Plant Molecular Biology Laboratory, Department of Botany, Raiganj University, Raiganj, West Bengal, India

Rabi Narayan Sahoo Division of Agricultural Physics, ICAR-Indian Agricultural Research Institute, New Delhi, India

Sirat Sandil Cooperative Research Centre of Environmental Sciences, Eötvös Loránd University, Budapest, Hungary

Muthusamy Sanjivkumar Department of Microbiology, K.R. College of Arts and Science, K.R. Nagar, Kovilpatti, Tamil Nadu, India

Anup Kumar Sarkar Department of Botany, Dukhulal Nibaran Chandra College, Aurangabad, Murshidabad, West Bengal, India;
Plant Molecular Biology Laboratory, Department of Botany, Raiganj University, Raiganj, West Bengal, India

R. Seenivasagan Department of Biotechnology, Arulmigu Kalasalingam College of Arts and Science, Krishnankoil, Tamil Nadu, India;
Food Security and Safety Niche Area, Faculty of Natural and Agricultural Sciences, North-West University, Mmabatho, South Africa

Balasundaram Sendilkumar Department of Microbiology, School of Allied Health Sciences, Vinayaka Missions Research Foundation (Deemed to be University), Salem, Tamil Nadu, India

Bakhtawar Shafique Institute of Food Science and Nutrition, University of Sargodha, Sargodha, Pakistan

Ghulam Mujtaba Shah Department of Earth and Environmental Sciences, Hazara University, Mansehra, Pakistan

Jyoti Sharma School of Basic and Applied Sciences, Shobhit Institute of Engineering and Technology (Deemed to-be-University), Meerut, Uttar Pradesh, India

Vishnu Sharma Department of Biotechnology, Biyani Girls College, Jaipur, Rajasthan, India

K. S. Shreenidhi Department of Biotechnology, Rajalakshmi Engineering College (Autonomous), (Affiliated to Anna University), Thandalam, Chennai, Tamilnadu, India

Ali Raza Siddiqui Pakistan Cricket Board, Lahore, Pakistan

Tamil Selvan Silambarasan Department of Microbiology, School of Allied Health Sciences, Vinayaka Missions Research Foundation (Deemed to be University), Salem, Tamil Nadu, India;
Fermentation Technology Laboratory, Department of Microbiology, School of Biosciences, Periyar University, Salem, Tamil Nadu, India

Reshma Soman School of Biosciences, Mahatma Gandhi University, Kottayam, Kerala, India

Shankarappa Sridhara Center for Climate Resilient Agriculture, University of Agricultural and Horticultural Sciences, Shivamogga, Karnataka, India

Ravichandran Subramaniam Regeneration and Stem Cell Biology Lab, Centre for Molecular and Nanomedical Sciences, International Research Centre, Sathyabama Institute of Science and Technology (Deemed to be University), Chennai, Tamil Nadu, India

Abhilash Surendran Department of Microbiology, Majlis Arts and Science College, Malappuram, Kerala, India

Subramaniyam Suresh Department of Biotechnology, College of Science and Humanities, SRM Institute of Science and Technology, Ramapuram Campus, Chennai, Tamil Nadu, India

Hafiz Muhammad Tauqeer Department of Environmental Science, Hafiz Hayat Campus, University of Gujrat, Gujrat, Pakistan

Veysel Turan Department of Soil Science and Plant Nutrition, Faculty of Agriculture, Bingöl University, Bingöl, Turkey

Muhammad Irfan Ullah Department of Entomology, College of Agriculture, University of Sargodha, Sargodha, Pakistan

V. L. Vasantha Department of Life Sciences, CHRIST (Deemed To Be University), Bangalore, Karnataka, India

Vyomesh Vibhaw Department of Biochemistry, Patna University, Patna, Bihar, India

Lufeng Wang College of Food Science and Technology, Huazhong Agricultural University, Wuhan, Hubei, China

Yojana Waychal Amity Institute of Biotechnology, Amity University, Panvel, Mumbai, Maharashtra, India

Ivan Wilson Department of Life Sciences, Graphic Era University, Dehradun, India

Niti Yashwardhini Department of Microbiology, Patna Women's College, Patna, Bihar, India

Beryl Vedha Yesudhason Regeneration and Stem Cell Biology Lab, Centre for Molecular and Nanomedical Sciences, International Research Centre, Sathyabama Institute of Science and Technology (Deemed to be University), Chennai, Tamil Nadu, India

Bioremediation of Soil: An Overview

1

Shafeeqa Irfan, Muhammad Modassar
Ali Nawaz Ranjha, Bakhtawar Shafique,
Muhammad Irfan Ullah, Ali Raza Siddiqui,
and Lufeng Wang

Abstract

The bioaccumulation of contaminants in soil causes toxicity to human, animals, microorganisms, and plants. Environmental biotechnology such as composting and wastewater treatment is not a new field, yet the recent fields are molecular biology and ecology. Bioremediation uses microorganisms (which may be indigenous or isolated from any other site), naturally occurring bacteria, fungi and plants to degrade or detoxify the contaminants (hazardous to human health and environment) into less toxic forms. Bioaugmentation process is used when microorganisms are imported to a contaminated site to enhance the detoxification. Public considers it more efficient than other technologies because bioremediation is based on natural attenuation. Bioremediation has certain limits such as high aromatic hydrocarbons are resistant to microbial attack. Bioremediation system mostly runs under aerobic conditions. Important factors include availability of contaminants to the microbial population and the environmental factors (pH, temperature, soil type, nutrients and presence of oxygen). Recent strategies for bioremediation include in situ bioremediation (these techniques are applied to soil at the site with minimal disturbance) and ex situ bioremediation (these techniques are applied to soil at the site which has been removed from the site through excavation). Bioremediation is a natural process and therefore has certain advantages along with some disadvantages. Phytoremediation, on the other hand, is the use of plants and their associated microbes for cleaning up the environment. This chapter develops the better understanding of bioremediation of soil, bioremediation strategies, especially the phytoremediation mechanisms.

Keywords

Bioaugmentation · Bioremediation · Biostimulation · Environmental biotechnology · In situ bioremediation · Ex situ bioremediation · Phytoremediation

S. Irfan · M. M. A. N. Ranjha · B. Shafique
Institute of Food Science and Nutrition,
University of Sargodha, Sargodha, Pakistan

M. I. Ullah
Department of Entomology, College of Agriculture,
University of Sargodha, Sargodha, Pakistan

A. R. Siddiqui
Pakistan Cricket Board, Lahore, Pakistan

L. Wang (✉)
College of Food Science and Technology,
Huazhong Agricultural University,
Wuhan 430070, Hubei, China
e-mail: pipilu001@foxmail.com

J. A. Malik (ed.), *Advances in Bioremediation and Phytoremediation for Sustainable Soil Management*,
https://doi.org/10.1007/978-3-030-89984-4_1

1.1 Introduction

The overall quality of environment inextricably decides the quality of life on Earth. In earlier times, our planet had immeasurable quality land and resources, today, which has been turned into contaminated land, in greater or less degree, due to carelessness and negligence of human practices and industrial activities (Cairney 1993). Earth's limited natural resources are facing decline due to urbanization, industrialization and population pressure which have been arising drastic changes and diverse environmental problems which can be specific to space and time. These environmental problems are becoming adversative with change in the magnitude and nature of environmental hazards and risks, generating bigger challenges and demanding appropriate and constant solution or technologies. In this framework, biotechnology has been proved potential to hold hope and provide protection, sustainability and management to environment in the form of (bio)treatment and/or (bio)remediation (Azadi and Ho 2010; Hatti-Kaul et al. 2007). Biotechnology offers two major applications: direct (bioremediation) and indirect (waste treatment, pollution prevention, and environmental remediation) (Juwarkar et al. 2010).

Products from natural sources are being used from centuries (Ranjha et al., 2020a, b). Bioremediation field is part of environmental biotechnology realm and in no means can be confused with biodegradation, which intercept mostly bacterial bases to metabolize the unmanageable or unusual compounds. Figure 1.1 shows the four interventions of environmental biotechnology. Bioremediation sorely deals with the biological interventions aiming at lessening pollution and assessment of environmental contamination. Generally, bioremediation includes natural attenuation or bioaugmentation or biostimulation depending on degree of interventions. Natural attenuation is the simplest method which means no human action involvement. Biostimulation involves the requirement of nutrients and electron acceptor or donor's addition to metabolize or promote the growth of certain microorganisms. Bioaugmentation is the conscious addition of microorganisms having desired catalytic capabilities either natural or engineered (Prasad et al. 2010).

Ecologically, bioremediation is the safe and natural process involving most effective metabolism of organic wastes through the natural bacterial strains as bacteria are available in nature and multiply in number in favorable conditions/food source such as in the presence of wastes. The strains of bacteria detonate and metabolize the hazardous wastes to complete the process of bioremediation resulting in harmless residues, as soon as the contaminant degrades, microbial population get declined. It is less expensive than other cleanup technologies (Sen and Chakrabarti 2009).

Bioremediation has been proved as the most promising new technology in effective treatment of effluent heavy metal mining wastes, municipal/urban sewage wastes, solid or liquid industrial wastes, hazardous wastes, and/or chemical spills, etc., from soil, water, and other environments. It is used to clean up environmental problems and contaminated groundwater. Bioremediation can be used as ex situ/in situ technology wherein the degradation of microbes can be enhanced by altering nutrients, type of available microorganisms, and climatic conditions such as moisture, temperature, pH, and oxygen levels (Sen and Chakrabarti 2009).

1.2 Principles of Bioremediation

Bioremediation is the natural process which involves microorganisms or plants to metabolize the persistent and toxic pollutants and break into carbon dioxide, water, microbial biomass and less hazardous by-products (metabolites) and eliminate them from the environment (Fig. 1.2). In bioremediation process, microorganisms can be native to contaminated site or brought to contaminated soil after isolation.

As explained earlier, bioremediation involves microbes and bacteria to metabolize the contaminants and ultimately depends on the growth and reproduction of microorganisms which

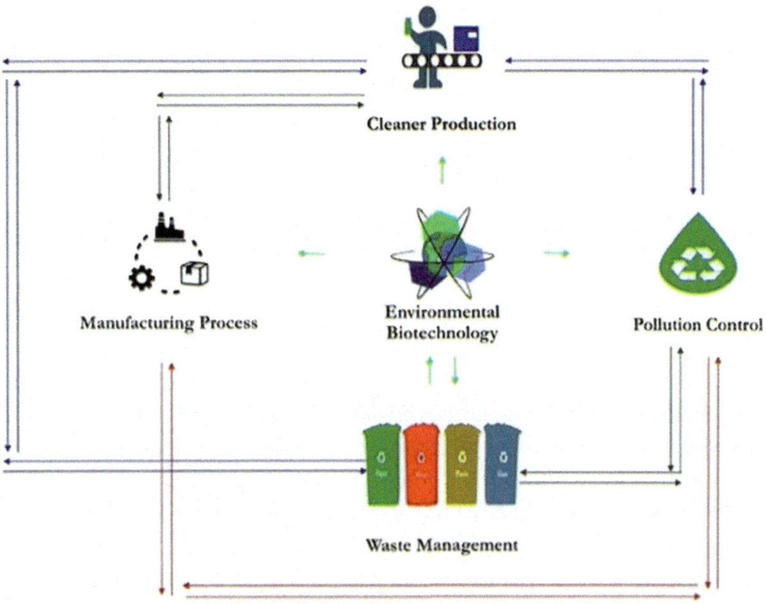

Fig. 1.1 Schematic diagram of four key interventions of environmental biotechnology. Adapted from Gavrilescu (2010)

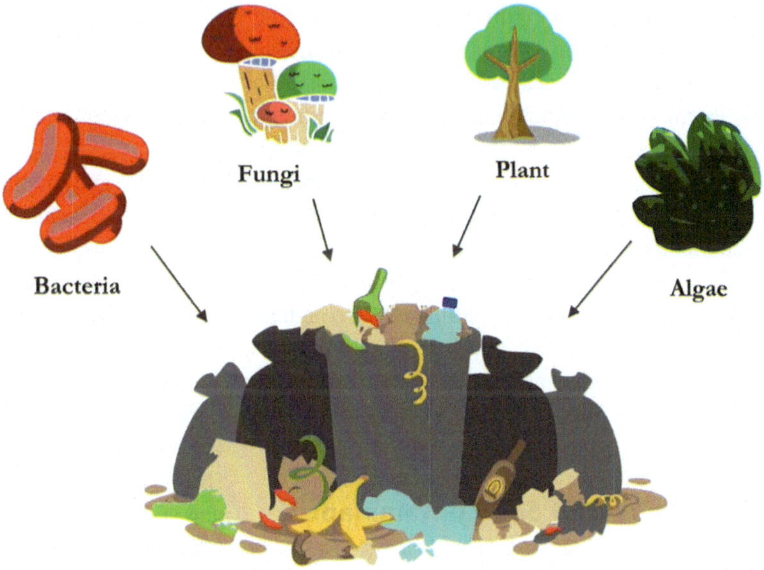

Fig. 1.2 Principle of bioremediation. Adapted from Tyagi and Kumar (2021)

require a source of energy and extraction of developed energy through suitable terminal electron acceptor (TEA). In order to sustain the metabolic functions (growth and reproduction), microorganisms utilize carbon as energy source. However, a terminal TEA is required by the metabolic processes to oxidize the carbon source (organic matter to CO2) enzymatically in bacteria (Sen and Chakrabarti 2009).

Organic matter + oxygen + biomass → $CO_2 + H_2O + \Delta H_f$.

where

ΔH_f = energy produced by reaction to fuel other metabolic processes (growth and reproduction).

This chapter develops a better understanding of bioremediation of soil, bioremediation strategies, especially the phytoremediation mechanisms.

Microorganisms have the potential to inhibit diverse environments like oceans, saline lakes, glaciers, desert, and thermal springs. Contaminated environments can be the sources to isolate potential degrading microbes to clean up the environment from variety of pollutants. The environments are polluted due to wastewater treatment plant and agricultural practices and lead to pesticide contaminated sites, petroleum-contaminated sites, landfills, and heavy metal-polluted sites. Under aerobic or anaerobic conditions, microorganisms utilize carbon source and hazardous contaminants to breakdown and convert them in less toxic metabolites to sustain metabolic activity (Tiwari and Singh 2014).

Pollutants need contact with desired microbes or application of surfactants for their effective degradation in soil as pollutants and microbes are not distributed uniformly. *Pseudomonas*, *Sphingomonas*, *Alcaligens*, and *Mycobacterium* (aerobic bacterial species) are efficient in degrading polycyclic aromatic hydrocarbons, alkanes, and pesticides. However, some of the aerobic methylotrophs are known in efficient degradation of dichloroethane and chlorinated aliphatics. On the other hand, few anaerobic bacterial species are efficient to degrade the chloroform, PCBs, and trichloroethylene/chlorinated solvents. Along with it, *Phanerochaete chrysosporium* (fungal species) are reported to efficiently remediate variability of persistent and hazardous contaminants (Ntougias et al. 2015; Harekrushna and Kumar 2012).

1.3 Sources of Soil Contaminants

1.3.1 Biological Contaminants

Different municipal wastes, sewage water, and sewage sludge contain number of different biological contaminants (Demirbas et al. 2017). These biological contaminants could include viruses (rotavirus, polio), *Salmonella* species, *Escherichia coli*, *Clostridium perfringens*, *Shigella* spp., *Campylobacter* species, Helminths (*Trichuris* species, *Ancylostoma* species, and *Ascaris* species), and parasitic protozoa (*Giardia lamblia* cysts, *Entamoeba histolytica*, and *Cryptosporidium parvum* oocysts). Furthermore, in addition to such microbiocidal contaminants, hookworms could also be present in such wastes and especially in the sewage sludge. These biological contaminants could lead to a number of different medical conditions like tuberculosis, gastroenteritis, bacillary, anemia, diarrhea, vomiting, ascariasis, typhoid, and different types of dysentery like amoebic and bacillary (Ünüvar 2018). Any of these illnesses could lead to an outbreak in any part of the world (Heymann 2020). However, the most harmful and widely reported microbes in human excreta that cause dysentery as well as typhoid, respectively, are *Salmonella* bacteria (Lee et al. 2020) and *Shigella dysenteriae* (Arianzad et al. 2020). Microbe-oriented pollution is typically very harmful in developed countries, because all kinds of wastes are borne by one drainage/sewerage system (Murtaza et al. 2014).

1.3.2 Organic Contaminants

Aromatic and aliphatic halogenated hydrocarbons that include benzene and petroleum hydrocarbons, phthalate esters, organochlorine pesticides, polynuclear hydrocarbons, polychlorinated biphenyls, grease, and oil are found in the main category of chemicals (Moore and Ramamoorthy 2012). The manufacturing of organic chemicals, synthetically, has been tremendously increased in the past 50 years (Bernhardt et al. 2017). As a result, the frequency and concentration of organic pollutants have also increased in effluents, waste, and biosolids. The occurrence and amount of organic pollutants are highly dependent on physicochemical properties, various local point sources of unique organic compounds, composition of the wastewater, and the operating parameters of the wastewater treatment plant. Organic contaminant concentrations are typically higher in industrial waste and considerably low in domestic waste (Trinh et al. 2020). More than 300 organic chemicals have been known with the varied concentrations (Smith 2000; Jacobs et al. 1987). Water treatment plants were mainly designed and manufactured for the removal of organic matters from contaminated water (Song et al. 2020), and the degradation mechanisms for degrading the bulk organic components have been well studied and understood; however, there were relatively few analysis to understand the processes for degrading synthetic organics. A variety of processes could be performed by an organic contaminant, including (I) sorption of solid surfaces, (II) volatilization, (III) chemical degradation, and (IV) biodegradation. In general, more the hydrophobic a compound is, the more it is vulnerable to collect sewage sludge particles.

Due to the unavailability of data to understand, only few generalizations can be drawn for the organic pollutants' behavior in the process of water treatment and sludge assimilation (Li et al. 2020). In bioremediation, the biodegradation of hydrocarbons varies from molecule to molecule; like highly branched and long chains are easily degraded in comparison with unbranched and short chains, respectively. Typically, unsaturated aliphatic compounds show more sensitiveness toward degradation comparing to saturated analogs. As organic contaminants have high lipophilicity and low water solubility, they are supposed to divide in sludges in the process of sedimentation (Bhandari and Xia 2005). The end result of organic pollutants is not fully explained during wastewater treatment process, as most of them require aerobic conditions for degradation and due to unavailability remain in anaerobic digestion accompanied by biosolids. Literature has seldom, in agricultural soils, details on the fate and behavior of biosolids-borne organic pollutants. In available literature, it can be seen that organic pollutants rapidly degrade under aerobic conditions of soil and are lost during land applications such as photolysis or volatilization, few are rarely degraded such as PAHs of high molecular mass, brominated flame retardants, and organotins while some are highly persistent such as organochlorines and dioxins (Haynes et al. 2009).

Characteristically, plants have poor absorption power for organic pollutants from soil and the only key route is through soil surface volatilization, accompanied by uptake of polluted air by vegetation above ground. Grazing animals can be affected by accumulation of organic pollutants in them due to direct ingestion and vegetation adhered biosolids. It has been studied that most organic pollutants have the power to break down in animals. However, others such as furan, polychlorinated dibenzo-p-dioxins, chlorinated hydrocarbons, pesticides and halogenated biphenyls show the bioconcentration in fat of animals and their products (milk) because these pollutants show resistance to degradation. The true worth of organic components is not described in literature; however, European Union has regulated certain limits for specific biosolid contaminants.

1.3.3 Inorganic Contaminants

This category encompasses the toxic gases, complexes of ions or pair of ions, heavy metals (loids), and their products resulted from interaction with other constituents of soil, from a number of pollutants from point and non-point sources. The major category of inorganic pollutants includes leach of Zn from domestic belongings, leach from Pb and Cu pipes, derivation of Cr, Ni, Pb, Cd, Zn, Cu from domestic wastewater discharge and industrial heavy metals or metalloids. In the process of sedimentation, heavy metals become the part of sludge in primary treatment due to their accumulation in bacterial cells and or their adsorption in bacterial cells in secondary treatment. In several parts of the world, concerns about loads of heavy metals present in sewage sludge/biosolids have developed guidelines and regulations that are typically focused on the highest permissible metal content limits in biosolids and/or the permissible loading limits of metals applied to the soil in biosolids. A drawback of these methods is that they consider heavy metal concentrations in soils to be absolute rather than biologically active (extractable).

Heavy metal toxicities restricting growth of the crop in soils amended with biosolid are seldom when agronomic rates (2–8 Mg ha-1) are used (Murtaza et al. 2011; Krogmann et al. 1999). The major concern about these pollutants is their accumulation in edible plant parts leading to food chain contamination and affecting grazing animals and humans toxically (Cobb et al. 2000).

It should be carefully noted that translocation of heavy metals in the above ground edible parts of crops is not easy, so contamination from food crop ingestion is not likely to occur under current regulations. The key possible mechanism for heavy metal accumulation in grazing animals' meat is through direct soil/biosolids ingestion. Their accumulation has a negative impact on the soil microbial and biochemical function, and their importance for soil is not explained fully (Haynes et al. 2009).

1.4 Factors Affecting Bioremediation

Via the action of bacteria, fungi and plants, bioremediation is a process of degrading, eliminating, modifying, detoxifying, or immobilizing different physical and chemicals waste from the setting. Through their enzymatic pathways, microorganisms are involved as biocatalysts and promote the biochemical reactions progression which evades the preferred pollutant. Eventually, the products of the process can contain simpler substances, such as carbon dioxide or water, and also cell biomass. Therefore, a risk for potential risks in the handling and storage of hazardous material is virtually removed. The efficacy of bioremediation is based on a variety of factors, including the chemical nature and concentration of contaminants, the physicochemical characteristics of the ecosystem, and the availability of microorganisms (Das et al. 2020; El Fantroussi and Agathos 2005; Lacalle et al. 2020). Since bacteria and contaminants do not touch each other, the rate of degradation is pretentious. In addition, in the atmosphere, bacteria and toxins do not disperse uniformly, and for this reason, the optimizing and monitoring of bioremediation procedures are a dynamic system. The presence of a microbial community capable of degrading pollutants, the availability of contaminants to the microbial population, and environmental factors are included here (type of soil, temperature, pH, the presence of oxygen or other electron acceptors, and nutrients).

1.4.1 Biological Factors

Biotic factors influence the degradation of organic compounds through rivalry between carbon-limited microorganisms, antagonistic connections between microorganisms, or protozoal and bacteriophageal microorganism's predation. The degradation rate of the contaminant is also dependent on the concentration of contaminant and the quantity of "catalyst" available. In this sense, the "catalyst" quantity reflects the amount of species capable of metabolizing the contaminant, as well as the quantity of enzymes that each cell releases. The cells' expression of unique enzymes will enhance or reduce the contaminant degradation rate. In addition, particular enzymes should be involved in the degree of contaminant metabolism, and their "affinity" is largely required for the contaminant and also for the contaminant availability. The main biological factors affecting bioremediation process include enzyme activity, mutation, interaction (succession, predation, and competition), self-growth, horizontal gene transfer, critical biomass, and the composition and population size (Madhavi and Mohini 2012; Boopathy 2000).

1.4.2 Environmental Factors

Possible interaction during the process is determined by the microorganisms' metabolic appearances and targeted pollutants' physicochemical properties. Nevertheless, like any chemical reaction, the site must be at the required temperature for the bioremediation process to be successful and nutrients must also be available, or the microbes will thrive too slowly or die.

Development and activity of microorganisms are influenced by pH, soil structure, temperature, humidity, water solubility, site characteristics, oxygen content, nutrients and redox potential, the absence of human resources trained in this field, and the physicochemical pollutant bioavailability (contaminant concentration, solubility, type, toxicity and chemical structure). The factors described above are determined by the degradation kinetics (Khudhaier et al. 2020; Madhavi and Mohini 2012; Adams et al. 2015; Hou et al. 2020). Biodegradation can occur in a broad pH range, but in most aquatic and terrestrial environments, a pH of 6.5–8.5 is typically perfect for biodegradation. The rate of contaminant metabolism is influenced by moisture because it determines the form and quantity of soluble materials available as well as the osmotic pressure and pH of terrestrial and aquatic systems (Cases and Lorenzo 2005).

1.4.3 Availability of Nutrients

The nutrients are added to cause changes in the necessary equilibrium of nutrients for the reproduction and growth of microbes and have an effect on the rate and efficacy of biodegradation. By optimizing the bacterial C: N: P ratio, balancing of nutrient especially the essential nutrients such as P and N supply may increase the efficiency of biodegradation. Microorganisms require a variety of nutrients, such as carbon, phosphorus, and nitrogen, to thrive and continue their microbial activities. The level of hydrocarbon degradation is also limited at low concentrations. A favorable stratagem for improving the microorganisms' metabolic activity and thus the rate of biodegradation in cold environments is the addition of a sufficient quantity of nutrients (Couto et al. 2014; Phulia et al. 2013). Biodegradation is limited by the nutrient availability in the aquatic environment (Thavasi et al. 2011). Oil-eating microbes, similar to the nutritional requirements of other species, often need nutrients for optimal development and growth. In the natural world, these nutrients are available but occur in low quantities (Macaulay 2014).

1.4.4 Temperature

One of the most important physical considerations for determining the microorganism's survival and the hydrocarbons composition is temperature (Das and Chandran 2011). Degradation of oil through natural procedures in cold environments is very slow such as the Arctic and places the microbes under more pressure to remove the spilled petroleum. The sub-zero water temperature in this area causes the channels of transport inside the microbial cells to freeze or even shut down the entire cytoplasm, making it metabolically inactive for most oleophilic microbes (Macaulay 2014; Yang et al. 2009). Biological enzymes are involved in the optimum temperature of the degradation process and may not have the same turnover for metabolism for any temperature. In addition, the process of degradation involves specific temperatures for specific compounds. Since microbial physiological properties are strongly affected, temperature often speeds up or slows down the process of bioremediation. The rate of microbial activity increases with temperature and at an optimal temperature reaches its maximum level. With further rise or reduction in temperature, it abruptly declined and finally stopped after hitting a certain temperature.

1.4.5 Concentration of Oxygen

Different species often need oxygen, some do not need oxygen depending on their need to better facilitate the process of biodegradation. For biological degradation which is only possible under favorable conditions such as aerobic and anaerobic environments, gaseous oxygen is vital and required by most living organisms to degrade. Oxygen presence will boost hydrocarbons breakdown in most cases (Macaulay 2014).

1.4.6 Moisture Content

Microorganisms need enough water to maintain their growth, function, and progression. In soil bioremediation, the moisture content of the soil plays a significant part. There should be ample water in the soil to sustain microbial activities. Microbial activity is hindered by small quantities of water in the soil, while excess water may fill pores in the soil and create resistance to oxygen transport toward microorganisms. Therefore, the optimum importance of the water content of the soil for bioremediation is very significant. Previous research has shown that when the moisture content of the soil is 60% of its water holding capacity, bacterial growth is optimum in the soil (Bahmani et al. 2018).

1.4.7 PH

Chemical pH, which is the essence of the compound's acidity, basicity, and alkalinity, shows its own effect on metabolic activity of microbes and therefore increases and decreases the process of removal. Soil pH measurements can indicate microbial growth potential (Asira 2013). Inferior findings were observed at low and high pH values; metabolic processes possess high sensitiveness to even minor changes in pH (Wang et al. 2011).

1.4.8 Site Characterization and Selection

Until proposing a bioremediation solution to sufficiently define the degree and nature of pollution, appropriate remedial investigative work must be conducted. At a minimum, this work should address the following factors: fully define the vertical and horizontal degree of contamination, list the criteria and locations to be sampled and the rationale for their selection, and explain the procedures to be utilized for the collection and analysis of the sample to be carried out.

1.4.9 Metal Ions

Metals play a vital role for bacteria and fungi in small quantities, but in large quantities, they possess the power to inhibit cells' metabolic activity. Metal compounds directly and indirectly influence the degradation rate.

1.4.9.1 Toxic Compounds

The higher concentration of certain pollutants has the ability to slow down the decontamination rate and produce toxic effects on microbes. The toxicity of pollutants varies depending on mechanism of toxicity, degree of toxicity, concentration, type, and exposed microorganism. However, the inorganic and organic compounds have the toxic impact on the targeted life forms (Madhavi and Mohini 2012).

1.5 Bioremediation Strategies

Bioremediation process has been broadly classified into ex situ and in situ remediation on the basis of pollutants' elimination from soil, transportation and origin.

1.5.1 Ex situ Bioremediation

It is based on the excavation, transport, and elimination of pollutants from native contaminated site to other site using wide range techniques of bioremediation. Ex situ has further subheaded, on the basis of degree of pollution and its depth, geographical features, geological features of contaminated site, type of pollutant, into different category like biopiling, landfarming, bioreactors, and biofilters (Atlas and Philp 2005). After ex situ remediation treatment, decontaminated soil becomes suitable for the landscaping purposes.

1.5.1.1 Biopiling

In biopiling or heap technique, both landfarming and compositing methods are collectively applied including laboratory testing in order to check soil's mechanical separation for soil homogenization, excavated polluted soil piling, degradation potential, forced aeration to increase degradation of microbes, soil improvement with nutrients and addition of potent microbes to biopile for pollutant's effective remediation. The arrangement consists of aeration system, treatment bed, nutrient, soil buried irrigation system, and a leachate collection system (Azubuike et al. 2016). For long time, biopiling has been effectively used to treat petroleum hydrocarbons (BTEX, PAHs, phenols) and low-molecular weight pollutant contaminated soil as well as an enormously polluted cold region environment (Dias et al. 2015; Ossai et al. 2020). This technique is widely used and efficient because of its cheap nature and controllable pH, temperature, and nutrient conditions (Ossai et al. 2020; Whelan et al. 2015; Mohammadi et al. 2020).

1.5.1.2 Landfarming

It utilizes autochthonous microbes for the aerobic degradation of excavated, spread, and timely tilled polluted soil over the support of fixed soil layer on ground. During landfarming, the major benefit of tilling the contaminated soil is that it increases the microbial activity through aeration, nutrients, and irrigation because of tillage (Nandy et al. 2021; Volpe et al. 2012; Skinder et al. 2020). It has been seen in studies that landfarming has a limitation to treat 10–35 cm of surface soil only (Kumar et al. 2018). Agricultural lime can be helpful to maintain the neutral pH. Although landfarming is a category of ex situ remediation, it can also play a role in in situ remediation in few cases. Mostly, landfarming plays beneficial and effective role to treat the polycyclic aromatic hydrocarbon, PCBs, and aliphatic polluted sites (Nandy et al. 2021; Silva-Castro et al. 2012; Shahzad et al. 2020). Landfarming is getting most attention due to cost effectiveness, low equipment, and maintenance operation and is also the best alternative to dumping method (Williams 2006).

1.5.1.3 Bioreactors

Bioreactor method involves the use of reactor or container to metabolize the pollutants under specific and controlled conditions. The system is based on engineered containments to treat the contaminated water, soil, sludge and sediments (Vidali 2001). Efficiency of the bioreactors method depends on some factors such as optimized reaction conditions, mass transfer, pollutants bioavailability, and nutrients supplementation which makes it bioreactor-centered bioremediation technique (Santillan et al. 2020; Mohan et al. 2004; Srivastava et al. 2021). Bioreactor has been found proficient in treating the soil or water polluted with volatile organic pollutants, i.e., xylene, ethylbenzene, toluene, and benzene. Bioreactor is extremely efficient in the biodegradation of pollutants than other techniques because it can manage, manipulate, and control process parameters according to biological reactions (Azubuike et al. 2016).

1.5.1.4 Biofilters

Biofilters technique involves the utilization of microbes embedded columns to eliminate gaseous pollutants. Biofilters technique has the advantage of only being used for gaseous pollutant removal (Meagher 2000).

1.5.2 In Situ Bioremediation

In situ bioremediation process is most favorable and trusted option because it does not involve any transportation or excavation and degrade the pollutants in the actual contaminated site (Adesipo et al. 2020; Kumar et al. 2018; Sharma, 2020). In situ remediation is further divided into two classes: (i) intrinsic in situ bioremediation in which no further enhancement practice is required and (ii) enhanced in situ bioremediation which involves phytoremediation, biosparging and bioventing (Tyagi and Kumar 2021; Azubuike et al. 2016). In situ bioremediation system does not spend any extra money except cost required to design and install equipment in order to increase the microbial activity; therefore, it is an economically feasible method. However, it is less effective and controllable in comparison with ex situ bioremediation method. This method is an effective treatment against hydrocarbons, dyes, heavy metals, and chlorinated solvents contaminated sites (da Silva et al. 2020; Kim et al. 2014).

1.5.2.1 Intrinsic in Situ Bioremediation

Intrinsic in situ bioremediation is also known as natural attenuation and this is one of the well-known in situ bioremediation techniques. This technique comprises no human practice involvement, with passive and unaided contaminated site remediation. Both aerobic and anaerobic processes can be followed to metabolize hazardous pollutants. However, this technique requires regular monitoring to attain sustainable and successful process and can also be termed as monitored natural attenuation (Tyagi and Kumar 2021; Azubuike et al. 2016).

1.5.2.2 Enhanced In Situ Bioremediation

As the name suggests, enhanced in situ bioremediation involves the addition of microbes, nutrients, and air to increase the microbial activity or excavation for efficient biodegradation of polluted sites. Following techniques are the examples of the enhanced in situ bioremediation method (Ghangrekar et al. 2020; Tiwari and Singh 2014);

Bioaugmentation

Bioaugmentation process involves addition of enzymatically active group of native or non-native or genetically engineered microorganisms to achieve the efficient biodegradation process. It is widely used to biodegrade sites contaminated with chlorinated and aromatic hydrocarbons and municipal wastewater. Proper management of the land treatment unit is crucial as it has been reported that non-native microorganisms compete with native microorganisms for their population growth (Sharma 2012).

Biostimulation

Biostimulation encompasses on the provision of desired nutrients to stimulate native microbial population of polluted sites or groundwater for the efficient and effective degradation of hazardous pollutants. This method has shown beneficial results in the treatment of sites contaminated with hydrocarbons and metals (Kao et al. 2008).

Bioventing

Bioventing bioremediation method is based on the controlled provision of oxygen and nutrients through wells to stimulate microbial activity in the contaminated sites. This process is operated in such a way that very low oxygen is introduced at low air flow rates to the polluted sites which is effective for both efficient degradation of microbes and minimized volatilization and pollutants release to environment (Atlas and Philp 2005). This method has shown beneficial results to remove pollutants from the vadose zone and effectively useful for biodegradation of soils polluted with volatile pollutants, spilled light petroleum, absorbed fuel residuals, and low-molecular weight hydrocarbons (Ho ̇hener and Ponsin 2014).

Bioslurping

Bioslurping bioremediation method is the collective use of an indirect supply of oxygen, bioventing, vacuum enhanced pumping, and soil vapor extraction to stimulate microbial activity for the cleanup of the soil and groundwater pollutants. This method has a few limitations that hinder its performance which includes low soil permeability that further declines activity of microbes and oxygen transfer rate. This method is suitable for biodegradation of semi-volatile and volatile organic contaminants available in both solid and liquid (Vidali 2001).

Biosparging

Biosparging mode of bioremediation has two major benefits based on the supply of air to the subsurface/saturated zone: (i) degradation of subsurface organic volatile pollutants by moving to surface zone and (ii) stimulate the microbial activity to boost pollutant elimination from polluted site. This method works on two core factors: biodegradability of pollutant and permeability of soil (Kumar et al. 2018). This method is broadly used to clean up aquifers polluted with petroleum products and hydrocarbons such as ethylbenzene, toluene, xylene, and benzene (Kao et al. 2008).

1.6 Advantages of Bioremediation

Bioremediation is a natural and safe process which has numerous advantages over conventional remediation methods such as incineration and landfilling as it can be beneficial to eliminate pollutants permanently, coupled with chemical and physical waste treatment techniques, operated at large scale, long-term viability elimination, and applied on actual polluted sites (Van Aken 2009). It has ample of benefits including greater public acceptance, friendly to environment, green technology, and cost effectiveness due to reduced installation and maintenance cost (Ali et al. 2013). Bioremediation technique possesses voluminous advantages such as prevention of leaching and metal erosions, precious metals recovery, and enhanced soil fertility (Mench et al. 2009).

1.7 Disadvantages of Bioremediation

Various drawbacks of bioremediation have also been reported in the previous studies. Microbial process is not sufficient for some compounds' biodegradation, e.g., radionuclides, chlorinated organic pollutants, and few metals. Some microorganisms during biodegradation produce toxic metabolites. Bioremediation process should be tailored to site-specific conditions as it is a scientifically intensive process and requires pilot treatment study before actual treatment of contaminated site (Mishra et al. 2021; Boopathy 2000; Tyagi and Kumar 2021). Phytoremediation method has some limitations including concentration of pollutant, toxicity and bioavailability to plants, its longer time treatment, and inability to degrade organic contaminants owing to the lack of degrading enzymes (Ali et al. 2013; Kuiper et al. 2004).

1.8 Phytoremediation

Phytoremediation follows the methodology of bioremediation which remediates polluted soil and water through plants and their roots. The studies have explained the mechanism followed by plants to remove pollutants from contaminated site, these employ passive process to up take pollutants, xylem flow to translocate pollutant from root to shoot, and pollutants get accrued in the shoot (Nandy et al. 2021; Miguel et al. 2013). Vegetation-based bioremediation methods act as biofilters and possess great potential to metabolize/degrade, transform, immobilize, and accumulate the pollutants (Meagher 2000). Phytoremediation is an emerging, innovative, and less costly replacement to other bioremediation methods to clean up the perilous pollutants from contaminated site. The phytoremediation encompasses a variety of techniques conditional to the type of pollutant either metal or organic (Fig. 1.3) (radionuclides, toxic heavy metals, chlorinated compounds, and hydrocarbons) and their fate (filtration, transformation, volatilization, stabilization, degradation, accumulation, and in combination) (Kuiper et al. 2004). Plants possess the ability to take up required components, i.e., heavy metals and nutrients from polluted sites. Some plants are known as hyperaccumulators which can up take and store large amounts of pollutants (heavy metals) without involving them in their metabolic functions. However, some plants also possess the ability to utilize several organic pollutants in their physiological processes by degrading them (Ozyigit et al. 2020; Vouillamoz and Milke 2001; Khan et al. 2004).

1.8.1 Phytoextraction

Phytoextraction bioremediation technique is the focus of research nowadays and is also called as rhizoaccumulation or phytoaccumulation. This process relies on plants to assist metal removal from polluted soil (Niazi et al. 2012). This method of phytoremediation recruits plants containing metals (pollutants) due to accrue of pollutants in the roots, shoots, and leaves which are further translocated to dispose and recycle (Meagher 2000).

1.8.2 Phytodegradation or Rhizodegradation

As the name suggests, this method involves the breakdown of pollutants due to soil microbes

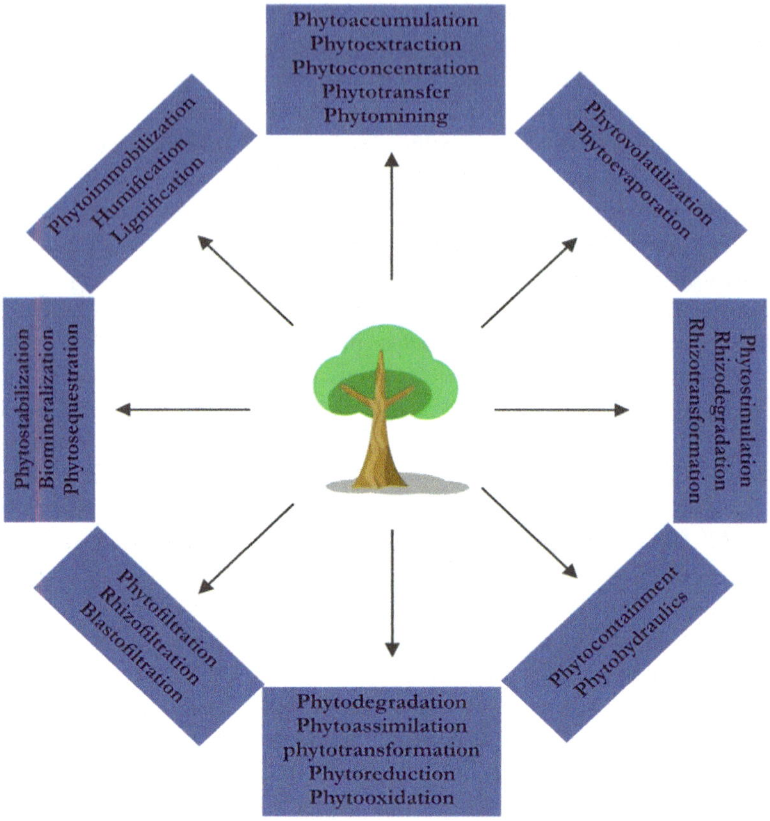

Fig. 1.3 Processes of phytoremediation involving wide scale of biodiversity. Adapted from Prasad et al. (2010)

present in the rhizosphere which produce proteins and enzymes. This method is an example of symbiotic relationship between plants and microbes in which plants are responsible for the delivery of nutrients to carry out biodegradation process. On the other hand, microbes are responsible for providing a suitable biodegradative environment (Miguel et al. 2013).

1.8.3 Phytostabilization

Phytostabilization is a technique in which plants develop a stable mass containing pollutants inside to prevent the mobility of pollutants in soil and water. Thus, this technique restricts further movement of contaminants to the atmosphere (Meagher 2000). This method, unlike phytoextraction, focuses on reducing the risk of pollutants to human health and environment than removal from the contaminated sites (Murtaza et al. 2014).

1.8.4 Phytotransformation

As the name indicates, phytotransformation technique transforms the hazardous organic pollutants into non-toxic derivatives by taking up from water, soil, and sediments (Kuiper et al. 2004). Plants up take the organic and inorganic pollutants from contaminated soil which get absorbed in the roots and become biochemically bound to cellular tissues in biologically inert or less active forms (Sinha et al. 2007; Watanabe 1997; Vara Prasad and de Oliveira Freitas 2003).

1.8.5 Rhizofiltration

Precipitation and flocculation are the conventional methods to treat the industrial wastewater and groundwater contaminated with metals, later sedimentation and then resulted sludge is disposed of (Ensley 2000). Rhizofiltration is the best alternate to the conventional methods. This method is quite similar to phytoextraction method, but rhizofilteration is a treatment method for contaminated groundwater in spite of contaminated soil. The mechanism involves the uptake of contaminant by plant roots, and then contaminant is either absorbed by the roots or get adsorbed on root surface. The plant is firstly acclimatized in the presence of contaminant for utilization in rhizofiltration method. Once the plant is acclimatized, it is planted in the contaminated groundwater to up take contaminant along with polluted water, until the roots of plant get saturated. After saturation, plants are harvested and disposed of safely. The cycle is continued on the contaminated site until or unless the complete removal of contaminants. Generally, this is a cost-competitive method beneficial for treating the groundwater, estuaries, and natural wetlands contaminated with significantly low metal loads such as Zn, Pb, and Cr (Verma et al. 2006; Shivalkar et al. 2020; Kumar et al. 1995; Miguel et al. 2013; Ensley 2000).

1.9 Conclusion

Bioremediation is a green technology which has been proved as the most promising new technology in effective treatment of effluent heavy metals mining wastes, municipal/urban sewage wastes, solid or liquid industrial wastes, hazardous wastes, and/or chemical spills, etc., from soil, water, and other environments. The process has been classified into ex situ and in situ remediation on the basis of pollutant elimination from soil, transportation, and origin. Sources of soil contaminants can be biological, organic, and/or inorganic. List of factors that can affect bioremediation process includes environmental and biological factors, temperature, metal ions, oxygen concentrations, moisture content, pH and toxic compounds, selection and characterization of soil and nutrient availability.

References

Adams GO, Fufeyin PT, Okoro SE, Ehinomen I (2015) Bioremediation, biostimulation and bioaugmention: a review. Int J Environ Bioremediation Biodegradation 3:28–39

Adesipo AA, Freese D, Nwadinigwe AO (2020) Prospects of In-situ remediation of crude oil contaminated lands in Nigeria. Scientific African e00403

Ali H, Khan E, Sajad MA (2013) Phytoremediation of heavy metals—concepts and applications. Chemosphere 91(7):869–881

Arianzad SA, Zeinoddini M, Haddadi A, Nazarian S, Sajedi RH (2020) In silico design of chimeric and immunogenic protein-containing ipab and ipad as a vaccine candidate against shigella dysenteriae. Curr Proteomics 17(4):333–341

Asira EE (2013) Factors that determine bioremediation of organic compounds in the soil. Acad J Interdisciplinary Stud 2:125–128

Atlas RM, Philp J (eds) (2005) Bioremediation: applied microbial solutions for real-world environmental cleanup. ASM Press, Washington

Azadi H, Ho P (2010) Genetically modified and organic crops in developing countries: a review of options for food security. Biotechnol Adv 28(1):160–168

Azubuike CC, Chikere CB, Okpokwasili GC (2016) Bioremediation techniques-classification based on site of application: principles, advantages, limitations and prospects. World J Microbiol Biotechnol 32(11):180

Bahmani F, Ataei SA, Mikaili MA (2018) The effect of moisture content variation on the bioremediation of hydrocarbon contaminated soils: modeling and experimental investigation. J Environ Anal Chem 5(2):236

Bernhardt ES, Rosi EJ, Gessner MO (2017) Synthetic chemicals as agents of global change. Front Ecol Environ 15(2):84–90

Bhandari A, Xia K (2005) Hazardous organic chemicals in biosolids recycled as soil amendments. Water pollution. Springer, Berlin, pp 217–239

Boopathy R (2000) Factors limiting bioremediation technologies. Biores Technol 74(1):63–67

Cairney T (1993) Contaminated land. Blackie, London

Cases I, de Lorenzo V (2005) Genetically modified organisms for the environment: stories of success and failure and what we have learned from them. Int Microbiol 8:213–222

Cobb GP, Sands K, Waters M, Wixson BG, Dorward E (2000) Accumulation of heavy metals by vegetables grown in mine wastes. Environ Toxicol Chem 19:600–607

Couto N, Fritt J, Jensen PE, Højrup M, Rodrigo AP (2014) Suitability of oil bioremediation in an artic soil using surplus heating from an incineration facility. Environ Sci Pollut Res 21(9):6221–6227 da Silva S, Gonçalves I, Gomes de Almeida FC, Padilha da Rocha e Silva NM, Casazza AA, Converti A, Asfora Sarubbo L (2020) Soil bioremediation: overview of technologies and trends. Energies 13(18):4664

Das MT, Kumar SS, Ghosh P, Shah G, Malyan SK, Bajar S, Singh L (2020) Remediation strategies for mitigation of phthalate pollution: challenges and future perspectives. J Hazardous Mater 124496

Das N, Chandran P (2011) Microbial degradation of petroleum hydrocarbon contaminants: an overview. Biotechnol Res Int

Demirbas A, Edris G, Alalayah WM (2017) Sludge production from municipal wastewater treatment in sewage treatment plant. Energy Sources Part a: Recovery Utilization Environ Effects 39(10):999–1006

Dias RL, Ruberto L, Calabro A, Balbo AL, Del Panno MT, Mac Cormack WP (2015) hydrocarbon removal and bacterial community structure in on-site biostimulated biopile systems designed for bioremediation of diesel-contaminated antarctic soil. Polar Biol 38(5):677–687

El Fantroussi S, Agathos SN (2005) Is bioaugmentation a feasible strategy for pollutant removal and site remediation? Curr Opin Microbiol 8(3):268–275

Ensley BD (2000) Rationale for use of phytoremediation. In: Raskin I, Ensley BD (eds) Phytoremediation of toxic metals using plants to clean up the environment. Wiley, New York, p 11

Gavrilescu M (2010) Environmental biotechnology: achievements, opportunities and challenges. Dyn Biochem Process Biotechnol Mol Biol 4(1):1–36

Ghangrekar MM, Sathe SM, Chakraborty I (2020) In situ bioremediation techniques for the removal of emerging contaminants and heavy metals using hybrid microbial electrochemical technologies. In: Emerging technologies in environmental bioremediation. Elsevier, pp 233–255

Harekrushna S, Kumar DC (2012) A review on: bioremediation. Int J Res Chem Environ 2(1):13–21

Hatti-Kaul R, Törnvall U, Gustafsson L, Börjesson P (2007) Industrial biotechnology for the production of bio-based chemicals—a cradle-to-grave perspective. Trends Biotechnol 25(3):119–124

Haynes R, Murtaza G, Naidu R (2009) Inorganic and organic constituents and contaminants of biosolids: implications for land application. Adv Agron 104:165–267

Höhener P, Ponsin V (2014) In situ vadose zone bioremediation. Curr Opin Biotechnol 27:1–7

Hou D, Bolan NS, Tsang DC, Kirkham MB, O'Connor D (2020) Sustainable soil use and management: an interdisciplinary and systematic approach. Sci Total Environ 138961

Jacobs LW, O'Connor G, Overcash M, Zabik M, Rygiewicz P (1987) Effects of trace organics in sewage sludges on soil-plant systems and assessing their risk to humans. Lewis, Chelsea

Juwarkar AA, Singh SK, Mudhoo AA (2010) Comprehensive overview of elements in bioremediation. Rev Environ Sci Biotechnol 9(3):215–288

Kao CM, Chen CY, Chen SC, Chien HY, Chen YL (2008) Application of in situ biosparging to remediate a petroleum-hydrocarbon spill site: field and microbial evaluation. Chemosphere 70(8):1492–1499

Khan FI, Husain T, Hejazi R (2004) An overview and analysis of site remediation technologies. J Environ Manage 71(2):95–122

Khudhaier SR, Al- AMA, Abbas RF (2020) A review article-technology of bioremediation. Int J Res Appl Sci Biotechnol 7(5):349–353

Kim S, Krajmalnik R, Kim JO, Chung J (2014) Remediation of petroleum hydrocarbon-contaminated sites by dna diagnosis-based bioslurping technology. Sci Total Environ 497:250–259

Krogmann U, Boyles LS, Bamka WJ, Chaiprapat S, Martel CJ (1999) Biosolids and sludge management. Water Environ Res 71(5):692–714

Kuiper I, Lagendijk EL, Bloemberg GV, Lugtenberg BJ (2004) Rhizoremediation: a beneficial plant-microbe interaction. Mol Plant-Microbe Interact 17(1):6–15

Kumar PN, Dushenkov V, Motto H, Raskin I (1995) Phytoextraction: the use of plants to remove heavy metals from soils. Environ Sci Technol 29(5):1232–1238

Kumar V, Shahi SK, Singh S (2018) Bioremediation: an eco-sustainable approach for restoration of contaminated sites. In: Singh J, Sharma D, Kumar G, Sharma N (eds) Microbial bioprospecting for sustainable development. Springer, Singapore, pp 115–136

Lacalle RG, Aparicio JD, Artetxe U, Urionabarrenetxea E, Polti MA, Soto M, Becerril JM (2020) Gentle remediation options for soil with mixed chromium (VI) and lindane pollution: biostimulation, bioaugmentation, phytoremediation and vermiremediation. Heliyon 6(8):e04550

Lee S, Yang YA, Milano SK, Nguyen T, Ahn C, Sim JH, Song J (2020) salmonella typhoid toxin pltb subunit and its non-typhoidal salmonella ortholog confer differential host adaptation and virulence. Cell Host Microbe 27(6):937–949

Li Y, Bräunig J, Angelica GC, Thai, PK, Mueller JF, Yuan Z (2020) Formation and partitioning behaviour of perfluoroalkyl acids (PFAAs) in waste activated sludge during anaerobic digestion. Water Res 189:116583

Macaulay BM (2014) Understanding the behavior of oil-degrading microorganisms to enhance the microbial remediation of spilled petroleum. Appl Ecol Environ Res 13:247–262

Madhavi GN, Mohini DD (2012) Review paper on–parameters affecting bioremediation. Int J Life Sci Pharma Res 2(3):77–80

Meagher RB (2000) Phytoremediation of toxic elemental and organic pollutants. Curr Opin Plant Biol 3(2):153–162

Mench M, Schwitzgue ́bel JP, Schroeder P, Bert V, Gawronski S, Gupta S (2009) Assessment of successful experiments and limitations of phytotechnologies: contaminant uptake, detoxification and sequestration, and consequences for food safety. Environ Sci Pollut Res 16(7):876

Miguel AS, Ravanel P, Raveton M (2013) A comparative study on the uptake and translocation of organochlorines by phragmites australis. J Hazard Mater 244:60–69

Mishra M, Singh SK, Kumar A (2021) Environmental factors affecting the bioremediation potential of microbes. In: Singh VK, Singh P, Mishra VK (eds) Kumar A. Microbe mediated remediation of environmental contaminants, Woodhead Publishing, pp 47–58

Mohammadi L, Rahdar A, Bazrafshan E, Dahmardeh H, Susan M, Hasan AB, Kyzas GZ (2020) Petroleum hydrocarbon removal from wastewaters: a review. Processes 8(4):447

Mohan SV, Sirisha K, Rao NC, Sarma PN, Reddy SJ (2004) Degradation of chlorpyrifos contaminated soil by bioslurry reactor operated in sequencing batch mode: bioprocess monitoring. J Hazard Mater 116(1–2):39–48

Moore JW, Ramamoorthy S (2012) Organic chemicals in natural waters: applied monitoring and impact assessment. Springer Science & Business Media

Murtaza G, Haynes RJ, Naidu R, Belyaeva ON, Kim K-R, Lamb DT, Bolan NS (2011) Natural attenuation of Zn, Cu, Pb and Cd in three biosolids-amended soils of contrasting pH measured using rhizon pore water samplers. Water Air Soil Pollut 221:351–363

Murtaza G, Murtaza B, Niazi NK, Sabir M (2014) Soil contaminants: sources, effects, and approaches for remediation. In: Improvement of crops in the era of climatic changes. Springer, New York, pp 171–196

Nandy S, Andraskar J, Lanjewar K, Kapley A (2021) Challenges in bioremediation: from lab to land. In: Kumar V, Shah MP (eds) Saxena G. Bioremediation for environmental sustainability, Elsevier, pp 561–583

Niazi NK, Singh B, Van Zwieten L, Kachenko AG (2012) Phytoremediation of an arsenic-contaminated site using pteris vittata l. and pityrogramma calomelanos var. austroamericana: a long-term study. Environ Sci Pollut Res 19(8):3506–3515

Ntougias S, Baldrian P, Ehaliotis C, Nerud F, Merhautová V, Zervakis GI (2015) Olive mill wastewater biodegradation potential of white-rot fungi–mode of action of fungal culture extracts and effects of ligninolytic enzymes. Bioresour Technol 189:121–130

Ossai IC, Ahmed A, Hassan A, Hamid FS (2020) Remediation of soil and water contaminated with petroleum hydrocarbon: a review. Environ Technol Innov 17:100526

Ozyigit II, Can H, Dogan I (2020) Phytoremediation using genetically engineered plants to remove metals: a review. Environ Chem Lett 1–30

Phulia V, Jamwal A, Saxena N, Chadha NK, Muralidhar (2013) Technologies in aquatic bioremediation, pp 65–91

Prasad MNV, Freitas H, Fraenzle S, Wuenschmann S, Markert B (2010) Knowledge explosion in phytotechnologies for environmental solutions. Environ Pollut 158:18–23

Ranjha MMAN, Amjad S, Ashraf S et al (2020) Extraction of polyphenols from apple and pomegranate peels employing different extraction techniques for the development of functional date bars. Int J Fruit Sci. https://doi.org/10.1080/15538362.2020.1782804

Ranjha MMAN, Irfan S, Nadeem M, Mahmood S (2020) A comprehensive review on nutritional value, medicinal uses, and processing of banana. Food Rev Int. https://doi.org/10.1080/87559129.2020.1725890

Sen R, Chakrabarti S (2009) Biotechnology—applications to environmental remediation in resource exploitation. Curr Sci 97(6):768–775

Shahzad A, Siddiqui S, Bano A, Sattar S, Hashmi MZ, Qin M, Shakoor A (2020) Hydrocarbon degradation in oily sludge by bacterial consortium assisted with alfalfa (*Medicago sativa* L.) and maize (*Zea mays* L.). Arab J Geosci 13(17): 1–12

Santillan JY, Rojas NL, Ghiringhelli PD, Nóbile ML, Lewkowicz ES, Iribarren AM (2020) Organophosphorus compounds biodegradation by novel bacterial isolates and their potential application in bioremediation of contaminated water. Bioresource Technol 317:124003

Sharma S. Bioremediation: features, strategies and applications. Asian J Pharm Life Sci 2(2):202–213

Sharma I (2020) Bioremediation techniques for polluted environment: concept, advantages, limitations, and prospects. In: Trace metals in the environment-new approaches and recent advances. IntechOpen

Shivalkar S, Singh V, Sahoo AK, Samanta SK, Gautam PK (2020) Bioremediation: a potential ecological tool for waste management. In: bioremediation for environmental sustainability. Elsevier, pp 1–21

Sinha RK, Herat S, Tandon PK (2007) Phytoremediation: role of plants in contaminated site management. In: Environmental bioremediation technologies. Springer, Berlin, pp 315–330

Silva-Castro GA, Uad I, Gónzalez-López J, Fandinõ CG, Toledo FL, Calvo C (2012) Application of selected microbial consortia combined with inorganic and oleophilic fertilizers to recuperate oil-polluted soil using land farming technology. Clean Technol Environ Pol 14:719–726

Skinder BM, Uqab B, Ganai, BA (2020) Bioremediation: a sustainable and emerging tool for restoration of polluted aquatic ecosystem. In: Fresh water pollution dynamics and remediation. Springer, Singapore, pp 143–165

Smith S (2000) Are controls on organic contaminants necessary to protect the environment when sewage sludge is used in agriculture? Prog Environ Sci 2:129–146

Song ZM, Xu YL, Liang JK, Peng L, Zhang XY, Du Y, Guan YT (2020) Surrogates for on-line monitoring of the attenuation of trace organic contaminants during advanced oxidation processes for water reuse. Water Res 116733

Srivastava AK, Singh RK, Singh D (2021) Microbe-based bioreactor system for bioremediation of organic contaminants: present and future perspective. In: Microbe mediated remediation of environmental contaminants. Woodhead Publishing, pp 241–253

Thavasi R, Jayalakshmi S, Banat IM (2011) Application of biosurfactant produced from peanut oil cake by Lactobacillus delbrueckii in biodegradation of crude oil. Bioresour Technol 102:3366–3372

Tiwari G, Singh SP (2014) Application of bioremediation on solid waste management: a review. J Bioremed Biodeg 5(6):248–1172

Trinh HT, Duong HT, Strobel BW, Le GT (2020) Comprehensive study of organic micro-pollutants in flooded paddy soils in central Vietnam: levels, pollution pathways and sources. Bull Environ Contamination Toxicol 105(4):572–581

Tyagi B, Kumar N (2021) Bioremediation: principles and applications in environmental management. In: Kumar V, Shah MP (eds) Saxena G. Bioremediation for environmental sustainability, Elsevier, pp 3–28

Ünüvar S (2018) Microbial foodborne diseases. In: Grumezescu AM (ed) Holban AM. Foodborne diseases, Academic Press, pp 1–31

Van Aken B (2009) Transgenic plants for enhanced phytoremediation of toxic explosives. Curr Opin Biotechnol 20(2):231–236

Vara MN, de Oliveira Freitas HM (2003) Hyperaccumulation in plants: biodiversity prospecting for phytoremediation technology. Electron J Biotechnol 6 (3):285–332

Verma P, George K, Singh H, Singh S, Juwarkar A, Singh R (2006) Modeling rhizofiltration: heavy-metal uptake by plant roots. Environ Model Assess 11 (4):387–394

Vidali M (2001) Bioremediation: an overview. Pure Appl Chem 73(7):1163–1172

Volpe A, D'Arpa S, Del Moro G, Rossetti S, Tandoi V, Uricchio VF (2012) Fingerprinting hydrocarbons in a contaminated soil from an italian natural reserve and assessment of the performance of a low-impact bioremediation approach. Water Air Soil Pollution 223(4):1773–1782

Vouillamoz J, Milke M (2001) Effect of compost in phytoremediation of diesel-contaminated soils. Water Sci Technol 43:291–295

Wang Q, Zhang S, Li Y, Klassen W (2011) Potential Approaches to improving biodegradation of hydrocarbons for bioremediation of crude oil pollution. Environ Protection J 2:47–55

Watanabe ME (1997) Phytoremediation on the brink of commericialization. Environ Sci Technol 31:182A-186A

Whelan MJ, Coulon F, Hince G, Rayner J, McWatters R, Spedding T, Snape I (2015) Fate and transport of petroleum hydrocarbons in engineered biopiles in polar regions. Chemosphere 131:232–240

Williams J (2006) Bioremediation of contaminated soils: a comparison of in situ and ex situ techniques, pages 12; Recuperado de/paper/Bioremediation-of-Contaminated-Soils-% 3A-A-Comparison-Williams/4c6afc722040e0d4807a744b7f89a5e7b9dac97f. Accessed 19 Dec 2020

Yang SZ, Jin HJ, Wei Z, He RX, Ji YJ (2009) Bioremediation of oil spills in cold environments: a review. Pedosphere 19:371–381

Current Soil Bioremediation Technologies: An Assessment

2

Arunima Nayak, Brij Bhushan, and Ivan Wilson

Abstract

Increasing global industrialization and modern agricultural practices has resulted in a rise in the incidences of soil pollutants like hydrocarbons, pesticides and heavy metals; all of which are linked to adverse health issues. Remediation of contaminated soil is essential not only for restoration of ecosystem but also for urban development. Physical, chemical and biological methods have been applied for soil remediation. Bioremediation or processes involving biological processes are fast picking up as effective treatment technologies not only because of their efficiency but also because of their environmental friendliness and cost effectiveness. The process is capable of degrading diverse types of pollutants including the persistent aromatic hydrocarbons; hence, bioremediation is a viable and effective technology for mitigation of soil pollutants. Choice of appropriate and feasible bioremediation technology for example phytoremediation, mycoremediation, bioventing, biopiles, composting etc. depends on the environmental conditions, type of pollutant, composition of soil, incurring treatment costs and available treatment time. Thus, detailed characterization and analysis of the contaminated site is a vital step toward successful bioremediation. More recently, use of surfactants/biosurfactants, nanomaterials along with genetically engineered biocatalysts has helped in enhancing the rate of removal/degradation of contaminants in the contaminated site. Use of more than one remediation technology has been preferred for remediation of complex sites. The aim of this chapter is to address the source, type and toxicity effects of soil contaminants followed by a detailed discussion on the different types of in situ and ex situ remediation processes applied till date along with their advantages and disadvantages. In addition, current scenario of new technologies vis-a-vis soil bioremediation is detailed. The objective of the chapter is to provide an updated information on the fundamentals, classification of bioremediation as a treatment technology along with affecting factors for the reclamation of contaminated soil, which will aid readers in selection of the appropriate technology. Future and current research prospects are addressed to assist researchers and scientists for the ongoing work in related field.

A. Nayak (✉) · B. Bhushan
Department of Chemistry, Graphic Era University, Dehradun 248002, India

I. Wilson
Department of Life Sciences, Graphic Era University, Dehradun 248002, India

J. A. Malik (ed.), *Advances in Bioremediation and Phytoremediation for Sustainable Soil Management*,
https://doi.org/10.1007/978-3-030-89984-4_2

Keywords

Biopile · Bioremediation · Biosparging · Bioventing · Ex situ · In situ · Phytoremediation

2.1 Introduction

Soil is a natural habitat to plants and microbial species; a source of nutrients, a carbon sink and a regulator for maintaining the quality and quantity of water. Intensive industrialization, agricultural activities as well as increasing modern man-made activities have resulted in introduction of unwanted species in soil. Soil pollution problem is aggravating because the contaminants are accumulating to unprecedented levels with increasing man-made activities (Mirsal 2008; Morillo et al. 2014, 2020). The major sources of soil pollution are man-made activities from agriculture, industry and urban lifestyle. Different pollutants identified in soil are heavy metals, nuclear wastes, pesticides, polycyclic aromatic hydrocarbons (PAHs), polychlorinated biphenyls (PCBs), dichlorodiphenyltrichloroethane (DDT), total petroleum hydrocarbon (TPH), plastics and sewage etc. But the major causes of concern are those contaminants that are recalcitrant to standard treatment technologies and hence have greater persistence in the environment. Such organic contaminants nomenclatured by regulatory bodies as persistent organic contaminants (POCs) are identified as petroleum-based hydrocarbons (e.g., alkanes, alkenes, cycloalkanes), chlorinated compounds (e.g., PCBs, PCDDs and PCDFs), monomeric aromatic hydrocarbons (e.g., benzene, toluene, ethylbenzene and xylene, collectively known as BTEX), polycyclic aromatic hydrocarbons (PAHs), organochlorine pesticides, etc. Such contaminants have diverse physical and chemical characteristics and induce permanent damage to the soil characteristics. Because of their capacity to accumulate in the food chain, such soil contaminants have adverse effect on the plants, animals, humans and the entire ecosystem. As per Weber et al. (2015), because of the toxicity, the POCs are considered as carcinogenic and are endocrine disrupters. Besides the POCs, urbanization and advanced lifestyle has resulted in release of various emerging contaminants (ECs) into the soil environment (Snow et al. 2017; Caliman and Gavrilescu 2009). Examples include pharmaceuticals, plasticizers, personal care products (cosmetics, soaps, detergents), perfluoroalkyl substances (PFAS), etc. The eco-toxicity effects of ECs have been less studied and have not yet been regulated. Soil has the capacity to degrade the contaminants. But excessive accumulation of contaminants in soil not only leads to degradation of soil quality but also has a damaging effect on the ecosystem (Cecchin et al. 2017).

Remediation of contaminants to reclaim the soil is essential to recover the ecosystem. An effective technology requires high level of efficiency for removal of diverse contaminants including recalcitrant contaminants like POCs and ECs, cost effectiveness of the treatment of contaminated site, environmental friendliness, etc. Among the physical, chemical and biological methods, remediation using biological methods is fast gaining in prominence mainly on account of its environmental friendliness and cost effectiveness for the treatment of diverse contaminant types present in contaminated sites. Remediation using biological species is known as bioremediation (Azubuike et al. 2016; Crawford and Crawford 1996). Such techniques involve microorganisms or plants which in turn use contaminants as a source of energy for their metabolism, development and growth. Because of the involvement of natural processes for the breakdown of toxic pollutants to non-toxic by-products, bioremediation is a sustainable and effective technique for mitigation of soil pollution. When biological species inhabiting the contaminated site are used for the treatment process, bioremediation is named as in situ. Ex situ process involves exogenous species for the treatment of a contaminated site. For a bioremediation method to be effective, characteristics of the pollutants, soil composition and environmental factors are vital factors. Because of the involvement of biological species, treatment may require longer period of time and depending on the area of contaminated site, treatment costs may be higher.

Based on the above considerations, the present chapter has a focus on a discussion of different soil pollutants along with their origin. This is followed by overview of the bioremediation technique and its classification. Future prospects in research on related area will be addressed. The chapter will provide a deeper insight to scientists, technocrats and students working on soil remediation.

2.2 Major Soil Pollutants, Their Sources and Toxicity Effects

The soil has always been the reservoir of organic and inorganic contaminants that are produced naturally or are incorporated by man-made activities (Duarte et al. 2018). Man-made activities are agricultural activities, urban lifestyle or industrial activities, etc. Organic chemicals present in septic effluents, animal or biowastes also cause soil pollution. The frequently occurring organic contaminants found in soils are polychlorinated biphenyls (PCBs), polybrominated biphenyls (PBB), polychlorinated dibenzofurans, polycyclic aromatic hydrocarbons (PAHs), organophosphorus and carbamate insecticides, herbicides and organic fuels, especially gasoline and diesel. Oil hydrocarbons-based contaminants like alkanes, alkenes and cycloalkanes have their source from oil industrial activities; chlorinated compounds like PCP, PCBs, PCDD/Fs result from manufacturing of pesticides/herbicides; monomeric aromatic hydrocarbons (BTEX) originate from oil industries, gasoline stations, etc. PAHs are known to originate from various activities related to incomplete burning of wood, municipal waste, fossil fuels etc. and from automobile exhaust (Achten and Hofmann). Heavy metals found in soil like Cu, Zn, Cd, Pb, Hg, Cr have been sourced from application of animal manure, gasoline stations, sawmills and wood preservation sites, mining and metallurgical industry, etc. (van-Camp et al., 2004). Nitroaromatics-based soil contaminants like TNT, nitrobenzene, nitrophenols and atrazine have resulted from industrial manufacture of explosives, aniline, dyes, drugs, pesticides and herbicides (Ye et al. 2004).

The adverse effect of the presence of such contaminants is irreparable change in soil texture, makeup and microbial community, reduced soil fertility, reduced nitrogen fixation, increased tendency for soil erosion, loss of soil nutrients, reduced crop yield and imbalances in soil flora and fauna (Stroud et al. 2007). Contaminated soil thus may cause a change in the entire ecosystem (Griffiths and Philippot 2013; Prosser). Soil contaminants have exhibited toxic effects on living organisms. Among the contaminants identified as priority pollutants requiring special attention for their handling and removal are the persistent organic pollutants (POPs) (e.g., DDT, dichloro-diphenyl-dichloro-ethylene (DDE), PCBs, PCDD/Fs and other halogenated compounds, polycyclic aromatic hydrocarbons (PAHs) (e.g., benzo-a-pyrene, naphthalene etc.), BTEX (e.g., benzene), emerging contaminants (e.g., bisphenol-A, polybrominated diphenyl ethers, perfluorinated compounds) (IPCS 1995; Clarke and Smith 2011). Various toxicity studies conducted on animals have demonstrated that such priority contaminants get accumulated in living tissues causing impairment of endocrine system, various abnormalities related to vitamin, thyroid, skin, reproduction, central nervous system, immune system, etc. and may cause chronic diseases like cancer, tumors (Vos et al. 2000; Sikka and Wang 2008; Rhind 2005). While low level exposure may result in blindness, blisters, rashes, nausea, diarrhea, etc., exposure at high concentrations may be fatal to both humans and animals.

2.3 Bioremediation: Technique and Affecting Environmental Factors

Bioremediation is the use of plants, algae and other microbial community for the natural breakdown of different organic and inorganic contaminants to lesser toxic forms that can be easily assimilated by living beings. The technology involves bioaugmentation and biostimulation of the microorganisms which in turn use the contaminant as food and source of energy. The success of bioremediation depends on healthy growth and development of the microbial community. Thus, factors related to physicochemical properties of contaminated soil and chemistry along with concentration of contaminant hold important aspect for the adaptation, growth of the microbial community. Soil porosity and soil permeability can influence the efficiency of bioremediation. Higher particle size of the soil cause higher contaminant adsorption leading to less availability of the contaminant to bioremediation (Guerin and Boyd 1992; Rarms 1996). Soil hydrophobicity can also reduce bioavailability of the contaminant to bioremediation (Nam and Alexander 1998). Nature of the contaminant also plays a major role in the bioremediation process. Majority of the soil contaminants are organic and are easily degraded by the microbial community. But there are certain chemical species that do not get degraded and persist in the soil environment. Many of these persistent contaminants are plastics which are not toxic to the microorganisms. Examples of recalcitrant toxic contaminants are chlordane, DDT, heptachlor, dieldrin, lindane and simazine, dichloro-diphenyl-dichloro-ethylene (DDE), PCBs, PCDD/Fs and other halogenated compounds, polycyclic aromatic hydrocarbons (PAHs), BTEX (e.g., benzene), emerging contaminants (e.g., bisphenol-A, polybrominated diphenyl ethers, perfluorinated compounds), etc. There are many contaminants which are water insoluble (e.g., benzo[a]pyrene, anthracene) and hence are bound to soil particles and are not accessible to microbes for degradation (Mahro 2000).

Other environmental factors like availability of moisture, nutrients, temperature and presence of low molecular weight organic acids, etc. hold a crucial role in the effectiveness of the bioremediation technique. Microbial species can affect their activity for degradation of soil contaminants.

Soil pH affects the functioning of the microorganisms as different species have an optimum operational pH. Soil pH can also alter the solubility, speciation and redox capability for different contaminant especially heavy metals (Brito et al. 2015). Metal ions can cause toxicity to the growing microbes and hence affect the bioremediation performance.

Temperature has a profound effect on the bioremediation efficiency (Bandowe et al. 2014). Extreme temperatures cause impairment to the healthy functioning of the microbes. Also, solubility of contaminants like the PAH and heavy metals increases with a rise in temperature; this in turn increases their bioavailability. Temperature induces high adsorption of contaminants onto the microbial surface; thereby causing enhanced interaction and better bioremediation.

Organic acids in soils having low molecular weight can alter the bioremediation efficiency. Soil organic compounds can bind to contaminants like heavy metal ions or PAHs thereby decreasing their binding to soil particles and increasing their bioavailability for enhanced degradation by microbes (Gao et al. 2015).

Soil contaminants can affect and inhibit the growth of microbes thereby affecting bioremediation. Also, new microbial species can develop for adapting to the changed environmental conditions. As per a study by Momose et al. 2008; the newly developed microbial strain has better adaptability and bioremediation capacity. The DNA of such resistant strain of microbes contain resistant genes; which when extracted and recombined have exhibited enhanced recombination efficiency (Haritash and Kaushik 2009; Mahmoudi et al. 2011). For example, the genes responsible for coding of the enzymes (ring-hydroxylating dioxygenase, RHDase) and the gene responsible for coding of enzyme (1-hydroxyl-2-naphthoate dioxygenase, 1H2Nase) were extracted from *Arthrobacter* sp. SA02, demonstrated high capacity for degrading phenanthrene (Li et al. 2015).

2.4 Classification of Bioremediation

The technology of bioremediation involves various ways and means for operation.

2.4.1 In Situ Bioremediation

2.4.1.1 Bioventing

Bioventing involves injecting of air under controlled conditions into the subsurface of contaminated site to promote microbial activity. The soil contaminants adsorbed to the soil are degraded to innocuous state (Philp and Atlas 2005). This form of in situ biodegradation has gained in popularity for reclamation of site contaminated with spilled petroleum products (Höhener and Ponsin 2014). In a study conducted by Sui and Li (2011), it was demonstrated that neither a high airflow rate nor a lower airflow rate brought about significant toluene degradation rate for the entire study period of 200 days. However, at an earlier study period of 100 days, it was observed that high airflow rate brought about enhanced toluene degradation rate as compared to a low airflow rate. The study demonstrated that in bioventing, airflow rate is a vital parameter for soil contaminant dispersal and redistribution. A similar study demonstrated by Frutos et al. (2010) showed that bioventing technology helped in remediation of soil contaminated by phenanthrene. The efficiency of phenanthrene removal was recorded as >93% after a period of seven months. Thomé et al. (2014) showed that a longer airflow time and a low airflow rate caused higher remediation efficiency for diesel-contaminated clayey soil. Besides the air injection time and airflow rate, the number of air injection points is crucial for ensuring uniform distribution of air flow to bring about efficient biodegradation. Increasing the number of air injection points helped in the removal of hydrocarbons from a sub-Antarctic hydrocarbon-polluted site, as demonstrated in a study by Rayner et al. (2007). Bioventing process has also helped in bioremediation of chlorinated compounds under anaerobic conditions, where, in place of air or oxygen, mixture of nitrogen, carbon dioxide and hydrogen were injected (Mihopoulos et al. 2000; Shah et al. 2001). Similarly, ozone injection helped in the oxidation of recalcitrant compounds and hence enabled enhanced biodegradation process (Philp and Atlas 2005).

The advantage of bioventing is its easy setup and installation, requiring shorter treatment times and is effective for the treatment of a wide variety of petroleum products. But certain limitations are there, e.g., ineffectiveness in the treatment of highly contaminated sites. Sites with clayey soil or exhibiting low permeability or saturation may exhibit difficulty in the treatment by bioventing (Khan et al. 2004).

Soil vapor extraction technique ensures a higher airflow rate as compared to the bioventing process, thereby maximizing the volatilization of volatile organic compounds (Magalhães et al. 2009).

2.4.1.2 Bioslurping

Bioslurping technique combines bioventing, soil vapor extraction and vacuum enhanced pumping for achieving remediation of soil contaminants as well as for recovery of products like light non-aqueous phase liquids (Gidarakos and Aivalioti 2007). The technique can also be used to remediate sites contaminated with volatile and semi-volatile organic compounds. Presence of excessive moisture in soil reduces microbial activity. While the disadvantage of this technique is its inability to remediate soils with low permeability, the merits are its low cost due to less generation of groundwater which in turn helps in minimizing storage, treatment and disposal costs (Philp and Atlas 2005).

2.4.1.3 Biosparging

Biosparging technique has an operating principle like bioventing which involves injection of air to promote microbial activity. But unlike bioventing, air is injected into the saturated zone of the contaminated soil. This causes the volatile organic contaminants to move upwards to the unsaturated zone, thereby promoting biodegradation. The technique has found wide application in sites contaminated with petroleum products like diesel and kerosene. Demonstrated that biosparging of BTEX contaminated site resulted in approximately >70% reduction of BTEX compounds. Increased dissolved oxygen levels revealed a shift to aerobic conditions. While biosparging has demonstrated improved bioremediation yet predicting the direction to apply the air injection onto the contaminated site is a major limitation of this technique.

2.4.1.4 Phytoremediation

Phytoremediation is an in situ process of bioremediation in which plants are used to remove the soil contaminants. The mechanism of removal of contaminants depends on the type of contaminant. Toxic metal ions are removed via the process of phyto-extraction, transformation and sequestration. Phytoremediation of organic contaminants involve enhanced rhizosphere biodegradation, phyto-accumulation, phyto-degradation and phyto-stabilization (Meagher 2000; Kuiper et al. 2004). Various factors affect the efficiency of the phytoremediation process. Some of the factors identified are physicochemical characteristics of soil, type of contaminant, root system of the plant involved in the phytoremediation and the plant species (Lee 2013). Longer and fibrous root systems allow greater contact with the soil contaminants; hence favoring greater degradation efficiency. For soil contaminants present on surface soil, grasses are favored while the deep penetrating roots of poplar help in degradation of contaminants found in greater depth below the soil. Selection of a plant species is vital for efficient phytoremediation. The characteristics desirable for selection of a plant species are (a) ability to sustain the climatic conditions of the contaminated site, (b) high biomass (c) long penetrating roots. Native plants are preferred as they have greater adaptability to the environmental conditions but also, such plants can help in restoration of habitat after the treatment process (Mench et al.). Soil properties like organic matter and clay content have a direct relationship to efficient phytoremediation. Phyto-degradation of soil contaminants in soils having higher organic content is less due to their lesser bioavailability for plants (Christman and Pfaender 2006). Organic contaminants tending to bind to roots more strongly have a higher potential to be phytoremediated (Briggs et al. 1983). It was reported that soils having greater clay content showed greater binding of contaminants like pyrene (Karickhoff et al. 1979). A similar study by Gardner et al., (1979) revealed that clayey soil showed four times greater binding of anthracene and two times greater binding for fluoranthene, benz[a] anthracene and benzo[a]pyrene as compared to sandy soil. Studies have shown that plants showing phytoremediation potential are *Phragmites australis* for PAH (Gregorio et al. 2014), *Dracaena reflexa* for diesel (Dadrasnia and Agamuthu 2013), *Amaranthus paniculatus* for Ni^{2+} (Iori et al. 2013), *Sparganium* sp. for polychlorinated biphenyls (PCBs) (Gregorio et al. 2013), *Rizophora* sp. for total petroleum hydrocarbon (TPH) (Moreira et al. 2013), grass for phenanthrene and pyrene (Xu et al. 2006).

The advantages of phytoremediation as cited in literature studies are many. Phytoremediation technique requires less environmental intervention and can be used for treatment of a diversity of soil contaminants. It is environmentally friendly due to minimal generation of secondary wastes (Khan et al. 2004). Also, the technology of using plants for remediation is aesthetically pleasing and is driven by solar energy. It is cost effective when applied on a large scale. Despite the advantages, limitations are the longer treatment periods. Climatic factors may affect the growth of plants and hence may be a limiting factor for the remediation efficiency. The plants may require specialized disposal options after the treatment of the contaminated site.

2.4.1.5 Ex situ Bioremediation

Depending on the intensity of contamination and nature of contaminants, ex situ bioremediation is employed. The process involves removal of the contaminants to specialized facilities for their subsequent treatment. The ex situ techniques thus have a better control over environmental conditions leading to enhanced biodegradation efficiency. But because of the lengthy excavation procedures and subsequent bioremediation treatment, the technique involves higher cost as compared to the in situ techniques.

2.4.1.6 Biopile and Windrows

In the biopile or windrows mediated ex situ bioremediation technique, the contaminated soil is excavated and is either stockpiled (biopile) or placed in rows (windrows) above the ground with subsequent treatment involving nutrient addition, irrigation and aeration to enhance the microbial activity and hence enhance the process of bioremediation. In the windrows technique, the contaminated soil is periodically turned. This ensures uniform aeration, distribution of contaminants, nutrients as well as microbial degradative activities, thereby enhancing the bioremediation rate (Barr et al. 2002). In a comparative study by Coulon et al. (2010), it was found that windrow treatment demonstrated higher hydrocarbon removal rates as compared to biopiling treatment. The study also demonstrated that use of friable soil was responsible for the better performance in windrows treatment. Windrows/biopile bioremediation technique is favorable and widely accepted due to better control of environmental factors like temperature, nutrient and aeration (Whelan et al. 2015). Microbial consortium and mature compost were added to the biopiles at low temperature conditions; the study showed that at the end of 94 days of treatment, 90.7% total petroleum hydrocarbons (TPHs) were achieved as compared to 48% TPH removal in control setups (Gomez and Sartaj 2014). Results revealed that action of both biostimulation and bioaugmentation brought about high reduction levels of TPH. A similar successful case of biopile remediation was achieved by Dias et al. (2015) who used pretreated contaminated soil for biopiling with subsequent biostimulation with fishmeal. Results reported were 71% reduction in total hydrocarbon concentration at the end of 50 day treatment period. Aislabie et al. (2006) incorporated heating into the biopile setup which helped to increase microbial activities and contaminant availability, thereby enhancing the rate of biodegradation. The advantage of biopile system was increased flexibility in the design which helped to bring about desired results in the bioremediation of contaminated sites. Delille et al. (2008) undertook aeration and sieving of contaminated soil prior to biopiling. Rodríguez-Rodríguez et al. (2010) used bulking agents like sawdust, straw, bark or wood chips to enhance bioremediation process in a biopile setup. Various studies have reported that biopile can treat large volumes of contaminated soil in limited space. Chemlal et al. (2013) was able to achieve similar results during biopiling in scale-up studies as compared to lab-scale studies.

Although the technology is easy to design and implement, requiring short treatment times, yet the requirement for large areas of land is a limitation. Other limitations are difficulty in achieving reduction in contaminant levels of over 95%; presence of high concentrations of heavy metals and petroleum products may be inhibitory to the microbial growth. Degradation of volatile compounds may require treatment emission control equipment (Khan et al. 2004). The overall cost of the technology is dependent on the contaminant type, procedure used for treatment (e.g., pre or posttreatment), requirement of emission control equipment, etc.

2.4.1.7 Bioreactors

The bioreactor system encompasses a closed cylindrical container in which the contaminated soil undergoes treatment under controlled environmental conditions of temperature, pH, agitation and aeration rates as well as nutrient addition. The contaminant and microbial levels can also be controlled within the reactor. Bioremediation can be conducted either aerobically or anaerobically, in which the contaminated soil can be placed either in a dry form or under suspension (Mohan et al. 2004). Bioreactors offer more flexibility and better control as compared to other ex situ bioremediation techniques like excellent control and manipulation of environmental factors and can treat a wide range of contaminants like heavy metal ions, pesticides, PAHs, BTEX compounds, etc. Better control of operating parameters ensures enhanced biological reactions within the bioreactor and reduced treatment times. Further, the system allows amendments in form of biostimulant or bioaugmenting agents for enhanced bioremediation (Chikere et al. 2012). Another advantage was that the bioreactor could be operated for short or long-term using crude oil-polluted slurry, and this helped to monitor changes in the dynamics of the microbial population (Zangi-Kotler et al. 2015). This was beneficial for characterization of microbial work dynamics involved in the bioremediation process. In a study by Plangklang and Alissara Reungsang (2010), a sequencing batch reactor involving carbofuran contaminated soil exhibited degradation efficiency of 88–97%. Fuller et al. (2003) used a glass jar paddle-type reactor for studying the removal efficiency of 2, 4, 6-trinitrophenylmethylnitramine contaminated soil, and the efficiency was demonstrated to be 99%. A packed bed biofilter was used for degradation of xylene in a pharmaceutical sludge and the removal % was demonstrated to be 95–99% (Saravanan et al. 2015). Similarly, Mustafa et al. (2015) showed a 95% removal efficiency of 2,4-dichlorophenoxyacetic acid contaminated soil using a roller slurry bioreactor.

Bioreactor applications for large scale treatment require more capital cost and are not cost effective. Failure of effective control of the various biological processes occurring simultaneously in the bioreactor may reduce microbial activity and finally make the bioreactor ineffective (Philp and Atlas 2005). Different contaminants may respond differently to different bioreactors; hence, appropriate design of the bioreactor is vital for effective bioremediation.

2.4.1.8 Landfarming

Landfarming involves simple operation, is of low cost, requires less capital and energy investment and can be used to treat large volume of contaminated soil with minimal environmental impact, thereby making the technology attractive (Maila and Colete 2004). It may be regarded as ex situ or in situ depending on the type of contaminated site and depth of contaminant penetration below the ground level (Nikolopoulou et al. 2013; Maila and Colete 2004). Maximum biodegradation results are obtained by control of soil conditions which is ensured via irrigation or spraying for enabling moisture content, via tilling for enabling aeration and via adding of acid or alkali agents for enabling neutral pH conditions. The technology has been successful for treatment of petroleum contaminated soils. In a study by Elis (1994), landfarming was conducted on a petroleum contaminated soil in a refinery site where the soil was tilled at frequent intervals for ensuring aeration and was treated with nutrients, surfactants, microbial inoculants. Results showed that at the end of 34 days of treatment, total petroleum hydrocarbon was reduced by 90%. Landfarming of kerosene contaminated soil in wheat fields was conducted by Dibbles and Bartha (1979) in which soil was excavated to a depth of 45 cm with subsequent addition of lime and nutrients. With frequent tilling in a period of two months, contaminant was reduced from 8,700 ppm to 30–3000 ppm. Wheat production was resumed one year after the landfarming operation.

Despite the various advantages, some of the limitations cited are requirement of large space for treatment and reduced efficiency for treatment of inorganic contaminants as well as for the heavier fractions of petroleum. For treatment of volatile contaminants, landfarming technique is not suitable especially for tropical climates and requires pretreatments. Heavier components of petroleum are less degraded by this technique. Reduction in the concentration level to less than 0.1 ppm is difficult to achieve (Khan et al. 2004).

2.5 Future Prospects

Both lab-scale and real-time site remediation cases have demonstrated the success of the bioremediation technique in the treatment of diverse contaminants as well as in the reclamation of soil ecosystem. Various factors have been found to affect the bioremediation efficiency and kinetics. Availability of nutrients, optimum levels of temperature, pH, aeration, etc., bioavailability of contaminants and microbial diversity, their abundance and their community structure play crucial role in the bioremediation efficiency.

Studies on molecular techniques (genomics, metabolomics, proteomics and transcriptomics) have helped in the identification of species, functions, metabolic pathways, which in turn has helped to overcome the limitations associated with microbial culture-dependent methods.

Biostimulation and bioaugmentation are two approaches that have been adopted for enhancing the microbial metabolism during the bioremediation of contaminated site. Enhancing microbial activity via addition of nutrients to the contaminated site is biostimulation. But, in a study by Wang et al. (2012), it was found that microbial activity and diversity was affected by excessive use of substrates and nutrients. Bioaugmentation is aimed to increase microbial population that are involved in degradation activities. Microbes when applied in consortium have resulted in more efficient degradation of soil contaminants as compared to pure isolated strains. The combined effect of biostimulation and bioaugmentation helped in effective degradation of both low as well as high molecular weight polycyclic aromatic hydrocarbons (PAHs), as revealed from a study by Sun et al. The degradation efficiency for low molecular weight PAH's and high molecular weight PAH's was 43.9% and 55% respectively. Although bioaugmentation has proved effective in the overall reduction in the contaminants, yet addition of exogenous microbes to the site may initiate competition between the endogenous and exogenous microbes may put risk on the survival endogenous microbes. Tyagi et al. (2011) suggested the use of agar, agarose, alginate, gelatin and polyurethane as carrier materials that may help in solving the problems associated with bioaugmentation. Use of biosurfactants has helped in improving contaminant bioavailability via initiating desorption and solubilization, hence increasing their mass transfer. The reasons cited for their mass acceptance for use in bioremediation purposes are because of their non-toxicity, biodegradability and are cheap. Addition of a biosurfactant that was isolated from a bacterial strain *Bacillus subtilis* MTCC1427 helped in enhancing the rate of biodegradation of a chlorinated pesticide (endosulfan) by 30–45% (Awasthi et al. 1999).

Studies have also shown that use of more than one remediation technique can help in minimizing or curtailing the shortfall/limitation of a single technology. Cassidy et al. (2015) used technologies like chemical oxidation, stabilization and anaerobic bioremediation in a single operation to reduce contaminant levels. Amendment with biochar followed by mycoremediation was used by García-Delgado et al. (2015) for immobilization and biodegradation of polycyclic aromatic hydrocarbons in creosote-contaminated soil. A similar study by Martínez-Pascual et al. (2015) was made by coupling chemical oxidation and biostimulation, and their effects were discussed on the natural attenuation capacity and resilience of the native microbial community in alkylbenzene-polluted soil. More recently, the use of nanotechnology followed by use of bioremediation in a sequential manner (nanobioremediation) has ensured reclamation of contaminated soil in less time, has demonstrated more efficiency and is environmentally friendly as compared to individual remediation technologies. Work on the techno-feasibility of nanobioremediation by Cecchin et al. was carried out on Brazilian residual clays contaminated with chlorinated organic contaminants (Cecchin et al. 2017). Work carried out thus far reveals the efficacy and sustainability of this technology. The use of different stabilized nanomaterials (nZVI, carbon-based nanoparticles etc.) has led to safe and effective reduction of the contaminants to by-products which on subsequent bioremediation are converted to innocuous forms (Koenig et al. 2016).

Use of genetically engineered microrganisms (GEMs) to biodegrade and biomineralize certain class of persistent contaminants (e.g., alachlor, benthiocarb, propanil, metachlor etc.) was also studied by various researchers as discussed in a review by Paul et al., (2005). Studies have also demonstrated that the newly designed GEMs helped in multiple degradation pathways for target contaminants (Timmis and Pieper 1999; Bohac et al. 2002; Chaloupkova et al. 2003; Okuta 2004).

2.6 Conclusion

Bioremediation is a clean and green technology for reclamation of contaminated soil. The chapter has put forth a variety of bioremediation technologies along with their lab-scale/scale-up applications, their advantages and their limitations. Each technology has its own limitations and advantages. Also, no single technology is appropriate for a particular contaminant type or for a particular site type. Thus, the environmental soil conditions, contaminant type, etc. along with the potential implication of the remediation type on the environment, on the cost effectiveness and on the efficiency ultimately determine the choice of a bioremediation technology. Because a contaminated site may be affected by more than one type of contaminant, and because of the complexity of soil type, more than one bioremediation technology may need to be adopted for effective reclamation of the soil.

Thus, elaborate geological characterization studies of not only the contaminated soil but also characterization studies of the contaminant are vital for effective selection of a bioremediation technology. Cost of treatment is another vital factor affecting the selection process. In comparison to in situ bioremediation, although ex situ bioremediation involves higher costs due to lengthy excavations and transportations to the treatment site, yet they have the potential to treat diverse contaminant type under controlled environmental conditions.

References

Achten C, Hofmann T (2009) Native polycyclic aromatic hydrocarbons (PAH) in coals—a hardly recognized source of environmental contamination. Sci Total Environ 407:2461–3247

Aislabie J, Saul DJ, Foght JM (2006) Bioremediation of hydrocarbon-contaminated polar soils. Extremophiles 10(3):171–179

Awasthi N, Kumar A, Makkar R, Cameotra SS (1999) Biodegradation of soil-applied endosulfan in the presence of a biosurfactant. J Environ Sci Health B 34(5):793–803

Azubuike CC, Chikere CB, Okpokwasili GC (2016) Bioremediation techniques–classification based on site of application: principles, advantages, limitations and prospects. World J Microbiol Biotechnol 32:1–18

Bandowe BAM, Bigalke M, Boamah L, Nyarko E, Saalia FK, Wilcke W (2014) Polycyclic aromatic compounds (PAHs and oxygenated PAHs) and trace metals in fish species from Ghana (West Africa): bioaccumulation and health risk assessment. Environ Int 65:135–146

Barr D, Finnamore JR, Bardos RP, Weeks JM, Nathanail CP (2002) Biological methods for assessment and remediation of contaminated land: case studies. Construction Industry Research and Information Association, London

Bohac M, Nagata Y, Prokop Z, Prokop M, Monincová M, Tsuda M, Koča J, Damborský J (2002) Halide-stabilizing residues of haloalkane dehalogenases studied by quantum mechanics calculations and site directed mutagenesis. Biochemistry 41:14272–14280

Briggs GG, Bromilow RH, Evans AA (1983) Relationships between lipophilicity and the distribution of non-ionised chemicals in barley shoots following uptake by the roots. Pestic Sci 14:492–500

Brito EM, De la Cruz BM, Caretta CA, Goñi-Urriza M, Andrade LH, Cuevas-Rodríguez G, Malm O, Torres JP, Simon M, Guyoneaud R (2015) Impact of hydrocarbons, PCBs and heavy metals on bacterial communities in Lerma River, Salamanca, Mexico: investigation of hydrocarbon degradation potential. Sci Total Environ 521:1–10

Caliman FA, Gavrilescu M (2009) Pharmaceuticals, personal care products and endocrine disrupting agents in the environment—A Review. Clean – Soil Air Water 37:277–303

Cassidy DP, Srivastava VJ, Dombrowski FJ, Lingle JW (2015) Combining in situ chemical oxidation, stabilization, and anaerobic bioremediation in a single application to reduce contaminant mass and leachability in soil. J Hazard Mater 297:347–355

Cecchin I, Reddy KR, Thomé A, Tessaro EF, Schnaid F (2017) Nano-bioremediation: Integration of nanoparticles and bioremediation for sustainable remediation of chlorinated organic contaminants in soils. Int Biodeterior Biodegradation 119:419–428

Chaloupková R, Sýkorová J, Prokop Z, Jesenská A, Monincová M, Pavlová M, Tsuda M, Nagata Y, Damborský J (2003) Modification of activity and specificity of haloalkane dehalogenase from *Sphingomonas paucimobilis* UT26 by engineering of its entrance tunnel. J Biol Chem 278:52622–52628

Chemlal R, Abdi N, Lounici H, Drouiche N, Pauss A, Mameri N (2013) Modeling and qualitative study of diesel biodegradation using biopile process in sandy soil. Int Biodeterior Biodegradation 78:43–48

Chikere CB, Chikere BO, Okpokwasili GC (2012) Bioreactor-based bioremediation of hydrocarbon-polluted Niger Delta marine sediment. Nigeria Biotech 2(1):53–66

Christman RF, Pfaender FK (2006) Molecular implications of hydrophobic organic partitioning theory. Acta Hydrochim Hydrobiol 34:367–374

Clarke BO, Smith SR (2011) Review of emerging organic contaminants in biosolids and assessment of international research priorities for the agricultural use of biosolids. Environ Int 37:226

Coulon F, Al Awadi M, Cowie W, Mardlin D, Pollard S, Cunningham C, Risdon G, Arthur P, Semple KT,

Paton GI (2010) When is a soil remediated? Comparison of biopiled and windrowed soils contaminated with bunker-fuel in a full-scale trial. Environ Pollut 158(10):3032–3040

Crawford RL, Crawford DL (1996) Bioremediation: principles and applications. Cambridge University Press, Cambridge, United Kingdom

Dadrasnia A, Agamuthu P (2013) Diesel fuel degradation from contaminated soil by *Dracaena reflexa* using organic waste supplementation. J Jpn Petrol Inst 56:236–243

Delille D, Duval A, Pelletier E (2008) Highly efficient pilot biopiles for on-site fertilization treatment of diesel oil-contaminated sub-antarctic soil. Cold Reg Sci Technol 54:7–18

Dias RL, Ruberto L, Calabró A, Balbo AL, Del Panno MT, Mac Cormack WP (2015) Hydrocarbon removal and bacterial community structure in on-site biostimulated biopile systems designed for bioremediation of diesel-contaminated Antarctic soil. Polar Biol 38:677–687

Dibble JT, Bartha R (1979) Rehabilitation of oil-inundated agricultural land: a case history. Soil Sci 128:56–60

Duarte RMBO, Matos JTV, Senesi N (2018) Organic Pollutants in Soils. Soil Pollution 103–126

Frutos FJG, Escolano O, García S, Mar Babín M, Fernández MD (2010) Bioventing remediation and ecotoxicity evaluation of phenanthrene-contaminated soil. J Hazard Mater 183:806–813

Fuller ME, Kruczek J, Schuster RL, Sheehan PL, Arienti PM (2003) Bioslurry treatment for soils contaminated with very high concentrations of 2,4,6-trinitrophenylmethylnitramine (tetryl). J Hazard Mater 100:245–257

Gao Y, Yuan X, Lin X, Sun B, Zhao Z (2015) Low-molecular-weight organic acids enhance the release of bound PAH residues in soils. Soil till Res 145:103–110

García-Delgado C, Alfaro-Barta I, Eymar E (2015) Combination of biochar amendment and mycoremediation for polycyclic aromatic hydrocarbons immobilization and biodegradation in creosote-contaminated soil. J Hazard Mater 285:259–266

Gardner WS, Lee RF, Tenore KR, Smith LW (1979) Degradation of selected polycyclic aromatic hydrocarbons in coastal sediments: importance of microbes and polychaete worms. Water Air Soil Pollut 11:339–347

Gidarakos E, Aivalioti M (2007) Large scale and long-term application of bioslurping: the case of a Greek petroleum refinery site. J Hazard Mater 149(3):574–581

Gomez F, Sartaj M (2014) Optimization of field scale biopiles for bioremediation of petroleum hydrocarbon contaminated soil at low temperature conditions by response surface methodology (RSM). Int Biodeterior Biodegradation 89:103–109

Gregorio SD, Azaizeh H, Lorenzi R (2013) Biostimulation of the autochthonous microbial community for the depletion of polychlorinated biphenyls (PCBs) in contaminated sediments. Environ Sci Pollut Res 20:3989–3999

Gregorio SD, Gentini A, Siracusa G, Becarelli S, Azaizeh H, Lorenzi, R (2014) Phytomediated biostimulation of the autochthonous bacterial community for the acceleration of the depletion of polycyclic aromatic hydrocarbons in contaminated sediments. BioMed Res Int 1–11

Griffiths BS, Philippot L (2013) Insights into the resistance and resilience of the soil microbial community. FEMS Microbiol Rev 37(2):112–129

Guerin WF, Boyd SA (1992) Differential availability of soil-sorbed naphthalene to two bacteria. Appl Environ Microbiol 58:1142–1152

Haritash AK, Kaushik CP (2009) Biodegradation aspects of polycyclic aromatic hydrocarbons (PAHs): a review. J Hazard Mater 169:1–15

Harms H (1996) Bacterial growth on distant naphthalene diffusing through water, air and water-saturated and non-saturated porous media. Appl Environ Microbiol 62:2286–2293

Höhener P, Ponsin V (2014) In situ vadose zone bioremediation. Curr Opin Biotechnol 27:1–7

Iori V, Pietrini F, Cheremisina A, Shevyakova NI, Radyukina N, Kuznetsov VV, Zacchini M (2013) Growth responses, metal accumulation and phytoremoval capability in Amaranthus plants exposed to nickel under hydroponics. Water Air Soil Pollut 224:1–10

IPCS, International Programme on Chemical Safety (1995) A review of selected persistent organic pollutants

Karickhoff SW, Brown DS, Scott TA (1979) Sorption of hydrophobic pollutants on natural sediments. Water Res 13:241–248

Khan FI, Husain T, Hejazi R (2004) An overview and analysis of site remediation technologies. J Environ Manage 71:95–122

Koenig JC, Boparai HK, Lee MJ, O'Carroll DM, Barnes RJ, Manefield MJ (2016) Particles and enzymes: combining nanoscale zero valent iron and organochlorine respiring bacteria for the detoxification of chloroethane mixtures. J Hazard Mater 308:106–112

Kuiper I, Lagendijk EL, Bloemberg GV, Lugtenberg BJ (2004) Rhizoremediation: a beneficial plant-microbe interaction. Mol Plant Microbe Interact 17(1):6–15

Lee JH (2013) An overview of phytoremediation as a potentially promising technology for environmental pollution control. Biotechnol Bioprocess Eng 18:431–439

Li F, Zhu L, Wang L, Zhan Y (2015) Gene expression of an arthrobacter in surfactant-enhanced biodegradation of a hydrophobic organic compound. Environ Sci Technol 49:3698–3704

Magalhães SM, Ferreira Jorge RM, Castro PM (2009) Investigations into the application of a combination of bioventing and biotrickling filter technologies for soil decontamination processes-a transition regime

between bioventing and soil vapour extraction. J Hazard Mater 170(2–3):711–715
Mahmoudi N, Slater GF, Fulthorpe RR (2011) Comparison of commercial DNA extraction kits for isolation and purification of bacterial and eukaryotic DNA from PAH-contaminated soils. Can J Microbiol 57:623–628
Mahro B (2000) Bioavailability of contaminants. In: Rehm H-J, Reed G (eds) Biotechnology, vol 11 b, 2nd edn. Wiley-VCH, Weinheim, pp 61–88
Maila MP, Colete TE (2004) Bioremediation of petroleum hydrocarbons through land farming: are simplicity and cost-effectiveness the only advantages? Rev Environ Sci Bio/technol 3:349–360
Martínez-Pascual E, Grotenhuis T, Solanas AM, Viñas M (2015) Coupling chemical oxidation and biostimulation: Effects on the natural attenuation capacity and resilience of the native microbial community in alkylbenzene-polluted soil. J Hazard Mater 300:135–143
Meagher RB (2000) Phytoremediation of toxic elemental and organic pollutants. Curr Opin Plant Biol 3(2):153–162
Mench M, Lepp N, Bert V, Schwitzguébel J-P, Gawronski S, Schröder P, Vangronsveld J (2010) Successes and limitations of phytotechnologies at field scale: outcomes, assessment and outlook from COST Action 859. J Soils Sediments 10:1039–1070
Mihopoulos PG, Suidan MT, Sayles GD (2000) Vapor phase treatment of PCE by lab-scale anaerobic bioventing. Water Res 34:3231–3237
Mirsal IA (2008) Soil pollution – origin, monitoring and remediation. Springer, Berlin, p 312
Mohan SV, Sirisha K, Rao NC, Sarma PN, Reddy SJ (2004) Degradation of chlorpyrifos contaminated soil by bioslurry reactor operated in sequencing batch mode: bioprocess monitoring. J Hazard Mater 116(1–2):39–48
Momose Y, Hirayama K, Itoh K (2008) Competition for proline between indigenous Escherichia coli and E. coli O157:H7 in gnotobiotic mice associated with infant intestinal microbiota and its contribution to the colonization resistance against E. coli O157:H7. Antonie Van Leeuwenhoek 94:165–171
Moreira ITA, Oliveira OMC, Triguis JA, Queiroz AFS, Barbosa RM, Anjos JASA, Reyes CY, Silva CS, Trindade MCLF, Rios MC (2013) Evaluation of the effects of metals on biodegradation of total petroleum hydrocarbons. Microchem J 110:215–220
Morillo E, Sánchez-Trujillo MA, Villaverde J, Madrid F, Undabeytia T (2014) Effect of contact time and the use of hydroxypropyl-β-cyclodextrin in the removal of fluorene and fluoranthene from contaminated soils. Sci Total Environ 496:144–154
Morillo E, Madrid F, Lara-Moreno A, Villaverde J (2020) Soil bioremediation by cyclodextrins. a review. Int J Pharm 591:119943
Mustafa YA, Abdul-Hameed HM, Razak ZA (2015) Biodegradation of 2,4-dichlorophenoxyacetic acid contaminated soil in a roller slurry bioreactor. Clean-Soil Air Water 43:1115–1266
Nam K, Alexander M (1998) Role of nanoporosity and hydrophobicity in sequestration and bioavailability tests with model solids. Environ Sci Technol 32:71–74
Nikolopoulou M, Pasadakis N, Norf H, Kalogerakis N (2013) Enhanced ex situ bioremediation of crude oil contaminated beach sand by supplementation with nutrients and rhamnolipids. Mar Pollut Bull 77(1–2):37–44
Okuta A (2004) Construction of chimeric catechol 2,3-dioxygenase exhibiting improved activity against the suicide inhibitor 4-methylcatechol. Appl Environ Microbiol 70:1804–1810
Paul D, Pandey G, Pandey J, Jain RK (2005) Accessing microbial diversity for bioremediation and environmental restoration. Trends Biotechnol 23(3):135–142
Philp JC, Atlas RM (2005) Bioremediation of contaminated soils and aquifers. In: Atlas RM, Philp JC (eds) Bioremediation: applied microbial solutions for real-world environmental cleanup. American Society for Microbiology (ASM), Washington Press, pp 139–236
Plangklang P, Alissara Reungsang A (2010) Bioaugmentation of carbofuran by *Burkholderia cepacia* PCL3 in a bioslurry phase sequencing batch reactor. Process Chem 45:230–238
Prosser JI (2012) Ecosystem processes and interactions in a morass of diversity. FEMS Microbiol Ecol 81:507–519
Rayner JL, Snape I, Walworth JL, Harvey PM, Ferguson SH (2007) Petroleum–hydrocarbon contamination and remediation by microbioventing at sub-antarctic macquarie Island. Cold Reg Sci Technol 48:139–153
Rhind S (2005) Are endocrine disrupting compounds a threat to farm animal health, welfare and productivity? Reprod Domest Anim 40:282
Rodríguez-Rodríguez CE, Marco-Urrea E, Caminal G (2010) Degradation of naproxen and carbamazepine in spiked sludge by slurry and solid-phase Trametes versicolor systems. Bioresour Technol 101(7):2259–2266
Saravanan V, Rajasimman M, Rajamohan N (2015) Performance of packed bed biofilter during transient operating conditions on removal of xylene vapour. Int J Environ Sci Technol 12:1625–1634
Shah JK, Sayles GD, Suidan MT, Mihopoulos P, Kaskassian S (2001) Anaerobic bioventing of unsaturated zone contaminated with DDT and DNT. Water Sci Technol 43(2):35–42
Sikka SC, Wang R (2008) Endocrine disruptors and estrogenic effects on male reproductive axis. Asian J Androl 10:134
Snow DD, Cassada DA, Larsen ML, Mware NA, Li X, D'Alessio M, Zhang Y, Sallach JB (2017) Detection, occurrence and fate of emerging contaminants in agricultural environments. Water Environ Res 89:897–920
Stroud JL, Paton GI, Semple KT (2007) Microbe-aliphatic hydrocarbon interactions in soil: implications for biodegradation and bioremediation. J Appl Microbiol 102:1239–1253

Sui H, Li X (2011) Modeling for volatilization and bioremediation of toluene-contaminated soil by bioventing. Chin J Chem Eng 19:340–348

Thomé A, Reginatto C, Cecchin I, Colla LM (2014) Bioventing in a residual clayey soil contaminated with a blend of biodiesel and diesel oil. J Environ Eng 140:1–6

Timmis KN, Pieper DH (1999) Bacteria designed for bioremediation. Trends Biotechnol 17:200–204

Tyagi M, da Fonseca MMR, de Carvalho CC (2011) Bioaugmentation and biostimulation strategies to improve the effectiveness of bioremediation processes. Biodegradation 22(2):231–241

van-Camp L, Bujarrabal B, Gentile AR, Jones RJA, Montanarella L, Olazabal C, Selvaradjou SK (2004) Reports of the technical working groups established under the thematic strategy for soil protection, vol 4. Contamination and land management. EUR 21319 EN/4, Office for Official Publications of the European Communities. Luxembourg

Vos JG, Dybing E, Greim HA, Ladefoged O, Lambre′ C, Tarazona JV, Brandt I, Vethaak AD (2000) Health effects of endocrine-disrupting chemicals on wildlife, with special reference to the European situation. Crit Rev Toxicol 30:71–133

Wang X, Wang Q, Wang S, Li F, Guo G (2012) Effect of biostimulation on community level physiological profiles of microorganisms in field-scale biopiles composed of aged oil sludge. Bioresour Technol 11:308–315

Weber R, Schlumpf M, Nakano T, Vijgen J (2015) The need for better management and control of POPs stockpiles. Environ Sci Pollut Res 22:14385–14390

Whelan MJ, Coulon F, Hince G, Rayner J, McWatters R, Spedding T, Snape I (2015) Fate and transport of petroleum hydrocarbons in engineered biopiles in polar regions. Chemosphere 131:232–240

Xu SY, Chen YX, Wu WX, Wang KX, Lin Q, Liang XQ (2006) Enhanced dissipation of phenanthrene and pyrene in spiked soils by combined plants cultivation. Sci Total Environ 363(1–3):206–215

Ye J, Singh A, Ward OP (2004) Biodegradation of nitroaromatics and other nitrogen-containing xenobiotics. World J Microbiol Biotechnol 20:117–135

Zangi-Kotler M, Ben-Dov E, Tiehm A, Kushmaro A (2015) Microbial community structure and dynamics in a membrane bioreactor supplemented with the flame retardant dibromoneopentyl glycol. Environ Sci Pollut Res Int 22(22):17615–17624

3 Phytoremediation of Soils Contaminated with Heavy Metals: Techniques and Strategies

Trinath Biswal

Abstract

The existence of heavy metals in the soil of agricultural lands is a vital environmental concern throughout the world and a threat to the biological community on our globe because of its entry into the food chain and food web. The presence of heavy metals in the soil is associated with a number of important human health hazard risks and causes of degradation of our ecosystem. Although different biological, physical, and chemical technologies with advanced nanotechnology are used for removal of heavy metals from the contaminated soil, but phytoremediation is more popular and potential strategy for removal of heavy metals from the soil in a cost-effective and sustainable way. The traditional technology of phytoremediation for removal of metals is less beneficial and associated with a number of limitations if applied at a larger scale; therefore, there is an urgent need for modification of this technique by application of nanotechnology, advanced chemical, genetic, and biological engineering tools. The hyperaccumulators of plants utilize their internal complex system of extremely effective homeostatic mechanisms in order to prevent the uptake, trafficking, accumulation, and detoxification of metals present in the soil. The process of phytoremediation includes a wide range of techniques, which can be effectively applied for the remediation of soil through stabilization, accumulation, volatilization, and sequestration of toxic heavy metals. Although more than four hundred kinds of plant species have the capability of remediating soils, but among these *Thlaspi*, *Brassica*, *Arabidopsis* and *Sedum alfredii* H. are comparatively more effective. Hence the use phytoremediation in a modified way for the removal of heavy metals from the soil is sustainable, green technology and highly beneficial to our society.

Keywords

Agricultural soils · Contaminated soil · Heavy metals · Nanotechnology · Phytodegradation · Sustainability

3.1 Introduction

The degradation of the quality of the environment and food security is due to excessive accumulation of heavy metals (HMs) because of massive agriculture, industrial and mining activities and is now becoming a serious issue throughout the world. It is the cause of damaging our normal natural ecosystem by disturbing the

T. Biswal (✉)
Department of Chemistry, VSS University of Technology, Burla, 768018, India

J. A. Malik (ed.), *Advances in Bioremediation and Phytoremediation for Sustainable Soil Management*,
https://doi.org/10.1007/978-3-030-89984-4_3

biodiversity. The continuous accumulation of HMs in water bodies and agricultural lands is the cause of potential threat to human and animal health because of their access into the food chain and food web. The HMs normally reaches into the ecosystem via anthropogenic and natural processes. The exposure of HMs for a longer time may even cause death. The various kinds of HMs show different levels of toxicity towards human health (Prieto et al. 2018; Liphadzi et al. 2005). The pollution load on the environment due to heavy metals is drastically increased because of massive industrial development and the concentration of pollutants has been found to be more in industrial locations as compared to the residential areas. The anthropogenic sources of HM contamination include sewage sludge, phosphate fertilizers and emission of pollutants from urban traffic, cement industries, power stations, and metal processing industries. Plants take up HMs from soil via cortical tissues of the roots because of their resemblance with more or less essential micronutrients present in the soil and find itself an appropriate path to reach the xylem transport system (Rizwan et al. 2016; Hasan et al. 2019). The presence of HMs is a major issue towards environmental contamination because it cannot be degraded and persist within the environment for a longer time period. HMs are responsible for inhibiting the growth rate of plants through the reduction in % of chlorophyll and the rate of photosynthesis. Metals may be the cause of water stress in a few plants via reduction in the rate of transpiration, stomatal conductance, and % of leaf relative water because of reduction in the number and dimension of chloroplasts, cell enlargement, and xylem vessels. The contaminated soils can be remediated by adopting different methods such as acid leaching, phytoremediation, soil washing, mechanical or physical separation of the pollutants, pyrometalurgical separation, electrokinetics, chemical treatment, biochemical methods, and electro-chemical treatment (Sall et al. 2020; Rai et al. 2019). The HMs are divided into two kinds, non-essential and essential metals. The essential metals are Cu, Mn, Zn, Ni, Fe, which have a significant role in various biological procedures, including cofactors of many enzymes and transferring electrons in proteins. The non-essential metals have no necessary biological activities, which include Cd, Pb, Hg, As, Sb. Among these different methods, phytoremediation is the cost-effective, environmental friendly, and sustainable method for remediation of HMs from polluted soil. Since plants are the key recipients of HMs, therefore the remediation of xenobiotic metals in the present scenario is found to be more attractive and interesting (Tchounwou et al. 2012; Zwolak et al. 2019; Arif et al. 2016).

3.2 Existence of HMs in Agroecosystems

The metals having atomic number >20 are termed as heavy metals, these are naturally occurring elements having density minimum of 5 times more than water. The toxicity and heaviness both are interrelated with each other, therefore all HMs and the metalloid (arsenic) possesses damaging property on exposure even at low concentrations. The metals like Co, Cr, Mg, Cu, Fe, Mn, Mo, Zn, Se, etc. are considered as essential nutrients, which are necessary for various physiological and biochemical activities and inadequate quantity of it may be the cause of numerous kinds of syndromes or diseases to human beings. On the basis of plant growth HMs are divided into two classes. Metals which are exclusively necessary for the growth of the plants (Cu, Fe, Zn, Ni, and Mn) at low concentrations, whereas at high concentration these metals are hazardous to plants and retard the plant growth, cause soil degradation, lessen % yield, poor quality of the food production along with significant health hazard risk to human beings, animals, and plants, are called essential micronutrients (Srivastava et al. 2017; Kumar et al. 2019). There are some other metalloids and metals (Ag, As, Sb, Cr, Cd, Pb, Hg, Se, etc.) having no biological functions to the plants and animals are referred to as non-essential micronutrients. The quantity of metals added into the ecosystem via anthropogenic sources are gradually increasing and are considered as one of the potential public health problems associated with environmental pollution. The HMs are used in a vast range of applications in our day to day life and easily enter into the environment predominantly via anthropogenic activities mainly mining, smelting, metal processing, agricultural and industrial activities. HMs are considered as a significant class of environmental pollutants which are the cause of alternation in the quality of the soil, composition of the atmosphere, water and pose toxic effects on human beings, plants, and animals. Due to extreme toxicity the introduction of HMs into the ecosystem, food chain, food web at different trophic levels is a threat to existence of the biotic community in the whole world (Kuerban et al. 2020; Jia et al. 2018). The HM contamination in agroecosystems is established both from natural and anthropogenic sources. The natural sources are volcanic eruption, weathering of some metal enriched rocks, and continental dust. The anthropogenic sources are mainly sewage sludge, effluents from agricultural land, combustion, metal containing fertilizers, pesticides, manufacturing units, electronic appliances, etc. (Jin et al. 2019). The soil is mainly contaminated by HMs because of mining, industrial and agricultural activity, and nowadays the contaminated soil is a serious concern to human health because of its bioaccumulation in the food chain and their adverse impact on the ecological system. The existences of both essential and non-essential heavy metals are harmful to both terrestrial and aquatic life (Sul et al. 2014).

3.2.1 Natural Sources

The primary and basic natural sources of HM contamination are the rocks or original material from which soil is produced. The total earth crust consists of about 5% sedimentary rocks and 95% ingenious rocks. The basaltic ingenious rocks mainly contain higher concentrations of HMs such as Cd, Co, Zn, Ni, and Cu, whereas the sedimentary rocks are mainly obtained from the fine sediments of inorganic and organic origin with high concentration of metals such as Zn, Mn, Cd, Cu and Pb. The heavy metals are present in the soil in the combined state in the form of different minerals such as carbonates, oxides, sulphides or their corresponding salts. The concentration of various kinds of metals and their minerals depends upon the type of soil and climatic condition of the region (Jaishankar et al. 2014; Selvi et al. 2019).

3.2.2 Anthropogenic Sources

Presently rapid development in the agriculture, industry, mining and smelting are the major cause of the increase in concentration of HMs in water and soil. Some metal enriched materials, biosolids and synthetic fertilizers play a significant role in the present scenario. Fertilizers used in agricultural farms are particularly rich in Zn metals. The phosphatic fertilizers are also containing heavy metals which are applied to plants to fulfill the necessary micronutrient availability. Different pesticides, fungicides, herbicides and insecticides used for disease management, controlling pests and insects in agricultural fields are the significant sources of many metals such as Cu, Zn, As and Fe, etc. (Zhang et al. 2008; Zhang et al. 2019). The phosphatic fertilizers mainly $Ca_3(PO_4)_2$ and triple super-phosphate contains a high concentration of Cd and Zn which depend upon the source of rock phosphate. It was found that some sources of rock phosphate contain Cd > 50 mg kg^{-1} and due to such a high concentration of Cd, it was banned in most of the countries in the world for agricultural applications (Roberts et al. 2014). The mineral solution containing metals like Zn, Fe, B, Cu, and Mn is sprayed onto plants and crops for protecting diseases, proper growth, and higher % of yield. Not only fertilizers but also some other organic substances, including composts, biosolids, and farm yard manures are also applied in agricultural fields to enhance the fertility of the soil, which is rich in heavy metal concentration. The level of soil contamination with HMs due to irrigation changes on the basis of location which again depends upon the extent of toxicity of the water used for irrigation. Normally industrial wastewaters and sewage water comprise higher quantities of HMs. The frequent use of untreated wastewater or sewage water in agricultural land is the cause of increasing metal concentration in the soil. The wet and dry deposition from various point sources such as the steel industry, metal refineries, foundries, metal smelters, and cement industries offer a great contribution of HMs in the soils. The extensive mining operation is also another cause of HMs in soil basically confined to a small area (Mar et al. 2012; Lwin et al. 2018). The factors affecting uptake mechanism of HMs are represented in Fig. 3.1.

3.3 Phytoremediation Strategies for Heavy Metal Remediation

According to United Nations Environment Programme (UNEP), phytoremediation may be defined as the use of green plants for in-situ elimination, degradation, control of pollutants in soils, groundwater and surface water. It uses green plants along with related microorganisms to reduce the toxic effect of hazardous pollutants from our environment. The term "phytoremediation" is obtained from the Greek word *Phyto* (means plant) and a Latin word *Remedium* (removal or correcting an evil). This method is

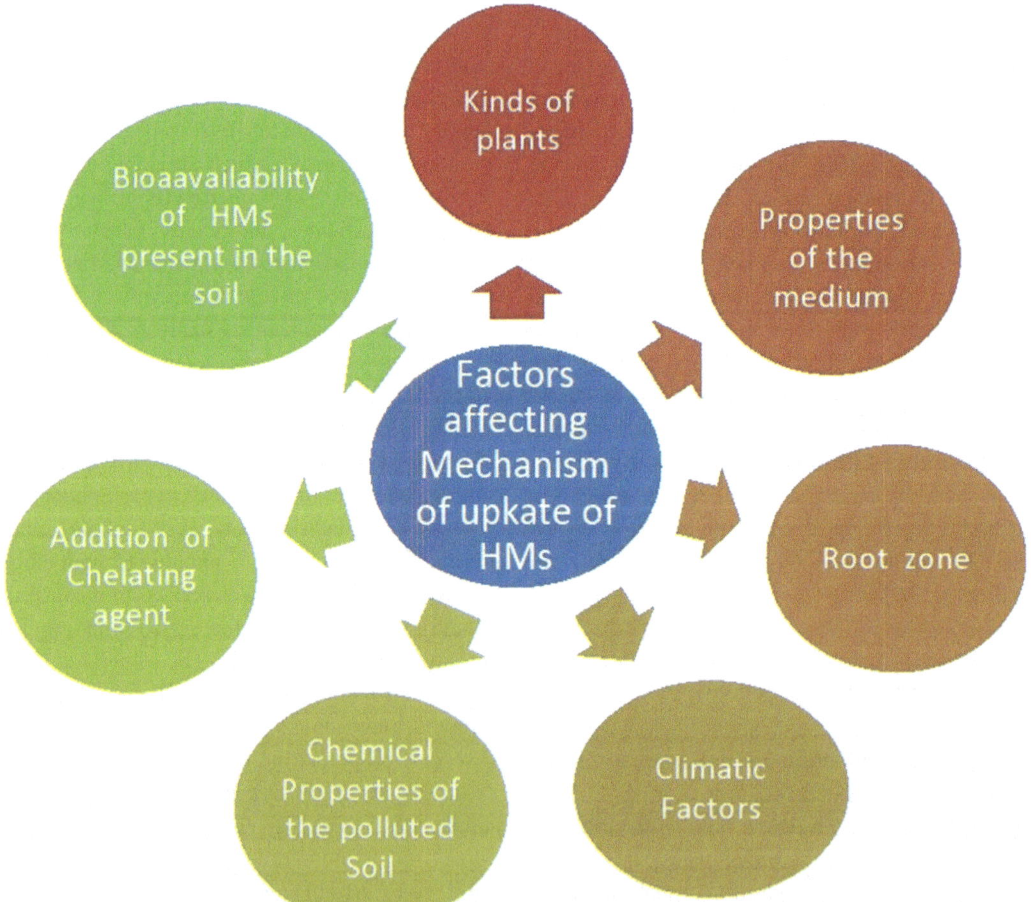

Fig. 3.1 Factors affecting the mechanism of uptake of HMs

successfully applied for remediation of HMs and metalloids from various polluted soils and is a highly effective and economical technique than any other engineering technique such as excavation, soil washing, flushing, soil incineration, solidification, etc. The concept of phytoremediation at first was suggested by Chaney in 1983. The plantation of green plants on polluted soil is cost effective in many ways such as;

- Phytostabilization
- Phytoextraction of valuable metals such as Hg, Ni, and Ag
- Proper land management.

The method of HM remediation was found to be highly popular and widely accepted because it is a sustainable, eco-friendly, and cost-effective process for remediation of many toxic metals and metalloids as compared to chemical and physical methods of remediation (Teng and Chen 2019; Ławniczak et al. 2020; Jawarkar et al. 2010).

3.3.1 Ideal Plants for Phytoremediation

There are many plants tolerable to metal contamination called hyperaccumulating plants which are abundantly found in nature in the soil enriched with metals. But these plant species are not appropriate for ideal phytoremediation because these plants are normally smaller in size and incapable of producing sufficient biomass. Otherwise, the plant species having appropriate growth rate exhibits lower ability of metal accumulation along with high intolerance to HMs. The plant species appropriate for phytoremediation must possess the following properties (Ansari et al. 2020);

- The capability of accumulating targeted HMs for removal must be preferably in the above land surface, nonetheless, the plant species which are unable to translocate HMs above the land surface can be useful for landscape recreation and phytostabilization.
- The plants must possess adequate tolerance capacity for accumulating any concentration of HMs.
- The plants should be able to grow rapidly with production of more biomass of metal accumulating capacity.
- The plants must be easily harvestable in an economical way.

It was estimated that hyperaccumulation and tolerance towards metals is more significant than the phytoremediation assisted by the production of high biomass (Ali et al. 2013; Tariq and Ashraf 2016). The different traditional and modified phytoremediation technique is represented in Fig. 3.2.

3.3.2 Traditional or Conventional Techniques

3.3.2.1 Phytostabilization

Phytostabilization is also termed as phytoimmobilization and is defined as the capability to retard the bioavailability and mobility of the metals either to inhibit its entry into the food web and food chain or leaching into the soil and mixed on the ground water by specified mechanisms such as complexation in the zone area of roots, precipitation, and adsorption by the roots. The immobilization of metals in phytostabilization is accomplished by reducing wind-blown dust, derceasing soil erosion and preventing the bioavailability in the food web, and reducing the solubility of the pollutants. The technique of phytostabilization can be used in the soil of varying concentrations of HMs with soil of different texture such as level of metals, salinity, soil pH, soil moisture, etc. and varies from location to location. Therefore the efficiency of phytostabilization can be increased by proper selection of suitable plant species and amendments applied to the soil of the selected field. Hence two key factors of phytostabilization are plant species and amendments applied (Cristaldi et al. 2017; Alkonta et al. 2010).

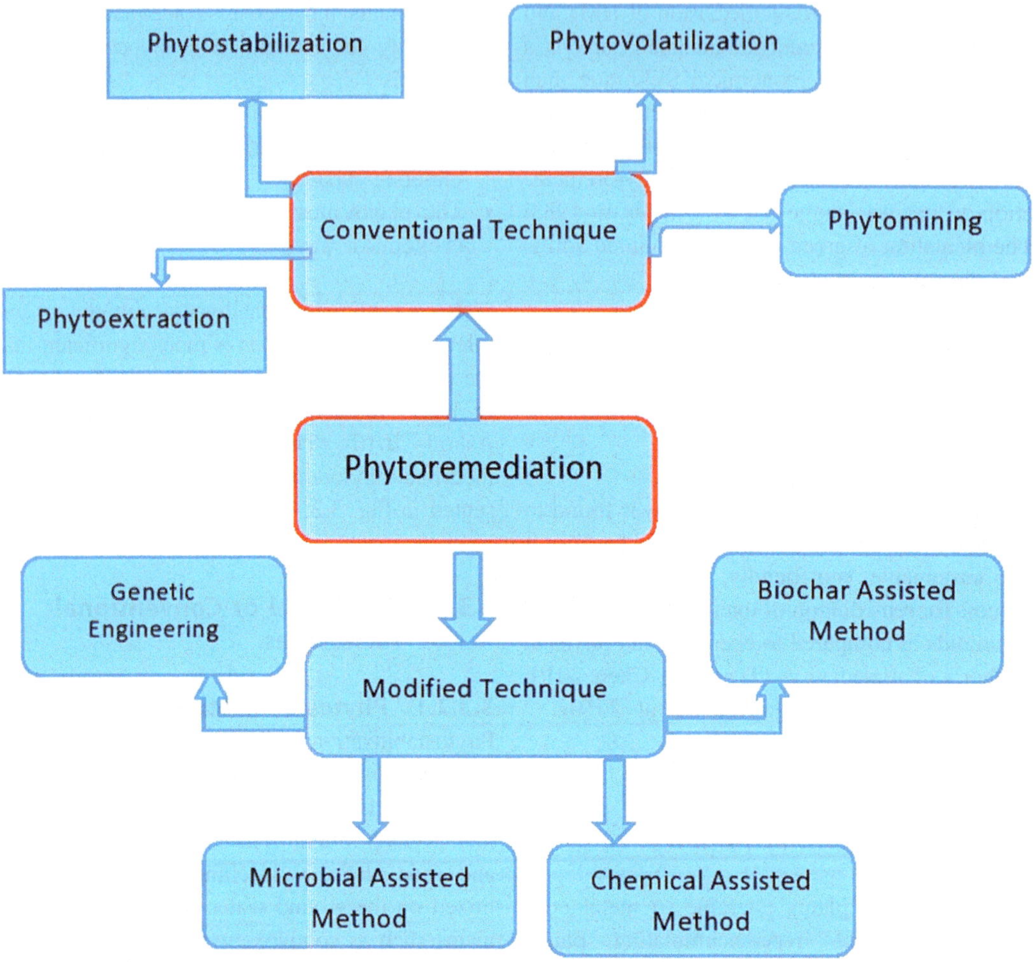

Fig. 3.2 Different methods of phytoremediation of heavy metals

The selection of plant species is a significant factor for phytostabilization. The plant species must be available in the native contaminated area and capable to tolerate environmental stresses such as drought, metals, and salts, which may restrict the accumulation of metals in the shoot. The plant species having bioconcentration factor of >1 with translocation factor <1 are selected as the promising plant species for the technique of phytostabilization. Again, another important criteria for selection of plants are its tolerance capacity for various metalloids and metals existing in the sediments (Cetinkaya et al. 2011). Hence the plant species used for effective phytostabilization of HMs must possess the following properties;

- The used plant species must have to be tolerable to the soil conditions.
- The plant species must grow rapidly and cover the ground surface.
- The root system of the used plants must be dense and strong.
- The plant must establish itself in local climatic condition.
- The used plant species must be having comparatively long life and can propagate itself.

During the process of phytostabilization the plant species can accumulate metals in its root tissue or nearby the rhizosphere and also have the capability of inhibiting the bioavailability or mobility of the metals by stabilizing and accumulating in the plant roots. The higher accumulation of the metals in the roots of the plant species is the cause for increase in the stabilization of metal and decrease in the mobility within the sediment (Radziemska et al. 2017; Zgorelec et al. 2020). Hence the primary mechanism behind phytostabilization mainly depends upon a number of factors such as the presence of microbes in the rhizosphere, binding of metal ions to the cell wall, root exudates, chelation of the metal ions via molecules of metal binding and their ultimate sequestration into the vacuoles. The process of phytostabilization is controlled by various factors of the soil such as pH, soil texture, redox potential, temperature, organic matter, and population of microbes. The process of phytostabilization is carried out by sorption, complexation, precipitation, reduction in valence of the metal (Muthusaravanan et al. 2018).

3.3.2.2 Phytovolatilization

Phytovolatalization is another method where the contaminants (HMs) are converted into volatile compounds and finally released into the earth's atmosphere via minute openings. This technique is particularly successful for Hg because the mercuric ion can be converted into a comparatively less toxic state. Phytovolatilization is the transpiration and acceptance of the pollutants by the plant species with simultaneous release of an improved form of pollutants into the troposphere from the plants. Phytovolatilization is normally carried out when the growing plants and trees consume water in addition to the contaminants. Some of the toxic pollutants reached at the leaves of the plants, through the roots, volatilize into the troposphere at relatively low concentrations.

Every plant and tree plays a crucial secondary role in stabilization of the soil with the help of their root system, protecting the surface of the soil, preventing erosion, and reducing the effect of heavy rain. Simultaneously the root system of the plants release nutrients which assist in sustaining a huge microbial community within the rhizosphere. The composition of the microorganisms in the rhizosphere is completely influenced by the complex interactions between the plant species, kinds of soil, and location of the root zone. The population of the microbes is generally found to be higher in rhizosphere relative to the soil without plant roots. This is due to the interdependent relationship between plants and soil microorganisms which increases the process of bioremediation.

The depth of plant roots plays a vital role in phytovolatilization especially in underground waters; the deep root system is preferred. The most significant aspect of this technique is the conversion of unnecessary toxic materials into less toxic substances, but the release of these less toxic materials to the environment is still a disadvantage. The roots take water along with toxic HMs from soil and supply it to different parts of the plant and finally to the leaves through the vascular system, where the pollutants are released to the atmosphere via volatilization or evaporation (Sarwar et al. 2017; Limmer and Burken 2016).

Example The plant species like *Arabidopsis thaliana* and *Brassica juncea* release HMs to the troposphere through phytovolatilization by the simultaneous transformation and absorption of contaminants into gaseous state. Similarly the plants like *Salix* and *Populus* are frequently used in phytovolatilization due to their capability to carry the harmful pollutants with phytoremediation (Nabi et al. 2019).

3.3.2.3 Phytoextraction

The technique of phytoextraction is considered the most significant phytoremediation technique for remediation of metalloids and metals from the polluted soil, biosolids, sediments, and water. There are various factors that influence the efficiency of phytoextraction which includes metal availability to plants, the properties of soil, speciation of metals, properties of the used plants. The plants chosen for successful phytoextraction must have to possess the following properties;

- The higher rate of growth.
- Production of more biomass products.
- Adequate hyperaccumulation of HMs.
- Widely distributed throughout the other plant species.
- Translocation of metals from the root system to shoot.
- Higher tolerance capacity towards the toxic effect of the metals.
- Higher resistance towards pests and pathogens.
- Must be possessing well adoption capability to usual atmospheric conditions.
- Must have to be easily cultivated and harvested.
- Must possess adequate capability to enter into the food chain through herbivores.

Normally shoot biomass and the concentration of shoot metal predominantly determines the appropriateness of the plants in the process of phytoextraction of metals. On the basis of these parameters, usually, two kinds of phytoextraction process have been used;

1. Use of hyperaccumulator plant species with comparatively less production of biomass.
2. Use of plants with comparable production of higher aboveground biomass along with less accumulation of metals (*Brassica juncea*).

Nevertheless, metal tolerance and hyperaccumulation is more preferred because the plant species have high levels of metal accumulation and less biomass production is easier to dispose for the separation of HMs. It was projected that the use of multi-cut plant species were found to be more effective due to their capability of extracting high concentration of metals as compared to the mono-harvest plant species. Therefore grasses, having high rate of growth, high tolerance level against abiotic stresses, short life cycle, and production of more biomass is normally favored over bushes and trees.

The use of non–hyperaccumulator plants including barley, maize, etc. for conventional phytoextraction needs different cropping seasons for removal of HMs to satisfactory level without adequate risk for contamination of food chain and food web. This technique is utilized for the absorption of inorganic and organic pollutants via roots and branches of plants. It is an appropriate process for retrieval of the polluted areas where the cultivated plant species have the capability of absorbing metals thereby removed from the contaminated soil (Asgari et al. 2019; Suman et al. 2018; Pulford 2003). Table 3.1 represents the plant species used for remediation of different kinds of heavy metals.

3.3.2.4 Phytomining

The technique of phytomining is the manipulation of sub-economic ore by using plant tissues. The method of phytomining is the use of plant species which have the capability of accumulating metals and metalloids in above ground parts. The plant species used for phytomining are called as hyperaccumulators and their ability in accumulating metals is commonly named hyperaccumulation. The bioharvesting of HMs from the crops of high biomass grown soil is specifically associated with the sub-economic mineralization and is termed phytomining. The technique of phytomining has the capability of recovering precious product materials from less valuable resource materials such as mining waste, industrial wastes, contaminated soil, etc.

Example Indian corn and mustard are successfully used to extract gold from an oxidized ore pile in induced hyperaccumulation (after thiocynate and cyanide treatment).

Phytomining is an innovative technology of phytoremediation which is used for production of less volume sulphide free bio-ore, if the targeted metals have to be recovered with appropriate profitable value, smelted, and recovered. The technology of phytomining is basically applied in the mineral processing industry for marketable production of metals through cropping of a variety of plants in an economic way. There are many areas of our globe which are enriched with precious metals and can potentially be phytomined. The plant species *Brassica juncea* (Indian mustard) can accumulate Au in the range of 0.6–39 mg/kg. Ni is also considered as a potential trace metal which can be successfully recovered through phytomining (Sheoran et al.

Table 3.1 Plant species used for phytoremediation of HMs from soil (Sumiahadi and Acar 2018)

Plant species	Heavy metals remediated
Allium schoenoprasum L. (Chive)	Ni, Cd, Co
Zea mays L. (corn)	Cd, Zn, Pb, Cu
Brassica juncea L. (Indian mustard)	Cd, Zn, Pb, Cu
Sorghum bicolor L. (sorghum)	Cd, Zn, Cu, Fe
Brassica napus L. (canola)	Cd, Zn, Cu, Pb
Solanum nigrum L. (black nightshade)	Cd
Cajanus Cajan (L.) Milsp. (pigeon pea)	As, Cd
Spinacia oleracea L. (spinach)	Cd, Cu, Fe, Ni, Zn, Pb, Cr
Cicer aeritinum L. (chickpea)	Cd, Cr, Pb, Cu
Rapanus sativus L. (radish)	As, Cd, Pb, Fe, Cu
Jatropha curcas L. (purging nut, physic nut)	Fe, Cu, Al, As, Mn, Zn, Cr, Hg
Pisum sativum L. (pea)	Pb, Cu, Fe, Zn,, Ni, Cd, As, Cr
Lantana camara L. (lantana)	Pb
Oryza sativa L. (rice)	Cu, Cd
Lens culinaris Medic. (lentil)	Pb
Medicago sativa L. (alfalfa)	Cd
Lepidium sativum L. (cress)	As, Cd, Pb, Fe, Hg
Lactuca sativa L. (lettuce)	Cu, Ni, Fe, Pb, Mn, Zn, Co, Cd, As

2009; Wang et al. 2019). The phytomining possesses following unique features;

- It provides the opportunity of exploiting mineralized soils or ores that can be cost effective and sustainable in the modified technology.
- 'Bio-ores' are almost free from sulphur and smelting of these kinds of ores needed less energy as compared to sulphide ores.
- The bio-ore usually contains higher % of metals than normal ores and requires less volume and space, regardless of lesser density of a bio-ore.
- Phytomining is considered as a green technology for recovering targeted metals from lower grade ore obtained from open-cast mining.

Till date phytomining technique has very limited applications because from the economic point of view the price of the metal obtained is higher than actual cost. The technique of phytomining is normally more attractive on the basis of the economic point of view if combined with phytoremediation and production of biofuel (Chaney and Baklanov 2017; Li et al. 2019, 2020).

3.3.3 Modified Techniques for Phytoremediation of HMs

For remediation of HMs from the polluted soil by using green plant species of various kinds, phytoremediation is a promising approach for minimization of metals and makes the soil fit for agricultural use. Although it is an attractive technology, still the conventional technique of phytoremediation possesses many demerits. The naturally growing hyperaccumulator species of plants either grow slowly with subsequent production of less above ground plant biomass or may not be adapted well to varying climatic conditions. Hence the modification of the technology is needed (Li and Luo 2019).

3.3.3.1 Limitations of Conventional Phytoremediation Technique

The different limitations faced by the traditional technique of phytoremediation are as follows.

- This process needed more time for remediation of polluted soil.
- The ability of phytoextraction by hyperaccumulator plant species is restricted because of the production of less above ground biomass.
- Only a small % of metals present in the soil are bioavailable and the concentration of bioavailability changes with the change in different soil parameters such as soil pH, competitive cations, calcareous materials, organic matter, etc.
- This technique is only suitable for an area having moderate or low contamination.
- It does not provide knowledge regarding agronomy, insect pests, disease spectrum and breeding potential.
- In this technique, the inaccuracy or mismanagement may be the cause of contamination of the food chain and food web.

These unfavorable disadvantages forced the researchers to develop and modify the conventional process of phytoremediation to overcome these demerits and confirm the technique for huge application for the removal of metals from the soil (Yaqoob et al. 2019; Sharma et al. 2018).

3.3.3.2 Chemical Assisted Phytoremediation with Non–hyperaccumulator Plants

The selection of appropriate plant species for the process of phytoremediation of HMs is the most important factor. Usually, the plant species, which have high capability of accumulating HMs is assumed to be suitable for phytoremediation and are commonly termed as hyperaccumulator plants. The ratio of the metal concentration of shoot to root is treated as the criteria to select the plants for hyper accumulation. If the ratio is >1.0, then the plant species specifies that metals normally accumulated more in shoot than in roots. This kind of plant species is called hyperaccumulator and is appropriate for phytoextraction. The concentration of HMs in aerial portions of a hyperaccumulator plant changes from 1000 to 10,000 mg/ kg depending upon the toxicity of the metals which is less than 100 mg/ kg in the case of highly hazardous metal like Cd.

The major demerits of the hyperaccumulator plants are very low growth rate and production of less aboveground biomass. Because of these demerits, the plants of this kind are not feasible for the method of phytoremediation. Hence, in order to overcome these demerits genetically modified (GM) plants were developed which have the capability of accumulating metals of high concentration and production of more above ground biomass.

The non–hyperaccumulator plant species have the capability of creating more above ground biomass are considered as suitable plants for potential phytoremediation. These kinds of plants remediate comparatively less amount of HMs, therefore for increasing the extraction ability, the organic or synthetic chelating agents are added and this method is called induced phytoextraction or chelate–assisted phytoextraction. There are various synthetic chelating agents, including ethylene diamine tetra–acetic acid (EDTA), ethylene glycol tetra–acetic acid (AGTA), and diethylene triamine penta–acetic acid (DTPA) which are effectively used to increase the bioavailability of metals and thus accepted by plants. But the major disadvantages of this process are high mobility of HMs which is a great risk towards environmental contamination. The uptake capability of chelated HMs is the major cause of increase in toxicity of the plants. Hence, further research work is needed to improve the induced phytoextraction by using different soil parameters such as temperature, redox potential, soil pH, organic matter, inter species struggle, status of soil fertility and plant morphology, which potentially influences the availability of metals (Souza et al. 2013; Zhuang et al. 2005).

Although the metal chelating agents have high potential to lead even stumpy accumulator plants for uptaking the excessive metals via enhancing the bioavailability of metals in soil, but such kind of plants by producing high biomass product with less accumulator capacity are genetically less tolerant towards the toxicity of metals. Mitigation of toxicity is obligatory for adopting such kind of innovative and new techniques of phytoremediation for their inexpensive nature and is capable of maximum removal of heavy metals from the polluted area (Dipu et al. 2012).

Salicylic acid (SA) is a phenolic compound and is treated as a famous signaling molecule in plants under abiotic and biotic stress situations. The use of SA in the pretreatment process of rice seeds at 0.1 mM for 24 h potentially enhances the parameters of seed germination and also seedling growth in the soil containing high concentration of Cd. The exogenous use of SA potentially improves the negative impact of Cd on proline levels, chlorophyll content and leaf growth comparable to the water content of maize. Therefore the pretreatment of soil by SA associated with chemical supported phytoremediation may improve the effectiveness of toxicity of HMs on metal extractor plants leading to the enhancement in biomass generation and significant removal of HMs from the soil for potential growth of plants (Isah 2019).

The most critical demerits of this process are the high cost of the synthetic chelating agents, because if phytoextraction is carried out on a large scale, it becomes highly expensive. Besides these costly chemicals, the low molecular weight organic acids like acetic acid, oxalic acid, citric acid, and malic acid can also be successfully used as chelating agents for extraction of HMs. These organic acid forms complexes with the metals present in the soil and can be easily removed. Other benefits of using these simple organic acids are their easy biodegradability nature in soil as compared to the synthetic chelating agents and impact least risk of environmental contamination (Koptsik 2014).

3.3.3.3 Genetic Engineering for Phytoremediation

The removal of HMs from soil through genetic engineering is a promising method for improving the capability of phytoremediation of the plant species. The GM plants are normally created by introducing the suitable genes from foreign sources/organisms, mainly plant species, animals or bacteria into the genome of the target plant species. After the recombination of DNA, foreign improved gene develops hereditary traits and confers explicit characters to the plant species. The genetic engineering is more beneficial than conventional breeding to modify the plant species with required characters for phytoremediation in a less time interval. Furthermore, the use of genetic engineering is the cause of transferring required genes from hyperaccumulator to the sexually unsuited plant species, which is almost impossible to attain via traditional breeding processes such as crossing. Hence application of genetic engineering for developing transgenic plants with necessary properties exhibits attractive prospects towards the field of phytoremediation. The technically developing plant species with high biomass, rapid growth to archive the high capability of accumulating HMs with high tolerance capacity is comparatively more suitable compared to engineering hyperaccumulators to acquire production of high biomass (Koźmińska et al. 2018; Ozyigit et al. 2020). Therefore, in most cases the high biomass, fast-growing plants are developed either to enhance the accumulation capability of HMs or increase the tolerance level against the HMs, which are treated as the vital characteristics of hyperaccumulators. Hence the gene selection for genetic engineering is based upon the knowledge of the mechanisms of accumulation and tolerance level of HMs by the plants.

This method is based on the over expression of definite genes that are involved in uptake, sequestration, translocation and tolerance of plant species towards xenobiotic substances in transgenic plants. The introduction of definite genes from plants, microbes or animals can be attained by using either direct DNA methods of transfer of genes or *Agrobacterium tumefaciens* –mediated transformation in order to develop/ synthesize transgenic plants (Fasani et al. 2017; Kawahigashi 2009).

Example The transgenic plant (*Arabidopsis thaliana*) possessing over–expression of gene might be accountable for expressing of Hg^{2+} ion reductase to enhance the tolerance level of Hg. Another plant, *Nicotiana tabaccum* with a yeast metallothionein expressing gene possesses high tolerance capacity against Cd and was developed first time to remediate metals present in the contaminated soil.

Hence transgenic plants are usually developed/ synthesized either to enhance the tolerance level of the plants against heavy metals or to increase immobilization, which facilitates the more accumulation and translocation of HMs in the plant biomass in their aboveground parts. The transport and metal uptake through the plasma membrane is normally the function of the transport proteins and greater affinity for the binding sites. The transgenic plants may be developed with a modified transport mechanism with higher levels of tolerance to sequester a high % of metals within the cellular organelles having decreased metabolic activities such as vacuoles (Yadab 2010).

Example The heavy metal (Arsenic) synthesized complexes with glutathione (–GSH) and phytochelatins (PCs) in vacuoles.

The transgenic plant species (*Thlaspi caerulescens*) is industrialized by applying bacterial gene ArsC from *E. coli*. The decrease of arsenate is generally considered a significant detoxification mechanism for As. The bacterial gene merA is accountable for encoding Hg^{2+} ion reductase and merB encoding organo–mercuial lyase within the transgenic plants and tolerance against Hg. Therefore transgenic plants having good tolerance capability and ability towards metal sequestration can be effectively used for phytoremediation of soil contaminated with HMs. However, in order to develop and synthesize transgenic hyperaccumulator plants the knowledge regarding detoxification mechanisms and tolerance level of metal is very significant (Maestri and Marmiroli 2011). The various steps of mechanism of phytoremediation are shown in Fig. 3.3.

The HMs are responsible for the generation of excessive generation of refractory organic substances and cause of oxidative stress, therefore the tolerance level of HMs is normally established via the strength of oxidative stress protection system. The most popular strategy for enhancing tolerance level of HMs is to increase the antioxidant activity which can be attained via overexpression of genes associated with antioxidant machinery. To enhance the accumulation of HMs by genetic engineering, the most popular strategy is use of overexpress genes which are associated with uptake, sequestration and translocation of HMs (Seth 2011). Therefore the gene encoding HM transporters might be overexpressed and transferred in the target plants to enhance the accumulation of HMs. Since the metal chelators function as metal binding ligands to enhance the bioavailability of HMs, stimulating the uptake capacity of metals, translocation through root-to-shoot and facilitate the intracellular sequestration of HM ions in organelles. Therefore, it is treated as an encouraging strategy to accelerate the accumulation of HMs by stimulating the formation of metal chelators through genetic engineering.

Another difficulty faced to develop GM plants is to get authorization for field testing in some regions of the world because of risk towards ecosystem safety and food chain. Hence, alternative procedure has to be developed to increase the performance of the plant species for phytoextraction, if genetic engineering becomes unviable (Yan et al. 2020). Consequently, it was suggested that the use of outside genes in plant species is an economically, ecologically, and environmentally viable process for phytoremediation of soil contaminated by HMs. However, there are some limitations of this process such as low biomass production especially above the earth's surface, low efficiency for phytoextraction of the used plant species.

Besides these conventional methods of phytoremediation for the removal of HMs from the soil, there are another two popular methods of phytoremediation termed rhizoremediation and phytodegradation, which are used for remediation of complex organic matter. Figure 3.4 shows the positive impact of introduction of genes into the plant species for phytoremediation of soil (Jabben et al. 2009; Emamverdian et al. 2015).

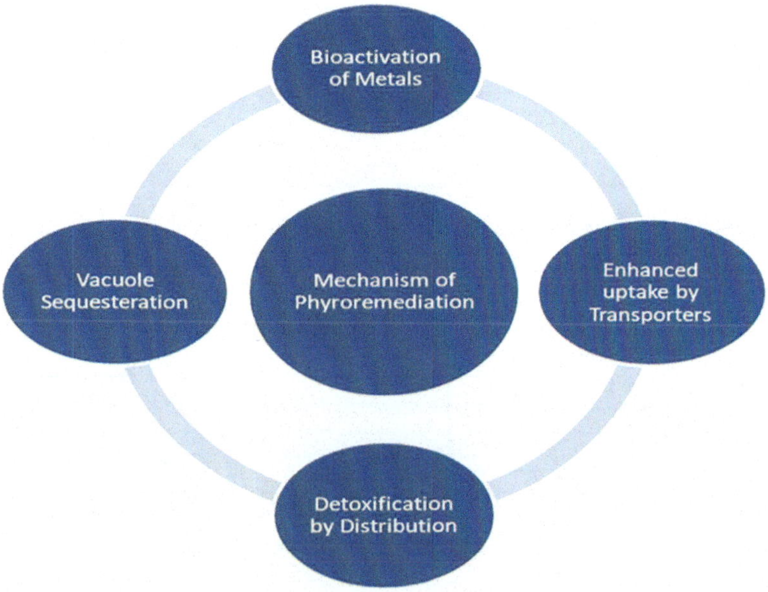

Fig. 3.3 Various steps of mechanism of phytoremediation

3.3.3.4 Biochar Assisted Method of Phytoremediation

Biochar is a carbonaceous, porous material synthesized due to the pyrolysis of organic compounds. The wood biochar (charcoal) is a popular and widely available biochar used since from long before. Biochar possesses some of its own chemical and physical properties such as high pH, alkaline nature, huge surface area for sorption of most of the metals, presence of ash, a higher ratio of carbon, and capability to immobilize toxic HMs. Therefore, it can be used as a successful remediator for various HMs.

Naturally biochars possess a huge surface area accessible for sorption which promotes the formation of HM-biochar complexes either through the exchange of cations of HMs or cations of some other metals such as Na^+, Ca^{2+}, K^+, Mg^{2+} with the different functional groups existing in biochar or through physical adsorption. The high alkalinity and pH of biochar might be the cause of reduction in the bioavailability of the metals and enhances the rate of precipitation in the soil on modification with biochar process (Evangelou et al. 2014; Paz-Ferreiro et al. 2013).

Biochar possesses many significant advantages such as enhancement of biological activity of the soil, decreasing emissions of greenhouse gas from various agricultural residues with subsequent increase in carbon sequestration in soil owing to its large % of recalcitrant carbon. It was found that the pH of biochar enhances with the pyrolysis temperature probably because of ash content in biochar. Biochar, if combined with a conventional phytoremediation method, the effectiveness of the remediation of HMs increases considerably as biochar enhances due to increase in biomass production and plant growth up to 10%. The increase in production of plant biomass products is credited to water holding capacity, cation exchange capacity, high pH of biochar and high nutrient availability, which influences nutrient cycling and increases nutrient turnover of the plant species. Biochar also affects the microbial community present in the soil, possibly favor towards useful microorganisms with suppressing pathogens (not useful). Biochar improves the % of yield and the rate of growth of the plant species and in combination with hyperaccumulators, EDTA and non–accumulators, can effectively remediate the toxic HMs present in the soil (Wu et al. 2020; Khalid et al. 2017).

Mechanism of Interaction Between Heavy Metals and Biochar

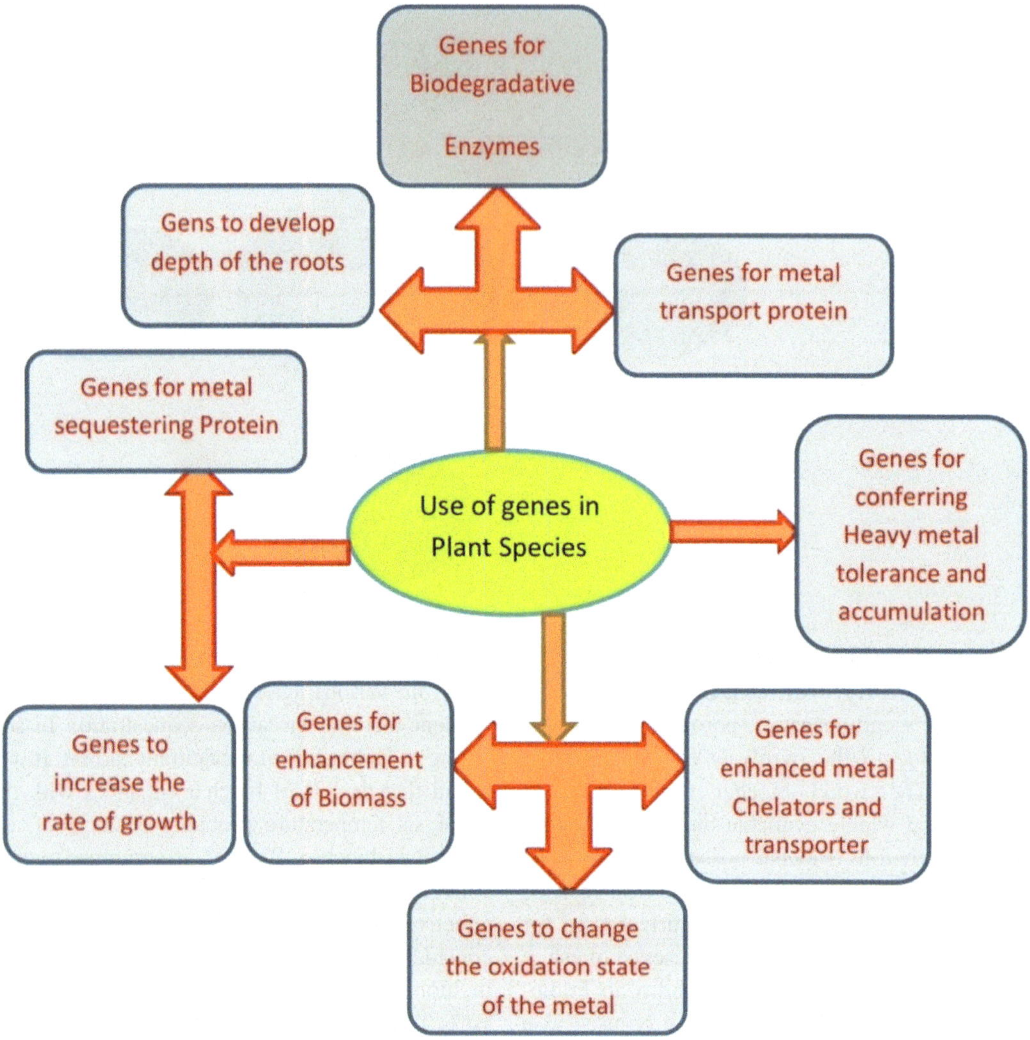

Fig. 3.4 Impact of introduction of genes into the plant species for improving phytoremediation

The properties of biochar are a function of many factors such as kinds of feedstock, temperature, the size of the particles in the feedstock and pyrolysis condition. The wide range of properties of biochar of some specific compounds makes it a successful candidate material for remediation of HMs than other processes. Therefore, for phytoremediation by using biochar, we have to be aware of not only the properties and kinds of soil, but also the properties of the biochar used in this process. There are some vital properties of the biochar mainly surface area, % of ash, pH and % of carbon which enhances the capability of biochar for immobilization of HMs. Figure 3.5 indicates the various functions of biochar in phytoremediation.

The biochar functions as the fraction of bioavailability of HMs and cause of reduction of their leachability. Another important property of biochar is large surface area, which facilitates the sorption of HM complexes on their surface. The sorption is owing to the formation of complexes with various functional groups existing in the biochar, which exchanges HMs with the cations related to biochar (Ca^{+2}, K^{+}, Na^{+}, Mg^{+2} and S) and partly because of physical adsorption. The

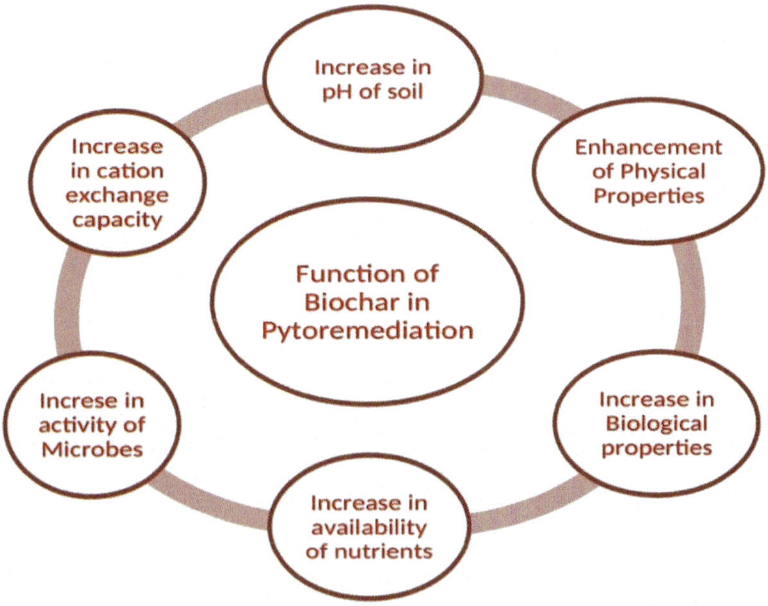

Fig. 3.5 The various functions of biochar in phytoremediation

functional groups containing O_2 can stabilize the HMs in the surface of biochar specifically for softer acids such as Cu^{+2} and Pb^{+2}. The sorption of Cu^{+2} was associated with the higher oxygenated surface groups along with more average pore diameter, enhanced superficial charge density, and the high exchange capacity of Mg^{+2} and Ca^{+2} of biochar. Probably the mechanism based on sorption significantly depends upon the presence of cations on the surface of both soil and biochar. The other components such as CO_3^{2-}, SO_4^{2-} and PO_4^{3-} also support the HM stabilization through the precipitation of these compounds with the contaminants (Gorovtsov et al. 2019; Sohi et al. 2010).

The alkaline property of the biochar can be partly accountable for the less % of HMs available in biochar amended soils. The high value of pH after the addition of biochar may be causing the precipitation of HMs in the soil. The pH value of the biochar enhances with the rise in pyrolysis temperature and related to the higher content of ash. The biochar can be able to decrease the mobility of HMs with changes in the redox state of the metal.

Example The addition of biochar can change Cr^{+6} states to the comparatively less mobile Cr^{+3}. The relative impacts of the various mechanisms on the immobilization of HMs through various kinds of biochar are still unknown.

The sorption of metals significantly varies with the variation in pH of the solution and provides significant impact to both charges of the biochar surface and speciation of metal. The alternation of the pH in soil potentially influences complexation properties of the functional groups such as carboxyl, amino, and hydroxyl groups.

Example The effective ionization property of the carboxyl group is in the pH range of 3–4, but the increase in pH is the cause of deprotonation of the carboxyl group which reduces the capability of complex formation with positively charged metal ions. If the pH rises from 3 to 7, the straw biochars surface becomes more negatively charged because of increase in the rate of deprotonation of the functional groups.

The five different mechanisms such as complexation, precipitation, electrostatic interactions, reduction, and cation exchange are considered effective for phytoremediation governing metal sorption through biochar from aqueous solutions. Figure 3.6 shows the five different mechanisms of metal sorption by biochar.

The role of every mechanism plays a specific role depending upon the kinds of metal and varies noticeably depending on the target metals.

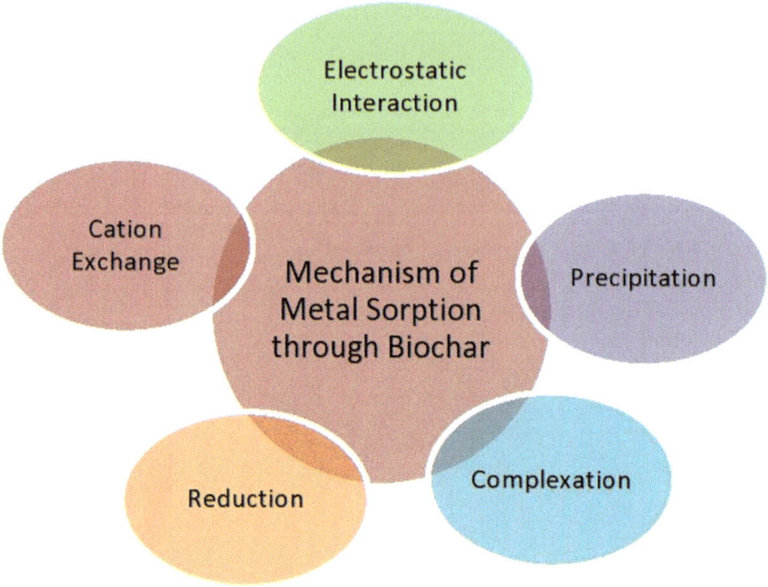

Fig. 3.6 Five mechanisms of metal sorption through biochar

Recently modified biochar is used to increase the capability of metal sorption by increasing the porosity, surface area, pH and enhancing the reactivity of functional groups (Tomczyk et al. 2020; Vijayaraghavan 2019).

3.3.3.5 Phytoremediation Assisted by Microbial Community

Plants associated with the microbes present in the soil significantly affect the availability of metals and its uptake by plants within the rhizosphere. Some of the microorganisms indecently exist, whereas some others require plant roots as support for their existence and growth. Mycorrhizal fungi are considered as the key constituents of living organisms in the zone of the root system and survive as interlinked with many of the higher plants in various forms such as ectomycorrhizas, ericaceous mycorrhizas, arbuscular mycorrhizas and orchid mycorrhizas along with arbuscular mycorrhizal fungi in coordination with root systems of terrestrial plants, which are more popular and widespread.

The coordination in between plant roots and fungi is advantageous for plants in various ways, such as plant nutrients, including N, P, Ca, K, S, Co, Zn, Ni, and Cu availability. These fungal species with plant roots are able to modify and develop composition of the root exudates, soil pH and accumulation in *Kummerowia striata*, *Echinochloa crusgalli* and *lxeris denticulate*, if added with arbuscular mycorrhizal fungi inoculums (Yang et al. 2020; Xu et al. 2020).

The plant growth stimulating bacteria are treated as another significant community of microorganisms that are supportive to plants for remediation HMs from the soil. These kinds of bacteria may enhance the growth of the plant species by different mechanisms such as nitrogen fixation, C_2H_2 creation under stress and explicit enzyme activity. These classes of microorganisms are able to alleviate the HM toxicity from plant species.

Example The inoculation of *Pseudomonas puteda* on application to *Brassica napus* is known to alleviate Cu toxicity.

Hence the use of appropriate microbial inoculum might support the plants for remediation of HMs from soil successfully (Mourato et al. 2015).

Mechanisms of Microbial Remediation

It was found that various kinds of microorganisms are used for treatment of polluted soil to degrade the complex organic substances into less toxic simple compounds. Specifically, the microbial degradation essentially means the degradation of PAHs and HMs through microbes and the microbial species effective for this purpose are *Bacillus*, *Mycobacterium* and *Escherichia*. *Mycobacterium* present in the natural soil exhibits strong capability to degrade the most dangerous organic compound PAHs. *Streptomycetaceae* obtained from PAH-contaminated soils can be able to remove 98.25% of diesel, 17.5% of phenanthrene and 99.14% of naphthalene within one week. In comparison to that some fungi can be able to degrade whole PAHs.

Example *Basidiomycetes*

, *Deuteromycetes* and white-rot fungi are able to degrade PAHs having four or more rings (Liu et al. 2017).

It was observed that usually fungi exhibit higher resistance to the degradation of HMs than actinomycetes and bacterial species. The filamentous kinds of fungi (*Gibberella*, *Phellinus*, *Aureobasidium* and *Saccharomyces*) are highly resistant to HM ions and prevent their absorption in significant amounts. The processes and mechanism behind the remediation of HMs by the microbial community includes absorption, precipitation and oxidation-reduction, however the decontamination of toxic complex organic materials are little known and involve different enzymatic pathways for transformation.

Normally the interactions between the various kinds of contaminants, climatic factors and activity of microbial community determine the effectiveness of phytoremediation assisted by the microbial community. The cations of metals bind to the negatively charged surfaces of microbial cells by the process of ionic exchange. The adsorption or binding of metal ions generally occurs outside or inside the microbial cells with the support of intracellular adsorption, which facilitates through binding to proteins. The complexation of HM ions to the extracellular polymeric substances (EPS) mainly includes nucleotides, carbohydrates, and proteins, which are able to measure the effectiveness of the extracellular adsorption technique, however, various biomolecules are found to be taking part in the complexation. Yet again it was recommended that peptidoglycan, chitin, and phosphoryl are the basic binding molecules responsible for metal complexation within the gram-positive bacteria, fungi and gram-negative bacteria (Jin et al. 2018; Malla et al. 2018).

Oxidation-reduction is treated as another potential mechanism of remediation assisted by the microbial community along with the enzymatic transformations by altering the oxidation state of the polyvalent metal cations to comparatively less toxic metal ions and other species. The popular microorganisms effective towards the change in oxidation state of the metal ions are *Arthrobacter*, *Bacillus*, *Phanerochaete*, *Saccharomyces*, *Pseudomonas*, *Cymodocea* and *Rhizopus* (Gill et al. 2014).

Example The bacterial strain changes selenate compounds into the colloidal form of Se and Pb (II) to Pb(I), therefore reduces the toxicity of Se and Pb. Similarly the toxic metals such as As (III), Fe(II), and Se(IV) are reduced or oxidized during detoxification.

Such kind of technique is carried out through the mechanism of indirect reductive metal precipitation, such as Cr(VI), which is being reduced by indirect oxidation of Fe^{2+} ion and S^{2-} ion. Lastly, some microbial community generate organic acids and amino acids which dissolve the metallic substances and their solution form and transform or translocate the metallic species through leaching, reduction and alkylation methods (Gadd 2004). In all these cases the chemistry of the metals changes with subsequent reduction of toxicity of these metals.

Example *Clostridium* can synthesize the organic acids such as butyric, lactic, and acetic acid which are able to dissolve the oxides of the metals Cu, Mn, Zn, Fe, and Cr, however the mechanism of interaction between the HMs and organic materials is still not clear (Gheju 2018).

3.4 Rhizoremediation

Rhizoremediation is very useful to degrade the toxic organic contaminants present in the soils. Rhizoremediation is a technique which disintegrates the contaminants present in the soil in the rhizosphere. The organic pollutants present in the soil are remediated through this method which cannot enter inside the plant species because of their extremely high hydrophobicity. Plants are not treated as a major approach of remediation for the pollutants by this method, but plant forms a niche for rhizosphere microbes for degradation of soil pollutants.

The microbe's selection for rhizosphere is very useful for remediation of PAHs and other toxic organic materials present in the soil. The transgenic plants can increase the activity of rhizosphere microbial communities and stimulate the efficiency of remediation. Rhizospheric microbes are successful for depredating majority of the pollutants present in the soil and the process of degradation stops when the microorganisms are deprived of the soil environment. The population of microbes is responsible for degradation of different kinds of complex organic contaminants such as polycyclic aromatic hydrocarbons (PAHs), lindane, total petroleum hydrocarbons (TPHs) and polychlorinated biphenyls (PCBs). To explain the process of rhizodegradation, also commonly called phytostimulation, different possible mechanisms are accountable for stimulating the microorganisms such as elevated oxygen level, habitat opportunity for microbes, and increase in availability of soil organic carbon (Saravanan et al. 2019; Yaashikaa et al. 2020).

Example The microorganism *Withania somnifera* is used for stimulating the degradation of lindane. Although the use of this technology may reduce the growth of the plants, the concentration of lindane in the soil is still decreased by 73% (from 20 to 5.38 mg/kg). The increase in microbial biomass carbon (MBC) specifies that the stimulated microbes generally are accountable for remediation of lindane (Boudh et al. 2017).

The presence of higher % of metal in the soils and their consumption by plant species are hazardous and provide negative impacts to symbiosis, growth, and consequently the % of yields of the crops through the degradation of cell organelles, disrupting the cell membranes, serving as genotoxic materials which disrupt the physiological methods such as photosynthesis, protein synthesis, deactivating respiration and carbohydrate metabolism.

The microorganism *Pseudomonas putida* is a potential root colonizer for carrying out rhizoremediation of contaminants with biological control and management of pests. If appropriate rhizosphere strain can be inoculated with a proper plant species (coated microorganism on the bacteria on plant seed), the microorganisms are settled on the root system of the plants along with the usual indigenous population of microbes stimulating the rate of bioremediation.

The land surface polluted soil having long periods of aging is normally found to be much less effective towards rhizodegradation as compared with newly spiked soil. It was concluded that less aging is one of the major reasons for the failure of rhizodegradation in the polluted field and aged spiky soils. This is a significant consequence for the usability of rhizodegradation along with evaluation of data measured on freshly or recently aged, spiked soil. There is another strategy for increasing the rate of rhizodegradation is inoculation of the degrader strains, which is normally not successful, where less bioavailability is the main limitation.

The controversial and unsatisfactory outcomes of conventional inoculation are the cause of increasing the requirement of rhizodegradation in a more sophisticated way. The effectiveness of the increase in the rate of degradation of inoculated microbes can be formed through initiation of nutritional favoritism in the direction of the inoculated strains. An effective rhizoremediation method mainly depends upon the extreme branched root system present, where the huge microbial population is found, secondary and primary metabolism can be established, the existence and ecological relations with other organisms. Therefore the root system of the plant species serve as an alternative for the digging of soil for incorporation of nutrients and for improving/ enhancing aeration in the soil.

Plant releases a number of photo-synthetically derived root exudates, which supports the degradation of contaminants. The root exudates contain many water insoluble, soluble and volatile substances such as sugars, proteins, alcohols, amino acids, nucleotides, phenolic compounds, organic acids, flavonones and some other specific enzymes (Boudh et al. 2017; Monoj et al. 2020). There is an interdependent relationship established in between microbes of plants and soil in the rhizosphere, in which plant delivers required nutrients to flourish the microorganisms, whereas the microbial community offers healthier soil conditions for proper growth of plant roots. Particularly plant species release soil and supplies H_2O and O_2 to the rhizosphere. Again the plant species provide some particular phytochemicals such as alcohols, sugars, carbohydrates, etc., which serve as a basic food source for some essential soil microbes that are responsible for creating healthier soil. Further the generated phytochemical may function as an allelopathic agent and suppresses the growth of other plants in the same soil and the required plant species are protected from the interference of other plants, toxin substances, soil pathogens, and some unwanted chemicals, which are originally present in the soil.

Rhizodegradation may be termed as rhizosphere biodegradation, phytostimulation, or plant assisted degradation/bioremediation, which can increase the rate of degradation of the pollutants through enhancement of bioactivity by utilizing the environment of the plant rhizosphere in order to promote the increase in population of microbial communities (Barea et al. 2005; Khan et al. 2005; Mishra et al. 2017).

3.5 Phytodegradation

Phytodegradation, also popularly called as phytotransformation which is the detoxification and breaking of toxic organic contaminants mainly due to plant metabolism, without the support of microorganisms. During this method the enzymes released by the plant species play a crucial role in the process of decomposition. Hence phytodegradation is also the process of degradation of organic contaminants that are generated by the plant species through metabolic activity (Lee 2013).

Example The metabolic activity of ibuprofen from plant species of wetland ecosystems such as cytochrome P450 monooxygenase and *Phragmites australis* on combination shows the property of catalysts for the conversion of ibuprofen. Hence the plant species are treated as the heart of the pedosphere (He et al. 2017).

Trinitrotoluene (TNT) is treated as the most harmful and persistent explosive worldwide. The presence of this dangerous explosive in soil and its remediation is highly difficult and expensive. It was found that there are a number of plants which are capable of disintegrating TNT, however, this technique tremendously influences the development and growth of plant species. This is the cause of prevention of this technique in large scale applications.

Entereo cloaca is a kind of soil bacterium, the enzymes synthesized by this bacterium species such as nitroreductase and PETN reductase are able to break TNT into comparatively less toxic products. The genes expressing synthesis of these two enzymes are presented into the tobacco plant named as *Nicotiana tabacum*.

In an experiment carried out by using a wild variety of plant and toxic plant; both are exposed to less concentration (0.25 mM) of TNT, it was found that the wild variety of plant species lose mass and become chlorotic, whereas the transgenic plant species grow comfortably. Hence the enzymes which are exposed to the transgenic plants supports the metabolization of TNT at a quicker rate as compared to the control plants. Hence transgenic plant species generally are highly resistant than non-transformed plant species which are strongly influenced in their modification and development (Zhu et al. 2012; Singh et al. 2013).

The phytodegradation process can be potentially used in the remediation of soil, underground waters, clay, and sediment. The most beneficial aspect of this process is that the degradation and reduction are generally carried out inside the plant species in physiological method and never depends upon the population and activity of microbial communities. The disadvantages of this process are detection of the appearance of intermediate toxic products and end products. The effectiveness of absorption of the toxic organic substance of the plant species depends upon the kinds of plants, residence time of the polluted metal in the soil, chemical and physical properties of the soil. The substances, which are easily dissolved in water, are challenging to absorb. The enzymes produced by the plants can capably disintegrate hazardous materials such as munitions wastes, herbicides, pesticides, chlorinated solvents mainly trichloroethane ($C_2H_3Cl_3$) (Laghlimi et al. 2015; Kumar et al. 2017).

3.6 Benefits and Limitations of Phytoremediation

In addition to several advantages of phytoremediation, there are also some limitations which restrict the application of this technique. The different benefits and demerits of phytoremediation are presented in Table 3.2.

3.7 Conclusion

The contamination due to the presence of HMs in agricultural soils is presently a significant environmental issue worldwide because of its quick accumulation, causing potential health hazard risks owing to the contamination of the food chain and food web. Hence, in order to remediate HM contamination and again make the land suitable for agricultural use, different kinds of techniques are investigated, however the phytoremediation technique proved as a significant tool to overcome this problem in a sustainable, environmental friendly, and cost-effective manner as compared to other different chemical and physical methods. The hyperaccumulator plant species can be efficiently used to extract and remove huge concentrations of harmful metals along with the other harmful organic and

Table 3.2 Benefits and limitations of phytoremediation (Paulo et al. 2014)

Benefits	Demerits
Passive and In-situ technique	Restricted to shallow soils or where the pollution is limited to the surface (<5 m)
The use of plant species can be easily controlled and managed compared to microorganisms	Bioavailability and toxicity of the products obtained after degradation are unidentified
Low cost and solar energy is used to make it a green technology process	This method is still under the undeveloped stage and therefore not recognized by most of the regulatory agencies
The biomass obtained can be sustainable and economically valuable	The location which has to be decontaminated must be in the large surface area to permit the application of farming techniques
The negative impact towards the environment is reduced and potentially contributes to the development of the landscape	There are few facts about genetics, farming, diseases and reproduction of phytoremediation plants
The collection of the organs or plants that have accumulated HMs can be easy to achieve with current technology	The plants discharge substances to enhance the mobility of the HMs present in the soil and, finally leached into groundwater
This technique was highly accepted by the people	The presence of high concentration of HMs in the soil might be toxic and deadly to plant species
Inhibits mobilization and leaching of the pollutants present in the soil	The plants selected for efficient phytoremediation may not adapt to environmental and climatic conditions at polluted locations
Provides habitat for animal life	Generally, plants are selective in metal remediation
Reduction of surface runoff	Contamination may spread through the food chain, if accumulator plants are ingested by animals
Reduction in dispersal of dust and contaminants by wind	Treatment rate is slower than the traditional physico-chemical techniques

inorganic pollutants present in the soils. Nevertheless, the conventional methods of phytoremediation are economically not viable to apply on large scales because hyperaccumulators found naturally are usually having low growth rate and synthesize comparatively less biomass product above the earth surface. The use of HM hyperaccumulators was found to be the most straightforward method relating to phytoremediation, and around hundreds of hyperaccumulator plant species are so far recognized. The phytoremediation associated with natural hyperaccumulators is a time consuming method and is still found to have many drawbacks. Therefore genetic engineering technique was developed with transgenic plants, which have the capability of producing high biomass above the ground surface, comparatively higher rate of HM accumulation, and higher tolerance capability against the toxicity of metals, can be well adapted to varying climatic conditions and might be sustainable, economical, and beneficial in many respects. There are some advanced phytoremediation techniques such as microbial assisted phytoremediation and chemical assisted phytoextraction which is now being used to remove HMs from contaminated soils on a large scale basis. In spite of so many advantages still, further research is needed to modify the field of genetic engineering to develop phytoremediation capability of transgenic plant species and to get knowledge regarding the effectiveness and mechanisms of phytoremediation to make this technology more efficient, economical, environmental friendly, sustainable and feasible.

References

Ali H, Khan E, Sajad MA (2013) Phytoremediation of heavy metals—Concepts and applications. Chemosphere 91(7):869–881

Alkorta I, Becerril JM, Garbisu C (2010) Phytostabilization of metal montaminated soils. Rev Environ Health 25(2):135–146

Ansari AA, Naeem M, Gill SS, AlZuaibr FM (2020) Phytoremediation of contaminated waters: an eco-friendly technology based on aquatic macrophytes application. Egyptian J Aquatic Res, in Press. https://doi.org/10.1016/j.ejar.2020.03.002

Arif N, Yadav V, Singh S, Singh S, Ahmad P, Mishra RK, Sharma S, Tripathi DK, Dubey NK, Chauhan DK (2016) Influence of high and low levels of plant-beneficial heavy metal ions on plant growth and development. Front Environ Sci 4:69–79

Asgari Lajayer B, Khadem Moghadam N, Maghsoodi MR (2019) Phytoextraction of heavy metals from contaminated soil, water and atmosphere using ornamental plants: mechanisms and efficiency improvement strategies. Environ Sci Pollut Res 26:8468–8484. https://doi.org/10.1007/s11356-019-04241-y

Barea JM, Pozo MJ, Azcón R, Azcón-Aguilar C (2005) Microbial co-operation in the rhizosphere. J Exp Bot 56(417):1761–1778

Boudh S, Tiwari S, Singh JS (2017) Microbial-mediated lindane bioremediation. Agro-Environ Sustainability 213–233. https://doi.org/10.1007/978-3-319-49727-3_11

Cetinkaya G, Sozen N (2011) Plant species potentially useful in the phytostabilization process for the abandoned CMC mining site in northern cyprus. Int J Phytorem 13(7):681–691. https://doi.org/10.1080/15226514.2010.500155

Chaney RL (1983) Plant Uptake of Inorganic Waste Constituents. In: Parr JF, Marsh PB, Kla JM (eds) Land treatment of hazardous wastes. Noyes Data Corporation, Park Ridge, pp 50–76

Chaney RL, Baklanov IA (2017) Phytoremediation and phytomining. Adv Botanical Res 189–221. https://doi.org/10.1016/bs.abr.2016.12.006

Cristaldi A, Conti GO, Jho EH, Zuccarello P, Grasso A, Copat C, Ferrante M (2017) Phytoremediation of contaminated soils by heavy metals and PAHs: A brief review. Environ Technol Innov 8:309–326. https://doi.org/10.1016/j.eti.2017.08.002

Dipu S, Kumar AA, Thanga SG (2012) Effect of chelating agents in phytoremediation of heavy metals. Remediat J 22(2):133–146. https://doi.org/10.1002/rem.21304

Emamverdian A, Ding Y, Mokhberdoran F, Xie Y (2015) Heavy metal stress and some mechanisms of plant defense response. Sci World J 2015:1–18. https://doi.org/10.1155/2015/756120

Evangelou MWH, Fellet G, Ji R, Schulin R (2014) Phytoremediation and biochar application as an amendment. Phytoremediation 253–263. https://doi.org/10.1007/978-3-319-10395-2_17

Fasani E, Manara A, Martini F, Furini A, DalCorso G (2017) The potential of genetic engineering of plants for the remediation of soils contaminated with heavy metals. Plant, Cell Environ 41(5):1201–1232. https://doi.org/10.1111/pce.12963

Gadd GM (2004) Microbial influence on metal mobility and application for bioremediation. Geoderma 122(2–4):109–119. https://doi.org/10.1016/j.geoderma.2004.01.002

Gheju M (2018) Progress in Understanding the Mechanism of CrVI removal in Fe0-based filtration systems water 10(5):651. https://doi.org/10.3390/w10050651

Gill RT, Harbottle MJ, Smith JWN, Thornton SF (2014) Electrokinetic-enhanced bioremediation of organic contaminants: a review of processes and environmental applications. Chemosphere 107:31–42. https://doi.org/10.1016/j.chemosphere.2014.03.019

Gorovtsov AV, Minkina TM, Mandzhieva SS, Perelomov LV, Soja G, Zamulina IV, Yao J (2019) The mechanisms of biochar interactions with microorganisms in soil. Environ Geochem Health 1–24.https://doi.org/10.1007/s10653-019-00412-5

Hasan MM, Uddin MN, Ara-Sharmeen I, Alharby FH, Alzahrani Y, Hakeem KR, Zhang L (2019) Assisting phytoremediation of heavy metals using chemical amendments. Plants 8(9):295–309. https://doi.org/10.3390/plants8090295

He Y, Langenhoff AAM, Sutton NB, Rijnaarts HHM, Blokland MH, Chen F, Schröder P (2017) Metabolism of Ibuprofen by *Phragmites australis*: Uptake and Phytodegradation. Environ Sci Technol 51(8):4576–4584. https://doi.org/10.1021/acs.est.7b00458

Isah T (2019) Stress and defense responses in plant secondary metabolites production. Biol Res 52(1):39–64. https://doi.org/10.1186/s40659-019-0246-3

Jabeen R, Ahmad A, Iqbal M (2009) Phytoremediation of Heavy Metals: Physiological and Molecular Mechanisms. Bot Rev 75:339–364. https://doi.org/10.1007/s12229-009-9036-x

Jaishankar M, Tseten T, Anbalagan N, Mathew BB, Beeregowda KN (2014) Toxicity, mechanism and health effects of some heavy metals. Interdiscip Toxicol 7(2):60–72. https://doi.org/10.2478/intox-2014-0009

Jia Z, Li S, Wang L (2018) Assessment of soil heavy metals for eco-environment and human health in a rapidly urbanization area of the upper Yangtze Basin. Sci Rep 8(1):3256–3270. https://doi.org/10.1038/s41598-018-21569-6

Jin Y, Luan Y, Ning Y, Wang L (2018) Effects and mechanisms of microbial remediation of heavy metals in soil: a critical review. Appl Sci 8(8):1336–1352. https://doi.org/10.3390/app8081336

Jin Y, O'Connor D, Ok YS, Tsang DCW, Liu A, Hou D (2019) Assessment of sources of heavy metals in soil and dust at children's playgrounds in Beijing using GIS and multivariate statistical analysis. Environ Int 124:320–328. https://doi.org/10.1016/j.envint.2019.01.024

Juwarkar AA, Singh SK, Mudhoo AA (2010) Comprehensive overview of elements in bioremediation. Rev Environ Sci Biotechnol 9:215–288. https://doi.org/10.1007/s11157-010-9215-6

Kawahigashi H (2009) Transgenic plants for phytoremediation of herbicides. Curr Opin Biotechnol 20 (2):225–230. https://doi.org/10.1016/j.copbio.2009.01.010

Khalid S, Shahid M, Niazi NK, Murtaza B, Bibi I, Dumat C (2017) A comparison of technologies for remediation of heavy metal contaminated soils. J Geochem Explor 182:247–268. https://doi.org/10.1016/j.gexplo.2016.11.021

Khan AG (2005) Role of soil microbes in the rhizospheres of plants growing on trace metal contaminated soils in phytoremediation. J Trace Elem Med Biol 18(4):355–364. https://doi.org/10.1016/j.jtemb.2005.02.006

Koptsik GN (2014) Problems and prospects concerning the phytoremediation of heavy metal polluted soils: a review. Eurasian Soil Sc 47:923–939

Koźmińska A, Wiszniewska A, Hanus-Fajerska E (2018) Recent strategies of increasing metal tolerance and phytoremediation potential using genetic transformation of plants. Plant Biotechnol Rep 12:1–14. https://doi.org/10.1007/s11816-017-0467-2

Kuerban M, Maihemuti B, Waili Y, Tuerhong T (2020) Ecological risk assessment and source identification of heavy metal pollution in vegetable bases of Urumqi, China, using the positive matrix factorization (PMF) method. PLoS ONE 15(4):1–19. https://doi.org/10.1371/journal.pone.0230191

Kumar B, Smita K, Cumbal Flores L (2017) Plant mediated detoxification of mercury and lead. Arab J Chem 10:S2335–S2342

Kumar V, Pandita S, Sharma A, Bakshi P, Sharma P, Karaouzas I, Cerda A (2019) Ecological and human health risks appraisal of metal(loid)s in agricultural soils: a review. Geology, Ecol Landscapes 1–13. https://doi.org/10.1080/24749508.2019.1701310

Laghlimi M, Baghdad B, El Hadi H, Bouabdli A (2015) Phytoremediation mechanisms of heavy metal contaminated soils: a review. Open J Ecol 5:375–388

Ławniczak Ł, zniak-Karczewska MW, Loibner AP, Heipieper HJ, Chrzanowski Ł (2020) Microbial degradation of hydrocarbons—basic principles for bioremediation: a review. Molecules 25(4):856–874

Lee JH (2013) An overview of phytoremediation as a potentially promising technology for environmental pollution control. Biotechnol Bioproc E 18:431–439. https://doi.org/10.1007/s12257-013-0193-8

Li C, Ji X, Luo X (2019) Phytoremediation of heavy metal pollution: a bibliometric and scientometric analysis from 1989 to 2018. Int J Environ Res Public Health 16(23):4755. https://doi.org/10.3390/ijerph16234755

Li C, Ji X, Luo X (2020) Visualizing Hotspots and future trends in phytomining research through scientometrics. Sustainability 12(11):4593–4609

Limmer M, Burken J (2016) Phytovolatilization of organic contaminants. Environ Sci Technol 50 (13):6632–6643. https://doi.org/10.1021/acs.est.5b04113

Liphadzi MS, Kirkham MB, Musil CF (2005) Phytoremediation of soil contaminated with heavy metals: a technology for rehabilitation of the environment. S Afr J Bot 71(1):24–37. https://doi.org/10.1016/s0254

Liu SH, Zeng GM, Niu QY, Liu Y, Zhou L, Jiang LH, Cheng M (2017) Bioremediation mechanisms of combined pollution of PAHs and heavy metals by bacteria and fungi: a mini review. Biores Technol 224:25–33. https://doi.org/10.1016/j.biortech.2016.11.095

Lwin CS, Seo BH, Kim HU, Owens G, Kim KR (2018) Application of soil amendments to contaminated soils for heavy metal immobilization and improved soil quality—a critical review. Soil Sci Plant Nutrit 64 (2):156–167. https://doi.org/10.1080/00380768.2018.1440938

Maestri E, Marmiroli N (2011) Transgenic plants for phytoremediation. Int J Phytorem 13(supl):264–279. https://doi.org/10.1080/15226514.2011.568549

Malla MA, Dubey A, Yadav S, Kumar A, Hashem A, Abd_Allah EF (2018) Understanding and designing the strategies for the microbe-mediated remediation of environmental contaminants using omics approaches. Front Microbiol 9:1132–1149. https://doi.org/10.3389/fmicb.2018.01132

Manoj SR, Karthik C, Kadirvelu K, Arulselvi PI, Shanmugasundaram T, Bruno B, Rajkumar M (2020) Understanding the molecular mechanisms for the enhanced phytoremediation of heavy metals through plant growth promoting rhizobacteria: a review. J Environ Manage 254:1–14

Mar SS, Okazaki M, Motobayashi T (2012) The influence of phosphate fertilizer application levels and cultivars on cadmium uptake by Komatsuna (*Brassica rapa* L. var. perviridis). Soil Sci Plant Nutrit 58(4):492–502. https://doi.org/10.1080/00380768.2012.704394

Mishra J, Singh R, Arora NK (2017) Alleviation of heavy metal stress in plants and remediation of soil by rhizosphere microorganisms. Front Microbiol 8:1706–1713. https://doi.org/10.3389/fmicb.2017.01706

Mourato M, Moreira I, Leitão I, Pinto F, Sales J, Martins L (2015) Effect of heavy metals in plants of the genus brassica. Int J Mol Sci 16(8):17975–17998. https://doi.org/10.3390/ijms160817975

Muthusaravanan S, Sivarajasekar N, Vivek JS, Paramasivan T, Naushad M, Prakashmaran J, Al-Duaij OK (2018) Phytoremediation of heavy metals: mechanisms, methods and enhancements. Environ Chem Lett 1–25. https://doi.org/10.1007/s10311-018-0762-3

Nabi RBS, Tayade R, Hussain A, Kulkarni KP, Imran QM, Mun BG, Yun BW (2019) Nitric oxide regulates plant responses to drought, salinity, and heavy metal stress. Environ Exp Bot 161:120–133

Ozyigit II, Can H, Dogan I (2020) Phytoremediation using genetically engineered plants to remove metals: a review. Environ Chem Lett 1–30.https://doi.org/10.1007/s10311-020-01095-6

Paulo JC, Pratas J, Varun M, DSouza R, Poul SM (2014) Phytoremediation of soils contaminated with metals and metalloids at mining areas: potential of native flora. In: Hernandez Soriano MC (ed) Environmental risk assessment of soil contamination. IntechOpen Limited, London, UK, pp 485–511

Paz-Ferreiro J, Lu H, Fu S, Méndez A, Gascó G (2013) Use of phytoremediation and biochar to remediate heavy metal polluted soils: a review. Solid Earth Discuss 5(2):2155–2179. https://doi.org/10.5194/sed-5-2155-2013

Prieto MJ, Acevedo SOA, Prieto GF (2018) Phytoremediation of soils contaminated with heavy metals. Biodiversity Int J 2(4):362–376. https://doi.org/10.15406/bij.2018.02.00088

Pulford I (2003) Phytoremediation of heavy metal-contaminated land by trees—a review. Environ Int 29(4):529–540. https://doi.org/10.1016/s0160-4120(02)00152-6

Radziemska M, Koda E, Bilgin A, Vaverková M (2017) Concept of Aided Phytostabilization of Contaminated Soils in Postindustrial Areas. Int J Environ Res Public Health 15(1):24–39. https://doi.org/10.3390/ijerph15010024

Rai PK, Lee SS, Zhang M, Tsang YF, Kim KH (2019) Heavy metals in food crops: Health risks, fate, mechanisms, and management. Environ Int 125:365–385. https://doi.org/10.1016/j.envint.2019.01.067

Rizwan M, Ali S, Rizvi H, Rinklebe J, Tsang DCW, Meers E, Ok YS, Ishaque W (2016) Phytomanagement of heavy metals in contaminated soils using sunflower: a review. Crit Rev Environ Sci Technol 46 (18):1498–1528. https://doi.org/10.1080/10643389.2016.1248199

Roberts TL (2014) Cadmium and phosphorous fertilizers: the issues and the science. Proc Eng 83:52–59. https://doi.org/10.1016/j.proeng.2014.09.012

Sall ML, Diaw AKD, Gningue-Sall D (2020) Toxic heavy metals: impact on the environment and human health, and treatment with conducting organic polymers, a review. Environ Sci Pollut Res 27:29927–29942. https://doi.org/10.1007/s11356-020-09354-3

Saravanan A, Jeevanantham S, Narayanan VA, Kumar PS, Yaashikaa P, Muthu CM (2019) Rhizoremediation—a promising tool for the removal of soil contaminants: a review. J Environ Chem Eng 8(2):1–41

Sarwar N, Imran M, Shaheen MR, Ishaque W, Kamran MA, Matloob A, Hussain S (2017) Phytoremediation strategies for soils contaminated with heavy metals: modifications and future perspectives. Chemosphere 171:710–721

Selvi A, Rajasekar A, Theerthagiri J, Ananthaselvam A, Sathishkumar K, Madhavan J, Rahman PKSM (2019) Integrated remediation processes toward heavy metal removal/recovery from various environments-a review. Front Environ Sci 7:66–81. https://doi.org/10.3389/fenvs.2019.00066

Seth CS (2011) A review on mechanisms of plant tolerance and role of transgenic plants in environmental clean-up. Bot Rev 78(1):32–62. https://doi.org/10.1007/s12229-011-9092-x

Sharma S, Tiwari S, Hasan A, Saxena V, Pandey LM (2018) Recent advances in conventional and contemporary methods for remediation of heavy metal-contaminated soils. 3 Biotech 8(4):216–233. https://doi.org/10.1007/s13205-018-1237-8

Sheoran V, Sheoran AS, Poonia P (2009) Phytomining: a review. Miner Eng 22(12):1007–1019. https://doi.org/10.1016/j.mineng.2009.04.001

Singh SN, Mishra S (2013) Phytoremediation of TNT and RDX. Biological Remediation of Explosive Residues 371–392. https://doi.org/10.1007/978-3-319-01083-0_16

Sohi SP, Krull E, Lopez-Capel E, Bol R (2010) A review of biochar and its use and function in soil. Adv Agronomy 47–82.https://doi.org/10.1016/s0065-2113(10)05002-9
Souza LA, Piotto FA, Nogueirol RC, Azevedo RA (2013) Use of non-hyperaccumulator plant species for the phytoextraction of heavy metals using chelating agents. Scientia Agricola 70(4):290–295. https://doi.org/10.1590/s0103-90162013000400010
Srivastava V, Sarkar A, Singh S, Singh P, de Araujo ASF, Singh RP (2017) Agroecological responses of heavy metal pollution with special emphasis on soil health and plant performances. Front Environ Sci 5:64–82. https://doi.org/10.3389/fenvs.2017.00064
Su C, Jiang L, Zhang WJ (2014) A review on heavy metal contamination in the soil worldwide: Situation, impact and remediation techniques. Environ Skeptics Critics 3(2):24–38
Suman J, Uhlik O, Viktorova J, Macek T (2018) Phytoextraction of heavy metals: a promising tool for clean-up of polluted environment? Front Plant Sci 9:1476–1490. https://doi.org/10.3389/fpls.2018.01476
Sumiahadi A, Acar R (2018) A review of phytoremediation technology: heavy metals uptake by plants. IOP Conf Ser: Earth Environ Sci 142–151:012023. https://doi.org/10.1088/1755-1315/142/1/012023
Tariq SR, Ashraf A (2016) Comparative evaluation of phytoremediation of metal contaminated soil of firing range by four different plant species. Arab J Chem 9(6):806–814. https://doi.org/10.1016/j.arabjc.2013.09.024
Tchounwou PB, Yedjou CG, Patlolla AK, Sutton DJ (2012) Heavy metal toxicity and the environment. Molecular, Clin Environ Toxicol 133–164https://doi.org/10.1007/978-3-7643-8340-4_6
Teng Y, Chen W (2019) Soil microbiomes—a promising strategy for contaminated soil remediation: a review. Pedosphere 29(3):283–297. https://doi.org/10.1016/s1002-0160(18)60061-x
Tomczyk A, Sokołowska Z, Boguta P (2020) Biochar physicochemical properties: pyrolysis temperature and feedstock kind effects. Rev Environ Sci Biotechnol 19:191–215. https://doi.org/10.1007/s11157-020-09523-3
Vijayaraghavan K (2019) Recent advancements in biochar preparation, feedstocks, modification, characterization and future applications. Environ Technol Rev 8(1):47–64. https://doi.org/10.1080/21622515.2019.1631393
Wang L, Hou D, Shen Z, Zhu J, Jia X, Ok YS, Rinklebe J (2019) Field trials of phytomining and phytoremediation: A critical review of influencing factors and effects of additives. Critical Rev Environ Sci Technol 1–51. https://doi.org/10.1080/10643389.2019.1705724
Wu P, Wang Z, Wang H (2020) Visualizing the emerging trends of biochar research and applications in 2019: a scientometric analysis and review. Biochar 2:135–150. https://doi.org/10.1007/s42773-020-00055-1
Xu Z, Wang D, Tang W, Wang L, Li Q, Lu Z, Guo S (2020) Phytoremediation of cadmium-polluted soil assisted by D-gluconate-enhanced *Enterobacter cloacae* colonization in the *Solanum nigrum* L. rhizosphere. Sci Total Environ 732:1–12. https://doi.org/10.1016/j.scitotenv.2020.139265
Yaashikaa PR, Senthil Kumar P, Varjani S, Saravanan A (2020) Rhizoremediation of Cu(II) ions from contaminated soil using plant growth promoting bacteria: an outlook on pyrolysis conditions on plant residues for methylene orange dye biosorption. Bioengineered 11(1):175–187. https://doi.org/10.1080/21655979.2020.1728034
Yadav SK (2010) Heavy metals toxicity in plants: an overview on the role of glutathione and phytochelatins in heavy metal stress tolerance of plants. S Afr J Bot 76(2):167–179. https://doi.org/10.1016/j.sajb.2009.10.007
Yan A, Wang Y, Tan SN, Mohd Yusof ML, Ghosh S, Chen Z (2020) Phytoremediation: a promising approach for revegetation of heavy metal-polluted land. Front Plant Sci 11:359–373. https://doi.org/10.3389/fpls.2020.00359
Yang Y, Liu Y, Li Z, Wang Z, Li C, Wei H (2020) Significance of soil microbe in microbial-assisted phytoremediation: an effective way to enhance phytoremediation of contaminated soil. Int J Environ Sci Technol 17:2477–2484. https://doi.org/10.1007/s13762-020-02668-2
Yaqoob A, Hassan NF, Sumreen A, Munawar N, Zia MA, Choudhary MS, Ashraf M (2019) Current scenario of phytoremediation: progresses and limitations. Int J Biosci 14(3):191–206. https://doi.org/10.12692/ijb/14.3.191-206
Zgorelec Z, Bilandzija N, Knez K, Galic M, Zuzul S (2020) Cadmium and Mercury phytostabilization from soil using *Miscanthus giganteus*. Sci Rep 10(1):6685–6694. https://doi.org/10.1038/s41598-020-63488-5
Zhang XY, Lin FF, Wong MTF, Feng XL, Wang K (2008) Identification of soil heavy metal sources from anthropogenic activities and pollution assessment of Fuyang County China. Environ Monitor Assess 154(1–4):439–449. https://doi.org/10.1007/s10661-008-0410-7
Zhang Q, Yu R, Fu S, Wu Z, Chen HYH, Liu H (2019) Spatial heterogeneity of heavy metal contamination in soils and plants in Hefei, China. Sci Rep 9(1):1049–1056. https://doi.org/10.1038/s41598-018-36582-y
Zhu B, Peng RH, Fu XY, Jin XF, Zhao W, Xu J, Yao QH (2012) Enhanced transformation of TNT by *Arabidopsis* Plants expressing an old yellow enzyme.

PLoS ONE 7(7):1–7. https://doi.org/10.1371/journal.pone.0039861

Zhuang P, Ye ZH, Lan CY, Xie ZW, Shu WS (2005) Chemically assisted phytoextraction of heavy metal contaminated soils using three plant species. Plant Soil 276(1–2):153–162. https://doi.org/10.1007/s11104-005-3901-0

Zwolak A, Sarzyńska M, Szpyrka E, Stawarczyk K (2019) Sources of soil pollution by heavy metals and their accumulation in vegetables: a review. Water Air Soil Pollut 230(7):164–172. https://doi.org/10.1007/s11270-019-4221-y

Bioremediation of Polluted Aquatic Ecosystems Using Macrophytes

4

R. Seenivasagan, A. Karthika, R. Kalidoss, and Junaid Ahmad Malik

Abstract

In the current situation, the aquatic ecosystem is heavily contaminated with household wastewater, industrial wastewater, agricultural fertilizer, pesticides, heavy metals, acid rain, organic, and inorganic compounds. Water bodies are the key targets for disposing of waste materials either directly or indirectly. Pollution in aquatic ecosystem leads to major threat to all living organisms in aquatic environment, also poses risk to human health. The traditional environmental cleaning methods are expensive to practice and also can lead to secondary contamination, without being shielded from environmental pollution. So, we need to remediate the problem by an eco-friendly method through phytoremediation (aquatic plant macrophytes). Plants in aquatic environment are most suitable for handling of such pollutants present in aquatic water bodies. As they have incredible capacity to absorb nutrients and other pollutants, it is considered to be one of the best eco-friendly ways to practice. Macrophytes are the primary producers of ecosystem that can provide habitats for periphytes, zooplanktons, and other species of invertebrates, fish, and frogs. Various types of macrophytes, like free-floating macrophytes (*Eichhornia crassipes, Salvinia, Ludwigia* sp., etc.), submerged macrophytes (*Egeria, Hydrilla* sp., etc.), emergent macrophytes (*Phragmites* sp., *Typha* sp.), viz. rooted shoreline, play a significant role of the ecosystem and function as biofilters. Now under the biological methods of water remediation, macrophytes play an important role as these have the potential to enhance water quality by extracting phosphorus, nitrogen, COD, BOD, suspended solids, organic carbon, phenols, heavy metals, pesticides, etc., from wastewater and also inhibit the algal bloom. This review summarizes the overall information reported recently about the bioremediation of contaminated water ecosystem through macrophyte plant species.

R. Seenivasagan
Department of Biotechnology, Arulmigu Kalasalingam College of Arts and Science, Krishnankoil, Tamil Nadu 626126, India

R. Seenivasagan
Food Security and Safety Niche Area, Faculty of Natural and Agricultural Sciences, North-West University, Mmabatho 2735, South Africa

A. Karthika (✉)
Department of Microbiology, The Standard Fireworks Rajarathnam College for Women, Sivakasi, Tamil Nadu 626123, India

R. Kalidoss
Department of Applied Biology, College of Sciences, University of Sharjah, University City, Sharjah, United Arab Emirates

J. A. Malik
Department of Zoology, Government Degree College, Bijbehara, Kashmir, J&K 192124, India

J. A. Malik (ed.), *Advances in Bioremediation and Phytoremediation for Sustainable Soil Management*, https://doi.org/10.1007/978-3-030-89984-4_4

Keywords

Aquatic ecosystem · Bioremediation · Macrophytes · Pesticides · Pollution

4.1 Introduction

Underdeveloped countries face serious environmental problems regarding water pollution (Bhat et al. 2017). Due to a lack of information, financial resources and eco-friendly policies, this issue has become a more dangerous and serious worldwide problem (Akpor and Muchie 2010; Eid et al. 2020). The main reason for water contamination is because of wastewater from industry, household, agricultural fertilizers, oil, heavy metals, acid rain, organic, inorganic compounds, etc., (Verla et al. 2018). Among these, water contamination due to inorganic pollutants and heavy metals causes serious concern due to the high level of toxicity. Also, it will not get easily disintegrated or degraded instead it gets accumulated in the land itself and thereby affects the food chain and food Web (Zhang et al. 2017; Jiang et al. 2018). To overcome this problem, there is a need for a proper wastewater management system for treating both industrial and municipal wastewater (Malakootian 2009; Bhat et al. 2018a, b). Comparing the wastewater management practices, phytoremediation is found to be a more advantageous and eco-friendly technique for clearing the water pollution while other traditional methods are more expensive and non-eco-friendly leading to the secondary pollutant formation (Rai 2009). Problems caused due to polyaromatic hydrocarbons, pesticides, chlorinated solvents may also be treated using phytoremediation (Khandare and Govindwar 2015). In phytoremediation, several categories of plants are cultivated to transfer, stabilize, destroy, and remove contaminants in the groundwater and soil. The cultivated plant roots absorb the contaminants existing in the wastewater, accumulates it, and transfers into harmless or low harmful forms. Recently, several works have been carried out regarding wastewater treatment using phytoremediation techniques (Favas et al. 2018; Vidal et al. 2019). Among them, the macrophytes involved in the phytoremediation process not only absorb the pollutant but also influences the better water quality in the surrounding aquatic ecosystem (Rai 2009). Though the toxic contaminants do not get easily destroyed, it can be directly transformed into a low lethal form (Zhang et al. 2017; Jiang et al. 2018).

Macrophytes are aquatic plants and are important component of the aquatic ecosystem due to their roles in nutrient recycling, sediment stabilization, controlling water quality, oxygen production, and providing shelter for aquatic life (Ravena 2001). Macrophytes are the strong phytoremediators encompass the mechanisms like phytovolatilization, phytostabilization, rhizofiltration, phytoextraction, and phytodegradation or phytotransformation, in which each process plays a different role in the remediation and accumulation of metals (Rai 2008, 2009, 2011, 2012). Rai et al. (2010) state the phytoremediation of polluted water is successfully used by several wetland macrophyte species. In comparison with terrestrial plants, aquatic macrophytes are reported to be more successful in wastewater treatment due to their greater production of biomass, faster growth, relatively higher pollutant uptake ability, and better purification effects because of direct connection with polluted water (Wickramasinghe and Jayawardana 2018). Moshiri (1993) states that the most promising application for wastewater treatment is the use of submerged macrophyte species such as *Ceratophyllum demersum, Egeria densa, Hydrilla verticillata, Elodea canadensis,* and *Elodea nuttallii.* The objective of this review is to summarize the overall information reported recently about the bioremediation of contaminated water ecosystem through macrophyte plant species.

4.2 Macrophytes

Macrophytes are photosynthetic organisms that are periodically or permanently grown in an aquatic environment by rooting in shallow water or growing up by emerging the vegetative parts above the water surface (Hasan and Chakrabarti 2009). According to Carpenter and Lodge (1986), Madsen et al. (2001), macrophytes help in nutrient cycling by contributing to general fitness and a healthy aquatic ecosystem. It also provides a habitat for insects, fish and produces food for other aquatic organisms (Madsen et al. 1996). Pompeo et al. (2017) state that the classification of macrophytes is mainly based on types of interactions as emergent, immersed, floating, submerged rooted, submerged free, amphiphytes submerged with floating leaves. Dordio and Carvalho (2013) constitute that macrophytes are essential compounds of wetland. Because of its capacity of large biomass production and high ability of absorbing pollutants, marcrophytes are nowadays more frequently utilized in the biochemical process of water treatment by directly connecting the root system and contaminated water. Sometimes macrophytes absorbing ability get decreased. According to Sood et al. (2012), macrophytes lose their decontamination and removal efficiency due to the interactions of different types of chemicals. So macrophyte polyculture is more preferable than monoculture in order to increase the efficiency of removing contaminants. Licata et al. (2019) stated that various macrophytes prefer various phytoremediation actions by biofilm establishment above the water surface when it gets exposed to contaminants. In response to the nature of contaminants, macrophytes can alter their growing pattern. According to Sand-Jensen (1998) and Wharton et al. (2006), macrophytes are considered to be the biological engineers of aquatic environment because of their capability of changing the affecting velocities, sediment patterns, and water depth in connection to the contaminants present. The absorbed toxic metals can be transferred from root to shoot or to different parts of macrophytes (Kassaye et al. 2016; Fawzy et al. 2012). Several authors like Champion and Tanner (2000), Baattrup-Pedersen et al. (2002) stated that the main roles played by aquatic macrophytes are emphasizing the knowledge of both physical and biological factors regulating them. Besides, macrophytes also provide shelter for many predators and also act as a habitat for their reproduction (Walker et al. 2013).

Macrophytes, such as *Salvinia rotundifolia* (Reddy and Tucker 1985), *Polugonum punctatum, Ludwiga helminthorriza* (Nunez et al. 2011), *Commelina cyanea, E. crassipes, Phragmites australis,* (Ajibade and Adewumi 2017), *Typha latifolia* (Vidayanti and Choesin 2017), *Colocasia esculenta* (Obi and Woke 2014), *Alocasia macrorrhiza, Pistia stratiotes, Alocasia puber* (Thani et al. 2019), *Lemna minor, Elodea Canadensis, Leptodictyum riparium* (Basile et al. 2012), *Wolffia borealis, Spirodela polyrhiza* (Lemon et al. 2001), *A. micrphylla* (Mishra et al. 2016), *Trapa natans* (Tsuchiya and Iwakuma 1993), *Nymphaea tetragona* (Kunii and Aramaki 1992), and *Hydrocharis dubia* (Tsuchiya 1989), were utilized for the wastewater treatment in constructed wetlands. According to Abdul Waheed et al. (2014), aquatic plants such as *Elodea Canadensis, Lemna minor,* and *Eichhornia crassipes* are photosynthetic plants having a high growth rate and efficient in absorbing pollutants. Macrophytes are well versed in absorbing toxic pollutants like pesticides, heavy metals, etc. They get hostile effects like transient storage, sorption, phytoremediation, and degradation of pesticides while exposed to the pollutant (Dosnon-Olette et al. 2009; Schulz 2004; Brogan and Relyea 2013a, b; Thomas and Hand 2011). According to Bhaskaran et al. (2013), macrophytes like *Salvinia, Lemna, Eichornia,* and *Pistia* have high phytoremediation capacity to purify the organic contaminants. Phytoremediation of dyes, phosphate, heavy metals, and radioactive isotopes can be achieved by common reed (*Phragmites* sp.) and cattail (*Typha* sp.); vetiver grass (*Vetiveria zizanioides*) (Bwire et al. 2011; Nyomora et al. 2012); ryegrass (*Lolium multiflorium*), duckweed (water Lemna), water hyacinth (*E. crassipes*), etc., (Zhang et al. 2010; Thapa et al. 2016; Nie et al. 2015; Priya and Selvan 2017). The studies of *Elodea canadensis, Spirodela polyrhiza,* and *Lemna minor* (Wahaab et al. 1995; Kahkonen and Manninen 1998; Miretzky et al. 2004) showed high capacities of absorbing heavy metal contaminated water.

Though, only a few reports showed the effectiveness of phytoremediation against pesticides (Olette et al. 2008; Gao et al. 2000) (Fig. 4.1).

4.3 Phytoremediation Through Macrophytes

4.3.1 Textile Industry Dyes

Textile effluents are considered to be heavily contaminated with pollutants such as dyes, suspended solids, total dissolved solids, toxic metals, common salt, and glauber salt. It gets discharged into the environment without any proper treatment (Sivakumar et al. 2013). Many macrophytes can be used in treating textile effluents. Pavithra and Kousar (2016) investigated the efficiency of *P. stratiotes* in textile effluent treatment for removing phosphate and nitrate within 7 days. According to Manjunath and Kousar (2016), *Pistia*, a kind of aquatic weed, can be successfully utilized for the degradation of textile industry waste. Sivakumar (2014) stated that *Lemna minor* L. effectively removes COD present in textile effluent at pH 8 within 4 days. Also, it is able to decolorize triphenylmethane dyes like malachite green and crystal violet (Torok et al. 2015). Shanmugam et al. (2020) stated that within 2 weeks, *Bacopa monnieri* (L.) Pennell can degrade about 90–100% of 14 azo dyes with the concentration of 0.04 g/L. *A. philoxeroides* and *Klebsiella* sp., have been found to be able to degrade reactive green dye (3 g/L) in 60 days (Sinha et al. 2019). Chandanshive et al. (2020) reported that Remazol red can be removed by *Vetiveria zizanioides*, a perennial bunchgrass up to 93% within 40 h. Roy et al. (2018) reported the *Eichhornia crassipes* is capable of removing 58% of color from textile wastewater. The macrophytes involved in removing contaminants in textile effluents are summarized in Fig. 4.2.

According to Ferdes et al. (2019), alga and macrophyte plants are effective in decolorizing the color in wastewater up to 90% for the blue dye and 70% for the black. *Chlorella* has a capacity of only 23% for the blue dye and 21% for the black dye. Keskinkan and Lugal Goksu (2007) reported that *Myriophyllum spicatum* and *Ceratophyllum demersum* aquaria have high removal capacity of basic blue 41 up to 94.8 and 94.1%. Patel and Adhvaryu (2015) studied the textile effluent treatment using *Eichornia* spp. and *Pistia* spp. by aquatic macrophytes and observed a high-color reduction of solutions containing royal blue dye and HD blue dye. According to Shehzadi et al. (2014) and Ong et al. (2010), aquatic plants like *Typha domingensis, Phragmites australis,* and *Chrysopogon zizanioides* were effective to treat textile wastewater. Rane et al. (2016) and Kadam et al. (2017) specified that wetlands were constructed with plants like *Fimbristylis dichotoma, Typha angustifolia, Ipomoea aquatica,* and *Salvinia molesta* and were utilized for dye wastewater

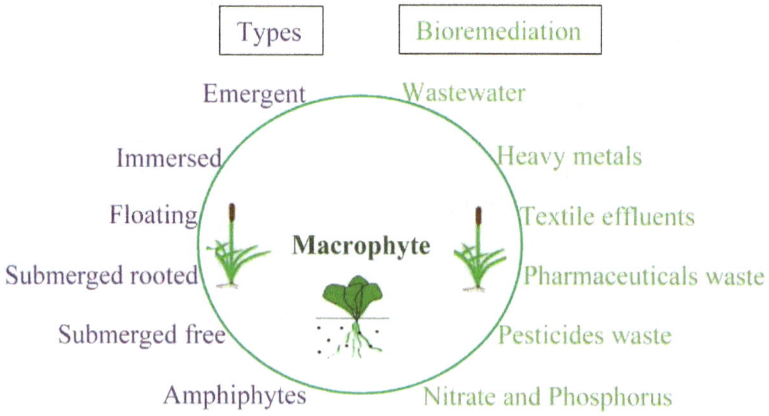

Fig. 4.1 Types of macrophytes utilized for bioremediation in aquatic environment

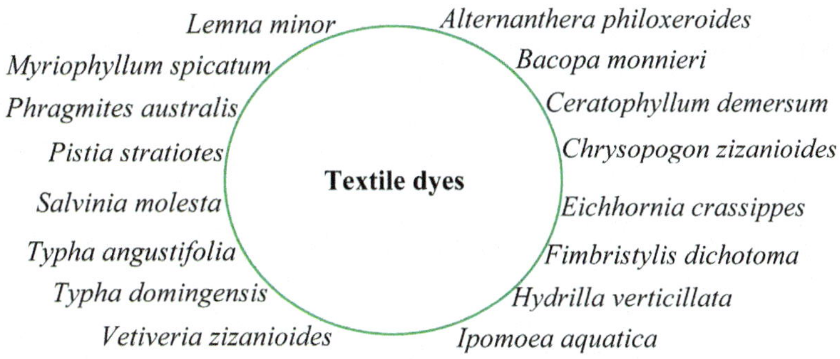

Fig. 4.2 Types of macrophytes remediating textile dye effluent pollutants

management. According to Binisha and Harsha (2019), macrophytes combine with water hyacinth (*Eichhornia crassippes*) are able to reduce the pollutants of textile industry effluent, conductivity, alkalinity, BOD, and COD. The phytoremediation treatment is also done using tank-based method. Yeruva et al. (2019) designed three tanks for the removal of various contaminants using floating plants; 1st tank for azo dye with aquatic macrophytes (*Eichhornia crassipes*), 2nd tank destined for oxidation of carbon compounds with nitrification by *Hydrilla verticillate,* and 3rd tank for removing TSS, carbon, color, etc.

4.3.2 Chemical Oxygen Demand (COD) and Biological Oxygen Demand (BOD)

COD analysis is the total quantification of all chemicals of wastewater (inorganics and organics), BOD measures the quantity of oxygen that is utilized to degrade organic components in the wastewater by the bacterium. The effluents from textile industries, both dye manufacturing and consuming, contain mixed amount of colored compounds which contain more amount of COD, BOD, and TSS (Aksu 2005). According to Tripathi and Shukla (1991), algae and *Eichhornia crassipes* are able to reduce 96.9% BOD from sewage wastewater. During wastewater treatment, about 67% of increased DO concentration is as a result of the reduction in COD and BOD. Suhendrayatna et al. (2012) stated that utilizing *Typha latifolia* for phyto-reduction of municipal wastewater has resulted into the reduction of the concentration of TSS, BOD, and COD upto 88.83, 56.72, and 50.15%, respectively. *Saccharum spontaneum* may also reduce the concentration of TSS, BOD, and COD upto 97.96, 37.31, and 56.41%, respectively, in the municipal wastewater. Shah et al. (2014) explained the treatment of wastewater on water hyacinth-based system and found the removal ability of upto 46.38% for COD, 40.34% for nitrogen, 18.76% for phosphorus, and 50.61% for BOD_5. For duckweed-based system, the efficiencies were 26.37% for COD, 17.59% for nitrogen, 15.25% for phosphorus, and 33.43% for BOD_5. Similarly, for water lettuce, 15.25% for phosphorus, 17.59% for nitrogen, 26.37% for COD, and 33.43% for BOD_5. According to Chandanshive et al. (2020), *Ipomoea aquatica, Vetiveria zizanioides*, and its consortium-VI are known to decrease the TSS by 34, 31, and 51%, TDS by 77, 75, and 83%, BOD by 73, 71, and 84%, COD by 75, 74, and 79%, ADMI by 68, 61, and 76%, respectively. Ugya et al. (2019) stated that *Lemna minor* is able to reduce COD, BOD, grease, and oil effectively, while *Eicchornia crassipes* and *Pistia stratiotes* have the maximum reduction efficiency of TDS, TSS, and Cl^-. Figure 4.3 shows the efficiency of marcrophytes involved in wastewater and textile wastewater system.

According to Tambunan et al. (2018), *Chrysopogon zizanioides* treated wastewater from the textile industry shows 89.05% COD and 98.47% BOD removal. *Eichhornia crassipes* is found to be the most effective plant in reducing COD, TDS, TSS, and BOD in textile wastewater

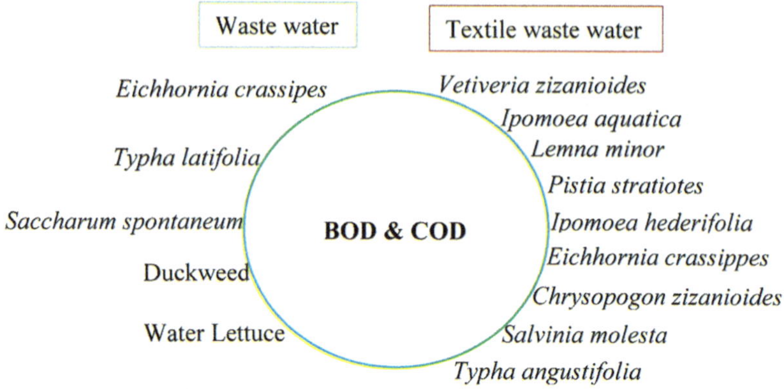

Fig. 4.3 Types of macrophytes involved in removal of BOD and COD in wastewater

(Roy et al. 2018). While treating textile wastewater, plants like *Typha angustifolia*, *Ipomoea hederifolia,* and *Salvinia molesta* were found to significantly reduce TDS, TSS, COD, BOD, ADMI, and electric conductivity (Rane et al. 2016; Chandanshive et al. 2016). Roy et al. (2010) stated that *Eichhornia crassipes, Pistia stratiotes,* and *Nostoc* are known to reduce chemical oxygen demand up to 69% in textile effluent.

4.3.3 Nitrate and Phosphorus

The effluent discharge from the industries containing nitrogen and phosphorus contaminants has become a serious problem nowadays (Li et al. 2009). According to Sangeeta and Savita (2007), the main source of these contaminants is from domestic and municipal wastewater, industrial effluents, agriculture drainage, and urban water runoff and is mainly responsible for eutrophication. Likewise, wastewater with an excess level of phosphorus is very harmful, leading to toxic algal blooms with low dissolved oxygen. It also affects aquatic biodiversity by affecting both vertebrates and invertebrates. It leads to serious effects on human well-being. This can be overcome by the action of microalga, sediment bacteria, aquatic macrophytes (Powell et al. 2009; Khoshmanesh et al. 2002; Reddy and DeBusk 1987) by taking the phosphorus contaminants.

Wetlands buildup the vegetation biomass of macrophytes by uptaking the nitrates and phosphates from wastewater as nutrients (Mitsch et al. 2001). Britto and Kronzucker (2002) stated that ammonia contaminated water also serves as a nutrient to plants but high concentration leads to stress. Fraser et al. (2004) stated that diverse vegetation of various macrophytes is more effective in nutrient removal compared to single species. In both natural and semi-natural system, aquatic macrophytes like *Cyperus papyrus* (Kyambadde et al. 2009), *Eichhornia crassipes* (Lin et al. 2004; Jayaweera and Kasturiarachchi 2010), *Typha latifolia* (Fraser et al. 2004), *Phragmites communis* (Vaillant et al. 2004), and *Pisitia stratiotes* (Sooknah and Wilkie 2004) are effective in removing inorganic nutrients N and P. According to Petrucio and Esteves (2000), high phosphorus and nitrogen concentrations were detected effectively by aquatic macrophytes such as *Salvinia auriculata* and *Eichhornia crassipes.* The total nitrogen and phosphorous absorption rates are found to be 91.2, 97.7, 96.4, 94.9% by *Alisma orientale, Acorus calamus, Monochoria korsakowii, Lythrum salicaria,* and 96.1, 83.4% by *A. calamus, L. salicaria,* respectively, (Fu and He 2015). Xu et al. (2020) state that *Ludwigia adscendens* and *Trapa natans* show a promising effect on removing total phosphorus and nitrogen from wastewater in subtropical areas. The macrophytes involved in nitrate and phosphorus removal are summarized in Fig. 4.4.

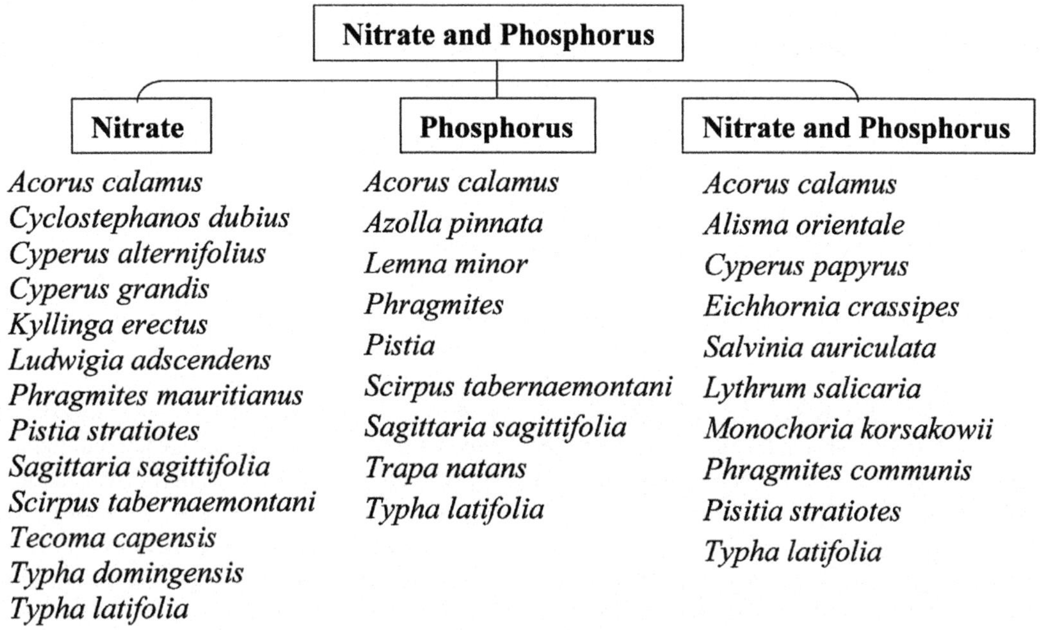

Fig. 4.4 Types of macrophyte remediating nitrate and phosphorus contaminants

Sangeeta (2007), Vymazal (2010), and Uka and Chukwuka (2011) reported that in tropical countries like India, Uganda, Kenya, Brazil, and Colombia, macrophytes such as *Cyclostephanos dubius, Kyllinga erectus, Typha domingensis, Tecoma capensis, Phragmites mauritianus,* and *Cyperus grandis* were used in nitrogen removal. Haule et al. (2013) in a study stated that *Kyllinga erectus* is a good candidate for removing nitrogen from wastewater. *Phragmites mauritianus* is supposed to be the best macrophyte because of its capacity of growing for longest period. According to Tao et al. (2015), *Cyperus alternifolius* and *Pistia stratiotes* efficiently removed total nitrogen up to 30.0–96.8%. On comparing several wetland species like *Phragmites*, *Typha,* etc., *Pistia* is found to be an efficient P accumulator with a percentage of up to 43.3 within 14 days (Shardendu et al. 2012). Muvea et al. (2019) stated that two macrophytes such as *Azolla pinnata* and *Lemna minor* are better in assimilating the nutrients mainly phosphorous in wastewater.

He et al. (2012) proposed that nitrate removal was efficiently carried out best by *Sagittaria sagittifolia* than *Scirpus tabernaemontani* and *Typha latifolia.* Similarly, for TP, the efficient removal was carried out by *Typha latifolia* than *Scirpus tabernaemontani, Acorus calamus*, and *Sagittaria sagittifolia*. In case of floating plants, the order of TP and NO_3^--N removal capacities was more in *Eichhornia crassipes* than in *Pistia stratiotes*. The native plants like *Ludwigia peploides* were found more effective than *E. crassipes* for purifying wastewater (Zhang 2007). Forni et al. (2001) recommend *Azolla pinnata* as the best macrophyte to remove nitrogen and phosphorous nutrients from wastewater and prevent it from eutrophication.

4.3.4 Heavy Metals

The drastic increase in science and technology increased the metal contaminants, pesticides, and fertilizers in wastewater (Harguinteguy et al. 2014; Romero Nunez et al. 2011). Worldwide heavy metal contamination is a common environmental problem and is a serious risk to aquatic ecosystems, human health, and agriculture (Garbisu and Alkorta 2003; Gupta et al. 2010; Ashraf et al. 2019). Due to anthropogenic activities like industrialization and urbanization aquatic ecosystems, these are highly contaminated with heavy metals and lead to global problems. Further, these contaminants reach the food chain, aquatic ecosystems, and pose significant health risks to human beings (Singh et al. 2018).

Mungur et al. (1997) state that macrophytes such as *Phragmites australis, Typha latifolia, Iris pseudacorus,* and *Schoenoplectus lacustris* are effective in removing heavy metals from industrial wastewater. This kind of removal can be done by surface adsorption or absorption by which the heavy metals incorporate in storage-bound form by aquatic macrophytes (Rai et al. 1995). According to Samecka-Cymermann and Kempers (1996), various plant species like *Elodea canadensis, Cabomba aquatic,* and *Lemna minor* are found to have good accumulative capacities of absorbing heavy metals. Hence, it is considered to be an efficient agent for phytoremediation of heavy metals and nutrient-contaminated water (Olette et al. 2008; Kahkonen and Manninen 1998; Wahaab et al. 1995).

According to Bonanno and Giudice (2010) and Sawidis et al. (1995), roots and rhizomes of *T. domingensis* and *P. australis* have cortex parenchyma with large intercellular air spaces that are able to accumulate a lot of heavy metals. El-Amier et al. (2018) demonstrated that heavy metals like Fe, Ni, Co, Cd, Pb, Cr, and Ni can get accumulated in hydrophytes *E. crassipes, E. stagnina, P. australis,* and *T. domingensis*. Plechonska and Klink (2014) in a study proved that various parts of reed canary grass (*Phalaris arundinace*) have phytoremediation abilities to remove trace metals such as Cr, Co, Cd, Ni, Cu, Pb, Mn, Zn, and Fe by which it gets accumulated or sedimented in tissues of plant parts. Sima et al. (2019) reported that *Phragmites australis* has the ability to remove As, Cr, and Cu effectively up to 36, 64, and 70%, respectively. Gudisa (2019) demonstrated that Mn, Zn, and Cu heavy metals can be absorbed by *Shoenoplectus lacustris* to a considerable level.

The phytoremediation activity of *Lemna minor* in a stream wastewater polluted with refinery and petrochemical compounds was investigated by Ugya (2015). The results showed that about 95.4, 100, 94.3, 99.3, 93.3, and 99.6% of Ag, Pb, Mn, Zn, Hg, and Cd were removed, respectively. Emergent macrophytes such as *Phragmites australis* and *Typha latifolia* effectively removed Cr, Fe, and Zn up to 66.2, 70.6, and 71.6%, respectively, from industrial effluents within 14 days (Lema et al. 2014; Kumari and Tripathi 2015). Nyquist and Greger (2007) studied the phytoremediation activity of *E. canadensis* in removing Zn, Cu, and Cd by accumulating the contaminants in plant tissues. But in long exposure, the activity of absorption got decreased. The removal of various heavy metals by various macrophytes is summarized in Fig. 4.5.

4.3.4.1 Chromium

In recent decades, the contamination of the biosphere has raised dramatically, mainly due to industrial wastewater spills. High levels of chromium (heavy metals) are typically combined with aromatic organic contaminants such as phenols (Ontanon et al. 2015). Chromium can stimulate the growth of many species at low concentrations in plants, while it is not an important component. At a high concentration, chromium alters both physiological and biochemical factors like germination inhibition, reduced seedlings development, necrosis, and reduced leaf growth (Shanker et al. 2009).

In constructed wetlands (CWs), the treatment of tannery effluent was done with *Penisetum purpureum, Iris pseudacorus, Cannabis indica, Brachiaria decumbens* (Mant et al. 2006), *Scirpus americanus, Typha latifolia* (Aguilar et al. 2008), *Juncus eusus* (Gruber et al. 2008), *Marselia quadrifolia, Cyperus kylinga, Cyperus rotundus, Ludwigia parvifloria* (Sundaramoorthy et al. 2010), *Leersia hexandra* (Liu et al. 2014), *Typha angustifolia, Vetiveria nemoralis, Vetiveria zizanioides, Cyperus esculentus* (Yadav and Chandra 2011), *Cyperus alternifolius, Parawaldeckia karaka, Typha*

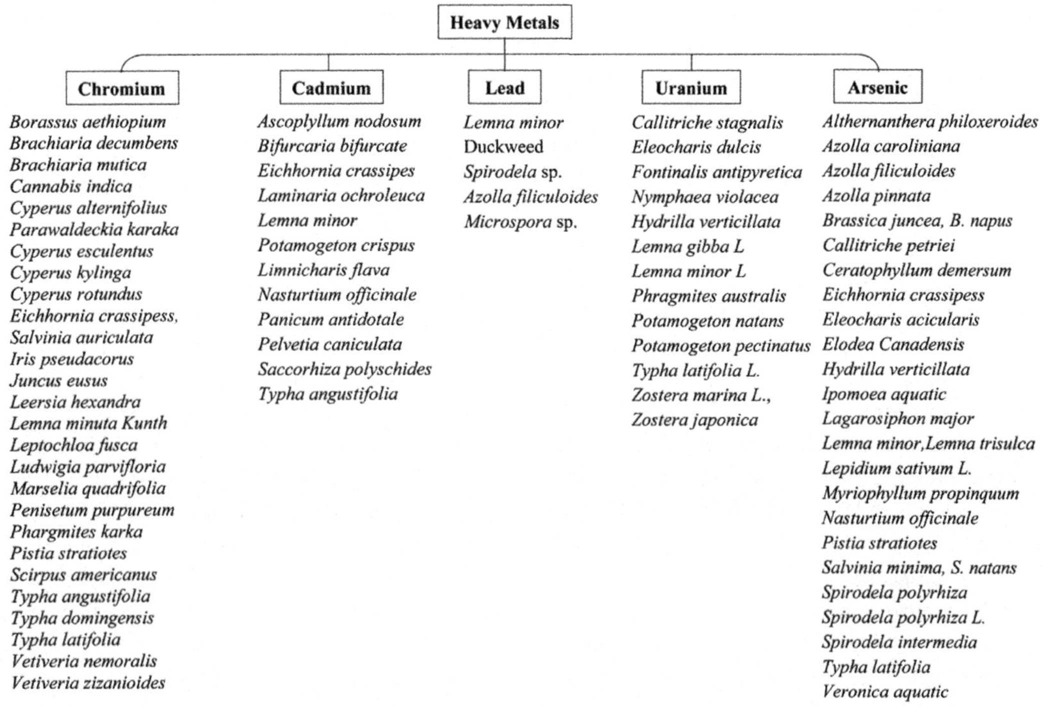

Fig. 4.5 Types of macrophytes remediating heavy metals in aquatic environment

domingensis, Borassus aethiopum, (Terfie and Asfaw 2015), *Leptochloa fusca, Typha domingensis*, and *B. mutica* (Ashraf et al. 2018). The wetland macrophyte plants like *Borassus aethiopium, Typha domingensis, Cyperus alternifolius*, and *Phargmites karka* were effective to remove 97% of chromium (Cr) (Tadesse and Seyoum 2015). Ashraf et al. (2020) reported that the removal of contaminants by *T. domingensis* and *Leptochloa fusca* performed better along with higher growth rates and biomass than other tested macrophytes. Hence, this species is considered to be more favored for tannery effluent treatment at CWs. Espinoza-Quinones et al. (2009) studied about the roots of *Eichhornia crassipess, Salvinia auriculata* and *Pistia stratiotes*, during metal biosorption, and reported the successful conversion of Cr (VI) to Cr(III). Paisio et al. (2017) reported that *Lemna minuta Kunth* has a high potential to treat Cr(VI) and phenol contaminants (Fig. 4.5).

4.3.4.2 Cadmium

In phosphate fertilizers, cadmium is a trace element and does not pose any ecological risk upto a certain limit. Cadmium contamination is also an important source of atmospheric deposition (ATSDR 2008). Cd enhancement in soil arises from both anthropogenic and natural sources (Pan et al. 2016). Geologically, rock weathering is the main usual cause of Cd pollutants (Khan et al. 2010; Liu et al. 2013), while primary anthropogenic sources of Cd include manufacturing (electronics components, automobile radiators, photography, and batteries), agrochemicals, mining, smelting, irrigation wastewater, and vehicular emission (Nawab et al. 2016; Naja and Volesky 2009; Khan et al. 2016). Both morphological and physiological levels of plant growth are affected by Cd (Shanying et al. 2017). Comparatively, Cd toxicity includes leaf chlorosis, photosynthesis, inhibition of respiration, and a delay in the growth rate (Navarro-Leon et al. 2019) decreased ability to take up nutrients and increased oxidative damage (Mohamed et al. 2012). It is documented that aquatic plants accumulate substantial amounts of Cd from polluted water such as *Nasturtium officinale*, when exposed to Cd metal for 14 days (Aslan et al. 2003). Demirezen and Aksoy (2004) recorded that *Typha angustifolia* can absorb certain level of cadmium. Kabata-Pendias and Pendias (1992) described that Cd has been absorbed effectively by both root and leaf. Arshad et al. (2015) reported that *Nasturtium officinale* shows phytoremediation activity in Cd contaminated waters. According to Zaidi et al. (2017), *E. crassipes* and *Chara* spp. are the best candidates for the remediation treatment of polluted water bodies containing Pb and Cd. Lodeiro et al. (2005) stated that *Laminaria ochroleuca, Ascoplyllum nodosum, Pelvetia caniculata, Saccorhiza polyschides,* and *Bifurcaria bifurcata* act as effective biosorbents in removing cadmium and other contaminants up to 50% within 3 h. Abhilash et al. (2009) increased the phytoremediation activity of *Limnicharis flava* by introducing the phytofiltration technique for Cd uptake. According to Singh et al. (2017), while comparing terrestrial land plants, aquatic wetland plants like *Eichhornia crassipes, Panicum antidotale, Lemna minor,* and *Potamogeton crispus* are suitable for uptaking more concentrations of Cd (Fig. 4.5).

4.3.4.3 Lead (Pb)

Lead (Pb) is considered to be highly toxic metal throughout the world. Its excessive usage can cause major environmental pollution and severe heath issues (Jaishankar et al. 2014). The main sources of lead pollution are metal finishing and plating, pesticides, excessive emissions from automobiles and fertilizers, battery industry effluents, ores smelting processes, wastes from gasoline, urban soil and dye additives, and chimney manufacturing units (Eick et al. 1999). According to Miretzky et al. (2005), *Pistia stratiotes* is consider to be an efficient and low-cost way to treat industrial effluent containing heavy metal ions like Cd and Pb. Uysal and Taner (2009) proved that *Lemna minor* is able to remove soluble lead at alkaline pH values at room temperature. Rahmani et al. (1999) reported that viable duckweed removes lead up to 70–80%. According to Srivastav et al. (1993), it was found that *Spirodela* is more effective for removing zinc and lead than *Salvinia*. Aquatic plants like *Azolla filiculoides, Microspora,* and *Lemna minor* show distinct remediation activity against Ni and Pb (Axtell et al. 2003) (Fig. 4.5).

4.3.4.4 Uranium (U)

Uranium is a ubiquitous heavy metal in nature, causing several pollutions to the environment (Palmer and Edmond 1993). Pratas et al. (2014) performed an experiment by growing three plant species *Potamogeton natans, Potamogeton pectinatus,* and *Callitriche stagnalis* in the laboratory phytofiltration system to reduce the uranium concentration in contaminated water up to 85.5%. The absorption occurred ranged from 0.98 to 1567 mg/kg, by *P. pectinatus* from 2.63 to 1588 mg/kg, and by *P. natans* from 3.46 to 271 mg/kg, respectively. Pratas et al. (2012) proved the evidence of deposition of arsenic and uranium content in some aquatic plants like *Lemna minor* L., *Lemna gibba* L., *Fontinalis antipyretica,* and *C. stagnalis* (Alvarado et al. 2008; Mkandawire and Dudel 2005). Several studies supported the evidence of utilizing aquatic plants in treatment of uranium contaminated water by phytofiltration technique, e.g., *Nymphaea violacea* (Pettersson et al. 1993), *Fontinalis antipyretica, Callitriche stagnalis* (Pratas et al. 2012), *Hydrilla verticillata* (Srivastava et al. 2010), *Typha latifolia, Phragmites australis*, (Carvalho et al. 2011; Soudek et al. 2007), *Lemna gibba* (Mkandawire and Dudel 2005), *Eleocharis dulcis* (Overall and Parry 2004), *Zostera marina*, and *Zostera japonica* (Kondo et al. 2003) are used for removing uranium pollutants (Fig. 4.5).

4.3.4.5 Arsenic (As)

The main cause of arsenic pollution is due to utilizing arsenic chemicals in agriculture, burning of fossil fuels, and mining (Bissen and Frimmel 2003). Mandal and Suzuki (2002) reported that worldwide large number of sites get contaminated by arsenic through geogenic, anthropogenic, and natural sources. Kadlec and

Zmarthie (2010) observed that *Typha latifolia* removes arsenic up to 29% in a leachate treatment system. Arsenic contamination in wastewater could be removed with the help of duckweeds like *Lemna, Spirodela, Wolffia*, and *Wolfiella* (Mkandawire and Dudel 2005; Mkandawire et al. 2004a, b; Zhang et al. 2009; Rahman et al. 2007, 2008a, b; Alvarado et al. 2008).

According to Rahman et al. (2008a, b) and Hoffmann et al. (2004), arsenic remediation can be achieved by plant species such as *Salvinia minima* and *Salvinia natans.* Alvarado et al. (2008) reported that arsenic can be effectively removed by *Lemna minor* up to 5%. Mishra et al. (2008) demonstrated that on comparing *Spirodela polyrhiza,* L. minor, and *E. crassipes* for the treatment of coal mine effluent, *E. crassipes* was observed to be the premier candidate for the removal efficiency up to 80% within 25 days. Upadhyay et al. (2017) and Niazi et al. (2017) studied the efficiency of *Ceratophyllum demersum, Lemna minor, Hydrilla verticillata, Brassica juncea,* and *B. napus* in arsenic removal. Many reports have supported the evidence of arsenic uptake from water by aquatic macrophytes such as *Nasturtium officinale, Lepidium sativum L., Ceratophyllum demersum, Elodea Canadensis, Callitriche petriei, Lagarosiphon major, Veronica aquatic, Myriophyllum propinquum, Eleocharis acicularis* (Robinson et al. 2003, 2005; Ha et al. 2009), *Pistia stratiotes* (Odjegba and Fasidi 2004; Lee et al. 1991), *Althernanthera philoxeroides* (Elayan 1999), *Azolla caroliniana, A. filiculoides, A. pinnata* (Rahman et al. 2008c; Zhang et al. 2008), *Ipomoea aquatic, Hydrilla verticillata* (Lee et al. 1991), *Spirodela polyrhiza L., Spirodela intermedia* (Rahman et al. 2007; Mishra et al. 2008; Rahman et al. 2008b, d), *Salvinia natans, Salvinia minima* (Hoffmann et al. 2004; Rahman et al. 2008b), *Eichhornia crassipes, and L. minor* L. (Mishra et al. 2008) (Fig. 4.5).

4.3.5 Pharmaceuticals

The pharmaceuticals industrial waste rooted in the land through the municipal wastewater. Several metabolite pharmaceuticals are not metabolized and excreted into the sewage system with urine and feces (Jones et al. 2002; Carballa et al. 2004; Zhang et al. 2008). Through wastewater treatment plant, several types of pharmaceutical effluents and their metabolites are released into coastal water, surface water, drinking and groundwater (Kim et al. 2007; Jelic et al. 2011; Vidal-Dorsch et al. 2012; Uslu et al. 2013).

Redshaw et al. (2008) stated that individual species of macrophytes are efficient in removing pharmaceutical contaminants in wastewater. Among them, *Typha angustifolia, Typha latifolia, Typha* spp., and *Phragmites australis* are considered to be most effective (Li et al. 2014). Sulfadimethoxin may also be removed from an aqueous medium by the aquatic fern *Azolla filiculoides.* In Forni et al. (2002) experiment, about 56−86% of pharmaceutical contaminants with the concentration of 0.05−0.45 g/L aqueous could be removed. Dordio et al. (2009) stated that 50−80% of clofibric acid was removed using *Typha* spp. within 48 h, in hydroponic systems. According to Gujarathi et al. (2005), *Pistia stratiotes* and *Myriophyllum aquaticum* completely removed oxytetracycline and tetracycline in 6 and 15d, respectively. Hwang et al. (2020) planted *Canna flaccida* in superficial wetland systems to demonstrate the efficiency in removing carbamazepine and acetaminophen contaminants up to 81–100%. Similarly, superficial wetland systems planted with *Eichornia crassipes* and *Heliconia rostrata* have been found to be efficient in removing phosphorylated compounds, nitrogen-containing, caffeine, ibuprofen, as well as organic compounds up to 80% (De Oliveira et al. 2019). In Chen et al. (2016) experiment, the wetland was constructed for removing the antibiotics sulfapyridine, sulfamethazine, trimethoprim, sulfamethoxazole, leucomycin, clarithromycin, azithromycin, and monensin, as well as antibiotic resistance genes with the help of *Iris tectorum* and *Thalia dealbata* up to 75.8–98.6% and 63.9–84%, respectively. Figure 4.6 demonstrates the various plant species involved in removing pharmaceutical pollutants.

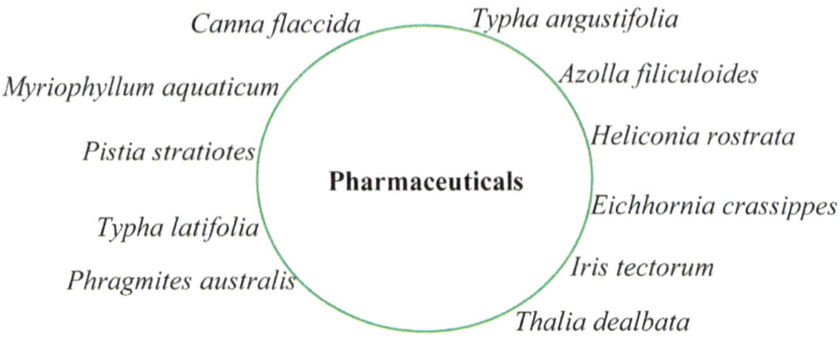

Fig. 4.6 Types of macrophytes remediating pharmaceuticals waste in aquatic environment

4.3.6 Pesticides

Pesticides used for agricultural crops may enter the aquatic environment through surface runoff leaching, erosion, drain flows, and atmospheric deposition (Reichenberger et al. 2007). Insecticides also pose a significant threat to the aquatic environment by affecting their habitats and decrease biodiversity (Geiger et al. 2010; Bradford et al. 2011). According to Lema et al. (2014), *Cyperus papyrus* and *Typha latifolia* are very effective in removing endosulfan, permethrin, and L-Cyhalothrin, respectively. Brogan and Relyea (2013a) proved an increase in *Elodea canadensis* density in the water column which was able to decrease malathion toxicity. Garcinuno et al. (2006) stated that *Lupinus angustifolius* displays the mass recoveries of carbaryl (57%), linuron (53%), and permethrin (55%) in a hydroponic system. Macrophytes such as *Anemone caroliniana* and *Lemna gibba* have greater potential in removing atrazine (Guimaraes et al. 2011). According to Dosnon-Olette et al. (2009), duckweeds (*Spirodela polyrhiza* and *Lemna minor*) are very effective in dimethomorph and pyrimethanil removal. Also macrophytes such as *Potamogeton crispus, Ceratophyllum demersum,* and *Elodea nuttallii* have greater ability to remove isoproturon and bifenox within 2 h of exposure (Stang et al. 2016). The high concentration of imazalil and tebuconazole could be phytoremediated by aquatic macrophytes like *E. nuttallii* (Elsaesser et al. 2013), *Elodea nuttallii* (Stang et al. 2013), *Sparganium americanum, Leersia oryzoides,* and *Typha latifolia* (Moore et al. 2013), respectively.

Potential removal of agrochemicals including carbamates, pyrethroids, organophosphates, and chlorhydrates can be done by macrophytes such as *Typha latifolia, Nymphaea lotus, Phragmites australis, Elodea canadensis, Spirodela oligorrhiza, Myriophyllum aquaticum,* and *Lemna minor* through bioaccumulation or metabolism (Gomes and Juneau 2016; Gomes et al. 2019; Dhir 2020). Zhao et al. (2011) stated that PCP removal from contaminated sediments can be achieved by planting *Scirpus validus* (99%), *Theileria orientalis* (99%), and *Phragmites communis Trin* (90%). Olette et al. (2008) specified that *Elodea canadensis* shows inordinate phytoremediation activity for the removal of pesticides, fungicides, and herbicides such as dimethomorph, flazasulfuron, copper sulfate, respectively. *Elodea canadensis, Lemna minor,* and *Eichhornia crassipes* shows translocation rates, volatilization, and diverse absorption of pesticides. Dosnon-Olette et al. (2010) stated that efficient phytoremediation could be achieved when pesticide mobility is directly proportional to surface adsorption on parts of the plant. Riaz et al. (2017) reported that organochlorine and pyrethroids can be removed effectively by *Pistia stratiotes, Eichhornia crassipes, Chaetomorpha sutoria, Sirogonium sticticum,* and *Zygnema* sp. De Souza et al. (2017) tested *Cynodon* spp., *Polygonum punctatum*, and *Mentha aquatic* for their capacity to remove chlorpyrifos pesticides in wastewater. Mahabali and Spanoghe (2014) stated that *Nymphaea amazonum* and *Eleocharis mutata* plants have the ability to absorb imidacloprid and Cyhalothrin pesticide, respectively, up to 79%. Figure 4.7 summarizes the aquatic

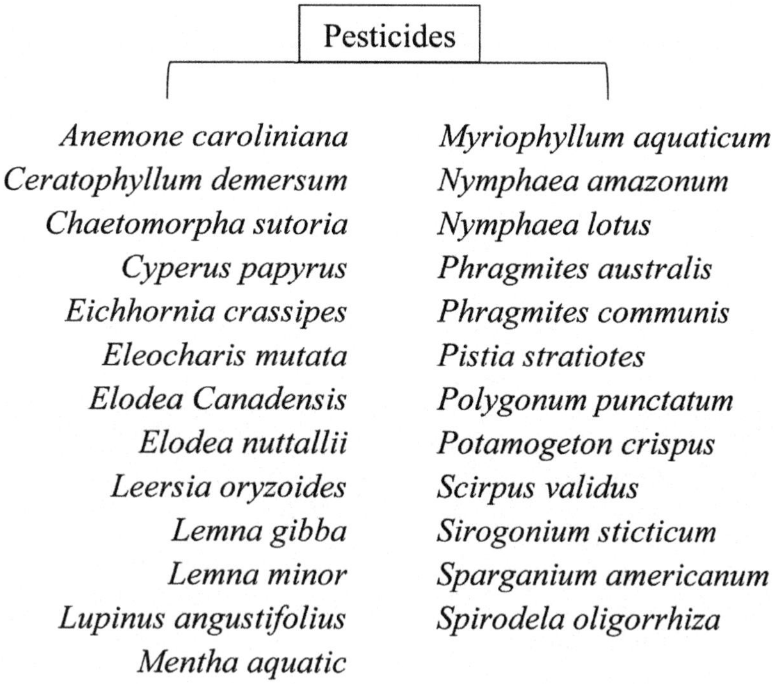

Fig. 4.7 Types of macrophytes involved in removal of pesticides waste in aquatic environment

plants involved in the removal of pesticides from the wastewater.

4.3.7 Phenols

Phenol and phenolic compound contaminants from petrochemicals, textiles, dyeing, phenolic resin manufacturing, and steel plant effluent lead to serious threats to microorganisms, plants, fishes, and other animals as well as can cause considerable effects in the environment. Metcalf and Eddy (2003) and Paisio et al. (2009) stated that phenol contaminants cause serious harm to anuran amphibians, fishes, crustaceans, and several other aquatic organisms. Jha et al. (2013) and Zhou et al. (2013) described that phenol in wastewater restricts the plant by inhibiting germination, development, and seedling growth, and also leads to chlorosis. According to Brezinova and Vymazal (2018), the highest concentrations of phenolics were found to be detected in *Scirpus sylvaticus* and *Carex nigra*, while the minimal observation were found in stems of *Phalaris arundinacea* and *Phragmites australis*. Macrophyte species such as *P. australis* sub sp. *americanus* and *Phragmites* spp., and *T. angustifolia* were found to absorb a high concentration of phenol through their roots and rhizomes (Ma and Havelka 2009; San Miguel et al. 2013). Ugya et al. (2019) reported that *Salvinia molesta* shows highest reduction efficiency of ammonical nitrogen and phenol in wastewater. The results of an experiment performed by Nafea (2019) revealed that *Lemna gibba* L. and *Pistia stratiotes* L. have higher removal efficiency of upto 88 and 83% in phenol polluted wastewater. Danh et al. (2009) state that *Vetiveria zizanioides* also degrade the benzo[a]pyren, atrazine, 2,4,6-trinitroluene, and phenol.

4.3.8 Sewage

Municipal wastewater is a mixture of utilized wastewater from households and industries. Such heterogeneous wastewater can cause serious challenges including water-borne diseases leading and other health hazards. It contains partially decomposed materials (inorganic and organic) and trace elements including iron (Fe), chromium (Cr), nickel (Ni), manganese (Mn), lead (Pb), copper (Cu), zinc (Zn), and cadmium (Cd). According to Clements and Newman (2002), Reddy and Kumar (2001), Vyas et al. (2008) and El-Gendy et al. (2004), *Pistia stratiotes, Eichhornia crassipes,* and *Lemna minor* show remediation to municipal sewage contaminants. According to Ekperusi et al. (2019), several biochemical processes such as total suspended solids, phosphate, ammonium nitrate, and BOD; ammonia and total nitrogen can be remediated by *Lemna gibb.* Zhou et al. (2018), Liu and Wu, (2018), and Liu et al. (2019) reported that alkylbenzenesulfonate was potentially removed by *Potamogeton perfoliatus, Myriophyllum spicatum, Lemna minor,* and *Chara vulgaris.*

4.4 Conclusion

Water is very important for survival for all living animals, plants, and humans. But continuously, water gets contaminated through various sources like industrial wastewater, heavy metals, agricultural fertilizers, acid rain, pesticides, household wastewater, organic and inorganic compounds, etc. In this review paper, wastewater management and pollution remediation by means of natural methods were discussed. As traditional methods are very expensive and create secondary waste, hence, phytoremediation using aquatic plant (macrophytes) was chosen to remediate the problem. Phytoremediation using macrophytes is one of the best cost-effective and eco-friendly technologies for pollutant removal from aquatic ecosystems. Many of the researchers reported that macrophytes, through adsorption/absorption techniques, remediate the pollutants from wastewater. From this review, we discuss major pollutants like phosphorus, nitrogen, COD, BOD, phenols, heavy metals (Chromium, Cadmium, Lead, Uranium, Cadmium, and Arsenic,) pesticides, textile effluent waste, pharmaceuticals waste, and their remediation through macrophytes. Specifically, some of the macrophytes such as *Pistia stratiotes, Eichhornia crassipes, Typha latifolia,* and *Lemna minor* were most involved in all pollutant remediation process. Among them, *Eichhornia crassipes* is the best species to treat all contamination types.

References

Abdul Waheed A, Awang S, Sarva P (2014) The comparison of phytoremediation abilities of water Mimosa and water Hyacinth. ARPN J Sci Techn 4 (12):722–731

Abhilash P, Pandey VC, Srivastava P, Rakesh P, Chandran S, Singh N, Thomas A (2009) Phytofiltration of cadmium from water by *Limnocharis flava* (L.) Buchenau grown in free-floating culture system. J Hazard Mater 170:791–797

Aguilar JRP, Cabriales JJP, Vega MM (2008) Identification and characterization of sulfur-oxidizing bacteria in an artificial wetland that treats wastewater from a tannery. Int J Phytorem 10:359–370

Ajibade FO, Adewumi JR (2017) Performance evaluation of aquatic macrophytes in a constructed wetland for municipal wastewater treatment. FUTA J Eng Eng Technol 11(1):1–11

Akpor OB, Muchie M (2010) Remediation of heavy metals in drinking water and wastewater treatment systems: processes and applications. Inter J Phys Sci 5:1807–1817

Aksu Z (2005) Application of bio sorption for the removal of organic pollutants: a review. Process Biochem 40:997–1026

Alvarado S, Guedez M, Lue-Meru MP, Nelson G, Alvaro A, Jesus AC, Gyula Z (2008) Arsenic removal from waters by bioremediation with the aquatic plants water hyacinth (*Eichhornia crassipes*) and lesser duckweed (*Lemna minor*). Bioresour Technol 99:8436–8440

Arshad M, Khalid R, Hina K, Ullah S, Ali MA (2015) Cadmium removal by aquatic macrophyte (*Nasturtium officinale*) and potential for oil production from the biomass. Pak J Agri Agril Engg Vet Sci 31:1–13

Ashraf S, Afzal M, Rehman K, Tahseen R, Naveed M, Zahir ZA (2018) Enhanced remediation of tannery effluent in constructed wetlands augmented with endophytic bacteria. Desalin Water Treat 102:93–100

Ashraf S, Ali Q, Ahmad Z, Ashraf S, Naeem H (2019) Phytoremediation: environmentally sustainable way for reclamation of heavy metal polluted soils. Ecotoxicol Environ Saf 174:714–727

Ashraf S, Naveed M, Afzal M, Seleiman MF, Al-Suhaibani NA, Zahir ZA, Mustafa A, Refay Y, Alhammad BA, Ashraf S, Alotaibi M, Abdella KA (2020) Unveiling the potential of novel macrophytes for the treatment of tannery effluent in vertical flow pilot constructed wetlands. Water 12(2):549

Aslan M, Unlu YU, Turkmen N, Yilmaz Z (2003) Sorption of cadmium and effects on growth, protein content, and photosynthetic pigment composition of

Nasturtium officinale R. Br. and *Mentha aquatica L.* Bull Environ Contam Toxicol 71:323–329

ATSDR (2008) Draft toxicological profile for cadmium Atlanta: US Department of Health and Human Services, Agency for Toxic Substances and Disease Registry. Available from: http://www.atsdr.cdc.gov/toxprofiles/tp5-p.pdf

Axtell NR, Sternberg SP, Claussen K. (2003) Lead and nickel removal using *Microspora* and *Lemna minor*. Biores Techn 89(1):41–48

Baattrup-Pedersen A, Larsen SE, Riis T (2002) Long-term effects of stream management on plant communities in two Danish lowland streams. Hydrobiologia 481:33–45

Basile A, Sorbo S, Conte B, Cobianchi RC, Trinchella F, Capasso C, Carginale V (2012) Toxicity, accumulation, and removal of heavy metals by three aquatic macrophytes. Int J Phytoremediation 14(4):374–387

Bhaskaran K, Vijaya Nadaraja A, Tumbath S, Babu Shah L, Gangadharan Puthiya Veetil P (2013) Phytoremediation of perchlorate by free floating macrophytes. J Hazard Mater 260:901–906

Bhat RA, Dervash MA, Mehmood MA, Hakeem KR (2018) Municipal solid waste generation and its management, a growing threat to fragile ecosystem in Kashmir Himalaya. Am J Envir Sci 13(6):388–397

Bhat RA, Dervash MA, Qadri H, Mushtaq N, Dar GH (2018) Macrophytes, the natural cleaners of toxic heavy metal (THM) pollution from aquatic ecosystems. Environmental contamination and remediation. Cambridge Scholars Publishing, Newcastle upon Tyne, pp 189–209

Bhat RA, Shafiq-ur-Rehman MMA, Dervash MA, Mushtaq N, Bhat JIA, Dar GH (2017) Current status of nutrient load in Dal Lake of Kashmir Himalaya. J Pharmacogn Phytochem 6(6):165–169

Binisha P, Harsha P (2019) Macrophytic Treatment of Textile Wastewater Together with Root Zone Technique 6(4):4940–4944

Bissen M, Frimmel FH (2003) Arsenic—a review. Part I: occurrence, toxicity, speciation, mobility. Acta Hydrochim Hydrobiol 31:9–18

Bonanno G, Giudice RL (2010) Heavy metal bioaccumulation by the organs of *Phragmites australis* (commonreed) and their potential use as contamination indicators. Ecol Ind 10:639–645

Bradford DF, Knapp RA, Sparling DW, Nash MS, Stanley KA, Tallent-Halsell NG, McConnell LL, Simonich SM (2011) Pesticide distributions and population declines of California, USA, alpine frogs, Rana muscosa, and Rana sierrae. Environ Toxicol Chem 30:682–691

Brezinova TD, Vymazal J (2018) Phenolic compounds in wetland macrophytes. Sci Agric Bohem 49(1):1–8

Britto D, Kronzucker H (2002) NH_4^+ toxicity in higher plants: a critical review. Plant Physiol 159(2):567–584

Brogan WR, Relyea RA (2013) Mitigating with macrophytes: submersed plants reduce the toxicity of pesticide-contaminated water to zooplankton. Environ Toxicol Chem 32(3):699–706

Brogan WR, Relyea RA (2013) Mitigation of malathion's acute toxicity by four submersed macrophyte species. Environ Toxicol Chem 32:1535–1543

Bwire KM, Njau KN, Minja RJ (2011) Use of vetiver grass constructed wetland for treatment of leachate. Water Sci Technol 63(5):924–930

Carballa M, Omil F, Lema JM, Llompart M, Garcia-Jares C, Rodriguez I, Gomez M, Ternes T (2004) Behavior of pharmaceuticals, cosmetics and hormones in a sewage treatment plant. Water Res 38(12):2918–2926

Carpenter SR, Lodge DM (1986) Effects of submersed macrophytes on ecosystem processes. Aquat Bot 26:341–370

Carvalho FP, Oliveira JM, Malta M (2011) Radionuclides in plants growing on sludge and water from uranium mine water treatment. Ecol Eng 37:1058–1063

Champion PD, Tanner CC (2000) Seasonality of macrophytes and interaction with flow in a New Zealand lowland stream. Hydrobiologia 441:1–12

Chandanshive V, Kadam S, Rane N, Jeon BH, Jadhav J, Govindwar S (2020) In situ textile wastewater treatment in high rate transpiration system furrows planted with aquatic macrophytes and floating phytobeds. Chemosphere 126513:1–9

Chandanshive VV, Rane NR, Gholave AR, Patil SM, Jeon BH, Govindwar SP (2016) Efficient decolorization and detoxification of textile industry effluent by *Salvinia molesta* in lagoon treatment. Environ Res 150:88–96

Chen J, Ying GG, Wei XD, Liu YS, Liu SS, Hu LX, He LY, Chen ZF, Chen FR, Yang YQ (2016) Removal of antibiotics and antibiotic resistance genes from domestic sewage by constructed wetlands: effect of flow configuration and plant species. Sci Total Environ 571:974–982

Clements WH, Newman MC (2002) Community ecotoxicology. Wiley, Hoboken, NJ, p 354

Danh LT, Truong P, Mammucari R, Tran T, Foster N (2009) *Vetiver grass, Vetiveria zizanioides*: a choice plant for phytoremediation of heavy metals and organic wastes. Int J Phytoremediation 11:664–691

De Oliveira M, Atalla AA, Frihling BEF, Cavalheri PS, Migliolo L, Filho FJCM (2019) Ibuprofen and caffeine removal in vertical flow and free-floating macrophyte constructed wetlands with *Heliconia Rostrata* and *Eichornia Crassipes*. Chem Eng J 373:458–467

De Souza TD, Borges AC, de Matos AT, Mounteer AH, de Queiroz MELR (2017) Removal of Chlorpyrifos insecticide in constructed wetlands with different plant species. Rev Bras Eng Agricola Ambient 21:878–883

Demirezen D, Aksoy A (2004) Accumulation of heavy metals in *Typha angustifolia* (L.) and *Potamogeton pectinatus* (L.) living in Sultan Marsh (Kayseri Turkey). Chemosphere 56:685–696

Dhir B (2020) Green technologies for the removal of agrochemicals by aquatic plants. Agrochemicals detection, treatment and remediation: pesticides and chemical fertilizers; Butterworth Heinemann. Elsevier, Oxford, UK, pp 569–591

Dordio AV, Duarte C, Barreiros M, Carvalho AP, Pinto A, da Costa CT (2009) Toxicity and removal efficiency of pharmaceutical metabolite Clofibric acid by *Tylpha* Spp.-Potential use for phytoremediation? Bioresour Technol 100:1156–1161

Dordio AV, Carvalho AJP (2013) Organic xenobiotics removal in constructed wetlands, with emphasis on the importance of the support matrix. J Hazard Mater 252–253:272–292

Dosnon-Olette R, Couderchet M, El Arfaoui A, Sayen S, Eullaffroy P (2010) Influence of initial pesticide concentrations and plant population density on dimethomorph toxicity and removal by two duckweed species. Sci Total Environ 408(10):2254–2259

Dosnon-Olette R, Couderchet M, Eullaffroy P (2009) Phytoremediation of fungicides by aquatic macrophytes: toxicity and removal rate. Ecotoxicol Environ Saf 72:2096–2101

Eick MJ, Peak JD, Bady PV, Pesek JD (1999) Kinetics of lead absorption and desorption on goethite: residence time effect. Soil Sci 164:28–39

Eid ME, Gala TM, Sewelam NA, Talha NI, Abdallah SM (2020) Phytoremediation of heavy metals by four aquatic macrophytes and their potential use as a contamination indicator: a comparative assessment. Env Sci Pollu Resea 27:12138–12151

Ekperusi AO, Sikoki FD, Nwachukwu EO (2019) Application of common duckweed (*Lemna minor*) in phytoremediation of chemicals in the environment: state and future perspective. Chemosphere 223:285–309

El-Amier YA, El-Alfy MA, Nofal MM (2018) Macrophytes potential for removal of heavy metals from aquatic ecosystem, Egypt: using metal accumulation index (MAI). Plant Arch 18:2131–2144

Elayan NMS (1999) Phytoremediation of arsenic and lead from contaminated waters by the emergent aquatic Macrophyte *Althernanthera philoxeroides* (alligatorweed). Southern University, Louisiana, USA

El-Gendy OSN, Biswas BJK (2004) Growth of water hyacinth in municipal landfill leachate with different pH. Environ Technol 25:833–840

Elsaesser D, Stang C, Bakanov N, Schulz R (2013) The Landau stream mesocosm facility: pesticide mitigation in vegetated flow-through streams. Bull Environ Contam Toxicol 90(6):640–645

Espinoza-Quinones FR, Martin N, Stutz G, Tirao G, Palacio SM, Rizzutto MA, Modenes AN, Silva FG Jr, Szymansky N, Kroumov AD (2009) Root uptake and reduction of hexavalent chromium by aquatic macrophytes as assessed by high-resolution X-ray emission. Water Res 43:4159–4166

Favas PJC, Pratas J, Rodrigues N, D'Souza R, Varun M, Paul MS (2018) Metal (loid) accumulation in aquatic plants of a mining area: Potential for water quality biomonitoring and biogeochemical prospecting. Chemosphere 194:158–170

Fawzy MA, Badr NE, El-Khatib A, Abo-El-Kassem A (2012) Heavy metal biomonitoring and phytoremediation potentialities of aquatic macrophytes in River Nile. Environ Monit Assess 184:1753–1771

Ferdes M, Voicu G, Dinca M (2019) Comparative study of water discoloration by phytoremediation using *Pistia* stratiotes and *Chlorella vulgaris*. Rom Biotechnol Lett 24(6):953–960

Forni C, Cascone A, Fiori M, Migliore L (2002) Sulphadimethoxine and *Azolla Filiculoides* Lam.: a model for drug remediation. Water Res 36:3398–3403

Forni C, Chen J, Tancioni L, Caiola MG (2001) Evaluation of the fern *Azolla* for growth, nitrogen and phosphorus removal from wastewater. Water Res 35(6):1592–1598

Fraser LH, Carty SM, Steer D (2004) A test of four plant species to reduce total nitrogen and total phosphorus from soil leachate in subsurface wetland microcosms. Bioresour Technol 94(2):185–192

Fu X, He X (2015) Nitrogen and phosphorus removal from contaminated water by five aquatic plants. In: International Conference on Mechatronics, Electronic, Industrial and Control Engineering (MEIC-15), pp 1274–1277

Gao J, Wayne Garrison A, Hoehamer C, Mazur CS, Lee Wolfe N (2000) Uptake and phytotransformation of organophosphorus pesticides by axenically cultivated aquatic plants. J Agric Food Chem 48:6114–6120

Garbisu C, Alkorta I (2003) Basic concepts on heavy metal soil bioremediation. Eur J Min Proc Environ Protect 13:58–66

Garcinuno RM, Fernandez Hernando P, Camara C (2006) Removal of carbaryl, linuron, and permethrin by *Lupinus angustifolius* under hydroponic conditions. J Agric Food Chem 54:5034–5039

Geiger F, Bengtsson J, Berendse F, Weisser WW, Emmerson M, Morales MB, Ceryngier P, Liira J, Tscharntke T, Winqvist C, Eggers S, Bommarco R, Part T, Bretagnolle V, Plantegenest M, Clement LW, Dennis C, Palmer C, Onate JJ, Guerrero I, Hawro V, Aavik T, Thies C, Flohre A, Hanke S, Fischer C, Goedhart PW, Inchausti P (2010) Persistent negative effects of pesticides on biodiversity and biological control potential on European farmland. Basic Appl Ecol 11:97–105

Gomes MP, Juneau P (2016) Oxidative stress in duckweed (*Lemna Minor* L.) induced by Glyphosate: Is the mitochondrial electron transport chain a target of this herbicide? Environ Pollut 218:402–409

Gomes MP, Tavares DS, Richardi VS, Marques RZ, Wistuba N, de Brito JCM, Soffatti P, Sant'Anna-Santos BF, da Silva MAN, Juneau P (2019) Enrofloxacin and Roundup® interactive effects on the aquatic macrophyte *Elodea canadensis* physiology. Environ Pollut 249:453–462

Gruber H, Wiessner A, Kuschk P, Kaestner M, Appenroth KJ (2008) Physiological responses of *Juncus eusus (rush)* to chromium and relevance for wastewater treatment in constructed wetlands. Int J Phytorem 10:79–90

Gudisa I (2019) Removal of heavy metals from wastewater in Hawassa textile factory treatment ponds by *Shoenoplectus Lacustris*: Opportunity of phytoremediation by aquatic macrophyte. J Envir Earth Sci 9 (4):9–14

Guimaraes FP, Aguiar R, Karam D, Oliveira JA, Silva J, Santos CL, Sant'Anna-Santos BF, Lizieri-Santos C (2011) Potential of macrophytes for removing atrazine from aqueous solution. Planta Daninha 29 (SPE):1137–1147

Gujarathi NP, Haney BJ, Linden JC (2005) Phytoremediation Potential of *Myriophyllum Aquaticum* and *Pistia Stratiotes* to modify antibiotic growth promoters, tetracycline, and oxytetracycline, in aqueous wastewater systems. Int J Phytorem 7:99–112

Gupta S, Satpati S, Nayek S, Garai D (2010) Effect of wastewater irrigation on vegetables in relation to bioaccumulation of heavy metals and biochemical changes. Environ Monit Assess 165:169–177

Ha NTH, Sakakibara M, Sano S (2009) Phytoremediation of Sb, As, Cu, and Zn from contaminated water by the aquatic macrophyte *Eleocharis acicularis*. Clean: Soil, Air, Water 37:720–725

Harguinteguy CA, Cirelli AF, Pignata ML (2014) Heavy metal accumulation in leaves of aquatic plant *Stuckenia filiformis* and its relations with sediment and water in the Suquia river (Argentina). Microchem J 114:111–118

Hasan MR, Chakrabarti R (2009) Use of algae and aquatic macrophytes as feed in small scale aquaculture: a review. Food and Agriculture Organization of the United Nations, pp 1–135

Haule AT, Pratap HB, Katima HJY, Mugittu K, Mbwette TSA (2013) Potential macrophytes for nitrogen removal from domestic wastewater in horizontal subsurface flow constructed wetlands in Tanzania. Open Enviro Eng J 6:14–21

He N, Sun ZX, Zhang YL, Liu MD (2012) Nitrogen and phosphorus removal from simulated wastewater with aquatic macrophytes. Advan Materi Resea 518–523:2597–2603

Hoffmann T, Kutter C, Santamaria J (2004) Capacity of *Salvinia minima* baker to tolerate and accumulate As and Pb. Eng Life Sci 4:61–65

Hwang JI, Li Z, Andreacchio N, Ordonez Hinz F, Wilson PC (2020) Potential use of floating treatment wetlands established with *Canna Flaccida* for removing organic contaminants from surface water. Int J Phytoremediat 22:1304–1312

Jaishankar M, Tseten T, Anbalagan N, Mathew BB, Beeregowda KN (2014) Toxicity, mechanism and health effects of some heavy metals. Interdiscip Toxicol 7(2):60–72

Jayaweera MW, Kasturiarachchi JC (2010) Removal of nitrogen and phosphorus from industrial wastewaters by phytoremediation using water hyacinth (*Eichhornia crassipes*). Water Sci Techn 50:217–225

Jelic A, Gros M, Ginebreda A, Cespedes-Sanchez R, Ventura F, Petrovic M, Barcelo D (2011) Occurrence, partition and removal of pharmaceuticals in sewage water and sludge during wastewater treatment. Water Res 45:1165–1176

Jha P, Jobby R, Kudale S, Modi N, Dhaneshwar A, Desai N (2013) Biodegradation of phenol using hairy roots of *Helianthus annuus* L. Int Biodeterior Biodegrad 77:106–113

Jiang C, Chen H, Zhang Y, Feng H, Shehzad MA, Wang Y, Xu T (2018) Complexation electro dialysis as a general method to simultaneously treat wastewaters with metal and organic matter. Chem Eng J 348:952–959

Jones OAH, Voulvoulis N, Lester JN (2002) Aquatic environmental assessment of the top 25 English prescription pharmaceuticals. Water Res 36:5013–5022

Kabata-Pendias A, Pendias H (1992) Trace elements in soils and plants, 2nd edn. CRC Press, Boca Raton

Kadam SK, Chandanshive VV, Rane NR, Patil SM, Gholave AR, Khandare RV, Bhosale AR, Jeaon BH, Govindwar SP (2017) Phytobeds with *Fimbristylis dichotoma* and *Ammannia baccifera* for treatment of real textile effluent: an insitu treatment, anatomical studies and toxicity evaluation. Environ Res 160:1–11

Kadlec RH, Zmarthie LA (2010) Wetland treatment of leachate from a closed landfill. Ecol Eng 36:946–957

Kahkonen MA, Manninen PKG (1998) The uptake of nickel and chromium from water by *Elodea canadensis* at different nickel and chromium exposure levels. Chemosphere 36:1381–1390

Kassaye YA, Skipperud L, Einset J, Salbu B (2016) Aquatic macrophytes in Ethiopian RiftValley lakes: their trace elements concentration and use as pollution indicators. Aquat Bot 134:18–25

Keskinkan O, Lugal Goksu MZ (2007) Assessment of the dye removal capability of submersed aquatic plants in a laboratory-scale wetland system using anova. Brazilian J Chem Eng 24(02):193–202

Khan S, Munir S, Sajjad M, Li G (2016) Urban park soil contamination by potentially harmful elements and human health risk in Peshawar City, Khyber Pakhtunkhwa, Pakistan. J Geochem Explor 165:102–110

Khan S, Rehman S, Khan AZ, Khan MA, Shah MT (2010) Soil and vegetables enrichment with heavy metals from geological sources in Gilgit, northern Pakistan. Ecotoxicol Environ Saf 73:1820–1827

Khandare RV, Govindwar SP (2015) Phytoremediation of textile dyes and effluents: current scenario and future prospects. Biotechnol Adv 33:1697–1714

Khoshmanesh A, Hart BT, Duncan A, Beckett R (2002) Luxury uptake of phosphorus by sediment bacteria. Water Res 36(3):774–778

Kim SD, Cho J, Kim IS, Vanderford BJ, Snyder SA (2007) Occurrence and removal of pharmaceuticals and endocrine disruptors in South Korean surface, drinking, and wastewaters. Water Res 41:1013–1021

Kondo K, Kawabata H, Ueda S, Hasegawa H, Inaba J, Mitamura O, Seike Y, Ohmomo Y (2003) Distribution of aquatic plants and absorption of radionuclides by plants through the leaf surface in brackish Lake Obuchi, Japan, bordered by nuclear fuel cycle facilities. J Radioanal Nucl Chem 257:305–312

Kumari M, Tripathi BD (2015) Effect of *Phragmites Australis* and *Typha latifolia* on bifiltration of heavy metals from secondary treated effluent. Int J Environ Sci Tech 12:1029–1038

Kunii H, Aramaki M (1992) Annual net production and life span of floating leaves in *Nymphaea tetragona* Georgi: a comparison with other floating-leaved macrophytes. Hydrobiologia 242:185–193

Kyambadde J, Kansiime F, Gumaelius L, Dalhammar G (2009) A comparative study of *Cyperus papyrus* and *Miscanthidium violaceum*-based constructed wetlands for wastewater treatment in a tropical climate. Water Res 38:475–485

Lee CK, Low KS, Hew NS (1991) Accumulation of arsenic by aquatic plants. Sci Total Environ 103:215–227

Lema E, Machunda R, Nicholas NJAU (2014) Influence of macrophyte types towards agrochemical phytoremediation in a tropical environment. Inter J Eng Resea General Sci 2(5):441–455

Lemon GD, Posluszny U, Husband BC (2001) Potential and realized rates of vegetative reproduction in *Spirodela polyrhiza, Lemna minor*, and *Wolffia borealis*. Aquat Bot 70:79–87

Li M, Wu YJ, Yu ZL, Sheng GP, Yu HQ (2009) Enhanced nitrogen and phosphorus removal from eutrophic lake water by *Ipomoea aquatica* with low-energy ion implantation. Water Res 43(5):1247–1256

Li Y, Zhu G, Ng WJ, Tan SK (2014) A review on removing pharmaceutical contaminants from wastewater by constructed wetlands: Design, performance and mechanism. Sci the Total Envir 468–469:908–932

Licata M, Gennaro MC, Tuttolomondo T, Leto C, La Bella S (2019) Research focusing on plant performance in constructed wetlands and agronomic application of treated wastewater—a set of experimental studies in Sicily (Italy). PLoS ONE 14(7):e0219445

Lin DJ, Tang SHJ, He JW (2004) Purification of wastewater by *Ipomoea aquatica* and *Eichhornia crassipes* in a floating culture system. J South Agric China 25:14–17

Liu J, Zhang XH, You SH, Wu QX, Chen SM, Zhou KN (2014) Cr (VI) removal and detoxification in constructed wetlands planted with *Leersia hexandra Swartz*. Ecol Eng 71:36–40

Liu N, Wu Z (2018) Toxic effects of linear alkylbenzenesulfonate on *Chara vulgaris L.* Envir Sci Pollu Res 25:4934–4941

Liu Y, Liu N, Zhou Y, Wang F, Zhang Y, Wu Z (2019) Growth and physiological responses in *Myriophyllum spicatum L.* exposed to linear Alkylbenzene Sulfonate. Envir Toxico Chemis 38(9):2073–2081

Liu Y, Xiao T, Ning Z, Li H, Tang J, Zhou G (2013) High cadmium concentration in soil in the three gorges region: geogenic source and potential bioavailability. Appl Geochem 37:149–156

Lodeiro P, Cordeiro B, Barriada JL, Herrero R, Sastre de Vicente ME (2005) Biosorption of cadmium by biomass of brown marine macroalgae. Bioresour Technol 96:1796–1803

Ma X, Havelka MM (2009) Phytotoxicity of chlorinated benzenes to *Typha angustifolia* and *Phragmites communis*. Environ Toxicol 24:43–48

Madsen JD, Bloomfield JA, Sutherland JW, Eichler LW, Boylen CW (1996) The aquatic macrophyte community of Onondaga Lake: field survey and plant growth bioassays of lake sediments. Lake Reservoir Manage 12:73–79

Madsen JD, Chambers PA, James WF, Koch EW, Westlake DF (2001) The interactions between water movement, sediment dynamics and submersed macrophytes. Hydrobiologia 444:71–84

Mahabali S, Spanoghe P (2014) Mitigation of two insecticides by wetland plants: feasibility study for the treatment of agricultural runo in Suriname (South America). Water Air Soil Pollut 225:1771

Mandal BK, Suzuki KT (2002) Arsenic round the world: a review. Talanta 58(1):201–235

Manjunath S, Kousar H (2016) Phytoremediation of textile industry effluent using *Pistia stratiote*. Int J Environ Sci 5(2):75–81

Mant C, Costa S, Williams J, Tambourgi E (2006) Phytoremediation of chromium by model constructed wetland. Bioresour Technol 97:1767–1772

Metcalf and Eddy (2003) Wastewater engineering treatment and reuse. Mc GrawHill International Edition, New York

Miretzky P, Saralegui A, Cirelli AF (2004) Aquatic macrophytes potential for the simultaneous removal of heavy metal. Chemosphere 57:997–1005

Miretzky P, Saralegui A, Cirelli AF (2005) Simultaneous heavy metal removal mechanism by dead macrophytes. Chemosphere 62:247–254

Mishra M, Mohapatra A, Satapathy KB (2016) A comparative study of vermicompost prepared from phytoremediated and naturally grown aquatic weeds on growth and yield of green gram (*Vigna radiata* (L.) Wilczek). Int J Curr Res Biosci Plant Biol 3:104–109

Mishra V, Upadhyay A, Pathak V, Tripathi B (2008) Phytoremediation of mercury and arsenic from tropical opencast coalmine effluent through naturally occurring aquatic macrophytes. Water Air Soil Pollut 192:303–314

Mitsch WJ, Day JW, Gilliam JW, Groffman PM, Hey DL, Randall GW, Wang N (2001) Reducing nitrogen loading to the Gulf of Mexico from the Mississippi River Basin: strategies to counter a persistent ecological problem: ecotechnology—the use of natural ecosystems to solve environmental problems—should be a part of efforts to shrink the zone of hypoxia in the Gulf of Mexico. J Biosci 51(5):373–388

Mkandawire M, Dudel EG (2005) Accumulation of arsenic in *Lemna gibba* L. (duckweed) in tailing waters of two abandoned uranium mining sites in Saxony, Germany. Sci Total Environ 336:81–89

Mkandawire M, Lyubun YV, Kosterin PV, Dudel EG (2004) Toxicity of arsenic species to *Lemna gibba* L. and the influence of phosphate on arsenic bioavailability. Environ Toxicol 19:26–35

Mkandawire M, Taubert B, Dudel EG (2004) Capacity of *Lemna gibba* L. (Duckweed) for uranium and arsenic phytoremediation in mine tailing waters. Int J Phytoremediation 6:347–362

Mohamed AA, Castagna A, Ranieri A, di Toppi LS (2012) Cadmium tolerance in *Brassica juncea* roots and shoots is affected by antioxidant status and phytochelatin biosynthesis. Plant Physiol Biochem 57:15–22

Moore MT, Tyler HL, Locke MA (2013) Aqueous pesticide mitigation efficiency of *Typha latifolia (L.)*, *Leersia oryzoides* (L.) Sw., and *Sparganium americanum* Nutt Chemosphere 92(10):1307–1313

Moshiri GA (1993) Constructed wetlands for water quality improvement. Lewis Publishers, CRC Press Inc, Florida

Mungur AS, Shutes RBE, Revitt DM, House MA (1997) An assessment of metal removal by a laboratory scale wetland. Water Sci Techn 35(5):125–133

Muvea FM, Ogendi GM, Omondi SO (2019) Nutrient removal efficiency by floating macrophytes; *Lemna minor* and *Azolla pinnata* in a constructed wetland. Global J Environ Sci Manage 5(4):415–430

Nafea EM (2019) Floating macrophytes efficiency for removing of heavy metals and phenol from wastewaters. Egyptian J Aqua Biolo Fisher 23:1–9

Naja GM, Volesky B (2009) Toxicity and sources of Pb, Cd, Hg, Cr, As, and radionuclides in the environment. In: Wang LK, Chen JP, Hung YT, Shammas NK (eds) Heavy metals in the environment. CRC Press, Bota Racon, pp 13–61

Navarro-Leon E, Oviedo-Silva J, Ruiz JM, Blasco B (2019) Possible role of HMA4a TILLING mutants of *Brassica rapa* in cadmium phytoremediation programs. Ecotoxicol Environ Saf 180:88–94

Nawab J, Khan S, Aamir M, Shamshad I, Qamar Z, Din I, Huang Q (2016) Organic amendments impact the availability of heavy metal (loid) s in mine-impacted soil and their phytoremediation by *Penisitum americanum* and *Sorghum bicolor*. Environ Sci Pollut Res 23:2381–2390

Niazi NK, Bibi I, Fatimah A, Shahid M, Javed MT, Wang H, Ok YS, Bashir S, Murtaza B, Saqib ZA, Shakoor MB (2017) Phosphate-assisted phytoremediation of arsenic by *Brassica napus* and *Brassica juncea*: Morphological and physiological response. Int J Phytoremediation 19(7):670–678

Nie X, Dong F, Ding DX, Liu N, Li GY, Zhang D, Liu M (2015) Ability of *Pistia stratiotes L.* and *Eichhornia crassipes* for remediation of uranium-contaminated waste water. Yuanzineng Kexue Jishu/Atomic Energy Sci Technol 49(11):1946–1953

Nunez SER, Negrete JLM, Rios JAH, Hadad R, Maine M (2011) Hg, Cu, Pb, Cd, and Zn accumulation in macrophytes growing in tropical wetlands. Water Air Soil Pollut 216:361–373

Nyomora AMS, Njau KN, Mligo L (2012) Establishment and growth of vetiver grass exposed to landfill leachate. J Solid Waste Techn Manag 38(2):82–92

Nyquist J, Greger M (2007) Uptake of Zn, Cu and Cd in metal loaded *Elodea canadensis*. Environ Exp Pot 60:219–226

Obi C, Woke J (2014) The removal of phenol from aqueous solution by Colocasia *Esculenta Araesia Linn* Schott. Sky J Soil Sci Environ Manag 3(6):59–66

Odjegba VJ, Fasidi IO (2004) Accumulation of trace elements by *Pistia stratiotes*: implications for phytoremediation. Ecotoxicology 13:637–646

Olette R, Couderchet M, Biagianti S, Eullaffroy P (2008) Toxicity and removal of pesticides by selected aquatic plants. Chemosphere 70:1414–1421

Ong S, Uchiyama K, Inadama D, Ishida Y, Yamagiwa K (2010) Treatment of azo dye Acid Orange 7 containing wastewater using up-flow constructed wetland with and without supplementary aeration. Bioresour Technol 101:9049–9057

Ontanon O, Gonzalez P, Agostini E (2015) Biochemical and molecular mechanisms involved in simultaneous phenol and Cr(VI) removal by *Acinetobacter guillouiae SFC 500-1A*. Environ Sci Pollut Res 22 (17):13014–13023

Overall RA, Parry DL (2004) The uptake of uranium by *Eleocharis dulcis* (Chinese water chestnut) in the ranger uranium mine constructed wetland filter. Environ Pollut 132:307–320

Paisio CE, Agostini E, Gonzalez PS, Bertuzzi ML (2009) Lethal and teratogenic effects of phenol on Bufo arenarum embryos. J Hazard Mat 167:64–68

Paisio CE, Fernandez M, Gonzalez PS, Talano MA, Medina MI, Agostini E (2017) Simultaneous phytoremediation of chromium and phenol by *Lemna minuta Kunth*: a promising biotechnological tool. Inter J Envir Sci Techn 15(1):37–48

Palmer MR, Edmond JM (1993) Uranium in river water. Geochim Cosmochim Acta 57(20):4947–4955

Pan LB, Ma J, Wang XL, Hou H (2016) Heavy metals in soils from a typical county in Shanxi Province, China: levels, sources and spatial distribution. Chemosphere 148:248–254

Patel VS, Adhvaryu MR (2015) Removal of textile dye by using *Eichhornia* spp. and *Pistia* spp. by aquatic macrophytes treatment systems (AMTS)—an eco-friendly technique. Inter J Sci Eng Res 4(7):62–64

Pavithra M, Kousar H (2016) Utilization of *Pistia* stratiotes for textile wastewater nitrates and phosphates removal SSRG. Inter J Agric Envir Sci 3 (6):31–34

Petrucio MM, Esteves FA (2000) Uptake rates of nitrogen and phosphorus in the water by *Eichhornia azurea* and *Salvinia auriculata*. Rev Bras Biol 60(2):229–236

Pettersson HBL, Johnston HA, Murray AS (1993) Uptake of uranium and thorium series radionuclides by the waterlily, Nymphaea violacea. J Environ Radioact 19:85–108

Plechonska L, Klink A (2014) Trace metal bioindication and phytoremediation potentialities of *Phalaris arundinacea L.* (red canary grass). J Geochem Explor 146:27–33

Pompeo M (2017) Monitoring and management of aquatic macrophytes in Brazilian tropical reservoirs
Powell N, Shilton A, Chisti Y, Pratt S (2009) Towards a luxury uptake process via microalgae defining the polyphosphate dynamics. Water Res 43(17):4207–4213
Pratas J, Favas PJC, Paulo C, Rodrigues N, Prasad MNV (2012) Uranium accumulation by aquatic plants from uranium-contaminated water in Central Portugal. Int J Phytoremediation 14:221–234
Pratas J, Paulo C, Favas PJC, Venkatachalam P (2014) Potential of aquatic plants for phytofiltration of uranium-contaminated waters in laboratory conditions. Ecol Eng 69:170–176
Priya ES, Selvan PS (2017) Water hyacinth (*Eichhornia crassipes*)—an efficient and economic adsorbent for textile effluent treatment—a review. Arab J Chem 10 (2):S3548–S3558
Rahman MA, Hasegawa H, Kitahara K, Maki T, Ueda K, Rahman MM (2008) The effects of phosphorous on the accumulation of arsenic in water fern (*Azolla pinnata* L.). J Ecotechnol Res 14:21–24
Rahman MA, Hasegawa H, Ueda K, Maki T, Okumura C, Rahman MM (2007) Arsenic accumulation in duckweed (*Spirodela polyrhiza*): a good option for phytoremediation. Chemosphere 69:493–499
Rahman MA, Hasegawa H, Ueda K, Maki T, Rahman MM (2008) Arsenic uptake by aquatic macrophyte *Spirodela polyrhiza L.:* interactions with phosphate and iron. J Hazard Mater 160:356–361
Rahman MA, Hasegawa H, Ueda K, Maki T, Rahman MM (2008) Influence of phosphate and iron ions in selective uptake of arsenic species by water fern (*Salvinia natans L.*). Chem Eng J 145:179–184
Rahman MA, Hasegawa H, Ueda K, Maki T, Rahman MM (2008) Influence of EDTA and chemical species on arsenic accumulation in *Spirodela polyrhiza L.* (duckweed). Ecotoxicol Environ Saf 70:311–318
Rahmani GNH, Sternberg SPK (1999) Bioremoval of lead from water using *Lemna minor*. Biores Technol 70:225–230
Rai PK (2008) Heavy-metal pollution in aquatic ecosystems and its phytoremediation using wetland plants: an eco-sustainable approach. Inter J Phytore 10:133–160
Rai PK (2009) Heavy metal phytoremediation from aquatic ecosystems with special reference to macrophytes. Critical Revie Envir Sci Techn 39:697–753
Rai PK (2011) Heavy metal pollution and its phytoremediation though wetland plants. Nova Science Publisher, New York, p 196
Rai PK (2012) An eco-sustainable green approach for heavy metals management: two case studies of developing industrial region. Envirl Monit Assess 184:421–448
Rai PK, Mishra A, Tripathi BD (2010) Heavy metals and microbial pollution of river Ganga: a case study on water quality at Varanasi. Aqua Ecos Healt Manag 13:352–361
Rai UN, Sinha S, Tripathi RD, Chandra P (1995) Wastewater treatability potential of some aquatic macrophytes: removal of heavy metals. Ecol Eng 5:5–12
Rane NR, Patil SM, Chandanshive VV, Kadam SK, Khandare RV, Jadhav JP, Govindwar SP (2016) *Ipomoea hederifolia* rooted soil bed and Ipomoea aquatic rhizofiltration coupled phytoreactors for efficient treatment of textile wastewater. Water Res 96:1–11
Ravena O (2001) Ecological monitoring for water body management. In: Timmerman JG (ed) Proceedings of the international workshop on information for sustainable water management (25–28 Sept 2000), pp 157–167
Reddy KR, DeBusk WF (1987) Nutrient storage capabilities of aquatic and wetland plants. In: Reddy KR, Smith WH (eds) Aquatic plants for waste treatment and resource recovery. Magnolia Publishing, Pineville, LA, pp 337–335
Reddy KR, Tucker JC (1985) Growth and nutrient uptake of pennyworth (*Hydrocotyle umbellate* L.) as influenced by the nitrogen concentration of the water. J Aquat Plant Manag 23:35–40
Reddy MV, Kumar AV (2001) Effects of Ganesh-idol immersion on some water quality parameters of Hussainsagar Lake. Curr Sci 81(11):1412–1413
Redshaw CH, Wootton VG, Rowland SJ (2008) Uptake of the pharmaceutical fluoxetine hydrochloride from growth medium by *Brassicaceae*. Phytochemistry 69:2510–2516
Reichenberger S, Bach M, Skitschak A, Frede H-G (2007) Mitigation strategies to reduce pesticide inputs into ground and surface water and their effectiveness; a review. Sci Total Environ 384:1–35
Riaz G, Tabinda AB, Iqbal S, Yasar A, Abbas M, Khan AM, Mahfooz Y, Baqar M (2017) Phytoremediation of organochlorine and pyrethroid pesticides by aquatic macrophytes and algae in freshwater systems. Int J Phytoremediat 19(10):894–898
Robinson B, Duwing C, Bolan N, Kannathasan M, Saravanan A (2003) Uptake of arsenic by New Zealand watercress (*Lepidium sativum L.*). Sci Total Environ 301:67–73
Robinson B, Marchetti M, Moni C, Schroeter L, van den Dijssel C, Milne G, Bolan N, Mahimairaja S (2005) Arsenic accumulation by aquatic and terrestrial plants. In: Naidu R, Smith E, Owens G, Bhattacharya P, Nadebaum P (eds) Managing arsenic in the environment: from soil to human health. CSIRO, Collingwood, Victoria, pp 235–247
Romero Nunez SE, Marrugo-Negrete JL, Arians-Rios JE, Hernan R, Maine MA (2011) Hg, Cu, Pb, Cd, and Zn accumulation in macrophytes growing in Tropical Wetlands. Water Air Soil Poll 216:361–373
Roy CK, Jahan MAA, Rahman SS (2018) Characterization and treatment of textile wastewater by aquatic plants (Macrophytes) and algae. Euro J Susta Develop Res 2(3):29

Roy R, Fakhruddin ANM, Khatun R, Islam MS, Ahsan MA, Neger AJMT (2010) Characterization of textile industrial effluents and its effects on aquatic macrophytes and algae. Bangladesh J Sci Ind Res 45 (1):79–84

Samecka-Cymermann A, Kempers AJ (1996) Bioaccumulation of heavy metals by aquatic macrophytes around Wrocław, Poland. Ecotox Envir Safety 35 (3):242–247

San Miguel A, Ravanel P, Raveton MA (2013) Comparative study on the uptake and translocation of organochlorines by *Phragmites australis*. J Hazard Mater 244–245:60–69

Sand-Jensen K (1998) Influence of submerged macrophytes on sediment composition and near-bed flow in lowland streams. Fresh Wat Biol 39:663–679

Sangeeta D (2007) Role of Macrophytes in improving water quality of an aquatic eco-system. J Appl Sci Environ Manag 11(4):131–135

Sangeeta D, Savita D (2007) Water quality improvement through Macrophytes: a case study. Asian J Exp Sci 21(2):427–430

Sawidis T, Chettri MK, Zachariadis GA, Stratis JA (1995) Heavy metals in aquatic plants and sediments from water systems in Macedonia, Greece. Ecotoxicol Environ Saf 32:73–80

Schulz R (2004) Field studies on exposure, effects, and risk mitigation of aquatic nonpoint-source insecticide pollution: a review. J Environ Qual 33:419–448

Shah M, Hashmi HN, Ali A, Ghumman AR (2014) Performance assessment of aquatic macrophytes for treatment of municipal wastewater. J Environ Health Sci Eng 12(1):106

Shanker AK, Djanaguiraman M, Venkateswarlu B (2009) Chromium interactions in plants: current status and future strategies. Metallomics 1:375–383

Shanmugam L, Ahire M, Nikam T (2020) *Bacopa monnieri* (L.) Pennell, a potential plant species for degradation of textile azo dyes. Environ Sci Pollut Res 27:9349–9363

Shanying H, Xiaoe Y, Zhenli H, Baligar VC (2017) Morphological and physiological responses of plants to cadmium toxicity: a review. Pedosphere 27:421–438

Shardendu S, Sayantan D, Sharma D, Irfan S (2012) Luxury uptake and removal of phosphorus from water column by representative aquatic plants and its implication for wetland management. ISRN Soil Sci 2012:1–9

Shehzadi M, Afzal M, Khan MU, Islam E, Mobin A, Anwar S, Khan QM (2014) Enhanced degradation of textile effluent in constructed wetland system using *Typha domingensis* and textile effluent-degrading endophytic bacteria. Water Res 58:152–159

Sima J, Svoboda L, Seda M, Krejsa J, Jahodova J (2019) The fate of selected heavy metals and arsenic in a constructed wetland. J Environ Sci Health Part A 54:56–64

Singh DV, Bhat JIA, Bhat RA, Dervash MA, Ganei SA (2018) Vehicular stress a cause for heavy metal accumulation and change in physico-chemical characteristics of road side soils in Pahalgam. Environ Monit Assess 190:353

Singh N, Kaur M, Katnoria JK (2017) Analysis on bioaccumulation of metals in aquatic environment of Beas River Basin: a case study from Kanjli wetland. GeoHealth 1:93–105

Sinha A, Lulu S, Vino S, Jabez Osborne W (2019) Reactive green dye remediation by Alternanthera philoxeroides in association with plant growth promoting *Klebsiella* sp. VITAJ23: a pot culture study. Microbiol Res 220:42–52

Sivakumar D (2014) Role of *Lemna minor* lin. in treating the textile industry wastewater. Int J Environ Earth Sci Eng 8(3):55–59

Sivakumar D, Shankar D, Vijaya Prathima AJR, Valarmathi M (2013) Constructed wetland treatment of textile industry wastewater using aquatic macrophytes. Inter J Env Sci 3(4):1223–1232

Sood A, Uniyal PL, Prasanna R, Ahluwalia AS (2012) Phytoremediation potential of aquatic macrophyte, Azolla. Ambio 41(2):122–137

Sooknah RD, Wilkie AC (2004) Nutrient removal by floating aquatic macrophytes cultured in anaerobically digested flushed dairy manure wastewater. Ecol Eng 22:27–42

Soudek P, Valenova S, Benesova D, Vanek T (2007) From laboratory experiments to large scale application: an example of the phytoremediation of radionuclides. In: Marmiroli N, Samotokin B, Marmiroli M (eds) Advanced science and technology for biological decontamination of sites affected by chemical and radiological nuclear agents, vol 75. NATO Science Series IV Earth and Environmental Sciences. Springer, Netherlands, pp 139–158

Srivastav RK, Gupta SK, Nigam KDP, Vasudevan P (1993) Use of aquatic plants for the removal of heavy metals from wastewater. Int J Environ Stud 45(1):43–50

Srivastava S, Bhainsa KC, D'Souza SF (2010) Investigation of uranium accumulation potential and biochemical responses of an aquatic weed *Hydrilla Verticillata* (Lf). Royle Bioresour Technol 101:2573–2579

Stang C, Bakanov N, Schulz R (2016) Experiments in water-macrophyte systems to uncover the dynamics of pesticide mitigation processes in vegetated surface waters/streams. Environ Sci Pollut Res Int 23(1):673–682

Stang C, Elsaesser D, Bundschuh M, Ternes TA, Schulz R (2013) Mitigation of biocide and fungicide concentrations in flow-through vegetated stream mesocosms. J Environ Qual 42(6):1889–1895

Suhendrayatna M, Andriani R, Fajriana Y, Elvitriana (2012) Removal of municipal wastewater BOD, COD, and TSS by phytoreduction: a laboratory-scale comparison of aquatic plant at different species *Typha latifolia* and *Saccharum spontaneum*. Int J Innov Res Sci Eng Technol 2:333–337

Sundaramoorthy P, Chidambaram A, Ganesh KS, Unnikannan P, Baskaran L (2010) Chromium stress in paddy: (i) Nutrient status of paddy under chromium stress;

(ii) Phytoremediation of chromium by aquatic and terrestrial weeds. Comptes Rendus Biol 333:597–607

Tadesse AT, Seyoum LA (2015) Evaluation of selected wetland plants for removal of chromium from tannery wastewater in constructed wetlands, Ethiopia. Afr J Environ Sci Technol 9:420–427

Tambunan JAM, Effendi H, Krisanti M (2018) Phytoremediating Batik wastewater using vetiver *Chrysopogon zizanioides* (L). Pol J Environ Stud 27:1281–1288

Tao W, Han Y, Li H (2015) Investigation into ammonia stress on *Cyperus alternifolius* and its impact on nutrient removal in microcosm experiments. Environ Manage 163:254–261

Terfie TA, Asfaw SL (2015) Evaluation of selected wetland plants for removal of chromium from tannery wastewater in constructed wetlands, Ethiopia. Afr J Environ Sci Technol 9:420–427

Thani NSM, Ghazi RM, Amin MFM, Hamzah Z (2019) Phytoremediaton of heavy metals from wastewater by constructed wetland microcosm planted with *Alocasia puber*. Jurnal Teknologi 81:5

Thapa G, Das D, Gunupuru LR, Tang B (2016) Endurance assessment of *Eichhornia crassipes* (Mart.) Solms, in heavy metal contaminated site—a case study. Cogent Environ Sci 2(1):1215280

Thomas KA, Hand LH (2011) Assessing the potential for algae and macrophytes to degrade crop protection products in aquatic ecosystems. Environ Toxicol Chem 30:622–631

Torok A, Buta E, Indolean C, Tonk S, Silaghi-Dumitrescu L, Majdik C (2015) Biological removal of triphenylmethane dyes from aqueous solution by *Lemna minor*. Acta Chim Slov 62:452–461

Tripathi BD, Shukla SC (1991) Biological treatment of wastewater by selected aquatic plants. Environ Pollut 69:69–78

Tsuchiya T (1989) Growth and biomass turnover of *Hydrocharis dubia L.* cultured under different nutrient conditions. Ecol Res 4:157–166

Tsuchiya T, Iwakuma T (1993) Growth and leaf life-span of a floating-leaved plant, *Trapa natans L.,* as influenced by nitrogen flux. Aquat Bot 46:317–324

Ugya A (2015) The efficiency of *Lemna minor L.* In the phytoremediation of Rani Stream: a case study of Kaduna Refinery and Petrochemical Company polluted stream. J App Biol Biotech 1:011–014

Ugya AY, Hua X, Ma J (2019) Phytoremediation as a tool for the remediation of wastewater resulting from dyeing activities. Appl Ecol Environ Res 17:3723–3735

Uka UN, Chukwuka KS (2011) Utilization of aquatic macrophytes in Nigerian freshwater ecosystem. J Fish Aquat Sci 6(5):490–498

Upadhyay MK, Gautam A, Mallick S, Srivastava S (2017) A successive application approach for effective utilization of three aquatic plants in Arsenic removal. Water Air Soil Pollut 228(2):54

Uslu MO, Jasim S, Arvai A, Bewtra J, Biswas N (2013) A survey of occurrence and risk assessment of pharmaceutical substances in the Great Lakes Basin. Ozone: Sci Eng 35(4):249–262

Uysal Y, Taner F (2009) Effect of pH, temperature and Lead concentration on the bioremoval of Lead from water using *L. minor*. Int J Phytoremed 11:591–608

Vaillant N, Fabien M, Huguette S (2004) Use of commercial plant species in a hydroponic system to treat domestic wastewaters. J Envir Qual 33:695–702

Verla AW, Verla EN, Amaobi CE, Enyoh CE (2018) Water pollution scenario at river Uramurukwa flowing through Owerri Metropolis, Imo State, Nigeria. Inter J Sci Resea 3:40–46

Vidal CF, Oliveira JA, da Silva AA, Ribeiro C, Farnese FDS (2019) Phytoremediation of arsenite-contaminated environments: Is *Pistia stratiotes* L. a useful tool? Ecol Ind 104:794–801

Vidal-Dorsch DE, Bay SM, Maruya K, Snyder SA, Trenholm RA, Vanderford BJ (2012) Contaminants of emerging concern in municipal wastewater effluents and marine receiving water. Envir Toxico Chemis 31 (12):2674–2682

Vidayanti V, Choesin DN (2017) Phytoremediation of chromium: distribution and speciation of chromium in Typha angustifolia. Int J Plant Biol 8(6870):14–18

Vyas A, Bajpai A, Verma N (2008) Water quality improvement after shifting of idol immersion site: a case study of Upper Lake, Bhopal, India. Environ Monit Assess 145(1–3):437–443

Vymazal J (2010) Constructed wetlands for wastewater treatment. Water 2:530–549

Wahaab AR, Lubberding HJ, Alaerts GJ (1995) Copper and chromium (III) uptake by duckweed. Water Sci Technol 32:105–110

Walker PD, Wijnhoven S, van der Velde G (2013) Macrophyte presence and growth form influence macroinvertebrate community structure. Aquat Bot 104:80–87

Wharton G, Cotton JA, Wotton RS, Bass JAB, Heppell CM, Trimmer M, Sanders IA, Warren LL (2006) Macrophytes and suspension-feeding invertebrates modify flows and fine sediments in the Frome and Piddle catchments, Dorset (UK). J Hydrol 330:171–184

Wickramasinghe S, Jayawardana CK (2018) Potential of aquatic macrophytes *Eichhornia crassipes, Pistia stratiotes* and *Salvinia molesta* in phytoremediation of textile wastewater. J Water Security 4:1–8

Xu L, Cheng S, Zhuang P, Xie D, Li S, Liu D, Li Z, Wang F, Xing F (2020) Assessment of the nutrient removal potential of floating native and exotic aquatic macrophytes cultured in Swine manure wastewater. Int J Environ Res Public Health 17:1103

Yadav S, Chandra R (2011) Heavy metals accumulation and eco physiological effect on *Typha angustifolia L.* and *Cyperus esculentus* L. growing in distillery and tannery effluent polluted natural wetland site, Unnao, India. Environ Earth Sci 62:1235–1243

Yeruva DK, Ranadheer P, Kumar AK, Mohan SV (2019) Tri-phasic engineered wetland system for effective treatment of azo dye-based wastewater. NPJ Clean Water 2(1):1–8

Zaidi J, Khan AH, Pal A (2017) Some aquatic macrophytes and their metal accumulation potentiality. J Ecophysiol Occupational Health 17(3/4):93–100

Zhang T, Lu Q, Su C, Yang Y, Hu D, Xu Q (2017) Mercury induced oxidative stress, DNA damage, and activation of antioxidative system and Hsp70 induction in duckweed (*Lemna minor*). Ecotox Envir Safety 143:46–56

Zhang WM (2007) Purifying effect and mechanism of hydrophytes in eutrophic water. PhD Thesis, Hehai University, Nanjing, China, pp 11–32

Zhang X, Lin AJ, Zhao FJ, Xu GZ, Duan GL, Zhu YG (2008) Arsenic accumulation by the aquatic fern Azolla: Comparison of arsenate uptake, speciation and efflux by A. *caroliniana and A. filiculoides*. Envir Pollu 156(3):1149–1155

Zhang X, Xia H, Li Z, Zhang P, Gao B (2010) Potential of four forage grasses in remediation of Cd and Zn contaminated soils. Bioresour Technol 101:2063–2066

Zhang X, Zhao FJ, Huang Q, Williams PN, Sun GX, Zhu YG (2009) Arsenic uptake and speciation in the rootless duckweed *Wolffia globosa*. New Phytol 182:421–428

Zhao L, Zhu C, Gao C, Jiang J, Yang J, Yang S (2011) Phytoremediation of pentachlorophenol-contaminated sediments by aquatic macrophytes. Environ Earth Sci 64:581

Zhou J, Wu Z, Yu D, Pang Y, Cai H, Liu Y (2018) Toxicity of linear alkylbenzene sulfonate to aquatic plant *Potamogeton perfoliatus L.* Envir Sci Pollu Resea 25:32303–32311

Zhou ZY, Liu WX, Pei G, Ren H, Wang J, Xu QL, Xie HH, Wan FH, Tan JW (2013) Phenolics from *Ageratina adenophora* roots and their phytotoxic effects on *Arabidopsis thaliana* seed germination and seedling growth. J Agric Food Chem 61(48):11792–11799

Bioremediation of Salt-Affected Soil Through Plant-Based Strategies

5

Anup Kumar Sarkar and Sanjoy Sadhukhan

Abstract

Soil salinity is a rising human concern and a significant challenge to biodiversity, since high salinity makes the soil inappropriate for most plants. The contamination of soils due to salinity damages almost one-fourth of agricultural land. Soil salinization is a natural process and is amplified by several anthropogenic practices. Saline soil affects the growth and development of the majority of the plants. The remediation of soil salinity is an economically expensive challenge of the present era. Since the last few decades, several approaches for amelioration of salt-affected soil have been used but among these techniques, few are less expensive. Though most of the plants are severely affected by soil salinization, some plants develop many tolerance mechanisms and detoxification strategies to remove excess salt from the soil. Plants are utilized to remove excess salt is nowadays considered as one of the effective and less expensive useful options. Such kind of remediation is also called as phytoremediation or green remediation. Plant-based remediation now has gained much attention as it facilitates benefit in various types of salt-affected habitats worldwide. This chapter reviews various plant-based strategies of bioremediation of salt-affected soil.

Keywords

Halophyte · Hyperaccumulation · Phytoextraction · Phytopumping · Phytoremediation · Phytostimulation · Rhizosphere · Saline soil · Sodic soil

5.1 Introduction

Since ancient times, human activities become a threat to the stability of the earth ecosystem. During evolution, human has been modifying the topographical texture and environment around them through agriculture, domestication, travel, urbanization and commercial purposes. With the advancement of technology, such changes gradually become a severe threat to sustainability and stability of the globe. Among the recognized threat to the earth, the most decisive one is soil salinity. Soil salinization is a natural process and also induced by anthropogenic activities. Normal salinity causes include rock erosion, capillary elevation from deeper brackish groundwater, seawater accumulation in the coastal area, salt-laden sand blown by sea wind, impeded

A. K. Sarkar
Department of Botany, Dukhulal Nibaran Chandra College, Aurangabad, Murshidabad, West Bengal 742201, India

A. K. Sarkar · S. Sadhukhan (✉)
Plant Molecular Biology Laboratory, Department of Botany, Raiganj University, Uttar Dinajpur, Raiganj, West Bengal 733134, India

J. A. Malik (ed.), *Advances in Bioremediation and Phytoremediation for Sustainable Soil Management*,
https://doi.org/10.1007/978-3-030-89984-4_5

drainage, etc., whereas anthropogenic factors include irrigation without a proper drainage system, industrial effluents, aberrant and unscientific use of fertilizers and pesticides, erosion, salt water flooding, high water level, low quality groundwater use for irrigation, etc. (Rasool et al.). Salinity turns out to be a problem while enough salts concentrate in the root system to adversely affect plant growth and metabolism. The problem of salinity affects almost all countries around the globe, and it causes a high ecological and economic cost and poses a challenge to global food security through agriculture (Fig. 5.1).

The salinity of the soil damages the land and thereby deteriorates soil fertility, soil stability, and biodiversity. A mild increase in salt is beneficial for soil as it helps in flocculation, i.e., binding of soil particles into aggregates, which in turn facilitates soil aeration, root penetration and root growth. Yet high salinity levels have detrimental and possibly fatal effects on plants. Thus, soil should regularly undergo thorough remediation processes at least for the food security and stability of the ecosystem. Remediation is a technical term used to refer to various strategies of soil decontamination. There are several techniques for soil remediation, each employing a distinct mechanism for removing contaminants like salts, heavy metals, etc. from the soil. Soil remediation techniques can be categorized into two groups—non-biological remediation and biological remediation (Fig. 5.2). Non-biological remediation includes physical–thermal treatment, manipulating water, structural engineering options (physical separation) and chemical-leaching of salts, applying chemical amendments (Bhuiyan et al. 2017). Although such tactics are scientifically approved and useful in some manner, due to high expenses and less eco-friendly approach, these are not widely utilized in all corners of the world. Alternatively, biological remediation or bioremediation has gained huge

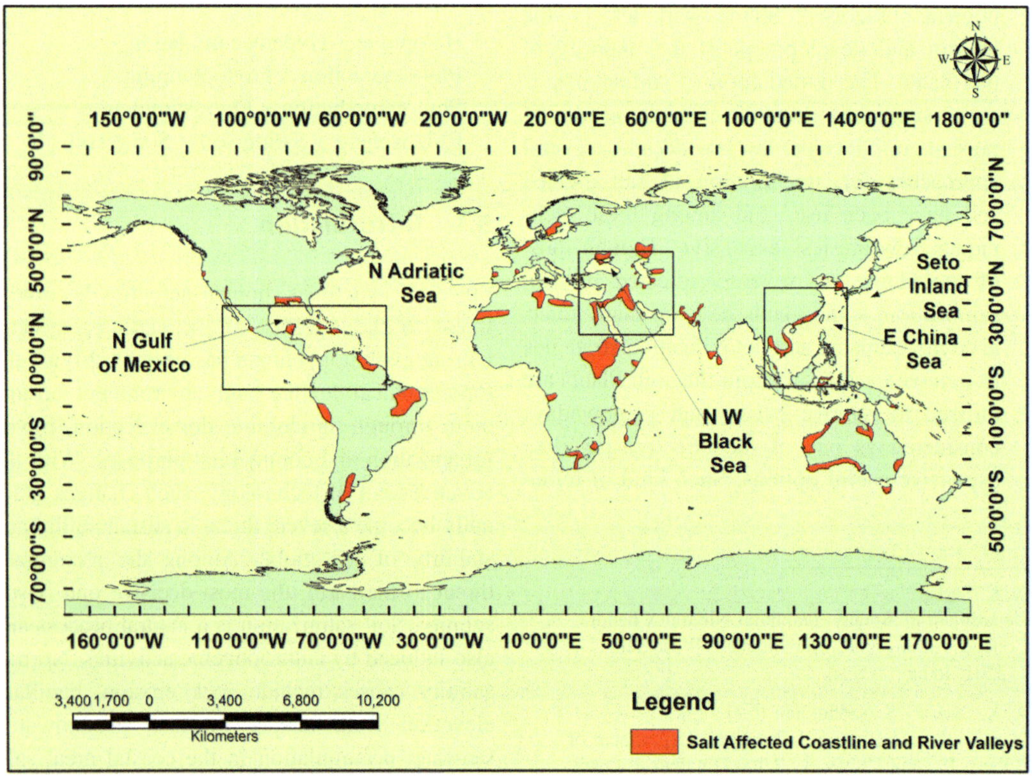

Fig. 5.1 World map of salt-affected soil. (Li et al. 2014)

acceptance due to its low expense and eco-friendly approach. The use of living species such as plants, microorganisms, animals, etc. is bioremediation by definition (Fig. 5.2) to degrade the environmental contaminants into less toxic forms (Kensa 2011). The technology involving bacterial, fungal, plants, animals or the enzymes secreted by them in the environment to clean the contaminants is included in bioremediation. Plants and/or plant-assisted remediation process is usually called phytoremediation.

5.2 Soil salinity—A Threat to Biodiversity of Earth

Soil salinity is a matter of environmental contention of the hour. The soil is theoretically known to be salty when the soil solution's electrical conductivity (EC) exceeds 4 dSm^{-1} (equivalent to 40 mM NaCl). The salty soil produces an osmotic pressure of around 0.2 MPa which dramatically decreases growth and productivity of many plants including agriculturally important crop plants (Munns and Tester 2008). Salts can be dissolved in water (soluble salts) or be present as solids. Based on their nature and properties, salt-affected soils are categorized into three main groups, viz. saline soil, sodic soil and saline-sodic soil. Saline soils contain sufficient neutral soluble salts, mainly chlorides and sulfates of sodium, calcium and magnesium, which adversely affect the plant growth. The structure, porosity and water content of the soil are affected by the salts resulting in the decrease in yield (McCauley et al. 2005). Sodic soils are distinguished by the presence of elevated sodium (Na^+) at levels which can negatively impact soil structure and the supply of some nutrients (Qadir et al. 2001). Sodic soils have a poor physical property, i.e., poor infiltration rate, and inadequate aeration and fertility problems that adversely affect the growth and yield of crops and difficult to cultivate (Qadir and Schubert 2002). Soil erosion due to salinity and/or sodium is a serious geo-environmental restriction with many detrimental consequences on plants and other components of biodiversity. Saline soils with sodium surface properties such as sodium abundance and impermeable structure are referred to as saline-sodic soils. It has characters intermediate between saline and sodic soils as following: EC > 4, ESP > 15 and pH <8.5 (Nouri et al. 2017). Recovery of saline-sodic soils is similar to sodic soils, i.e., by removal of excess salts and sodium.

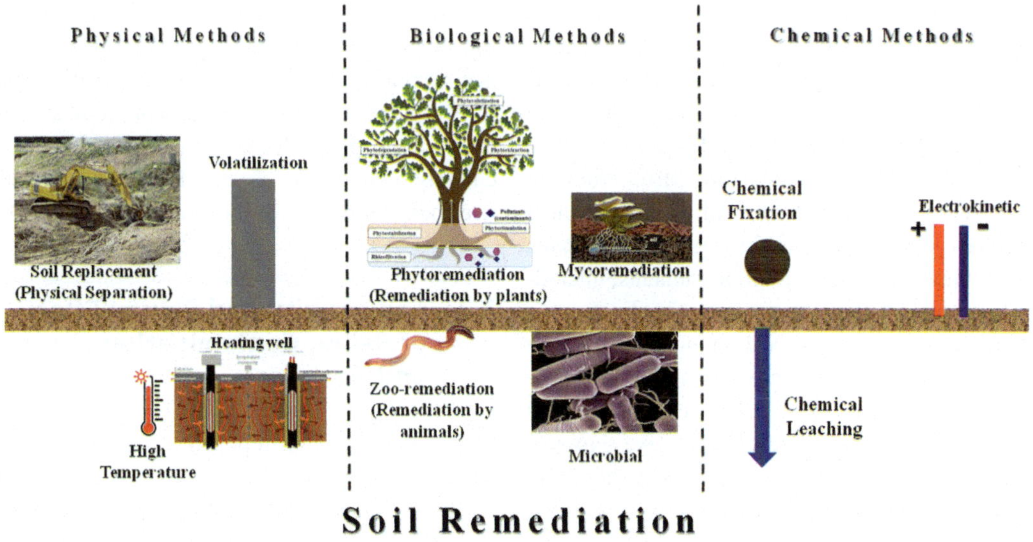

Fig. 5.2 Different methods of soil remediation

Around 831 Mha of land is salt-affected soil worldwide (Zhang et al. 2018), of which crop loss of above US$ 27 billion per year is due to irrigation-induced salinity (Qadir et al. 2014). Salts are present in the soil as natural components of the earth's crust and useful to plants up to a certain level but beyond this level, it exerts negative influence to plants. Degradation of salt-affected soil has triggered imbalances between the supply of food and plant-based natural resources and the demands of a society that exploit these (Qadir et al. 2007). Thus, salinity is a major threat to global food security, farming and livelihoods as salt-affected soils are not suitable for crops. Various remediation methods are applied to restore the quality of these soils, among which the application of plants for the extraction of salt appears to be sufficient for most circumstances.

In general, saline soil contains salts—table salt (NaCl), gypsum ($CaSO_4$), $CaCl_2$, magnesium sulfate, potassium chloride and sodium sulfate. The negative effects on soil due to sodium salts are counterbalanced by the high concentration of calcium and magnesium salt. Sodium (Na^+) is the key ion of concern in most cases of soil salinization due to its impact on soil structure and its interaction with potassium (K^+) in plant activity and metabolism (Karadag et al. 2016; Wakeel 2013). Since Na^+ and Cl^- usually stay together in the soil; therefore, toxicity due to chloride (Cl^-) is also linked with saline conditions. Multiple elements are found in hypersaline soil, such as silica (Si), fluorine (F), boron (B), aluminum (Al), rubidium (Rb), barium (Ba) lithium (Li), manganese (Mn), molybdenum (Mo), strontium (Sr) and selenium (Se); each of these will affect animals and plants. (Tanji 1990). In plants exhibiting physiological improvements, including stomatal closing, hyperosmotic shock, reticence of cell division and photosynthesis, the impacts of high salt concentrations in soils are noted (Jesus et al. 2015). The remediation and maintenance of salt-affected soils give expectation of land development and growth improvement for imminent nutrition safety under the scenario, where agricultural fields are declining due to rapid urbanization (Kumar and Sharma 2020). The process of remediation involves binding, buffering, immobilization, detoxification, filtering, or conversion of salts, sodium ions and chloride ions into a non-toxic soil mineral and not a consumption process. After desalinization, the reclamation zones can be utilized for other purposes like the cultivation of crops, fodder grasses, forest trees and aquaculture.

The United Nations Food and Agriculture Organization (FAO) reported that about 6.5 percent of the world's arable and marginal soils are either sodic or salty, based on soil surveys performed between 1970 and 1980 (FAO 2016). A pioneering effort in the comprehensive evaluation of salt-affected soils by means of this method was conducted by Szabolcs (1979), where salt-affected soils were categorized as saline soil, alkaline soil and potentially salt-affected soils. It was also clearly stated that due anthropogenic activities potentially salt-affected soils can be converted into saline soils. The 2018 Plenary Assembly of the Global Soil Partnership (GSP) looked at the global knowledge deficit for salt-affected soils and called for global information update mobilization (GSP-FAO 2018), and according to the resolution, for a first step, the GSP surveyed Member States to raise awareness of the state of saline challenges and to highlight problems for working on updated details. The survey concluded that over 70% of nations have various aspects of salt issues and evidence for salt-affected soil mapping. However, the soil map is not sufficient to get a clear idea about the status of salt-affected soil as electrical conductivity distribution, pH, soluble ions, etc. can be inaccurately defined or absent. Besides, soil map overlook some areas of the world map (Fig. 5.1).

5.3 Soil Salinity and Plants: Effects of Saline Contamination and Response

Soil is the crucial component of the ecosystem which performs the pivotal role for the existence of both natural and anthropogenic systems (Volchko et al. 2014). Presence of any kind of contaminants downgrades the quality of the soil.

Salts are the most common types of contaminants in soil, especially in agricultural soils. Salt contaminated soil is broadly categorized into three types: saline soil, sodic soil and saline-sodic soil. Saline soils contain high amounts of soluble salts such as chloride and carbonate of sodium (Na^+) calcium (Ca^{2+}), magnesium (Mg^{2+}), potassium (K^+), etc. Salinity altered various soil characteristics including swelling, dispersion and aggregation, all ending up with soil degradation (Tejada and Gonzalez 2005). Salt-affected soils normally contain soluble salts or their ions in at least one of their horizons above the toxicity level (the highest allowable concentration of salts that would not inhibit crop growth). Soil salinity can impact plant growth both physically (osmotic effect) and chemically (nutrient and/or toxicity effect), and it becomes more difficult for plants to take up water with the rise in salt content in the soil. Salinity influences several structural and functional changes in plants. Some of them are as follows;

- Lignification and thickening of the cell wall (Reinhardt and Rost 1995; Wahid et al. 1998).
- Reduced size of plasmalemma and contraction of plasmalemma away from the cell walls (Bliss et al. 1984).
- Intra-membranous particle disaggregation (Bliss et al. 1984).
- Reduced and damaged mitochondrial apparatus (Petruzzelli et al. 1991).
- Formation of small pro-vacuoles instead of the single large vacuole (Bliss et al. 1986).
- Dispersion and condensation of the substances with chromatin in embryo (Bliss et al. 1986; Petruzzelli et al. 1991).
- Abnormal size of cortical cell in mesocotyl (Wahid et al. 1998).
- Induction of endodermis with Casparian band (Reinhardt and Rost 1995).
- Suberin lamellae close to root base (Reinhardt and Rost 1995).
- Earlier development and differentiation of secondary xylem in hypocotyls (Reinhardt and Rost 1995).
- Constriction of cortical tissue of mesocotyl (Wahid et al. 1998).
- Increased lignification of secondary tissues (Valenti et al. 1992).
- Cessation of leaf blade expansion and early senescence of leaf as salinity intensifies (Abdul Qados 2011).
- Salinity decreases the number of nodes per plant, flowers per nod, number of fruits and seeds (Ghassemi-Golezani et al. 2018).
- By inhibiting microsporogenesis and stamen filament elongation, improving programmed cell death in certain tissue forms, ovule abortion and senescence of fertilized embryos, salinity adversely affects reproductive growth (Shrivastava and Kumar 2015).
- Salinity retards the rate of biosynthesis of various chlorophyll molecules, and thus, there is a reduction of photosynthetic substance (Chutipaijit et al. 2011).
- Salinity disturbs stomatal conductance rapidly and transiently due to interruption in water relations and sharply the local synthesis of short-lived ABA in roots (Fricke 2004), and as a consequence of the reduction in transpiration rate, leaf temperature shows significant increases (Mohammadian et al. 2005).
- Salinity causes a drop in the pyruvate carrier and pyruvate dehydrogenase subunit abundance, which decreases the ability of the TCA cycle's convective activity and provides respiratory reduction. (Che-Othman et al. 2020).
- Due to salinity most of the plant cells accumulate the measurable amount of reactive oxygen species (ROS) such as hydroxyl radicals, hydrogen peroxide and superoxide anions which can severely damage cell organelles and biomolecules like polysaccharides, proteins, lipids and nucleic acids (Shen et al. 1997; Zhu et al. 2007)

As a whole, a plant under salt-induced stress shows vast modification from ultra-cellular level to organ level. Most of the glycophytic plants are highly affected by the salt stress but halophytes and some salt-tolerant glycophyte adopt diverse morphological, biochemical and physiological modification to cope with this stressful situation.

5.4 Salt Elimination from Soil

Traditional approaches for the amelioration of saline soils include leaching the salts beneath the root area and drainage to lower the water table. There are various useful salt-affected land enhancement strategies, including water leaching, chemical recovery and phytoremediation (Sharma and Minhas 2005; Qadir et al. 2007). Different methods of soil remediation consist of physical, biological and chemical methods (Fig. 5.2). Physical methods include soil replacement—the contaminated soil is separated and replaced by good soil, high temperature is used to decontaminate the soil, volatilization—air is blown through a pipe which flows through the soil, activated carbon, is used to trap volatilized contaminants; high temperature—heating wells are used in high temperature is used. The use of microbes, algae, fungi, animals (earthworms) and plants is part of biological methods. Chemical methods include chemical fixation, chemical leaching and electrokinetic. Plant-assisted bioremediation or phytoremediation will provide soils and groundwater polluted with salt, metals, radionuclides and different forms of organic compounds with low-cost and eco-friendly methods of remediation than conventional remediation methods. This plant-based approach is of great significance, particularly for many developed countries, where as chemical modifications become more and more costly, vast portions of land are impacted by salt.

Phytoremediation is a word derived from the Greek prefix "*Phyto*" meaning plant, and the Latin suffix "*remedium*" meaning "remove the evil" or clean or restore (Cunningham et al. 1997). The term phytoremediation is technically defined as the cleaning of environmental contaminant by utilizing either naturally occurring plant or genetically engineered plants or plant-microbial association (Flathman and Lanza 1998). Globally phytoremediation is used at a broad scale to utilize the saline soil or saline-sodic soil. The aim of this technology is to enhance the physical, biological and chemical characteristics of the soil in areas influenced by salt. The technology completely exploits the natural hydraulic and metabolic processes of plants and/or associated microbes and thus, it is solar-driven. Qadir and Oster (2004) proposed that phytoremediation is a role of four main factors: (i) CO_2 partial pressure inside the rhizosphere, (ii) root proton release (in the case of N_2-fixing plants), (iii) improvement of soil porosity by root expansion and (iv) shoot sodium content (removed by the harvest). These authors concluded that salt-affected soil phytoremediation is largely dependent on leaching and shoots have a very little contribution in salt remediation process in comparison to roots. However, in partially drained salt-affected lands, particularly in arid and semi-arid regions where rainfall is very low for leaching salts from the rhizosphere, the important contribution of the shoot in the salt/sodium recovery process is noticed (Shiyab et al. 2003).

Various terms are used in the phytoremediation category which includes the use of soil-associated plants, microbiota to clean pollutants which are phytoextraction, phytostabilization, phytopumping or dendroremediation, rhizofiltration, phytotransformation, phytodetoxification, phytostimulation, etc. All of these processes are involved in the removal of salt or the salt ions from the rhizosphere but different ways. Phytoextraction corresponds to the isolation and aggregation of toxins by roots and surface shoots in harvestable plant parts. In this process, roots engross the pollutants together with sap and are stored within various above-ground organs like stems and leaves, as well as in storage modified roots. There is no detoxification process in this phytoremediation strategy. Phytostabilization is the use of plants to increase sequestration of contaminants (like salt or heavy metal) in the soil and/or the plant root. The contaminants are absorbed and precipitated by plants through phytostabilization, thereby reducing the mobility and inhibits the flow of contaminants to groundwater through leaching or transport through wind or even entry into the food chain (Miller 1996). Phytostabilization is absorption and accumulation of contaminants occurring in plant tissues, adsorption on roots or precipitation within the rhizosphere; thus, their migration in the soil is checked, even their transportation by erosion and deforestation. In phytopumping, certain plants use a hydraulic obstruction to either generate an upward movement of water in roots, thereby avoiding contaminants either percolating down or dispersing horizontally (Pilon-Smits 2005). Enormous underground web of roots formed by living plants operates as solar-driven organic pumps that remove and condense essential elements and contaminant from soil and water (Susarla et al. 2002). Rhizofiltration refers the absorption, concentration, and precipitation of salts, heavy metals etc. by plant roots. Plant root absorption and adsorption play a crucial role in this procedure, and large root surface areas are typically needed as a result (Peer et al. 2005). Growth of root along with the accumulation of organic matter to the root zone within the affected region would increase soil hydraulic conductivity, which raises the risk of natural salt leaching from the upper and lower horizons of the soil (Akhter et al. 2004; Qadir et al. 2007; Ammari et al. 2013). Phytotransformation or phytodegradation is considered to be the degradation of complex contaminant molecules into simple molecules and the incorporation of those molecules into plant tissues. It includes the decomposition of pollutants either internally through physiological activities or externally through the release of enzymes into the soil produced by plants. Phytostimulation refers to the stimulation of contaminant microbial degradation through exudate/enzyme release into the root zone (rhizosphere). In most cases, phytoextraction and phytostimulation deal with the cleaning of salts contaminant of soil. Previously, plants and microorganisms were utilized separately for bioremediation, but at present numerous studies reveal that synergistically, they can strive to improve the mitigation process of various matrices, such as water, air and soil (Khan et al. 2018). Many researchers have determined that perennial halophytes desalinate and fertilise the rhizosphere, supplying glycophytes with a beneficial microhabitat for improved growth and production (Fig. 5.3). Abdelly et al. (1995) described that after successful soil desalination in salt-affected soil, annual glycophytes like *Medicago* sp. grow well inside the bunch of perennial halophytes compared to outside.

Phytoremediation is the process of removing the contaminants of the environment using different plants to shift, remove, degrade or stabilize contaminants in the soil and/or groundwater. Effective phytoremediation requires three basic steps assessment of the salt-afflicted site, selection and plantation of correct species and implementation of a suitable crop management practice (Robinson et al. 2003). The achievement of phytoremediation is dependent on appropriate plant selection. Phytoremediator plants should have several advantageous features which not only make them fit to survive in salt-affected soil but also decrease the level of salinity in the habitat. Appropriate plant selection could help achieve a progressive improvement in salt-affected soil. Firstly, phytoremediator plants should be tolerant to salinity and drought. They should have a deep root organization which allows water entree and great ion buildup (Bell 1999). They can promote soil productivity through the enhancement of microbial biomass as well as organic matter.

It has been shown that many plant species effectively reduce electric conductivity and sodium absorption ratio (SAR) in salt-affected soils. Most halophytes are enormous phytoremediator plants and are able to obtain and accumulate considerably higher salt and Na^+ amounts in their shoots. Halophytes are just 1 percent of all plant species that can finish their development process in moderately high saline conditions, up to or over 200 mM NaCl (Flowers and Colmer 2008). Halophytes function as model plants; it is used in the study of different adaptive mechanisms, i.e., initiation of the antioxidant functions of enzymes, the aggregation of harmful ions in their vacuoles, preservation in response to cellular tension of compatible soluble compounds, etc. (Otlewska et al. 2020). Since halophytic plants are present naturally in the environment, they seek special interest because these plants are characterized by an excess of toxic ions, primarily Na^+ and Cl^- ions. Such

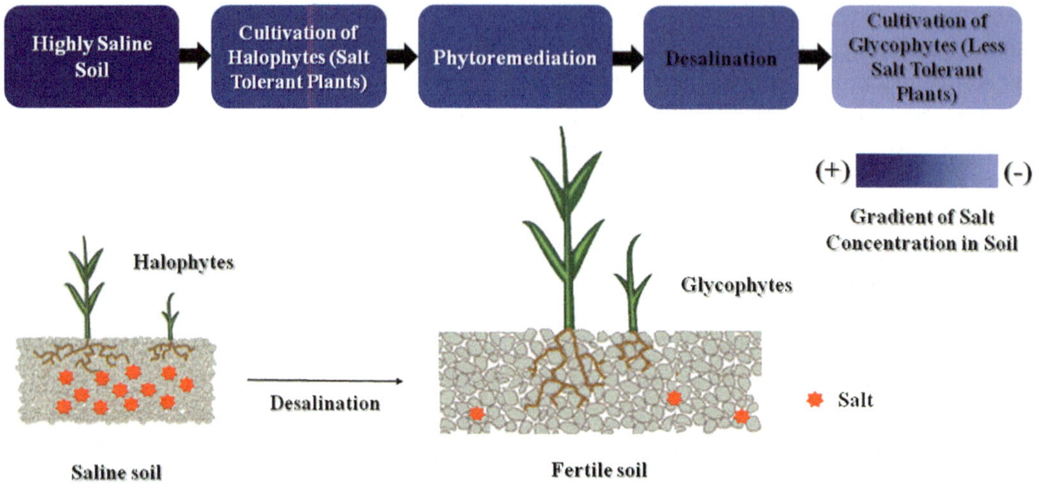

Fig. 5.3 Desalination of saline soil by halophytes turning it to fertile soil suitable for the growth of glycophytes

adaptive feature makes them ideal candidates for phytoremediation of salt-affected soils. The basic principle of remediation is the removal of excess salt to the desired level in the root zone. Most of them accumulate salt in their above-ground, aerial part and by removal of the aerial parts of those plants would remove salt from the soil (Manousaki and Kalogerakis 2011). Based on the different activities in salt remediation processes, halophytes are classified into three groups by Yensen and Biel (2006); excluder, accumulator and conductor plants. As a salt tolerance framework, excluder plants prevent salts from entering their tissues; accumulator plants take up and accumulate salts in their tissues while conducting plants, absorb salts and excrete them through salt glands, leading salts from the soil into the air (Jesus et al. 2015). Phytoremediators are not thought to be halophytes which excrete salt ions from modified leaf glands and drop aged organs with Na^+ and other stored ions because the salt ultimately returns to the soil within the plant (Gerhardt et al. 2017). Thus, the selected plants should be appropriate for repeated harvesting of foliar tissues and/or above-ground parts containing salts or Na^+ ions for effective phytoremediation. The use of salt accumulating species has become an option to remove salts from contaminated sites to remediate aqueous or soil environments. Though the identification of such species may be accomplished through screening, the search for constant, reliable markers has been thus far relatively unsuccessful (Garnett et al. 2002).

5.5 Cultivating Suitable Plant Species

Among the salt phytoremediation strategies, the most effective one is repeated cultivation and harvestation of suitable plants especially native plants of salt afflicted areas. Besides, some salt-tolerant glycophytes and genetically modified salt stress-tolerant plants are also cultivated to clear salt contaminants from soil. Such plants can accumulate higher concentration of Na^+ and Cl^- ions in their shoots (Manousaki and Kalogerakis 2011). The roots of these plants absorb Na^+ and Cl^- ions from the rhizosphere along with sap and translocate them to the aerial shoots where they accumulate. Via the passive symplastic pathway, plants mainly absorb Na^+ and Cl^- ions by potential gradient and transpiration cycles (White and Broadley 2001; Hedrich 2012). These ions are then transferred by the xylem circulatory system from the plant roots to the leaves and eventually deposited in the foliage as well as other organs (Zhu 2007; White and Broadley 2001; Hasegawa 2013). However, slight ion flow ensues via the phloem down to the roots (White and Broadley 2001; Hasegawa 2013). The aerial parts of the plants are harvested and cleared after sufficient development, resulting in the permanent removal of salt from contaminated soil. Some plants, in quantities 100–500 times larger than normal plants, can accumulate salt in their aerial components without any effect on their developmental performance and are often referred to as hyperaccumulator plants. While these strategies contribute to the complete eradication of salts from polluted soils, it is highly recommended for salt-contaminated soils at modest concentrations, as extreme salty soils cannot be decontaminated by most plant species.

5.5.1 Enhanced Calcium Levels in Soil Solution

The primary objective of salt abatement is to decrease the amount of soil distribution caused by the Na^+ ion, which is achieved by adding calcium ions (Oster et al. 1999; Qadir et al. 2001). Phytoremediation in saline soils is accomplished by increasing the dissolution rate of calcite through plant roots leading to improved levels of Ca^{2+} in the soil solution to practically displace Na^+ from the cation exchange complex (Qadir et al. 2007). Plant-assisted solubilization of $CaCO_3$ is due to root respiration for non-nitrogen fixer plant and/or due to H^+ in case of nitrogen fixer plants (Qadir et al. 2014). Hydrogen ions formed during calcite dissolution, in turn, react with calcite liberating more calcium ions (Qadir et al. 2003). Due to excessive irrigation, the substituted Na^+ is either percolated from the rhizosphere, a mechanism involving water infiltration and the possibility of a natural or artificial water supply or consumed by crops (Qadir et al. 2007). Balance of Na^+ and Ca^{2+} in the hyperaccumulator plants is genetically regulated. For example, PeSTZ1 in *Populus*, a positive zinc finger transcription factor, adjusts K^+/Na^+ homeostasis under salt stress to relieve salt toxicity in plant photosynthetic organs, improving photosynthetic activity under salinity stress (He et al. 2020).

5.5.2 Exploiting Plant Roots' Potential to Enhance Dissolution

Numerous halophytes as well as salt-tolerant glycophytes are utilized in phytoremediation of salt adulterated soil as they capable of retaining higher concentration of salt or salt ions in their roots without changing the salt concentration of above-ground parts. The chemical compounds emitted by plants prevent contaminants' migration and instead of the contaminants breakdown. Plants chosen for such clearing strategy tolerate salt and accumulate absorbed salts in roots but do not translocate it to their aerial parts. About 90% of Na^+ can be accumulated in the roots of these plants but the concentration remains low in the shoot (Imada et al. 2009). Some workers believed that these plants have an efficient salt management mechanism through which the pH and soil-moisture content around their roots are regulated (Tester and Davenport 2003). This process is facilitated by the increase in calcite dissolution in the rhizosphere, partial pressure of carbon dioxide in root zones and an increase of proton release from plant roots (Imadi et al. 2016). In this way, salt ions are coagulated by those plant roots and the availability of salts to other plants gradually reduced. Different plants have different capacities for storage of salts within their vacuoles, and more or less, it has been estimated that 500 mM Cl^- would be maximal for most plants (Cram 1973). Plants having such capability are sown or transplanted into the soil contaminated with salt, and other salt-sensitive plants are cultivated in association with them by appropriate agricultural practices.

5.5.3 Sodium Removal Through Soil-Cation Exchanger

Removal and replacement of Na^+ by Ca^{2+} through soil cation exchange sites cause a significant improvement in soil content noticed by phytoremediation. Roots crosstalk with neutral salts by the formation of exchange acidity and exhibit "suspension effect," thereby exhibiting the presence of a cation double-layer coupled with root surface (Williams and Coleman 1950). Usually, H-ions are associated with root surfaces in an exchangeable form and may be replaced by other cations like Na^+, K^+, Ca^+, etc. Through the cation exchange process, the plant absorbs almost 70% of the cations it needs, hence imbalances of cations on the clay colloids, also known as the "base"; in the plant, the absorption of adequate balance of vital nutrients is prevented. The high interchangeable sodium level and salt adsorption ratio correlated with the pH limits of these soils are the key factors regulating nutritional disparities and their deficiencies (Yadav 1980). The cation exchange ability of the roots varies significantly with the nature of the species, type and time of testing, crop age, conditions of growth, root zone, level of soil nutrients and form of soil (Chamuah and Dey 1987). The application of N and Zn has been shown in an experiment to greatly improve the cation exchange potential of rice and wheat roots in salt-affected soil (Srivastava and Srivastava 1992).

Inherent or precipitated sources of Ca^{2+}, usually calcite ($CaCO_3$), are found in many salt and saline sodium soils at various depths within the soil. Phytoremediation of sodium and saline-sodium soils, on the other hand, is accomplished not solely by the removal of Na + in plants, but by the ability of plant roots to increase the calcite dissolution rate (Qadir et al. 2005). Salt-affected soils are those soils having salts beyond the critical limit in the solution phase and/or Na^+ on the cation exchange sites.

5.5.4 Remove Salt by Natural Leaching

Leaching of salts from upper to lower soil depths has been a typical form of salt-affected soil amelioration among the different methods used. In the irrigated land, salts are accumulated in the root zone when soil water is vigorously absorbed and transpired by cultivated plant or lost by evaporation from the soil surface. Due to such water loss, upper soil layer dries in comparison to the deeper layer which enforces potential for the upward flow of water into the root zone. The inceptive salinity, texture of soil and plant species dictates the appropriate depth for leaching, which plays a vital role in desalinization and desodification programs (Konuku et al. 2006; Corwin et al. 2007). Salt builds up in the rhizosphere when upward salt migration due to evaporation surpasses the downward gravitational movement (Mondal et al. 2001). To improve the crop growth and supporting lowest amount of salinity in the rhizosphere, a substantial quantity of water is required to clear out salinity when the field is irrigated with saline water (Aktas et al. 2006). Some salt-tolerant, vigorously growing plants with large, fibrous root systems that contain substantial quantities of root hairs are required to create a large surface area for absorption of the salt contaminant (Dushenkov et al. 1995). Growth of root along with the accumulation of organic matter to the root zone within the affected region will raise hydraulic conductivity of the soil, which upsurges the probability for natural leaching of salt from upper to lower soil horizons (Akhter et al. 2004; Qadir et al. 2007; Ammari et al. 2013). Significant desalinization of salt-affected soils needs a downward movement of the zone of salt accumulation to a depth from where resalinization is not possible or extremely limited. If the soil is deep enough, then salts can leach out of the soil, and permeability is satisfactory and without any water table near the surface.

5.5.5 Plants-Microbe Association Mediated Remediation

Using plants to facilitate microbial degradation or biotransformation of pollutants through the rhizodegradation process is an integral part of phytoremediation. Phytostimulation or plant-facilitated bioremediation is described as the stimulation of microbial and fungal degradation by the secretion of exudates/enzymes in the rhizosphere. For plant growth and health, microbial populations associated with plant roots are essential and are therefore referred to as the second genome of the plant (Berendsen et al. 2012). Previously, bioremediation studies were performed through plants and microorganisms separately, whereas in recent time numerous studies reveal that harmonious effect has enhanced remediation process in several models such as water, soil, and air (Khan et al. 2018). Under salt stress condition, the rhizospheric microorganisms, directly and indirectly, support plant growth and yield (Dimkpa et al. 2009). The microbial decomposition process induces a reduction in pH by the processing of carbonic acid, which promotes the breakdown of carbonate minerals like $CaCO_3$ and raises the concentration of Ca^+ ions to displace the exchangeable Na^+ in saline soils. CO2 from plant root respiration dissolves in soil solution and increases partial CO_2 pressure (Wong et al. 2009; Qadir et al. 2014). Numerous researchers suggested the utilization of association of microbes, i.e., phosphate-solubilizing microbes present in the phytoremediation to obtain efficient remediation of contaminated soils (Jia et al. 2016).

In general, fungi will be much more essential for distinguishing between root zone microbiomes of different salt-tolerant plant species than bacteria and archaea (Wang et al. 2020). Promotion of nutrient absorption, the ability of water uptake and osmolyte pile-up under the situation of salinity stress are enhanced by Arbuscular Mycorrhizal Fungi (AMF) (Hanin et al. 2016). The absorptive surface of plant roots increases by AMF through the widespread hyphal web thus improves plant growth and support phytoremediation by producing phytohormones (Gohre and Paszkowski 2006; Vamerali et al. 2010). In salt marsh, plant species have revealed that AMF improves the nutrient supply and ameliorate water use and photosynthesis efficiency, whereas AMF colonization exceptionally declines at moderate salinities (Caravaca et al. 2005).

5.6 Improvement in Soil Fertility

In some cases, routine drainage, conventional remediation techniques and crop management practices do not reduce high salinity in soils over time. In such situation, amelioration of saline soils by phytoremediation is important for long term protection of soil and sustainable agriculture. In addition to collecting biologically active chemicals, plant roots actively generate and secrete compounds to promote soil fertility directly or indirectly in their immediate environment. Various studies of root exudates from diverse plant species reveal that during salt accumulation such plants secrete isoflavonoids, flavonoids, strigol, herbicidal allelochemicals, phytosiderophores, etc. (Peters and Long 1988; Maxwell and Phillips 1990; Inoue et al. 1992; Shepherd and Davies 1993; Yu and Matsui 1994). These substances provide several physiological and ecological advancements to the plants. Some salt phytoremediator plants also help in clearing of other soil pollutants like heavy metals. Moray et al. (2016) identified 60 species from a diverse range of angiospermic groups that capable of accumulating both salts and heavy metals.

5.7 Selection of Plants for Phytoremediation

Phytoremediation is the procedure of removing the contaminants of the environment using different plants to shift, remove, degrade or stabilize contaminants in the soil and/or groundwater. Since phytoremediation is a plant-based technology, the achievement naturally dependent upon the proper selection of plant. Phytoremediator plants should have several advantageous features which not only make them fit to survive in salt-affected soil but also decrease the level of salinity in the habitat. Appropriate plant selection could help achieve a progressive improvement in salt-affected soil. Firstly, phytoremediator plants should be tolerant to salinity and drought. They have a deep root structure that allows for water access and heavy ion deposition. Via enhancement of microbial biomass and organic matter, they will increase soil fertility.

In salt-affected soils, several plant species are also used to effectively decrease electrical conductivity and sodium absorption ratio. Most of the halophytes are marvelous phytoremediator plants and have the capability to obtain and accrue rather high levels of salts and Na^+ within their shoots (Table 5.1). Halophytes serve as model plants; it is used in the study of different adaptive mechanisms, i.e., activation of the antioxidant activities of the enzymes, the concentration of harmful ions in their vesicles, the preservation of appropriate soluble compounds in response to cellular tension, etc. (Otlewska et al. 2020). The special interest of halophytic plants is that they are naturally found in conditions marked by an abundance of radioactive ions, primarily sodium and chloride. Such adaptive feature makes them ideal candidates for phytoremediation of salt-affected soils. The basic principle of remediation is the removal of excess salt to the desired level in the root zone. Most of them accumulate salt in their above-ground aerial part and by removal of the aerial parts of those plants would remove salt from the soil (Manousaki and Kalogerakis 2011). Based on the different activities in salt remediation processes, halophytes are classified into three groups by Yensen and Biel (2006): excluder, accumulator and conductor plants. As a salinity resistance

Table 5.1 Phytoremediator plants used in the removal of soil salinity

Plants	Mode of remediation	References
Agropyron elongatum	Accumulation of salt	Weimberg (1986)
Apocynum lancifolium	Removal of chloride, sodium, magnesium and calcium ions from the salt-affected soils	Hamidov et al. (2007)
Arthrocnemum indicum	Remove sodium from the soil by accumulating it in shoots	Rabhi et al. (2009)
Arundo donax	Hyperaccumulation of salts	Devi et al. (2008)
Atriplex nummularia	Hyperaccumulation of Na^+ and Cl^- uptake	Silva et al. (2016)
Atriplex lentiformis	Hyperaccumulation of salts	Devi et al. (2008)
Atriplex amnicola	Hyperaccumulation of salts	Devi et al. (2008)
Bassia indica	Phytoextraction of Na^+ and K^+ ions	Shelef et al. (2012)
Chenopodium album	Removal of chloride, sodium, magnesium and calcium ions from the salt-affected soils	Hamidov et al. (2007)
Clerodendron inerme	Accumulation of salt in above-ground parts	Ravindran et al. (2007)
Cynodon dactylon	Accumulates Na^+ and Cl^- in glands in their leaves	Parthasarathy et al. (2015)
Glycyrrhiza glabra	Accumulate salts in their rapid growing biomass.	Kushiev et al. (2005)
Haloxylon recurvum	Hyperaccumulation of salts	Devi et al. (2008)
Helianthus annuus	Remove considerable quantities of chloride and sodium ions when grown on heavy clay soils with saline patches	Bhatt and Indirakutty (1973), Iwasaki (1987)

(continued)

Table 5.1 (continued)

Plants	Mode of remediation	References
Heliotropium curassavicum	Hyperaccumulation of salts	Ravindran et al. (2007)
Heliotropium eichwaldi	Hyperaccumulation of salts	Devi et al. (2008)
Ipomoea pes-caprae	Hyperaccumulation of salts	Ravindran et al. (2007)
Kalidium folium	Hyperaccumulation of salts	Zhao et al. (2005)
Kosteletzkya virginica	Phytostimulation of mycorrhiza	Qin et al. (2015)
Lactuca sativa	Absorbed more salt at a higher salt concentration	Ben Asher et al. (2012)
Leptochloa fusca	The above-ground part contains salt in the range of 40–80 g kg^{-1} when grown in salty soils	Malik et al. (1986)
Lotus corniculatus	Accumulation of salts	Aydemir and Sunger 2011)
Medicago sativa	Enhanced calcium level in soil solution by dissolving calcite in soil solutions and forming calcium ions	Qadir et al. (2003)
Oryza sativa	Accumulation of sodium ion	Iwasaki (1987)
Phragmites australis	Phytoextraction of chloride	McSorley et al. (2016), Yun et al. (2019)
Populus euphratica	Phytopumping	He et al. (2020)
Portulaca oleracea	Hyperaccumulation of salts	Devi et al. (2008), Ben Asher et al. (2012)
Puccinellia nuttalliana	Accumulation of Na^+ and Cl^- ions	McSorley et al. (2016)
Salicornia europaea	Accumulation of Na^+ and Cl^- at 20% and 25% of dry weight, respectively	Balnokin et al. (2010)
Salsola baryosma	Hyperaccumulation of salts	Devi et al. (2008)
Sesbania bispinosa	Hyperaccumulation of salts	Ilyas et al. (1993), Qadir et al. (2002)
Sesuvium portulacastrum	Hyperaccumulation of salts in their tissues as well as higher reduction of salts in the soil medium	Ravindran et al. (2007)
Sorghum sudanense	Leaching of Na^+ ion	Robbins (1986)
Spartina pectinata	Excretion of K^+ and Cl^- ions	McSorley et al. (2016)
Suaeda fruticosa	Desalinized rhizosphere by salt leaching.	Rabhi et al. (2009)
Suaeda maritime	Hyperaccumulation of salts in their tissues as well as higher reduction of salts in the soil medium	Ravindran et al. (2007), Devi et al. (2008)

(continued)

Table 5.1 (continued)

Plants	Mode of remediation	References
Suaeda nudiflora	Hyperaccumulation of salts	Devi et al. (2008)
Tamarix chinensis	Reduce the salt concentration in saline soils and increase the abundance of soil nutrients.	Zhang et al. (2008), Cao et al. (2014)
Tetragonia tetragonoides	Absorbed more salt at a higher salt concentration	Ben Asher et al. (2012)

system, excluder plants avoid salts from penetrating their tissues; accumulator plants pick up and accumulate salts in their tissues when conducting plants absorb salts and excrete them by salt glands, leading salts from the soil into the air (Jesus et al. 2015). Halophytes that excrete salt ions from specialized leaf glands and drop older organs that have Na^+ and other ions collected are not known to be phytoremediators since the salt ultimately returns to the soil near the plant (Gerhardt et al. 2017). Thus, the chosen plants should be ideal for repetitive harvesting of foliar tissues and/or overground sections carrying salts or Na^+ ions for successful phytoremediation. Use of salt absorbing organisms is becoming an alternative to extract salts from polluted areas to fix aqueous or soil conditions. Though the identification of such species may be accomplished through screening, the search for constant, reliable markers has been thus far relatively unsuccessful (Garnett et al. 2002).

5.8 Advantages of Using Phytoremediation for Saline Soil Amendments

In several areas, phytoremediation methods of salt recovery from the soil have been tested successfully. These assessments concluded that phytoremediation is not only a realistic alternative for environmental remediation but also has multiple benefits relative to other remediation strategies.

(i) Phytoremediation of moderately sodium and saline-sodium soils is an effective, low-cost approach to their improvement and is a viable option for poor farmers.
(ii) No need for investment to purchase instruments and chemical amendments.
(iii) It protects the topsoil, retains the topsoil productivity and encourages the use of crops to mitigate soil erosion as well as metal leaching.
(iv) There are some possibilities to get financial or other benefits by selection and cultivation of economically important phytoremediator plants in addition to amelioration.
(v) As a result of root operation, soil aggregate stabilization and porosity are enhanced and hydraulic soil is consequently strengthened.

5.9 Limitations of Using Phytoremediation for Saline Soil Amendments

Despite of being the most effective process for the remediation of salt ions from the contaminated soils, phytoremediation is also having some drawbacks while implementing the process on the same soil.

(i) Phytoremediation of salt is limited to the surface area and the depth occupied by the roots.
(ii) A long time is often required for effective phytoremediation of salt, and it may take up to several years or several decades to remediate a contaminated site.
(iii) With plant-based remediation schemes, mitigation of the leaching of salts into the groundwater cannot be achieved. The salt toxicity of the polluted land and the

general state of the soil also affect the survival of the plants.

(iv) Many findings suggested that soil salt leaching by plants are often disadvantageous as it removes soil nutrients such as N and K, disorganizes soil particles and causes flooding and other environmental issues (El-Haddad and Noaman 2001; Yurtseven et al. 2005; Kolahchi and Jalali 2007).

(v) Bioaccumulation of salt ions in plants entering the food chain, from customers at the primary level onwards, or the healthy disposal of the infected plant material.

(vi) Harvested biomass from hyperaccumulator plants may be appearing as hazardous waste and hence disposal should be proper.

5.10 Conclusion and Future Prospects

Most countries face immense challenges in the current decade, with extensive salinity-induced depletion of land resources and degradation of water quality. Due to numerous natural mechanisms and anthropogenic behaviors, salt concentration in the soil has risen exponentially. Due to physical and chemical influence on the plant, salt-affected soil has posed a serious threat to biodiversity, food security and the ecosystem. Since few decades, several plants-based remediation technologies have been discovered and implemented in several corners of the world especially in arid and semi-arid regions. Such plant-based remediation takes advantages of the fact that they have a high level of public acceptance and attractive for remote and sensitive locations. Plants are a special part of the earth ecosystem armed with excellent mechanical, metabolic and absorption capacities, as well as transport mechanisms that can selectively suck up nutrients or pollutants such as soil or water salt. Large volumes of soils are being explored by plants widespread root systems for the efficient remediation of different contaminants from various soil types. There is no doubt that the most effective eco-friendly approaches of removal of salt from impacted soils are vegetation and harvesting of phytoremediator plants. The challenge for plant researchers is therefore to improve the efficiency of plants in the reclamation of salts and other toxicants from the soil, which need more basic research and knowledge on the mechanisms of phytoremediator plants. Gradual research on salt-tolerant glycophytes and halophytes is very crucial because these plants can be exploited in the reclamation of the salt-affected area and overcoming the global problem of food scarcity and energy crises by saline agriculture and production of biomass industries.

Acknowledgements Authors would like to thank Mr Subhasish Sutradhar and Dr Prolay Mondal, Department of Geography, Raiganj University, for their help in the preparation of Fig. 1.

References

Abdelly C, Lachaâl M, Grignon C, Soltani A, Hajji M (1995) Episodic association of strict halophytes and glycophytes in a salt hydromorphic ecosystem in a semi-arid zone. Agronomy 15:557–568

Abdul Qados AMS (2011) Effect of salt stress on plant growth and metabolism of bean plant *Vicia faba* (L.). J Saudi Soc Agricult Sci 10(1):7–15 https://doi.org/10.1016/j.jssas.2010.06.002

Akhter J, Murray R, Mahmood K, Malik KA, Ahmed S (2004) Improvement of degraded physical properties of a saline-sodic soil by reclamation with kallar grass (*Leptochloa fusca*). Plant Soil 258:207–216. https://doi.org/10.1023/B:PLSO.0000016551.08880.6b

Aktas H, Abak K, Cakmak I (2006) Genotypic variation in the response of pepper to salinity. Sci Hortic-Amsterdam 110:260–266. https://doi.org/10.1016/j.scienta.2006.07.017

Ammari TG, Al-Hiary S, Al-Dabbas M (2013) Reclamation of saline calcareous soils using vegetative bioremediation as a potential approach. Arch Agron Soil Sci 59:367–375. https://doi.org/10.1080/03650340.2011.629813

Aydemir S, Sunger H (2011) Bioreclamation effect and growth of a leguminous forage plant (*Lotus corniculatus*) in calcareous saline-sodic soil. Afr J Biotechnol 10(69):115571–115577. https://doi.org/10.5897/AJB.9000083

Balnokin Y, Myasoedov N, Popova L, Tikhomirov A, Ushakova S, Lasseur C, Gros JB (2010) Use of halophytic plants for recycling NaCl in human liquid waste in a bioregenerative life support system. Adv Space Res 46:768–774

Bell DT (1999) Australian trees for the rehabilitation of waterlogged and salinity damaged soils. Aust J Bot 47:697–716

Ben Asher J, Beltrao J, Aksoy U, Anac D, Anac S (2012) Controlling and simulating the use of salt removing species. Int J Energy Environ 6(3):360–370

Berendsen RL, Pieterse CM, Bakker PA (2012) The rhizosphere microbiome and plant health. Trends Plant Sci 17:478–486. https://doi.org/10.1016/j.tplants.2012.04.001

Bhatt JG, Indirakutty KN (1973) Salt tolerance and salt uptake by sunflower. Plant Soil 39:457–460

Bhuiyan MS, Raman A, Hodgkons DS (2017) Plants in remediating salinity-affected agricultural landscapes. Proc Indian Natn Sci Acad 83(1):51–66 https://doi.org/10.16943/ptinsa/2016/48857

Bliss RD, Platt-Aloia KA, Thomson WW (1984) Changes in plasmalemma organization in cowpea radicle during imbibition in water and NaCl solutions. Plant Cell Environ 7:601–606. https://doi.org/10.1111/1365-3040.ep11592139

Bliss RD, Platt-Aloia KA, Thomson WW (1986) The inhibitory effect of NaCl on barley seed germination. Plant Cell Environ 9:727–733. https://doi.org/10.1111/j.1365-3040.1986.tb02105.x

Cao D, Shi F, Koike T, Lu Z, Sun J (2014) Halophyte plant communities affecting enzyme activity and microbes in saline soils of the Yellow River Delta in China. Clean: Soil, Air, Water 42:1433–1440. https://doi.org/10.1002/clen.201300007

Caravaca F, Alguacil MM, Torres P, Roldan A (2005) Plant type mediates rhizospheric microbial activities and soil aggregation in a semiarid Mediterranean salt marsh. Geoderma 124(3–4):375–382. https://doi.org/10.1016/j.geoderma.2004.05.010

Chamuah GS, Dey JK (1987) Root cation exchange capacity in relation to nutrient uptake of rice. J Indian Soc Soil Sci 35:113–114

Che-Othman MH, Jacoby RP, Millar AH, Taylor NL (2020) Wheat mitochondrial respiration shifts from the tricarboxylic acid cycle to the GABA shunt under salt stress. New Phytol 225(3):1166–1180. https://doi.org/10.1111/nph.15713

Chutipaijit S, Chaum S, Sompornpailin K (2011) High contents of proline and anthocyanin increase protective response to salinity in *Oryza sativa* L. spp. *indica*. Aust J Crop Sci 5(10):1191–1198

Corwin DL, Rhoades JD, Simunek J (2007) Leaching requirement for soil salinity control: Steady-state versus transient models. Agric Water Manag 90 (3):165–180

Cram WJ (1973) Internal factors regulating nitrate and chloride influx in plant cells. J Exp Bot 24:328–341

Cunningham SD, Shann JR, Crowley DE, Anderson TA (1997) Phytoremediation of contaminated water and soil. In: Kruger EL, Anderson TA, Coats JR (eds) Phytoremediation of soil and water contaminants. ACS Publications, Washington, pp 2–17

Devi S, Rani C, Datta KS, Bishnoi SK, Mahala SC, Angrish R (2008) Phytoremediation of soil salinity using salt hyperaccumulator plants. Indian J Plant Physiol 13(4):347–356

Dimkpa C, Weinand T, Ash F (2009) Plant-rhizobacteria interactions alleviate abiotic stress conditions. Plant Cell Environ 32:1682–1694. https://doi.org/10.1111/j.1365-3040.2009.02028.x

Dushenkov V, Nanda Kumar PBA, Motto H, Raskin I (1995) Rhizofiltration: the use of plants to remove heavy metals from aqueous streams. Environ Sci Technol 29(5):1239–1245. https://doi.org/10.1021/es00005a015

El-Haddad ESH, Noaman MM (2001) Leaching requirement and salinity threshold for the yield and agronomic characteristics of halophytes under salt stress. J Arid Environ 49(4):865–874. https://doi.org/10.1006/jare.2000.0783

FAO (2016) Saline Soils and their Management. Food and Agriculture Organization of the United Nations. Available online: http://www.fao.org/3/x5871e/x5871e04.htm

FAO (2020) Mapping of Salt Affected Soils. FAO, Rome

Flathman PE, Lanza GR (1998) Phytoremediation: Current views on an emerging green technology. J Soil Contam 296:415–432. https://doi.org/10.1080/10588339891334438

Flowers TJ, Colmer TD (2008) Salinity tolerance in halophytes. New Phytol 179(4):945–963. https://doi.org/10.1111/j.1469-8137.2008.02531.x

Fricke W (2004) Rapid and tissue-specific accumulation of solutes in the growth zone of barley leaves in response to salinity. Planta 219(3):515–525. https://doi.org/10.1007/s00425-004-1263-0

Garnett MR, Murch SJ, Krisnaraj S, Dixon MA, Saxena PK (2002) The Rhizofiltration of Sodium from hydroponic fluid using scented Geraniums. Water Air Soil Pollut 140:343–366

Gerhardt KE, MacNeill GJ, Gerwing PD, Greenberg BM (2017) Phytoremediation of salt-impacted soils and use of plant growth-promoting Rhizobacteria (PGPR) to enhance phytoremediation. In: Ansari et al (eds) Phytoremediation. Springer, Cham, pp 19–51. https://doi.org/10.1007/978-3-319-52381-1_2

Ghassemi-Golezani K, Farhangi-Abriz S, Bandehagh A (2018) Salicylic acid and jasmonic acid alter physiological performance, assimilate mobilization and seed filling of soybean under salt stress. Acta Agricult Slovenica 111(3):597–607. https://doi.org/10.14720/aas.2018.111.3.08

Gohre V, Paszkowski U (2006) Contribution of the arbuscular mycorrhizal symbiosis to heavy metal phytoremediation. Planta 223:1115–1122. https://doi.org/10.1007/s00425-006-0225-0

GSP-FAO (2018) Report of the Sixth Meeting of the Plenary Assembly of the Global Soil Partnership. GSPPAVI/18/Report. http://www.fao.org/3/CA0389EN/ca0389en.pdf

Hamidov A, Khaydarova V, Sharipova S, Costa C, Beltrao J (2007) Salt removal potential of *Portulaca oleracea* golden purslane. In: Proceedings of 3rd IASME/WSEAS international conference on energy, environment, ecosystem and sustainable development. Agios Nikolaos, Greece, July 24–26, pp 151–156

Hanin M, Ebel C, Ngom M, Laplaze L, Masmoudi K (2016) New insights on plant salt tolerance mechanisms and their potential use for breeding. Front Plant Sci 7:1787. https://doi.org/10.3389/fpls.2016.01787

Hasegawa PM (2013) Sodium (Na^+) homeostasis and salt tolerance of plants. Environ Exp Bot 92:19–31. https://doi.org/10.1016/j.envexpbot.2013.03.001

He F, Niu M-X, Feng CH, Li HG, Su Y, Su WL, Pang H, Yang Y, Yu X, Wang HL, Wang J, Liu C, Yin W, Xia X (2020) PeSTZ1 confers salt stress tolerance by scavenging the accumulation of ROS through regulating the expression of PeZAT12 and PeAPX2 in *Populus*. Tree Physiol 40(9):1292–1311. https://doi.org/10.1093/treephys/tpaa050

Hedrich R (2012) Ion channels in plants. Physiol Rev 92(4):1777–1811. https://doi.org/10.1152/physrev.00038.2011

Ilyas M, Miller RW, Qureshi RH (1993) Hydraulic conductivity of saline-sodic soil after gypsum application and cropping. Soil Sci Soc Am J 57:1580–1585

Imada S, Yamanaka N, Tamai S (2009) Effects of salinity on the growth, Na partitioning, and Na dynamics of a salt-tolerant tree *Populus Alba* L. J Arid Environ 73 (3):245–251. https://doi.org/10.1016/j.jaridenv.2008.10.006

Imadi SR, Shah SW, Kazi AG, Azooz M, Ahmad P (2016) Phytoremediation of saline soils for sustainable agricultural productivity. plant metal interaction. Elsevier, New York, pp 455–468

Inoue M, Nishimura H, Li HH, Mizutani J (1992) Allelochemicals from *Polygonum sachalinense* Fr. Schm. (Polygonaceae). J Chem Ecol 18(10):1833–1840

Iwasaki K (1987) The effectiveness of salt-accumulating plants in reclaiming salinized soils. Japan J Trop Agr 31(4):255–260

Jesus JM, Danko AS, Fiuza A, Borges M-T (2015) Phytoremediation of salt-affected soils: a review of processes, applicability, and the impact of climate change. Environ Sci Pollut Res 22(9):6511–6525. https://doi.org/10.1007/s11356-015-4205-4

Jia X, Liu C, Song H, Ding M, Du J, Ma Q, Yuan Y (2016) Design, analysis and application of synthetic microbial consortia. Synth Syst Biotechnol 1(2):109–117. https://doi.org/10.1016/j.synbio.2016.02.001

Karadag S, Eren E, Cetinkaya E, Ozen S, Deveci S (2016) Optimization of sodium extraction from soil by using a central composite design (CCD) and determination of soil sodium content by ion selective electrodes. Eurasian J Soil Sci 5(2):89–96

Kensa VM (2011) Bioremediation- an overview. J Indus Pollut Control 27(2):161–168

Khan N, Bano A, Zandi P (2018) Effects of exogenously applied plant growth regulators in combination with PGPR on the physiology and root growth of chickpea (*Cicer arietinum*) and their role in drought tolerance. J Plant Interact 13(1):239–247. https://doi.org/10.1080/17429145.2018.1471527

Kolahchi Z, Jalali M (2007) Effect of water quality on the leaching of potassium from sandy soil. J Arid Environ 68(4):624–639

Konuku F, Gowing GW, Rose DA (2006) Dry drainage: A sustainable solution to water logging and salinity problems in irrigation area. Agricul Water Manage 83(1–2):1–12. https://doi.org/10.1016/j.agwat.2005.09.003

Kumar P, Sharma PK (2020) Soil Salinity and Food Security in India. Front Sustain Food Syst 4:533781 https://doi.org/10.3389/fsufs.2020.533781

Kushiev H, Noble AD, Abdullaev I, Toshbekov U (2005) Remediation of abandoned saline soils using *Glycyrrhiza glabra*: a study from the hungry steppes of central Asia. Int J Agric Sustain 3(2):102–113. https://doi.org/10.1080/14735903.2005.9684748

Li P, Dai X, Xu Y, Liu L (2014) Evolution of soil properties following reclamation in coastal areas: A review. Geoderma 226–227(1):130–139. https://doi.org/10.1016/j.geoderma.2014.02.003

Malik KA, Aslam Z, Naqvi M (1986) Kallar grass: a plant for saline land. Nuclear Institute for Agriculture and Biology. Faisalabad (No. REP-3884 CIMMYT)

Manousaki E, Kalogerakis N (2011) Halophytes- an emerging trend in phytoremediation. Int J Phytorem 13(10):959–969. https://doi.org/10.1080/15226514.2010.532241

Maxwell CA, Phillips DA (1990) Concurrent synthesis and release of nod-gene-inducing flavonoids from alfalfa roots. Plant Physiol 93:1552–1558

McCauley A, Jones C, Jacobsen J (2005) Basic soil properties. Soil Water Manage Mod 1(1):1–12

McSorley KA, Rutter A, Cumming R, Zeeb BA (2016) Chloride accumulation vs chloride excretion: Phytoextraction potential of three halophytic grass species growing in a salinized landfill. Sci Total Environ 572:1132–1137. https://doi.org/10.1016/j.scitotenv.2016.08.023

Miller RR (1996) Phytoremediation, technology overview report. Ground-water remediation technologies analysis center, Pittsburgh, PA, USA

Mohammadian R, Moghaddam M, Rahimian H, Sadeghian SY (2005) Effect of early season drought stress on growth characteristics of sugar beet genotypes. Turk J Agric for 29:357–368

Mondal MK, Bhuiyan SI, Franco DT (2001) Soil salinity reduction and prediction of salt dynamics in the coastal ricelands of Bangladesh. Agric Water Manag 47(1):9–23. https://doi.org/10.1016/S0378-3774(00)00098-6

Moray C, Goolsby EW, Bromham L (2016) The phylogenetic association between salt tolerance and heavy metal hyperaccumulation in angiosperm. Evol Biol 43:119–130. https://doi.org/10.1007/s11692-015-9355-2

Munns R, Tester M (2008) Mechanisms of salinity tolerance. Annu Rev Plant Biol 59:651–681. https://doi.org/10.1146/annurev.arplant.59.032607.092911

Nouri H, Borujeni SC, Nirola R, Hassanli A, Beecham S, Alaghmand S, Saint C, Mulcahy D (2017) Application of green remediation on soil salinity treatment: a review on halophytoremediation. Process Saf Environ Prot 107:94–107. https://doi.org/10.1016/j.psep.2017.01.021

Oster JD, Shainberg I, Abrol IP (1999) Reclamation of salt-affected soils. Agricultural Drainage 38:659–691

Otlewska A, Migliore M, Dybka-Stepien K, Manfredini A, Struszczyk-Swita K, Napoli R, Białkowska A, Canfora L, Pinzari F (2020) When salt meddles between plant, soil, and microorganisms. Front Plant Sci 11:553087. https://doi.org/10.3389/fpls.2020.553087

Parthasarathy M, Pemaiah B, Natesan R, Padmavathy SR, Pachiappan J (2015) Real-time mapping of salt glands on the leaf surface of *Cynodon dactylon* L. using scanning electrochemical microscopy. Bioelectrochem 101:159–164. https://doi.org/10.1016/j.bioelechem.2014.10.004

Peer WA, Baxter IR, Richards EL, Freeman JL, Murphy AS (2005) Phytoremediation and hyperaccumulator plants. In: Tamas MJ, Martinoia E (eds) Molecular biology of metal homeostasis and detoxification. Topics in current genetics, vol 14, Springer, Berlin, Heidelberg, pp 299–340

Peters NK, Long SR (1988) Alfalfa root exudates and compounds which promote or inhibit induction of *Rhizobium meliloti* nodulation genes. Plant Physiol 88 (2):396–400. https://doi.org/10.1104/pp.88.2.396

Petruzzelli L, Melillo MT, Zache TB, Taranto G (1991) Physiological and ultrastructural changes in isolated wheat embryos during salt and osmotic shock. Ann Bot 69:25–31. https://doi.org/10.1093/oxfordjournals.aob.a088302

Pilon-Smits E (2005) Phytoremediation. Annu Rev Plant Biol 56:15–39. https://doi.org/10.1146/annurev.arplant.56.032604.144214

Qadir M, Noble AD, Oster JD, Schubert S, Ghafoor A (2005) Driving forces for sodium removal during phytoremediation of calcareous sodic and saline–sodic soils: a review. Soil Use Manag 21:173–180. https://doi.org/10.1111/j.1475-2743.2005.tb00122.x

Qadir M, Oster JD (2004) Crop and irrigation management strategies for saline-sodic soils and waters aimed at environmentally sustainable agriculture. Sci Total Environ 323(1–3):1–19. https://doi.org/10.1016/j.scitotenv.2003.10.012

Qadir M, Oster JD, Schubert S, Noble AD, Sahrawatk KL (2007) Phytoremediation of sodic and saline-sodic soils. Adv Agron 96:197–247. https://doi.org/10.1016/S0065-2113(07)96006-X

Qadir M, Quillérou E, Nangia V, Murtaza G, Singh M, Thomas RJ, Drechsel P, Noble AD (2014) Economics of salt-induced land degradation and restoration. Nat Res Forum 38(4):282–295. https://doi.org/10.1111/1477-8947.12054

Qadir M, Qureshi RH, Ahmad N (2002) Amelioration of calcareous saline sodic soils through phytoremediation and chemical strategies. Soil Use Manag 18:381–385. https://doi.org/10.1111/j.14752743.2002.tb00256.x

Qadir M, Schubert S (2002) Degradation processes and nutrient constraints in sodic soils. Land Degrad Dev 13(4):275–294. https://doi.org/10.1002/ldr.504

Qadir M, Schubert S, Ghafoor A, Murtaza G (2001) Amelioration strategies for sodic soils: a review. Land Degrad Dev 12:357–386. https://doi.org/10.1002/ldr.458

Qadir M, Steffens D, Yan F, Schubert S (2003) Proton release by N_2-fixing plant roots: a possible contribution to phytoremediation of calcareous sodic soils. J Plant Nutr Soil Sci 166(1):14–22. https://doi.org/10.1002/jpln.200390007

Qin P, Han R, Zhou M, Zhang H, Fan L, Seliskar DM, Gallagher JL (2015) Ecological engineering through the biosecure introduction of *Kosteletzkya virginica* (seashore mallow) to saline lands in China: a review of 20 years of activity. Ecol Eng 74:174–186. https://doi.org/10.1016/j.ecoleng.2014.10.021

Rabhi M, Hafsi C, Lakhdar A, Hajji S, Barhoumi Z, Hamrouni MH, Abdelly C, Smaoui A (2009) Evaluation of the capacity of three halophytes to desalinize their rhizosphere as grown on saline soils under nonleaching conditions. Afr J Ecol 47(4):463–468. https://doi.org/10.1111/j.1365-2028.2008.00989.x

Rasool S, Hameed A, Azooz MM, Muneeb-u-Rehman, Siddiqi TO, Parvaiz A (2013) Salt stress: causes, types and responses of plants. In: Ahmed et al (eds) Ecophysiology and responses of plants under salt stress. Springer, pp 1–24 https://doi.org/10.1007/978-1-4614-4747-4_1

Ravindran KC, Venkatesan K, Balakrishnan V, Chellappan KP, Balasubramanian T (2007) Restoration of saline land by halophytes for Indian soils. Soil Biol Biochem 39(10):2661–2664. https://doi.org/10.1016/j.soilbio.2007.02.005

Reinhardt DH, Rost TL (1995) Salinity accelerates endodermal development and induces an exodermis in cotton seedling roots. Environ Exper Botany 35:563–574

Robbins CW (1986) Sodic calcareous soil reclamation as affected by different amendments and crops. Agron J 78(5):916–920. https://doi.org/10.2134/agronj1986.00021962007800050034x

cRobinson B, Green S, Mills T, Clothier B, Fung L, Hurst S, et al. (2003) Assessment of phytoremediation as best management practice for degraded environments. In: Currie LD, Stewart RB, Anderson CWN (eds) Environmental management using soil-plant systems: proceedings of the 16th annual workshop held by the fertilizer and lime research centre, palmerston: fertilizer and lime research centre, pp 39–49

Sharma BR, Minhas PS (2005) Strategies for managing saline/alkali waters for sustainable agricultural production in South Asia. Agric Water Manag 78(1–2):136–151. https://doi.org/10.1016/j.agwat.2005.04.019

Shelef O, Gross A, Rachmilevitch S (2012) The use of *Bassia indica* for salt phytoremediation in constructed wetlands. Water Res 46(13):3967–3976. https://doi.org/10.1016/j.watres.2012.05.020

Shen B, Jensen RG, Bohnert H (1997) Increased resistance to oxidative stress in transgenic plants by targeting mannitol biosynthesis to chloroplasts. Plant Physiol 113(4):1177–1183. https://doi.org/10.1104/pp.113.4.1177

Shepherd T, Davies HV (1993) Carbon loss from the roots of forage rape (*Brassica napus* L.) seedlings following pulse-labeling with (CO_2)-C-14. Ann Bot-Lond 72:155–163

Shiyab S, Shibli R, Mohammad M (2003) Influence of sodium chloride salt stress on growth and nutrient acquisition of sour orange *in vitro*. J Plant Nutr 26 (5):985–996. https://doi.org/10.1081/PLN-120020070

Shrivastava P, Kumar R (2015) Soil salinity: a serious environmental issue and plant growth promoting bacteria as one of the tools for its alleviation. Saudi J Biol Sci 22(2):123–131. https://doi.org/10.1016/j.sjbs.2014.12.001

Silva YJAB, Silva YJAB, Freire MBGS, Lopes EAPL, Santos MA (2016) *Atriplex nummularia* Lindl. as alternative for improving salt-affected soils conditions in semiarid environments: a field experiment. Chilean J Agricult Res 76(3):343–348. https://doi.org/10.4067/S0718-58392016000300012

Srivastava AK, Srivastava OP (1992) Cation- exchange capacity of roots in relation to response of fertilizer nutrients in salt- affected soil. Indian J Agric Sci 62 (3):200–204

Susarla S, Medina VF, McCutcheon SC (2002) Phytoremediation: an ecological solution to organic chemical contamination. Ecol Engineer 18(5):647–658

Szabolcs I (1979) Review of research on salt-affected soils. UNESCO, France

Tanji KK (1990) Nature and extent of agricultural salinity. In: Tanji KK (ed) Agricultural salinity assessment and management, ASCE manuals and reports on engineering practice. No 71, New York, NY: ASCE, pp 1–17

Tejada M, Gonzalez JL (2005) Beet vinasse applied to wheat under dryland conditions affects soil properties and yield. Eur J Agron 23(4):336–347. https://doi.org/10.1016/j.eja.2005.02.005

Tester M, Davenport R (2003) Na^+ tolerance and Na^+ transport in higher plants. Ann Bot 91:503–527. https://doi.org/10.1093/aob/mcg058

Valenti GS, Melane L, Orsi O, Riveros F (1992) Anatomical changes in *Prosopis cineraria* (L.) Druce seedlings at different levels of NaCl salinity. Ann Bot 70:399–404. https://doi.org/10.1093/oxfordjournals.aob.a088494

Vamerali T, Bandiera M, Mosca G (2010) Field crops for phytoremediation of metal-contaminated land. A Rev Environ Chem Lett 8:1–17. https://doi.org/10.1007/s10311-009-0268-0

Volchko Y, Norrman J, Rosen L, Norberg T (2014) A minimum data set for evaluating the ecological soil functions in remediation projects. J Soils Sediments 14:1850–1860. https://doi.org/10.1007/s11368-014-0939

Wahid A, Javed I, Ali I, Baig A, Rasul E (1998) Short term incubation of sorghum caryopses in sodium chloride levels: changes in some pre and post germination physiological parameters. Plant Sci 139:223–232. https://doi.org/10.1016/S0168-9452(98)00124-1

Wakeel A (2013) Potassium-sodium interactions in soil and plant under saline-sodic conditions. J Plant Nutr Soil Sci 176(3):344–354. https://doi.org/10.1002/jpln.201200417

Wang X, Sun R, Tian Y, Guo K, Sun H, Liu X, Chu H, Liua B (2020) Long-term phytoremediation of coastal saline soil reveals plant species-specific patterns of microbial community recruitment. Ecol Evolut Sci 5 (2):e00741-e819. https://doi.org/10.1128/mSystems.00741-19

Weimberg R (1986) Growth and solute accumulation in 3-week-old seedlings of *Agropyron elongatiun* stressed with sodium and potassium salts. Physiol Plant 67(2):129–135. https://doi.org/10.1111/j.1399-3054.1986.tb02433.x

White PJ, Broadley MR (2001) Chloride in soils and its uptake and movement within the plant: a review. Ann Bot 88(6):967–988. https://doi.org/10.1006/anbo.2001.1540

Williams DE, Coleman NT (1950) Cation exchange properties of plant root surfaces. Plant Soil 2:243–256. https://doi.org/10.1007/BF01852352

Wong VNL, Dalal RC, Greene RSB (2009) Carbon dynamics of sodic and saline soils following gypsum and organic material additions: a laboratory incubation. Appl Soil Ecol 41:29–40. https://doi.org/10.1016/j.apsoil.2008.08.006

Yadav JSP (1980) Efficacy of fertilizers used in saline and alkali soils for crop production. Fertiliser News 25:19–27

Yensen NP, Biel KY (2006) Soil remediation via salt-conduction and the hypotheses of halosynthesis and photoprotection ecophysiology of high salinity tolerant plants. In: Khan MA, Weber DJ (eds), vol 40, Tasks for vegetation science 34. Springer Netherlands, pp 313–344. https://doi.org/10.1007/1-4020-4018-0_21

Yu JQ, Matsui Y (1994) Phytotoxic substances in root exudates of cucumber (*Cucumis sativus* L.). J Chem Ecol 20:21–31. https://doi.org/10.1007/BF02065988

Yun K, Rutter A, Zeeb BA (2019) Composting of the halophyte *Phragmites australis* following phytoaccumulation of chloride from a cement kiln dust (CKD)-contaminated landfill. Waste Manag 87:119–124. https://doi.org/10.1016/j.wasman.2019.01.030

Yurtseven E, Kesmez GD, Unlukara A (2005) The effects of water salinity and potassium levels on yield, fruit quality and water consumption of a native central Anatolian tomato species (*Lycopersicon esculantum*). Agricul Water Manage 78(1–2):128–135. https://doi.org/10.1016/j.agwat.2005.04.018

Zhang L, Song R, Wu X (2008) Salt tolerance capability of *Tamarix chinensis* and the effects on the improvement of coastal saline soil. J Anhui Agric Sci 36 (13):5424–5426

Zhang X, Zhong B, Shafi M, Guo J, Liu C, Guo H, Peng D, Wang Y, Liu D (2018) Effect of EDTA and citric acid on absorption of heavy metals and growth of moso bamboo. Environ Sci Pollut Res 25:18846–18852

Zhao K-F, Fan H, Song J, Sun M-X, Wang B-Z, Zhang S-Q, Ungar IA (2005) Two Na^+ and Cl^- hyperaccumulators of the chenopodiaceae. J Integr Plant Biol 47 (3):311–318. https://doi.org/10.1111/j.1744-7909.2005.00057.x

Zhu J, Fu X, Koo YD, Zhu JK, Jenney FE Jr, Adams MW, Zhu Y, Shi H, Yun DJ, Hasegawa PM, Bressan RA (2007) An enhancer mutant of Arabidopsis salt overly sensitive 3 mediates both ion homeostasis and the oxidative stress response. Mol Cell Biol 27(14):5214–24. https://doi.org/10.1128/MCB.01989-06

Zhu JK (2007) Plant salt stress. In: Encyclopedia of life sciences (eLS). Wiley, Chichester http://www.els.net. https://doi.org/10.1002/9780470015902.a0001300.pub2

Bioremediation of Waste Dumping Sites

6

Reshma Soman and Abhilash Surendran

Abstract

Human activities have resulted in the accumulation of enormous tons of waste products day by day in the form of raw sewage to nuclear wastes. These wastes are dumped everywhere and are needed to be treated efficiently to make the earth neat and clean. The wastes which can be bio-treated include heavy metal wastes, agricultural wastes, solid wastes, rubber wastes and xenobiotic wastes. Microorganisms, which are having profound effects on the day to day life of humans, can be effectively used for the remediation of these waste materials at their dumping sites. The use of microbes for waste treatment presents an eco-friendly as well as economical treatment method and a wide range of microbes can be used, based on the type of wastes, at the dumping sites. Heavy metals can be treated with heavy metal degrading bacteria in which the bacteria convert the harmful heavy metals to their less toxic forms. Other wastes, including solid and agricultural wastes, can be treated by bioremediation methods, including composting, biopiling or bioaugmentation methods. These wastes are transferred to big bioreactors, which can be further treated with microorganisms to detoxify it to less harmful products. This chapter emphasizes the various types of bioremediation methods used for the treatment of wastes at their dumping sites.

Keywords

Bioremediation · Health hazards · Municipal waste · Pollution · Waste dumping sites

6.1 Introduction

Our mother planet, the earth, is endowed with a wide range of natural resources comprising abiotic factors such as soil, air, water, wind, and biotic factors such as microorganisms, plants, and animals. Human life started as a nomadic type and they co-existed peacefully with the other life forms. However, with the advent of civilization, human life started to emerge in an unparalleled way as compared with other life forms. The progress in human life has eventually resulted in the misuse of natural resources, resulting in their depletion (Gosavi et al. 2004). The higher rate of growth in science, technology, and industry have resulted in the accumulation of different types of waste products, including

R. Soman
School of Biosciences, Mahatma Gandhi University, Kottayam, Kerala 686560, India

A. Surendran (✉)
Department of Microbiology, Majlis Arts and Science College, Malappuram, Kerala 676552, India

J. A. Malik (ed.), *Advances in Bioremediation and Phytoremediation for Sustainable Soil Management*,
https://doi.org/10.1007/978-3-030-89984-4_6

nuclear wastes, thereby presenting an alarming situation for the existence of not only humans but also of the earth itself (Karigar and Rao 2011). Hence proper care must be taken for the effective removal of these waste products.

6.2 How Waste is Produced?

Waste is mostly a byproduct of human activities. Tremendous and unregulated developments in human lifestyles have resulted in the accumulation of complex types of waste materials throughout the environment. Introduction of advanced technologies in industries, medical facilities and agricultural practices have contributed much to the modernization of human life. But at the same time, they have contributed significantly to the deposition of waste products in the environment. The wastes are mainly of three states: solid, liquid, and gas. Since most of these waste products are highly toxic to the environment and humans, proper remediation measures must be employed for ensuring the wellbeing of the environment.

6.3 Importance of Waste Dumping Sites

Waste dumping sites are isolated areas that are specifically used for the deposition and decomposition of waste materials. The deposition of wastes materials at dumping sites is an age-old process that is in practice right from the prehistoric periods. The deposition and decomposition of toxic waste materials in isolated areas prevent the risk of health hazards to humans. With the progress in human civilization, more waste materials are released into the environment on a daily basis and hence there arises the need for more waste dumping sites. Scientific waste dumping and decomposition processes can reduce the toxic impacts of waste materials on the environment. On the other hand, unscientific and careless management of such dumping sites may make them a source of several dangerous diseases that may affect the wellbeing of humans. Hence proper strategies much be applied for the remediation of waste materials.

6.4 Waste Remediation Strategies

Various strategies have been adopted for the removal of waste products. These include: (1) Incineration- application for higher temperature for the treatment of waste materials. (2) Sanitary landfills- landfills are the sites where waste materials are kept isolated from the human dwelling areas. It is generally considered the most common method with wide acceptance throughout the world. (3) Recycling of waste materials- Conversion of waste materials into some useful products which can be applied for human uses. (4) Reduction and reuse- Making reusable/ repairable products that may reduce the rate of waste deposition. Several advanced technologies have been put forth including the application of chemical decomposition (including base-catalyzed dechlorination, UV oxidation) and incineration at high temperature. Even though they can be effectively used for treating a large range of contaminants, they present a wide range of drawbacks also. They are much intricate, too costly and hence they lack appreciation from the common man (Karigar and Rao 2011).

6.5 Health Issues Associated with Waste Treatment

Unscientific measures at any step of waste treatment (handling, treatment, and disposal processes) may cause health problems. This can be either direct (affecting the laborers associated with waste treatment) or indirect (through the contact with contaminated water, soil or air). Ground water pollution due to the failure of waste treatment processes has been reported even in highly developed nations. Degradation of waste materials may pose direct health impacts on humans (Shukla et al. 2000). Improper use of gloves, uniforms, and safety equipment by the workers associated with waste remediation may result in health hazards. People living in close proximity to waste dumping sites are prone to infections by gastrointestinal parasitic worms and respiratory infections. Open type waste dumping sites may release large quantities of methane as a byproduct of the degradation process into the surrounding areas. This may often result in explosions thereby acting as a major contributing factor for global warming (Giusti 2009; Slagstad and Bratteb 2013). Such waste dumping sites may cause the migration of leachates to soil and water bodies (Dasgupta et al. 2013; Muhammad et al. 2020).

Improper degradation of tires and other waste materials may result in the storage of water in the waste dumping sites. This may promote mosquito growth resulting in malaria, dengue and West Nile fever, and several other diseases. Unscientific and uncontrolled burning of waste materials at their dumping sites may release smog, harmful gases, and other fine particles into the atmosphere. The effluents may contain CO, CO_2, arsenic, mercury, polycyclic aromatic hydrocarbon (PAHs) and other toxic components. Inhalation of such toxic air may result in profound health hazards including cancer in new born babies and adult population. Improper handling of such waste dumping sites may result in the leaching of microplastics into the ocean which may cause damages to the ocean food chain and thereby affecting the survival of fish and other living beings (Sridevi et al. 2012; Azar and Azar 2016; Ghosh 2016; Ahamed et al. 2020).

Even though several advanced techniques have been developed so far for the treatment of waste products, the adverse effects on humans and the environment is still not negligible (Viel et al. 2008). Emissions from incinerators of waste treatment plants are found to cause respiratory problems (acute and chronic diseases) in humans. Cancers of different organs such as the liver, larynx, lungs, bladder, stomach, and leukemias, as well as skin and gastrointestinal ailments have been reported at higher rates in individuals living close to the incinerator sites (Crowley et al. 2003). Hence advanced methods for the effective decomposition of the waste products, without causing hazards to humans and the environment, have to be developed.

6.6 Categorization of Waste Materials

According to the source of wastes, solid waste can be categorized into three. The household wastes which is otherwise known as the municipal wastes are produced tremendously from our houses are a major reason of waste dumping sites around the world. Second type of waste produced is industrial waste which is considered the most hazardous waste to society. The third type of waste that are produced in our ecosystem is the hospital wastes, also known as infectious wastes which are not treated properly, serve as the reason for most of the infectious diseases that are spreading in our society.

Waste dumping sites is the common form of municipal solid waste. Municipal solid waste (MSW) consists of the general wastes including the household wastes, plastics, agricultural wastes, food wastes, sanitation residues, street wastes, and so on. The rise in urbanization and modernization leads to changes in lifestyles and also the food habits of people. This leads to the change in waste generation and the composition of the waste. For example, the solid waste generated in India during 1947 was six million tonnes, and after 50 years by 1997, the estimated amount of waste produced was hiked to forty-eight million tonnes. Garbage wastes can be categorized into four types;

Organic wastes These are the wastes that are biodegradable wastes which include the kitchen wastes, flowers, leaves, fruits and vegetables.

Toxic wastes These are the wastes that are toxic to humans and living forms. Examples of these wastes include old medicines, spray cans, chemicals, bulbs, batteries, shoe polish, paints, fertilizer and pesticide containers.

Recyclable wastes These wastes can be recycled and made into another compound according to the need of human. These wastes include paper, glass, metals and plastics.

Soiled waste These wastes are being dirty because of various sources like hospital waste such as cloth soiled with blood and other body fluids.

6.7 Municipal Waste—Segregation

Municipal solid wastes that are segregated at a source can be of different types.

6.7.1 Recyclable Dry Waste

The recyclable dry wastes can be again categorized into two by the property of burning. Those which can be burned readily are grouped as combustibles including packing material, and paper which can be further used as fuel pellets. Those which cannot be burned are called non- combustibles which includes glass and metals. These types of wastes can be separated and recycled in industries.

6.7.2 Organic Fraction

These types of wastes are subjected to biological treatments. These wastes can further be treated and used for composting, vermicomposting, biogas and landfill gases.

6.7.3 Inert Debris

These types of wastes cannot be further treated. Hence, they can be used for low grade constructions and paving of roads.

6.7.4 Hazardous Wastes

These wastes are toxic to nature. They can be accumulated from hospital wastes and by other means. Hospital wastes are further incinerated to remove the toxicants while other hazardous wastes have to be planned according to the specific treatment that is required for the removal of waste.

A schematic diagram of the typical municipal waste management system is presented in Fig. 6.1.

6.8 Bioremediation

The first two decades of the twenty-first century witnessed tremendous advancements in the application of bioremediation techniques, with the main purpose of efficiently reducing environmental pollution. Bioremediation presents a cost-effective as well as an eco-friendly approach for the treatment of waste products. Several methods have been developed by researchers around the globe which comes under the bioremediation category. Due to the vast range of pollutants, a single bioremediation technique cannot offer cent percent efficiency in their treatment. The key factors for bioremediation are the microorganisms and hence the environmental conditions at the waste dumping sites must support their growth, survival, and metabolism (Verma and Jaiswal 2016). Eco-friendly natures, as well as cost-effectiveness, are the positive aspects of bioremediation over various chemical and physical methods of degradation.

6.9 Bioremediation Principles

Bioremediation can be defined as a process that involves the controlled biological degradation of organic wastes to an innocuous state or to concentrations below the standard levels mentioned by the regulatory authorities (Mueller et al. 1996). Bioremediation can be done either under aerobic or anaerobic conditions. The microorganisms employed for the biological degradation of waste materials maybe either indigenous to the waste dumping sites or they may be introduced from elsewhere. The degradation of pollutants is carried out by a mixture of different types of microorganisms and this occurs as a part of their normal metabolic processes. The process of introduction of microorganisms to waste dumping sites from other sites for enhancing the bioremediation process efficiency is called as bioaugmentation. Microorganisms act by secreting extracellular enzymes that may act on toxic materials and convert them into non-toxic forms, thereby, deteriorating the toxic potential of these substances on humans as well as to the environment. Environmental parameters must be ideal for the optimal growth and activity of microorganisms, which in turn, is essential for the effective decomposition of toxic waste products at a faster rate. Another important factor for proper

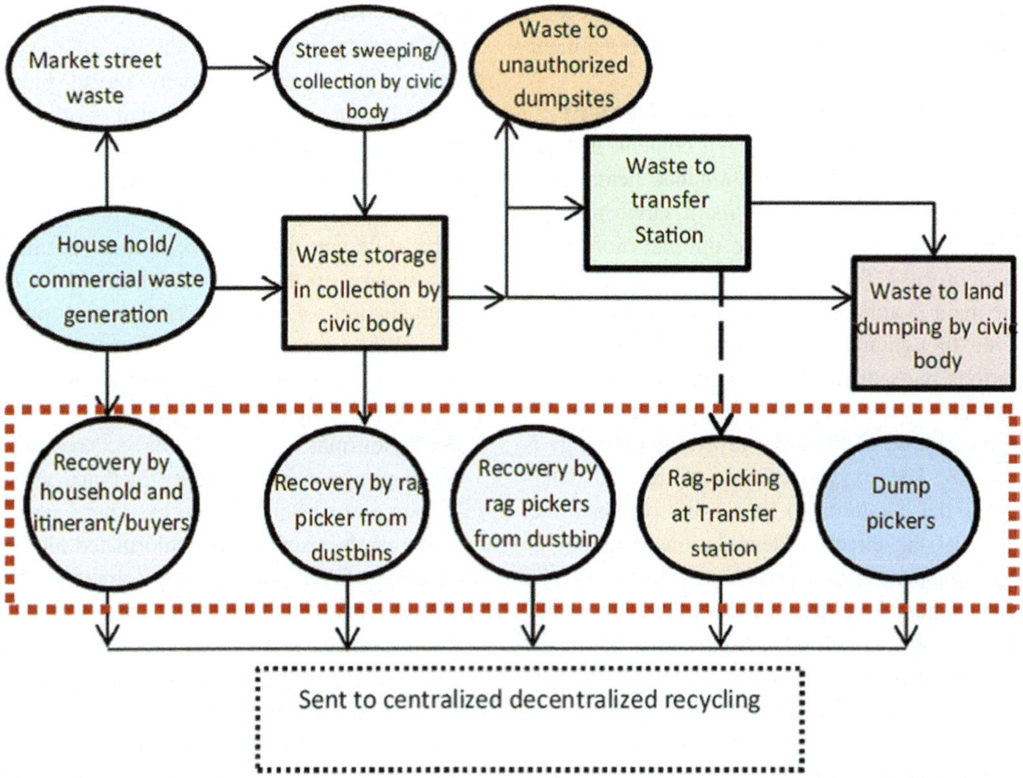

Fig. 6.1 Schematic diagram of the typical municipal waste management system (Modified after: Joseph 2002)

biodegradation is the contact between microorganisms and the pollutants and this is dependent on microbial mobility. Certain bacteria can sense the pollutant and move towards them by chemotaxis while other microbes including fungi grow towards the pollutants in a filamentous form. Surfactants such as sodium dodecyl sulphate (SDS) can serve as stimulating factors for microbial mobilization (Hetherington et al. 2006).

6.10 Different Factors Affecting the Bioremediation Process at the Waste Dumping Sites

Since microorganisms are an inevitable part of bioremediation, the control and optimization of bioremediation at waste dumping sites is a complex process. Bioremediation is affected by environmental and microbial factors (Vidali 2001). These parameters needed to be optimum and in adequate amounts for the growth of microbes thereby ensuring the effective degradation of contaminants or pollutants to less harmful products at a rapid rate (de la Cueva et al. 2016).

6.10.1 Environmental Factors

The environmental factors that affect the bioremediation process can be classified under two categories; the presence of nutrients and other environmental requirements.

6.10.1.1 Nutrients

There are an enormous amount of microorganisms present in the soil but the availability of the specific microorganisms might or can be lower in adequate amounts that are required for the enhancement of bioremediation. So, to attain a better treatment, nutrient supplementation including carbon, nitrogen, phosphorous, hydrogen and oxygen are added to the treatment.

6.10.1.2 Environmental Requirements

Soil moisture, soil pH, oxygen content, nutrient content, temperature, contaminants, heavy metals, types of soil are the main environmental requirements needed for the optimal bioremediation process. Temperature, pH, and moisture are the main factors that determine bacterial growth. Even though microbes had been isolated from various parts of the world including all adverse conditions, the optimum temperature is mesophilic (15–40 °C) and the favorable pH range is 5.5–8.8. The oxygen content of the soil determines the condition to be aerobic or anaerobic. Most of the effective degradation of hydrocarbons happens in the aerobic temperature while compounds like chlorurate can only be degraded in anaerobic conditions. Apart from these, the soil texture also holds an important role in bioremediation. The soil provides all these conditions that are required for the growth and establishment of microbes in the environment (Vidali 2001; Prescott et al. 2002).

6.10.2 *Microbial Factors*

Microorganisms are ubiquitous and can be profusely isolated from environmental samples that vary in temperature, pH, pressure, salinity, aerobic and anaerobic conditions. To survive in these conditions, the carbon source is used as an energy source. Because of all these characteristics, microbes became an inevitable part of the bioremediation process. These microorganisms can be characterized into aerobic, anaerobic, ligninolytic fungi and methylotrophs.

6.10.2.1 Aerobic

Aerobic bacteria are those bacteria which use oxygen for their growth. These microbes can use contaminants as their source of carbon and energy. Hence, they are extensively used for the degradation of hydrocarbons and pesticides. Examples of aerobic bacteria include *Pseudomonas, Alcaligenes, Mycobacterium, Sphingomonas and Rhodococcus.*

6.10.2.2 Anaerobic

These bacteria grow in the absence of oxygen which makes them extensively useful in the bioremediation of polychlorinated bisphenyls (mainly present in the river water), chloroform, and dechlorination of trichloroethylene.

Lignolytic Fungi

The different fungal species that normally cause diseases to the plants can be used for the bioremediation of toxic environmental pollutants. White rot fungus *Phanaerochaete chrysosporium* is a potent example of fungi that are extensively used for this process.

Methylotrophs.

This group of bacteria uses methane as the sole source of carbon and energy. These aerobic bacteria can be used against chlorinated aliphatics trichloroethylene and 1,2-dichloroethane.

For achieving a better remediation process, the bacteria need to be in contact with the contaminants and work effectively. Two strategies can be applied based on this aspect. First one is the case with mobile bacteria. Such bacteria can move towards the contaminants in a similar fashion to root nodule formation through chemotaxis. The bacteria through chemotactic response sense the presence of contaminants that are nearby, move and adhere to the contaminants and thus bioremediate it. The second strategy involves the utilization of filamentous fungus in bioremediation. These fungal species can move towards the contaminants by forming filaments and once they came in contact with the pollutant, can detoxify and bioremediate them. Apart from these, the use of surfactants like sodium dodecyl sulphate can be used to immobilize the contaminants and thereby enhancing the bioremediation process (Vidali 2001; Glazer and Nikaido 2007; Gupta and Prakash 2020).

6.11 Types of Bioremediation Strategies at the Waste Dumping Sites

Based on the site of operation, the bioremediation process can be classified into ex-situ bioremediation and in-situ bioremediation. Ex-situ bioremediation technique involves the gouging and transportation of pollutants from a place to another site for treatment. These treatment strategies are used according to the degree and depth of pollution, the pollutant type and treatment cost for it, geographical location, and geology of the contamination site (Philp and Atlas 2005). Biopile, land farming, bioreactors and windrows are the methods of ex-situ bioremediation techniques (Azubuike et al. 2016).

The in-situ bioremediation technique involves the treatment of contaminants in the pollution site itself without excavation and alteration to the soil texture. These techniques are more economical than ex-situ bioremediation as it rules out the costs of excavation and transportation. However, the major problem is the requirement of sophisticated instruments and their on-site installation. In-situ bioremediation techniques can be further divided into enhanced and non-enhanced treatments. Enhanced in-situ bioremediation includes bioventing, biosparging and phytoremediation. Non-enhanced in-situ bioremediation comprises intrinsic bioremediation and natural attenuation (Folch et al. 2013; Kim et al. 2014; Frascari et al. 2015; Roy et al. 2015).

6.11.1 Composting

Nearly 35–40% of municipal solid waste that is produced in India is composed of organic matter. The main method that is used to recycle these wastes is composting. Composting is a natural biological process where bacteria and fungi covert the organic waste into biodegradable humus like substances. This is excellent manure for the growth of plants that is rich in carbon and nitrogen. It shows that the waste produced in the kitchen (like leftovers, leaves, and other veggies) which are meant to throw away or rot can be recycled as nutrients and finally reach soil in the form of nutrients (Ramtek 2010).

6.11.2 Bioventing

In this method of in-situ bioremediation, air is supplied at slow rate for the slow provision of adequate amounts of oxygen and nutrients, for the controlled stimulation of microorganisms. This is an effective method for the bioremediation of hydrocarbons due to its property to devoid of volatile gases (Banerjee et al. 2016).

6.11.3 Biosparging

This method of bioremediation is widely used in contamination sites where there is an interaction between ground water and the soil occurs. It is done by providing aeration to the water table that is present below the remediation site. In addition, provision of low pressure is ensured which helps to increase the availability of oxygen for microbes. This also facilitates the better interaction of groundwater with the soil. It is a very simple technique and the air injectors are easy to handle, making it a widely used technique in contaminated sites (Gupta and Prakash 2020).

6.11.4 Bioreactors

This is an ex-situ bioremediation technique which is also widely used for bioremediation of contaminated sites. In this type of treatment, slurry reactors or aqueous reactors are used for the treatment of contaminated water or soil. The contaminants are first placed in a containment vessel. These contaminants are then subjected to the three-phase system (solid, liquid, and gas) mixing using different apparatus, and slurry is formed. The slurry so formed is rich in microbes and this ensures the efficient degradation of contaminants (Banerjee et al. 2016, Gupta and Prakash 2020).

6.12 Municipal Solid Waste—Operation Methods

Municipal solid waste (also known as urban solid waste) is the waste that is produced within a given area mainly constituting household wastes and commercial wastes. They can be solid or semi solid form. Solid waste management can be categorized into the following phases;

6.12.1 Waste Generation

Those materials or substances which are no longer valuable and meant to be thrown away are compiled together and considered wastes.

6.12.2 Waste Handling, Generation, Storage, and Processing at the Source

The wastes are collected in a container and these loaded waste containers are transported/moved to separation places. These steps are collectively called the handling and storage of solid waste.

6.12.3 Collection

These solid and recyclable wastes are gathered and transported to the location where they are collected. These locations where the wastes are being transferred should have the facility to process the material, or have a landfill or have a transfer station.

6.12.4 Separation, Processing, and Transformation of Solid Wastes

Curbiside collection, drop off and buy back centers are the means by which the recovery of wastes is normally done. Separation is done at the source and at places where material recovery facility is available, transfer stations, have combustion facilities and disposal sites.

6.12.5 Transfer and Transport

This step includes the transfer of the municipal waste from collection vehicles to the large fetch equipment, and transferring of wastes to disposal sites or processing sites. These steps use large transport equipments that can run long distances to the processing sites.

6.12.6 Disposal

Landfilling and landspreading are the destinations of all residential or commercial solid wastes. Modernization leads to more advanced and engineered sanitary landfills which have the facility to dispose wastes without creating hazards to public health like ground water contamination, breeding of rats, insects, and stray dogs.

6.13 Techniques Adopted for the Bioremediation of Different Types of Wastes

As mentioned above, the municipal solid wastes are separated and treated differently according to the component it is made up of. These bioremediation processes are:

6.13.1 Bioremediation of Hydrocarbons

Hydrocarbons refer to compounds that have hydrogen and carbon as the main constituents and these compounds are a major source of energy. Petroleum products are rich sources of hydrocarbons. Oil spills from petroleum products often occur in the ocean and this can eventually lead to the contamination of oceans. In addition to this, they may also promote global warming which is harmful to air, water and soil and may even affect the existence of various living species on earth. Hence the need for proper bioremediation of hydrocarbons is very crucial for the wellbeing of life on earth. Microorganisms utilize the hydrocarbons that are present on the dump sites and convert them into compounds that can be further degraded by other microorganisms which are present in the environment. These microorganisms again degrade the components and finally produce carbon dioxide, water and other inorganic compounds. Apart from microorganisms, plants also help to degrade hydrocarbons into less harmful compounds (de la Cueva et al. 2016).

6.13.2 Bioremediation of Plastics

Long ago, the materials were made of stones, glass, woods and metals. These materials had been used widely for the preparation of utensils, household items, and other daily use equipment. But when time goes on, the brittle nature and weight of these materials lead to the necessity of a new material that is lightweight, do not tend to change the shape on heating, durable, non-breakable, and can be used in adverse conditions. This leads to the accidental discovery of plastics. Later on, plastics replaced glass, metals, stones, and woods by its properties.

Due to the hydrophobic nature, plastics are widely used in situations where water contact should be avoided. It is free from microbial attacks and requires decades of decay. It is ubiquitous around the land and water and is one of the main reasons for the death of fishes (Fig. 6.2). Hence, plastics are a serious threat to human beings and nature. Still, the use of plastics cannot be ceased. The decay of plastics is less compared to the production of plastics. The need for the removal of this harmful waste is an important issue in our environment. Even though it is considered a major non-biodegradable contaminant, studies proved that certain microorganisms can breakdown plastics into less toxic compounds and thereby help for the bioremediation of plastics (Shahnawaz et al. 2019).

Biodegradation of plastics occurs in various ways. The main step of biodegradation is the biofragmentation of plastics. This can be achieved mainly by two methods; hydrolysis and oxidation. Hydrolysis is a simple process, and it is a method commonly adopted by different sets of microbes, while the oxidation process is limited to an aromatic ring containing compounds including polythene and polypropylene. *Ideonella sakaiensis* and *Pseudomonas spp.* are the two major microbial agents associated with the biodegradation of plastics. *Ideonella sakaiensis* refers to a Gram-negative rod-shaped bacterium categorized under the class Betaproteobacteri. This type of bacteria is endowed with the capability of utilizing Polyethylene terephthalate (PET) as a carbon source, thereby ensuring PET biodegradation. *I. sakaiensis* can attach with its substrate (PET) by using its tendrils. After making proper attachment with PET, *I. sakaiensis* releases two enzymes, polyethylene terephthalate hydrolase (PETase) and mono-2-hydroxyethyl terephthalate hydrolase (MHETase). PETase enzyme catalyzes the hydrolysis of PET plastic mainly into mono-2-hydroxyethyl terephthalate (MHET) and small fractions of bis (2-hydroxyethyl) terephthalic acid (BHET) and TPA (terephthalic acid). The other enzyme, MHETase, facilitates the hydrolysis of MHET and BHET to TPA and EG (ethylene glycol). TPA so formed undergoes degradation to pyruvate and oxaloacetate through several reactions involving the formation of intermediates such as protocatechuic acid (PCA) and EG is incorporated into the glyoxylate metabolism of these microbial agents. *Thermobifida fusca* is a thermophilic actinobacterium found in decaying organic matter. This bacterium also possesses the capability to degrade PET through a double enzyme system comprising a polyester hydrolase and carboxylesterase enzymes respectively (Jenkins et al. 2019).

6.13.3 Bioremediation of Food Wastes

Food is an inevitable part of our life. There are many food industries around the world such as the industries dealing with fruit and vegetable canning, frozen vegetables, vegetable dehydration, fruit and vegetable drying, fruit pulping, tomato juice concentrate, and fruit concentrate (Punnagaiarasi et al. 2017). Food wastes are tremendously produced from these food industries and are mainly dumped in landfills. These wastes are a major source of hydrocarbons (sugars, nitrogen, and cellulose fibers), salts, proteins, fats, and large amounts of moisture to the soil (Schaub and Leonard 1996; Mavropoulos 2011). The water that leaches out from the vegetables and the water wastes contain pesticides, herbicides, dissolved compounds and cleaning chemicals. Thus the food industry serves as a source of untrapped energy and also

Fig. 6.2 The plastic accumulation on the banks of Vembanad Lake in Kerala (India) after the 2018 floods

as a cause of emission of greenhouse gases to the atmosphere (Mavropoulos 2011).

The wastes that are produced from the food industry are of different types including solid and liquid wastes. Hence, separate waste treatments are required as per the nature of waste that is produced. The solid wastes can be treated by composting. The liquid containing solid produces slurry which can be treated with the help of big bioreactors and land farming. Since this treatment involves the use of microorganisms for detoxifying the wastes, pretreatments are required for the enhancement of microbial growth and removal of excess water (Schaub and Leonard 1996). Since wastes are produced enormously, the bulk density is also needed to be rectified. For improving the porosity of the sludge and to decrease the bulk density, bulking agents are added into the sludge. Once the porosity is enhanced, this can exert pressure on the sludge or by gravity which will eventually increase the drainage of water (Grobe 1994; Schaub and Leonard 1996). Another method used for the treatment of solid food wastes is by the use of aerated piles. This beneficial treatment technique helps for the best mixing of sludge (Nakata 1994).

Another type of food waste that is also the main reason for dumped waste is meat and poultry wastes. These wastes are produced from the slaughter houses and mainly contain the fats and residues from the intestine, blood, paunch grass, manure, and meat and bone pieces (Cournoyer 1996). The wastewater produced from slaughterhouses is obnoxious having a foul odor, high BOD, high nitrogen and moisture (90–95%) along with a rich supply of pathogens. Hence pretreatment of the wastewater is a prerequisite for the bioremediation process of these wastes. The wastes can be treated either aerobically or anaerobically. In aerobic treatment, bulking agents are added in order to increase the porosity and reduction of moisture, and this will eventually help to increase the carbon and aeration of the wastewater. Contact processes, sludge beds, filters, packed beds, and hybrid reactors are used as a part of aerobic treatment techniques (Punnagaiarasi et al. 2017).

6.14 Scope of the Bioremediation Process at the Waste Dumping Sites

Bioremediation of waste materials has received widespread appreciation from the public due to the fact that it is a harmless, natural process. The major concept of this process is the transformation of hazardous contaminants into eco-friendly products. The vital residues for bioremediation comprise carbon dioxide, water, and cell biomasses which are non-toxic to the environment. Bioremediation offers the advantage of treating waste materials at the source of origin thereby preventing the health and environmental hazards that can occur during the waste transportation process. Theoretically, bioremediation can offer complete decomposition of waste products with less expenditure as compared with the other waste treatment technologies.

The major drawback of bioremediation is that only biodegradable compounds can be treated using this process. Since bioremediation is a highly specific process, some toxic compounds cannot be completely degraded. Bioremediation is dependent on several factors such as sufficient microbial populations, appropriate microbial growth conditions, required amount of nutrients and favorable environmental conditions for microbial metabolism. If any of these factors are affected, microbial activity will be affected, thereby affecting the overall efficiency of the bioremediation process. Hence pilot scale studies cannot effectively predict the success of bioremediation process at a particular site (Dua et al. 2002). More field-level research is usually required for arriving at a final conclusion of the process efficiency at a particular waste dumping site. Bioremediation is comparatively a slower waste remediation process as compared with excavation and incineration processes. However, bioremediation ranks the best in terms of ensuring ecological safety. Innovations to improve the speed and decomposition of more waste products can make it a much more effective waste remediation process.

References

Ahamed A, Veksha A, Yin K, Weerachanchai P, Giannis A, Lisak G (2020) Environmental impact assessment of converting flexible packaging plastic waste to pyrolysis oil and multi-walled carbon nanotubes. J Hazard Mater 390:121449

Azar SK, Azar SS (2016) Waste related pollutions and their potential effect on cancer incidences in Lebanon. J Environ Prot 07(06):778–783

Azubuike CC, Chikere CB, Okpokwasili GC (2016) Bioremediation techniques–classification based on site of application: principles, advantages, limitations and prospects. World J Microbiol Biotechnol 32(11):180

Banerjee A, Roy A, Dutta S, Mondal S (2016) Bioremediation of hydrocarbon a review. Int J Adv Res 4 (6):1303–1313

Cournoyer, MS (1996) Sanitation and stabilization of slaughterhouse sludges through composting. In: Proceedings of the Canadian meat research institute technology symposium, Ontario, Canada, Canadian Meat Research Institute, Toronto, pp 1–7

Crowley D, Staines A, Collins C, Bracken J, Bruen M, Fry J, Hrymak V, Malone D, Magette B, Ryan M, Thunhurst C (2003) Health and Environmental Effects of Landfilling and Incineration of Waste - A Literature Review 153- 187

Dasgupta B, Yadav VL, Mondal MK (2013) Seasonal characterization and present status of municipal solid waste (MSW) management in Varanasi, India. Adv Environ Res 2(1):51–60

de la Cueva SC, Rodríguez CH, Cruz NOS, Contreras JAR, Miranda JL (2016) Changes in bacterial populations during bioremediation of soil contaminated with petroleum hydrocarbons. Water Air Soil Poll 227(3):91

Dua M, Sethunathan N, Johri AK (2002) Biotechnology bioremediation success and limitations. Appl Microbiol Biotechnol 59(2–3):143–152

Frascari D, Zanaroli G, Danko AS (2015) In situ aerobic cometabolism of chlorinated solvents: a review. J Hazard Mater 283:382–399

Ghosh SK (2016) Sustainable SWM in developing countries focusing on faster growing economies, India and China. Procedia Environ Sci 35:176–184

Giusti L (2009) A review of waste management practices and their impact on human health. J Waste Manag 29 (8):2227–2239

Glazer AN, Nikaido H (2007) Microbial biotechnology. Fundame Appl Microbiol 8:510–528

Gosavi K, Sammut J, Gifford S, Jankowski J (2004) Macroalgal biomonitors of trace metal contamination in acid sulfate soil aquaculture ponds. Sci Total Environ 324:25–39

Grobe K (1994) Composter links up with food processor. Biocycle 34(40):42–43

Gupta C, Prakash D (2020) Novel bioremediation methods in waste management: novel bioremediation methods. Waste management: concepts, methodologies, tools, and applications. IGI Global, USA, pp 1627–1643

Hetherington M, Moore M, Pepperman R, Shea TG, Stone L, Reimers R, Rom P (2006) Emerging Technologies for Biosolids Management. Office of Wastewater Management, US Environmental Protection Agency, Washington, DC

Jenkins S, Quer AMI, Fonseca C, Varrone C (2019) Microbial degradation of plastics: new plastic degraders, mixed cultures and engineering strategies. Soil Microenviron Bioremed Polym Prod 213–238
Joseph K (2002) Perspectives of solid waste management in India. In; International symposium on the technology and management of the treatment and reuse of the municipal solid waste, Shanghai, China, pp 1–14
Kim S, Krajmalnik-Brown R, Kim JO, Chung J (2014) Remediation of petroleum hydrocarbon-contaminated sites by DNA diagnosis-based bioslurping technology. Sci Total Environ 497:250–259
Mueller JG, Cerniglia CE, Pritchard PH (1996) Bioremediation of environments contaminated by polycyclic aromatic hydrocarbons. Biotechnol Res Series 6:125–194
Muhammad N, Nafees M, Khan MH, Ge L, Lisak G (2020) Effect of biochars on bioaccumulation and human health risks of potentially toxic elements in wheat (*Triticum aestivum* L.) cultivated on industrially contaminated soil. Environ Pollut 260:113887
Nakata B (1994) Recycling by-products on California vineyards. Biocycle 34:61
Philp JC, Atlas RM (2005) Bioremediation of contaminated soils and aquifers. In: Bioremediation. American Society of Microbiology, pp 139–236
Prescott LM, Harley JP, Klein DA (2002) Microbiology. Fundament Appl Microbiol 2:1012–1014
Punnagaiarasi A, Elango A, Rajarajan G, Prakash S (2017) Application of bioremediation on food waste management for cleaner environment. Bioremediation and sustainable technologies for cleaner environment. Springer, Cham, pp 51–56
Ramtek DS (2010) Reclamation and remediation of solid waste through bio-chemical process. In: Fulekar MH (ed) Bioremediation technology. Springer, Dordrecht, pp 285–314
Roy M, Giri AK, Dutta S, Mukherjee P (2015) Integrated phytobial remediation for sustainable management of arsenic in soil and water. Environ Int 75:180–198
Schaub SM, Leonard JJ (1996) Composting: An alternative waste management option for food processing industries. Trends Food Sci Technol 7(8):263–268
Shahnawaz M, Sangale MK, Ade AB (2019) General introduction. Bioremediation technology for plastic waste. Springer, Singapore, pp 1–19
Shukla SR, Akolkar AB, Bhide AD, Dhussa AK, Varshney AK, Acharya DB, Datta MM, Dutta M, Mazumdar NB, Asnani PU, Ramanathan R, Ramaprasad VB, Uppal SBB, Dhinadhayalan M (2000) Manual on municipal solid waste management. The Ministry of Urban Development, Government of India, p 663
Slagstad H, Bratteb H (2013) Influence of assumptions about household waste composition in waste management LCAs. Waste Manag 33(1):212–219
Sridevi V, Musalaiah M, Ch MVV, Lakshmi A, Kesavarao L (2012) A review on integrated solid waste management. Int J Eng Sci Adv Techno 2 (5):1491–1499
Verma JP, Jaiswal DK (2016) Book review: advances in biodegradation and bioremediation of industrial waste. Front Microbiol 6:1–2
Vidali M (2001) Bioremediation. An overview. Pure Appl Chemistry 73(7):1163–1172
Viel JF, Daniau C, Goria S, Fabre P, de Crouy-Chanel P, Sauleau EA, Empereur-Bissonnet P (2008) Risk of Hodgkin's lymphoma in the vicinity of French municipal solid waste incinerators. Envirol Health 7 (29):51

7 Plant-Based Bioadsorbents: An Eco-friendly Option for Decontamination of Heavy Metals from Soil

Shankarappa Sridhara, Nandini Ramesh, Shivanand Pradeep, and Manjunath K. Naik

Abstract

Rapid industrialization coupled with anthropogenic activities resulted in significant environmental pollution. Development in the establishment of industries, tanneries, mining, and paint industries paved way for heavy metals accumulation in the environment beyond acceptable limits. Even the heavy metal accumulation in soils, water, environment, and living organisms, including food crops, fruits, and vegetables, is being noticed in the recent past. Most of these heavy metals are known for bioaccumulation and are non-degradable in nature. This has resulted in manifestation of diseases related to nervous system, liver, bones, and disruption in the normal functioning of vital enzymes in the human body apart from carcinogenic effect. Various sustainable and eco-friendly technologies, viz., microbial phytoremediation, filtration through biomembranes, photocatalytic oxidation and reduction, and adsorption through adsorbents, were developed to decontaminate the heavy metals from the environment. This chapter focuses on different bio-based adsorbents especially the plant-derived, economic, and environmental feasibility, merits and demerits of their use. Besides, the chapter also focuses on the mechanism behind the use of bioadsorbents as well as their commercial application. Further, it also focuses on the best bet combination of microbial and bio-based adsorbents for sustainable reclamation of species-specific heavy metals from the soils and environment.

Keywords

Bioaccumulation · Bioadsorbents · Eco-friendly technologies · Heavy metals · Soil health

S. Sridhara (✉) · N. Ramesh
Center for Climate Resilient Agriculture, University of Agricultural and Horticultural Sciences, Shivamogga, Karnataka 577204, India

S. Pradeep
Organic Farming Research Center, University of Agricultural and Horticultural Sciences, Shivamogga, Karnataka 577204, India

M. K. Naik
University of Agricultural and Horticultural Sciences, Shivamogga, Karnataka 577204, India

7.1 Introduction

Modern life style, urbanization, industrialization have posed serious threat to whole biota, endangering the existence of life on earth. The air we breathe, water we drink, food we consume are contaminated due to intensive and unscientific agricultural practices, mining, improper disposal of industrial wastes creating accumulation of

J. A. Malik (ed.), *Advances in Bioremediation and Phytoremediation for Sustainable Soil Management*,
https://doi.org/10.1007/978-3-030-89984-4_7

heavy metals which are non-biodegradable and toxic. Even the minimal amounts of heavy metals can cause disastrous effects to humans, animals, and plants as well. Soil pollution due to heavy metals from natural and anthropogenic sources is of significant concern worldwide (Alloway, 1995; Song et al. 2018; Adimalla et al. 2019). Cadmium, mercury, chromium, copper, nickel, zinc, arsenic, selenium are the major heavy metals accumulated due to mining, waste water, and discharge of industrial effluents, agriculture runoffs, fossil fuel combustion, etc. leading to acute or chronic toxicity (Li et al. 2014). Heavy metals contaminating the agricultural soils are of prime concern as they directly affect the food chain and ground water regimes causing severe health hazards. Soil pollution due to heavy metals is an irreversible process, and therefore, reclaiming the toxicity levels is the most challenging task. Although various remedial measures are developed to reduce the toxicity level of heavy metals, their high-cost investment, heavy sludge discharge, cumbersome handling processes have made them ineffective and uneconomical. Hence, an attempt was made in this chapter to focus mainly on economically feasible, easily available, and environmental-friendly plant-derived adsorbents in remediating the heavy metal toxicity hazard.

7.2 Causes for Heavy Metal Contamination

Both natural as well as anthropogenic sources discharge heavy metals. The major source of metals like Zn, Pb, Al, Ni, Mn, Hg, and Cu accumulation is due to volcanic eruptions. Weathering of rocks leads to higher accumulation of Mn, Pb, Zn, Cr, Se, Co, Cu, Ni, Cd, and Hg, whereas forest fire havocs are the major source of Se and Hg (Nagajyoti 2010; Srivastava et al. 2017). Different sources of heavy metals leading to environmental pollution are defined in Tables 7.1, 7.2 and Fig. 7.1.

Table 7.1 Agricultural sources of heavy metals

Source	Heavy metal released
Pesticides	Copper (Cu), lead(Pb), zinc (Zn), arsenic (As), cadmium (Cd)
Fertilizers	Chromium (Cr), copper (Cu), manganese (Mn), zinc (Zn), cadmium (Cd), nickel (Ni), lead (Pb)
Biosolids and manures	Zinc (Zn), lead (Pb), arsenic (As), mercury (Hg), copper (Cu), nickel (Ni), cadmium (Cd), chromium (Cr)
Waste water	Zinc (Zn), nickel (Ni), lead (Pb), cadmium (Cd), chromium (Cr), copper (Cu), arsenic (As), mercury(Hg)

Table 7.2 Industrial sources of heavy metals in India (Central Water Commission 2019)

Heavy metal	Source
Chromium (Cr)	Metallurgic and chemical industries, mining, leather tanning, chromium salt manufacturing
Arsenic (As)	Smelters, thermal power plants, fuel burning
Zinc (Zn)	Electroplating, galvanizing process, smelting, brass manufacture, refineries
Lead (Pb)	Mining, paints, coal burning, smelters
Nickel (Ni)	Electroplating, battery manufacturing industries, metallurgical industries, smelting operations
Iron (Fe)	Alloys, cast iron, wrought iron, steel and machine manufacturing
Mercury (Hg)	Mining and refining, pesticides industries, dispensary wastes, electrical appliances
Cadmium (Cd)	Refining, nuclear fission plants, e-waste, batteries, smelting operations
Copper (Cu)	Disposal of municipal and industrial wastes, mining, smelters

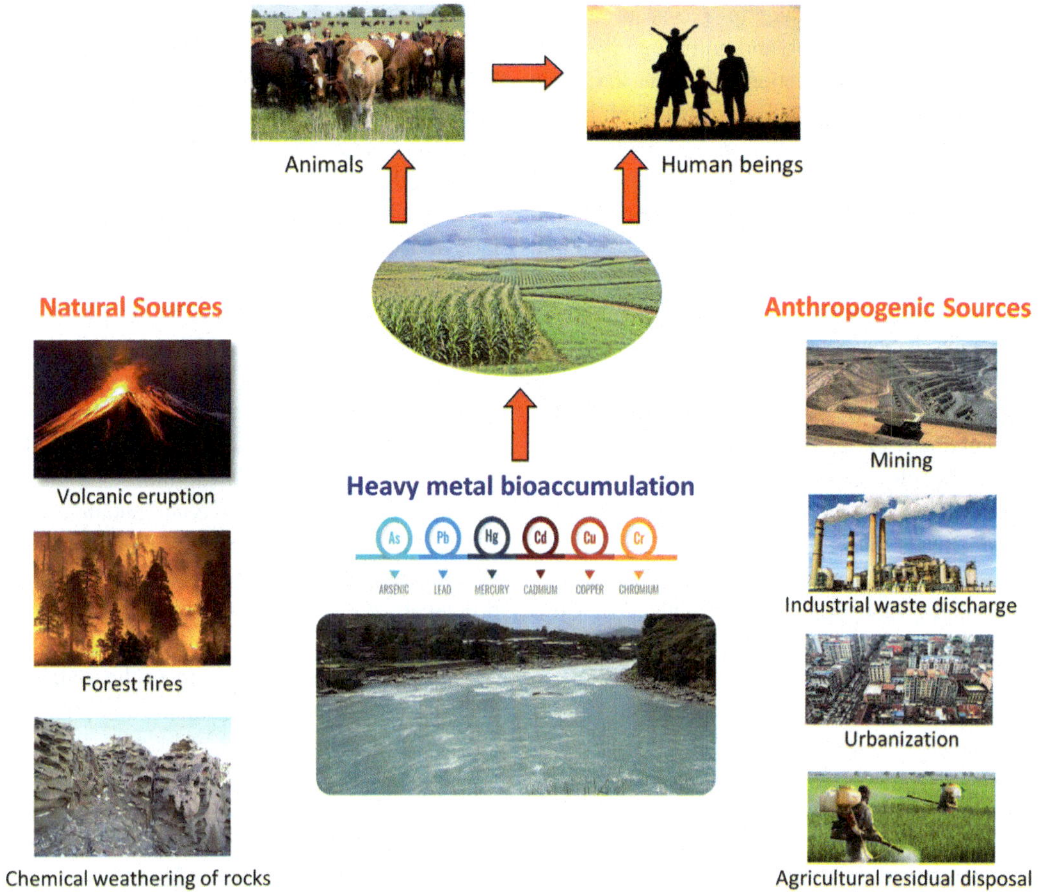

Fig. 7.1 Various processes responsible for absorption and entry of heavy metals into food chain from soil, water, and air

7.3 Status of Heavy Metal Toxification in India

A glance toward literature on status of heavy metals in India from 1991 to 2018 depicts that Indian soils have exceeded their limits in Zn and Pb accumulation. According to Indian soil guidelines, Zn and Pb were 22.1 and 13.1 $\mu g\ g^{-1}$, respectively, and as per the Canadian soil guidelines, Zn was found to be 200 $\mu g\ g^{-1}$. The contamination of Cd, As, and Cu assessed in all types of Indian soil exceeded the prescribed limit. Cluster and factor analysis, viz. potential contamination index, ecological risk index, enrichment factor, contamination factor, was employed to find out the important contaminants in Indian soils which revealed that Cd and As being the major culprits (Kumar et al. 2018).

With a view to study the level of heavy metal accumulation in Indian soils, different assessment studies have been carried out in vegetables/crops. Gupta et al. (2008) assessed the heavy metal contamination in vegetables in waste water irrigated areas of West Bengal and reported that the concentrations of Pb, Cr, Zn, Cd, and Ni were present at limits beyond the recommended levels in all the vegetables examined like mint, lettuce, cauliflower, spinach, and onion. The soils of Nuggihalli chromite mining areas in Karnataka were assessed for heavy metal concentration of barium, cobalt, chromium, copper, molybdenum, arsenic, nickel, lead, selenium, vanadium, zinc, and zirconium. According to geo-accumulation index, enrichment factor, and pollution index, Cr, Ni, and Co were the major pollutants loaded in the study area beyond the prescribed limit of soil quality guidelines (Krishna et al. 2013). Similarly, Yadav et al. (2013) assessed the status of heavy metals in soils and vegetables of Allahabad state of Uttar Pradesh which revealed that the bio-concentration of Zn, Cd, Fe, Cu, Ni, and Pb in edible parts of vegetables (tomato, coriander, cabbage, spinach, and radish) cultivated using waste water discharged from industries was more than the specified limits. Chabukdhara et al. (2015) evaluated the vegetable samples grown in Ghaziabad industrial areas and inferred that the concentration of cadmium, lead, nickel, and zinc in vegetables exceeded the FAO specified limits. The transfer factor (TF) value was in the sequence of Cd > Pb > Ni > Zn > Cu > Cr showing the movement of metals from soil to consumable parts of the vegetables.

Chandrasekaran et al. (2015) used multivariate statistical approach and spectrometry to assess the concentration of heavy metals in soils of Yelagiri hills of Tamil Nadu. The X-ray spectrometry results showed that aluminum was abundantly present in all the soil samples tested. Cr, Co, Ti, and Mn showed strong correlation with the anthropogenic sources prevailed in the sampling sites. Adimalla et al. (2019) analyzed the agricultural soils for heavy metals accumulation in Northern Telangana by using multivariate analysis considering pollution index, geo-accumulation index (GAI), enrichment factor (EF), and spatial distribution as factors. The results showed that Zn, Cu, barium (Ba), and vanadium (Va) were significantly higher than the prescribed values. Copper contamination reached to extreme level which was indicated by high geo-accumulation index (1.04).

Solid municipal waste was evaluated in Tiruchirappalli city for heavy metals accumulation which showed heavier concentrations of Cr and Pb than the Indian standards (Noorjahan et al. 2012). Agricultural fields, water, and the sediments of Roro Hill, Chaibasa, India, were assessed by Kumar et al. (2015) for contamination of heavy metals. The results showed higher levels of chromium (1148 mg kg^{-1}) and nickel (1120 mg kg^{-1}) far beyond threshold values in the contaminated agricultural fields, and the main source for this accumulation was found to be chromite-asbestos mine. Krishna and Govil (2004) analyzed the soils of Pali industrial area in Rajasthan for heavy metal toxicity, and the results showed elevated levels of Pb (293 mg kg^{-1}), Cr (240 mg kg^{-1}), Cu (298 mg kg^{-1}), V (377 mg kg^{-1}), Zn (1,364 mg kg^{-1}), and Sr (2,694 mg kg^{-1}) which were more than the threshold levels. Same group of scientists (Krishna and Govil 2008) in 2008 assessed the soils of Manali industrial area in Chennai for heavy metal contamination using GAI, EF, contamination factor as well as degree of contamination. They found higher concentration levels of Cr, Cu, Ni, Zn, and Mo which were beyond the limits of soil quality guidelines.

Many health risks arising from the toxicity of heavy metals in humans, animals, and plants are presented in Table 7.3.

7.4 Technologies Employed to Decontaminate Heavy Metals

For the treatment of heavy metals, different procedures have been used, such as filtration through membranes, oxidation, precipitation by chemical reaction, extraction by using solvents, adsorption, coagulation/flocculation, ozonation, ion exchange, electrodialysis, photochemical irradiation, and reverse osmosis (Meunier et al. 2006; Lee et al. 2006; Djedidi et al. 2009). The above methods are expensive and also found ineffective at metal ion concentration of $\leq$ 100 ppm (Inbaraj et al. 2009) (Table 7.4). This urged a need to develop an approach which is cost-effective, eco-friendly, and feasible; thus, one of the finest mechanisms to eliminate heavy metal toxicity was evolved, viz. "bioremediation/bioadsorption" which utilizes enzymes, microorganisms, plant biomass, agricultural by-products, etc. Thus, bioadsorbents are the most suitable option to decontaminate heavy metals from the soil and atmosphere due to their unique

Table 7.3 Health hazards owing to heavy metal toxicity

Heavy metal/toxicant	Health hazards	
	Human beings/animals	**Plant species**
Chromium (Cr)	Damages liver, kidney and central nervous system (CNS), cancer, skin ulceration Affects the immune system of fresh water fishes	Disrupts seed and seedling emergence, decreased photosynthetic rate and dry matter production, reduced growth
Cadmium (Cd)	Toxic to bones, liver, and blood vessels, and renal dysfunction causes lung cancer, Itai-itai disease	Stunted growth, chlorosis, necrosis, purple coloration, reddish petioles and veins
Arsenic (As)	Damages lungs, skin, gastrointestinal issues, breathing problems, colic pain, and nausea	Reduction in growth, red necrotic spots on older leaves
Lead (Pb)	Disorders in kidney, nervous system, reproductive organs, and brain and affects hemoglobin synthesis	Foliage will become stunted and dark green, and more number of shoots will be produced
Nickel (Ni)	Affects the nervous system and heart and causes insomnia, vomiting, nausea, and tachycardia	Stunting of plants, decreased leaf area and roots, chlorosis and necrosis
Copper (Cu)	Nausea, diarrhea, paralysis, Wilson's disease and damages kidney, spleen, liver. Poisons aquatic ecosystem	Purple and yellow coloration of midrib, inhibition of root growth, chlorosis
Mercury (Hg)	Affects CNS, brain and kidney, insomnia, vomiting, loss of smelling sense	Browning of leaf tips, severe stunting of roots and seedlings, reduced growth
Selenium (Se)	Colic pain, respiratory and heart disorders, drowsiness, reproduction problems, loss of appetite	Yellowing of younger leaves, interveinal chlorosis,, and pink discoloration on roots
Zinc (Zn)	Nausea and vomiting in children and anemia and cholesterol problems in adults	Reduced growth of leaves, roots and plant, chlorosis of younger leaves

physicochemical properties, metal ion binding functional groups, etc. (Taha et al. 2011; Inyang et al. 2012; Gupta et al. 2015; Singh et al. 2019).

7.5 Mechanisms of Biosorption Process

Biosorption involves various complex physicochemical processes for binding the metal ions onto the surface of bioadsorbents that includes complexation, chelation, coordination, ion exchange, precipitation, and reduction.

7.5.1 Complexation

The reaction between metal ions and the functional groups forms the complex. Active binding groups such as carboxyl, hydroxyl, phosphoric, sulfhydryl, amine, and phenolic form surface complexes with heavy metals (Wu et al. 2012). Such complexation mechanism has been observed in removal of Cd(II) using alkali-modified sewage sludge (Hu et al. 2012). Aloe vera wastes biosorbent used surface complexation process for detoxifying uranium and cadmium from the aqueous solution (Noli et al. 2019).

7.5.2 Chelation

This method involves attaching chelating agent to metal ion at more than one position, creating a material known as chelates that due to numerous binding activities are more stable than complexes. Rice straw used to remove Cd(II) and soybean meal waste to remove Cr(III) and Cu(II) utilized chelation mechanism forming chelates

Table 7.4 Pros and cons of employed methods used to mitigate heavy metal toxicity

Method	Treatment	Advantages	Disadvantages	Reference
Physical	Mechanical separation	Heavy metals volume will be reduced to a greater extent in contaminated soil	Homogenous distribution of pollutants in soil could not be remediated	Ottosen and Jensen (2005)
	Electro-kinetics	Different metal forms can be reclaimed from the contaminated soil	Heterogeneity in soil medium reduces the treatment efficiency Soil acidification will be a major disadvantage	Tahmasbian and Nasrazadani (2012)
Chemical	Ex situ soil washing	Effective method to remove inorganic contaminants, viz. heavy metals, toxic anions, radionuclides	Expensive method, construction, cleaning, and installation are quite cumbersome	Wuana and Okieimen (2011)
	Soil washing (in situ)	Invasive method to detoxify inorganic contaminants, viz. heavy metals, toxic anions, radionuclides to some extent	Heavy discharge of liquid as well as semi-liquid residues	Wuana and Okieimen (2011)
Soil amendments	Application of amendments like lime	Movement of lead, copper, nickel, zinc, and cadmium metal ions will be restricted	Physicochemical properties of soil will be changed	Guo et al. (2006)
	Applying of chelating agents	Reduces the mobility of copper (Cu) and lead (Pb) metals	Physicochemical properties of soil will be changed	Sukumara et al. (2012)

which bind with carboxyl and hydroxyl groups present on the surface of biosorbents (Ding et al. 2012; Witek-Krowiak and Reddy, 2013). Chelation was found to be the plausible mechanism in decontaminating lead(II) and cadmium(II) ions from the water sources using *Leucaena leucocephala* residues as bioadsorbent (Cimá-Mukul et al. 2019).

7.5.3 Coordination

The metal atom binds with neighboring one having lone pair of electrons from the non-metal atom which is known as coordinating/donor atom, while the principal metal ion is known as acceptor atom (Tsezos et al. 1995). Vanadium metal biosorption on carbohydrate biomass types indicated the coordination of metal ions with donors like $COO^-$and OH^- groups (García et al. 2013).

7.5.4 Ion Exchange

In the bioadsorption system, ion exchange is one of the most critical processes involving the exchange of binary metal ions with counter ions present on the bioadsorbent surface. It may be anion or cation exchange. Ion exchange relationship between the sequestration of nickel ions and calcium ions was reported by Williams and Edyvean (1997). *Spirulina* has been used to detoxify Cr(III), Cu(II), and Cd(II) and used cation exchange process with phosphate, carboxyl and hydroxyl groups (Chojnacka et al. 2005). Watermelon rind is reported to have biosorption capacity to remove copper(II), lead (II), and Zinc(II) by K^+, Na^+, Ca^+, and Mg^+ ion exchange (Liu et al. 2012). Similarly, Kanamarlapudi et al. (2018) studies supported the evidence for ion exchange process in adsorbing cadmium(II) ions using rice straw by exchanging K, Mg, Na, and Ca ions.

7.5.5 Precipitation

Metal ion precipitation is found with functional groups of the microbial cells which remain in contact or inside the cells. The formation of inorganic as well as metal precipitation occurs as metals are easily bound to extracellular polymeric substances (Tavares and Quintelas, 2014). Such kind of precipitation mechanism was reported by Witek-Krowiak and Reddy (2013) while using soybean meal to remove Copper (II) and Chromium (II) toxicity and also tomato husk sorption method removed Fe and Mn metal ions using precipitation (García-Mendieta et al., 2012).

7.5.6 Reduction

This process involves reduction of metal ions during interaction with functional carboxyl groups. Removal of hexavalent chromium toxic metal from the solution is due to reduction process and is one of the classic examples. Many micro-organisms detoxify Cr(VI) by reducing it to Cr(III) from waste water due to chemical reduction (Park et al. 2005; Quintelas et al. 2009). Microbes decontaminate heavy metals from tannery waste water by reducing metal ion from highly toxic to less toxic state (Igiri et al. 2018).

7.6 Factors Affecting Bioadsorption

pH, temperature, dose, initial concentration, contact time, particle size, and agitation rate are the key determining factors influencing the bioadsorption process.

7.6.1 Effect of pH

pH is one of the most influencing factors in biosorption because it strongly affects the degree of ionization, site dissociation of the adsorbent surface, and chemistry of heavy metals with the bioadsorbents (Memon et al. 2008). It also plays a vital role in complexation of organic/inorganic ligands, precipitation, reduction and oxidation reactions. The biosorptive capacity will be enhanced with increase in solution pH (Ofomaja and Ho 2008; Nejadshafiee & Islami, 2019).

7.6.2 Effect of Temperature

Most of the bioadsorption processes are endothermic in nature which increase number and size of the active pores on the adsorbent's surface attracting more metal ions toward it, thus enhancing the adsorptive capacity (Kirbiyik et al. 2016). In case of exothermic process, the efficiency of bioadsorption will be decreased with increase in temperature levels which was proved in case of peanut shell removing Pb(II) ions from aqueous solution (Taşar et al. 2014; Redha, 2020).

7.6.3 Effect of Contact Time

Bioadsorption capacity and removal efficiency increase with increase in contact time with the adsorbate. When the state of equilibrium is attained, saturation of biosorption process occurs (Din and Mirza, 2013; Kirbiyik et al. 2016). Increased contact time enhanced the biosorption capacity of *Aloe barbadensis* residual leaves powder to remove nickel(II) metal ions to an extent of 42.8 percent (Gupta and Kumar, 2019).

7.6.4 Effect of Metal Ion Concentration

Bioadsorption process is influenced to a larger extent by the concentration of initial metal ion. As the metal ion concentration increases, there will be higher uptake of heavy metals on to the bioadsorbent surface (Tavares and Quintelas, 2014). Khajavian et al. (2019) reported that brown algae (*Cystoseria indica*) adsorbed Ni and Cd metal ions up to 17.5 and 35 mg/L at an initial metal ion concentration of 30 mg/L.

7.6.5 Effect of Adsorbent Quantity

Quantity/dose of adsorbent is the most determining factor of bioadsorption capacity because as the amount of adsorbent increases, surface area and number of active binding sites will be increased, thus enhancing the adsorptive capacity of the bioadsorbents (Atar et al. 2012). The importance of adsorbent dosage in enhancing biosorption capacity was revealed by the studies carried out by Gaur et al. (2018) wherein the soybean used as biosorbent removed higher quantity of toxicants such as lead (Pb) and arsenic (As) ions from waste water at 3 g 100 ml^{-1} of soybean absorbent.

7.7 Low-cost Effective Technologies Using Plant-based Bioadsorbents to Decontaminate Heavy Metals

Green technology, *i.e.,* fruit and vegetable peels of pineapple, citrus, potato, orange, banana, pomegranate, tomato were tested to decontaminate heavy metals. The adsorption was due to first-order kinetics with endothermic process. Orange fruit peel removed Ni, Pb, Zn, Cu, and Cr, tomato waste adsorbed Co, banana and pumpkin removed Pb, pomegranate adsorption was with Fe, and pineapple decontaminated Sofranin. The efficiency of fruit and vegetable peels to decontaminate heavy metals ranged from 43 to 96 percent (Jain, 2015). Hazelnut shell (*Corylus avellna* L.), an inexpensive biological adsorbent, was used to remove heavy metals (Pb and Cd) from edible leafy vegetables in Tehran, Iran. Results revealed that hazelnut shell was potential enough to remove Pb and Cd to an extent of 87.7 and 100 percent, respectively, from parsley (Fatahi et al. 2020).

Fruit peels contain polar functional groups like –OH, –NH2, and –COOH which can attract metal ions on to their surface. Considering this simple mechanism, fruit peels of avocado (*Persea americana*), hami melon (*Cucumis melo*), and dragon fruits (*Hylocereus undatus*) were experimented to detoxify heavy metals, dyes, and dissolved compounds in water. Dragon fruit peel was found effective with higher extraction potential of 71.85 and 62.58 mg g^{-1} for alcian blue and methylene blue dyes individually. Hami melon removed Pb and Ni to an extent of 7.89 and 9.45 mg g^{-1}, correspondingly (Mallampati et al. 2015). Other economically feasible materials such as peanut shells and banana peels were utilized to detoxify heavy metals such as copper, lead, zinc, and cadmium from waste water. Atomic adsorption spectrophotometry (AAS), scanning electron microscopy (SEM), and X-ray diffraction techniques were employed to assess the adsorption of heavy metal ions. The adsorption process was found better in peanut shells with the order of lead>zinc > copper > cadmium, whereas banana peel removed heavy metals in the order of cadmium > copper > lead > zinc as assessed by AAS method (Jaishankar et al. 2014).

Balaji et al. (2014) used three species of Spirulina (*Arthospira*), *A. indica*, *A. maxima*, and *A. platensis*as bioremediation agent to treat Pb, Cd, and Cr toxicity from river water affected with tannery effluent. Atomic spectrometry showed maximum bioadsorption potential of all the three species in removing the aforesaid heavy metals. Similarly, *Arthospira platensis* was used to decontaminate heavy metals like Cr, Cd, and Pb from tannery effluent discharged from Ambur industrial area, Tamil Nadu. Atomic absorption spectrometry (AAS) technique was used to analyze the heavy metal accumulation. The study revealed the potentiality of *Arthospira platensis* in reducing heavy metals from tannery effluents by enhancing the efficacy of antioxidant enzymes such as super-oxidase dismutase and catalase (Balaji et al. 2015).

Wastes from agro-industries can be used as biosorbents to decontaminate heavy metals from waste water. Saxena et al. (2017) used castor leaves, rice husk, and sugarcane bagasse as bioadsorbents to treat Pb and Ni toxicity. Fourier transform infrared FT-IR) spectrophotometer was used to know the interactions of heavy metal ions and the biosorbents. The main mechanism of adsorption process was found to be compelxation of metal ions onto the surface of biosorbents, and the castor leaves emerged as the best bioadsorbent compared to rice husk and sugarcane bagasse.

Gasco et al. (2019) conducted an experiment through phytoremediation to reclaim soils contaminated with heavy metals in mining areas of Riotinto, Spain. Cu, Pb, Zn, and As were the main contaminants present. Rabbit manure was used as biochar prepared at 450 °C and 600 °C. *Brassica napus* was used as a tester plant. Results revealed that the combination of biochar with *B. napus* was successful in eliminating As, Co, Cr, Cu, Se, and Pb from the soil. Further, application of biochar increased the yield components attributing to improvement in soil nutrient status, organic carbon, and pH. Another phytoremediation study using *Prosopis juliflora* biochar (PJB) and rice husk ash (RHA) to detoxify Pb accumulation in castor plants was conducted by Kiran and Prasad (2019). The PJB applied @ 5% reduced Pb accumulation up to 59% in roots, 60% in shoots, and 62% in leaves, while RHA was found more efficient which decreased Pb to an extent of 87%, 71%, and 99% in roots, shoots, and leaves of castor plant, respectively. Further, the addition of PJB and RHA amendments improved plant growth, chlorophyll, and protein content by immobilizing the Pb content.

Kirbiyik et al. (2016) used three adsorbents such as sesamum stalk, biochar, and activated carbon to eliminate Fe and Cr toxicity. SEM and FT-IR spectrophotometry techniques were used to characterize adsorbents. Langmuir isotherm model and the kinetics of pseudo-second order proved better with higher correlation coefficients for metal ions adsorption onto the surface of above-quoted adsorbents. Among the three adsorbents, sesamum stalk which is economically cheaper and easy available exerted higher adsorption efficiency compared to biochar and activated carbon. Thus, it can be potentially utilized as a raw precursor to adsorb Fe and Cr metal ions.

Puschenreiter et al. (2005) suggested some reliable and low-cost agro-techniques to eradicate heavy metals. Selection of crops with low metal uptake capacity is one of the effective approaches. Cereals, beans, potatoes can resist heavy metal accumulation, whereas leafy vegetables especially spinach and lettuce have high transfer factor indicating high metal uptake. Crop rotation and growing of fiber crops/bioenergy crops (cotton, flex, and hemp) can also reclaim metal-polluted soils to greater extent. Application of organic (FYM) and inorganic (lime-zeolites and Fe-oxides) amendments is another effective method which reduced metal ions transfer to the edible parts of crops.

The agricultural residues (rice and wheat straw) and *Salvinia* biomass which are relatively cheaper, recyclable, and easily accessible were used to treat waste water contaminated with Cr, Cd, and Ni. *Salvinia* plants proved effective with higher potential to remove Cr, Ni, and Cd heavy metals compared to rice and wheat straw. Langmuir and Freundlich isotherm models performed better with equilibrium data (Dhir and Kumar, 2010). Tajik et al. (2020) used coffee bean waste as adsorbent to remove Ni from soil and coriander plants. Low-cost coffee waste was found effective in detoxifying the Ni metal ion in very short time due to commendable level of pH, dose of adsorbent, contact time, and initial metal ion concentration. Similarly, Arabian et al. (2020) discovered the potentiality of black tea and herbal tea (*Salvia officinalis*) residual wastes in detoxifying the pharmaceutical effluents loaded with heavy metal ions such as Cd, Co, and Ni. Contact time is the most effective factor which showed significant potentiality of bioadsorbents used.

Obayomi et al. (2019) developed agricultural waste composite activated carbon to adsorb Pb and As from water sources. Brunauer–Emmett–Teller (BET) was used to study the surface area and porosity which revealed adsorption capacity of agricultural waste composite activated carbon to be 200 mg g^{-1} for As and 250 mg g^{-1} for Pb. The adsorption process of agricultural waste composite activated carbon fitted well in Langmuir isotherm model following pseudo-second-order kinetics. Activated carbon obtained from cobs of maize and petai hull was experimented by Lestari et al. (2018) for the removal of lead contaminant. Adsorbent was grouped differently based on the concentration of corn cob and petai hull used. Native activated carbon of corn cob and petai hull was grouped as B(1:3) and C (1:1), while the modified activated carbon using KOH as D(3:1) and H(1:1). The results highlighted the adsorbent capacity of H group as more efficient in eliminating lead(II) with 2368 mg g^{-1} capacity at 300 ppm concentration compared to other group of adsorbents (D, C, and B). Another attempt was made using jackfruit wood sawdust as bioadsorbent which was further modified by treating with phosphoric acid to remove lead (Pb) metal ions. The adsorption capacity exerted was found to be 1.4382 mg g^{-1} at 0.33 concentration of phosphoric acid which fitted well with pseudo-second-order kinetics (Mutiara et al. 2018). Some easily and commonly available plant-based adsorbents to eliminate heavy metals are depicted in the table below (Table 7.5).

Table 7.5 Some of the plant-based biosorbents to decontaminate heavy metals

Biosorbent (plant derivatives)	Heavy metal removed	References
Rice husk	Cd(II), Cu(II), Se(IV), Hg(II), Pb(II)	Akhtar et al. (2010); Wong et al. (2003); El-Shafey (2007 and 2010); Kumar and Bandyopadhyay (2006)
Rice husk ash	Pb(II), Cd(II)	El-Shafey (2007); Naiya et al. 2009
Wheat straw	Cd(II), Cr(VI), Cu(II)	Dang et al. (2009); Farooq et al. (2011);Chen et al. (2010)
Wheat bran	Cd(II), Cr(VI), Cu(II), Pb(II), Zn(II), Hg(II)	Farajzadeh and Monji (2004); Dupont et al. (2005); Nouri et al. (2007); Singh et al. (2009); Wang et al. (2009); Naiya et al. (2009)
Coconut coir pith	Cd(II), Hg(II), Ni(II), Co(II), Pb(II)	Kadirvelu and Namasivayam, (2000); Parab et al. (2006); Anirudhan et al. (2008)
Orange peel	Cd(II), Cu(II), Ni(II)	Ajmal et al. (2000); Sha et al. (2009)
Mango peel	Pb(II), Cd(II), Cu(II), Zn(II), Ni(II)	Sha et al. (2009); Iqbal et al. (2009)
Banana peel	Pb(II), Cd(II), Cr(Vi)	Memon et al. (2009); Anwar et al. (2010a)
Potato peel	Cu(II)	Moreno-Piraján and Giraldo (2011)
Lemon peel	Co(II)	Bhatnagar et al. (2010)
Chestnut shell	Pb(II), Cu((II), Zn(II)	Vázquez et al. (2009)
Almond shell	Cr(VI), Pb(II)	Pehlivan and Altun, (2008); Pehlivan et al. (2009)
Hazelnut shell	Cr(VI), Pb(II)	Pehlivan and Altun, (2008); Pehlivan et al. (2009)
Walnut shell	Cr(VI), Hg(II)	Pehlivan and Altun, (2008); Zabihi et al. (2009)
Peanut shell	Cu	Zhu et al. (2009)
Shell carbon	Zn(II), Pb(II)	Amuda et al. (2007); Sekhar (2008)
Tea waste	Cd(II), Ni(II), Cu(II), Pb(II), Zn(II)	Cay et al. (2004); Choi and Yun (2004); Ahluwalia and Goyal (2005); Amarasinghe and Williams (2007)
Black tea and green tea	Pb(II)	Zuorro and Lavecchia (2010)
Peanut hull	Cu(II)	Zhu et al. (2009)
Groundnut husk	Cr(VI)	Dubey and Gopal (2007)
Coffee waste	Cd(II), Pb(II)	Azouaou et al. (2010); Reddy et al. (2010)
Sesamum stalk	Ni(II), Zn(II)	Kirbiyik et al. (2012)
Parthenium	Cd(II)	Ajmal et al. (2006)

(continued)

Table 7.5 (continued)

Biosorbent (plant derivatives)	Heavy metal removed	References
hysterophorous		
Bael fruit	Cr(VI)	Anandkumar and Mandal (2009)
Tomato waste and apple juice residue	Pb(II)	Heraldy et al. (2018)
Blackgram husk	Pb, Cd, Zn, Cu, Ni	Saeed et al. (2002)
Dal husk	Cr(VI), Fe(III)	Parate and Talib (2014)
Grape bagasse	Pb	Farinella et al. (2007)
Barley straw	Cu, Pb	Pehlivan et al. (2009)
Pumpkin waste	Cr(VI), Pb	Okoye et al. (2010)
Pea waste	Cr(VI)	Anwar et al. (2010b)
Sugarbeet pulp	Cu	Aksu and Isoglu (2005)
Coconut husk	Cr(III), Cr(II), Hg(II)	Ahmad et al. (2005)
Coconut copra meal	Cd(II)	Ofomaja and Ho (2007)
Coconut shell charcoal	Cr(VI)	Babel and Kurniawan, (2004)
Guava seeds	Cr(VI)	Abdelwahab et al. (2007)
Papaya seed	Cu(II)	Hadi et al. (2011)
Pine leaves	As(V)	Shafique et al. (2012)
Neem leaves	Cr(V)	Babu and Gupta (2008)
Moringa oleifera leaves	Cd(II), Cu(II), Ni(II)	Reddy et al. (2012)
Sunflower leaves	Cu(II)	Benaïssa and Elouchdi (2007)
Oil palm roots	Cu(II), Pb(II)	Bhaumik et al. (2014)
Eucalyptus bark	Cr(VI)	Sarin and Pant (2006)
Corn stalk	Cd(II)	Zheng et al.(2010)
Sunflower stalk	Pb(II), Cd(II)	Jalali and Aboulghazi (2013)
Bengal gram husk	Cr(III)	Ahalya et al. (2005)
Agave bagasse	Cd(II), Pb(II), Zn(II)	Velazquez-Jimenez et al. (2013)
Sugarcane bagasse	Pb(II), Hg(II)	Martín-Lara et al. (2010); Khoramzadeh et al. (2013)
Coconut fiber	Hg(II)	Johari et al. (2014)
Cashew nut shell	Pb(II)	Kumar et al. (2011)

7.8 Conclusion

Pollution of soil, water, and atmosphere due to heavy metal accumulation is of global concern as it is evolved from natural as well as due to anthropogenic activities. Remediation of heavy metal toxification is a very challenging due to their non-biodegradable and hazardous in nature. Chemical reclamation methods have greater limitations due to high-cost treatments, installation set up, and heavy residual discharge of sludge, liquid and semi-liquid wastes. Thus, a technique which is cost-effective, eco-friendly with zero residual wastes should be employed to decontaminate heavy metal accumulation. One such method is biosorption/bioadsorption which utilizes agricultural residues/by-products, plant-based adsorbents, micro-organisms which are easily available, recyclable, and low cost. Although many researchers have worked on low-cost plant-based bioadsorbents at *in vitro* level, further needs are to be explored at larger scale to industrial/pilot level, making commercially feasible to all the desired sectors.

References

Abdelwahab O, El Sikail A, Khaled A, El Nemr A (2007) Mass-transfer processes of chromium(VI) adsorptiononto guava seeds. Chem Ecol 23(1):73–85

Adimalla N, Qian H, Wang H (2019) Assessment of heavy metals (HM) contamination in agricultural soil lands in northern Telangana, India: an approach of spatial distribution and multivariate statistical analysis. Environ Monit Asess 191:246. https://doi.org/10.1007/s10661-019-7408-1

Ahalya N, Kanamadi RD, Ramachandra TV (2005) Biosorption of chromium(VI) from aqueous solutions by the husk of Bengal gram (*Cicer arientinum*). Electron J Biotechn 8:258–264

Ahluwalia SS, Goyal D (2005) Removal of heavy metals by waste tea leaves from aqueous solution. Eng Life Sci 5:158–162

Ahmad R, Hasany SM, Chaudhary MH (2005) Adsorption characteristics of Cr(III) ions onto coconut husk from aqueous solution. Adsorpt Sci Technol 23 (6):467–478

Ajmal M, Rao RAK, Ahmad R, Ahmad J (2000) Adsorption studies on *Citrus reticulate* (fruit peel of orange): removal and recovery of Ni (II) from electroplating wastewater. J Hazard Mater 79:117–131

Ajmal M, Rao RAK, Ahmad R, Khan MA (2006) Adsorption studies on *Parthenium hysterophorous* weed: removal and recovery of Cd (II) from wastewater. J Hazard Mater 135:242–248

Akhtar M, Iqbal S, Kausar A, Bhanger MI, Shaheen MA (2010) An economically viable method for the removal of selected divalent metal ions from aqueous solutions using activated rice husk. Colloids Surf 275:149–155

Aksu Z, Isoglu IA (2005) Removal of copper(II) ions from aqueous solution by biosorption onto agricultural waste sugar beet pulp. Process Biochem 40(9):3031–3044

Alloway BJ (1995) Heavy metals in soils. Blackie Academic & Professional, Chapman & Hall, London (UK)

Amarasinghe BMWPK, Williams RA (2007) Tea waste as a low cost adsorbent for the removal of Cu and Pb from wastewater. Chem Eng J 132:299–309

Amuda OS, Giwa AA, Bello IA (2007) Removal of heavy metal from industrial wastewater using modified activated coconut shell carbon. Bio Chem Eng J 36:174–181

Anandkumar J, Mandal B (2009) Removal of Cr(VI) from aqueous solution using Bael fruit (*Aegle marmeloscorrea*) shell as an adsorbent. J Hazard Mater 168:633–640

Anirudhan TS, Divya L, Ramachandran M (2008) Mercury(II) removal from aqueous solutions and wastewaters using a novel cation exchanger derived from coconut coir pith and its recovery. J Hazard Mater 157:620–626

Anwar J, Shafique U, Waheed-uz-Zaman (2010b) Removal of chromium from water using pea waste—a green approach. Green Chem Lett Rev 3(3):239–243

Anwar J, Shafique U, Waheed-uz-Zaman SM, Dar A, Anwar S (2010a) Removal of Pb(II) and Cd(II) from water by adsorption on peels of banana. Bioresour Technol 101:1752–1755

Arabian S, Parisa Ziarati P, Sawicka B (2020) Waste herbal and black tea as a novel adsorbent for detoxification of pharmaceutical effluent. Med Discov 5(3):jmd20040. https://doi.org/10.24262/jmd.5.3.20040

Atar N, Olgun A, Wang S (2012) Adsorption of cadmium (II) and zinc(II) on boron enrichment process waste in aqueous solutions: Batch and fixed-bed system studies. Chem Eng J 192:1–7. https://doi.org/10.1016/j.cej.2012.03.067

Azouaou N, Sadaoui Z, Djaafri A, Mokaddem H (2010) Adsorption of cadmium from aqueous solution onto untreated coffee grounds: equilibrium, kinetics and thermodynamics. J Hazard Mater 184:126–134

Babel S, Kurniawan TA (2004) Cr(VI) removal from synthetic waste water using coconut shell charcoal and commercial activated carbon modified with oxidizing agents and/or chitosan. Chemosphere 54:951–967

Babu BV, Gupta S (2008) Adsorption of Cr(VI) using activated neem leaves: kinetic studies. Adsorpt 14:85–92

Balaji S, Kalaivani T, Rajasekaran C, Shalini M, Siva R, Singh RK, Akthar MA (2014) *Arthrospira* (*Spirulina*) species as bioadsorbents for lead, chromium, and cadmium—a comparative study. Clean Soil Air Water 42(9999):1–8

Balaji S, Kalaivani T, Rajasekaran C, Shalini M, Vinodhini S, Priyadharshini SS, Vidya AG (2015) Removal of heavy metals from tannery effluents of Ambur industrial area, Tamil Nadu by *Arthrospira* (Spirulina) *platensis*. Environ Monit Assess 187:325. https://doi.org/10.1007/s10661-015-4440-7

Benaïssa H, Elouchdi MA (2007) Removal of copper ions from aqueous solutions by dried sunflower leaves. Chem Eng Process 46:614–622

Bhatnagar A, Minocha AK, Sillanpää M (2010) Adsorptive removal of cobalt from aqueous solution by utilizing lemon peel as biosorbent. Bio Chem Eng J 48:181–186

Bhaumik M, Choi HJ, Seopela MP, McCrindle RI, Maity A (2014) Highly effective removal of toxicCr (VI) from wastewater using sulfuric acid-modified avocado seeds. Indo Eng Chem Res 53:1214–1224

Cay S, Uyanik A, Özas A (2004) Single and binary component adsorption of copper(II) and cadmium(II) from aqueous solutions using tea-industry waste. Sep Purif Technol 38:273–280

Central Water Commission (CWC) India (2019) Report on heavy metal contamination problem in Indian rivers. www.cwc.gov.in

Chabukdhara M, Munjal A, Nema AK, Gupta SK, Kaushal R (2015) Heavy metal contamination in vegetables grown around peri-urban and urban-industrial clusters in Ghaziabad, India. Hum Ecol Risk Assess Int J 22(3):736–752. https://doi.org/10.1080/10807039.2015.110572

Chandrasekaran A, Ravisankar R, Harikrishnan N, Satapathy KK, Prasad MVR, Kanagasabapathy KV (2015) Multivariate statistical analysis of heavy metal concentration in soils of Yelagiri Hills, Tamil Nadu, India—Spectroscopical approach. Spectrochemica Acta Part A Molecular Biomol Spectros 137:589–600

Chen S, Yue Q, Gao B, Xu X (2010) Equilibrium and kinetic adsorption study of the adsorptive removal of Cr(VI) using modified wheat residue. J Coll Interf Sci 349:256–264

Choi SB, Yun YS (2004) Lead biosorption by waste biomass of *Corynebacterium glutamicum* generated from lysine fermentation process. Biotechnol Lett 26:331–336

Chojnacka K, Chojnacki A, Gorecka H (2005) Biosorption of Cr^{3+}, Cd^{2+} and Cu^{2+} ions by blue-green algae *Spirulina* sp.: kinetics, equilibrium and the mechanism of the process. Chemosphere 59:75–84

Cimá-Mukul CA, Abdellaoui Y, Abatal M, Vargas J, Santiago AA, Barrón-Zambrano JA (2019) Eco-efficient biosorbent based on *Leucaena leucocephala* residues for the simultaneous removal of Pb(II) and Cd(II) ions from water system: sorption and Mechanism. Bioinorg Chem. https://doi.org/10.1155/2019/2814047

Dang VBH, Doan HD, Dang-Vu T, Lohi A (2009) Equilibrium and kinetics of biosorption of cadmium (II) and copper(II) ions by wheat straw. Bioresour Technol 100:211–219

Dhir B, Kumar R (2010) Adsorption of heavy metals by *Salvinia* biomass and agricultural residues. Int J Environ Res 4(3):427–432

Din MI, Mirza ML (2013) Biosorption potentials of a novel green biosorbent *Saccharum bengalense* containing cellulose as carbohydrate polymer for removal of Ni(II) ions from aqueous solutions. Int J Biol Macromol 54:99–108

Ding Y, Jing D, Gong H, Zhou L, Yang X (2012) Biosorption of aquatic cadmium(II) by unmodified rice straw. Bioresour Technol 114:20–25

Djedidi Z, Boudaa M, Souissi MA, Cheikhb RB, Mercier G, Tyagi RD, Blais JF (2009) Metals removal from soil, fly ash and sewage sludge leachates by precipitation and dewatering properties of the generated sludge. J Hazard Mater 172:1372–1382

Dubey SP, Gopal K (2007) Adsorption of chromium(VI) on low cost adsorbents derived from agricultural waste material: A comparative study. J Hazard Mater 145:465–470

Dupont L, Bouanda J, Dumonceau J, Aplincourt M (2005) Biosorption of Cu(II) and Zn(II) onto a lignocellulosic substrate extracted from wheat bran. Environ Chem Lett 2:165–168

El-Shafey EI (2007) Sorption of Cd(II) and Se(IV) from aqueous solution using modified rice husk. J Hazard Mater 147:546–555

El-Shafey EI (2010) Removal of Zn(II) and Hg(II) from aqueous solution on a carbonaceous sorbent chemically prepared from rice husk. J Hazard Mater 175:319–327

Farajzadeh MA, Monji AB (2004) Adsorption characteristic of wheat bran towards heavymetal cations. Separ Pur Tech 38:197–207

Farinella NV, Matos GD, Arruda MAZ (2007) Grape bagasse as a potential biosorbent of metals in effluent treatments. Bioresour Technol 98(10):1940–1946

Farooq U, Khan MA, Athar M, Kozinski JA (2011) Effect of modification of environmentally friendly biosorbent wheat (*Triticum aestivum*) on the biosorptive removal of cadmium(II) ions from aqueous solution. Chem Eng J 171:400–410

Fatahi A, Ziarati P, Jafarpour A, Cruz-Rodriguez L (2020) Heavy metal removal from edible leafy vegetable by low cost novel adsorbents: Hazelnut Shell. J Sci Discov 4(2):jsd20039. https://doi.org/10.24262/jsd.4.2.20039

García J, Gonza'lez JC, Frascaroli MI, García S, Blanes P, Correia I, Pessoa JC, Sala LF, (2013) Spectroscopic studies of vanadium biosorption on different types of carbohydrate biomass. Can J Chem 91:186–195. https://doi.org/10.1139/cjc-2012-0208

García-Mendieta A, Olguín MT, Solache-Ríos M (2012) Biosorption properties of green tomato husk (*Physalis philadelphicalam*) for iron, manganese and iron–manganese from aqueous systems. Desalination 284:167–174

Gasco G, Alvarez ML, Paz-Ferreiro J, Mendez A (2019) Combining phytoextraction by *Brassica napus* and biochar amendmentfor the remediation of a mining soil in Riotinto (Spain). Chemosphere 231:562–570

Gaur N, Kukreja A, Yadav M, Tiwari A (2018) Adsorptive removal of lead and arsenic from aqueous solution using soya bean as a novel biosorbent: equilibrium isotherm and thermal stability studies. Appl Water Sci 8(98):1–12. https://doi.org/10.1007/s13201-018-0743-5

Guo G, Zhou Q, Ma LQ (2006) Availability and assessment of fixingadditives for the in situ remediation of heavy metal contaminated soils: a review. Environ Monit Assess 116:513–528

Gupta N, Khan DK, Santra SC (2008) An assessment of heavy metal contamination in vegetables grown in wastewater-irrigated areas of Titagarh, West Bengal. India. Bull Environ Contam Toxicol. 80:115–118. https://doi.org/10.1007/s00128-007-9327-z

Gupta S, Kumar A (2019) Removal of nickel (II) from aqueous solution by biosorption on *A. barbadensis*

Miller waste leaves powder. Appl Water Sci 9 (96):1–11 https://doi.org/10.1007/s13201-019-0973-1
Gupta VK, Nayak A, Agarwal S (2015) Bioadsorbents for remediation of heavy metals: current status and their future prospects. Environ Eng Res 20:1–18
Hadi NBA, Rohaizer NAB, Sien WC (2011) Removal of Cu(II) from water by adsorption on papaya seed. Asian Trans Eng 1(5):49–55
Heraldy E, Lestari WW, Permatasari D, Arimurti DD (2018) Biosorbent from tomato waste and apple juice residue for lead removal. J Environ Chem Eng 6:1201–1208
Hu JL, He XW, Wang CR, Li JW, Zhang CH (2012) Cadmium adsorption characteristic of alkali modified sewage sludge. Bioresour Technol 121:25–30
Igiri BE, Okoduwa SIR, Idoko, GO, Akabuogu EP, Adeyi AO, Ejiogu IK (2018) Toxicity and bioremediation of heavy metals contaminated ecosystem from tannery wastewater: a review. J Toxicol 1–16
Inbaraj BS, Wang JS, Lu JF, Siao FY (2009) Adsorption of toxic mercury(II) by an extracellular biopolymer poly (-glutamic acid). Bioresour Technol 100:200–207
Inyang M, Gao B, Xue Y, Zimmerman AR, Pullammanappallil P, Cao X (2012) Removal of heavy metals from aqueous solution by biochars derived from anaerobically digested biomass. Bioresour Technol 110:50–61
Iqbal M, Saeed A, Kalim I (2009) Characterization of adsorptive capacity and investigation of mechanism of Cu^{2+}, Ni^{2+} and Zn^{2+} adsorption on mango peel waste from constituted metal solution and genuine electroplating effluent. Sep Sci Technol 44:3770–3791
Jain N (2015) Removal of heavy metal by using different fruit peels, vegetable peels and organic waste—a review. Int J Adv Res 3(11):916–920
Jaishankar M, Mathew BB, Shah MS, Murthy KTP, Gowda SKR (2014) Biosorption of few heavy metal ions using agricultural wastes. J Environ Pollut and Human Health 2(1): 1–6. https://doi.org/10.12691/jephh-2-1-1
Jalali M, Aboulghazi F (2013) Sunflower stalk, an agricultural waste, as an adsorbent for the removal of lead andcadmium from aqueous solutions. J Mater Cycles Waste Managt 15:548–555
Johari K, Saman N, Song ST, Heng JYY, Mat H (2014) Study of Hg(II) removal from aqueous solution using lignocellulosic coconut fiber biosorbents: equilibrium and kinetic evaluation. Chem Eng Commun 201:1198–1220
Kadirvelu K, Namasivayam C (2000) Agricultural by-product as metal adsorbent: sorption of lead (II) from aqueous solution onto coir pith carbon. Environ Technol 21:1091–1097
Kanamarlapudi LRK, Chintalpudi VK, Muddada S (2018) Application of biosorption for removal of heavy metals from wastewater. Biosorption 18:69
Khajavian M, Wood DA, Hallajsani A, Majidian N (2019) Simultaneous biosorption of nickel and cadmium by the brown algae *Cystoseria indica* characterized by isotherm and kinetic models. Appl Biol Chem 62(69):1–12. https://doi.org/10.1186/s13765-019-0477-6
Khoramzadeh E, Nasernejadb B, Halladj R (2013) Mercury biosorption from aqueous solutions by Sugarcane bagasse. J Taiwan Inst Chem Eng 44:266–269
Kiran BR, Prasad MNV (2019) Biochar and rice husk ash assisted phytoremediation potentials of *Ricinus communis* L. for lead-spiked soils. Ecotoxicol Environ Saf 183:109574
Kirbiyik Ç, Kiliç M, Çepelioğullar Ö, Pütün AE (2012) Use of sesame stalk biomass for the removal of Ni(II) and Zn(II) from aqueous solutions. Water Sci Tech 66 (2):231–238
Kirbiyik C, Pütün AE, Pütün E (2016) Comparative studies on adsorptive removal of heavy metal ions by biosorbent, bio-char and activated carbon obtained from low cost agro-residue. Water Sci Technol 73 (2):423–436
Krishna AK, Govil PK (2004) Heavy metal contamination of soil around Pali industrial area, Rajasthan, India. Environ Geol 47:38–44. https://doi.org/10.1007/s00254-004-1124-y
Krishna AK, Govil PK (2008) Assessment of heavy metal contamination in soils around Manali industrial area, Chennai, Southern India. Environ Geol 54:1465–1472. https://doi.org/10.1007/s00254-007-0927-z
Krishna AK, Rama Mohan K, Murthy NN, Periasamy V, Bipinkumar G, Manohar K, Srinivas Rao S (2013) Assessment of heavy metal contamination in soils around chromite mining areas, Nuggihalli, Karnataka India. Environ Earth Sci 70(2):699–708. https://doi.org/10.1007/s12665-012-2153-6
Kumar PS, Ramalingam S, Abhinaya RV, Thiruvengadaravi KV, Baskaralingam P, Sivanesa S (2011) Lead(II)adsorption onto sulphuric acid treated cashew nut shell. Sep Sci Technol 46:2436–2449
Kumar U, Bandyopadhyay M (2006) Sorption of cadmium from aqueous solution using pretreated rice husk. Bioresour Technol 97:104–109
Kumar V, Sharma A, Kaur P, Singh SGP, Bali AS, Bhardwaj R, Thukral AK, Cerda A (2018) Pollution assessment of heavy metals in soils of India and ecological risk assessment: a state of the art. Chemosphere 216:449–462. https://doi.org/10.1016/j.chemosphere.2018.10.066
Lee J, Kwon T, Moon I (2006) Performance of polyamide reverse osmosis membranes for steel wastewater reuse. Desalination 189:309–322
Lestari AYD, Masruroh K, Widyastuti I (2018) Activated carbon from corn cob and petai hull as novel adsorbent for lead removal. Mat Sci Forum 934:165–169. https://doi.org/10.4028/www.scientific.net/MSF.934.165
Li Z, Ma Z, Van der Kuijp TJ, Yuan Z, Huang L (2014) A review of soil heavy metal pollution from mines in China: pollution and health risk assessment. Sci Total Environ 468–469 & 843–853

Liu C, Ngo HH, Guo W (2012) Watermelon rind: agro-waste or superior biosorbent? Applied Biochem Biotechnol 167:1699–1715

Mallampati R, Xuanjun L, Adin A, Valiyaveettil S (2015) Fruit peels as efficient renewable adsorbents for removal of dissolved heavy metals and dyes from water. ACS Sustain Chem Eng 3:1117–1124. https://doi.org/10.1021/acssuschemeng.5b00207

Martín-Lara MÁ, Rico ILR, Vicente ICA, García GB, de Hoces MC (2010) Modification of the sorptive characteristics of sugarcane bagasse for removing lead from aqueous solutions. Desalination 256:58–63

Memon JR, Memon SQ, Bhanger MI, El-Turki A, Hallam KR, Allen GC (2009) Banana peel: a green and economical sorbent for the selective removal of Cr (VI) from industrial wastewater. Colloids Surf B 70:232–237

Memon JR, Memon SQ, Bhanger MI, Memon GZ, El-Turki A, Allen GC (2008) Characterization of banana peel by scanning electron microscopy and FT-IRspectroscopy and its use for cadmium removal. Colloids Surf B 66(2):260–265

Meunier N, Drogui P, Montané C, Hausler R, Mercier G, Blais J-F (2006) Comparison between electro-coagulation and chemical precipitation for metals removal from acidic soil leachate. J Hazard Mater 137(1):581–590. https://doi.org/10.1016/j.jhazmat.2006.02.050

Moreno-Piraján JC, Giraldo L (2011) Activated carbon obtained by pyrolysis of potato peel for the removal of heavy metal copper (II) from aqueous solutions. J Anal Appl Pyrolysis 90:42–47

Mutiara T, Rofiki I, Al Ghifari MAD (2018) Bio-adsorbent from modified Jackfruit Wood Sawdust for removal of lead ions. Mat Sci Forum 934:159–164. https://doi.org/10.4028/www.scientific.net/MSF.934.159

Nagajyoti PC (2010) Heavy metals, occurrence and toxicity for plants: a review. Environ Chem Lett 8 (3):199–216

Naiya TK, Bhattacharya AK, Mandal S, Das SK (2009) The sorption of lead(II) ions on rice husk ash. J Hazard Mater 163:1254–1264

Nejadshafiee V, Islami MR (2019) Adsorption capacity of heavy metal ions using sultone-modified magnetic activated carbon as a bio-adsorbent. Mat Sci Eng C 101:42–52

Noli F, Kapashi E, Kapnisti M (2019) Biosorption of uranium and cadmium using sorbents based on aloe Vera wastes. J Environ Chem Eng 7(2):102985. https://doi.org/10.1016/j.jece.2019.102985

Noorjahan A, Dhanakumar S, Mohanraj R, Ravichandran M (2012) Status of heavy metals distribution in municipal solid waste in Tiruchirappalli city, India. Int J Appl Biol & Pharmaceut Technol 3(3):252–258

Nouri L, Ghodbane I, Hamdaoui O, Chiha M (2007) Batch sorption dynamics and equilibrium for the removal of cadmium ions from aqueous phase using wheat bran. J Hazard Mater 149:115–125

Obayomi KS, Bello JO, Nnoruka JS, Adediran AA, Olajide PO (2019) Development of low-cost bio-adsorbent from agricultural waste composite for Pb(II) andAs(III) sorption from aqueous solution. Cogent Eng 6(1):1687274. https://doi.org/10.1080/23311916.2019.1687274

Ofomaja A, Ho Y (2007) Effect of pH on cadmium biosorption by coconut copra meal. J Hazard Mater 139:356–362

Ofomaja A, Ho YS (2008) Effect of temperatures and pH on methyl violet biosorption by Mansonia wood sawdust. Bioresour Technol 99:5411–5417

Okoye AI, Ejikeme PM, Onukwuli OD (2010) Lead removal from wastewater using fluted pumpkin seed shell activated carbon: Adsorption modelling and kinetics. Int J Environ Sci Technol 7(4):793–800

Ottosen LM, Jensen PE (2005) Electro-remediation of contaminated soil. In: Lens P, Grotenhuis T, Malina G, Tabak H (eds) Soil and sediment remediation. IWA Publishing, London, pp 264–287

Parab H, Joshi S, Shenoy N, Lali A, Sarma US, Sudersanan M (2006) Determination of kinetic and equilibrium parameters of the batch adsorption of Co (II), Cr(III) and Ni (II) onto coir pith. Process Biochem 41:609–615

Parate VR, Talib MI (2014) Study of metal adsorbent prepared from tur dal (*Cajanus cajan*) husk: a value addition to agro-waste. IOSR J Environ Sci Toxicol Food Technol 8(9):43–54

Park D, Yun YS, Park JM (2005) Studies on hexavalent chromium biosorption by chemically treated biomass of *Ecklonia* sp. Chemosphere 60:1356–1364

Pehlivan E, Altun T (2008) Biosorption of chromium (VI) ion from aqueous solutions using walnut, hazelnut and almond shell. J Hazard Mater 155:378–384

Pehlivan E, Altun T, Cetin S, Bhanger MI (2009) Lead sorption by waste biomass of hazelnut and almond shell. J Hazard Mater 167:1203–1208

Puschenreiter M, Horak Friesl OW, Hartl W (2005) Low-cost agricultural measures to reduce heavy metal transfer into the food chain—a review. Plant Soil Environ 51(1):1–11

Quintelas C, Rocha Z, Silva B, Fonseca B, Figueiredo H, Tavares T (2009) Removal of Cd(II), Cr(VI), Fe(III) and Ni(II) from aqueous solutions by an *E. Coli* biofilm supported onkaolin. Chem Eng J 149:319–324

Reddy DHK, Seshaiah K, Reddy AVR, Lee SM (2012) Optimization of Cd(II), Cu(II) and Ni(II) biosorption by chemically modified *Moringa oleifera* leaves powder. Carbohydr Polym 88:1077–1086

Reddy DHK, Seshaiah K, Reddy AVR, Rao MM, Wang MC (2010) Biosorption of Pb^{2+} from aqueous solutions by *Moringa oleifera* bark: equilibrium and kinetic studies. J Hazard Mater 174:831–838

Redha AA (2020) Removal of heavy metals from aqueous media by biosorption. Arab J Basic Appl Sci 27 (1):183–193. https://doi.org/10.1080/25765299.2020.1756177

Saeed A, Iqbal M, Akhtar MW (2002) Application of bio-waste materials for the sorption of heavy metals in contaminated aqueous medium. Pakistan J Sci Industrial Res 45(3):206–211
Sarin V, Pant KK (2006) Removal of chromium from industrial waste by using eucalyptus bark. Bioresour Technol 97:15–20
Saxena A, Bhardwaj M, Allen T, Kumar S, Sahney R (2017) Adsorption of heavy metals from wastewater using agricultural–industrial wastes as biosorbents. Water Sci 31:189–197
Sekhar MC (2008) Removal of lead from aqueous effluents by adsorption on coconut shell carbon. J Environ Sci Eng 50:137–140
Sha L, Xueyi G, Ningchuan F, Qinghua T (2009) Adsorption of Cu^{2+} and Cd^{2+} from aqueous solution by mercapto-acetic acid modified orange peel. Colloids Surf B 73:10–14
Shafique U, Ijaz A, Salman M, ZumanWu Jamil NR, Javid A (2012) Removal of arsenic from water using pine leaves. J Taiwan Inst Chem Eng 43:256–263
Singh KK, Hasan HS, Talat M, Singh VK, Gangwar SK (2009) Removal of Cr(VI) from aqueous solutions using wheat bran. Chem Eng J 151:113–121
Singh S, Kumar V, Datta S, Singh D, Sharma K, Samuel J, Singh J (2019) Current advancement and future prospect of biosorbents for bioremediation. Sci Total Environ. https://doi.org/10.1016/j.scitotenv.2019.135895
Song H, Hu K, An Y, Chen C, Li G (2018) Spatial distribution and source apportionment of the heavy metals in the agricultural soil in a regional scale. J Soils Sediments 18(3):852–862
Srivastava V, Sarkar A, Singh S, Singh P, deAraujo ASF, Singh RP (2017) Agroecological responses of heavy metal pollution with special emphasis on soil health and plant performances. Front Environ Sci 5:64. https://doi.org/10.3389/fenvs.2017.00064
Sukumara D, Kumar AA, Thanga SG (2012) Effect of chelating agents in phytoremediation of heavy metals. Adv Agric Sci Eng Res 2(9):364–372
Taha MF, Kiat CF, Shaharun MS, Raml A (2011) Removal of Ni(II), Zn(II) and Pb(II) ions from single metal aqueous solution using activated carbon prepared from rice husk. World Academy of Sci Eng Technol 60
Tahmasbian I, Nasrazadani A (2012) Soil electerokinetic remediation and its effects on soil microbial activity-a review. African J Microbiol Res 6(10):2233–2238
Tajik S, Ziarati P, Cruz-Rodriguez L (2020) Coffee waste as novel bio-adsorbent: detoxification of nickel from contaminated soil and *Coriandrum Sativum*. J B Bio Engine 1(3):1–11
Taşar S, Kaya F, Ozer A (2014) Biosorption of lead(II) ions from aqueous solution by peanut shells: equilibrium, thermodynamic and kinetic studies. J Environ Chem Eng 2(2):1018–1026. https://doi.org/10.1016/j.jece.2014.03.015
Tavares T, Quintelas C (2014) Biosorption of metals-from the basics to high value catalysts production. In: Heavy metals in water: presence, removal and safety. Royal society of chemistry, pp 296–314 www.rsc.org
Tsezos M, Remoudaki E, Angelatou, V (1995) A systematic study on equilibrium and kinetics of biosorptive accumulation. The case of Ag and Ni Int Biodeter Biodegr 35(1–3):129–153
Vázquez G, Calvo M, Freire MS, González-Alvarez J, Antorrena G (2009) Chestnut shell as heavy metal adsorbent: optimization study of lead, copper and zinc cations removal. J Hazard Mater 172:1402–1414
Velazquez-Jimenez LH, Pavlick A, Rangel-Mendez JR (2013) Chemical characterization of raw and treated agave bagasse and its potential as adsorbent of metal cations from water. Ind Crop Prod 43:200–206
Wang XS, Li ZZ, Sun C (2009) A comparative study of removal of Cu (II) from aqueous solutions by locally low-cost materials: marine macroalgae and agricultural by-products. Desalination 235:146–159
Williams CJ, Edyvean RGJ (1997) Optimization of metal adsorption by seaweeds and seaweed derivatives. Trans Inst Chem Eng 75:19–26
Witek-Krowiak A, Reddy DHK (2013) Removal of micro-elemental Cr(III) and Cu(II) by usingsoybean meal waste–unusual isotherms and insights of binding mechanism. Bioresour Technol 127:350–357
Wong KK, Lee CK, Low KS, Haron MJ (2003) Removal of Cu and Pb by tartaric acid modified rice husk from aqueous solutions. Chemosphere 50:23–28
Wu Y, Li T, Yang L (2012) Mechanisms of removing pollutants from aqueous solutions by microorganisms and their aggregates: a review. Bioresour Technol 107:10–18
Wuana RA, Okieimen FE (2011) Heavy metals in contaminated soils: a review of sources, chemistry, risks and best available strategies for remediation. Int Scholarly Res Netw ISRN Ecol 1–20
Yadav A, Yadav PK, Shukla DN (2013) Investigation of Heavy metal status in soil and vegetables grown in urban area of Allahabad, Uttar Pradesh. India. Int J Sci Res Publications 3(9):1–7
Zabihi M, Ahmadpour A, Asl AH (2009) Removal of mercury from water by carbonaceous sorbents derived from walnut shell. J Hazard Mater 167:230–236
Zheng L, Dang Z, Yi X, Zhang H (2010) Removal of cadmium(II) from aqueous solution by corn stalk graft copolymers. Bioresour Technol 101:5820–5826
Zhu CS, Wang LP, Chen WB (2009) Removal of Cu(II) from aqueous solution by agricultural by-product: peanut hull. J Hazard Mater 168:739–746
Zuorro A, Lavecchia R (2010) Adsorption of Pb(II) on spent leaves of green and black tea. Am J Appl Sci 7:153–159

8 Aquatic Plants in Phytoextraction of Hexavalent Chromium and Other Metals from Electroplating Effluents

Suseela Lanka and Sowjanya Goud Murari

Abstract

Electroplating industry is one of the major industries associated with the release of toxic materials and heavy metals in the form of effluent wastewater. The effluent exhibits strong colour with high-biochemical oxygen demand, chemical oxygen demand, inorganic contaminants (calcium, sodium, sulfates, phosphates, nitrates, copper, chromium, iron, cyanide, and nickel), dissolved and suspended solids. Of these, the hexavalent chromium (Cr VI) is highly toxic and harmful to health if exposed for prolonged periods. The discharge of such effluent wastewater without proper treatment can as well contaminate and pollute the soil. As this wastewater is unsafe to human health and the environment, measures have to be taken to dispose it of safely. Among various remediation techniques, phytoremediation is a green approach, and is widely employed for treating polluted water bodies & soil by growing plants. Aquatic plants owing to their potential to absorb and accumulate heavy metals & other trace metals, play a pivotal role in the remediation of industrial effluents. The aim of the present paper is to look for various aquatic plants that play a crucial role in the remediation of toxic hexavalent chromium, and other metals from electroplating effluents to safeguard the soil from their toxic effects.

Keywords

Aquatic plants · Effluent wastewater · Electroplating industry · Heavy metals · Hexavalent chromium · Phytoremediation

8.1 Introduction

The industrial revolution has made most of the developing countries as 'hot spots' of heavy metal pollution. The industries that are mainly associated with heavy metal pollution include mining, chrome plating, electroplating, geothermal energy plants, pharmaceutical, pesticide manufacturing industries, automotive, agriculture, tannery industries, etc. (Sterrett et al. 1996). The heavy metals so released are toxic and non-biodegradable, and hence pose serious ecological consequences. The non-biodegradability also allows them to accumulate in different trophic levels, and this ultimately results in biomagnification (Gupta et al. 2012; Mudgal et al. 2010). Hence, remediation of effluent wastewaters polluted with heavy metals is extremely important to safeguard health of the people and the surroundings.

S. Lanka (✉) · S. G. Murari
Department of Bio-Sciences and Biotechnology, Krishna University, Machilipatnam, Andhra Pradesh 521004, India

J. A. Malik (ed.), *Advances in Bioremediation and Phytoremediation for Sustainable Soil Management*,
https://doi.org/10.1007/978-3-030-89984-4_8

Of various industries associated with heavy metal release, the electroplating industry occupies a prominent place. Most of the electroplating industries in India (over 300,000 units) operate as small scale industries (Chitraprabha and Sathyavathi 2018). The effluents from these industries possess several toxic pollutants and heavy metals (Iron, Chromium, Copper, Cyanide, and Nickel) released in the electroplating process. The discharge of such effluent wastewater without proper treatment can as well contaminate and pollute the soil. Of heavy metals, Cr(VI) is the most toxic and needs to be remediated. As per the available literature, the tannery industries of India release around 2.0 to 3.2 kt of Cr (VI) annually (Chandra et al. 1997). Electroplating industries are predominant sources of discharge of toxic chromium next to tannery industries. Chromium is a skin irritant and also a carcinogenic agent to humans (CPCB 2012; ATSDR 2012). Of the two oxidation states of chromium (Cr VI and Cr III), Cr (VI) is highly toxic upon prolonged exposures. In humans, lipid and carbohydrate metabolic pathways require some considerable amount of Cr (III) (ATSDR, 2012). The ill effects of hexavalent chromium inhalation or ingestion over a period of time include ulcers, nosebleeds, convulsions, liver and kidney damage, cancers of different types, and sometimes even lead to death (ATSDR 1998, 2012).

Heavy metal removal from effluent wastewaters can be done by several technologies. Most of these techniques are very much expensive and some are even dangerous to health. The available techniques for chromium removal from effluents include membrane filtration (Ndiaye et al. 2005), precipitation (Pathasarathy et al. 1986), nanofiltration (Simons 1993; Religa et al. 2010; Abhang et al. 2013), ion exchange (Ruixia et al. 2002), electro coagulation, flotation (Hu et al. 2005), and adsorption (Mohapatra et al. 2004). But each of these technologies have their own drawbacks like incomplete removal, the requirement of high-energy, treatment cost, and production of sludge which in turn can create an environmental burden. For developing countries like India, such treatments impose additional financial burden. Cost-effective and environmental-friendly methodologies offer good solutions to resolve the issues related to heavy metal remediation. The plants, nature's gifts offer sustainable solution, and technology of using plants in remediation called the phytoremediation, is also highly economical, esthetic and effective for heavy metal remediation from contaminated waters (Terry and Banuelos 2000; Mohanty et al. 2005; Mohanty and Patra 2011; Mohanty 2015). The technology also requires a little skill to operate. In phytoremediation, plants absorb pollutants through their root systems further accumulating them in their other body parts. The use of plants as heavy metal accumulators and other pollutants were traced back to 300 years. Different methods of phytoremediation include phytoextraction (Kumar et al. 1995; Tangahu et al. 2011), rhizofiltration (Dushenkov et al. 1995; Elias et al. 2014) phytostabilization (Salt et al. 1995), and phytotransformation/phytodegradation (Susarla et al. 2002). The growth rate and the photosynthetic activity of the plant mainly contribute to the success of phytoremediation technology.

Among the various phytoremediation methods, phytoextraction occupies a prominent place as it holds the heavy metals in different parts of the plants. It is also called phytoaccumulation and involves the uptake of contaminants through root systems and then their transport to the above ground level parts, the shoots. A plant is said to be ideal for phytoextraction if it possess certain characteristics like nativity, quick growth rate, ability to accumulate and tolerate a wide variety of heavy metals in its body parts, etc. The aquatic plants in phytoremediation offer cost-effective and resourceful cleanup technologies, and can be applied to remediate large polluted areas. They serve as natural absorbers and are proven to be most efficient in the remediation of toxic heavy metals owing to their extensive root system. Artificial wetlands constructed using aquatic plants are also in use worldwide for the treatment of effluent wastewaters. The choice of a suitable aquatic plant is very crucial for the effective removal of heavy metals using phytoremediation.

The present paper mainly discusses the efficacy of various aquatic plants viz., free-floating, submerged, and emergent in the remediation of hexavalent chromium and other metals from effluent wastewaters of electroplating industries in order to safeguard the mother soil from the accumulation of such potent and toxic heavy metals.

8.2 Hexavalent Chromium and Soil Contamination

The soil may be contaminated by chromium through anthropogenic deposition, dumping of chromium-containing industrial effluents, solid wastes in the form of chromium by-products (Kimbrough et al. 1999). Cr(VI) is the most toxic oxidative form of chromium, that is produced from various industries viz., Cr plating, chromite mining, chemical manufacturing, etc. Cr (VI) released from these industries contaminates surrounding water bodies and as well the soil. In the soil, chromium exists in combinations of Cr (III) and Cr (VI). Chromium undergoes various transformations once it enters the soil viz., oxidation, precipitation, reduction, dissolution, and sorption (Kimbrough et al. 1999). The soil oxidants like dissolved oxygen and manganese dioxide (MnO_2) can convert Cr (III) to hexavalent chromium, the highly toxic form (Fendorf and Zasoski 1992). Cr (VI) in the soil is as well reduced by iron, vanadium, organic materials, and sulfides (Cary 1982). Hexavalent chromium may remain for years in the soil when the reducing capacity of soil is low, and this is true with sandy soils and soils with low organic matter.

The effluent wastewater has to be remediated as per the norms of environment regulatory authorities before discharge into the surrounding environment. The chemical method is the routine method to remove hexavalent chromium from the effluent wastewaters but this method finally lefts with various residual chemicals that may further harm the environment. Especially, these residual chemicals accumulate in the soil and further damage the fertility of the soil. Researchers are trying hard to find out the best method to remove this toxic heavy metal. Phytoremediation offers an eco-friendly and economical method to achieve the statutory norms of environmental authorities. Aquatic plants owing to their fast growth rate and tolerance to heavy metals are widely employed to treat industrial effluents and contaminated water bodies.

8.3 Phytoremediation of Heavy Metals from Industrial Effluents

Industrial effluents if discharged without any treatment contaminates both soil and water bodies. This causes a serious menace to human and environmental health (Francová et al. 2017). Phytoremediation, coupled with some biological and engineering strategies found more effective in the remediation of heavy metals from effluent wastewaters through phytoextraction and phytostabilization (Cheraghi et al. 2011). Phytoremediation capacity of twelve different aquatic plants was tested for their potential to remove different heavy metals from the industrial effluents in Swabi district, Pakistan, and the results revealed the capability of these aquatic plants in the heavy metal remediation with the following efficacy: Cr (89%), Cd (90%), Pb (50%), Fe (74.1%), Ni (40.9%), and Cu (48.3%) (Khan et al. 2009). *T. Domingensis* showed better heavy metal (Zn, Fe, Ni, and Cr) removal efficiency from metallurgical plant effluents (Maine et al. 2017), and found to be a dominant plant having good toxic metal tolerance and heavy metal accumulation. Highest accumulation was found in roots rather than in leaves, and hence, rhizofiltration was found to be the superior phytoremediation mechanism in *T. Domingensis* (Hegazy et al. 2011).

The aquatic plant species viz., *Marsilea quadrifolia*, *Hydrilla verticillata*, and *Ipomea aquatica* showed good accumulation potential and TF value (translocation factor) for Cr, Fe, Zn, As, Hg, Cd, Pb, and Cu from the industrial effluents (Ahmad et al. 2011).

8.4 Aquatic Plants in Phytoremediation

Large contaminated areas can be cleaned by phytoremediation using aquatic ecosystems as it is a low cost and resourceful cleanup technique. The aquatic plants act as natural absorbers of heavy metal contaminants (Pratas et al. 2014), and the application of aquatic plants in heavy metal remediation is the most economical, profitable, and proficient method (Ali et al. 2013; Guittonny-Philippe et al. 2015). Aquatic plants in the form of constructed wetlands were the most widely employed technologies across the world for effluent wastewater treatment (Gorito et al. 2017; Mesa et al. 2015). The type of aquatic plant used in heavy metal accumulation is very important for effective phytoremediation (Galal et al. 2018; Fritio and Grege 2003). Aquatic plants owing to their ability in heavy metal absorption and accumulation have gained a good reputation in the cleanup of contaminated sites (Gorito et al. 2017; Gopal 2003). The extensive root system of aquatic plants makes them suitable options as the best accumulators of contaminants (Mays and Edwards 2001; Stoltz and Greger 2002). Though cultivating and growing aquatic plants is a time-consuming process (Said et al. 2015), the potentiality of these plants in wastewater treatment is innumerable (Kozminska et al. 2018; Syukor et al. 2014).

Aquatic plants remediate heavy metals via different processes like absorption, accumulation, surface adsorption, and integration into their various body parts (Rai et al. 1995; Sas-Nowosielska et al. 2008). Aquatic plants by their potential to remediate effluent wastewaters protect the surrounding environment.

8.4.1 Types of Aquatic Plants and Their Role in Heavy Metal Remediation

8.4.1.1 Free-Floating Plants

These are plants with floating leaves and submerged roots. These plants are studied for their potential to remove metals from polluted waters (Muthusaravanan et al. 2018; Maine et al. 2001; Olguín et al. 2002). Their root system accumulates heavy metals and transports them to different plant parts. The main free-floating aquatic plants involved in heavy metal remediation are water hyacinth, water lettuce, and duckweed (Anaokar et al. 2018; Chen et al. 2018; Hua et al. 2012; Singh et al. 2012).

Water Hyacinth

It is the most widely used aquatic plant in heavy metal remediation including hexavalent chromium because of its extensive growth rate (Xia and Ma 2005). The best species among the seven species of water hyacinth is the perennial *Eicchornia crassipes*. The enormous growth rate, ability to tolerate high pollution, the nutrient and heavy metal absorption capacities (Chanakya et al. 1993; Singhal and Rai 2003; Ingole and Bhole 2003; Liao and Chang 2004; Jayaweera and Kasturiarachchi 2004; Swarnalatha and Radhakrishnan 2015) make it the best choice to use in effluent water treatment. The ash and activated carbon derived from water hyacinth showed its good heavy metal (copper, nickel, zinc, and chromium) accumulation capacity. The production of minimal sludge and bio-sorbent nature facilitates easy metal recovery (Mahmood et al. 2010).

Water Lettuce

Water lettuce (*Pistia stratiotes L.*) is an aquatic plant commonly found in streams, lakes, and ponds (Quattrocchi 2017). This plant species possess extraordinary pH and temperature tolerance over an extensive range (Lima et al. 2013). The accumulation of heavy metals (Cu, Zn, Fe, Cd, and Cr) by this plant species does not show any harmful effect on it, making this a good choice as hyper accumulator plant for heavy metal removal from contaminated wastewater (Eloy et al. 2019; Mishra and Tripathi 2008).

Duckweed

It is mainly seen floating on the surface of still and slow-moving water. Water lens is the other name for duckweed. Duckweeds are smaller and very fast growing plants mainly grow in canals, ponds, ditches. These aquatic plants tolerate broad pH and temperature ranges (Radic et al.

2010), and hence found application in phytoremediation (Krishna and Polprasert 2008). The plant plays role in the removal of heavy metals, organic & inorganic pollutants, nutrients, pesticides of agricultural runoff, treatment of sewage, industrial effluents, and domestic wastewater (Daud et al. 2018; Chen et al. 2018; Mkandawire and Dudel 2007).

Water Fern

Salvinia auriculatais, a water fern, is a small aquatic plant that is extensively seen in aquatic ecosystems. Within no time this plant produces widespread colonies because of its substantial growth rate owing to its quick reproduction ability (Henry-Silva and Camargo 2006), and hence suitable for phytoremediation applications (Gardner and Al-Hamdani 1997). The rapid growth rate and ability to tolerate toxic metals make S. natans suitable for heavy metal removal applications (Dhir 2009; Dhir et al. 2011). The plant roots show higher metal accumulation potential. Among different *Salvinia* sp., *S. minima* exhibits high BCF (bioaccumulation factor) for cadmium and lead (Olguín et al. 2005).

8.4.1.2 Submerged Aquatic Plants

Leaves of these plants play an immense role in the uptake of metals. They can accumulate heavy metals present in sediments and water (Keskinkan et al. 2003; Rai et al. 2003; Saygıdeger and Dogan 2004). Some of the submerged plants well known for metal accumulation include American pondweed (*Potamogeton pectinatus*), parrot feather (*Myriophyllum spicatum*), pond weed (*Potamogeton crispus*), hornwort (*Ceratophyllum demersum*), water mint, *Mentha aquatica*, and *Vallisneria spiralis* (Brankovic et al. 2012; El-Khatib et al. 2014; Borisova et al. 2014; Peng et al. 2008; Casagrande et al. 2018).

8.4.1.3 Emergent Aquatic Plants

These plants are found on submerged soils with a water table present 0.5 m underneath the soil. Different plants belonging to this group accumulate heavy metals in different body parts viz., *Spartina alterniflora* (smooth cordgrass) accumulate heavy metals in leaves (Hempel et al. 2008), *Phragmites australis* (common reed) accumulate heavy metals in the roots (Ha and Anh 2016). *Typha latifolia* (Cattail), *Scirpus* sp. (bulrush), *Phragmites* (common reed) and *Polygonum hydropiperoides* (smartweed) are the best aquatic plants used for the remediation of metals such as Zn, Pb, Fe, Cr, Cd, Ni, Cu (Sasmaz et al. 2008; Kutty and Al-Mahaqeri 2016; Rudin et al. 2017). Table 8.1 shows the potential of different aquatic species (recent references) in the remediation of chromium and other metals.

8.5 Constructed Wetlands-Aquatic Plants

Wastewater remediation using constructed wetlands has been in wide use across the world and is a method of choice for wastewater treatment (Wang et al. 2017). These wetlands are mainly designed to treat effluent wastewaters that originated within the controlled environment. The wastewaters of municipal (Abou-Elela et al. 2013), agriculture (Vymazal and Brezinová 2015), storm water (Griffiths and Mitsch 2017), leachate from landfills (Madera-Parra et al. 2015), industrial effluents (Saeed et al. 2018), etc., can be treated using constructed wetlands. These artificial wetlands are highly economical, simple, and efficient in wastewater remediation without disturbing natural resources (Rizzo et al. 2020). The constructed wetlands (CW's) mainly use aquatic plants for remediation of polluted waters. Aquatic plants in these artificial wetlands serve two indirect but significant functions-firstly, the stems and leaves of these plants facilitate attachment of microbial communities by enhancing the surface area, and secondly, aquatic plants also tolerate anaerobic environment prevailed in the effluent wastewaters by transporting oxygen toward rhizosphere (Brix 1997). The rhizosphere microbial communities of aquatic plants are well supported to handle different tasks such as alteration of metallic ions, nutrients, and various other compounds, thus facilitating remediation of effluent wastewaters (Hammer 1989). CW's play an immense role in the heavy metal elimination from contaminated waters and this depends on

Table 8.1 Aquatic plants in chromium remediation

Name of the Aquatic plant	Heavy metal remediation	Removal rate	References
Water hyacinth	**Cr** and Cu from Tannery effluents	Cr-99.98%	Sarkar et al. (2017)
	Cr from aqueous solutions	Cr-63% on 3rd day and a removal rate of 80% on 9th day	Swarnalatha and Radhakrishnan (2015)
Water Lettuce	Mn, Fe, **Cr,** Cu, Pb, Zn, Co, Ni from Al-Sero drain of Giza	With an exception to Cr and Pb, all other heavy metals studied have a bio-concentration factor more than 1000. Likewise, the translocation factor with an exception of Pb and Cu, not exceeded one. The Rhizofiltration Potential for Cr, Pb, and Cu is 100 and it is higher than 1000 for iron.	Galal et al. (2018)
	Mn, Fe, Na, Pb, Ni, Cu, **Cr**, Zn, Ca, Al, Co, Cd, Mg, K from storm water detention ponds	Roots were found to absorb and accumulate Al, Ni, Cr, Cu, and Pb	Lu et al. (2011)
Duck weeds	Cr at a concentration of 0, 10, 100, 200 μM	Chromium uptake % increased with increased concentrations used with *L. minor*	Sallah-ud-Din et al. (2017)
	Cr and Pb at concentrations of 2, 4, 10 and 15 mg/L prepared using lab water	Observed a removal rate of 86.2–94.8% for Cr and 91.0–96.4% for Pb	Abdallah (2012)
	Co, Fe, Cu, Cd, Mo, Ni, Mn, **Cr**, Zn, Se from polluted mining water that is rich with selenium	A removal rate of 87% for Co, 55% for Se, and a removal rate of 35–60% for remaining heavy metals was observed	Flores-Miranda et al. (2014)
Salvinia (Water ferns)	Zn, Ni, Cu, and **Cr** at an initial concentration of 15 mg/L	Noticed a removal rate of 73.8% for Cu, 84.8% for Zn, 56.8% for Ni, 41.4% for Cr	Dhir et al. (2011)
	Cu, **Cr,** Cd Pb	Initial concentration Cu-(1.092 ± 0.026) Cr-(2.201 ± 0.0024) Pb-(2.974 ± 0.018) Cd-(0.251 ± 0.017) After treatment Cu-(2.035 ± 0.014) Cr-(1.052 ± 0.022) Pb-(1.924 ± 0.012) Cd-(0.018 ± 0.018)	Ranjitha et al. (2016)
	Pb, Cu, Ni, Mn, Zn, **Cr**, Fe, Cd metals from effluent waters of coal mines	Studied a removal rate of 97.01% for Ni, 96.96% for Pb, 96.77% for Cu, 96.22% for Mn, 96.38% for Zn, 94.12% for Fe, 80.99% for Cd and 92.85% for Cr	Lakra et al. (2017)

the type of metallic element, its ionic form, substrate, season, and plant species (Marchand et al. 2010). CW's with their enormous population of aquatic plants have already proved their effectiveness in the heavy metal remediation from effluent wastewaters (Baharudin and Shahrel 2008; Yeh et al. 2009). The circulation of essential components by the aquatic plants plays an important role in the maintenance of wetlands biochemistry (Brezinová and Vymazal 2015).

Research has clearly shown the role of aquatic plants in the removal of heavy metal contaminants using constructed wetlands (Maine et al. 2017, 2016; Baharudin and Shahrel 2008; Xu and Mills 2018). This technology could be a good choice to remediate electroplating industrial effluents. But proper remediation of contaminated water bodies ultimately depends upon the type of aquatic plant selected (Yeh 2009).

8.6 Conclusions and Future Prospects

Rapid industrialization is the leading cause of heavy metal load in the soil. These heavy metals are present as persistent pollutants in our environment. Among heavy metals, hexavalent chromium is highly toxic that severely impairs human health, and also the environment. The effluents from electroplating industries harbors various toxic heavy metals including chromium. The chemical method that is being currently used to treat electroplating effluent wastewaters is neither economical nor safe for the environment. Proper remediation methods are the need of the hour to safeguard both human health and our environment. The phytoremediation technology proves to be eco-friendly, economical, and sustainable cleanup technology. The application of suitable plant species plays a major role in phytoremediation. On par with other hyperaccumulator plants, aquatic plants are equally potent in the heavy metal removal from contaminated sites. Heavy metal pollution can be eradicated successfully by employing aquatic plants in either bioaccumulation or bio-sorption. The mode of interaction, chelator activities, and transport, control the accumulation and storage of heavy metals by these plants. Techniques of genetic engineering can be employed to engineer the plants to enhance their potential to accumulate and tolerate toxic heavy metals. Already engineered terrestrial plants showed good tolerance and metal accumulation capacity, but studies with regard to aquatic plants are in their preliminary stages. The use of aquatic plants in phytoremediation also eliminates post-filtration processes that are required in other conventional chemical & physical methods, and hence can be applied to remediate higher volumes of contaminated wastewaters. This inturn safeguards the soil. The plant biomass after treatment can be used as animal feed, bio-fertilizer, or can be used for biogas production. The technology of employing plants as bio accumulators is also esthetic, making contaminated sites visually pleasing in addition to their cost-effectiveness. The procedure also preserves the sustainability of whole ecosystems.

References

Abdallah MAM (2012) Phytoremediation of heavy metals from aqueous solutions by two aquatic macrophytes, *Ceratophyllum demersum* and *Lemna gibba* L. Environ Technol 33:1609–1614

Abhang RM, Wani KS, Patil VS, Pangarkar BL, Parjane SB (2013) Nanofiltration for recovery of heavy metal ions from wastewater—a review. Int J Res Environ Sci Technol 3(1):29–34

Abou-Elela SI, Golinielli G, Abou-Taleb EM, Hellal MS (2013) Municipal wastewater treatment in horizontal and vertical flows constructed wetlands. Ecol Eng 61:460–468

Ahmad A, Ghufran R, Zularisam A (2011) Phytosequestration of metals in selected plants growing on a contaminated Okhla industrial areas, Okhla, New Delhi, India. Water Air Soil Pollut 217:255–266

Ali H, Khan E, Sajad MA (2013) Phytoremediation of heavy metals—concepts and applications. Chemosphere 91:869–881

ATSDR S (2012) Toxicological profile for chromium. Agency for toxic substances and disease registry. Public Health Service, US Department of Health and Human Services. http://www.atsdr.cdc.gov/toxprofiles/tp.asp

ATSDR U (1998) Public Health Service. Toxicological profile for Barium and compounds

Anaokar G, Sutar T, Mali A, Waghchoure K, Jadhav R, Walunj S, Hyacinth J (2018) Low cost municipal wastewater treatment using water. Water Resour Pollut Stud 3:2

Baharudin B, Shahrel M (2008) Lead and Cadmium removal in synthetic wastewater using constructed wetland. Ph.D Thesis, Faculty of Chemical & Natural Resources Engineering University, Gambang, Malaysia

Borisova G, Chukina N, Maleva M, Prasad M (2014) *Ceratophyllum demersum* L. and *Potamogeton alpinus* Balb. from Iset'river, Ural region Russia differ in adaptive strategies to heavy metals exposure–a comparative study. Int J Phytoremediation 16:621–633

Brankovic S, Pavlovic-Muratspahic D, Topuzovic M, Glisic R, Milivojevic J, Dekic V (2012) Metals concentration and accumulation in several aquatic macrophytes. Biotechnol Biotechnol Equip 26:2731–2736

Brezinová TB, Vymazal J (2015) Evaluation of heavy metals seasonal accumulation in *Phalaris arundinacea* in a constructed treatment wetland. Ecol Eng 79:94–99

Brix H (1997) Do macrophytes play a role in constructed treatment wetlands. Water Sci Technol 35:11–17

Cary EE (1982) Chromium in air, soils, and natural waters. In: Langard S (ed) Biological and environmental aspects of chromium. Elsevier Biomedical, New York, pp 49–63

Casagrande GCR, dos Reis C, Arruda R, de Andrade RLT, Battirola LD (2018) Bioaccumulation and biosorption of Mercury by *Salvinia biloba Raddi* (Salviniaceae). Water Air Soil Pollut 229:166

Chandra P, Sinha S, Rai UN (1997) Bioremediation of chromium from water and soil by vascular aquatic plants. In: Kruger EL, Anderson TA, Coats JR (eds) Phytoremediation of soil and water contaminants. ACS Publications, Washington, DC, pp 274–282

Chanakya S, Borgaonkar G, Meena KS, Jagadish (1993) Solid-phase biogas production with garbage or water hyacinth. Bioresour Technol 46(3):227–231

Chitraprabha K, Sathyavathi S (2018) Phytoextraction of chromium from electroplating effluent by *Tagetes erecta* (L.). Sustain Environ Res 28(3):128–134

CPCB (2012) General standards for discharge of environmental pollutants (Part-A) effluents the environment (Protection) (Second Amendment) Rules, Central Pollution Control Board, New Delhi, India

Chen G, Huang J, Fang Y, Zhao Y, Tian X, Jin Y, Zhao H (2018) Microbial community succession and pollutants removal of a novel carriers enhanced duckweed treatment system for rural wastewater in Dianchi lake basin. Bioresour Technol 276:8–17

Cheraghi M, Lorestani B, Khorasani N, Yousefi N, Karami M (2011) Findings on the phytoextraction and phytostabilization of soils contaminated with heavy metals. Biol Trace Elem Res 144:1133–1141

Daud MK, Ali S, Abbas Z, Zaheer IE, Riaz MA, Malik A, Hussain A, Rizwan M, Rehman MZ, Zhu SJ (2018) Potential of duckweed (*Lemna minor*) for the phytoremediation of landfill leachate. J Chem 1–9

Dhir B (2009) Salvinia: an aquatic fern with potential use in phytoremediation. Environ Int J Sci Technol 4:23–27

Dhir B, Sharmila P, Saradhi PP, Sharma S, Kumar R, Mehta D (2011) Heavy metal induced physiological alterations in *Salvinia natans*. Ecotoxicol Environ Saf 74:1678–1684

Dushenkov V, Kumar PBAN, Motto H (1995) Rhizofiltration—the use of plants to remove heavy metals from aqueous streams. Environ Sci Technol 29 (5):1239–1245

Elias S, Mohamed M, Ankur AN, Muda K, Hassan MAHM, Othman MN, Chelliapan S (2014) Water hyacinth bioremediation for ceramic industry wastewater treatment application of rhizofiltration system. Sains Malaysiana 43(9):1397–1403

El-Khatib A, Hegaz A, Abo-El-Kassem AM (2014) Bioaccumulation potential and physiological responses of aquatic macrophytes to Pb pollution. Int J Phytoremediation 16:29–45

Eloy GG, Marta R, Gertjan M, Miquel C, Rosina G (2019) Quantitative risk assessment of norovirus and adenovirus for the use of reclaimed water to irrigate lettuce in Catalonia. Water Res 153:91–99

Fendorf SE, Zasoski RJ (1992) Chromium(III) oxidation by δ-MnO_2. 1. Characterization. Environ Sci Technol 26(1):79–85

Flores-Miranda DCM, Luna-Gonz alez A, Cort es-Espinosa DV, onica Escamilla-Montes R (2014) Bacterial fermentation of *Lemna* sp. as a potential substitute of fish meal in shrimp diets. Afr J Microbiol Res 8:1516–1526

Francová A, Chrastný V, Šillerová H, Vítková M, Kocourková J, Komárek M (2017) Evaluating the suitability of different environmental samples for tracing atmospheric pollution in industrial area. Environ Pollut 220:286–297

Fritio A, Grege M (2003) Aquatic and terrestrial plant species with potential to remove heavy metals from stormwater. Int J Phytoremediation 5:211–224

Galal TM, Eid EM, Dakhil MA, Hassan LM (2018) Bioaccumulation and rhizofiltration potential of *Pistia stratiotes* L for mitigating water pollution in the Egyptian wetlands. Int J Phytoremediation 20:440–447

Gardner JL, Al-Hamdani SH (1997) Interactive effects of aluminum and humic substances on Salvinia. J Aquat Plant Manag 35:30–34

Gopal B (2003) Perspectives on wetland science, application and policy. Hydrobiologia 490:1–10

Gorito AM, Ribeiro AR, Almeida CMR, Silva AM (2017) A review on the application of constructed wetlands for the removal of priority substances and contaminants of emerging concern listed in recently launched EU legislation. Environ Pollut 227:428–443

Griffiths LN, Mitsch WJ (2017) Removal of nutrients from urban storm water runoff by storm-pulsed and seasonally pulsed created wetlands in the subtropics. Ecol Eng 108:414–424

Guittonny-Philippe A, Petit ME, Masotti V, Monnier Y, Malleret L, Coulomb B, Combroux I, Baumberger T, Viglione J, Laffont-Schwob I (2015) Selection of wild macrophytes for use in constructed wetlands for phytoremediation of contaminant mixtures. J Environ Manag 147:108–123

Gupta K, Chatterjee C, Gupta B (2012) Isolation and characterization of heavy metal tolerant gram-positive bacteria with bioremedial properties from municipal waste rich soil of Kestopur canal (Kolkata), West Bengal, India. Biologia 67:827–836

Hammer DA (1989) Constructed wetlands for wastewater treatment municipal, industrial and agricultural. CRC Press, Boca Raton, FL

Ha H, Nguyen T, Anh BTK (2017) The removal of heavy metals by iron mine drainage sludge and *Phragmites australis*. In: Proceedings of the IOP conference series earth and environmental science, Bandung, Indonesia (20–22 September 2016), IOP Publishing, Bristol, UK, p 012022

Hegazy A, Abdel-Ghani N, El-Chaghaby G (2011) Phytoremediation of industrial wastewater potentiality by *Typha domingensis*. Int J Environ Sci Technol 8:639–648

Hempel M, Botté SE, Negrin VL, Chiarello MN, Marcovecchio JE (2008) The role of the smooth cordgrass *Spartina alterniflora* and associated sediments in the heavy metal biogeochemical cycle within Bahía Blanca estuary salt marshes. J Soils Sediments 8:289

Henry-Silv GG, Camargo AFM (2006) Efficiency of aquatic macrophytes to treat Nile tilapia pond effluents. Sci Agric 63:433–438

Hua J, Zhan C, Yin Y, Chen R, Wang X (2012) Phytoremediation potential of three aquatic macrophytes in manganese-contaminated water. Water Environ J 26:335–342

Hu CY, Lo SL, Kuan WH, Lee YD (2005) Removal of fluoride from semiconductor wastewater by electro coagulation-flotation. Water Res 39:895–901

Ingole NW, Bhole AG (2003) Removal of heavy metals from aqueous solution by water hyacinth (*Eichhornia crassipes*). J Water SRT–Aqua 52:119–128

Jayaweera MW, Kasturiarachchi JC (2004) Removal of nitrogen and phosphorus from industrial wastewaters by phytoremediation using water hyacinth (*Eichhornia crassipes* (Mart.) Solms). Water Sci Technol 50:217–225

Keskinkan O, Goksu M, Yuceer A, Basibuyuk M, Forster C (2003) Heavy metal adsorption characteristics of a submerged aquatic plant (*Myriophyllum spicatum*). Process Biochem 39:179–183

Khan S, Ahmad I, Shah MT, Rehman S, Khaliq A (2009) Use of constructed wetland for the removal of heavy metals from industrial wastewater. J Environ Manag 90:3451–3457

Kimbrough DE, Cohen Y, Winer AM, Creelman L, Mabuni C (1999) A critical assessment of chromium in the environment. Crit Rev Environ Sci Technol 29 (1):1–46

Krishna KB, Polprasert C (2008) An integrated kinetic model for organic and nutrient removal by duckweed-based wastewater treatment (DUBWAT) system. Ecol Eng 34:243–250

Kozminska A, Wiszniewska A, Hanus-Fajerska E, Muszynska E (2018) Recent strategies of increasing metal tolerance and phytoremediation potential using genetic transformation of plants. Plant Biotechnol Rep 12:1–14

Kumar PBAN, Dushenkov V, Motto H (1995) Phytoextraction—the use of plants to remove heavy metals from soils. Environ Sci Technol 29:1232–1238

Kutty AA, Al-Mahaqeri SA (2016) An investigation of the levels and distribution of selected heavy metals in sediments and plant species within the vicinity of Ex-Iron Mine in Bukit Besi. J Chem Article ID 2096147:1–9

Lakra KC, Lal B, Banerjee T (2017) Decontamination of coal mine effluent generated at the Rajrappa coal mine using phytoremediation technology. Int J Phytoremediat 19:530–536

Liao S, Chang W (2004) Heavy metal phytoremediation by water hyacinth at constructed wetlands in Taiwan. J Aqua Plant Manage 42:60–68

Lima L, Pelosi B, Silva M, Vieira M (2013) Lead and chromium biosorption by *Pistia stratiotes* biomass. Chem Eng Trans 32:1045–1050

Lu Q, He ZL, Graetz DA, Stoffella PJ, Yang X (2011) Uptake and distribution of metals by water lettuce (*Pistia stratiotes L.*). Environ Sci Pollut Res 18:978–986

Maine MA, Hadad HR, Sánchez G, Caffaratti S, Pedro MC (2016) Kinetics of Cr (III) and Cr (VI) removal from water by two floating macrophytes. Int J Phytoremediation 18:261–268

Marchand L, Mench M, Jacob DL, Otte ML (2010) Metal and metalloid removal in constructed wetlands, with emphasis on the importance of plants and standardized measurements: a review. Environ Pollut 158:3447–3461

Madera-Parra CA, Peña-Salamanca EJ, Peña MR, Rousseau DPL, Lens PNL (2015) Phytoremediation of Landfill Leachate with *Colocasia esculenta*, *Gynerum sagittatum* and *Heliconia psittacorum* in constructed wetlands. Int J Phytoremediation 17: 16–24

Mahmood T, Malik SA, Hussain ST (2010) Biosorption and recovery of heavy metals from aqueous solutions by *Eichhornia crassipes* (water hyacinth) ash. Biol Resour 5:1244–1256

Maine MA, Duarte MV, Suñé NL (2001) Cadmium uptake by floating macrophytes. Water Res 35:2629–2634

Maine M, Hadad H, Sánchez G, Di Luca G, Mufarrege M, Caffaratti S, Pedro M (2017) Long-term performance of two free-water surface wetlands for metallurgical effluent treatment. Ecol Eng 98:372–377

Mays P, Edwards G (2001) Comparison of heavy metal accumulation in a natural wetland and constructed wetlands receiving acid mine drainage. Ecol Eng 16:487-500

Mesa J, Mateos-Naranjo E, Caviedes M, Redondo-Gómez S, Pajuelo E, Rodríguez-Llorente I (2015) Heavy metal resistant and plant growth promoting rhizobacteria in the native metal rhizoaccumulator *Spartina maritima* Scouting contaminated estuaries. Mar Pollut Bull 90:150–159

Mishra VK, Tripathi B (2008) Concurrent removal and accumulation of heavy metals by the three aquatic macrophytes. Bioresour Technol 99:7091–7097

Mkandawire M, Dudel EG (2007) Are *Lemna* spp. effective phytoremediation agents. Biorem Biodiv Bioavail 1:56–71

Mohanty M, Pattnaik MM, Mishra AK, Patra HK (2005) Assessment of soil and water quality of chromite mine area of South Kaliapani, (Sukinda, Orissa). Bull Environ Sci 23(2):109–113

Mohanty M, Patra HK (2011) Attenuation of chromium toxicity by bioremediation technology. Rev Environ Contam Toxicol 210:1–34

Mohanty M (2015) Phytoremediation—an innovative approach for attenuation of chromium toxicity and rice cultivation in mining areas. J Rice Res 3:e116

Mohapatra D, Mishra D, Mishra SP, Chaudhury R, Das RP (2004) Use of oxide minerals to abate fluoride from water. J Colloid Interf Sci 275:355–359

Mudgal V, Madaan N, Mudgal A, Singh R, Mishra S (2010) Effect of toxic metals on human health. Open Nutraceuticals J 3:94–99

Muthusaravanan S, Sivarajasekar N, Vivek J, Paramasivan T, Naushad M, Prakashmaran J, Al-Duaij OK (2018) Phytoremediation of heavy metals mechanisms, methods and enhancements. Environ Chem Lett 16:1339–1359

Ndiaye PI, Moulin P, Dominguez L, Millet JC, Charbit F (2005) Removal of fluoride from electronic industrial effluent by RO membrane separation. Desalination 173:25–32

Olguín E, Hernández E, Ramos I (2002) The effect of both different light conditions and the pH value on the capacity of *Salvinia minima* Baker for removing cadmium, lead and chromium. Acta Biotechnol 22:121–131

Olguín EJ, Sánchez-Galván G, Pérez-Pérez T, Pérez-Orozco A (2005) Surface adsorption, intracellular accumulation and compartmentalization of Pb (II) in batch-operated lagoons with *Salvinia minima* as affected by environmental conditions, EDTA and nutrients. J Ind Microbiol Biotechnol 32:577–586

Pathasarathy N, Buffle J, Haerdi W (1986) Study of interaction of polymeric aluminium hydroxide with fluoride. Can J Chem 64:24–29

Peng K, Luo C, Lou L, Li X, Shen Z (2008) Bioaccumulation of heavy metals by the aquatic plants *Potamogeton pectinatus L.* and *Potamogeton malaianus* Miq and their potential use for contamination indicators and in wastewater treatment. Sci Total Environ 392:22–29

Pratas J, Paulo C, Favas PJ, Venkatachalam P (2014) Potential of aquatic plants for phytofiltration of uranium-contaminated waters in laboratory conditions. Ecol Eng 69:170–176

Quattrocchi U (2017) CRC world dictionary of plant names common names scientific names eponyms synonyms and etymology. CRC Press, Boca Raton, FL, p 728

Radic S, Stipanicev D, Cvjetko P, Mikelic IL, Rajcic MM, Širac S, Pavlica M (2010) Ecotoxicological assessment of industrial effluent using duckweed (*Lemna minor* L.) as a test organism. Ecotoxicology 19:216–222

Rai UN, Sinha S, Tripathi RD, Chandra P (1995) Wastewater treatability potential of some aquatic macrophytes: removal of heavy metals. Ecol Eng 5:5–12

Rai U, Tripathi R, Vajpayee P, Pandey N, Ali M, Gupta D (2003) Cadmium accumulation and its phytotoxicity in *Potamogeton pectinatus L*(Potamogetonaceae). Bull Environ Contam Toxicol 70:0566–0575

Ranjitha J, Raj A, Kashyap R, Vijayalakshmi S, Donatus M (2016) Removal of heavy metals from industrial effluent using *Salvinia molesta*. Int J Chem Technol Res 9:608–613

Religa P, Klimezak AK, Gierycz P (2010) Application of nanofiltration for chromium concentration in the tannery wastewater. J Hazard Mater 186(1):288–292

Rizzo A, Bresciani R, Martinuzzi N, Masi F (2020) Online monitoring of a long-term full-scale constructed wetland for the treatment of winery wastewater in Italy. Appl Sci 10:555

Rudin SM, Murray DW, Whitfeld TJ (2017) Retrospective analysis of heavy metal contamination in Rhode Island based on old and new herbarium specimens. Appl Plant Sci 5:1600108

Ruixia L, Jinlong G, Hongxiao T (2002) Adsorption of fluoride, phosphate, and arsenate ions on a new type of ion exchange fiber. J Colloid Interf Sci 248:268–274

Saeed T, Muntahaa S, Rashid M, Sun G, Hasna A (2018) Industrial wastewater treatment in constructed wetlands packed with construction materials and agricultural by-products. J Clean Prod 189:442–453

Said M, Cassayre L, Dirion J-L, Nzihou A, Joulia X (2015) Behavior of heavy metals during gasification of phytoextraction plants: thermochemical modelling computer aided. Chem Eng 37:341–346

Sallah-Ud-Din R, Farid M, Saeed R, Ali S, Rizwan M, Tauqeer HM, Bukhari SAH (2017) Citric acid enhanced the antioxidant defense system and chromium uptake by *Lemna minor* L. grown in hydroponics under Cr stress. Environ Sci Pollut Res 24:17669–17678

Salt DE, Blaylock M, Kumar PBAN (1995) Phytoremediation: a novel strategy for the removal of toxic elements from the environment using plants. Biotechnology 13:468–474

Sarkar M, Rahman A, Bhoumik N (2017) Remediation of chromium and copper on water hyacinth (*E. crassipes*) shoot powder. Water Resour Indust 17:1–6

Sasmaz A, Obek E, Hasar H (2008) The accumulation of heavy metals in *Typha latifolia L.* grown in a stream carrying secondary effluent. Ecol Eng 33:278–284

Sas-Nowosielska A, Galimska-Stypa R, Kucharski R, Zielonka U, Małkowski E, Gray L (2008) Remediation aspect of microbial changes of plant rhizosphere in mercury contaminated soil. Environ Monit Assess 137:101–109

Saygıdeger S, Dogan M (2004) Lead and Cadmium accumulation and toxicity in the presence of EDTA in *Lemna minor* L. and *Ceratophyllum demersum* L. Bull Environ Contam Toxicol 73:182–189

Simons R (1993) Trace element removal from ash dam waters by nanofiltration and diffusion dialysis. Desalination 89:325–341

Singhal V, Rai JPN (2003) Biogas production from water hyacinth and channel grass used for phytoremediation of industrial effluents. Bioresour Technol 86:221–225
Singh D, Gupta R, Tiwari A (2012) Potential of duckweed (*Lemna minor*) for removal of lead from wastewater by phytoremediation. J Pharm Res 5:1578–1582
Sterrett S, Chaney R, Gifford C, Mielke H (1996) Influence of fertilizer and sewage sludge compost on yield and heavy metal accumulation by lettuce grown in urban soils. Environ Geochem Health 18:135–142
Stoltz E, Greger M (2002) Accumulation properties of As, Cd, Cu, Pb and Zn by four wetland plant species growing on submerged mine tailings. Environ Exp Bot 47:271–280
Susarla S, Medina VF, McCutcheon SC (2002) Phytoremediation: an ecological solution to organic chemical contamination. Ecol Eng 18:647–658
Swarnalatha K, Radhakrishnan B (2015) Studies on removal of Zn & Cr from aqueous solutions using water hyacinth. Pollution 1(2):193–202
Syukor AA, Zularisam A, Ideris Z, Ismid MM, Nakmal H, Sulaiman S, Nasrullah M (2014) Performance of phytogreen zone for BOD5 and SS removal for refurbishment conventional oxidation pond in an integrated phytogreen system. World Acad Sci Eng Technol 8:159
Tangahu BV, Abdullah SRS, Basri H, Idris M, Anuar N, Mukhlisin M (2011) A review on heavy metals (As, Pb & Hg) uptake by plants through phytoremediation. Int J Chem Eng 2011:939161
Terry N, Banuelos G (2000) Phytoremediation of contaminated soil and water. Lewis Publishers, New York, p 389
Vymazal J, Brezinova T (2015) The use of constructed wetlands for removal of pesticides from agricultural runoff and drainage, a review. Environ Int 75:11–20
Wang M, Zhang DQ, Dong JW, Tan SK (2017) Constructed wetlands for wastewater treatment in cold climate, a review. J Environ Sci 57:293–311
Xia H, Ma X (2005) Phytoremediation of ethion by water hyacinth (*Eichhornia crassipes*) from water. Biores Technol 97:1050–1054
Xu X, Mills GL (2018) Do constructed wetlands remove metals or increase metal bioavailability. J Environ Manag 218:245–255
Yeh TY, Chou CC, Pan CT (2009) Heavy metal removal within pilot-scale constructed wetlands receiving river water contaminated by confined swine operations. Desalination 249:368–373
Yeh TY (2009) Removal of metals in constructed wetlands: Review. Pract Period Hazard Toxic Radioact Waste Manag 12:96–101

Phytoremediation of PAH-Contaminated Areas

9

Vijaya Geetha Bose, K. S. Shreenidhi, and Junaid Ahmad Malik

Abstract

Globalization has to lead to increased exploitation of our natural resources and thereby contaminating the environment equally. As the term remediation has been frequently used in the current scenario to reduce the contaminants level, phytoremediation has shown the efficient removal of the same. Phytoremediation is a kind of bioremediation process where the plants or parts of the plants are used for pollutant degradation and thereby complete removal. Restoration of ecosystems and maintaining ecological balance is trivial in the current scenario. Polyaromatic hydrocarbons (PAH) that are present in the effluents of various industries reach the environment and end up polluting the biota. Different PAHs can be removed by rhizofiltration, phytoextraction, and other actions of plants like Indian mustard, willow, etc. In this chapter, we will walk through various phytoremediation techniques to act on the PAH-contaminated areas.

Keywords

Bioremediation · Contaminants · Effluents · PAH · Phytoextraction · Phytoremediation · Rhizofiltration

9.1 Introduction

Drastic economic changes and intense technology advancements have led to the progress of mankind. Almost every industry working toward utilization of resources and a better product/service comes into the market daily. As the resources get used up, the ecological balance is disturbed. There are numerous NGOs and socio-activists who work toward achieving it. In spite of all these actions/activities carried out, the environment gets contaminated. The contaminants come in various forms, starting from solid wastes to the toxic gases emitted from the industries, reactors, and power plants. Polycyclic aromatic hydrocarbons (PAHs) constituting carbon and hydrogen have been found to be toxic to organisms and humans. Exposure to PAHs for a long time causes kidney- and liver-related diseases. The Center for Disease Control and Prevention claims that humans become susceptible to PAHs upon inhalation of motor vehicle exhaust and consumption of grilled foods.

V. G. Bose (✉) · K. S. Shreenidhi
Department of Biotechnology, Rajalakshmi Engineering College (Autonomous), (Affiliated to Anna University), Thandalam, Chennai, Tamilnadu 602105, India
e-mail: vijayageetha.b@rajalakshmi.edu.in

J. A. Malik
Department of Zoology, Government Degree College, Bijbehara, Kashmir (J&K) 192124, India

J. A. Malik (ed.), *Advances in Bioremediation and Phytoremediation for Sustainable Soil Management*,
https://doi.org/10.1007/978-3-030-89984-4_9

Though the PAHs are excreted through urine and feces, there is a possibility of micro-amount of PAHs being present in the body. In the longer run, the accumulation leads to diseases like cancer and life-threatening conditions. When humans come in contact with PAHs, be it in the form of inhaling them or consumption of chemicals with PAHs in them, they are found to cause cancer (skin cancer, stomach cancer, and lung cancer). This has been proved when tested with laboratory animals.

Polyaromatic hydrocarbon contaminants serve to be the principal characteristic to be subjected to taken control of; various restoratives could be incurred to have adventing a preclusion to the affecting contamination. PAHs can be classified into;

(a) **High Molecular Weight PAHs**—have four or more rings, can be dependent on soil and sediments; also found to be difficult to degrade by microbes owing to its hydrophobicity; can be noxious to bacterial cells.
(b) **Low Molecular Weight PAHs**—have two or three rings; volatile and soluble nature makes them highly degradable.

Inclusive of the fact to include the aspects of bioremediation into the contamination treatments, varied strategies have been developed to limit the intrusive effects (Sandeep et al. 2015). PAH are found in coal, crude oil, and also emitted when coal, garbage, or burnt tobacco; as natural reserves. Owing to their properties of low biodegradation and hydrophobicity properties, they are found to easily pile up in the environment thereby having a check on the environment is mandatory (Sun et al. 2008); the effect on the surroundings is something to be considered and dealt promptly as it might lead to serious impacts (Dua et al. 2002). Statistically, the contribution of PAHs to global pollution stands around 90% as it is the result of many combustion activities of organic materials on the soil surface layer (Andreoni and Gianfreda 2007).

Commonly found PAHs include anthracene, naphthalene, fluorine, and pyrene. Polycyclic aromatic hydrocarbons are used in plastics, dyes, and some medicines. They are seen in land (soil), water, and air. Soil contamination paves way for the PAH to enter the ground water table which in turn reaches the domestic land (Fig. 9.1). All these natural resources get contaminated leading to the onset of different diseases to the organisms. Power plants and industries have checkpoints to check the amount of the PAH being released to the atmosphere. The PAH contaminants in soil require remediation to prevent affecting the growth of trees and water tables. Remediation through biological methods/techniques is currently in trend and is indeed found to be efficient.

9.2 Modus Operandi of Remediation Types and PAH

Based on the method in which PAH gets degraded or metabolized into simpler complexes, there are a few types of remediation. Most common technique is phytodegradation or phytotransformation which involves plants metabolizing the organic contaminants by direct uptake from soil (Edwards et al. 2011). With the help of the roots and its nodules, the degrading ability of the microbes is enhanced in the rhizosphere. This type is called rhizodegradation/rhizoremediation (Corgie et al. 2003). The degradation action by microbes can be increased also by composting (Barker and Bryson 2002). The nonpolarity, hydrophobicity, and strong link with soil organic fraction limit the plant uptake and amassing of the contaminants (Kim et al. 2001). Soil properties, concentrations, environmental factors and contaminant properties are the key factors that influence the uptake, buildup, and absorption of PAHs in plants (Kapusta et al. 2004). Various methodologies are being developed to bring about the efficient bioremediation of PAHs contagion in the natural environment. The utilization of the microbes and genetically modified microorganisms that take up the hydrocarbons as their only source of carbon and energy has been found to be quite significant (Liste and Felgentreu 2006; Liang et al. 2011). Further literature survey claims and proves the fact that microbes

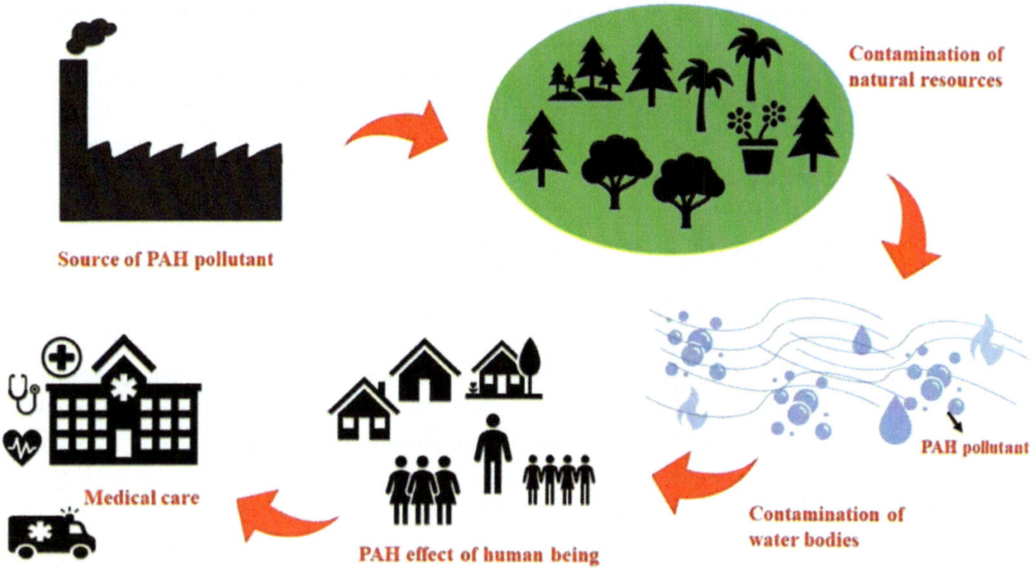

Fig. 9.1 Pictorial representation of the transmission of PAH within the environment

like bacteria and fungi along with the plants help in significant reduction of PAH contaminants in the environment (Fig. 9.2).

However, bacteria like *Alcaligenes*, *Arthrobacter*, *Bacillus*, *Micrococcus*, *Mycobacterium*, and *Pseudomonas* play a vital role (Tejeda-Agredano et al. 2013).Wet laboratory experiments carried out by the researchers have proved the effectiveness of bioremediation of soil contaminated with PAHs (Ciesielczuk et al. 2014). *Azospirillum* and combination of other bacterial species when used in inoculated plants

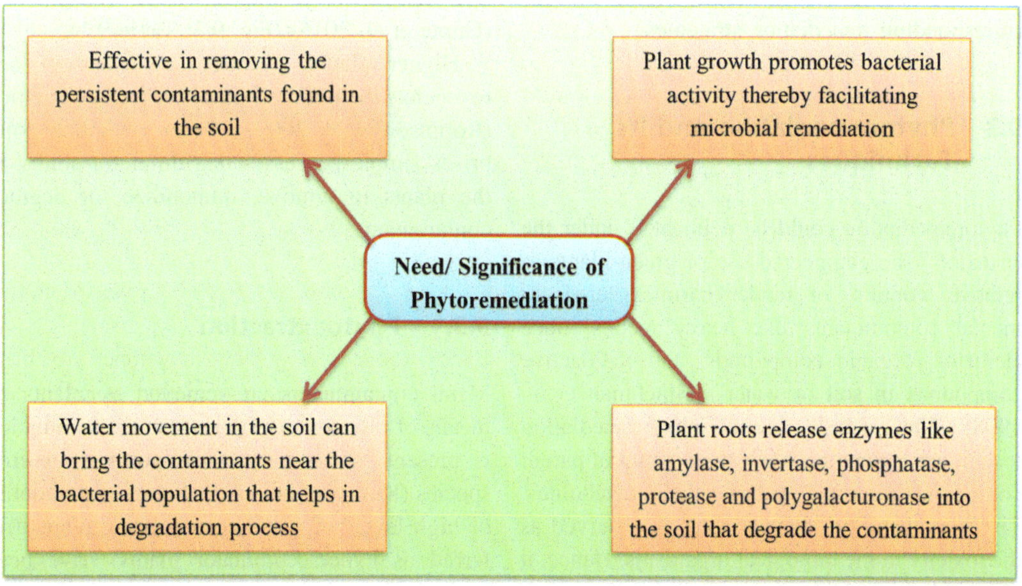

Fig. 9.2 Need for phytoremediation

have been found to enhance the plant growth thereby increasing the phytoremediation process (Couilleror et al. 2013).

9.2.1 Remediation and Its Types

Remediation refers to subsidence of the damage being caused or hindered in the due course of its action. The most commonly utilized environmental remediation is carried out through resources like water and plant growth related environmental factors. Organic contaminants are altered to environmentally safe compounds with threshold levels in the groundwater and soil. The conversion reactions involved are called biotransformation and biodegradation (Mueller et al. 1997).

The different participating environmental remediation types include remediation of soil, ground water, surface water, and sedimental regions. Furthermore depending on the materials used for remediation, we have microbial contribution and plants. Upon developmental studies, it is culminated that microbe-assisted remediation processes were found to be more efficient and whirlwind where plant-assisted remediation known as phytoremediation works on the longer run but only certain plants are found to slow the corresponding remediation strategies.

9.3 Phytoremediation and Its Techniques

Phytoremediation could be delineated under the terms of "The engineered use of green plants to remove, contain, or render harmless environmental contaminants like heavy metals, trace elements, organic compounds, and radioactive compounds in soil or water" (Hinchman et al. 1998). The success rate of phytoremediation techniques is determined by the amount of parent compounds removed ignoring their metabolites. The responsibility of metabolites is observed as the toxicity in the biological organisms. Hence, it is mandatory that the contaminants are detoxified preventing the high risk to organisms. Before declaring that the soil has been almost completely removed of the contaminants, it is essential to carry out appropriate toxicity assays to confirm the same (Mendonca and Picado 2002). Phytoremediation is one among the different types of remediation processes involving plants to exclude the contaminants from the environment especially soil. PAHs, one of the soil organic contaminants, can be removed either directly or indirectly. *Direct*—PAHs degraded by the action of enzymes. *Indirect*—Developing a favorable environment for the microbes that help in PAHs degradation at an increased rate (Rasmussen and Olsen 2004).

Phytoremediation, being an upcoming trending technology, has more advantages as it employs the plants to alter the contaminants' characteristics in an eco-friendly way. The various means of phytoremediation include volatilization, rhizoremediation, phytotransformation, phytostabilization, and hydraulic control. The incorporation of degraded simple molecules contaminants into the plant tissues is called phytodegradation (also known as phytotransformation) (Newman and Reynolds 2004). Using inorganic nutrients long with *E. crassipes* has been found to give combined degradation and phytodegradation which has become an innovative approach gaining a wide acceptability (Gupte et al. 2016) (Fig. 9.3; Table 9.1).

Higher plants are used for clear-out and revegetate the polluted sites by phytoremediation (Robinson et al. 2009). Phytoremediation comprises simple processes that differ on ability of the plants to remove, immobilize, or degrade contaminants.

9.3.1 Phytoextraction

Metal contaminants are removed by plants by means of phytoextraction and their accumulation is present in the harvestable portions of crop species (Kumar et al. 1995). Significant varieties of high levels of metal accumulating plants (referred as hyperaccumulator plants) have been known. Since their growth is slow and produces less quantity of biomass, this methodology faces

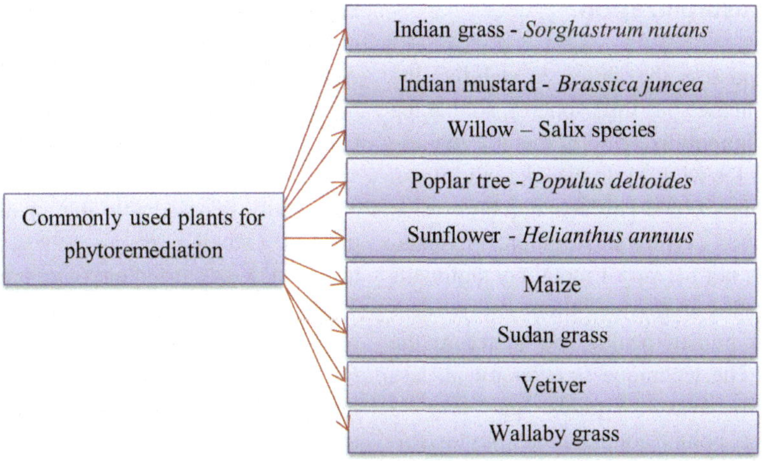

Fig. 9.3 Customary plants wielded for phytoremediation

Table 9.1 Different techniques of phytoremediation—an overview

Phytoremediation techniques	Description	References
Phytosequestration (or) Phytostabilization	Involves absorption and adsorption processes near the roots; biochemical reaction occurs that precipitates, sequester or immobilize the contaminants in the vicinity of the roots *Example*: Remediation of Ni using *Solanum nigrum*	Ferraz et al. (2012)
Rhizodegradation	Takes place in and around plant roots' vicinity (water/soil) *Example*: Perchlorate rhizodegradation carried out using *Salix nigra*	Yifru and Nzengung (2008)
Phytohydraulics	Contaminants that interact with the deep rooted plants are degraded/sequestered *Example*: Populus tree	Pivetz (2001)
Phytoextraction/ Phytoaccumulation	Contaminants are taken up, stored in the stem/tissues of the plants may or may not be degraded It is referred to as phytomining when metals are taken up by the plants *Example*: Phytoextraction of Sr, B, Zn, Cr(VI), As, Cu, Ni, Se, Pb, and Cd using *Brassica juncea*	Salido et al. (2003)
Phytodegradation	Contaminants are degraded/ biotransformed/ metabolized in the plants (roots/stems/leaves) *Example*: *Leucaena leucocephala*	Doty et al. (2003)
Phytovolatilization	Contaminants that are volatile in nature are utilized by the plants, metabolized, and released into the atmosphere via leaves *Example*: *Brassica juncea*	Bañuelos et al. (1998)

limitations. To overcome this, chelate-assisted phytoextraction is being carried out. This makes the plant fast growth and high biomass production. Run-off or leaching of solubilized metals into the surface and sub-surface water bodies must be taken into consideration. Introduction of modified genes into normal plants that has the ability to accumulate can bring about revolution in this technique. The normal plant will become capable of hyperaccumulation with high growth and increased biomass productions thereby accomplishing phytoextraction.

9.3.2 Phytostabilization

Some plants possess the ability to stabilize the contaminants in the soil (rhizosphere), so that the sequestration immobilizes them, thereby making it inaccessible for wildlife, livestock, and human exposure (Wong 2003). This technique is called phytostabilization or phytorestoration. Phytostabilization does not get rid of metal contaminants from a site, rather inactivates them and reduces the danger to human health and therefore the environment. The technique is economical, less environmentally evasive, quick, and easy to implant. These points confirm that phytostabilization is more beneficial than other soil-remediation practices (Berti and Cunningham 2000). In case of highly unclean soils, phytostabilization approach is employed as it tends to neutralize the affected area with plants that exhibit tolerance and have no danger of leaching and pollution of water bodies.

9.3.3 Rhizofiltration

Plants are initially propagated and grown in favorable conditions. Once they reach the development stage, they are transferred to sites where there is contamination of metals and PAHs (Henry 2000). In some cases, the plants are developed in the contaminated waters and remediation is carried out. This technique is called rhizofiltration. By this method, it is possible to treat storm waters, agricultural runoffs, diluted sludge, radionuclide-contaminated solutions, downwashes from power lines, industrial and residential effluents, acid mine drainage, and surface water and groundwater.

Plants fit for rhizofiltration applications can also competently get rid of toxic metals incurred through their rapid root growth systems. Several terrestrial plant species are established to effectually take out toxic metals like Ni^{2+}, Zn^{2+}, Cr^{6+}, Cu^{2+}, Pb^{2+}, and Cd^{2+} from aqueous solutions (Dushenkov et al. 1995).

9.3.4 Phytovolatilization

As the name phytovolatilization states, the technique wherein the organic and inorganic contaminants are converted to volatile gaseous species in the presence of water within the plants and finally released into the air at little concentrations (Mueller et al. 1999). Initially, this technique was carried out from transforming Hg^{2+} into less toxic Hg. It employs the insertion of bacterial Hg ion reductase gene into plants like tobacco (*Nicotiana tabacum* L.) and *Arabidopsis thaliana* L. This incorporation helped in larger range of phytovolatilization of mercury (Bizily et al. 1999). Unfortunately, it is still found that mercury released into the environment is recycled by precipitation and returns to the ecosystem (Henry 2000). Phytovolatilization advantages include minimal site disruption and minimal erosion.

9.3.5 Phytodegradation

Organic pollutants are degraded through several metabolic processes by plants and also by association plants with microbes (Burken and Schnoor 1997). Another technique under phytoremediation is phytodegradation. This method involves metabolizing the organic contaminants by the roots and additional plant parts in its tissues to lesser toxic substances. This technique was found to triumph in remedying the hydrophobic organic contaminants. In order to phytodegrade the toxic and recalcitrant organic compounds, most commonly the Poplar trees (*Populus* spp.) are used. Rhizodegradation occurs in the rhizosphere, wherein the soil contaminants are turned into fewer toxic levels. This procedure is facilitated by root exudates (organic molecules) that withstand populations of soil microbes. It can also be augmented by inoculation of selected bacteria in the contaminated sites/soils. The inoculum has the strains possessing the metabolic activity for degrading the specific contaminants. When genetically modified bacterial strains are integrated, significant results are observed. By biostimulation, the numbers of bacteria are amplified. The nutrient and the pH levels of the soils containing the contaminants are manipulated in this biostimulation system.

On a longer run, techniques like lagooning are not efficient. Phytoremediation technique serves as the "living cap" of plants and associated microorganisms which will diminish the pollution in the medium term (tens of years). The cost figuratives of such waste treatments are observed to be far below than those that are required for an impermeable cap restoration or the incineration of contaminated soil as accredited by the Centre of Disease Control and Prevention.

9.4 Environmental Prevalence of PAH

PAHs that are either released into the atmosphere or environment experience evaporation phenomenon. The common sources of emancipation of PAHs are air, soil, or water. PAHs assimilate dust particles from the atmosphere thereby undergoing photo-oxidation. The photo-oxidation process causes disruption of the chemical bonds, and this could prolong from days to weeks. The insoluble nature of the PAHs causes sedimentation at the bottom of lakes and rivers thereby contaminating the water. Degradation of PAHs may occur gradually over a period of months to weeks as sedimentation of soil and mixed microbial population are present (Prince 2015).

9.4.1 Ecotoxic Effects of PAH and Their Subsidiaries-PAH Toxicity

PAHs of various types that are prevalent have been ascertained as of substantial concern in regards to the likely contact and unfavorable health things on humans and are measured as a group. By which means, 17 predominant PAHs are levitated as most hazardous. Biological monitoring of such PAHs' exposures is of pivotal and prime interests, by virtue of which the widespread dispersal of these compounds to their toxicological relevance is understood.

Nonetheless, the health effects of individual PAHs are not as synonymous, as they are assumed to be. In record of the International Agency for Research on Cancer (IARC): the classification of PAHs may or may not be carcinogenic to humans (Group 1, 2A, or 2B). Among these certain figurative examples are benz[a]anthracene, chrysene, benzo[a]pyrene (Group 1), benzo[k]fluoranthene, benzo[b]fluoranthene (Group 2B), and naphthalene. Also few PAHs crossed significantly as carcinogens, teratogens, and mutagens, therefore constituting a severe ultimatum to human health and well-being. The gravest health consequence observed is likely from the inhalation subjection to PAHs causing an additional risk of lung cancer. (Hussein et al. 2016).

The aquatic organisms are affected by toxicity of such PAHs and the mechanisms of metabolism as well as by the effect of photo-oxidation. Specifically, they are observed to be toxic at high level in the prevalence of ultraviolet (UV) light. PAH possesses nearly reasonable to high acute toxicity in marine life and birds. However, the soilcontaining PAHs are improbable to have adverse effects on the earthbound invertebrates prevalence, except when the soil that they are prevalent in, is greatly contaminated. The furbished organisms have the antagonistic effects that include tumor, reproductive deficits and development, and also their immunity. Mammals could absorb potent toxic PAHs by various ways such as: skin contact, inhalation, and also ingestion (Dong et al. 2012). While on the subsidiary, plants absorb these PAHs from soils through their root nodules and translocation occurs to rest of the plant parts thereby accustoming with the PAHs. The uptake rates of these xeno-chemicals are wrangled by the factors of amount present, water solubility, and their physicochemical state, and also the soil nature and their various types. But, the PAH-induced phytotoxic effects are sparse in consideration.

The regarding full information and their databases are under development and under the scope of future research.

Contrarily, plants also accommodate certain substances that can protect themselves from the PAH effects. Plants also have the ability to orchestrate PAHs to act as growth hormones; thereby, they could assist in the remedial process in order to detoxify the effects of PAHs in the due course of their contamination. Thus, this aspect of plants could provide an ordeal as to why plants adapt themselves and produce betterment for bioremedial perspectives (Veltman et al. 2011). The contaminating PAHs were moderately unfluctuating in the subjective environment within. In terms of aquatic contaminations, the respective concentrations of PAHs in water solubility and fish water solubility were anticipated to be much greater than in their surrounding environment. Similar bioaccumulations were also witnessed in terrestrial invertebrates.

Subsequently, a wide range of these effects were observed in the divers' microbiota prevalent in the surroundings, most predominantly in aquatic biomes. In the studies undergone by Cerniglia (1992), PAHs accumulations had been documented to cause DNA damage causing mutations. Unlike lipophilic organic compounds, PAHs convert themselves to water soluble compounds, forming reactive intermediates in the due process. These react with the DNA of the cells forming adducts, preventing the gene in normal functionality.

9.4.2 Health Hazards Induced by PAH —Acute and Chronic Terms

In humans, as the accumulation of PAHs are observed by the characteristic of PAHs' diffusion with air particles, detrimental effects have been observed on human health. Based on the period of exposure to this xeno substance, the effects on health vary accordingly; a mixture of these compounds likely causes skin irritations and inflammation, with disclosure to certain PAHs such as naphthalene, benzo(a)pyrene, and anthracene acting as direct skin irritation. These are some significant acute effects that were conquered by the International Programme on Chemical Safety. Enduring exposure to PAHs resulted in declining immune responses, cataracts, kidney damage, respiratory issues, and lung abnormalities. Repeated exposure of the human skin to PAHs has resulted in various skin allergies and skin diseases (Diggs et al. 2011).

9.4.3 Genotoxicity and Teratogenicity of PAHs

Embryonic effects have been noticed in the animals chosen for the studies of experimental exposure to PAHs like benzo(a)pyrene, naphthalene, and benzo(a)anthracene. The wet laboratory studies carried out in mice showed that elevated levels of benzo(a)pyrene (on ingestion) diminished body weight in the offspring and in pregnancy period resulted in birth defects. In humans, their effects are not well known, yet it was put forward that the exposure of PAH subsequently alters and brings about adverse birth defects like premature delivery, less birth weight, and aberration in heart.

Furthermore, high prenatal exposure to PAH is also correlated with inferior IQ at a young age and heightened behavior problems at early ages, and childhood asthma. In the cord blood of babies exposed to PAHs, DNA damage was connected to cancer. The PAHs compounds undergo many metabolic modifications that lead to the development of electrophilic derivatives (e.g., diol epoxides, quinones, and conjugated hydroxyalkyl derivatives) competent of attesting covalent interaction with nucleophilic centers of macromolecules. Evenmore owing to the base pair substitutions, binding of PAHs to DNA bases could also prompt deletions, strand breakage, a variety of chromosomal alterations, S-phase arrest, and frameshift mutations, facilitating adverse genetic effects (Hussein et al. 2016).

9.5 Potential Possibilities of Phytoremediation of PAHs

The result of the work by Rui and colleagues (Liu et al. 2015) indicated that the alkaline phosphatase levels (ALPs) have a less influential role in the promotion of PAH hydrolysis, especially in the phytoremediation of PAH contaminated soil. Phytoremediation of such organic pollutants in soils were nearly pursued by the properties of the plant types; in the interim, the phytoremediation is also instigated by the catalysis of several enzymes (including hydrolases and oxidoreductase) in the plant rhizosphere. Studies done by Liu et al. (2016) signified that the growing of either perennial ryegrass or the white clover on creosote contaminated sediments along with subsequent presence and absence of a clean soil overlay having some specific observations. They concluded that the sediments with creosote contaminated exhibit a significant ($P = 0.007$) toxicity to the perennial ryegrass and even greater toxicity on the white clover. An inhibition of 94% was observed in the perennial ryegrass biomass production in the creosote contaminated sediments. By a subsequent assimilation of 2 cm clean soil intersection on the contaminated sediment periphery, significantly ($P = 0.002$) lessened the growth inhibition, with the perennial ryegrass plants display a perceptible drop of 53% in their biomass. In comparison with the plants cultivated in pristine soils, it had established a prospective benefactor of the remedial phenomenon. The chromosomal abnormalities monitored in the soil prior to phytoremediation confirmed that the soil with PAH contamination exert both eugenic as well as clastogenic effects before phytoremediation. The most shared irregularities in the unplanted control and zeroth-day soil were anaphase bridges, micronuclei, and also stickiness. The chromosome bridges results in chromatid break indicating the clastogenic effects of PAHs in plants. However, the micronuclei form a weighty alterations in the chromosomes, for example losses and breaks that were mistakenly repaired by the parent cells (Leme and Marin-Morales 2009). Therefore, the ordination of the micronuclei indicated a tough clastogenic effect by the inhibition of the spindle fibers (Cabaravdic 2010).

The affluence of plant species as efficacious rhizoremediation assemblages may depend on their extremely branched root system to harbor a considerable number of bacteria, primary and secondary metabolism, and setting up, survival, and also the ecological interfaces with other ambient organisms. Plant roots could act as a substitute soil tilling to include the additives (nutrients) and improving encompassing aeration. On this motive, the microbial availability of the targeted area also plays a principal factor in the remedial strata. A plenty of bacterial species are known to degrade PAHs which are secluded from tainted soil or sediments. The long-term petrochemical waste discharge harboring bacteria possess the ability of mortifying PAH to a sizeable extent. Among the PAH that rife in petrochemical waste, for example: Benzo (a) pyrene (BaP) are reflected to be the most carcinogenic and toxic by their nature. According to US-EPA, among all the PAHs known, Benzo (a) Pyrene is acknowledged as the maximum perilous pollutant constituting as a paramount constituent of smoke released from cigarettes (Renner 1993). On the note of judicious parlance of the remediation by plants, it has been shown that the fescue grass (*Festuca arundinacea*) and switchgrass (*Panicum virgatum*) degrade high 38% of pyrene contamination present in the soil in about 190 days (Chen et al. 2003). The thus observed intricacies would profit from the specific characterization by auxiliary studies, which would likely prove a productive field of inquiry. According to the work concorded by Anna et al., the study of *Azospirillum* sp. and *Pseudomonas stutzeri* was subjected to the judicious result of PAH bioremediation approach. By this protocol, it was observed that statistically significant results were upwent where the PAH were controlled with the roots of the legume plants scrutinized under the region of rhizosphere where the rhizoremediation was possibly observed. Under the reduction process, a significant result led to the captivation of pollutants and translocation to the vascular bundles of the plants, thereby transpirated by leaves of the plant subjected—meadow fescue. This drew an optimistic conclusion that alongside the effect of causing less pollution, these organisms also gave a possibility of fixing nitrogen in turn of aiding as a fertilization factor; and the aromatic hydrocarbons consumptive as the singular cause of carbon and energy suggest a potential possibility to use these strains for the bioremediation of PAH polluted soils at a limited habitat supplementation along nitrogen fertilizers (Anna et al. 2012). Generally, the range of PAHs removal was stated to subside upon appending the molecular weight as well as the PAH ring number (Haritash and Kaushik 2009). In the study by Rank and Nielsen (1998), the mitotic index (%) contemplation of the phytoremediation soils (observed at 60th and 120th day of the experiment) increased upon its comparison to the zeroth day and the unplanted control. This evidenced that the soil contaminated with PAHs before remediation expedited higher cytotoxicity and it is inference with the cell division by extending the S phase following the DNA and protein synthesis inhibition (Rank and Nielsen 1998). On the other hand, the multi-component remediation system had the boundless level for PAH eliminated from the contaminated soil, with an mediocre removal of nearly 16 PAHs at 80% efficiency, and the total material eliminated were 95% compared to the initial contamination observed. The superlative refinement attested was for the strongly soil bound by a pseudo-linear range, much longer than with any further single method (Huang et al. 2001).

9.5.1 Ecopiling

Ecopiling is an abatement of the convectional passive composting method but includes the addition of phytoremediation process. The process of ecopiling includes biostimulation of the original hydrocarbon degraders and the bio-augmentation process which involves inoculation of known hydrocarbon sources causing degradation of consortia and phytoremediation, the outcome of root growth and perforation through the soil. For optimizing the feasibility of the usage of perennial ryegrass and clover for phytoremediation, greenhouse trials were set up.

Significant reduction in biomass concentration was observed when the plants were cultivated in creosote polluted soil (Liste and Felgentreu 2006). The process of biostimulation of hydrocarbons polluted with cow dung in the Sudd wetland shows greater possibilities in restoring the hydrocarbon adulterated soil by the process of bioremediation. The presence of *Tithonia diversifolia, Oryza longistaminata, Hyparrhenia rufa,* and *Sorghum arundinaceum* plants act as natural phytoremediators resulting in greater bioremediation, though the above species exhibit exceptional phytoremediation property their growth gets reticent by high mass of TPH (Ruley et al. 2017).

PAHs contaminated soil exhibited both clastogenic and eugenic before phytoremediation thereby reflecting the chromosomal malformation present in the soil. The commonly observed defects in the unplanted control and zeroth-day soil are the anaphase bridges, stickiness, and micronuclei. The clastogenic effects of PAHs in plants are indicated by the chromosome bridges. Micronuclei defects are formed due to the damage of chromosomes and incorrect repair of the parent cells (Leme and Marin-Morales 2009). The strong clastogenic effect exhibited by PAHs contaminated soil lead to the inhibition of spindle fibers triggering the micronuclei (Cabaravdic 2010).

9.5.2 Methodologies for PAHs Estimation

The protocols endured to evaluate and study the toxic effects of PAHs by Altschul et al (1997) were plate counting method, isolation of the strains, and bacterial culture distinguishing. Alternative posit samples of slurry, 15 mL each were acquired for monitoring the corresponding microbial evolution. For the observed count of bacterial specimens, the moratoriums were first made to suffice for the distinctive separation of the sediment and the pellet. Following this, 0.1 mL of the clear supernatant was collected to implement a ten-fold serial dilution for the plates inoculated with a GAE (asparagine, glucose, yeast extract) medium (of the compositions 10 g/L glucose; 0.5 g/L yeast extract; 1 g/L L-asparagine; 20 g/L agar; 0.01 g/L $FeSO_4$; 0.5 g/L $MgSO_4.7H_2O$). Later, the annexed plates were incubated for nearly 48–72 h under a temperature of 30 °C. The colonies observed to be viable were then staunched and intimated as colony-forming units per ml (CFU/ml). Upon subsequent steps in the protocol underwent, the perceptible morphologies of the colonies were stipulated on their relative abundance basis. Thereafter, they were refined in the similar cataloged substratum to the point where their pureness was moderated. The purified bacteria obtained were consecutively bracketed by 16S rDNA gene sequencing following gene amplification via polymerase chain reaction (PCR) employing a MinicyclerTM (MJ research). Nearly a full-length 16S bacterial rDNA sequence (1500 bp) was amplified by enlisting the 27F and 1492R primers (Ikenaga et al. 2002), and purified with the GFXTM PCR DNA and gel band purification kit (Amersham Biosciences) for the automated DNA sequencing. Consequently, the 16S rDNA sequences expressed a similar profile arranged as a unique ribotype. An exclusive envoy of each ribotype was considered for sequencing, alongside the previous 16S rDNA sequences procured as well, exercising an ABI PrismTM 3100 Genetic Analyzer (Perking Elmer). Following this, comparison with the GenBank sequences was done by the aid of the USA National Center for Biotechnology Information NCBI's Basic Local Alignment Search Tool (BLAST) (Altschul et al. 1997). The subsequent denominations of the taxonomic identities were concluded following the RDP II (Cole et al. 2009).

In accordance to the work done by Guarino et al. (2019), it was inferred that microbe-assisted phytoremediation was a highly beneficial method in comparison with other strategies as this method resulted in higher rates of degradation. This method secondly demonstrated that *Lotus corniculatus, Piptatherum miliaceum,* and *Plantago lanceolata* symbiosis could be an eco-friendly method for the treatment of very old poly aromatic hydrocarboncontaminated soils. The work by Forján et al. (2020) revealed the application of bioaugmentation in the degradation of PAHs in slurry bioreactor experiments, and it also provided a clear understanding of the complexity in microbial relationships. The control surrounding conditions in the bioreactor expedited the bioaccessibility of the pollutants thereby exhibiting significant increase in the depletion of PAH. In a case study, there was about 40% degradation of PAHs in a period of 60 days in heap pile bioremediation, while on the contrary in this work, the yields were replicated for short duration of about 15 days, and it was inferred that the degeneration notably affected heavy PAHs and other organics. This was estimated using quantitative analyses and GC–MS qualitative. Bioslurry technique is quite simpler and economical in comparison with other strategies such as biopiles; therefore, designing should be carried out in such a way that it is inclusive of bioslurry, so that this method could be used for treatments of heavy PAHs (Dhir 2009) (Table 9.2).

The criterions that are considered while choosing the plants for remediation are availability of the soil the type of root nodule present. This approach provides the overall plants that are capable of being considered as plant models for the study.

A study by Jeelani et al. (2017) explored the phytoremediation ability of *A. calamus* in soil co-contaminated with cadmium and poly aromatic hydrocarbons. The results obtained concluded *A. calamus* prove to be operational for phytoremediation of soil polluted with PAHs and Cd. Results obtained from the work of Gałązka and Gałązka, 2005 concluded that the solicitation of grass inoculation with *P. stutzeri* and *Azospirillum* spp. manifested an effective response on the degradation procedures of polycyclic aromatic hydrocarbons. Various treatment methods yielded varied results; surfactant-aided process had a 90% eradication rate, whereas the compost modified phytoremediation evinced about 58–99% removal of pyrene. Chemical oxidation yielded in complete removal of PAHs (Saeid et al. 2018).

The phenomenon of chemotaxis discusses the maneuver of microorganisms under the effect of

Table 9.2 Legion of plants espoused for the PAH remediation facilitations

PAHs	Plant/s used
Naphthalene, acenaphthylene, fluorene, anthracene, and ascenaphthene	Fescue (*Festuca arundinacea*), annual ryegrass (*Lolium multiflorum*), and yellow sweet clover (*Melilotus officinalis*) (Parrish et al. 2004)
Anthracene, pyrene, and phenanthrene	Alfalfa cultivar crioula (Alves et al. 2018)
Mix of PAHs in soil	Ryegrass (*Lolium multiflorum* Lam.) along with *Ochrobactrum* sp. (PW) and alfalfa (*Medicago sativa* L.) (Xu et al. 2020)
Fluoranthene, acenaphthene, phenanthrene, anthracene, and naphthalene	Green ash (*Fraxinus pennsylvanica Marshall*) and hybrid poplar (*Populus deltoides* × *P. nigra* DN 34) (Spriggs et al.2005)
Naphthalene and phenanthrene	*Helianthus annuus* (sunflower) and *Avena sativa* (oat plant) (Reddy et al. 2020)
Phenantherene and pyrene	*Acorus calamus* (Jeelani et al. 2017)
Phenantherene and pyrene	Mixed cropping of rape and alfalfa; mixed cropping of rape and white clover (Lei and Wu 2014)
Anthracene, pyrene, and phenanthrene	Alfalfa cultivar *crioula* (Alves et al. 2018)
Anthracene, naphthalene, acenaphthene, phenanthrene, benzo(a)pyrene, chrysene, indeno(1,2,3-cd) pyrene + Benzo(g, h, i) perylene, fluoranthene, pyrene, benzo(alpha) anthracene, benzo(beta) fluoranthene, dibenz (a, h) anthracene, acenaphthylene, fluorene, benzo(k) fluoranthene	*Cymbopogon jwarancusa* (lemongrass) and *Helianthus annuus* (sunflower)—(D'Souza et al.2015)

the chemical gradient that aids in finding the finest conditions for growth and survival has been evidenced to promote the bacterial bioavailability isolated from the polluted rhizosphere degrading the PAHs. Another crucial reduction in the PAH contamination is the perspective of enzymatic degradation. Under enzymatic degradation, the enzymes involved in PAH deprivation are chiefly oxygenase, lignolytic, phosphatases, and dehydrogenase enzymes. Optimum temperatures are a necessity for their acme activity, and it is observed that most of the degradative enzymes work at mesophilic temperatures and their activity lessen with very low and high temperatures (Sandeep et al. 2015).

9.6 Conclusion

From the onset of the need and necessity of energy production for the proliferation and sustenance of human lifestyle, the demand and supply of energy reservoirs is an unfaltering challenge that requires the dire need of sustainable energy necessity. Fossil fuels discovery dawned the never imagined boom in the industrialization factor as well as laying the foundation for natural resource usage to enhance economy and exploiting the reserves for the benefactor of the human civilizational appraisals. But, as integrated and judicious the development of these resources had been observed by far, it is quite evident as to how the pollution of this production of energy has been noticed at minute levels. The aggregation of the pollutants affects the reservoirs that harness life forms from microbial, plant, aquatic, and animal biomes of the ecosystem. With delineated and scientifically substantiated methodologies, we can elucidate the future retrospectives to procure a better precluding fact of environmental sustenance. In this study as we discussed the variety of plants utilized for the remediation, this serves the importance of how resource recycling is acquired. Aspects of rhizoremediation, chemotaxis, and using genetically engineered organisms with extenuated modifications are evident examples as to how the processes differ from one plant to another in their own PAH degradation extents. Let alone the course of actions where restorations are rendered, and a possible understanding of plant ecosystems importance could be well furnished with optimal affirmation.

9.7 Future Perspectives

Bioremediation is an emerging technological advancement that has its own regime of expertise with ongoing pivotals of innovation. Approaches of incorporating the natural remedials alongside the essence of the regional natural reserves prevalent in the region. The notions of bioremediation as well as the context of traditional plants present can also be learned. With the aspect of phytoremediation, this could thereby incite the researchers to focus more on the availability of the solar energy prevalence being subjected to precise usage. In line with the recent upcoming field of biotechnology, this can serve as a note to various interdisciplinary ordeals to put forth a balanced study of renovation. In-silico surveys and studies on the remedials by regional plants observed, data analytics regarding the remediation extents and hence furbished results could be integrated to blend the technicalities of contrasting subjective fields. Academics of how the combinations are done and the incorporation of mathematical significance can also be studied and analyzed, as brooked in the study by (Geetha et al. 2018) on determining the histopathological and HPLC analysis. Regarding which, the approach of in-silico as well as in-vitro studies was concorded orderly to familiarize a greater understanding of the toxic-inductive phenomenon of the xeno-biotic component. Similar studies could be undertaken for the PAHs remediation for coherent conclusions. Unique coalitions of natural resources and synthetics would spawn new disciplines of expertise in remedials; hence, sustainable development of extinguishable resources can be assayed and applauded.

References

Altschul SF, Madden TL, Schäffer AA, Zhang J, Zhang Z, Miller W, Lipman DJ (1997) Gapped BLAST and PSI-BLAST: a new generation of protein database search programs. Nucleic Acids Res 25(17):3389–402. https://doi.org/10.1093/nar/25.17.3389

Alves WS, Manoel EA, Santos NS, Nunes RO, Domiciano GC, Soares MR (2018) Phytoremediation of polycyclic aromatic hydrocarbons (PAH) by cv. Crioula: A Brazilian alfalfa cultivar. Int J Phytoremediation 20(8):747–755. https://doi.org/10.1080/15226514.2018.1425663

Anna G, Maria K, Andrzej P (2012) The efficiency of rhizosphere bioremediation with *Azospirillum sp.* and *Pseudomonas stutzeri* in soils freshly contaminated with PAHs and diesel fuel. Pol J Environ Stud 21 (2):345–353

Bañuelos GS, Ajwa HA, Wu L, Zambrzuski S (1998) Selenium accumulation by *Brassica Napus* grown in Se-Laden soil from different depths of Kesterson reservoir. J Soil Contam 7(4):481–496. https://doi.org/10.1080/10588339891334393

Barker AV, Bryson GM (2002) Bioremediation of heavy metals and organic toxicants by composting. Sci World J 2:407–420. https://doi.org/10.1100/tsw.2002.91

Berti WR, Cunningham SD (2000) Phytostabilization of metals. In: Raskin I, Ensley BD (eds) Phytoremediation of toxic metals: using plants to clean-up the environment. Wiley, New York, pp 71–88

Bizily SP, Rugh CL, Summers AO, Meagher RB (1999) Phytoremediation of methylmercury pollution: mer B expression in Arabidopsis thaliana confers resistance to organomercurials. Proc Natl Acad Sci USA 96:6808–6813. https://doi.org/10.1073/pnas.96.12.6808

Burken JG, Schnoor JL (1997) Uptake and metabolism of atrazine by poplar trees. Environ Sci Technol 31:1399–1406. https://doi.org/10.1021/es960629v

Cabaravdic M (2010) Induction of chromosome aberrations in the *Allium cepa* test system caused by the exposure of cells to Benzo(a) pyrene. Med Arch 64:215

Cerniglia CE (1992) Biodegradation of polycyclic aromatic hydrocarbons. Biodegradation 3:351–368. https://doi.org/10.1016/0958-1669(93)90104-5

Chen YC, Banks MK, Schwab AP (2003) Pyrene degradation in the rhizosphere of tall fescue (*Festuca arundinacea*) and switchgrass (*Panicum virgatum*). Environ Sci Technol 37:5778–5782. https://doi.org/10.1021/es030400x

Ciesielczuk T, Kusza G, Poluszyńska J, Kochanowska K (2014) Pollution of flooded arable soils with heavy metals and poly-cyclic aromatic hydrocarbons (PAHs). Water Air Soil Pollution 225(10):2145. https://doi.org/10.1007/s11270-014-2145-0

Cole JR, Wang Q, Cardenas E, Fish J, Chai B, Farris R, Kulam-Syed-Mohideen AS, Mcgarrell D, Marsh T,

George G, Tiedje J (2009) The ribosomal database project: improved alignments and new tools for rRNA analysis. Nucleic Acids Res 37:141–145. https://doi.org/10.1093/nar/gkn879

Corgié S, Joner EJ, Leyval C (2003) Rhizospheric degradation of phenanthrene is a function of proximity to roots. Plant Soil 257:143–150. https://doi.org/10.1023/A:1026278424871

Couillerot O, Ramírez-Trujillo A, Walker V, Von Felten A, Jansa J, Monika M, Défago G, Prigent-Combaret C, Comte G, Caballero-Mellado J, Moënne-Loccoz Y (2013) Comparison of prominent *Azospirillum* strains in *Azospirillum-Pseudomonas-Glomus consortia* for promotion of maize growth. Appl Microbiol Biotechnol 97(10):4639–4649. https://doi.org/10.1007/s00253-012-4249-z

D'Souza R, Varun M, Lakhani A, Singla V, Paul MS (2015) PAH contamination of urban soils and phytoremediation. Int Phytoremediation Cham 219–241

Dhir B (2009) Potential of aquatic macrophytes for removing contaminants from the environment. Crit Rev Environ Sci Technol 39:754–781. https://doi.org/10.1080/10643380801977776

Diggs DL, Huderson AC, Harris KL, Myers JN, Banks LD, Rekhadevi PV, Niaz MS, Ramesh A (2011) Polycyclic aromatic hydrocarbons and digestive tract cancers: a perspective. J Environ Sci Health C Environ Carcinog Ecotoxicol 29(4):324–357. https://doi.org/10.1080/10590501.2011.629974

Dong CD, Chen CF, Chen CW (2012) Determination of polycyclic aromatic hydrocarbons in industrial harbor sediments by GC-MS. Int J Environ Res 9(6):2175–2188. https://doi.org/10.3390/ijerph9062175

Doty S, Shang T, Wilson A, Moore A, Newman L, Strand S, Gordon M (2003) Metabolism of the soil and groundwater contaminants, ethylene dibromide and trichloroethylene, by the tropical leguminous tree, *Leuceana leucocephala*. Water Res 37:441–449. https://doi.org/10.1016/s0043-1354(02)00291-9

Dua M, Singh A, Sethunathan N, Johri A (2002) Biotechnology and bioremediation: successes and limitations. Appl Microbiol Biotechnol 59:143–152. https://doi.org/10.1007/s00253-002-1024-6

Dushenkov V, Motto H, Raskin L, Kumar PBAN (1995) Rhizofiltration: the use of plants to remove heavy metals from aqueous streams. Environ Sci Technol 30:1239–1245. https://doi.org/10.1021/es00005a015

Edwards R, Dixon DP, Cummins I, Brazier-Hicks M, Skipsey M (2011) New perspectives on the metabolism and detoxification of synthetic compounds in plants. In: Schröder P, Collins C (eds) Organic xenobiotics and plants. Plant ecophysiology 8. Springer, Dordrecht. https://doi.org/10.1007/978-90-481-9852-8_7

Ferraz P, Fidalgo F, Almeida A, Teixeira J (2012) Phytostabilization of nickel by the zinc and cadmium hyperaccumulator *Solanum nigrum* L. Are metallothioneins involved? Plant Physiol Biochem 57:254–260. https://doi.org/10.1016/j.plaphy.2012.05.025

Forján R, Lores I, Sierra C, Baragaño D, Gallego JLR, Peláez AI (2020) Bioaugmentation treatment of a PAH-polluted soil in a slurry bioreactor. Appl Sci 10:2837. https://doi.org/10.3390/app10082837

Gałązka A, Gałązka R (2005) Phytoremediation of polycyclic aromatic hydrocarbons in soils artificially polluted using plant-associated-endophytic bacteria and *Dactylis glomerata* as the bioremediation plant. Pol J Microbiol 64(3):241–252. https://doi.org/10.5604/01.3001.0009.2119

Geetha V, Sujata R, Shreenidhi KS, Sundararaman T (2018) Histopathological and HPLC analysis in the hepatic tissue of *Pangasius* sp. exposed to diclofenac. Pol J Environ Stud 27(6):2493–2498. https://doi.org/10.15244/pjoes/75829

Gianfreda L (2007) Bioremediation and monitoring of aromatic-polluted habitats. Appl Microbiol Biotechnol 76:287–308. https://doi.org/10.1007/s00253-007-1018-5

Guarino C, Zuzolo D, Marziano M, Conte B, Baiamonte G, Lorenzo M, Benotti D, Gresia D, Stacul E, Cicchella D, Sciarrillo R (2019) Investigation and assessment for an effective approach to the reclamation of polycyclic aromatic hydrocarbon (PAHs) contaminated site: SIN Bagnoli, Italy. Sci Rep 9:11522. https://doi.org/10.1038/s41598-019-48005-7

Gupte A, Tripathi A, Rudakiya D, Patel H, Gupte S (2016) Bioremediation of polycyclic aromatic hydrocarbon (PAHs): a perspective. Open Biotechnol J 10:363–378. https://doi.org/10.2174/1874070701610010363

Haritash A, Kaushik C (2009) Biodegradation aspects of polycyclic aromatic hydrocarbons (PAHs): a review. J Hazard Mater 169:1–15. https://doi.org/10.1016/j.jhazmat.2009.03.137

Henry JR (2000) An overview of the phytoremediation of lead and mercury. National Network of Environmental Management Studies, US Environmental Protection Agency, pp 21–22

Hinchman RR, Negri MC, Gatliff EE (1998) Phytoremediation: using green plants to clean up contaminated soil, groundwater, and wastewater. Submitted to the US Department of Energy, Assistant Secretary for Energy Efficient and Renewable Energy under Contract W-31-109-Eng-38

Huang XD, Penrose D, Dixon GD, Glik BR, Greenberg BM (2001) An integrated multi-component phytoremediation system for removal of polycyclic aromatic hydrocarbon contaminants from soil. Department of Biology, University of Waterloo, Canada

Hussein I, Abdel S, Mona SM, Mansour (2016) A review on polycyclic aromatic hydrocarbons: source, environmental impact, effect on human health and remediation. Egypt J Pet 25(1):107–123.https://doi.org/10.1016/j.ejpe.2015.03.011

Ikenaga M, Muraoka Y, Toyota K, Kimura M (2002) Community structure of the microbiota associated with nodal roots of rice plants along with the growth stages: estimation by PCR-RFLP analysis. Biol Fertil

Soils 36:397–404. https://doi.org/10.1007/s00374-002-0543-8

Jeelani N, Yang W, Lingqian X, Qiao Y, An S, Leng X (2017) Phytoremediation potential of *Acorus calamus* in soils co-contaminated with cadmium and polycyclic aromatic hydrocarbons. Sci Rep 7:8028. https://doi.org/10.1038/s41598-017-07831-3

Kapusta P, Szarek-Lukaszewska G, Kiszka J (2004) Spatial analysis of lichen species richness in a disturbed ecosystem (Niepolomice Forest, S Poland). Lichenologist 36:249–260. https://doi.org/10.1017/S0024282904014112

Kim IS, Park JS, Kim KW (2001) Enhanced biodegradation of polycyclic aromatic hydrocarbons using nonionic surfactants in soil slurry. Appl Geochem 16:1419–1428. https://doi.org/10.1016/S0883-2927(01)00043-9

Kumar PDAN, Dushenkov V, Motto H, Raskin L (1995) Phytoextraction: the use of plants to remove heavy metals from soils. Environ Sci Technol 29(5):1232–1238. https://doi.org/10.1021/es00005a014

Lei ZH, Wu YX (2014) Phytoremediation on PAHs in soils and laboratorial technics. Appl Mech Mater 675:1012–1017. https://doi.org/10.4028/www.scientific.net/AMM.675-677.1012

Leme DM, Marin-Morales MA (2009) Allium cepa test in environmental monitoring: a review on its application. Mutat Res. 682(1):71–81. https://doi.org/10.1016/j.mrrev.2009.06.002. Epub 2009 Jul 2. PMID: 19577002

Liang Meng L, Qiao M, Peter H (2011) Phytoremediation efficiency of a PAH-contaminated industrial soil using ryegrass, white clover, and celery as mono- and mixed cultures. H Arp J Soils Sediments 11:482–490. https://doi.org/10.1007/s11368-010-0319-y

Liste HH, Felgentreu D (2006) Crop growth, culturable bacteria, and degradation of petrol hydrocarbons (PHCs) in a long-term contaminated field soil. Appl Soil Ecol 31:43–52. https://doi.org/10.1016/j.apsoil.2005.04.006

Liu R, Dai Y, Sun L (2015) Effect of rhizosphere enzymes on phytoremediation in PAH-contaminated soil using five plant species. PLoS ONE 10(3):369. https://doi.org/10.1371/journal.pone.0120369

Liu X, Kiwanuka S et al (2016) Use of ecopiling to remediate PAH-contaminated storm-water lagoon sediment. J Bioremediat Biodegrad 7:355. https://doi.org/10.4172/2155-6199.1000355

Mendonca E, Picado A (2002) Ecotoxicological monitoring of remediation in a coke oven soil. Environ Toxicol 17:74–79. https://doi.org/10.1002/tox.10034

Mueller B, Rock S, Gowswami D, Ensley D (1999) Phytoremediation decision tree. Prepared by—Interstate Technology and Regulatory Cooperation Work Group, pp 1–36

Mueller JG, Devereux R, Santavy DL, Lantz SE, Willis SG, Pritchard PH (1997) Phylogenetic and physiological comparisons of PAH-degrading bacteria from geographically diverse soils. Antonie Van Leeuwenhoek 71(4):329–343. https://doi.org/10.1023/a:1000277008064

Newman LA, Reynolds CM (2004) Phytodegradation of organic compounds. Curr Opin Biotechnol 15(3):225–230. https://doi.org/10.1016/j.copbio.2004.04.006

Parrish ZD, Banks MK, Schwab AP (2004) Effectiveness of phytoremediation as a secondary treatment for polycyclic aromatic hydrocarbons (PAHs) in composted soil. Int J Phytoremediation 6(2):119–137. https://doi.org/10.1080/16226510490454803

Pivetz BE (2001) Ground water issue: phytoremediation of contaminated soil and ground water at hazardous waste sites. United States Environmental Protection Agency, EPA 36

Prince U (2015) Environmental effects of polycyclic aromatic hydrocarbons. J Nat Sci 5

Rank J, Nielsen MH (1998) Genotoxicity testing of wastewater sludge using the *Allium cepa* anaphase-telophase chromosome aberration assay. Mutat Res Genet Tox En 418:113–119. https://doi.org/10.1016/s1383-5718(98)00118-1

Rasmussen G, Olsen RA (2004) Sorption and biological removal of creosote contaminants from groundwater in soil/sand vegetated with orchard grass (*Dactylis glomerata*). Adv in Environ Res 8:313–327. https://doi.org/10.1016/S1093-0191(02)00105-3

Reddy KR, Chirakkara RA, Martins Ribeiro LF (2020) Effects of elevated concentrations of co-existing heavy metals and PAHs in soil on phytoremediation. JHTRBP 24(4):04020035. https://doi.org/10.1061/(ASCE)HZ.2153-5515.0000538

Renner R (1993) EPA to strengthen persistent, bioaccumulative, and toxic pollutant controls-Mercury first to be targeted. Environ Sci Technol 33(3):62A. https://doi.org/10.1021/es992653p

Robinson BH, Banuelos G, Conesa HM, Evangelou MWH, Schulin R (2009) The phytomanagement of trace elements in soil. Crit Rev Plant Sci 28:240–266. https://doi.org/10.1080/07352680903035424

Ruley JA, Amoding A, Tumuhairwe JB, Basamba T, Oryem-Origa H (2017) Hydrocarbons in crude-oil production sites of Sudd-region South-Sudan: implication on soil fertility and plants species risk. Int J Curr Res 9(12):63070–63075. https://doi.org/10.1155/2020/4614286

Saeid G, George AS, Soroush G, Mahdeih B (2018) Treatment technologies for PAH-contaminated sites: a critical review. Environ Monit Assess 190:546.https://doi.org/10.1007/s10661-018-6936-4

Salido AL, Hasty KL, Lim J, Butcher DJ (2003) Phytoremediation of arsenic and lead in contaminated soil using Chinese brake ferns (*Pteris vittata*) and Indian Mustard (*Brassica juncea*). Int J Phytoremediation 5(2):89–103. https://doi.org/10.1080/713610173

Sandeep B, Piyush P, Bhavya B, Shivesh S, Vivek K, Krishan DS (2015) Bioremediation of polyaromatic hydrocarbons (PAHs) using rhizosphere technology. Braz J Microbiol 46(1):7–21. https://doi.org/10.1590/S1517-838246120131354

Spriggs T, Banks MK, Schwab P (2005) Phytoremediation of polycyclic aromatic hydrocarbons in manufactured gas plant–impacted soil. J Environ Qual 34(5):1755–1762. https://doi.org/10.2134/jeq2004.0399

Sun YB, Zhou QX, Diao CY (2008) Effects of cadmium and arsenic on growth and metal accumulation of Cdhyperaccumulator *Solanum nigrum* L. Bioresource Technol 99(5):1103–1110. https://doi.org/10.1016/j.biortech.2007.02.035

Tejeda-Agredano MC, Gallego JVS, Grifoll M, Ortega-Calvo JJ, Cantos M (2013) Influence of the sunflower rhizosphere on the biodegradation of PAHs in soil. Soil Biol Biochem 57:830–840.https://doi.org/10.1016/j.soilbio.2012.08.008

Veltman K, Huijbregts MA, Rye H, Hertwich EG (2011) Including impacts of particulate emissions on marine ecosystems in life cycle assessment: the case of offshore oil and gas production. Integr Environ Assess Manag 7(4):678–686. https://doi.org/10.1002/ieam.246

Wong MH (2003) Ecological restoration of mine degraded soils, with emphasis on metal contaminated soils. Chemosphere 50(6):775–780. https://doi.org/10.1016/s0045-6535(02)00232-1

Xu C, Yang W, Wei L, Huang Z, Wei W, Lin A (2020) Enhanced phytoremediation of PAHs-contaminated soil from an industrial relocation site by *Ochrobactrum* sp. Environ Sci Pollut Res Int 27(9):8991–8999. https://doi.org/10.1007/s11356-019-05830-7

Yifru D, Nzengung V (2008) Organic carbon biostimulates rapid rhizodegradation of perchlorate. Environ Toxicol Chem 27:2419–2426. https://doi.org/10.1897/08-008.1

Bioremediation of Petroleum-Contaminated Soil

10

Yojana Waychal, Shreya Gawas, and Sagar H. Barage

Abstract

Petroleum contamination of soil has always been a threat to ecosystem and human health. On average, contaminated soils across the globe are found to have the total petroleum hydrocarbon (TPH) concentrations in an alarming range of 1.17–236.7 g per kilogram of soil. The polycyclic and monocyclic aromatic hydrocarbons and heavy metals are the most polluting components of oils. These components have carcinogenic, teratogenic and mutagenic effects and heavy metals pose about 95% carcinogenic risks to humans. In addition, oil pollution significantly and adversely affect the moisture content, total nutrient composition, hydraulic conductivity, etc. of soil. These parameters ultimately affect the soil fertility leading to low crop yield and petroleum contamination in vegetables and fruits. The petroleum contamination causes adverse and hazardous effects on plants and animals. Thus, numerous physico-chemical methods are used for remediation of petroleum-contaminated soil but these methods are expensive and have some disadvantages. Hence, alternative technologies which are cost effective, eco-friendly and greater ease of practical application are expected. Bioremediation is more efficient and cost effective and less environmentally aggressive method for remediation of petroleum-contaminated soil. Researchers have been trying to harness the potential of microbial species for the bioremediation of contaminated soils. In present chapter, we will focus on recent advances in the field of microbial bioremediation along with its large scale application and associated downfall. On the other hand, the information on alteration of pH, temperature and other factors on microbial growth and metabolism will be useful for improvement of present strategies of microbial bioremediation. Overall, the present information will be useful to broaden our current understanding and help design alternative strategies for remediation of petroleum-contaminated soil.

Keywords

Bioremediation · Contaminated soil · Microbial bioremediation · Mycoremediation · Petroleum · Petroleum contamination · Phycoremediation

10.1 Introduction

Soil pollution is considered a global threat to the health and agricultural sectors. The major contribution of petroleum contaminants in soil pollution is increasing with time. Oil contamination

Y. Waychal · S. Gawas · S. H. Barage (✉)
Amity Institute of Biotechnology, Amity University, Mumbai—Pune Expressway, Bhatan Post—Somathne, Panvel, Mumbai, Maharashtra 410206, India

J. A. Malik (ed.), *Advances in Bioremediation and Phytoremediation for Sustainable Soil Management*,
https://doi.org/10.1007/978-3-030-89984-4_10

is common due to the activities like transportations, leakages, spillage, etc. Due to petroleum contamination, pollutants like pyrene, benzene, heavy metals, etc. are released in soils hampering the natural processes occurring in it (Yuniati 2018). Organisms living in soil are directly exposed to such toxicants are affected adversely. These toxicants get accumulated in the living cells of such organisms leading their way up in the food chain. Humans are the largest consumer species of the food web which leads to adverse teratogenic, carcinogenic and mutagenic effects of petroleum pollutant components (Gargouri et al. 2014). Thus, it is important to remediate such pollutants in order to reduce the toxic effects. The treatment of polluted soils is primarily done by physical methods like incineration and excavation and by chemical methods like oxidation. Although these conventional methods are quite cheap, fast and effective, they have certain drawbacks like secondary pollution and non-sustainable approach. Bioremediation is a major area of environmental biotechnology that has attracted a lot of attention in the past 2 decades due to its cheap and eco-friendly approach. Bioremediation involves degradation or decontamination of pollutants using plants and microbes and very rarely invertebrate animals as well. Various approaches and techniques like bioaugmentation, bioaccumulation, anaerobic digestion in bioreactors, etc. are widely used for remediating pollutants from different media. Bacteria like *Pseudomonas, Comamonas, Bacillus*, etc. are some representative genera that show effective results on remediating toxicants like toluene and naphthalene, While archaea and some eukaryotes like fungi and microalgae have excellent metabolic pathways that have the capability of accumulating and degrading petroleum pollutants like Polycyclic Aromatic Hydrocarbons (PAHs) and heavy metals (Cadmium, Nickel, Copper, Cobalt) (Devatha et al. 2019). The natural metabolism of prokaryotes and eukaryotic microorganisms is effective for degradation or decontamination of petroleum pollutants, but it is time consuming hence research is going on to find out ways to accelerate the process and enhance efficiency using genetic and molecular tools, controlling physical and chemical parameters, etc. This chapter focuses on recent findings associated with microbial bioremediation of petroleum compounds.

10.2 Petroleum Pollutants: Nemesis and Composition

Petroleum is one of the largest contributors to organic pollutants of soils and water. Petroleum hydrocarbons have accumulation capabilities and are hazardous for human health, crops, environment, etc. Petroleum products hamper the growth of plants and disturb the healthy population of native microflora in the soils (Borowik et al. 2019). Petroleum contaminants affect the ecosystem adversely as invertebrates come easily in contact with these toxicants and this furthers upwards to the predators in the food chain. Recent studies have shown that consumables like vegetables and fruits were also found to be contaminated by petroleum compounds (Zhang et al. 2014). The petroleum compounds, especially benzene and xylene, are cancer causing agents which also enhance the risks of pancreatic and aerodigestive dysfunctions (Khanna and Gharpure 2017). Naphthalene and associated compounds are also found to be teratogenic (malformation) (Varjani et al. 2017).

Crude oil and diesel contribute largely in the soil pollution. The petroleum pollution is caused by a mixture of aromatic and aliphatic hydrocarbons and heavy metals. Petroleum on combustion leaves behind Polycyclic Aromatic Hydrocarbons (PAHs) that cause adverse pollution effects anthracene, benzene (derivatives-benzo-pyrene), fluoranthene, naphthalene and pyrene is the most polluting PAHs found in soils all across the globe (Souza et al. 2013). Industrialization has added yet another threat which is heavy metal pollution. Heavy metals like Zn, Hg, Cd, Cu, Ni, Co, etc. show oligodynamic action towards native microbes and have tendency to accumulate in plant and animal tissues. Majorly found organic compounds in petroleum waste are listed in Table 10.1.

Table 10.1 Organic compounds in petroleum along with their Mol. Wt. (Varjani et al. 2017)

Compound	Molecular weight (g)
Benzene	78
Toluene	92
Xylene	106
Naphthalene	128.17
Anthracene	178.23
Fluorene	166.22
Pyrene	202.26

10.3 Analysis of Petroleum-Contaminated Soils

Aliphatic and aromatic hydrocarbons impart adverse effects on the environment as well as human health, some serious effects of these products can be classified as mutagenicity, toxicity and carcinogenicity (Tahseen et al. 2016). The oil degradation often causes the change in the soil moisture levels, limits of total carbon content, total nitrogen, Atterberg limits (plastic index of the soil which include liquid oil limit and plastic limits present in the soil) and hydraulic conductivity (Devatha et al. 2019). The oil contamination blocks soil aeration leads to shift in the microbial and ecological functions in the contaminated soil (Sutton et al. 2013). This leads to contamination of the ground water contamination (Rahman et al. 2010).

The soil organic matter plays an important role in the bioremediation of the soil. If the organic matter in the soil is high, the rate of bioremediation decreases in the soil. The biodegradation is also dependent on the size and nature of the components of the crude oil, for example, whether the oil is hydrophobic or amphibolic in nature and if the oil component has lower molecular weight or higher molecular weight (Scherr et al. 2007). The oil biodegradation in momentary polluted soil would contend with migration and adjustment of oil in non-available soil compartments. Differentiating biodegradation patterns in soils of varying structures are identified with disparate microenvironments influencing toxin accessibility also as well as abundance and variety of the microbial consortium (Amellal et al. 2001). Poor microbial expansion and diversity are ordinary for soils with sandy surfaces and low organic carbon, which is appended by lower corruption rates as contrasted with clay loam and soil. The hydrophobic organic chemicals are bordered by an increase in sorption to the soil and a decline in the rate and degree of biodegradation. But the biodegradability due to hydrophobic organic chemicals is not yet totally characterized.

10.4 Treatment Methods

10.4.1 Physico-Chemical Remediation

Conventionally remediation of contaminated sites was done by physical methods like land excavation followed by incineration or pyrolysis of the soil (Yuniati 2018). Chemicals and associated reactions like Fenton's reagent, chemical oxidation processes, etc. are the most commonly known treatments. Fenton's reagent is found to be applied in treating soils that are polluted by hazardous organic compounds. It is a solution, made of hydrogen peroxide and ferrous ions, that is catalyst using naturally present iron mineral, goethite, in order to carry out efficient oxidation of hazardous petroleum contaminants from soils with upto 72% efficiency (Ouriache et al. 2019). Some persistent compounds like naphthalene are difficult to oxidize for which photocatalytic technique brought into action. Chemicals like TiO_2 and ZnO is used as catalysts in the presence of UV radiations for a low cost and fast fix for hydroxylation of aromatic rings (Varjani et al. 2020). An even better efficiency (99%) in removal of total petroleum hydrocarbon (TPH) is achieved in a continuous rotating kiln reactor by subjecting the soil to pyrolysis at 420 °C for 15 min (Gao and Zygourakis 2019). Although impressive petroleum pollutant removal is witnessed in application of these chemical and physical methods, selection of the apt technology is equally dependent on the process energy consumption. To address this aspect, studies are being conducted to use electroremediation method where current is applied to contaminated soil samples. Such electrokinetic treatments have shown a significant drop in energy consumption by around 56% than the conventional physico-chemical methods (Streche et al. 2018). These methods, however, pose underlying problems like higher costs and secondary pollution due to formation of polychlorinated dibenzo-p-dioxins (PCDDs) and dibenzofurans, that makes them environmentally unsustainable (Kong et al. 1998).

10.4.2 Biological Remediation

Treatment of contaminated soil or water, using biological agents like plants and microorganisms have by far been the most environment friendly and cost-effective method used in the world. Such treatment is termed as bioremediation that occurs naturally in all habitats but biotechnological advancements of these processes are aimed towards accelerating the rate to enhance the impact. There are various bioremediation technologies (in situ *and* ex situ) depending on mechanism of action like bioreactors, bioventing, composting, land farming (Table 10.2). Bacteria, algae, fungi and plants are the most common and most suitable biological agents that are used for detoxifying contaminated soil. Bioremediation directed by plants is termed as phytoremediation which includes various treatment methods like phytotransformation, phytodegradation and phytovolatilization (Zouboulis and Moussas 2011). Usually, the mechanism of phytoremediation involves uptake and degradation of organic pollutants in the plant tissues making them less toxic or harmless. Microbial bioremediation is pollutant degradation using bacteria, fungi and microalgae which is discussed in detail in the following sections of this chapter.

10.5 Factors Influencing Petroleum-Contaminated Soil Bioremediation

The contaminated soil has many factors that lead to inefficient bioremediation. There are challenges related to bioremediation when the process is taking place in bulk and in situ conditions, various environmental factors affect the process. There are many parameters which ascend the bioremediation among which some are: oxygen content in the soil, nutrient availability, pH and temperature, while environmental factors affecting the degradation of the oil are the microbial consortium, the aggregation of the oil contaminants and type of soil for example if the soil is loamy, clay or black soil (Koshlaf and Ball 2017). The factors affecting the whole process are listed below.

10.5.1 Temperature

It is the most important factor affecting the whole process of oil degradation. The impact of temperature is observed on the structural complexity of the oil and the microbial consortium which degrades the oil. It influences gas solubilities, influences microbial development rate, soil lattice and metabolic activity of the microorganisms, physical and certain kinds of toxin produced by microorganisms (Varjani and Upasani 2017).

10.5.2 Nutrients

The whole microbial growth depends on the availability of the nutrients, where carbon and nitrogen are most important nutrients for a microbial cell. Since here the microorganisms are dealing with oil pollutants it is critical to find nutrients in the bioremediation. The phosphorus

Table 10.2 Common bioremediation methods (Zouboulis and Moussas 2011)

Bioaugmentation	Ex situ method; inoculation of microbial culture in contaminated soils
Bioreactors	Ex situ method; reactor set up with controlled conditions to stimulate microbial bioremediation of polluted media
Bioventing	In situ method; Oxygen is passed through contaminated soils in low flow rates to enhance growth of aerobic microbiota
Composting	Both in situ and ex situ; Piled organic waste is subjected to elevated temperature with provision of air and moisture to stimulate microbial growth
Land farming	Ex situ- Performed in biotreatment cells In situ-Performed on the upper soil zone

and nitrogen are important for the creation of biomass.(Varjani and Upasani 2017) but if there is excess amount of nitrogen, potassium and phosphorus present in the soil it may also, lead to negative impacts on the oil degrading microorganisms (Koshlaf and Ball 2017). Metabolic results obtained about carbon and nitrogen sources, during biodegradation restrained essentially the oxidation of oil when their focuses on the critical levels of oil degradation (Chaillan et al. 2006).

10.5.3 Bioavailability and Biosurfactants

10.5.3.1 Bioavailability

The degradation of the oil increases if the amount of the oil pollutant in the soil is biologically available to the microorganisms (Koshlaf and Ball 2017). The availability of substrate surface territory for cell connection would be the restricting component in substrate take-up for this situation (Goswami and Singh 1991). The availability of the pollutant also depends upon the oil-microorganism interaction, the state of the pollutant in which they are present in the soil. The chemical factors responsible for availability are hydrophobicity and volatility (Varjani and Upasani 2017).

10.5.3.2 Biosurfactants

The emerging category of environmental cleaners, these are secondary metabolites which reduce the opposite interaction between two phases of the pollutant for example: gases-liquid phase or liquid-solid phase. It helps in increment of bioavailability of the oil pollutants. These are amphibolic molecules especially helpful in microbial growth and are nontoxic and have higher biodegradability rate. These metabolites also help in reducing the surface tension of the pollutants (Varjani and Upasani 2017).

10.5.3.3 Oxygen Content and Movement

Oxygen content always depends on the soil type and the moisture levels of the soil. The significant amount of oxygen and oxygenases required for the respiration cycle in the ensuing debasement pathway of the hydrocarbons (Koshlaf and Ball 2017). The key to a proper and successful biodegradation is to restrict the contribution of oxygen adequately so as to forestall the combination of the high-impact biodegradative pathways for raw petroleum (Kristanti 2011).

10.5.3.4 Ecological Toxicity

The bioremediation has toxic effect on the microorganisms, combination of the bioassays with chemical monitoring for assessing the bioremediation effectiveness and evaluating the debased/remediated soils is recommended for preventing the eco-toxicity. Ecological toxicity increases the amount of organic matter in the soil affecting the microbial uptake of oxygen which leads to inefficient microbial degradation of the oil (Koshlaf and Ball 2017).

10.6 Petroleum Bioremediation Using Microorganism and Mechanism

10.6.1 Bacteria and Archaea

One of the primary mechanisms of petroleum decontamination is remediation by natural microflora, especially by bacteria. Bacteria have the capability of transforming hazardous petroleum parent compounds to much less toxic products like CO_2, water and salts, etc. by different metabolic pathways. Bacterial bioremediation has been a topic of interest amongst researchers for more than 2 decades and associated technological advances have been of absolute import in various industries in recent times. The most common bacterial strains that are used for petroleum bioremediation belong to *Comamonas*, *Pseudomonas*, *Sphingomonas*, *Streptomyces* and *Bacillus* genera (Xu et al. 2018). The organic compounds in petroleum contamination are usually degraded by aerobic chemo-organotrophic bacteria that are natural habitats of soils. These include *Xanthomonas* sp., *Pseudomonas* sp., *Corynebacterium* sp., etc. (Shokrollahzadeh et al. 2008). Bacteria can perform bioremediation as pure cultures, consortia and endophytic relation with plants. Some bacterial strains carry out efficient degradation of saturated and aromatic fractions of petroleum as isolates/pure cultures. The most common ones are *Stenotrophomonas acidaminiphila, Bacillus megaterium, Pseudomonas aeruginosa* and *Bacillus cerues* with 91.7%, 89%, 86.7% and 88.4% biodegradation efficiencies respectively (Cerqueira et al. 2011). Another bacterial strain, *Comamonas* sp. JB cells has shown great degrading action against Benzene, Toluene, Ethylbenzene, Xylene (BTEX) compounds present in petrochemical waste. In this approach, the *Comamonas* sp. JB cells strain is immobilized using Fe_2O_3 nanoparticles as biocatalysts that resulted in almost complete degradation of xylene and phenol (Jiang et al. 2015). Recently, the aerobic denitrifying bacteria isolated from sewage treatment pool was grown on heterotrophic nitrification media (HNM) to understand ammonia reduction potential and then tested to degradation ability of petroleum contamination. The bacteria isolated in this study were *Pseudomonas, Acinetobacter and Sphingomonas* out of which a novel strain of *Sphingomonas* (YY2) showed surprisingly excellent petroleum degradation properties (Lang et al. 2019). It has been observed that bacterial consortium show a synergistic effect in treating pollutants for example consortium of Actinobacteria and Proteobacteria along with rhamnolipid inoculation give total degradation of organic petroleum pollutants (Xue et al. 2020). Recent studies revealed that mixed culture of *Pseudomonas aeruginosa* and *Serratia marcescens* has the capability of flourishing in PAHs contaminated soil and can degrade upto 97% of phenanthrene and 98% of fluorine (Fathi and Ebrahimipour 2018). On the other hand, studies on endophytic bacteria, shown that they promote plant growth and act as remediation agents (Gupta et al. 2020). Endophyte strain of *Bacillus* sp. SBER3 with host plant *Populus deltoids* have been shown to degrade anthracene and benzene whereas *Staphylococcus* sp. BJ106 of host plant *Alopecurus* biodegrades toxic pyrene (Feng et al. 2017). Although bacteria are quite efficient in degrading contaminants, members of Archaea are known to remediate heavy metals and PAHs. *Haloarcula* sp., *Haloferax* sp. and *Halobacterium salinarum* have been found to degrade anthracene, benzene and crude oil respectively (Krzmarzick et al. 2018).

10.6.1.1 Mechanism of Action

Bacteria are tolerant to many stress conditions including increase in pollutant concentrations. One of the ways they do so is by increasing membrane lipids which results in lower membrane fluidity that thwarts the fluidizing state caused due to pollutants. The microbes also produce stress proteins, like in *Bacillus* and *Escherichia coli,* that help the cells remain viable under toxic environments (Katarína et al. 2018). The bacteria and archaea get acclimatized in such environments that start expression of genes that produce enzymes like oxygenases and hydrogenases which use compounds like pyrene and fluoranthene as carbon sources degrade them and ultimately thrive (Liu et al. 2017). Archaea bacteria like *Sulfolobus metallicus*, which is a chemolithotroph, has the ability to oxidize petroleum toxicants but the co-relation between Archaea and oil contamination is still unclear (Krzmarzick et al. 2018).

10.6.2 Fungi

The bioremediation using fungi is called mycoremediation. Mycoremediation is one of the cost-effective methods, it is also known as the green method, i.e. it is an ecologically friendly method. Various factors such as fast growth, tremendous hyphal network, secretion of lignolytic enzymes, flexibility to fluctuating pH, temperature and presence of metal-restricting proteins make fungi capable of remediation (Akhtar and Mannan 2020). Growths can make an assortment of natural surroundings with complex soil network filling in as the significant area for fungal colonization (Deshmukh et al. 2016). The bacterial degradation is a far slower process as compared to fungal degradation due to higher molecular weight of the crude oil (Akhtar and Mannan 2020). Hence, mycoremediation is preferable in case of crude oil contaminated soil. The fungal growth always leads to a better soil structure, providing better bioavailability (Li et al. 2020). Prominent fungal species used for crude oil contaminated soils among the studies were observed to be *Pleurotus ostreatus, Pleurotus eryngi, A. niger, Rhizopus* sp., *Candida* sp., *Penicillium* sp., *Mucor* sp. and *Ganoderma lucidum* (Pozdnyakova 2012; Deshmukh et al. 2016; Akhtar and Mannan 2020; Li et al. 2020) (Table 10.3). The above-mentioned fungal strains/species degrade aliphatic hydrocarbons, polycyclic aromatic hydrocarbons and chlorophenols. Fungal strains are fit for remediating oil hydrocarbons by secreting various enzymes for example laccases, tyrosinases, manganese peroxidases, cytochrome P450 monooxygenases, reductive dehalogenases (Li et al. 2020). The ligninolytic protein framework assumes a vital part in the beginning advance of PAH corruption by white-decay and litter decomposing fungi. The subsequent metabolites are more dissolvable and can be taken inside the cell, where extraordinary intracellular proteins (e.g. cytochrome P-450) can act. The principle inconsistency is that PAH degradation happens previously in extracellular protein creation. This logical inconsistency can be tackled if one considers the presence of a mycelia surface-bound LAC pool, which might be engaged with the beginning phases of PAH degradation (Pozdnyakova 2012).

10.6.2.1 Mechanism of Action

Fungal biodegradation with extracellular enzymes: hydrocarbons can sometimes provide carbon as energy source to fungal species in the surrounding area of oil contamination. Fungal cell films are penetrable to oil hydrocarbons or organic compounds oxidized by extracellular chemicals, which can go through additional metabolism including hydrolysis, dehalogenation, β-oxidation and passage into the TCA cycle (Li et al. 2020). In the recent discussion, it has been found that some non-lignolytic fungal strains are capable of degrading oil contaminants, *Dentipellis sp.* KUC8613 uses cytochrome P450 and other enzymes such as dehydrogenases, FAD-dependent monooxygenases and dioxygenases for biodegradation, these enzymes enlisted above are extracellular enzymes working on extracellular degradation and then these convert hydrocarbons into quinone, which further is broken and processed into ring fission (Fig. 10.1).

The limiting factor for mycoremediation is the reduced bioavailability, due to hydrophobic nature of the hydrocarbons, which leads to slow transportation across the membrane. This limitation can be solved by forming symbiotic associations with the bacteria present in the soil, the fungal strains can transport crude oil to these bacteria. A fungus secretes emulsifying agents, which increase the bioavailability and degradation by increment in the solubility. To increase the bioavailability chitin and cellulose can be added to the soil, but excess amounts of these

Table 10.3 Fungal strains and the compounds degraded

Fungi used in bioremediation	Compounds degraded by fungi	Reference
Penicillium sp.	Decane	Li et al. (2020)
Pleurotus ostreatus	Anthracene	Li et al. (2020)
Polyporus sp.	Chrysene	Li et al. (2020)
Coriolopsis gallica	2-methylanthracene; Acenaphthene, carbazole N-ethyl carbazole	Pozdnyakova (2012)
Aspergillus spp.	Benzo[a]pyrene, Pyrene, Phenanthrene, Naphthalene, Crude oil	Al-Hawash et al. (2018a, 2018b)
Ganoderma lucidum	Acenaphthene; Acenaphthylene	Pozdnyakova (2012)
Exophiala xenobiotica	Gasoline	Deshmukh et al. (2016)
T. versicolor	benzo[b]fluoranthene; benzo[k]fluoranthene	Pozdnyakova (2012)

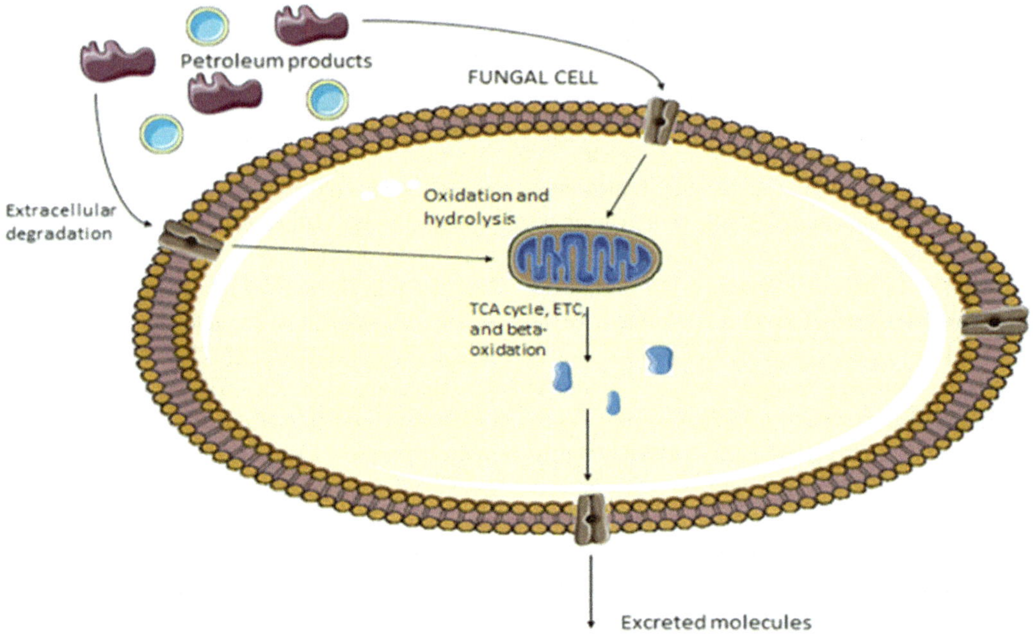

Fig. 10.1 Fungal mechanism of oil degradation

agents can lead to reduced uptake of carbon source from hydrocarbons (Akhtar and Mannan 2020).

10.6.3 Microalgae

The remediation of polluted media using algae or phycoremediation is being studied across the globe not just for waste treatment but also for procuring value added products from the process. The photosynthetic eukaryotic organisms called algae, are primarily found in two forms namely microalgae (microscopic; a few microns long) like *Chlorella* sp. and *Spirulina* sp. and macroalgae like seaweeds (Khan et al. 2018). This chapter will be primarily focusing on bioremediation of petroleum contaminated soil mediated by microalgae and their mechanism of action.

Various species of microalgae like *Selenastrum capricornutum*, *Chlamydomonas* sp., *Scenedesmus* sp. is used for degradation of polycyclic aromatic hydrocarbons (PAHs) like pyrene. In over 7 days, *S. capricornutum* was found to have degraded about 78% of total PAHs including fluoranthene and pyrene (Mondal et al. 2019). *Chlorella vulgaris*, when treated with 10–20 g/L crude oil/water concentrations for a fortnight, shows about 94% and 88% remediation of light and heavy crude oil hydrocarbons respectively. This can be credited to the resistance build-up towards these toxic hydrocarbons which also resulted in increased dry weight of the species (Xaaldi Kalhor et al. 2017). A comparative study between natural attenuation and biostimulation to increase remediation rates of diesel contaminated soils was conducted using *Spirulina platensis* biomass extracts. Around 88% diesel removal was recorded in the contaminated soil that was biostimulated using phycocyanin (emulsifier from *S. platensis*) (Decesaro et al. 2017). Though phycoremediation is environment friendly and economically feasible, there are a few setbacks. Sometimes the pollutant and intermediate compounds show a toxic effect on some microalgal species, for example, *Chlamydomonas angulosa* is sensitive to naphthalene molecules that result in death of around 90% of the algal cells within a span of 24 h. Efficiency of microalgal remediation is species specific which often results in poor removal rates, for example, *C. vulgaris* efficiently removes other petroleum compounds but tends to remove just 48% of PAHs (Mondal et al. 2019).

10.6.3.1 Mechanism of Action

Some cellular mechanisms of microalgae have developed over the years and have made some species tolerant to toxicants like PAHs, heavy metals, toluene, etc. Two of the most common remediation mechanisms used by microalgae are bioconcentration and biosorption. Biosorption is an adsorption phenomenon in which dead algal biomass adsorbs toxicants like metal ions whereas live biomass accumulates pollutants in case of bioconcentration. The mechanism is associated with presence of surface proteins in the cell membrane that has high affinity towards heavy metal ions which occurs by chelation, ion-exchange or complex formation (Ahemad and Kibret 2013). Some strains like *Chlamydomonas reinhardtii*, *Synechococcus leopoliensis* and *Cyanidioschyzon merolae* possess metabolic mechanism that converts toxic heavy metal cadmium to its sulphide form which is less harmful due to its low solubility in water. Some autotrophic microalgal species possess pigments that have photon-capturing machinery in which generated ATP to use in dark reactions in which CO_2 is digested. Toluene gets converted to acetylCo-A/pyruvate via cleavage pathways (Fig. 10.2) (Gupta et al. 2019). These compounds further take part in TCA and thus toluene is biodegraded (Hammed et al. 2016). Algae also facilitate bacterial remediation of phenolics by providing oxygen that is produced as a result of algal photosynthesis. There are many more phycoremediation phenomena that are under study and some of the recent advances are discussed further in this chapter.

10.7 Limitation of Bioremediation

Along with crude oil the soil is contaminated with heavy metals like mercury which hampers the soil texture as well as the efficiency of the microorganisms to bioremediate. When it comes to the use of biosurfactants, they inhibit the ability of the pollutant to transform from one phase to another leading to biosurfactant toxicity. The enzyme-soil interactions, can affect the enzyme activity of the enzymes secreted by microorganisms during remediation. The consequence of this can be a shift in the pH of the optimum catalytic activity of the enzymes participating in the biodegradation, leading to decreased efficiency of the bioremediation. This problem can be solved by keeping the track of the organic composition of the polluted soil and then optimizing them. Sometimes enzymes used for bioremediation can undergo post translational modification, which obviously lowers the bioremediation ability of the microorganisms (Mougin et al. 2009). Anaerobic remediation requires

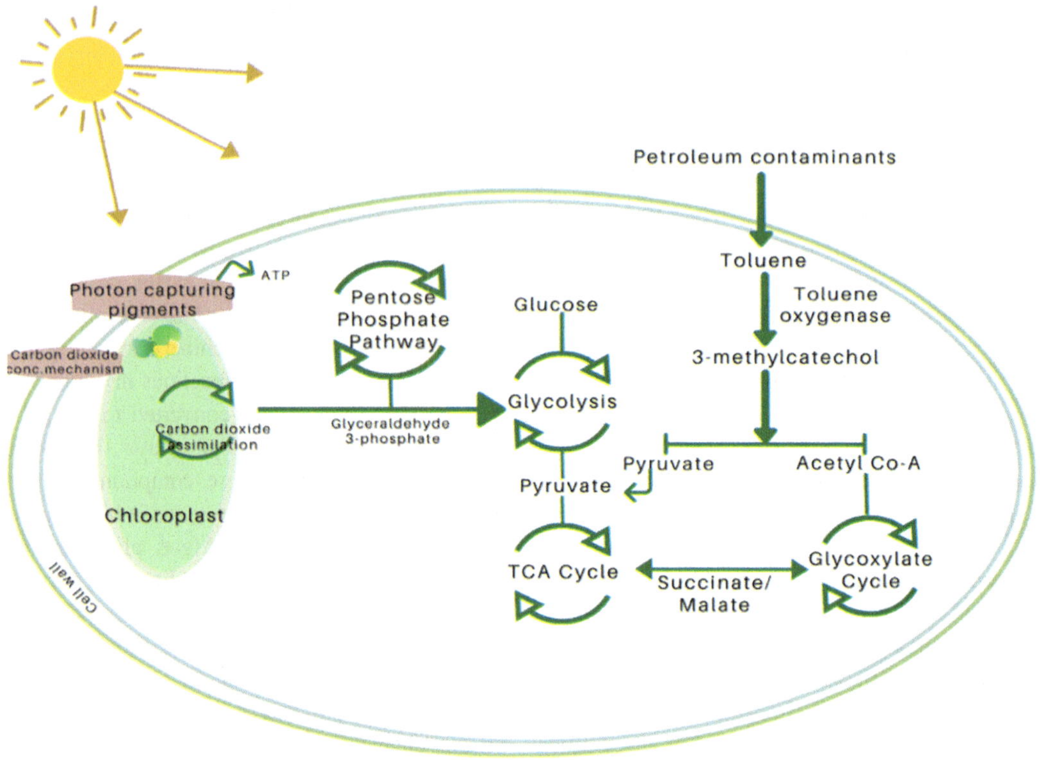

Fig. 10.2 Toluene biodegradation pathway in microalgal cell

longer start up time. These are more susceptible to toxic substance. The anaerobic remediation has the ability to produce bad odour and corrosive agents. The removal of nitrogen and phosphorus is inhibited in anaerobic respiration. To remove these elements, the soil needs to undergo the process of alkylation.

The biological remediation may predict different results when performed in a lab, the nature of remediation differs in situ process of remediation. In situ bioremediation requires a continuous monitoring system, with maintenance of all the environmental factors, which is a cumbersome task for the developers of bioremediation (Perelo 2010). Bioremediation may lead to degradation of the oil pollutants, but certain amount of pollutants cannot be degraded hence, there is no guarantee of the soil to be completely clean. Sometimes microorganism has a tendency to release secondary metabolites which leads to more toxicity than the pollutant present in the soil. Thus, when it comes to the microbial degradation of volatile compounds like oil, the bioremediation process is not so effective. This is because microorganisms can degrade the pollutants present in one phase. For example, if all the pollutants in the soil are present in liquid phase the mechanism of biodegrading used by microorganisms will be for liquid phase. But in the soil, the pollutants are present in all the phases in mixture, i.e. solid, liquid and gases, which hampers the mechanism used by microorganisms. There are no such techniques developed yet to solve this limitation (Vidali 2001).

10.8 Recent Advances and Rising Technologies

The very focus of all the studies related to the bioremediation of oil contaminated soil has always been on the bacterial consortium used in degradation of the oil. But there is a void in the field of research on microbial matter.

The parameters mainly observed for efficiency of bioremediation are; (1) nutrient content of the soil for example nitrogen phosphorus and the amount of dissolved oxygen in the soil, (2) the augmented oil degrading microorganisms and (3) the impact of other environmental factors. Hence, the current remediation techniques mainly focus on the oil degrading methods which are based on optimization of the parameters (Lim et al. 2016). According to a study by (Martínez Álvarez et al. 2015) the level of phosphorus, carbon and nitrogen was adjusted to develop a pilot scale study of the diesel contaminated soil. The result stated that the carbon, nitrogen and phosphorus ratio when adjusted to 100:17.6:1.73, gave the diesel degradation efficiency of 54.9% while unadjusted C:N:P of 100:10:1 only had efficiency of 27.8% which apparently substantiated the point of parameter adjustment. The study also focuses on the parameter of the temperature stating that incubation temperature of the microorganism if adjusted can lead to more effective oil degradation.

Anaerobic bioremediation is a technique which has recently been in focus of several researchers. This technique of remediation basically focuses on the usage of sulphate, nitrate, CO_2 and iron as substitution to oxygen by microorganisms. The method being cost effective prevent the oxygen to be present at the site of remediation (Lim et al. 2016). Other new strategies that could quicken the oil degradation rate incorporate the investigation of a novel supplement application system for the disguising of supplements that could boost supplement home time which could be utilized for all oil contaminated seashores if the seashore has the oil contamination. Maximizing the living arrangement season of supplements in this zone is the key to achieve quick and practical remediation. The utilization of the nutrients arrangement should start following the elevated tide and should keep going for a large portion of a flowing cycle, for efficient remediation (Li et al. 2007).

Since the microbial consortia used in the bioremediation is congruent it is okay to use the combination of the different microorganisms to enhance the efficiency of the bioremediation the utilization of individual microbial consortia and developed compost indicated comparable evacuation efficiencies of 55% and 52% separately, yet the mix of microbial consortia and developed compost gave a critical increment in evacuation productivity up to 82% by diminishing the amount of oil in the soil. But further research with various sorts of soil networks at field scale still needs to be concentrated to evaluate use of these methodologies under cold weather. In this manner, rather than landfill and destructive treatment techniques, for example, cremation, the utilization of mature compost and consortium immunization for bioremediation in cold weather advances soil manageability and re-utilization of oil hydrocarbons sullied soil (Gomez and Sartaj 2013).

10.9 Conclusion

Petroleum pollution of soil has become a global issue due to rapid growth in industrialization and reckless waste disposal. Petroleum pollutants can easily enter human body through topical or ingestion routes resulting in adverse health conditions. Due to secondary pollution associated with conventional physical and chemical treatments, decontaminating should be rely upon usage of biological reagents for remediation of toxicants like PAHs and heavy metals. Microorganisms are versatile mini-factories that can ingest, digest and transform pollutants like toluene, benzene, pyrene, etc. to their nontoxic forms just by the usage of their natural metabolic machinery. Although microbial bioremediation is eco-friendly and cost effective it is quite a time consuming and requires technological advancements for increased efficiency of the process. Recent advances involving genetic engineering, technological upgradation, use of nanotechnology, etc. have shown hope that microbial bioremediation can completely fix petroleum associated pollution problems on a large scale in the near future.

References

Ahemad M, Kibret M (2013) Recent trends in microbial biosorption of heavy metals: a review. Biochem Mol Biol J 1:19–26. https://doi.org/10.12966/bmb.06.02.2013

Akhtar N, Mannan MA (2020) Mycoremediation: expunging environmental pollutants. Biotechnol Rep (amst) 9(26):e00452. https://doi.org/10.1016/j.btre.2020.e00452

Al-Hawash AB, Alkooranee JT, Zhang X, Ma F (2018) Fungal degradation of polycyclic aromatic hydrocarbons. Int J Pure Appl Biosci 6(2):8–24

Al-Hawash AB, Alkooranee JT, Zhang X, Ma F (2018) Fungal degradation of polycyclic aromatic hydrocarbons. Int J Pure App Biosci 6:8–24. https://doi.org/10.18782/2320-7051.6302

Amellal N, Portal JM, Berthelin J (2001) Effect of soil structure on the bioavailability of polycyclic aromatic hydrocarbons within aggregates of a contaminated soil. Appl Geochem 16:1611–1619. https://doi.org/10.1016/S0883-2927(01)00034-8

Borowik A, Wyszkowska J, Kucharski M, Kucharski J (2019) Implications of soil pollution with diesel oil and BP petroleum with ACTIVE technology for soil health. Int J Environ Res Public Health 16(14):2474. https://doi.org/10.3390/ijerph16142474

Cerqueira VS, Hollenbach EB, Maboni F et al (2011) Biodegradation potential of oily sludge by pure and mixed bacterial cultures. Bioresour Technol 102(23):11003–11010. https://doi.org/10.1016/j.biortech.2011.09.074

Chaillan F, Chaîneau CH, Point V et al (2006) Factors inhibiting bioremediation of soil contaminated with weathered oils and drill cuttings. Environ Pollut 144(1):255–265. https://doi.org/10.1016/j.envpol.2005.12.016

Decesaro A, Rampel A, Machado TS et al (2017) Bioremediation of soil contaminated with diesel and biodiesel fuel using biostimulation with microalgae biomass. J Environ Eng 143:04016091. https://doi.org/10.1061/(ASCE)EE.1943-7870.0001165

Deshmukh R, Khardenavis AA, Purohit HJ (2016) Diverse metabolic capacities of fungi for bioremediation. Indian J Microbiol 56:247–264. https://doi.org/10.1007/s12088-016-0584-6

Devatha CP, Vishnu Vishal A, Purna Chandra Rao J (2019) Investigation of physical and chemical characteristics on soil due to crude oil contamination and its remediation. Appl Water Sci 9:89. https://doi.org/10.1007/s13201-019-0970-4

Fathi Z, Ebrahimipour G (2018) Isolation and identification of polycyclic aromatic hydrocarbons (pahs) degrading bacteria from arak petrochemical wastewater. J Microbiol Biotechnol Food Sci 7:454–459. https://doi.org/10.15414/jmbfs.2018.7.5.454-459

Feng NX, Yu J, Zhao HM, Cheng YT, Mo CH, Cai QY, Li YW, Li H, Wong MH (2017) Efficient phytoremediation of organic contaminants in soils using plant-endophyte partnerships. Sci Total Environ 583:352–368. https://doi.org/10.1016/j.scitotenv.2017.01.075

Gao Y, Zygourakis K (2019) Kinetic study of the pyrolytic treatment of petroleum-contaminated soils. Ind Eng Chem Res 58:10829–10843. https://doi.org/10.1021/acs.iecr.9b01494

Gargouri B, Karray F, Mhiri N et al (2014) Bioremediation of petroleum hydrocarbons-contaminated soil by bacterial consortium isolated from an industrial wastewater treatment plant: bioremediation of petroleum hydrocarbons-contaminated soil. J Chem Technol Biotechnol 89:978–987. https://doi.org/10.1002/jctb.4188

Gomez F, Sartaj M (2013) Field scale ex-situ bioremediation of petroleum contaminated soil under cold climate conditions. Int Biodeterior Biodegradation 85:375–382. https://doi.org/10.1016/j.ibiod.2013.08.003

Goswami P, Singh HD (1991) Different modes of hydrocarbon uptake by two *Pseudomonas* species. Biotechnol Bioeng 37:1–11. https://doi.org/10.1002/bit.260370103

Gupta A, Singh SK, Singh VK et al (2020) Endophytic microbe approaches in bioremediation of organic pollutants. In: Microbial endophytes. Elsevier, pp 157–174

Gupta P, Ranjan S, Gupta S (2019) Phycoremediation of petroleum hydrocarbon-polluted sites: application, challenges and future prospects. In: Application of microalgae in wastewater treatment, vol 1. Springer, Cham, pp 145–162

Hammed AM, Prajapati SK, Simsek S et al (2016) Growth regime and environmental remediation of microalgae. Algae 31(3):189–204. https://doi.org/10.4490/algae.2016.31.8.28

Jiang B, Zhou Z, Dong Y et al (2015) Bioremediation of petrochemical wastewater containing BTEX compounds by a new immobilized bacterium *Comamonas* sp. JB in magnetic gellan gum. Appl Biochem Biotechnol 176:572–581. https://doi.org/10.1007/s12010-015-1596-0

Katarína D, Slavomíra M, Hana D et al (2018) The adaptation mechanisms of bacteria applied in bioremediation of hydrophobic toxic environmental pollutants: how indigenous and introduced bacteria can respond to persistent organic pollutants-induced stress? In: Donyinah SK (ed) Persistent organic pollutants. IntechOpen, London, pp 71–97. https://doi.org/10.5772/intechopen.79646

Khan MI, Shin JH, Kim JD (2018) The promising future of microalgae: current status, challenges, and optimization of a sustainable and renewable industry for biofuels, feed, and other products. Microb Cell Fact 17:36. https://doi.org/10.1186/s12934-018-0879-x

Khanna S, Gharpure AS (2017) Petroleum carcinogenicity and aerodigestive tract in context of developing nations. Cureus 9(4):e1202. https://doi.org/10.7759/cureus.1202

Kong SH, Watts RJ, Choi JH (1998) Treatment of petroleum-contaminated soils using iron mineral. Chemosphere 37(8):1473–1482

Koshlaf E, Ball AS (2017) Soil bioremediation approaches for petroleum hydrocarbon polluted environments. AIMS Microbiol 3(1):25–49. https://doi.org/10.3934/microbiol.2017.1.25

Kristanti RA (2011) Bioremediation of crude oil by white rot fungi Polyporus sp. S133. J Microbiol Biotechnol 21:995–1000. https://doi.org/10.4014/jmb.1105.05047

Krzmarzick M, Taylor D, Fu X, McCutchan A (2018) Diversity and niche of archaea in bioremediation. Archaea 2018:3194108. https://doi.org/10.1155/2018/3194108

Lang X, Li Q, Xu Y et al (2019) Aerobic denitrifiers with petroleum metabolizing ability isolated from caprolactam sewage treatment pool. Biores Technol 290:121719. https://doi.org/10.1016/j.biortech.2019.121719

Li H, Zhao Q, Boufadel MC, Venosa AD (2007) A universal nutrient application strategy for the bioremediation of oil-polluted beaches. Mar Pollut Bull 54 (8):1146–1161. https://doi.org/10.1016/j.marpolbul.2007.04.015

Li Q, Liu J, Gadd GM (2020) Fungal bioremediation of soil co-contaminated with petroleum hydrocarbons and toxic metals. Appl Microbiol Biotechnol 104:8999–9008. https://doi.org/10.1007/s00253-020-10854-y

Lim MW, Lau EV, Poh PE (2016) A comprehensive guide of remediation technologies for oil contaminated soil—present works and future directions. Mar Pollut Bull 109:14–45. https://doi.org/10.1016/j.marpolbul.2016.04.023

Liu SH, Zeng GM, Niu QY et al (2017) Bioremediation mechanisms of combined pollution of PAHs and heavy metals by bacteria and fungi: a mini review. Bioresour Technol 224:25–33. https://doi.org/10.1016/j.biortech.2016.11.095

Martínez Álvarez LM, Lo Balbo A, Mac Cormack WP, Ruberto LAM (2015) Bioremediation of a petroleum hydrocarbon-contaminated Antarctic soil: Optimization of a biostimulation strategy using response-surface methodology (RSM). Cold Reg Sci Technol 119:61–67. https://doi.org/10.1016/j.coldregions.2015.07.005

Mondal M, Halder G, Oinam G et al (2019) Bioremediation of organic and inorganic pollutants using microalgae. In: New and future developments in microbial biotechnology and bioengineering. Elsevier, pp 223–235

Mougin C, Boukcim H, Jolivalt C (2009) Soil bioremediation strategies based on the use of fungal enzymes. In: Singh A, Kuhad RC, Ward OP (eds) Advances in applied bioremediation. Springer, Berlin Heidelberg, Berlin, Heidelberg, pp 123–149

Ouriache H, Arrar J, Namane A, Bentahar F (2019) Treatment of petroleum hydrocarbons contaminated soil by Fenton like oxidation. Chemosphere 232:377–386. https://doi.org/10.1016/j.chemosphere.2019.05.060

Perelo LW (2010) Review: in situ and bioremediation of organic pollutants in aquatic sediments. J Hazard Mater 177(1–3):81–89. https://doi.org/10.1016/j.jhazmat.2009.12.090

Pozdnyakova NN (2012) Involvement of the ligninolytic system of white-rot and litter-decomposing fungi in the degradation of polycyclic aromatic hydrocarbons. Biotechnol Res Int 2012:1–20. https://doi.org/10.1155/2012/243217

Rahman ZA, Hamzah U, Ahmad N (2010) Geotechnical characteristics of oil-contaminated granitic and metasedimentary soils. Asian J Appl Sci 3:237–249. https://doi.org/10.3923/ajaps.2010.237.249

Scherr K, Aichberger H, Braun R, Loibner AP (2007) Influence of soil fractions on microbial degradation behavior of mineral hydrocarbons. Eur J Soil Biol 43:341–350. https://doi.org/10.1016/j.ejsobi.2007.03.009

Shokrollahzadeh S, Azizmohseni F, Golmohammad F et al (2008) Biodegradation potential and bacterial diversity of a petrochemical wastewater treatment plant in Iran. Bioresour Technol 99(14):6127–6133. https://doi.org/10.1016/j.biortech.2007.12.034

Souza RB de, Maziviero TG, Christofoletti CA et al (2013) Soil contamination with heavy metals and petroleum derivates: impact on edaphic fauna and remediation strategies. In: Soriano (ed) Soil processes and current trends in quality assessment. IntechOpen. https://doi.org/10.5772/52868

Streche C, Cocârţă DM, Istrate I-A, Badea AA (2018) Decontamination of petroleum-contaminated soils using the electrochemical technique: remediation degree and energy consumption. Sci Rep 8(1):3272. https://doi.org/10.1038/s41598-018-21606-4

Sutton NB, Maphosa F, Morillo JA et al (2013) Impact of long-term diesel contamination on soil microbial community structure. Appl Environ Microbiol 79:619–630. https://doi.org/10.1128/AEM.02747-12

Tahseen R, Afzal M, Iqbal S et al (2016) Rhamnolipids and nutrients boost remediation of crude oil-contaminated soil by enhancing bacterial colonization and metabolic activities. Int Biodeterior Biodegradation 115:192–198. https://doi.org/10.1016/j.ibiod.2016.08.010

Varjani S, Joshi R, Srivastava VK et al (2020) Treatment of wastewater from petroleum industry: current practices and perspectives. Environ Sci Pollut Res 27:27172–27180. https://doi.org/10.1007/s11356-019-04725-x

Varjani SJ, Gnansounou E, Pandey A (2017) Comprehensive review on toxicity of persistent organic pollutants from petroleum refinery waste and their degradation by microorganisms. Chemosphere 188:280–291. https://doi.org/10.1016/j.chemosphere.2017.09.005

Varjani SJ, Upasani VN (2017) A new look on factors affecting microbial degradation of petroleum hydrocarbon pollutants. Int Biodeterior Biodegradation 120:71–83. https://doi.org/10.1016/j.ibiod.2017.02.006

Vidali M (2001) Bioremediation: an overview. Pure Appl Chem 73(7):1163–1172. https://doi.org/10.1351/pac200173071163

Xaaldi Kalhor A, Movafeghi A, Mohammadi-Nassab AD et al (2017) Potential of the green alga *Chlorella vulgaris* for biodegradation of crude oil hydrocarbons. Mar Pollut Bull 123(1–2):286–290. https://doi.org/10.1016/j.marpolbul.2017.08.045

Xu X, Liu W, Tian S et al (2018) Petroleum hydrocarbon-degrading bacteria for the remediation of oil pollution under aerobic conditions: a perspective analysis. Front Microbiol 9:2885. https://doi.org/10.3389/fmicb.2018.02885

Xue S, Huang C, Tian Y et al (2020) Synergistic effect of rhamnolipids and inoculation on the bioremediation of petroleum-contaminated soils by bacterial consortia. Curr Microbiol 77:997–1005. https://doi.org/10.1007/s00284-020-01899-3

Yuniati MD (2018) Bioremediation of petroleum-contaminated soil: a review. IOP Conf Ser: Earth Environ Sci 118:012063. https://doi.org/10.1088/1755-1315/118/1/012063

Zhang J, Fan S, Yang J et al (2014) Petroleum contamination of soil and water, and their effects on vegetables by statistically analyzing entire data set. Sci Total Environ 476–477:258–265. https://doi.org/10.1016/j.scitotenv.2014.01.023

Zouboulis AI, Moussas PA (2011) Groundwater and soil pollution: bioremediation. In: Nriagu JO (ed) Encyclopedia of environmental health. Elsevier, Burlington, pp 1037–1044

Phytoremediation of Radioactive Contaminated Sites

11

Iram Gul, Imran Ahmad, and Ghulam Mujtaba Shah

Abstract

Radioactive substances are widely used in different sectors including industries, agriculture, energy sector, food industries, and medical sector for the improvement of mankind's lifestyle. The major sources of radioactive substances are nuclear weapon production and testing, industrial processes like smelting of metals, mining, and research laboratories. Due to the excessive use of these substances in industries and other sectors, they contaminate the soil as well as water resources. The management of these contaminants is challenging as radioactive substance required certain time for decaying. Therefore, it is necessary to use some environment friendly methods for the removal of these contaminants from soil for the betterment of human. Among various techniques, phytoremediation is proven to be eco-friendly and realistic technique for the decontamination of radioactive contaminated sites. In this chapter, major radioactive substances, their sources, and exposure pathway in environment and major impacts on ecosystem is discussed. Further, the different remediation techniques are also highlighted, but the focus is on phytoremediation.

Keywords

Chelating agents · Contaminated sites · Mobilization · Phytoremediation · Radioactive substances

11.1 Introduction

Due to the increase in industrialization and urbanization, technological advancement has also been increased to fulfill the community demands. The unplanned development in all sectors, resulted in the release of large amount of environmental contaminants including antibiotics, hydrocarbons, heavy metals, and radioactive substances (Mukhtar et al. 2020; Arshad et al. 2020; Gul et al. 2020). All the contaminants from different sources affect the normal functioning of ecosystem, and most of them are either carcinogenic, mutagenic, or genotoxic. Among these contaminants, radioactive substances are of serious concern due to their long half-life, carcinogenic, and genotoxic nature (Yan et al. 2020; Gupta et al. 2016). The radioactive substances are released from the natural sources including

I. Gul (✉) · G. M. Shah
Department of Earth and Environmental Sciences, Hazara University, Mansehra, Pakistan
e-mail: iram.k87@gmail.com; iramgul.es@hu.edu.pk

I. Ahmad (✉)
Department of Geology, University of Malakand, Dir (Lower) Chakddara, Pakistan
e-mail: imran_geo@uom.edu.pk; iramgul.es@hu.edu.pk

J. A. Malik (ed.), *Advances in Bioremediation and Phytoremediation for Sustainable Soil Management*,
https://doi.org/10.1007/978-3-030-89984-4_11

cosmic rays and natural reservoirs of these substances. However, large amount of these radioactive substances in environment is resulted from the anthropogenic sources including mining of ores, nuclear weapon testing, nuclear accidents, and use of these substances in research sectors (Schlumberger et al. 2012).

Radioactive substances from different sources enter into air through emission and water through effluents from nuclear powerplants. From the contaminated water and air, they are accumulated in soil, and through food chain, they are consumed by humans. The presence of radioactive substances cause air, water, and soil pollution. The presence of dangerous radioactive substances in environment affects the soil microorganism by damaging their cell membrane, DNA and may also cause death. They also affect the plants by causing necrosis, reduce the plant height, biomass, and chlorophyll content by disturbing the photosynthetic activity. Furthermore, they also cause genotoxicity by damaging the DNA of plant, and long-term exposure to these dangerous substances causes death (Ogwu et al. 2019; Koonin and Wolf 2008). Humans are exposed to these substances directly by inhaling the contaminated air and drinking the contaminated water and through food. The presence of these substances in the body affects the nervous and cardiovascular system and also causes skeletal cancer (Tawalbeh et al. 2013). Therefore, the remediation of radioactive contaminated sites is important. For this purpose, different remediation techniques have been used including soil excavation, washing, and landfill (Dushenkov 2003; Khan et al. 2000). These conventional techniques have some disadvantages; therefore, there is a need for the environment friendly technique for the removal of contaminants from the environment.

Phytoremediation is a green method in which plants are used for the removal of contaminants from the environment. This technique is economically feasible and permanently removes the contaminants from soil and water. The contaminants accumulated in shoots can be handled and disposed of easily as compared to soil. In phytoremediation, different mechanisms such as phytoextraction, phytodegradation, phytostabilization, and phytovolatilization are used. The phytoextraction and phytostabilization are the suitable techniques for the radioactive contaminated sites. The efficient removal of radioactive substances depends on: (i) selection of plants—having capability to accumulate high level of contaminants without phytotoxicity; produced high biomass, (ii) availability of contaminants for plants' uptake. All techniques have some advantages and disadvantages, and therefore, one method is not suitable for all types of contaminants. In this chapter, the sources along with their exposure routes and impact on ecosystem are discussed. Further the phytoremediation technique for the decontamination of radioactive substances is highlighted.

11.2 Sources of Radioactive Substances

Radioactive substances are present naturally, and the amount is increased due to the various anthropogenic activities. The natural sources include: (a) cosmic rays from the outer space—these are the atomic nuclei and high energy proton travel through the space. The cosmic rays are generally composed of nuclei of well-known atoms (99%) and solitary electron (1%). The nuclei atoms are composed of nuclei of heavy elements (1%), alpha particles (9%), and simple protons such as hydrogen nuclei (90%); (b) the emission from the radioactive substance present in the earth crust—large amount of radioactive substances are naturally found in the earth's crust. Uranium (U), thorium (Th), radon (Rn), polonium (Po), radium (Ra), and plumbum (Pb) are the most abundant naturally found substances. The radium and radon gases are released from the decay of uranium and thorium. These naturally occurring radioactive substances are known as naturally occurring radioactive materials (NORM). Apart from these NORM, technologically enhanced naturally occurring radioactive materials (TENORM) are also released in the environment due to anthropogenic activities.

The high concentration of radioactive substances is resulted from the anthropogenic activities. From different anthropogenic activities, radioactive substances enter into environment such as air, water, and soil and cause radioactive contamination. The radioactive contamination also known as radiological contamination is the presence or deposition of undesirable radioactive substances in/on the surface of soil/water/air. The major anthropogenic sources include nuclear weapons, nuclear power plants, transportation and disposal of nuclear waste, mining and processing of radioactive ores, leakage from nuclear powerplants, nuclear accidents, fossil fuel consumption, phosphate fertilizer, use of radioactive substance in medical technology, and research facilities (Yan et al. 2020; Gupta et al. 2016; Jagetiya et al. 2014). In agricultural research, ^{14}C has been used for tracing the changes in substances (Valldor et al. 2015), and ^{131}I has been used for treating the thyroid cancer (Schlumberger et al. 2012). Generally, more than two hundred radioactive substances have been released from the nuclear power plants, and they decay naturally at low level. The large amount of radioactive substances released into atmosphere from the Fukushima and Chernobyl incident highly contaminated the aquatic and terrestrial environment (Yoschenko et al. 2018; Hu et al. 2010). It has been reported that 1.2 × 107 TBq of radioactive substances were released in the atmosphere due to the Chernobyl incident (Hu et al. 2010), and release of radioactive materials due to the Fukushima incident contaminated the lands and oceans (Yan et al. 2020; Steinhauser et al. 2014).

11.3 Major Radioactive Substances

As discussed above, large amount of radioactive substances are present in the environment. Radioactive substances are unstable substances that are decayed naturally and release either gamma radiations or alpha, beta particles. The radiations produced from the decaying of radioactive substances are dangerous and invisible. The alpha particles are slow moving and unable to penetrate in skin but cause damages upon swallowing. Beta particles are the high energy particles, able to penetrate, but cause less damages. The gamma radiations are the extremely high energy radiations and cause serious damage. Large number of radioactive substances are present in environment including Cs, Sr, Th, U, Rn, Pb, and I. However, in this chapter, the main focus is on the common radioactive substances like Cs, Sr, U, and Th present in the environment.

Cesium (Cs) is a silvery-gold and soft metal. This metal has thirty-nine known isotopes, and among these isotopes, only one isotope, i.e., ^{133}Cs, is stable (Moogouei et al. 2017). ^{135}Cs has the longest half-life of 2.3 million years. The ^{134}Cs and ^{137}Cs have half-life of 2 and 30 years, respectively. The presence of these radioactive substances affects the ecosystem because the behavior of ^{133}Cs and ^{137}Cs is similar to the potassium. Therefore, these isotopes are easily taken up by the plants (Tsukada et al. 2002).

Uranium (U) is a silvery-gray metal. Three isotopes, i.e., ^{238}U, ^{235}U, and ^{234}U, are naturally present in the earth's crust, and all isotopes are radioactive. The most stable and abundant isotope is ^{238}U. Naturally, 0.3–1.0 mg kg^{-1} of uranium is present in the earth crust, but the concentration increased to 100 mg kg^{-1} due to the anthropogenic activities (Kabata-Pendias 2011).

Thorium (Th) is a tarnishes black and silvery radioactive metallic chemical. It is naturally occurring substance and is three times more abundant than the uranium. It has total of six unstable isotopes, and among these, only one isotope, i.e., ^{232}Th, is relatively stable with half-life of fourteen billion years. Other thorium isotopes, i.e., ^{230}Th, ^{227}Th, ^{234}Th, ^{229}Th, ^{228Th}, are the trace thorium radioisotopes and produced due to the decay of ^{238}U, ^{232}Th, and ^{235}U.

Strontium (Sr) is a silvery-gray, soft metal with seventeen different isotopes. Among these seventeen isotopes, four isotopes are stable like ^{88}Sr, ^{87}Sr, ^{86}Sr, and ^{84}Sr (Gupta et al. 2018). The ^{88}Sr is mostly found isotope of strontium, which is mostly (83%) composed of natural strontium.

The isotopes ^{89}Sr and ^{90}Sr are widely used in the radioecology (Amano et al. 2016). Among different isotopes of strontium, ^{90}Sr is considered as the most dangerous isotope because of its high half-life period, affects the human health and ecosystem (Gupta et al. 2018). Strontium released from the nuclear power plants or different sources affects the soil and water due to its high solubility and affects humans through food chain (Ogawa et al. 2016).

Agency for toxic substance and disease registry categorized cesium, uranium, thorium, and strontium as the 217, 97, 102, and 123 most hazardous substances, respectively (ATSDR 2019). The common isotopes of cesium, uranium, strontium, and thorium are ^{137}Cs, ^{134}Cs, ^{235}U, ^{238}U, ^{89}Sr, ^{90}Sr, and ^{232}Th, respectively (Table 11.1). The principle radiation released from the decay of cesium isotopes are β and γ radiation, the decay of uranium release α, β, and γ radiations (Dragović et al. 2015; Hu et al. 2010; White and Broadley 2000). These radiations affect the soil microorganisms, plants, and humans.

11.4 Exposure Pathway of Radioactive Substances

Radioactive substance from all sources, i.e., natural and anthropogenic sources, enter into water, soil, and air through discharges of effluents and emissions. Human and other living organisms are exposed to these dangerous radioactive substances through various ways like inhalation, ingestion, and dermal contact (Fig. 11.1). The exposure pathway of radioactive substances could be explained as:

(a) Generally, the radioactive substances contaminate the surface water as they are deposited in the rivers and ultimately affect the aquatic life. The plants and aquatic animals get directly exposed to these substances and consume them. Further, the freshwater in also used for drinking purpose, and humans get directly exposed to the radioactive substances.
(b) The other source through which radioactive substances are introduced in the environment is the anthropogenic activity such as nuclear powerplant. From the nuclear powerplant, substances are directly entered into atmosphere through emission and cause air pollution. The aquatic life (freshwater and marine water) is contaminated through effluents discharged from the plant and causes the water pollution. Furthermore, radioactive substances released into the atmosphere get deposited into water and soil. The humans are directly exposed to radioactive substances through inhalation. The contaminated water is generally used for the irrigation purpose due to shortage of freshwater; therefore, these substances are

Table 11.1 Summary of the common radioactive substances in environment

Radioactive substances	Isotopes	Sources	Half-life (years)	Radiations	References
Cesium (Cs)	^{137}Cs	Fuel processing, nuclear accidents, weapon testing	30.17	β and γ radiation	White and Broadley (2000)
	^{134}Cs	Nuclear accidents, weapon testing	2.06	β and γ radiation	Hu et al. (2010)
Uranium (U)	^{235}U	Mining, nuclear weapon, nuclear waste and natural sources	7×10^8	α, β, and γ radiation	Dragović et al. (2015)
	^{238}U	Mining, nuclear weapon, nuclear waste and natural sources	4.5×10^9	α, β, and γ radiation	
Thorium (Th)	^{232}Th	Mining, nuclear weapon, nuclear waste and natural sources	1.4×10^{10}	α and β radiation	Jagetiya et al. (2014)
Strontium (Sr)	^{89}Sr	Nuclear accidents, weapon testing	50.52 d	β radiation	Burger and Lichtscheidl (2019)
	^{90}Sr	Nuclear accidents, weapon testing	29.1	β radiation	

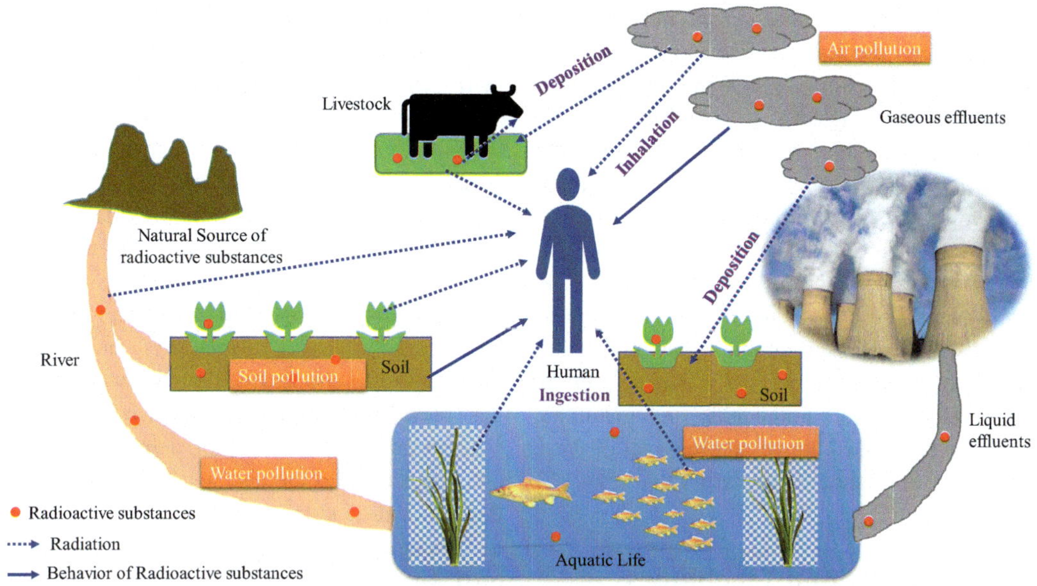

Fig. 11.1 Exposure pathway of radioactive substances in environment

taken up by the edible plants. Animal and human feed on these plants and are exposed to radioactive substances through ingestion.

11.5 Impact of Radioactive Substances on Ecosystem

Radioactive substance accumulated in the soil affects the soil microorganisms and their enzymatic activities. These contaminants affect the cell membrane of microorganisms and disturb the structure and function of proteins (Hollosy 2002; Cox and Battista 2005; Kujawa et al. 2004). The radioactive substances are genotoxic affect the genetic material of soil microorganisms by disturbing the DNA structure and functioning (Ogwu et al. 2019; Koonin and Wolf 2008). The susceptibility of microorganisms to the radioactive substances is different and depends on the species. Some microorganisms are susceptible, and other are resistant to these dangerous substances and can survive at high concentrations. The susceptibility and resistance of microorganisms depend on the genetic and biochemical processes (Narumi 2003; Jolivet et al. 2003).

Radioactive substances also affect the growth and development of plants. The uptake and accumulation of dangerous substances cause the water and nutritional imbalance. Due to the similar characteristics of radioactive substances like cesium and strontium with essential nutrients such as potassium and calcium, these radioactive substances are taken up by plants. The presence of these substances in plants causes the necrosis and chlorosis. They also cause the oxidative stress, damage the leaf membrane; reduce the plant biomass and chlorophyll content by affecting the photosynthetic pigments (Yoschenko et al. 2016). These radioactive substances also reduce the seed germination and plant height. Due to the genotoxic nature of radioactive substances, they damage the genetic material by causing alteration in the DNA (Stojanović et al. 2010). The Fukushima nuclear accident polluted the coastal ecosystem due to the spread of radioactive substances (Yoschenko et al. 2018; Hashimoto et al. 2012; Garnier-Laplace et al. 2011).

Radioactive substances are taken up by humans through plants, water, and inhalation, and these are toxic to humans due to the carcinogenic and genotoxic nature, and they have a long half-life (Tawalbeh et al. 2013). Radiations released from the radioactive substances induce DNA breakage, and the reactive free radicals are also released which ultimately enhance the genomic lesions and carcinogenesis (Moysich et al. 2002) (Fig. 11.2). Uranium (U) like ^{234}U, ^{238}U, and ^{235}U accumulate in different body parts like liver, kidney, nervous, and cardiovascular system (Ten Hoeve and Jacobson 2012). The ^{90}Sr has similar characteristics to calcium hence accumulates in human teeth and bones, due to their half-life it is the most toxic element among other isotopes of strontium (Comar et al. 1957; Leggett et al. 2003). Due to the carcinogenic and genotoxic effects as mentioned above, the removal of these radioactive substances is necessary.

11.6 Remediation Techniques

Different physical and chemical techniques have been intensively used for the removal of radioactive substances from the environment. Some of the techniques are:

(i) contaminated solid wastes are excavated, dumped in the landfill, and allowed for the decomposition, use of site for recreation or other purposes;
(ii) separation of contaminants from the mineral ore;
(iii) high temperature thermal treatment;
(iv) use of chemicals for the removal of contaminants from the solution (Dushenkov 2003; Khan et al. 2000).

The above-mentioned physical and chemical techniques are very expensive. In physical methods, e.g., soil excavation, high machinery has been used which is expensive and also requires much labor. Further, the soil structure and characteristics are also getting disturbed. For the removal of radioactive substances by chemical methods, e.g., soil washing, chemicals are used. The use of chemical affects the soil fertility, characteristics, and production of secondary pollutant may be more toxic which affects the soil microorganisms and functioning. These techniques are only suitable for small contaminated sites (Khan et al 2000). Therefore, for the removal of these contaminants, environment-friendly technique is required.

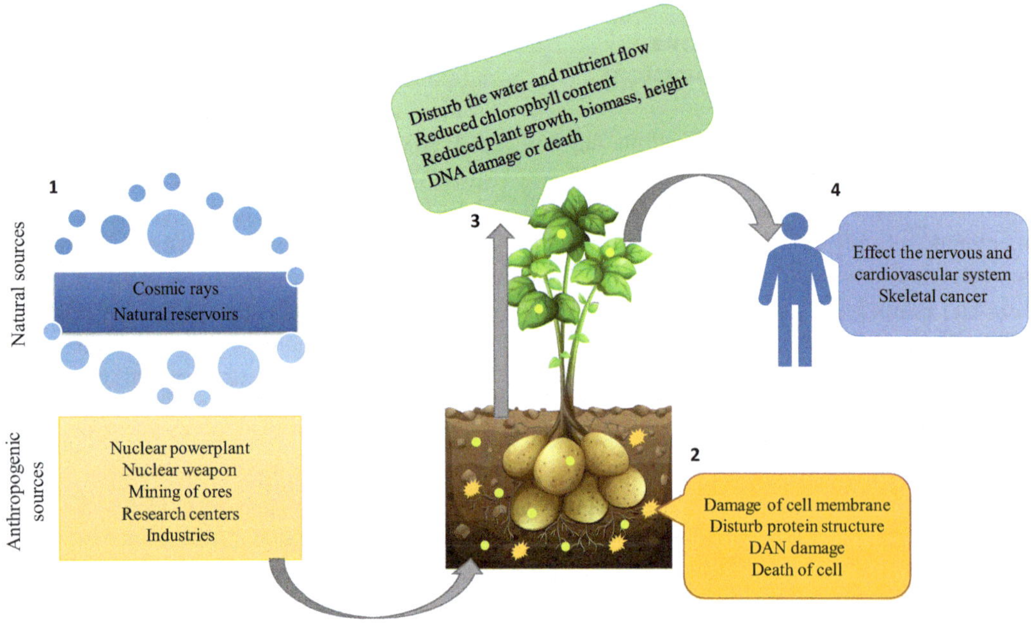

Fig. 11.2 Sources and effect of radioactive substances on environment. (1) The natural and anthropogenic sources of radioactive substances. (2) Effect on soil microorganisms. (3) Effect on plant. (4) Effect on human if consumed through food

11.6.1 Phytoremediation

Phytoremediation is a green technology in which different contaminants such as heavy metals, antibiotics, and radioactive substances have been removed from soil by using plants (Gul et al. 2020; Arshad et al. 2020; Gul et al. 2019a). The phytoremediation is environment-friendly method because the plants and microorganisms are used for the decontamination of sites without harming the surrounding conditions. Further, this technique is economically feasible and does not affect the soil characteristics (Gul et al. 2019b; Arshad et al. 2016; Chaney et al. 2014). By using plants, contaminants are completely removed from site as they are accumulated in plants. In phytoremediation, plant responds to contaminants in different way;

(i) **Contaminant Excluders**—restrict the translocation of contaminants from roots to shoot;
(ii) **Contaminant Accumulators**—accumulate and concentrate the contaminants in aerial parts and survive at high level of contaminants without causing phytotoxicity;
(iii) **Contaminants Indicators**—control the contaminants translocation from roots to the above ground parts (Khan 2020).

Phytoremediation has various techniques including:

Phytoextraction—the contaminants are taken up by plants, transferred to the shoot and accumulated in the aerial parts and can be easily harvested for further processing;
Phytostabilization—the contaminants are stabilized into soil or plant root tissue;
Phytodegradation—plants and associated microflora convert the toxic contaminant into less or non-toxic toxic form;
Phytovolatilization—volatilization of contaminants;
Rhizofiltration—absorption of contaminants by plant roots and is the combination of phytostabilization and phytoextraction (Gul et al. 2019c; Khan 2000, 2005, 2009).

Among all different phytoremediation techniques, phytoextraction and phytostabilization are commonly used techniques for the remediation of heavy metal and radioactive substance contaminated sites. Plants with high biomass hyperaccumulate the radioactive substances and help in their accumulation and concentration into shoots. Other plants known as non-accumulators could uptake the metals but do not translocate them to the shoots (Ogar et al. 2015).

For the phytoremediation of radioactive substances, both terrestrial and aquatic plants are used for the removal of contaminants from soil and water (Pilon-Smits 2005). Further, some plants are used to stabilize the contaminants in soil, reduce their phytoavailability, mobility in soil, and prevent the movement of pollutants in deep soil. But the phytostabilization did not remove the contaminant from soil permanently but reduced the contaminant migration in environment (Dushenkov 2003).

11.6.1.1 Plants Used for the Phytoremediation of Radioactive Substances

The success of phytoremediation is mainly depending on the capability of plants to uptake radioactive substances, transferring, and accumulating in the aerial parts. Therefore, the selection of plant is very important for efficient removal of radioactive substances. Generally, plants having ability to tolerate and survive at high level of radioactive substance without causing phytotoxicity and producing high biomass are suitable for the remediation of contaminated sites (Manzoor et al. 2018; Gupta et al. 2016). The response of plants to tolerate and accumulate radioactive substances involves complex processes including synthesis of phytochelation, organic acid, and metallothioneins (Dalvi and Bhalero 2013). Translocation factor (TF) and bioconcentration factor (BCF) are used for the selection of hyperaccumulator plants. TF is the ratio of radioactive substance concentration in shoot to root, and BCF is the ratio of concentration between soil and root (Bitterli et al. 2010).

Studies have shown that *Helianthus annuus* has ability to remove different radioactive substances such as ^{137}C, U, and ^{90}Sr from water and soil. However, *Amaranthus retroflexus* efficiently accumulates ^{90}Sr and ^{137}C in shoots (Alsabbagh and Abuqudaira 2017; Roongtanakiat et al. 2010). *Zea mays*, *Callitriche hamulate*, *Callitriche lusitanica,* and *Typha latifolia* efficiently remove the uranium from soil and water (Stojanović et al. 2010; Favas et al. 2014). *Vetiveria zizanoides*, *Helianthus annuus,* and *Triticum vulgare* were found to be efficient in the uptake of cesium, strontium, and thorium (Pentyala and Eapen 2020; Shtangeeva and Ayrault 2004; Alsabbagh and Abuqudaira 2017). Different plant species have ability to accumulate high amount of radioactive substances in the aerial parts and are used for the phytoextraction of radioactive substances (Table 11.2).

Another important factor involved in the phytoextraction is the phytoavailability of radioactive substances for plant uptake. The use of organic acid and synthetic chelating agents could be used to increase the solubilization of

Table 11.2 Plant species used for the phytoremediation of cesium and strontium

Radioactive material	Concentration	Plant species	Effects	References
Cesium	20, 40, 60, 80 ppm	*Cannabis sativa*	Not absorbed by plant	Hoseini et al. (2012)
	9.87, 13.4, 11.27 Bq g^{-1}	*Amaranthus retroflexus* L *Brassica juncea* *Phaseolus acutifolius* A. Gray	7.02 Bq g^{-1} 35.9 Bq g^{-1} 1.95 Bq g^{-1}	Fuhrmann et al. (2002)
	0.002 and 20 mM	*Arabidopsis halleri*	2140 mg kg^{-1} of Cs in the leaves at higher Cs level	Burger et al. (2019)
	100 and 400 mg kg^{-1}	*Sorghum bicolor*	TF and BCF greater than 1 86–92% removal of Cs 1147, 2473, and 2393 mg kg^{-1} in roots, stem, and leaves	Wang et al. (2017)
Strontium	20, 40, 60, 80 ppm	*Cannabis sative*	45%, 40%, and 15% absorption in root, stem, and leaves, respectively	Hoseini et al. (2012)
	0.05, 0.069, 0.059 Bq g^{-1}	*Amaranthus retroflexus* L *Brassica juncea* *Phaseolus acutifolius* A. Gray	0.330 Bq g^{-1} 0.400 Bq g^{-1} 0.792 Bq g^{-1}	Fuhrmann et al. (2002)
	0.001 and 100 mM	*Arabidopsis halleri*	122.59 mg kg^{-1} of Sr in leaves ate higher Sr level	Burger et al. (2019)
Uranium	1–11,900 ppm	*Vetiveria zizanioides* L	Plant survive at high level of uranium 90–95% uranium was recovered below 200 ppm	Pentyala and Eapen (2020)
	0–300 mg kg^{-1}	*Sebania rostrata*	23.3% has been extracted from plant	Ren et al. (2019)
	0–318 µg g^{-1}	*Helianthus annuus*	3% U accumulated in shoots Most U accumulated in roots	Alsabbagh and Abuqudaira (2017)

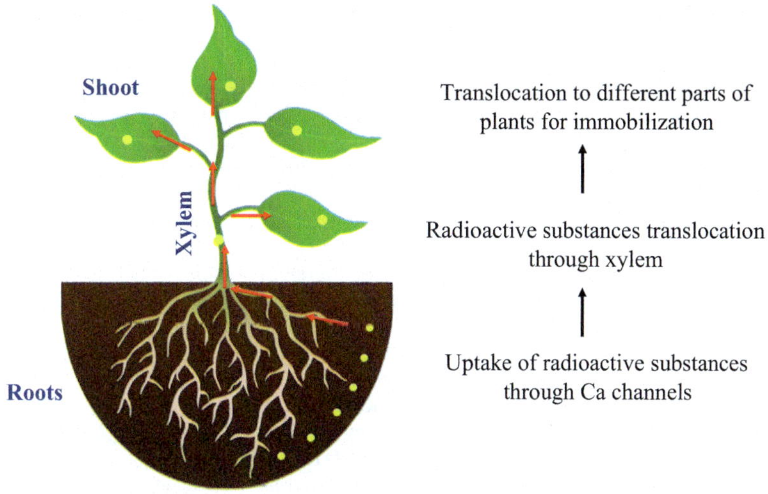

Fig. 11.3 Overview of mechanism of radioactive substances uptake in plants

contaminants for plant uptake (McGrath and Zhao 2003). Among different amendments used to enhance the phytoextraction, citric acid was more efficient in increasing the mobility and accumulation of radioactive substances in *Brassica chinensis* and *Brassica juncea* (Huang et al. 1998). Oxalic acid and citric acid were proven to be efficient in enhancing the uranium phytoextraction (Shahandeh and Hossner 2002). The addition of poultry litter increases the cesium and strontium accumulation in *Sorghum halpense* (Entry et al. 2001).

11.6.1.2 Mechanism of Radioactive Substances Uptake in Plants

Plants uptake available portion of radioactive substances by roots through symplastic and apoplastic movement along with nutrient and are transported to shoot through xylem. The movement of radioactive substances from soil to shoot is generally carried out in following steps: (i) transportation and uptake of radioactive substances from outer environment to roots; (ii) movement/translocation of substances from root to shoot through xylem loading; (iii) sequestration of accumulation of radioactive substances in different parts of plants (Fig. 11.3). Generally, the available amount of radioactive substances is taken up by roots. The strontium is taken up through calcium channel due to the similar analogy, than these are transported to the shoot through xylem, and from xylem vessels, they are translocated to the other parts of plants, i.e., shoot and leaves (Gupta et al. 2018; Scotti and Carini 2000; White et al. 2002).

11.7 Conclusions

Technological advancement increased the use of radioactive substance in different field like medicines and research facilities. These radioactive substances enter into environment from natural and anthropogenic sources and affect the microorganism, plants, and human through food chain. Due to the carcinogenicity and genotoxicity of these substance, different physical and chemical remediation techniques are used but have some disadvantages. Therefore, phytoremediation technique is used, which is an eco-friendly method. Different plants have been used for the extraction of radioactive substances from soil and water. Further the process has been enhanced by using chelating agents. The phytoremediation of radioactive contaminated sites is eco-friendly and economically feasible technique. However, to enhance the efficiency and reduce the removal time, integrated method (plants along with soil microorganism and natural chelating agents) should be used.

References

Alsabbagh AH, Abuqudaira TM (2017) Phytoremediation of Jordanian uranium-rich soil using Sunflower. Water Air Soil Pollut 228(6):219

Amano H, Sakamoto H, Shiga N, Suzuki K (2016) Method for rapid screening analysis of Sr-90 in edible plant samples collected near Fukushima, Japan. Appl Radiat Isotopes 112:131–135

Arshad M, Merlina G, Uzu G, Sobanska S, Sarret G, Dumat C, Silvestre J, Pinelli E, Kallerhoff J (2016) Phytoavailability of lead altered by two Pelargonium cultivars grown on contrasting lead-spiked soils. J Soils Sedim 16(2):581–591

Arshad M, Naqvi N, Gul I, Yaqoob K, Bilal M, Kallerhoff J (2020) Lead phytoextraction by *Pelargonium hortorum*: comparative assessment of EDTA and DIPA for Pb mobility and toxicity. Sci Total Environ 748:141496

ATSDR (2019) Toxicological profile for lead. US Department of Health and Human Services, Public Health Service, Atlanta, GA

Bitterli C, Bañuelos GS, Schulin R (2010) Use of transfer factors to characterize uptake of selenium by plants. J Geochem Explor 107(2):206–216

Burger A, Weidinger M, Adlassnig W, Puschenreiter M, Lichtscheidl I (2019) Response of *Arabidopsis halleri* to cesium and strontium in hydroponics: extraction potential and effects on morphology and physiology. Ecotoxicol Environ Saf 184:109625

Chaney RL, Reeves RD, Baklanov IA, Centofanti T, Broadhurst CL, Baker AJ, Roseberg RJ (2014) Phytoremediation and phytomining: using plants to remediate contaminated or mineralized environments. In: Rajakaruna N, Boyd RS, Harris T (eds) Plant ecology and evolution in harsh environments. Nova Science Publishers Inc., Hauppauge, NY, USA, pp 365–392

Comar CL, Russell RS, Wasserman RH (1957) Strontium-calcium movement from soil to man. Science 126(3272):485–492

Cox MM, Battista JR (2005) *Deinococcus radiodurans*—the consummate survivor. Nat Rev Microbiol 3 (11):882–892

Dalvi AA, Bhalerao SA (2013) Response of plants towards heavy metal toxicity: an overview of avoidance, tolerance and uptake mechanism. Ann Plant Sci 2(9):362–368

Dragović S, Petrović J, Dragović R, Đorđević M, Đokić M, Gajić B (2015) The influence of edaphic factors on spatial and vertical distribution of radionuclides in soil. In: Walther C, Gupta DK (eds) Radionuclides in the environment, Springer, Cham

Dushenkov S (2003) Trends in phytoremediation of radionuclides. Plant Soil 249(1):167–175

Entry JA, Watrud LS, Reeves M (2001) Influence of organic amendments on the accumulation of ^{137}Cs and ^{90}Sr from contaminated soil by three grass species. Water Air Soil Pollut 126(3–4):385–398

Favas PJ, Pratas J, Varun M, D'Souza R, Paul MS (2014) Accumulation of uranium by aquatic plants in field conditions: prospects for phytoremediation. Sci Total Environ 470:993–1002

Fuhrmann M, Lasat MM, Ebbs SD, Kochian LV, Cornish J (2002) Plant and environment interactions. J Environ Qual 31:904–909

Garnier-Laplace J, Beaugelin-Seiller K, Hinton TG (2011) Fukushima wildlife dose reconstruction signals ecological consequences 5077–5078

Gul I, Manzoor M, Silvestre J, Rizwan M, Hina K, Kallerhoff J, Arshad M (2019a) EDTA-assisted phytoextraction of lead and cadmium by Pelargonium cultivars grown on spiked soil. Int J Phytoremediat 21 (2):101–110

Gul I, Manzoor M, Hashim N, Yaqoob K, Kallerhoff J, Arshad M (2019b) Comparative effectiveness of organic and inorganic amendments on cadmium bioavailability and uptake by *Pelargonium hortorum*. J Soils Sediments 19(5):2346–2356

Gul I, Manzoor M, Hashmi I, Bhatti MF, Kallerhoff J, Arshad M (2019c) Plant uptake and leaching potential upon application of amendments in soils spiked with heavy metals (Cd and Pb). J Environ Manage 249:109408

Gul I, Manzoor M, Kallerhoff J, Arshad M (2020) Enhanced phytoremediation of lead by soil applied organic and inorganic amendments: Pb phytoavailability, accumulation and metal recovery. Chemosphere 258:127405

Gupta DK, Chatterjee S, Datta S, Voronina AV, Walther C (2016) Radionuclides: accumulation and transport in plants. Rev Environ Contam Toxicol 241:139–160

Gupta DK, Schulz W, Steinhauser G, Walther C (2018) Radiostrontium transport in plants and phytoremediation. Environ Sci Pollut Res 25(30):29996–30008

Hashimoto S, Ugawa S, Nanko K, Shichi K (2012) The total amounts of radioactively contaminated materials in forests in Fukushima. Japan. Sci Rep 2:416

Hollósy F (2002) Effects of ultraviolet radiation on plant cells. Micron 33(2):179–197

Hoseini PS, Poursafa P, Moattar F, Amin MM, Rezaei AH (2012) Ability of phytoremediation for absorption of strontium and cesium from soils using *Cannabis sativa*. Int J Environ Health Eng 1(1):17

Hu QH, Weng JQ, Wang JS (2010) Sources of anthropogenic radionuclides in the environment: a review. J Environ Radioact 101(6):426–437

Huang JW, Blaylock MJ, Kapulnik Y, Ensley BD (1998) Phytoremediation of uranium-contaminated soils: role of organic acids in triggering uranium hyperaccumulation in plants. Environ Sci Technol 32(13):2004–2008

Jagetiya B, Sharma A, Soni A, Khatik UK (2014) Phytoremediation of radionuclides: a report on the state of the art. Radionuclide contamination and remediation through plants. Springer, Cham, pp 1–31

Jolivet E, I'Haridon S, Corre E, Forterre P, Prieur D (2003) Thermococcus gammatolerans sp. Nov., a

hyperthermophilic archaeon from a deep-sea hydrothermal vent that resists ionizing radiation. Int J Syst Evol Microbiol 53(3):847–851
Kabata-Pendias A (2011) Trace elements in soils and plants/fourth editions. CRC Taylor and Francis Group, Boca Raton 505
Khan AG (2005) Role of soil microbes in the rhizospheres of plants growing on trace metal contaminated soils in phytoremediation. J Trace Elem Med Biol 18(4):355–364
Khan AG (2009) Role of vetiver grass and arbuscular mycorrhizal fungi in improving crops against abiotic stresses. Salinity and water stress. Springer, Dordrecht, pp 111–116
Khan AG (2020) In situ phytoremediation of uranium contaminated soils. Phytoremediation. Springer, Cham, pp 123–151
Khan AG, Kuek C, Chaudhry TM, Khoo CS, Hayes WJ (2000) Role of plants, mycorrhizae and phytochelators in heavy metal contaminated land remediation. Chemosphere 41(1–2):197–207
Koonin EV, Wolf YI (2008) Genomics of bacteria and archaea: the emerging dynamic view of the prokaryotic world. Nucleic Acid Res 36(21):6688–6719
Kujawa J, Zavodnik L, Zavodnik I, Buko V, Lapshyna A, Bryszewska M (2004) Effect of low-intensity (3.75–25 J/cm2) near-infrared (810 nm) laser radiation on red blood cell ATPase activities and membrane structure. J Clin Laser Med Surg 22(2):111–117
Leggett RW, Williams LR, Melo DR, Lipsztein JL (2003) A physiologically based biokinetic model for cesium in the human body. Sci Total Environ 317(1–3):235–255
Manzoor M, Gul I, Silvestre J, Kallerhoff J, Arshad M (2018) Screening of indigenous ornamental species from different plant families for Pb accumulation potential exposed to metal gradient in spiked soils. Soil Sediment Contam 27(5):439–453
McGrath SP, Zhao FJ (2003) Phytoextraction of metals and metalloids from contaminated soils. Curr Opin Biotechnol 14(3):277–282
Moogouei R, Borghei M, Tajadod G (2017) Analysis of transfer factor, anatomical changes and growth of plants during phytoremediation of cesium contaminated solutions. Impact of cesium on plants and the environment. Springer, Cham, pp 253–264
Moysich KB, Menezes RJ, Michalek AM (2002) Chernobyl-related ionising radiation exposure and cancer risk: an epidemiological review. Lancet Oncology 3(5):269–279
Mukhtar A, Manzoor M, Gul I, Zafar R, Jamil HI, Niazi AK, Arshad M (2020) Phytotoxicity of different antibiotics to rice and stress alleviation upon application of organic amendments. Chemosphere 258:127353
Narumi I (2003) Unlocking radiation resistance mechanisms: still a long way to go. Trends Microbio 11 (9):422–425
Ogar A, Sjöberg V, Karlsson S (2015) Phytostabilization of uranium-containing shale residues using *Hieracium pilosella*. Uranium-past and future challenges. Springer, Cham, pp 425–432
Ogawa K, Fukuda T, Han J, Kitamura Y, Shiba K, Odani A (2016) Evaluation of *Chlorella* as a decorporation agent to enhance the elimination of radioactive strontium from body. PLoS ONE 11(2):e0148080
Ogwu MC, Kerfahi D, Song H, Dong K, Seo H, Lim S, Adams JM (2019) Changes in soil taxonomic and functional diversity resulting from gamma irradiation. Sci Rep 9(1):1–13
Pentyala VB, Eapen S (2020) High efficiency phytoextraction of uranium using *Vetiveria zizanioides* L. Nash. Inter J Phytoremed 22.1–10
Pilon-Smits E (2005) Phytoremediation. Anul Rev Plant Biol 56:15–39
Ren CG, Kong CC, Wang SX, Xie ZH (2019) Enhanced phytoremediation of uranium-contaminated soils by arbuscular mycorrhiza and rhizobium. Chemosphere 217:773–779
Roongtanakiat N, Sudsawad P, Ngernvijit N (2010) Uranium absorption ability of sunflower, vetiver and purple guinea grass. Kasetsart J Nat Sci 44:182–190
Schlumberger M, Catargi B, Borget I, Deandreis D, Zerdoud S, Bridji B, Vera P (2012) Strategies of radioiodine ablation in patients with low-risk thyroid cancer. New Engl J Med 366(18):1663–1673
Scotti IA, Carini F (2000) Heavy metal effect on uptake and translocation of 134Cs and 85Sr in aubergine plants. J Environ Radioact 48(2):183–190
Shahandeh H, Hossner LR (2002) Enhancement of uranium phytoaccumulation from contaminated soils. Soil Sci 167(4):269–280
Shtangeeva I, Ayrault S (2004) Phytoextraction of thorium from soil and water media. Water Air Soil Pollut 154:19–35
Steinhauser G, Brandl A, Johnson TE (2014) Comparison of the Chernobyl and Fukushima nuclear accidents: a review of the environmental impacts. Sci Total Environ 470:800–817
Stojanović MD, Stevanović DR, Milojković JV, Grubišić MS, Ileš DA (2010) Phytotoxic effect of the uranium on the growing up and development the plant of corn. Water Air Soil Pollut 209(1–4):401–410
Tawalbeh AA, Samat SB, Yasir MS (2013) Radionuclides level and its radiation hazard index in some drinks consumed in the central zone of Malaysia. Sains Malaysiana 42(3):319–323
Ten Hoeve JE, Jacobson MZ (2012) Worldwide health effects of the Fukushima Daiichi nuclear accident. Energy Environ Sci 5(9):8743–8757
Tsukada H, Hasegawa H, Hisamatsu SI, Yamasaki SI (2002) Transfer of 137Cs and stable Cs from paddy soil to polished rice in Aomori, Japan. J Environ Radioact 59(3):351–363
Valldor P, Miethling-Graff R, Martens R, Tebbe CC (2015) Fate of the insecticidal Cry1Ab protein of GM crops in two agricultural soils as revealed by 14 C-tracer studies. Appl Microbiol Biotechnol 99 (17):7333–7734

Wang X, Chen C, Wang J (2017) Cs phytoremediation by Sorghum bicolor cultivated in soil and in hydroponic system. Int J Phytoremed 19(4):402–412

White P, Broadley M (2000) Mechanisms of caesium uptake by plants. New Phytol 147:241–256

White PJ, Bowen HC, Demidchik V, Nichols C, Davies JM (2002) Genes for calcium-permeable channels in the plasma membrane of plant root cells. Biochim Biophys Acta Biomembr 1564(2):299–309

Yan L, Van Le Q, Sonne C, Yang Y, Yang H, Gu H, Peng W (2020) Phytoremediation of radionuclides in soil, sediments and water. J Hazard Mater 407:124771

Yoschenko V, Nanba K, Yoshida S, Watanabe Y, Takase T, Sato N, Keitoku K (2016) Morphological abnormalities in Japanese red pine (*Pinus densiflora*) at the territories contaminated as a result of the accident at Fukushima Dai-Ichi Nuclear Power Plant. J Environ Radioact 165:60–67

Yoschenko V, Ohkubo T, Kashparov V (2018) Radioactive contaminated forests in Fukushima and Chernobyl. J Forest Res 23(1):3–14

Willows: Cost-Effective Tools for Bioremediation of Contaminated Soils

12

Sirat Sandil and Nandini Gowala

Abstract

Soil degradation due to contamination with toxic substances is an environmental concern worldwide. Various anthropogenic activities like improper industrial waste disposal, run-offs from waste disposal sites, and irrigation with wastewater or water containing geogenic contaminants could lead to elevated concentration of toxic substances in the soil. These contaminants can further be transported in the food chain by consumption of edible plants cultivated in such soils. Mechanisms like phytoremediation and bioremediation have thus been used to remove these contaminants. Bioremediation is the process of removing contaminants from the environment by converting them into less hazardous forms. It traditionally employs plants and naturally occurring bacteria and fungi to remove the contaminants by uptaking and metabolizing them. Several plant species are known to flourish on contaminated soils despite bioaccumulating high amounts of toxic substances. Willows (*Salix* species) have been used for bioremediation studies due to their ability to sequester heavy metals in above-ground mass and high biomass productivity. *Salix* is a cost-effective plant which can grow in a range of environments and absorb both organic and inorganic pollutants. In this chapter, we elaborate on the various bioremediation studies involving *Salix* species and describe the physiological characteristics and ecology of *Salix* which makes it an efficient tool for bioremediation.

Keywords

Bioremediation · Cost-Effective · Heavy metals · Organic pollutants · Phytoremediation · *Salix* · Willows

S. Sandil (✉)
Cooperative Research Centre of Environmental Sciences, Eötvös Loránd University, Pázmány Péter sétány 1/A, 1117 Budapest, Hungary

N. Gowala
Department of Environmental Science, Tezpur University, Napaam, Tezpur, Assam 784028, India

12.1 Introduction

Soil pollution due to heavy metals and persistent organic pollutants is a serious concern worldwide. The chief cause of this is the rapid change in land use, industrialization, and urbanization. Pollutants like oils, hydrocarbons, heavy metals, and excessive nutrients cause soil pollution. Healthy soil has the ability to filter and convert pollutants, which is lost due to alteration of soil quality caused by the presence of pollutants (Zalesny et al. 2021). Heavy metals find their way into the soil from both anthropogenic and geogenic sources like mining, smelting, industrial and atmospheric deposition,

J. A. Malik (ed.), *Advances in Bioremediation and Phytoremediation for Sustainable Soil Management*,
https://doi.org/10.1007/978-3-030-89984-4_12

fossil fuel combustion, irrigation with heavy metal contaminated water, use of pesticides and fertilizers, disposal of municipal waste, weathering, and erosion (Ali et al. 2013; Jadia and Fulekar 2009). Plants require certain essential heavy metals for normal growth (Cu, Fe, Mn, Zn) but high concentration of these trace metals can affect the plant negatively. Increased concentration of these metals leads to soil toxicity, which impacts photosynthesis, growth, and mineral nutrition in plants (Ojuederie and Babalola 2017; Jadia and Fulekar 2009). Plants when cultivated on contaminated soil can easily uptake the heavy metals, which then enter the food chain and pose a potential risk to human health (Ojuederie and Babalola 2017).

In order to mitigate soil pollution, there is a need for inexpensive and efficient methods. Conventional remediation techniques are expensive and deteriorate the soil quality so alternative techniques like phytoremediation have gained prominence (Pulford and Watson 2003). Plants have been used for water treatment and sludge disposal for a long time. They also consume carbon dioxide and decelerate global warming. Their role in maintaining balance in the environment has led to the focus on using plants for remediation purposes as well. Phytoremediation is the use of plants with high accumulation capacities for uptaking contaminants, and cleaning the contaminated sites. Plants when applied for phytoremediation either uptake and sequester the contaminant in the plant cells, or breakdown and stabilize the contaminant in the soil. Plants suitable for phytoremediation, especially phytoextraction, should exhibit a fast growth rate, high biomass, high accumulation of contaminants and should be easily harvested. Using plants for phytoextraction has several advantages like low expenditure, application to a wide range of pollutants, minimal disturbance to environment, elimination of secondary products, and easy public acceptance. When plants are grown for a few rotation cycles in the soil, they either remove all the contaminants or breakdown the contaminants into non-toxic forms (Kacálková et al. 2015; Fosso-Kankeu and Mulaba-Bafubiandi 2014; Cunningham and Ow 1996).

Willows are suitable for phytoremediation of contaminated sites due to their high growth and evapotranspiration rates, easy rejuvenation from vegetative part, high tolerance and translocation of contaminants (Beauchamp et al. 2018). They are pioneer species in many environments and are adaptable to a wide range of environments (Kuzovkina and Quigley 2005). Willows (*Salix* species) are found in the temperate and subtropical regions in form of shrubs or trees, adapted to a range of soil types and climatic conditions. Since they can be grown easily in a short period they play an important role in agriculture, forestry, and industry (Ball et al. 2005). *Salix* species have been found to be efficient remediators of organic pollutants, heavy metals, landfill leachate, and sewage sludge (Mleczek et al. 2018; Justin and Zupančič 2009; Cunha et al. 2012). Willows cultivated in short rotation forestry for phytoremediation could play an important role in energy supply and help meet renewable energy targets. The willows harvested after phytoextraction can also be used as materials for metal enrichment (Gomes 2012). Willows are also effective in conserving soil from being eroded, in restoration and rehabilitation of ecosystems, and in carbon sequestration (Ball et al. 2005).

12.2 Distribution of *Salix* Species Around the World

There are over 300 species of the genus *Salix* worldwide (Lindegaard and Barker 1996), dominant mostly in the Northern Hemisphere. They are found in every continent except Antarctica and Australia. The variation of distribution of this genus is noteworthy. Willows can be in form of small shrubs or trees present in many topographic variations ranging from mountains to lowlands to deserts (Kuzovkina and Quigley 2005). *Salix* are pioneer species near water bodies, as they can survive in anaerobic flooded areas (McLeod and McPherson 1973). Willows belong to the family Salicaceae, genus *Salix* and are mainly of two types, alluvial (growing on bank of rivers and streams) and wetland (thriving on flooded soil). Some of the *Salix* species prefer mineral soil over organic soil (Kuzovkina and Quigley 2005). In Europe, 32 species of willows and their hybrids are found. *Salix alba* is the most common willow species found along with *Salix triandra*, *Salix caprea*, *Salix phylicifolia* and *Salix myrsinifolia*. Among the hybrid species the cross between *Salix purpurea* and *Salix viminalis* is common in Romania, Hungary, and Poland. The cross of *Salix triandra* and *Salix viminalis*, and of *Salix phylicifolia* and *Salix myrsinifolia* is also common in European countries (Cronk et al. 2015). At least 130 species of willows are found in South America. In Eastern Asia, 313 species of *Salix* are found. *Salix* is more abundantly found in areas with abundant water. They are also found in mountain slopes and lowlands. However, in Eastern Asia *Salix* is dominant in mountain areas (Wang et al. 2017). Willows have been divided into the sub-genera, *Protitea*, *Salix*, *Longifoliae*, *Chamaetia*, and *Vertix* by Canadian biologist George W. Argus (Abdollahzadeh et al. 2011).

12.3 Economic Importance of Willows

Willows can grow rapidly under favorable conditions. They are easy to grow and are hybridized for their numerous economic values. Willows are grown in short rotation forestry which has the advantage of high biomass yield, short harvest cycles of 3–5 years, low costs, and ecological benefits (Kahle et al. 2007). Short rotation forestry is important for the bio-energy industry as it can supply the biomass needed for fossil fuel replacement. The plants are either converted to biofuels or combusted for heat and electricity. Willows are also being touted as future energy crops as 54 species have been identified for producing very high above-ground biomass in a quick growth cycle. Short rotation forestry also plays an important role in carbon sequestration (Lindegaard and Barker 1996; Laureysens et al. 2004; Marmiroli et al. 2011).

Willows have been employed in several countries like the USA, China, UK, Canada, New Zealand, Sweden, and India for the conservation of soil and water. Willows act as riparian buffers, prevent soil erosion, provide high amount of wood biomass, help in nitrogen removal from leachates and livestock farming, enable enhanced carbon storage, are good phytoremediators, and are used in agroforestry industry (Ball et al. 2005). *Salix* species growing in flood plains help in stabilizing sediments on the riverbanks and in breaking river currents. Their high rate of vegetative propagation makes them tolerant of any disturbance in the environment. They also improve the nutrients level in soils, C/N ration, and enhance the biodiversity (Markus-Michalczyk et al. 2016). Some cultivars produce straight timber which is used for construction purposes, furniture, and matchstick production. The foliage has a high nutritive value and is used as fodder (Wilkinson 1999). Weeping willow (*Salix babylonica*) is the most preferred ornamental willow species worldwide (Isebrands and Richardson 2014).

Willows host a rich biodiversity of insects, birds, and animals in their natural environments. They are a source of food and shelter for animals like rabbits, beavers, moose, reindeer, elk, and mice. Beavers use young willow trees for building dams (Charles 2014). Willows act as a breeding ground for at least sixty bird species. Many insects, integral part of food webs, feed on willows. Their large overgrown aerial parts provide shade to animals and act as viewing grounds for animals, thus serving recreational purposes (Kuzovkina and Quigley 2005).

12.4 Bioremediation

Soil remediation is the return of soil to the condition it existed in, prior to any disturbance (Jadia and Fulekar 2009). Conventional physico-chemical remediation processes like ion exchange, chemical reduction, and reverse osmosis are problematic due to their low effectiveness in removing contaminants, and the high energy and expense incurred (Mani and Kumar 2014). Due to this, there is a need to apply methods which are effective as well as low on cost. Bioremediation is the process of transforming environmental contaminants into benign forms with the help of plants, fungi, or bacteria (Vidali 2001). It helps in dealing with persistent environmental contaminants by applying natural biological functions which breakdown the contaminants. Plants with the help of their photosynthetic pathways can uptake and accumulate pollutants (Gu 2018), while microbes (aerobic and anaerobic) and fungi carry out the degradation of persistent pollutants through their metabolic pathways (Juwarkar et al. 2010). Bioremediation is a cost-effective method as compared to other conventional methods (Mani and Kumar 2014), and has thus gained importance over the past few decades.

Bioremediation techniques are mainly classified into two broad categories, i.e., *Ex-situ* and *In-situ*. *Ex-situ* methods are applied after the removal of the contaminated medium from the site, while *in-situ* methods are applied at the site of contamination. *Ex-situ* techniques include biopiles, composting, bioreactors, and landfarming, and *in-situ* techniques are bioslurping, bioventing, biosparging, and phytoremediation (Vidali 2001). Bioremediation has several benefits over conventional processes, like low cost, no disturbance to the contamination site, and no additional operational costs. It is simpler than any other technology and has fewer requirements for successful operation. But there are some limitations to the procedure because heavy metals and highly chlorinated compounds are not susceptible to degradation. The procedure is also slower if lower amount of substrate is available to the microorganisms for functioning (Juwarkar et al. 2010; Bollag et al. 1994).

12.5 Phytoremediation

Phytoremediation is the use of plants for remediation of contaminated sites (soil, sediment, sludge, leachate, groundwater). Plants using solar energy clean-up sites with low or moderate contamination by degrading, containing or removing the contaminants *in-situ* (USEPA 1999). Plants can remove the contaminants by absorption and translocation into their above-ground mass where the contaminants are mineralized by the plant metabolic processes (Bollag et al. 1994). Phytoremediation has been applied for the removal of heavy metals, ammonia, pesticides, oil, explosives, or any contaminating substance (Rhodes 2013). Phytoremediation is applied at contaminated sites where other methods are not practically feasible or are too expensive. When low amount of contamination is spread over a large area, and when contaminated site needs to be closed then phytoremediation is the favored method (Vidali 2001). The remediation of trace metals is dependent on the metal concentration in soil, soil properties, soil pH, and type of plant employed for phytoremediation (Jensen et al. 2009). The uptake of metals by plants depends on the bioavailable fraction and not on the total fraction of the contaminant in the soil (Vamerali et al. 2010). Phytoremediation has several advantages like removal of both organic and inorganic pollutants, low cost, and no loss of fertility in the soil due to its non-invasiveness (Ojuederie and Babalola 2017). Since phytoremediation is a natural solar driven process it is environment friendly and readily accepted (Wenzel 2009).

Phytoremediation can be of different types based on the uptake and removal method and the element involved (USEPA 1999; Gonzaga et al. 2006; Wenzel 2009):

- Phytoextraction: the removal of contaminants from the site by plants which uptake and store large concentrations of the contaminants in their above-ground mass. Upon completion of plant growth, the aerial parts can be harvested thus removing the contaminants.
- Phytostabilization: roots contain the toxicity of contaminants by decreasing their mobility in the soil. Helps in restricting the movement of contaminants.
- Rhizo-filtration: roots absorb and immobilize contaminants.
- Phytodegradation: plants in association with soil microorganisms cause the breakdown of contaminants by releasing exudates.

- Phytovolatilization: volatilization of some compounds in the leaves and their release into the atmosphere.

For a plant to be an effective phytoremediator it should bear the following characteristics (Chaney et al. 1997; Ali et al. 2013):

1. The plant should accumulate high concentrations of the targeted pollutant.
2. The amount of above-ground biomass should be higher, so that more pollutants can be absorbed from the soil.
3. The plant should be able to tolerate high concentrations of the pollutant. This tolerance could be due to the presence of chelating agents like metallothioneins and phytochelatins, and compartmentalization in vacuoles.
4. The plant should have a high root to shoot translocation rate.
5. The plant should be easily cultivable and harvestable.
6. The plant should be resistant to pests and grazers.

Plants are classified as hyperaccumulator when they can accumulate high concentrations of an element in their aerial tissues. The efficiency of phytoremediation is based on both the soil metal concentration and the aerial part metal concentration. The bioaccumulation factor (BAF) and translocation factor (TF) are used to classify the plant as a good accumulator or hyperaccumulator. The bioaccumulation factor is the ratio of the element's concentration in the aerial part to that in the soil and represents the ability of plant in removing the element from the soil. The translocation factor is the ratio of the element's concentration in the aerial part to that in the roots and represents the translocation ability of the plant. Plants which have a BAF and TF greater than 1 have a good uptake efficiency and translocation, and can be used for phytoremediation studies. Hyperaccumulators like Brassicaceae or good accumulators like *Salix* are an example of this (Gonzaga et al. 2006). Through phytoextraction, plants are able to accumulate high concentrations of contaminants in the aerial part, which is then harvested and ashed to remove the contaminant from the site (Krämer and Chardonnens 2001).

Phytoextraction studies generally utilize hyperaccumulating plants. Hyperaccumulators are plants with the capacity to uptake and accumulate a very high concentration of contaminants in their aerial parts without displaying any toxicity symptoms. Hyperaccumulators have an enhanced contaminant uptake rate, increased translocation, and detoxification (Verbruggen et al. 2009). However, most hyperaccumulating plants have a low growth rate, low biomass production, and shallow roots, which makes them unsuitable for phytoremediation studies (Laureysens et al. 2004; Bissonnette et al. 2010). Willows on the other hand, due to their high biomass, are an ideal candidate for phytoremediation. In willows, phytoextraction and phytostabilization are the most explored phytoremediation types due to their (1) high shoot biomass production, (2) ability to accumulate contaminants in the shoot, (3) vast root network, (4) high evaporation and transpiration rates which helps in increased root to shoot translocation, and (5) association with soil microorganisms, both bacteria and fungi (Krämer and Chardonnens 2001; Złoch et al. 2017; Bissonnette et al. 2010). *Salix* species have been suggested to be good agents for phytoremediation due to their deep roots which can penetrate further into the ground and cover a large area, thus stabilizing large areas of soil and also absorbing high amount of contaminants from them (Tőzsér et al. 2018).

12.6 *Salix*: A Potential Candidate for Bioremediation

Plants that grow rapidly and accumulate a high amount of heavy metals in their leaves or above-ground mass are ideal candidates for phytoremediation (Fig. 12.1). Approximately 20% of *Salix* species are essential for both biomass production and contaminants uptake (Mleczek et al. 2010). *Salix* is a cost-effective tool for bioremediation because new plants are regenerated from

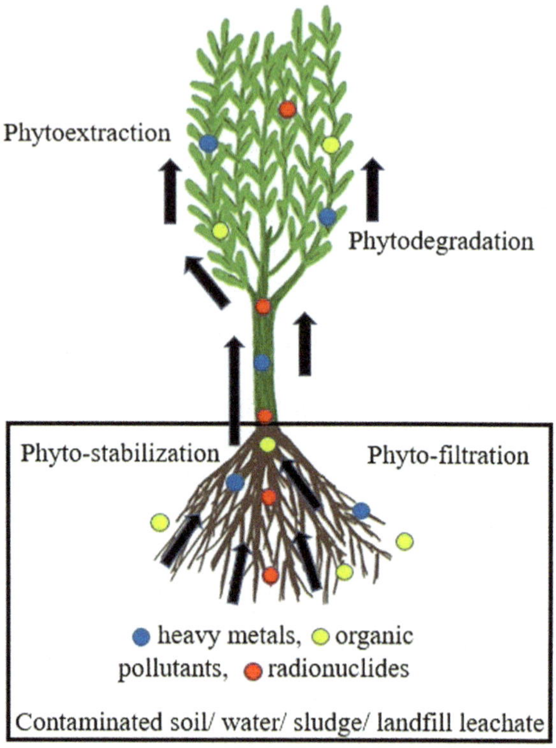

Fig. 12.1 Graphical representation of phytoremediation by *Salix* plant

residual roots whenever destroyed by any environmental factor or livestock (Wilkinson 1999). Some species are also tolerant to salinity, drought, and fire. Willows as pioneer species can establish themselves in all types of environments. They have light weight seeds which can travel long distances and establish themselves in sunlight (Kuzovkina and Quigley 2005).

Salix is an ideal plant for phytoremediation studies due to the following characteristics (Meers et al. 2005; Wilkinson 1999; Marmiroli et al. 2011):

1. Rapid Growth: *Salix* species grow rapidly and can survive in different environmental conditions. Their juvenile growth rate is faster (1–4 m per year) as compared to any other tree species of the temperate climes. They can also be grown from unrooted cuttings.
2. Easily cultivable: Most of the willows and their clones can be grown easily through vegetative propagation from poles. Willows can regrow quickly after being harvested, due to which they are preferred for short term extraction of pollutants. Plants can regrow from coppiced stems. Willows species such as *Salix euxina*, *Salix ragiles,* and *Salix nigara* are suitable for vegetative propagation.
3. Large above-ground biomass: Willows produce a very large amount of aerial biomass. Under favorable conditions they produce yearly biomass of about 10–12 oven dry tons per hectare. Due to this hefty aerial biomass willows are able to accumulate large amounts of pollutants within a short time.
4. Tolerance: *Salix* species can grow in highly polluted environments and accumulate large amounts of pollutants. They also exhibit tolerance to anaerobic conditions and grow well in saturated or flooded soils.
5. High evapotranspiration rates during growth period increase the translocation of pollutants to above-ground part.
6. Extensive network of fine fibrous roots: the root system holds soil together preventing soil erosion. Many species have tap root systems

which can penetrate deep into the soil and find a water source. This helps them to survive in all sorts of environments.

7. The large root system and high evapotranspiration helps in increased uptake of nutrients and nitrogen.
8. Resistance to pests and grazers: The presence of salicin in the *Salix* species makes the leaves of the plant bitter to some herbivores like Opossum (Palo 1984).

Willows have been successfully used for phytoremediation in many experiments with different contaminated mediums (Tőzsér et al. 2018; Salam et al. 2016; Dimitriou and Aronsson 2010; Landberg and Greger 1996).

12.7 *Salix* Root and Soil Microorganisms

The efficiency of phytoremediation by *Salix* can be increased by its association with microorganisms. Plants can remove contaminants from the soil by increasing interactions with microorganisms in the rhizosphere, creating aerobic conditions in soil for nitrifying bacteria which helps improve plant growth, and by providing leaf litter for the microorganisms to thrive on. The leaf litter is converted into humic substances in the soil which can bind to heavy metals and prevent their mobility, thus helping in phytostabilization (Bollag et al. 1994). The rhizosphere of the plants play an important role in the uptake of contaminants. Heavy metals exhibit a greater movement in the rhizosphere of metal-accumulating plants like *Salix* due to the presence of an active bacterial community which enhances the heavy metals movement and availability (Kuffner et al. 2008). Root density, length, depth are essential morphological characters which help in stress adaptation. Roots can decrease or increase the bioavailability of metals depending on the plant type (accumulator/excluder). Plant roots can modify the pH and redox potential, and release organic acids and chelating agents (Vamerali et al. 2010; Wenzel 2009). *Salix* roots release low molecular weight, easily biodegradable organic acids like citric acid, oxalic acid, and malic acid during oxidative stress. These organic acids are good chelators, which act as a carbon source for microorganisms, thus helping in mobilization of heavy metals in the soil (Salam et al. 2019). Metals uptaken by plants are sequestered in the vacuoles with the help of metallothionines and phytochelatins which enables the plant to cope with the high concentration of metals (Fosso-Kankeu and Mulaba-Bafubiandi 2014). Soil bacteria and fungi play an important role in phytoremediation. They are present in association with roots and make the contaminants available for uptake by plants. In soils with less nutrients the roots develop a symbiotic relationship with mycorrhizal fungi which ensures a continuous flow of nutrients to the plant (Kuzovkina and Quigley 2005).

The availability of contaminants in the soil depends on the microorganisms associated with the roots, and root exudates like organic acids and siderophores (Verbruggen et al. 2009). Soil microbes can promote growth through many mechanisms like nitrogen fixation, enhancement of nutrient uptake, and production of growth regulators. They can increase the uptake of heavy metals by the plant by bringing about a change in the soil pH, redox conditions or by causing release of organic acids and siderophores which help in chelation of heavy metals (Złoch et al. 2015). They also help in growth and adaptation of willows in highly contaminated soils (Kuffner et al. 2008; Guarino et al. 2018). Willows grown on constructed wetlands are effective in lowering the total nitrogen, nitrate, and chemical oxygen demand of landfill leachate. Willow roots release oxygen creating aerobic conditions for nitrifying bacteria thus enhancing ammonia conversion and removal. Uptake of nitrogen by plant also plays an important role in heavy metal uptake since willows are very fast growing plants (Białowiec et al. 2012).

Plant association with bacteria greatly improves the heavy metal uptake efficiency. The metal uptake efficiency is dependent on the bacterial siderophores, which are compounds released by bacteria for iron acquisition. They could help in the mobilization of other metals as well in the soil, and improve plant growth (Kuffner et al. 2008). A similar observation was made by Złoch et al. 2017 in *Salix dasyclados* inoculated with three metal tolerant strains of the bacteria *Streptomyces*. The overall phytoextraction efficiency increased in the plants and the maximum phytoextraction efficiency was observed for the strain which produced the highest siderophore. They used metalliferous soils from areas near a mining plant and observed that inoculation with bacteria increased the extractable fraction of the heavy metals in soil. The bacteria also increased plant biomass growth, uptake of Zn and Cd by the plant, and decreased oxidative stress in the plant.

Salix plant displays dual mycorrhizal association, with both arbuscular mycorrhizal and ectomycorrhizal fungi. This association can exist separately or together and benefit both partners. The mycelium of the fungi helps the plant in spreading far and absorbing more water and nutrients and the fungi in return can get a supply of energy from the plant (Dagher et al. 2020). Arbuscular mycorrhizal fungi are common in almost all terrestrial ecosystems. They form obligatory symbiotic relationships with tree roots. They improve plant growth and resistance by acting as extended roots which cover a large area. This helps increase their nutrient, and water supply. Ectomycorrhizal fungal associations on the other hand are rarer, and they do not colonize the cortex. They increase the phytoextraction efficiency of plants by increasing the nutrient supply to plant, because their hyphae extend to much longer distances. They increase plant tolerance to pollutants by storing them in vacuoles and on mycelial surfaces. Both mycorrhizal associations can either increase the heavy metals absorption by plant or decrease it, depending upon the soil heavy metal concentration. At low heavy metal concentration in soil the translocation efficiency is higher, but at high soil concentration the fungi bind the heavy metals in the soil and reduces uptake by plant (Bissonnette et al. 2010; Baum et al. 2006). Few studies on growth and phytoextraction efficiency of *Salix* in association with bacteria and fungi are given in Table 12.1.

12.8 Physiological Response of Willows to Contaminants

In response to metal pollution plants release low molecular weight organic acid in the rhizosphere. These acids lower the pH and enhance the metal bioavailability and mobility in soil. These root exudates help in metabolic transformation and degradation of contaminants (Mleczek et al. 2018; Suresh and Ravishankar 2004). All *Salix* species have a different reaction to contaminants, depending on the type and concentration of the contaminant. This was confirmed in a study by Beauchamp et al. 2018 on ten different cultivars of willow grown in a site contaminated by petrochemical contaminants. They reported that different cultivars accumulated varying levels of heavy metals, and cultivars procured from a contaminated environment had a higher biomass. Despite not being a hyperaccumulator, *Salix* species have a fast growth rate which is helpful in the removal of contaminants. Biomass yield is the most important characteristic for phytoextraction by *Salix* as older plants were found to accumulate higher concentrations of contaminants. Phytoextraction also depends on the concentration of essential elements in the tissues, as the tissues with higher essential elements had lower biomass and lower accumulation of contaminants (Zárubová et al. 2015). Willows are also sensitive to high concentrations of contaminants as shown by Lunáčková et al. 2003, in four species of willows (*Salix viminalis* L., *Salix alba* L., *Salix purpurea* L. and *Salix cinerea* L.). Willow roots were found to be more sensitive to Cd as compared to the shoots, and root growth was suppressed by 66–99.7% in all the species. *Salix borealis* displayed a similar stress response to terrestrial pollution of sulfur and heavy metals in the boreal zone. The leaves exhibited fluctuating asymmetry in response to foliar damage caused by pollution (Zvereva et al. 2017). Yu et al. 2010 reported the temperature dependence of *Salix matsudana Koidz* × *alba* L. in the removal of chromium. They observed that the uptake rate of chromium increased linearly with temperature, suggesting that contaminant uptake in *Salix* is also affected by temperature.

Table 12.1 Influence of bacteria and fungi on pollutant uptake and growth in a few *Salix* species

Salix species	Microorganism involved	Source	Effect on plant	Reference
Salix dasyclados	Ectomycorrhizal fungus *Paxillus involutus*, one strain from a contaminated site, another from uncontaminated site	Contaminated soil with NH_4NO_3-extractable metals	Both strains increased the plant biomass. Extractable metal fraction impacted the mycorrhiza and heavy metal uptake The total metal mobilization for Pb, Zn and Cd increased The origin of the strain is important as it significantly affects the density of the mycorrhizal colonization and the dependence of *Salix* on it	Baum et al. (2006)
Salix caprea	Rhizosphere isolates of *Pseudomonas, Janthinobacterium, Serratia, Flavobacterium, Streptomyces, Agromyces*	Soil from lead mine ontaminate with Zn, Cd, Pb	*Streptomyces* AR17 improved Zn and Cd uptake in plant *Agromyces* AR33 increased plant growth and augmented Zn and Cd uptake Bacterial siderophores had no effect on Zn and Cd uptake	Kuffner et al. (2008)
Salix viminalis	Arbuscular mycorrhizal Fungus *Glomus intraradices*	Slightly ontaminate soil	Association with fungi resulted in high aerial part biomass production High concentration of Cu and Pb in roots, and Cd and Zn in shoot	Bissonnette et al. (2010)
Salix viminalis	Bacteria (*Massilia* species and *Pseudomonas* species), saprophytic fungus (*Clitocybe* species)	Medium containing 1 mM of Cd	Bacterial strains improved phytoremediation, both in terms of Cd quantity and bioaccumulation factor and transfer factor Inoculation with fungus increased biomass production	Złoch et al. (2015)
Salix purpurea subspecies *lambertiana*	Microorganism comprising of mycorrhizae and plant growth promoting rhizobacteria	Topsoil ontaminate with As, Cd, Pb, Zn	Inoculation with microorganisms caused an increase in heavy metal concentration in roots, contributed to scavenging of reactive oxygen species, improved the plant's resistance and adaptation	Guarino et al. (2018)
Salix miyabeana	Arbuscular mycorrhizal fungus *Rhizophagus irregularis*, and an ectomycorrhizal fungus *Sphaerosporella brunnea*	Industrial landfill	Ectomycorrhizal fungus: Higher biomass in willows, lower concentration of Cu, Sn, and Pb in the soil. And increased phytoextraction of Cd, Zn, and Ba Arbuscular mycorrhizal fungi: no significant changes in plant or soil.	Dagher et al. (2020)

12.9 *Salix* in Remediation of Sewage Sludge and Leachate

Landfill leachate arises from the decomposition of organic wastes in municipal landfills. This leachate contains a high nitrogen content and toxic elements due to which it is not treated in wastewater treatment plants (Justin and Zupančič 2009). Landfill leachate is generally treated with processes like activated sludge, reverse osmosis, or ozonation. These processes are expensive and difficult to maintain, especially in rural settings (Białowiec et al. 2012). Covering landfills with fast growing and deep-rooted plants like *Salix* helps in leachate remediation. The leachate is uptaken by the plants as fertigation (Zalesny and Bauer 2007). Leachates contain a high amount of macro and micro elements which can be used as fertilizers in growth of willows (Justin and Zupančič 2009). In Sweden, large-scale cultivation of *Salix* in short rotation forestry is being carried out with municipal wastewater and sewage sludge. This is beneficial as the waste is being used to produce biomass and also to remediate the waste, at minimal expense (Dimitriou and Aronsson 2011). Usage of sludge as a medium for growth of willows helps in plant growth and metabolism by acting as a fertilizer. It improves the soil quality, increases humus content and total organic carbon content, and helps increase the microbial population in soil. It also results in increased growth of willows, improved chlorophyll content, and enhanced leaf length and surface area (Urbaniak et al. 2017).

Dimitriou and Aronsson (2010) studied the effect of leachate irrigation on willows grown in clay lysimeter. They observed that the growth of willows was enhanced at the higher leachate treatment and willows showed better nitrogen and phosphate retention than poplars. Białowiec et al. 2012 in an experiment with landfill leachate observed that leachate tanks covered with willows had lower chemical oxygen demand, nitrate and total nitrogen concentration, than untreated leachate tanks. This is because the plant actively uptakes nitrogen for its metabolism and releases oxygen in the rhizosphere, which decreases the chemical oxygen demand. In an experiment conducted with wastewater sludge treatment in *Salix viminalis* and *Salix discolor* to check the accumulation of pollutants in the species, it was found that the plants receiving the maximum sludge treatment had the highest growth. The increase in plant height was more in case of *Salix viminalis* as compared to *Salix discolor*. The accumulation of Zn and Pb was also higher in case of *Salix viminalis*. Both the species accumulated a good amount of Cd and Zn, but a lesser amount of Cu, Pb, Ni, and Hg. The authors concluded that *Salix viminalis* was a good filter for purifying wastewater sludge (Labrecque et al. 1995).

12.10 *Salix* and Organic Pollutants

For the remediation of organic contaminants, the use of plants in association with microorganisms is a very effective method. Soil microorganisms play an important role in the degradation of organic contaminants. Microorganisms associated with plant roots can break down xenobiotic compounds and pesticides like trichloroethylene (Bollag et al. 1994). The microorganisms can enhance remediation of organic pollutants by volatilizing the pollutants or by using them to produce humic substances (Alkorta and Garbisu 2001). Plants carry out remediation of organic pollutants in three ways: (a) absorb and store the pollutants in plant cells by converting them into non-toxic form with the help of metabolic pathways inside plant, (b) release exudates like organic acids and phenols in the rhizosphere, which help in increasing the microbial activity and transforming the pollutant, (c) enhance mineralization of pollutant in soil with help of bacterial and fungal associations (Schnoor et al. 1995). Root exudates contain substances like organic acids, phenols, and proteins, which can serve as sources of nutrition for the microorganisms which help in the breakdown of organic pollutants (Alkorta and Garbisu 2001).

A number of studies have been carried out on the remediation of organic pollutants by willows: ethylene glycol (Carnegie and Ramsay 2009), methyl tert-butyl ether (MTBE) (Yu and Gu 2006), trichloroethylene (Miller and Khan 2011), polycyclic aromatic hydrocarbons (PAHs) (Rentz et al. 2005), and polychlorinated biphenyls (PCBs) (Slater et al. 2011), and pharmaceuticals like ibuprofen (Iori et al. 2013). Details of a few studies are summarized in Table 12.2. Vervaeke et al. 2003 studied the effect of willows (*Salix viminalis* L. 'Orm') on dredged sediments. Dredged sediments usually contain a large amount of organic and inorganic contaminants like heavy metals, PAHs, and oil products. This sediment when disposed on land can act as a substrate for willows grown in short rotation forestry. Willow plantation on the sediment resulted in increased breakdown of the oil products but did not significantly affect the PAH concentration. This was unexpected because soils covered by vegetation have a higher microbial activity which should have degraded the PAH concentration. In contrast to this Cunha et al. 2012, observed a significant decrease in PAH content in soil contaminated with petroleum derived hydrocarbons (total hydrocarbons and PAHs). They used the willows *Salix rubens* and *Salix triandra*, for phytoremediation of the contaminated soil and found the plants to be resistant to the contamination. The species were very good remediators as they decreased the total hydrocarbon concentration by almost 98%. Concentrations of pyrene, chrysene, benzo(a)pyrene all decreased to below detectable limit. The removal efficiency of PAH is further improved when Salix is grown in association with microorganisms. Ma et al. 2020 observed very high removal efficiency (80%) of PAHs from PAH-contaminated soil when *Salix viminalis* was inoculated with *Crucibulum leave*.

Table 12.2 Literature review of *Salix* species involved in remediation of organic pollutants

Species name	Medium	Findings of the study	Reference
Salix babylonica	Groundwater contaminated with ethanol-gasoline mixture	Willow tree cutting decreased ethanol and benzene concentration by more than 99% in a week The removal was dependent on the plant transpiration rate and on sorption by roots	Corseuil and Moreno (2001)
Salix babylonica L	Hydroponic solution containing methyl tert-butyl ether (MTBE)	Plant removed MTBE through phytovolatilization, because a large amount of MTBE was found in the air, and not in plant tissues. Willows could tolerate MTBE concentrations up to 200 mg/L. Beyond that the plant displayed toxicity symptoms *Salix* was very efficient in removing MTBE from the solution	Yu and Gu (2006)
Salix viminalis	Hydroponics with 4-Chlorobenzoic acid at concentration 5 and 50 mg/L	4- chlorobenzoic acid was removed by uptake with water and sorbed to plant tissues 4-CBA was toxic to willow at 50 mg/L. Willow could remove 10–30% of 4-CBA	Deavers et al. (2010)
Salix alaxensis	Polychlorinated biohenyl (PCBs) contaminated soil	Soil treated with willow root had a greater PCB loss Willow treated soil had less toxins in the environment *Salix alaxensis* is effective for rhizoremediation	Slater et al. (2011)
Twelve *Salix species*	Toxicity, uptake and degradation of trichloroethylene was tested	*Salix* discolor had the highest mass gain A wild *Salix* species was best suited for trichloroethylene degradation	Miller and Khan (2011)

12.11 Phytoremediation of Heavy Metals by *Salix* Species

Plants involved in remediation processes for heavy metals need to convert the heavy metals into non-toxic forms or remove them by uptaking them into their tissues. This is because heavy metals are not susceptible to degradation (Cunningham and Ow 1996). Heavy metals toxicity in soil is dependent on a few factors like the total and extractable metal fraction, pH, soil properties, organic matter content and redox potential. Increase in soil contamination does not increase the phytoextraction efficiency of plants. This is because the phytoextraction efficiency is based on the metal concentration in the aerial part to that in the soil, thus, soils with less contamination have better heavy metal removal capacity (Złoch et al. 2017). Several studies have established that certain species of *Salix* have high bioremediation potential for heavy metals like Cd, Mn, Zn, and Pb (Table 12.3). Uptake of these elements is aided by high translocation and bioconcentration factors.

Heavy metal accumulation in plants also depends on the rate of plant growth and the duration of exposure (Guarino et al. 2018). Several hybrid species of willows have been employed in phytoremediation practices. This is because natural tolerance or hyperaccumulation of a species is generally restricted to one trace element, but there are multiple contaminants in the polluted soil. Hybrid species aid in the removal of multiple contaminants at once (Krämer and Chardonnens 2001).

Recent studies have focused on growing willows on highly contaminated sites like landfills (Justin and Zupančič 2009), sewage sludge disposal sites (Urbaniak et al. 2017), and sediment dredge disposal sites (Vervaeke et al. 2003). Willows have been grown on these mediums to contain the movement of contaminants from these mediums into the surroundings (Fig. 12.2). Willows due to their physiological characteristics like high transpiration rate and large root network can aid in contaminant removal. Willows can phytoextract and translocate a large amount of contaminants from the soil into their above-ground mass, which is then harvested (Gonzaga et al. 2006). In addition, willows are perennial plants, which can be grown easily through vegetative propagation and are very fast growing, which makes them suitable candidates for phytoremediation studies (Mertens et al. 2006). Phytoextraction is the absorption and translocation of contaminants by plants. Plants accumulate the contaminants in their plant parts, usually the aerial parts. This procedure when applied several times to the soil decreases the soil contamination levels (Meers et al. 2005).

Tlustoš et al. (2007) studied the phytoextraction efficiency of seven clones of *Salix* namely: *S. smithiana* Willd. (hybrid of *S. viminalis* L. and *S. caprea* L.), *S. smithiana* Willd., *S. rubens* Schr. (natural hybrid of *S. alba* L. and *S. ragiles* L.), *S. dasyclados* Vimm. (natural hybrid of *S. viminalis* L. and *S. cinerea* L.), *S. alba* L. (white willow), *S. alba* L.'Pyramidalis', and *S. viminalis* L. (basket willow). All plants were cultivated in pots with three soil types: heavily polluted fluvisol contaminated by waste from smelters, moderately contaminated cambisol, and unpolluted chernozem. All clones demonstrated a high concentration of Cd and Zn in the aerial part, and As and Pb in the roots. In all clones, Cd and Zn accumulation were more in the leaves and less in the twigs. *S. smithiana* and *S. rubens* accumulated the highest Cd while *S. rubens* accumulated the maximum Zn. The Cd and Zn concentrations in aerial parts were higher in plants grown with contaminated soil, indicating that *Salix* clones could be used as bio-monitors for Cd and Zn pollution in soil. The maximum plant biomass was observed in the cambisol, while in the heavily polluted fluvisol toxicity symptoms could be observed. The authors concluded that the difference in metal uptake was primarily due to the properties of the plants.

Concentration of heavy metals in aerial part is dependent on their extractable fraction in soil. This was observed by Meers et al. (2005) in a phytoremediation experiment with *Salix viminalis* "Orm" to assess heavy metal uptake on a moderately contaminated dredged sediment

Table 12.3 Literature review of *Salix* species involved in remediation of heavy metals

Species name	Medium	Findings of the study	Reference
Salix viminalis, Salix dasyclados, Salix daphnioides, Salix purpurea, S. triandra, and *Salix. Viminalis*	Heavy metal polluted and unpolluted areas	Heavy metal fraction in aerial part was not correlated with the extractable metal fraction in soil in both areas The *Salix* species collected from polluted areas (*S. viminalis* and *S. dasyclados* contained a higher concentrations of Cd, Cu, and Zn in their roots, due to higher tolerance The metals translocation is *Salix* species from polluted areas were lower, probably as a plant protection mechanism	Landberg and Greger (1996)
Salix dasyclados, Salix triandra, Salix ragiles, Salix purpurea × *Salix daphnoides, Salix schwerinii*	Pot experiment with three soil types: moderately contaminated dredge sediment-surface soil, heavily polluted sediment, sandy soil with moderate contamination	*Salix dasyclados, S. ragiles*, and *S. schwerinii* accumulated the maximum Cd and Zn concentration in the aerial part An annual amount of 5–27 kg per hectare Zn and 0.25–0.65 kg per hectare Cd can be harvested from these species Metal uptake is higher in pot experiments due to small surface area. If cultivated in field the metal uptake would be different Addition to ethylene diamine disuccinate, a chelating agent, improved the metal removal in the heavily polluted sediment and in sandy soil but not in the dredge sediment derived soil	Meers et al. (2007)
Salix viminalis	Field and growth chamber trials on contaminated soil obtained from construction-demolition sites	Metal uptake was 2–10 times higher in the field Leaves contained 80 mg/kg Cd and 3000 mg/kg Zn when grown a highly polluted soil, but the biomass production was poor Similar biomass was obtained for unpolluted control soil Willows are good remediators of moderately polluted soil Despite poor biomass the removal of a high concentration of heavy metal from soil helps in phytostabilization	Jensen et al. (2009)
Salix viminalis and *Salix alba* cultivars	Fertile brown soil and clay	*Salix viminalis* cultivars accumulated higher concentration of all heavy metals (Cu, Hg, Pb, Zn, Cd) Different cultivars also had different cellulosic content	Mleczek et al. (2010)

(continued)

Table 12.3 (continued)

Species name	Medium	Findings of the study	Reference
Salix × *smithiana,* Willd	Pot experiment to monitor Pb, Cd, and Zn in leachate To determine the effect of ectomycorrhizal inoculum on plant growth and metal uptake	Cd, Zn accumulated in large amounts in leaves. Pb was mostly in roots Cd and Zn decreased in leachate Inoculation improved plant growth but did not impact metal uptake. It reduced plant stress caused by heavy metals.	Trakal et al. (2013)
Salix viminalis L. clone 'SV068'	Pot experiment for two seasons with sediment containing high concentrations of Cr, Cu, Zn, Ni, Cd, Pb	Sediment containing heavy metals decreased plant growth in first season but improved robustness in terms of plant height and leaf area in 2015 Zn and Cd showed higher accumulation in aerial part (phytoextraction), while the other heavy metals, Pb, Cu, Ni, and Cr were accumulated in roots (phytostablization) Negatively impacted plant pigments	Pilipović et al. (2019)

Fig. 12.2 A view of *Salix* tree in Hungary

disposal site. The plant biomass production was high and ranged between 13.2–17.8 tons DW per hectares per year. But the plant did not uptake a large amount of pollutants. The concentration of Zn, Cd, and Pb in the bark and leaves were correlated well with their respective concentrations in the soil, but for the other elements (Cu, Ni, Cr) there was no difference between uptakes at the different sampling sites. The authors observed that increase in Zn and Cd uptake by plant depended on the increase in the extractable fraction of Zn and Cd in the soil, and not on the total fraction of these elements. The extractable fraction of Zn and Cd in the soil was high, indicating higher mobility of these elements in soil. On the other hand, Pb was poorly available in the extractable fraction. They concluded that using *Salix viminalis* for phytoremediation of such sites was both economically and environmentally profitable. Dredged sediments contain high amount of nutrition but have low economic value due to their contamination with heavy metals. *Salix viminalis* could thrive well on these polluted sites and produce good biomass, helping in remediation of such sites by improving the soil quality. In addition, they absorbed a high amount of Cd and Zn in their aerial part which could be harvested and used for metal recovery, while the biomass could be useful in providing energy.

Metal uptake by plant is dependent on the age of the plant and on seasonal change. Mertens et al. 2006 grew *Salix ragiles* L. and *Salix triandra* L. in a field with contaminated dredge sediments. They observed the heavy metal accumulation in the plant parts was in the following order: leaves > bark > wood. The accumulation pattern of heavy metals in the plant parts changed with the age of the plant. Old plants had the lowest metal concentration in their plant parts while young plants had the highest metal concentration. This phenomenon is explained by the fact that young tissues exhibit active growth which helps in higher uptake of the metals. Older plants further display a dilution effect of metals in their tissues, because they have a larger biomass and comparatively lower metal uptake. Metal concentrations in the plant parts showed a seasonal variation. This could be due to either a growth-dilution effect in the plant tissues or due to change in the available fraction of metals in the soil. Leaves displayed a larger fluctuation with season. Young growing leaves had lower concentrations of Cu, Cr, Ni, and Zn, which increased at the time of senescence. This change in concentration is promising for phytoremediation because plants harvested at the end of growth period will have the maximum biomass as well as the maximum metal concentration. In contrast to this Mleczek et al. (2009) observed that older plants (2–3 years) had higher metal accumulation than younger plants (1 year). They studied the cuttings of *Salix viminalis* to understand the effect of age on heavy metal uptake. All cuttings came from the same rootstock and had no leaves. The experiment was carried out in hydroponic pots to which seven heavy metals (Cd, Co, Cr, Ni, Pb, Zn, Cu) were added at 0.1, 0.5, 1, and 1.5 mmol concentrations. They postulated that the higher uptake of heavy metals by older willow was probably due to a well-developed root system. All plants accumulated high levels of metals which indicated the adaptation of the plant to the toxic surroundings, and a high root to shoot translocation rate.

The uptake of heavy metals by willows can be enhanced by the presence of chelating substances like ethylenediaminetetraacetic acid (EDTA), glyphosate, and organic acids like citric acid. Chelating agents like EDTA, citric acid, oxalic acid, and malic acid improves phytoextraction efficiency and plant growth. The addition of EDTA is known to increase the presence of Pb in the soil solution by a 1000-fold. Inorganic amendments like sulfur, ammonium sulfate and chloride salts also caused a similar increase in Cd and Zn concentration (Schmidt 2003; Salam et al. 2019). In pot studies with non-accumulator plants like pea and corn, addition of chelate increased the Pb concentration in shoot to 1% DW, which was unsustainable and led to plant death. Soil amendments are chosen based on the desired outcome of the experiment, whether phytoextraction or phytostabilization (Cunningham and Ow 1996). Salam et al. 2019 studied the effect of lime and bisphosphonates on the growth of a willow variety (*Salix viminalis* × *S.*

schwerinii × *S. dasyclados*) cultivated in soil with heavy contamination of Ni, Zn, and Cu, and irrigated with wastewater from a mine. They found that the soil amendments when individually applied greatly improved plant growth, photosynthesis, and phytoextraction. When the amendments were applied together the improvement in plant growth and phytoextraction efficiency was much higher.

12.12 Limitations in Using *Salix* for Bioremediation

Despite willows being very efficient bioremediators, they have some limitations (Ali et al. 2013; Davison 2005; Wenzel 2009):

- Plants used for phytoremediation could pose a risk in the food chain if consumed by grazing wild animals. The high elemental concentration of the leaves may cause harm to the animals and find its way higher up into the food chain.
- Since most of the pollutant is accumulated in the leaves, there is a need to collect the leaves in time, to avoid the re-release of the pollutant into the environment.
- There is a risk of cross-pollination and gene flow between cultivated clones and wild varieties which could result in the extinction of the wild varieties.
- Willows accumulate lower amount of pollutants as compared to their high biomasses. Despite the ability of willows to accumulate a large variety of pollutants, they can be used only in low to moderately contaminated sites. Jensen et al. 2009, observed that although willows accumulated high amount of Cd and Zn in their leaves (80 mg/kg and 3000 mg/kg, respectively) they were not suited for remediation in highly contaminated soils, due to their low biomass production.
- The remediation is constrained in the soil up to the zone where the roots reach. If the contamination has aged and reached into deeper layers, phytoremediation will not be very responsive.

12.13 Conclusion

Ùapid increase in urbanization and industrialization there is an ever-increasing burden of environmental pollution. Soil pollution, both agricultural and otherwise, impacts the food chain. Crops cultivated in contaminated soil accumulate the contaminants in their edible parts which reach human beings. There is thus an urgency to implement techniques which can help in remediation of the soil, without impacting soil fertility levels. Using plants to mitigate soil contamination is a low-cost and non-invasive method which helps in dealing with contamination at the site. Willows are an ideal candidate for use in phytoremediation studies because of their fast growth rate and their adaptability in all environments. They have an extensive root system and high evapotranspiration rate which allows them to transfer and store the contaminants in their above-ground biomass. The large number of diverse willow species are capable of remediating a wide range of contaminants. Along with being an able phytoremediator willows play a secondary role in preventing soil erosion by holding erosion prone soil or sediments with the help of their extensive network of fibrous roots. Willow roots release oxygen and nutrients thereby sustaining a thriving microbial population. This association with microbes further helps the plant in degrading and absorbing the contaminants. Bioremediation with willows is a cost-effective procedure because willows can propagate vegetatively and thrive in a range of environmental conditions. Further studies are needed to gain an in-depth understanding of the remediation potential of willows to implement them for large-scale bioremediation purposes in field.

References

Abdollahzadeh A, Osaloo SK, Maassoumi A (2011) Molecular Phylogeny of the Genus *Salix* (Salicaceae) with an emphasize to its species in Iran. Iran J bot

Ali H, Khan E, Sajad MA (2013) Phytoremediation of heavy metals-concepts and applications. Chemosphere 91:869–881. https://doi.org/10.1016/j.chemosphere.2013.01.075

Alkorta I, Garbisu C (2001) Phytoremediation of organic contaminants in soils. Bioresour Technol. https://doi.org/10.1016/S0960-8524(01)00016-5

Ball J, Carle J, Del Lungo A (2005) Contribution of poplars and willows to sustainable forestry and rural development. Unasylva 56:3–9

Baum C, Hrynkiewicz K, Leinweber P, Meißner R (2006) Heavy-metal mobilization and uptake by mycorrhizal and nonmycorrhizal willows (*Salix* x *dasyclados*). J Plant Nutr Soil Sci 169:516–522. https://doi.org/10.1002/jpln.200521925

Beauchamp S, Jerbi A, Frenette-Dussault C et al (2018) Does the origin of cuttings influence yield and phytoextraction potential of willow in a contaminated soil? Ecol Eng 111:125–133. https://doi.org/10.1016/j.ecoleng.2017.11.019

Białowiec A, Davies L, Albuquerque A, Randerson PF (2012) Nitrogen removal from landfill leachate in constructed wetlands with reed and willow: Redox potential in the root zone. J Environ Manage 97:22–27. https://doi.org/10.1016/j.jenvman.2011.11.014

Bissonnette L, St-Arnaud M, Labrecque M (2010) Phytoextraction of heavy metals by two Salicaceae clones in symbiosis with arbuscular mycorrhizal fungi during the second year of a field trial. Plant Soil 332:55–67. https://doi.org/10.1007/s11104-009-0273-x

Bollag J-M, Mertz T, Otjen L (1994) Role of microorganisms in soil Bioremediation. 2–10. https://doi.org/10.1021/bk-1994-0563.ch001

Carnegie D, Ramsay JA (2009) Anaerobic ethylene glycol degradation by microorganisms in poplar and willow rhizospheres. Biodegradation 20:551–558. https://doi.org/10.1007/s10532-008-9244-9

Chaney RL, Malik M, Li YM et al (1997) Phytoremediation of soil metals. Curr Opin Biotechnol 8:279–284. https://doi.org/10.1016/S0958-1669(97)80004-3

Charles JG (2014) Poplars and willows: trees for society and the environment

Corseuil HX, Moreno FN (2001) Phytoremediation potential of willow trees for aquifers contaminated with ethanol-blended gasoline. Water Res 35:3013–3017. https://doi.org/10.1016/S0043-1354(00)00588-1

Cronk Q, Ruzzier E, Belyaeva I, Percy D (2015) Salix transect of Europe: Latitudinal patterns in willow diversity from Greece to arctic Norway. Biodivers Data J. https://doi.org/10.3897/BDJ.3.e6258

Cunningham SD, Ow DW (1996) Promises and prospects of phytoremediation. Plant Physiol 110:715–719. https://doi.org/10.1104/pp.110.3.715

Da Cunha ACB, Sabedot S, Sampaio CH et al (2012) Salix rubens and salix triandra species as phytoremediators of soil contaminated with petroleum-derived hydrocarbons. Water Air Soil Pollut 223:4723–4731. https://doi.org/10.1007/s11270-012-1228-z

Dagher DJ, Pitre FE, Hijri M (2020) Ectomycorrhizal fungal inoculation of sphaerosporella brunnea significantly increased stem biomass of salix miyabeana and decreased lead, tin, and zinc, soil concentrations during the phytoremediation of an industrial landfill. J Fungi 6:1–12. https://doi.org/10.3390/jof6020087

Davison J (2005) Risk mitigation of genetically modified bacteria and plants designed for bioremediation. J Ind Microbiol Biotechnol 32:639–650. https://doi.org/10.1007/s10295-005-0242-1

Deavers K, Macek T, Karlson UG, Trapp S (2010) Removal of 4-chlorobenzoic acid from spiked hydroponic solution by willow trees (Salix viminalis). Environ Sci Pollut Res 17:1355–1361. https://doi.org/10.1007/s11356-010-0321-3

Dimitriou I, Aronsson P (2010) Landfill leachate treatment with willows and poplars—efficiency and plant response. Waste Manag 30:2137–2145. https://doi.org/10.1016/j.wasman.2010.06.013

Dimitriou I, Aronsson P (2011) Wastewater and sewage sludge application to willows and poplars grown in lysimeters-plant response and treatment efficiency. Biomass Bioenerg 35:161–170. https://doi.org/10.1016/j.biombioe.2010.08.019

Fosso-Kankeu E, Mulaba-Bafubiandi AF (2014) Implication of plants and microbial metalloproteins in the bioremediation of polluted waters: a review. Phys Chem Earth 67–69:242–252. https://doi.org/10.1016/j.pce.2013.09.018

Gomes HI (2012) Phytoremediation for bioenergy: challenges and opportunities. Environ Technol Rev 1:59–66. https://doi.org/10.1080/09593330.2012.696715

Gu JD (2018) Bioremediation of toxic metals and metalloids for cleaning up from soils and sediments. Appl Environ Biotechnol **3**:48–51. https://doi.org/10.26789/AEB.2018.02.006

Guarino C, Paura B, Sciarrillo R (2018) Enhancing phytoextraction of HMs at real scale, by combining salicaceae trees with microbial consortia. Front Environ Sci 6:1–11. https://doi.org/10.3389/fenvs.2018.00137

Iori V, Zacchini M, Pietrini F (2013) Growth, physiological response and phytoremoval capability of two willow clones exposed to ibuprofen under hydroponic culture. J Hazard Mater 262:796–804. https://doi.org/10.1016/j.jhazmat.2013.09.017

Isebrands JG, Richardson J (2014) Introduction. In: Poplars and willows: trees for society and the environment, CABI

Jadia CD, Fulekar MH (2009) Phytoremediation of heavy metals: recent techniques. African J Biotechnol 8:921–928. https://doi.org/10.5897/AJB2009.000-9152

Jensen JK, Holm PE, Nejrup J et al (2009) The potential of willow for remediation of heavy metal polluted calcareous urban soils. Environ Pollut 157:931–937. https://doi.org/10.1016/j.envpol.2008.10.024

Justin MZ, Zupančič M (2009) Combined purification and reuse of landfill leachate by constructed wetland and irrigation of grass and willows. Desalination 246:157–168. https://doi.org/10.1016/j.desal.2008.03.049

Justin MZ, Pajk N, Zupanc V, Zupančič M (2010) Phytoremediation of landfill leachate and compost wastewater by irrigation of Populus and Salix: biomass and growth response. Waste Manag 30:1032–1042. https://doi.org/10.1016/j.wasman.2010.02.013

Juwarkar AA, Singh SK, Mudhoo A (2010) A comprehensive overview of elements in bioremediation. Rev Environ Sci Biotechnol 9:215–288. https://doi.org/10.1007/s11157-010-9215-6

Kacálková L, Tlustoš P, Száková J (2015) Phytoextraction of risk elements by Willow and Poplar Trees. Int J Phytoremediation 17:414–421. https://doi.org/10.1080/15226514.2014.910171

Kahle P, Hildebrand E, Baum C, Boelcke B (2007) Long-term effects of short rotation forestry with willows and poplar on soil properties. Arch Agron Soil Sci 53:673–682. https://doi.org/10.1080/03650340701648484

Krämer U, Chardonnens AN (2001) The use of transgenic plants in the bioremediation of soils contaminated with trace elements. Appl Microbiol Biotechnol 55:661–672. https://doi.org/10.1007/s002530100631

Kuffner M, Puschenreiter M, Wieshammer G et al (2008) Rhizosphere bacteria affect growth and metal uptake of heavy metal accumulating willows. Plant Soil 304:35–44. https://doi.org/10.1007/s11104-007-9517-9

Kuzovkina YA, Quigley MF (2005) Willows beyond wetlands: uses of salix L. species for environmental projects. Water Air Soil Pollut 162:183–204. https://doi.org/10.1007/s11270-005-6272-5

Labrecque M, Teodorescu TI, Daigle S (1995) Effect of wastewater sludge on growth and heavy metal bioaccumulation of two Salix species. Plant Soil 171:303–316. https://doi.org/10.1007/BF00010286

Landberg T, Greger M (1996) Differences in uptake and tolerance to heavy metals in Salix from unpolluted and polluted areas. Appl Geochemistry 11:175–180. https://doi.org/10.1016/0883-2927(95)00082-8

Laureysens I, Blust R, De Temmerman L et al (2004) Clonal variation in heavy metal accumulation and biomass production in a poplar coppice culture: I. seasonal variation in leaf, wood and bark concentrations. Environ Pollut 131:485–494. https://doi.org/10.1016/j.envpol.2004.02.009

Lindegaard KN, Barker JHA (1996) Breeding willows for biomass. Asp Appl Biol Biomass Energy Crop 49:1–3

Lunáčková L, Šottníková A, Masarovičová E et al (2003) Comparison of cadmium effect on willow and poplar in response to different cultivation conditions. Biol Plant 47:403–411

Ma X, Li X, Liu J et al (2020) Enhancing Salix viminalis L.–mediated phytoremediation of polycyclic aromatic hydrocarbon–contaminated soil by inoculation with Crucibulum laeve (white-rot fungus). Environ Sci Pollut Res 27:41326–41341. https://doi.org/10.1007/s11356-020-10125-3

Mani D, Kumar C (2014) Biotechnological advances in bioremediation of heavy metals contaminated ecosystems: An overview with special reference to phytoremediation. Int J Environ Sci Technol 11:843–872. https://doi.org/10.1007/s13762-013-0299-8

Markus-Michalczyk H, Hanelt D, Jensen K (2016) Effects of Tidal Flooding on Juvenile Willows. Estuaries Coasts 39:397–405. https://doi.org/10.1007/s12237-015-0014-8

Marmiroli M, Pietrini F, Maestri E et al (2011) Growth, physiological and molecular traits in Salicaceae trees investigated for phytoremediation of heavy metals and organics. Tree Physiol 31:1319–1334. https://doi.org/10.1093/treephys/tpr090

McLeod KW, McPherson JK (1973) Factors limiting the distribution of *Salix nigra*. Bull Torrey Bot Club 102–110. https://doi.org/10.2307/2484398

Meers E, Lamsal S, Vervaeke P et al (2005) Availability of heavy metals for uptake by Salix viminalis on a moderately contaminated dredged sediment disposal site. Environ Pollut. https://doi.org/10.1016/j.envpol.2004.12.019

Meers E, Vandecasteele B, Ruttens A et al (2007) Potential of five willow species (Salix spp.) for phytoextraction of heavy metals. Environ Exp Bot 60:57–68. https://doi.org/10.1016/j.envexpbot.2006.06.008

Mertens J, Vervaeke P, Meers E, Tack FMG (2006) Seasonal changes of metals in willow (Salix sp.) stands for phytoremediation on dredged sediment. Environ Sci Technol 40:1962–1968. https://doi.org/10.1021/es051225i

Miller RS, Khan Z (2011) Comparison of Trichloroethylene Toxicity, Removal, and Degradation by Varieties of Populus and Salix for Improved Phytoremediation Applications. J Bioremediation Biodegrad S **7**. https://doi.org/10.4172/2155-6199.s7-001

Mleczek M, Łukaszewski M, Kaczmarek Z et al (2009) Efficiency of selected heavy metals accumulation by Salix viminalis roots. Environ Exp Bot 65:48–53. https://doi.org/10.1016/j.envexpbot.2008.03.003

Mleczek M, Rutkowski P, Rissmann I et al (2010) Biomass productivity and phytoremediation potential of Salix alba and Salix viminalis. Biomass Bioenerg 34:1410–1418. https://doi.org/10.1016/j.biombioe.2010.04.012

Mleczek M, Gąsecka M, Waliszewska B et al (2018) Salix viminalis L.—a highly effective plant in phytoextraction of elements. Chemosphere 212:67–78. https://doi.org/10.1016/j.chemosphere.2018.08.055

Ojuederie OB, Babalola OO (2017) Microbial and plant-assisted bioremediation of heavy metal polluted environments: a review. Int J Environ Res Public Health **14**. https://doi.org/10.3390/ijerph14121504

Palo RT (1984) Distribution of birch (Betula SPP.), willow (Salix SPP.), and poplar (Populus SPP.) secondary metabolites and their potential role as chemical defense against herbivores. J Chem Ecol. https://doi.org/10.1007/BF00988096

Pilipović A, Zalesny RS, Rončević S et al (2019) Growth, physiology, and phytoextraction potential of poplar and willow established in soils amended with heavy-metal contaminated, dredged river sediments. J Environ Manage 239:352–365. https://doi.org/10.1016/j.jenvman.2019.03.072

Pulford ID, Watson C (2003) Phytoremediation of heavy metal-contaminated land by trees—a review. Environ Int 29:529–540. https://doi.org/10.1016/S0160-4120(02)00152-6

Rentz JA, Alvarez PJJ, Schnoor JL (2005) Benzo[a]pyrene co-metabolism in the presence of plant root extracts and exudates: Implications for phytoremediation. Environ Pollut 136:477–484. https://doi.org/10.1016/j.envpol.2004.12.034

Rhodes CJ (2013) Applications of bioremediation and phytoremediation. Sci Prog 96:417–427. https://doi.org/10.3184/003685013X13818570960538

Salam MMA, Kaipiainen E, Mohsin M et al (2016) Effects of contaminated soil on the growth performance of young Salix (Salix schwerinii E. L. Wolf) and the potential for phytoremediation of heavy metals. J Environ Manage 183:467–477. https://doi.org/10.1016/j.jenvman.2016.08.082

Salam MMA, Mohsin M, Pulkkinen P et al (2019) Effects of soil amendments on the growth response and phytoextraction capability of a willow variety (S. viminalis × S. schwerinii × S. dasyclados) grown in contaminated soils. Ecotoxicol Environ Saf 171:753–770. https://doi.org/10.1016/j.ecoenv.2019.01.045

Schmidt U (2003) Enhancing phytoextraction : the effect of chemical soil manipulation on mobili ... Science (80) 1939–1954

Schnoor JL, Licht LA, McCutcheon SC et al (1995) Phytoremediation of organic and nutrient contaminants: pilot and full-scale studies are demonstrating the promise and limitations of using vegetation for remediating hazardous wastes in soils and sediments. Environ Sci Technol 29:318. https://doi.org/10.1021/es00007a747

Silva Gonzaga MI, Gonzaga Santos JA, Ma LQ (2006) Arsenic phytoextraction and hyperaccumulation by fern species. Sci Agric 63:90–101. https://doi.org/10.1590/s0103-90162006000100015

Slater H, Gouin T, Leigh MB (2011) Assessing the potential for rhizoremediation of PCB contaminated soils in northern regions using native tree species. Chemosphere 84:199–206. https://doi.org/10.1016/j.chemosphere.2011.04.058

Suresh B, Ravishankar GA (2004) Phytoremediation—a novel and promising approach for environmental clean-up. Crit Rev Biotechnol 24:97–124. https://doi.org/10.1080/07388550490493627

Tlustoš P, Száková J, Vysloužilová M et al (2007) Variation in the uptake of Arsenic, Cadmium, Lead, and Zinc by different species of willows Salix spp. grown in contaminated soils. Cent Eur J Biol 2:254–275. https://doi.org/10.2478/s11535-007-0012-3

Tőzsér D, Harangi S, Baranyai E et al (2018) Phytoextraction with Salix viminalis in a moderately to strongly contaminated area. Environ Sci Pollut Res 25:3275–3290. https://doi.org/10.1007/s11356-017-0699-2

Trakal L, Neuberg M, Száková J et al (2013) Phytoextraction and assisted phytoextraction of metals from agriculture used soil. Commun Soil Sci Plant Anal 44:1862–1872. https://doi.org/10.1080/00103624.2013.790403

United States Environmental Protection Agency (1999) https://www.epa.gov/sites/production/files/2015-04/documents/phytoresgude.pdf

Urbaniak M, Wyrwicka A, Tołoczko W et al (2017) The effect of sewage sludge application on soil properties and willow (Salix sp.) cultivation. Sci Total Environ 586:66–75. https://doi.org/10.1016/j.scitotenv.2017.02.012

Vamerali T, Bandiera M, Mosca G (2010) Field crops for phytoremediation of metal-contaminated land. A Review. Environ Chem Lett 8:1–17. https://doi.org/10.1007/s10311-009-0268-0

Verbruggen N, Hermans C, Schat H (2009) Molecular mechanisms of metal hyperaccumulation in plants. New Phytol

Vervaeke P, Luyssaert S, Mertens J et al (2003) Phytoremediation prospects of willow stands on contaminated sediment: A field trial. Environ Pollut 126:275–282. https://doi.org/10.1016/S0269-7491(03)00189-1

Vidali M (2001) Bioremediation. An overview. In: Pure and Applied Chemistry

Wang Q, Su X, Shrestha N et al (2017) Historical factors shaped species diversity and composition of Salix in eastern Asia. Sci Rep. https://doi.org/10.1038/srep42038

Wenzel WW (2009) Rhizosphere processes and management in plant-assisted bioremediation (phytoremediation) of soils. Plant Soil 321:385–408. https://doi.org/10.1007/s11104-008-9686-1

Wilkinson AG (1999) Poplars and willows for soil erosion control in New Zealand. Biomass Bioenerg 16:263–274. https://doi.org/10.1016/S0961-9534(99)00007-0

Yu XZ, Gu JD (2006) Uptake, metabolism, and toxicity of methyl tert-butyl ether (MTBE) in weeping willows. J Hazard Mater 137:1417–1423. https://doi.org/10.1016/j.jhazmat.2006.04.024

Yu XZ, Peng XY, Xing LQ (2010) Effect of temperature on phytoextraction of hexavalent and trivalent chromium by hybrid willows. Ecotoxicology 19:61–68. https://doi.org/10.1007/s10646-009-0386-2

Zalesny RS, Bauer EO (2007) Selecting and utilizing Populus and Salix for landfill covers: implications for leachate irrigation. Int J Phytoremediation 9:497–511. https://doi.org/10.1080/15226510701709689

Zalesny RS, Casler MD, Hallett RA, et al (2021) Bioremediation and soils

Zárubová P, Hejcman M, Vondráčková S et al (2015) Distribution of P, K, Ca, Mg, Cd, Cu, Fe, Mn, Pb and Zn in wood and bark age classes of willows and poplars used for phytoextraction on soils contaminated by risk elements. Environ Sci Pollut Res 22:18801–18813. https://doi.org/10.1007/s11356-015-5043-0

Złoch M, Tyburski J, Hrynkiewicz K (2015) Analysis of microbiologically stimulated biomass of Salix viminalis L. In the presence of CD2+ under in vitro conditions—implications for phytoremediation. Acta Biol Cracoviensia Ser Bot 57:67–78. https://doi.org/10.1515/abcsb-2015-0024

Złoch M, Kowalkowski T, Tyburski J, Hrynkiewicz K (2017) Modeling of phytoextraction efficiency of microbially stimulated Salix dasyclados L. in the soils with different speciation of heavy metals. Int J Phytoremediation 19:1150–1164. https://doi.org/10.1080/15226514.2017.1328396

Zvereva EL, Kozlov MV, Haukioja E (1997) Stress responses of *Salix borealis* to pollution and defoliation. J Appl Ecol 1387–1396

13 Bioremediation of Arsenic Contaminated Soil

Deepak Kumar Jha, Niti Yashwardhini, Vyomesh Vibhaw, and Kumar Pranay

Abstract

Arsenic (As) is a trace metalloid element present in many water ecosystems and soils of the world. The source of arsenic is natural as well as anthropogenic. Most arsenic is present in two ways in the environment: one is arsenite [As(III)], and the other is arsenate [As(V)]. Arsenite is more toxic as its show high affinity for thiol-containing glutathione and thiol-groups of cysteines in many proteins and thus inhibit important biochemical reactions. Arsenate functions as a phosphate analogue which, by the production of arsenylated derivatives, can alter biological reactions. Groundwater is polluted with arsenic (As) in more than 100 nations in the world at a rate greater than the permissible level prescribed by the World Health Organisation (10 μg/L). This represents a serious health hazard to around 150 million people living in these areas. Arsenic toxicity can cause melanosis, keratosis, skin and bladder cancer. The largest route of arsenic contamination has been proven to be groundwater used for drinking purposes polluted with arsenic. However, recently, several studies pointed out that arsenic can enter food chain from soils irrigated with groundwater contaminated with arsenic. Arsenic removal from groundwater includes adsorption, reverse osmosis, membrane filtration and coagulation. These methods are expensive and ineffective in the removal of arsenic at low concentration. Microorganisms have shown ability of bioremediation, i.e. cleaning of pollutants that are harmful to environment and human health. Microorganisms have also exhibited the ability of converting [As(III)] to [As(V)]. These organisms harbour As-resistance genes and exhibit greater genetic diversity in arsenic contaminated regions. These organisms have intrinsic ability of detoxifying, transforming, mobilising or immobilising arsenic through redox reactions, complexation, biomethylation and sorption. Hence, characterisation of microorganisms that can withstand high concentration of arsenic are very potent candidates for bioremediation of arsenic present in groundwater and agricultural fields.

D. K. Jha
P.C. Vigyan Mahavidyala, Chapra, Bihar 841301, India

N. Yashwardhini
Department of Microbiology, Patna Women's College, Patna, Bihar, India

V. Vibhaw
Department of Biochemistry, Patna University, Patna, Bihar 800005, India

K. Pranay (✉)
Department of Biochemistry, Indira Gandhi Institute of Medical Sciences, PatnaBihar 800014, India

J. A. Malik (ed.), *Advances in Bioremediation and Phytoremediation for Sustainable Soil Management*,
https://doi.org/10.1007/978-3-030-89984-4_13

Keywords

Arsenic · Arsenite · Bioremediation · Microorganisms · Toxicity

13.1 Introduction

The last 50 years has witnessed a rapid industrial growth across the world which has improved standard of life in many sectors. However, industrial growth has also contaminated our environment by releasing toxic compounds including heavy metals (Ali et al. 2013). Heavy metals and metalloids are toxic chemical entities which can hinder normal functioning of the cell in a number of ways, e.g. due to the release of reactive oxygen species, and they can adhere to respiratory components and cause oxidative damage (Canovas et al. 2003). Prominent heavy metals present in industrial waste include cadmium (Cd), chromium (Cr), nickel (Ni), zinc (Zn), cobalt (Co), mercury (Hg), lead (Pb), arsenic and iron (Fe). Amongst heavy metals, arsenic is extremely important as it is one of the most toxic and carcinogenic element in nature. Arsenic is a metalloid having atomic number and atomic weight as 33 and 74.92, respectively. Arsenic is a tasteless and fragrance free element with a half-life of 10 h. From health point of view, arsenic and its related compounds can be subdivided into three groups, namely, the elemental, inorganic and organic forms (Wu et al. 1989; NRC 2001). In environment, arsenic is mainly found in two oxidation states, one is arsenate [As(V)], and other is arsenite [As(III)]. Arsenite is more toxic as its shows high affinity for thiol-containing glutathione and thiol-groups of cysteines in many proteins and thus inhibits many important biochemical reactions (Shen et al. 2013). Arsenate behaves as a phosphate analogue and through the production of arsenylated substances, and it can modify biological processes (Gottesman and Mustaev 2018). Arsenic is chemically similar to essential element phosphate and hence can replace it in many biomolecules like ATP/ADP and DNA–protein complexes. Toxicity largely depends on chemical species as well as on dose and exposure time. Groundwater is polluted with arsenic (As) in more than 100 countries in the world at a rate greater than the acceptable level suggested by the World Health Organisation (WHO, 2006) (10 μg/L) (Singh and Stern 2017). According to USA Environmental Protection Agency's Superfund list, arsenic is the most prevalent environmental pollutant. Arsenic toxicity affects multiple organ systems including cardiovascular, auditory, developmental, hepatic, renal, nervous and respiratory systems. Arsenic is also a potent co-carcinogen and has been implicated in many types of cancers (lung, liver, skin and bladder). Arsenate [As(V)] and arsenite [As(III)] after entering the food chain are biotransformed by the process of methylation to less toxic organic form monomethylarsonic acid (MMA). MMA undergoes further methylation to form dimethylarsinic acid (DMA). Initially, methylation was thought to be a process that reduces toxicity (Kitchin 2001). However, recent studies suggest that both, MMA (III) and DMA (III) can cause cell toxicity, enzyme inhibition and genotoxicity (Styblo et al. 2002). In addition, DMA (III) induces oncogenes activation and produces bladder cancer in rats (Wei et al. 2002).

Arsenic removal from groundwater includes adsorption, reverse osmosis, membrane filtration and coagulation. These methods are expensive and ineffective in the removal of arsenic at low concentration. Bioremediation is a biotic procedure in which living entities are used to detoxify and degrade harmful environmental pollutants. In comparison with physical and chemical processes, bioremediation is slow but its effects are sustainable in nature. Bioremediation can be done in several ways and depends on the organism of choice which can be microorganism, plant or both. Biological byproducts are also used in bioremediation. Bioremediation can be done through several mechanisms such as phytoremediation, biosorption, bioaccumulation, biotransformation, biomineralization and bioleaching.

13.2 Sources of Arsenic

The occurrence of arsenic and its cycling has gained extreme importance because of the toxicity associated with it (Gihring and Banfield 2001). Sources of arsenic can be grouped into two categories; natural and anthropogenic (Table 13.1).

13.2.1 Natural Source of Arsenic

Arsenic, though in trace quantity, is pervasive in nature. Arsenic accumulation in our atmosphere is enhanced by geochemical processes such as leaching, habitat loss, rock weathering, nuclear explosion and forest fire (Naureen and Rehman 2016). Oxidation of arsenic containing minerals like arsenopyrite (FeAsS) in water has raised the accumulation of arsenic in water resources in recent decades (Shakoor et al. 2016).

13.2.2 Anthropogenic Sources of Arsenic

Tremendous industrial growth and economic activities have increased the concentration of arsenic manifolds in the earth's crust. In a study, it was found that every year 152,000–1120,000 tonnes of excess arsenic was being released into environment because of anthropogenic activities (Sarkar and Paul 2016). Another research confirmed that 35 g/kg of arsenic was added into environment due to industrial activities (Chappell et al. 2001).

13.2.3 Arsenic in the Food Chain

Living organisms accumulate heavy metals like arsenic and transfer them to next trophic level leading to biomagnification (Ali et al. 2019). Water, soil and sediments, components of abiotic environment are sources of heavy metals for living organisms. Arsenic (As) is a pollutant of the ecosystem and the food chain (Ali et al. 2019). Paddy crops like rice accumulate arsenic very easily which is a major health risk. Arsenate is easily reduced by plant species to arsenite, which is neutralised by thiol-rich peptide complex formation such as phytochelatins and/or vacuolar sequestration. A variety of prevention approaches, from agronomic and plant breeding steps to genetic engineering, may be used to minimise the arsenic intake by food crops.

13.3 Bioremediation of Arsenic

Bioremediation utilises microorganisms to eliminate or transform toxic environmental pollutants into less toxic or non-toxic forms. Microorganisms have acquired the ability of bioremediation through complex interactions with toxic metals. Metal ions can't be destroyed nonetheless microorganisms act upon them to reduce their toxicity by following mechanisms:

Table 13.1 Sources of arsenic pollution

Natural sources	Anthropogenic sources
Arsenic-rich sediments	Semiconductor industries
Volcanic eruptions	Insecticides
Disintegration of complexes of arsenic into pyrite ores	Pesticides
Weathering of arsenic bearing minerals	Mining and smelting activities
Marine aerosols	Coal combustion
Swampy soils and marshy areas	Phosphate fertilisers

(a) By increasing their solubility, microbes extract harmful metals from polluted areas.
(b) Microorganisms precipitate toxic metals ions out of soil solution and thus immobilising them.
(c) Microorganisms modify the chemical properties of metals by changing their oxidation states through redox action, e.g. arsenic oxidation (As(III) to As (V) and its reduction (As(V) to As (III).

Plethora of studies has focussed on the testing and certification of potential arsenic resistant or remediator microorganisms from arsenic contaminated sites as well as other sources. Metal ions are sequestered by microorganisms through the mechanisms like bioaccumulation, biotransformation and biosorption.

Bioaccumulation is a two-step ATP-dependent process in which, firstly metal ion binds to the cell wall of bacteria and then enters into the cell through specific metal-binding proteins such as metallothioneins (MTs). It is an energy dependent slow and irreversible process. In a study, it was reported that a strain of *Corynebacterium glutamicum* can accumulate arsenic (Satypal et al. 2016). *Brevibacillus brevis* can tolerate upto 1000 mg/L of arsenite and 500 mg/L of arsenate (Banerjee et al. 2013). Another study from Taiwan reported that a novel As(III)-oxidising bacterium strain can oxidise 2300 mg/L of arsenite. Similarly, *Marinomonas communis* has shown potential of accumulating 2290 μg/g of arsenic.

Biotransformation is another apparatus used by microorganisms by which they transform or modify a toxic form of metal ion into its less toxic form. This mechanism essentially involves oxidation–reduction process. Biotransformation of arsenic means conversion of arsenate (V) into arsenite (III) and methylation of As to MetAs compounds. Many bacterial strains capable of oxidising arsenic have been isolated and characterised (Elahi et al. 2019; Naureen and Rehman et al. 2016; Khan et al. 2015; Sarkar et al. 2013). In another study under aerobic conditions, As (III) was oxidised to less toxic form [As(V)] (Suhadolnik et al. 2017).

Biosorption is an energy independent process which involves passive uptake of metals by microorganisms which results in the formation of metalorganic complex with extracellular polymers or capsule synthesised by microorganisms. Biosorption is a slow and reversible process and has no toxic effect. Biosorption involves mechanism like Van der Waals forces, electrostatic interaction, redox interaction and chelation (Dhankhar and Hooda 2010). A sulphate-reducing bacterium has shown potential of removing 6.6% arsenite and 10.5% arsenate (Teclu et al. 2008). *Pseudomonas aeruginosa* has exhibited capacity of 98% biosorption (Tariq et al. 2019). *Bacillus* sp., *Thiothrix* sp. And *Pseudomonas* sp. Are routinely used for biosorption.

13.4 Phytoremediation of Arsenic

Plant-based detoxification or removal of environment pollutants is known as phytoremediation (Pulford and Watson 2003). Phytoremediation is sustainable in nature, and hence is widely employed in reducing arsenic from contaminated sites (Mesa et al. 2017). However, it is a slow process and often involves erticillate so plants used are either hyperaccumulators or tolerant species. Plants synthesise metabolites such as cysteine, glutathione, phytochelatins and non-protein thiols against arsenic induced stress (Dixit et al. 2016). Plant-based arsenic removal is more efficient, more expensive and requires less time in comparison of other remediation techniques (Wan et al. 2016). Plants involved in arsenic bioremediation should have high accumulation factor, high arsenic tolerance, short life cycle, high propagation rate and wide distribution (Visoottiviseth et al. 2002). Phytoremediation involves processes like phytoextraction, phytoaccumulation, phytostabilisation, phytofiltration and phytovolatilization.

Phytoaccumulation is primarily carried out by weeds or non-edible plants since it requires the ingestion of arsenic within the roots and its concentration. *Pteris vittata* represents the best example of phytoaccumulation for arsenic.

P. vittata in co-cultivation with rice not only removed arsenic but also resulted in significant decline in DMA level (Ye et al. 2011). In another study, *Arabidopsis thaliana* was engineered to harbour and code two bacterial genes *arsC* and γ-*ECS*. *arsC* codes for arsenite reductase that allowed conversion of arsenate into erticil, whilst γ-*ECS* codes for glutamyl cysteine synthase which leads to formation of thiols which helps in detoxification of As (III) by forming As-protein thiols which are sequestered in vacuoles (Doucleff and Terry 2002). To extract arsenic from polluted soil and water, terrestrial and marine plants may be used (Rahman and Hasegawa 2011). Some important aquatic plants used for arsenic removal from water bodies are *Lemna gibba*, *Lemina minor* and *Hydrilla erticillate* (Rahman and Hasegawa 2011).

Phytostabilisation is another important way of arsenic removal. It involves holding of contaminated sediments, water and soils in place by metallophytes and hyperaccumulators which results in immobilisation of contaminants in rhizosphere. This mechanism reduces the accumulation of metals in plants. *Retama sphaerocarpa* has shown efficient hyperstabilisation ability of heavy metals like As, Cd, Zn and Al (Moreno-Jimenez et al. 2011). This approach, however, only offers a temporary solution as only the mobility of metals is restricted in soil (Vangronsveld et al. 2009).

Another important method of arsenic phytoremediation is phytoextraction. The process involves accumulation of metals in the harvestable regions of hyperaccumulator plants. *Sedum alfredi*, *Rumex crispus* and *Viola baoshamesis* have been used for arsenic phytoextraction (Zhuang et al. 2007).

Phytovolatilization is a process in which plants uptake metals from soil and water, convert them into less toxic organic forms which are extruded into the air through stomatal openings. *Arundo donax* can volatilise about 7–22% of arsenic at 300–1000 μg/L (Mirza et al. 2011). In another study, transgenic rice was engineered to harbour a gene *arsM* from *Rhodopseudomonas palustris*. *arsM* codes for arsenic methyltransferase which converts arsenic to trimethylarsine (TMA) and can volatilise significant quantity of arsenic (Meng et al. 2011). The applicability of phytovolatilzation is under scrutiny as it causes release of metals in to air which could have hazardous health consequences (Padmavathiamma and Li 2007).

13.5 Mycoremediation

Cleaning of environment with the help of fungi is termed as mycoremediation. Like bacteria, fungi also use biosoprtion, biotransformation and bioaccumulation to transform metalloids and heavy metals (Igiri et al. 2018; Boriova et al. 2014). *Aspergillus, Fusarium, Emericella and Rhizomucor* have been isolated from agriculture fields contaminated with arsenic. In addition, these fungi improve the growth of plants. *Glomus mosseae*, an arbuscular mycorrhizal fungi has successfully decreased the arsenic content in plants (Wu et al. 2009). *Saccharomyces cerevisiae* has removed arsenic up to 90% from contaminated water (Roy et al. 2013).

13.6 Phycoremediation

Bioremediation based on algae and microalgae is called as phycoremediation. Microalgae decrease the toxicity caused due to inorganic arsenic by As (III) oxidation, complex formation with glutathione and phytochelations, extracellular adsportion, excretion from the cell, biotransformation into arsenosugars/arsenolipids and As (V) reduction (Papry et al. 2019). Phosphate has a similar outer electronic configuration as arsenate [As(V)] and so it plays a key role in bioaccumulation and detoxification of arsenic in microalgae (Sun et al. 2015). Low level of phosphate induces efficient uptake of arsenic from environment by high-affinity phosphate transporters like PIT and PST. On the other hand, high concentration of phosphate inhibits arensic uptake by microalgae such as *Chlorella* sp. (Bahar et al. 2016). Cultures of *Dunaliella salina* in association with symbiotic bacteria and under phosphate limiting condition induce the reduction of As (V) and excretion of As (III) and DMA (Wang et al. 2016). Microalgae use the mechanism of adsportion to remove arsenic at concentration below 100 mg/L (Jiang et al. 2011).

13.7 Phytobial Remediation

Phytoremediation assisted by endophytes is a new emerging approach of cleaning environment contaminated with heavy metals. In this process, higher plants work in close association with microorganisms and fungi for the remediation of polluted sites. For metal sequestration, this correlation is very beneficial as it can facilitate chelation, extracellular metal absorption redox balance and helps to increase metal concentrations. Studies have shown that arsenic speciation and bioavailability in soil–plant environments are impaired by microbial behaviour, and it has been investigated that arsenic toxicity can be minimised by plant-microbial induced remediation (Roy et al. 2015). Alternatively, previous studies have also reported that microbial action in soil can increase or decrease phytoremediation of arsenic (Khalid et al. 2017).

13.8 Metagenomics and Bioremediation of Arsenic

Metagenomics is a technique which utilises ecology, molecular biology and microbiology skills to analyse microbial communities at genomic level. It provides a picture of the uncultured microbiota. Metagenomic studies are extremely important with regard to bioremediation of arsenic as detailed knowledge of microbial flora controlling the biogeochemical cycle of arsenic in environment is difficult to obtain (Das et al. 2017). Metagenomic studies have been carried out for arsenic contaminated soils (Luo et al. 2014; Layton et al. 2014). Studies have shown that proteobacteria are the dominant phylum in arsenic contaminated soils (Layton et al. 2014). At the genus level, *Nitrosomonas* is the most prominent, followed by *Pirellula* (White et al. 2012). The *Pirellula* sp. Strain 1 has a complex array of arsenic metabolising and transporting genes (Glockner 2003). Metagenomic profiling has revealed that bacterial harbouring arsenic metabolising genes play a prominent role in biogeochemical cycle of arsenic.

13.9 Conclusion

Arsenic is a toxic metalloid found ubiquitously. Arsenic concentration in environment increases mainly from two sources; one is the natural source, whilst the other is anthropogenic activities. Arsenic toxicity is a major health hazard throughout the world particularly in South Asia. The presence of arsenic in terrestrial food chain and in aquatic waterbodies represents serious health implications. Scientific studies have extensively focussed on developing ways or tools of mitigating the menace of arsenic in which various chemical methods like membrane filtration, physical approach, biosorption, etc. However, these methods are expensive and incapable of removing arsenic at low concentration. So at present, the technique of bioremediation has gained importance because of its sustainable nature. The technique of bioremediation is inexpensive and eco-friendly. Microbial diversity present at sites contaminated with arsenic offers a promising tool of arsenic removal from the environment.

References

Ali H, Khan E, Ilahi I (2019) Environmental chemistry and ecotoxicology of hazardous heavy metals: environmental persistence, toxicity, and bioaccumulation. J Chem 2019:1–14 Article ID 6730305

Ali H, Khan E, Sajad MA (2013) Phytoremediation of heavy metals-concepts and applications. Chemosphere 91(7):869–881

Bahar MM, Megharaj M, Naidu R (2016) Influence of phosphate on toxicity and bioaccumulation of arsenic in a soil isolate of microalga *Chlorella* sp. Environ Sci Pollut Res 23(3):2663–2668

Banerjee S, Majumdar J, Samal AL, Bhattachariya P, Santra SC (2013) Biotransformation and bioaccumulation of arsenic by *Brevibacillus brevis* isolated from arsenic contaminated region of West Bengal. J Environ Sci Toxicol Food Technol 3(1):1–10

Boriova K, Cemansky S, Matus P, Bujdos M, Simonovicova A (2014) Bioaccumulation and biovolatilization of various elements using filamentous fungus *Scopulariopsis brevicaulis*. Lett Appl Microbiol 59(2):217–223. https://doi.org/10.1111/lam.12266

Canovas D, Cases I, De Lorenzo V (2003) Heavy metal tolerance and metal homeostasis in *Pseudomonas putida* as revealed by complete genome analysis. Environ Microbiol 5(12):1242–1256

Chappell WR, Abernathy CO, Calderon RL (eds) (2001) Arsenic exposure and health effects IV. Elsevier, New York, p 492

Das S, Bora SS, Yadav RNS, Barooah M (2017) A Metagenomic approach to decipher the indigenous microbial communities of arsenic contaminated groundwater of Assam. Genom Dat 12:89–96

Dhankhar R, Hooda A (2010) Fungal biosorption-an alternative to meet the challenges of heavy metal pollution in aqueous solutions. Environ Technol 32 (5):467–491

Dixit G, Singh AP, Kumar A, Mishra S, Dwivedi S, Kumar S, Trivedi PK, Pandey V, Tripathi RD (2016) Reduced arsenic accumulation in rice (*Oryza sativa* L.) shoot involves sulfur mediated improved thiol metabolism, antioxidant system and altered arsenic transporters. Plant Physiol Biochem 99:86–96

Doucleff M, Terry N (2002) Pumping out the arsenic. Nat Biotechnol 20:1094–1096

Elahi A, Rehman A (2019) Comparative behavior of two gram positive Cr^{6+} resistant bacterial strains *Bacillus aerius* S1 and *Brevibacterium iodinum* S2 under hexavalent chromium stress. Biotechnol Rep **21**: e00309

Gihring TM, Banfield JF (2001) Arsenite oxidation and arsenate respiration by a new Thermus isolate. FEMS Microbiol Lett 204(2):335–340

Glockner FO, Kube M, Bauer M, Teeling H, Lombardot T, Ludwig W, Gade D, Beck A, Borzym K, Heitmann K, Rabus R, Schesner H, Amann R, Reinhardt R (2003) Complete genome sequence of the marine planctomycete *Pirellula* sp. Strain 1. Proc Natl Sci USA 100:8298–8303

Gottesman ME, Mustaev A (2018) Inorganic phosphate, arsenate, and vanadate enhance exonuclease transcript cleavage by RNA polymerase by 2000- fold. Proc Natl Acad Sci USA 13:2746–2751. https://doi.org/10.1073/pnas.1720370115

Igiri BE, Okoduwa SIR, Idoko GO, Akabuogu EP, Adeyi AO, Ejiogu IK (2018) Toxicity and bioremediation of heavy metals contaminated ecosystem from tannery wastewater: a review. J Toxicol. https://doi.org/10.1155/2018/2568038

Jiang Y, Purchase D, Jones H, Garelick H (2011) Effects of arsenate (As^{5+}) on growth and production of glutathione (GSH) and phytochelatins (PCS) in *Chlorella vulgaris*. Int J Phytoremediat 13(8):834–844

Khalid S, Shahid M, Niazi NK, Rafiq M, Bakhat HF, Imran M, Abbas T, Bibi I, Dumat C (2017) Arsenic behaviour in soil-plant system: biogeochemical reactions and chemical speciation influences. In: Enhancing cleanup of environmental pollutants. Springer, Cham, pp 97–140. https://doi.org/10.1007/978-3-319

Khan Z, Nisar MA, Hussain SZ, Arshad MN, Rehman A (2015) Cadmium resistance mechanism in *Escherichia coli* P4 and its potential use to bioremediate environmental cadmium. Appl Microbiol Biotechnol 99:10745–10757

Kitchin KT (2001) Recent advances in arsenic carcinogenesis: modes of action, animal model systems, and methylated arsenic metabolites. Toxicol Appl Pharmacol 172:249–261

Layton AC, Chauhan A, Williams DE, Mailloux B, Knappett PSK, Ferguson AS, Mckay LD, Alam MJ, Ahmed KM, Geen AV, Sayler GS (2014) Metagenomes of microbial communities in arsenic-and pathogen-contaminated well and surface water from Bangladesh. Genome Announc **2**(6):e01170

Luo J, Bai J, Liang J, Qu (2014) Metagenomic approach reveals variation in microbes with arsenic and antimony metabolism genes from highly contaminated soil. PLoS One **9**:e108185

Meng XY, Qin J, Wang LH, Duan GL, Sun GX, Wu HL, Chu CC, Ling HQ, Barry PR, Zhu YG (2011) Arsenic biotransformation and volatilization in transgenic rice. New Phytol 191(1):49–56

Mesa V, Navazas A, González-Gil R, González A, Weyens N, Lauga B, Gallego JLR, Sánchez J, Peláez AI (2017) Use of endophytic and rhizosphere bacteria to improve phytoremediation of arsenic contaminated industrial soils by autochthonous *Betula celtiberica*. Appl Environ Microbiol 83(8):e03411–e03416

Mirza N, Arshid PA, Qaisar MQ, Shah MM, Shafqat MN (2011) Ecological restoration of arsenic contaminated soil by *Arundo donax*. Ecol Eng 37 (12):1949–1956

Moreno-Jimenez EE, Vázquez S, Carpena-Ruiz RO, Esteban E, Penalosa JM (2011) Using Mediterranean shrubs for the phytoremediation of a soil impacted by pyritic wastes in Southern Spain: A field experiment. J Environ Manag 92:1584–1590

Naureen A, Rehman A (2016) Arsenite oxidizing multiple metal resistant bacteria isolated from industrial effluent: their potential use in wastewater treatment. World J Microbiol Biotechnol 32(8):133

NRC (National Research Council) (2001) Arsenic in drinking water: update. DC, National Academy Press, Washington

Padmavathiamma PK, Li LY (2007) Phytoremediation technology: hyperaccumulation metals in plants. Water Air Soil Pollut 184:105–126

Papry RI, Ishii K, Mamun MAA, Miah S, Naito K, Mashio AS, Maki T, Hasegawa H (2019) Arsenic biotransformation potential of six marine diatom species: effect of temperature and salinity. Sci Rep 9 (1):10226

Pulford ID, Watson C (2003) Phytoremediation of heavy metal contaminated land by trees-a review. Environ Int 29(4):529–540

Rahman MA, Hasegawa H (2011) Aquatic arsenic: phytoremediation using floating macrophytes. Chemosphere 83(5):633–646
Roy D, Gaur P, Verma N, Pathireddy M, Singh KP (2013) Bioremediation of Arsenic (III) from water using baker yeast *Sacchromyces cerevisiae*. Int J Environ Bioremed Biodegrad 1(1):14–19
Roy M, Giri AK, Dutta S, Mukherjee P (2015) Integrated phytobial remediation for sustainable management of arsenic in soil and water. Environ Int 75:180–198. https://doi.org/10.1016/j.envint.2014.11.010
Sarkar A, Kazy SK, Sar P (2013) Characterization of arsenic resistant bacteria from arsenic rich groundwater of West Bengal. India. Ecotoxicology 22(2):363–376
Sarkar A, Paul B (2016) The global menace of arsenic and its conventional remediation—a critical review. Chemosphere **1**(158):37–49. https://doi.org/10.1016/j.chemosphere.2016.05.043
Satyapal GK, Rani S, Kumar M, Kumar N (2016) Potential role of arsenic resistant bacteria in bioremediation: current status and future prospects. J Microb Biochem Technol 8(3):256–258
Shakoor MB, Niazi NK, Bibi I, Murtaza G, Kunhikrishnan A, Seshadri B, Shahid M, Ali S, Bolan NS, Ok YS, Abid M, Ali F (2016) Remediation of arsenic-contaminated water using agricultural wastes as biosorbents. Crit Rev Environ Sci Technol 46(5):467–499
Shen S, Li XF, Cullen WR, Weinfeld M, Le XC (2013) Arsenic binding to proteins. Chem Rev 113:7769–7792. https://doi.org/10.1021/cr300015c
Singh SK, Stern EA (2017) Global arsenic contamination: living with the poison nectar. Environment 59(2):24–28
Styblo M, Drobna Z, Jaspers I, Lin S, Thomas DJ (2002) The role of biomethylation in toxicity and carcinogenicity of arsenic: a research update. Environ Health Perspect 110(5):767–771
Suhadolnik ML, Salgado AP, Scholte LL, Bleicher L, Costa PS, Reis MP, Dias MF, Ávila MP, Barbosa FA, Chartone-Souza E (2017) Novel arsenic-transforming bacteria and the diversity of their arsenic-related genes and enzymes arising from arsenic polluted freshwater sediment. Sci Rep **7**(1):11231. https://doi.org/10.1038/s41598-017-11548-8
Sun J, Cheng J, Yang Z, Li K, Zhou J, Cen K (2015) Microstructures and functional groups of *Nannochloropsis* sp. cells with arsenic adsorption and lipid accumulation. Bioresour Technol 194:305–311
Tariq A, Ullah U, Asif M, Sadiq I (2019) Biosorption of arsenic through bacteria isolated from Pakistan. Int Microbiol 22(1):59–68
Teclu D, Tivchev G, Laing M, Wallis M (2008) Bioremoval of arsenic species from contaminated waters by sulphate-reducing bacteria. Water Res 42 (19):4885–4893
Vangronsveld J, Herzig R, Weyens N, Boulet J, Adriaensen K, Ruttens A, Thewys T, Vassilev A, Meers E, Nehnevojova E, Lelie DVD, Mench M (2009) Phytoremediation of contaminated soils and groundwater: lessons from the field. Environ Sci Pollut Res 16:765–794
Visoottiviseth P, Francesconi K, Sridokchan W (2002) The potential of Thai indigenous plant species for the phytoremediation of arsenic contaminated land. Environ Pollut 118(3):453–461
Wan X, Lei M, Chen T (2016) Cost-benefit calculation of phytoremediation technology for heavy-metal-contaminated soil. Sci Total Environ 563:796–802
Wang Y, Zhang CH, Lin MM, Ge Y (2016) A symbiotic bacterium differentially influences arsenate absorption and transformation in *Dunaliella salina* under different phosphate regimes. J Hazard Mater 318:443–451
Wei M, Wanibuchi H, Morimura K, Iwai S, Yoshida K, Endo G, Nakae D, Fukushima S (2002) Carcinogenicity of dimethylarsinic acid in male F344 rats and genetic alterations in induced urinary bladder tumors. Carcino 23(8):1387–1397
White CP, Debry RW, Lytle DA (2012) Microbial survey of a full scale, biologically active filter for treatment of drinking water. Appl Environ Microbiol 78:6390–6394
Wu FY, Ye ZH, Wong MH (2009) Intraspecific differences of arbuscular mycorrhizal fungi in their impacts on arsenic accumulation by *Pteris vittata* L. Chemosphere 76(9):1258–1264
Wu MM, Kuo TL, Hwang YH (1989) Dose-response relation between arsenic concentration in well water and mortality from cancers and vascular diseases. Am J Epidemiol 130:1123–1132
Ye WL, Khan MA, McGrath SP, Zhao FJ (2011) Phytoremediation of arsenic contaminated paddy soils with *Pteris vittata* markedly reduces arsenic uptake by rice. Environ Pollut 159(12):3739–3743
Zhuang P, Yang QW, Wang HB, Shu WS (2007) Phytoextraction of heavy metals by eight plant species in the field. Water Air Soil Pollut 184:235–242

Bioremediation and Detoxification of Asbestos from Soil

14

T. Gopishankar, H. Baraiya Divyeksha, V. L. Vasantha, and N. Praveen

Abstract

Asbestos is referred to as magic mineral and used as excellent building material. It finds its application in wide range of products such as floor tiles, pipes, paper, rope, cloth, insulated partition board, etc. On average, India uses 3,50,000 tons of asbestos annually and asbestos fibers readily undergo weathering releasing them into soil, water and air. Occupational and environmental exposure to this asbestos is leading to asbestosis (asbestos-related disease), lung cancer, and heart failure. Considering the serious health risk, countries like Australia, Brazil, and Canada had banned the use of asbestos. As asbestos is extensively used in construction of buildings, the demolished materials are dumped in the soil and thus it finds its route in soil as pollutant. Soil borne microbes like bacteria, fungi and lichens are found to be best means to reduce the toxicity of asbestos. These microorganisms remove iron from asbestos and reduce its toxicity. Another most effective bioremediation approach is phytoremediation to clean up the soil wherein vegetative cover on contaminated soil can remove iron and breaks down asbestos as source of inorganic nutrient. The main advantage of phytoremediation is that it can be extended to any geographical area where plants can grow. This chapter emphasizes various means of use and disposal of asbestos, followed by various means of bioremediation using microbes and plants and as an alternate for the sustainable soil condition.

Keywords

Asbestos · Asbestosis · Bioremediation · Detoxification · Phytoremediation · Pollution · Mesothelioma

14.1 Introduction

The word 'asbestos' is derived from a Greek word which means 'inextinguishable' or 'unquenchable', and is being used for its high tensile strength, flexibility, large surface area, heat and chemical resistance, and non-conductive property. Asbestos actually occurs in every region in the world (King 2020). The functionality and application of mineral was known centuries ago which brought an industrial revolution. Archaeologists uncovered asbestos fibers in rubble dated back to the Stone Age, about 750,000 years ago. By the end of the first century, cremation garments, sheets, and wicks for temple lights were crafted by chrysotile asbestos from Cyprus and

T. Gopishankar · H. Baraiya Divyeksha ·
V. L. Vasantha · N. Praveen (✉)
Department of Life Sciences, CHRIST (Deemed To Be University), Bangalore, Karnataka 560029, India
e-mail: praveen.n@christuniversity.in

J. A. Malik (ed.), *Advances in Bioremediation and Phytoremediation for Sustainable Soil Management*,
https://doi.org/10.1007/978-3-030-89984-4_14

tremolite asbestos from Italy. There are clay pots in Finland dating back to 2500 B.C. made of asbestos fibers, which are thought to fortify the pots besides making them resilient to fire (King 2020). In ancient times in Europe and Asia, asbestos was given the name "magic mineral", from the idea of mineral wool which can be spun and woven which was commended itself to the Romans (Murray 1990). During early 1700s, paper made out of asbestos was discovered in Italy. Global demand for asbestos-based goods has exploded since the late 1800s. Asbestos ores were in great demand and gave more than 3,000 different products (King 2020). The term asbestos is given to the category of six naturally existing minerals consisting of smooth, versatile fibrous structures dependent on silicate (Wallis et al. 2020). There are two configurations according to its chemistry and morphology: serpentine and amphibole. Here serpentine mineral group includes chrysotile asbestos and amphibole mineral group includes actinolite, amosite (brown), anthophyllite, remolite, and crocidolite (blue) asbestos. Chrysotile (white asbestos) accounts for about 95% of usage of asbestos in twentieth century (Virta 2005). Physical properties of serpentine mineral, chrysotile is heat resistant, fibrillary structure, large surface area, and resistant to higher temperature. This is a commercial mineral consisting of hydrated magnesium silicates. Furthermore, moderate amount of iron and aluminum can substitute magnesium and silicon respectively (Radetzki 2010). Amphibole asbestos is much more brittle, resistant to chemical attack, hardness, has high tensile strength, and greater resistance towards high temperatures. The amphibole asbestos is hydrated silicate mineral and due to its crystalline nature, considerable amount of elemental substitution occurs in crystal lattice (Virta 2005). Thousands of commercial products, such as insulation and fireproofing materials, car brakes and textile products, water supply lines, fire blankets, gaskets, cement, and wallboard materials, have been using these minerals for decades (CDC 2012; WHO 2018). Ever since the industrial revolution, its low cost has further extended its applicability (Bhattacharya et al. 2015). The United States imports asbestos and produces cement products, paper, footwear, yarn and thread, mill board, gaskets, civil crafts, fabricated asbestos fibers, clothing pads, friction materials, insulating materials, cement pipes, plastic flooring and thermal insulation (Fig. 14.1) (Kumar et al. 2016). The latest report by the United States Geological Survey (USGS) showed gradual decrease in usage of asbestos from year 2013–2018, in the United States 100 tons of asbestos was imported from Russia (a top worldwide mineral producer) and consumed. In 2019, total 1.1 million tons of waste were produced worldwide where Russia, Kazakhstan, and China were the top producers. Due to adverse effects of asbestos on health, more than 60 countries have banned the usage of asbestos including Australia and the UK. Recently, in June 2019 Environmental Protection Agency (EPA) passed a Significant New Use Rule (SNUR) which can prohibit the use of asbestos until EPA grants permission (USGS 2020).

Asbestos is responsible for causing cancer, mesothelioma, pneumoconiosis, other respiratory disorders, soil, water, and air pollution. Exposure and inhalation of asbestos fibers result in pleural calcification, lung disorders, peritoneal mesothelioma, pulmonary fibrosis, and several cancers. Approximately 125 million persons are subjected to asbestos in their workplaces, leading to asbestos-related diseases (ARD) and non-occupational asbestos exposures (WHO 2014). During the weathering of natural deposits, extraction of minerals, refining, wearing down of processed asbestos materials, and excavation of asbestos-containing rocks, these fibers are present in the atmosphere and their internalization can contribute to the deposition of fibers in the lungs (Kilburn 2000). After inhalation, the fibers get deposited in lung parenchyma cells leading to redox chemical reactions (Daghino et al. 2006) and also generation of free radicals and reactive oxygen species (ROS) (David et al. 2021), which can lead to development of severe lung disorders and cancer. The meta-analysis was performed to explain the relationship between asbestos exposure and mortality of cardiovascular-related disease, resulting in a substantially increased risk of

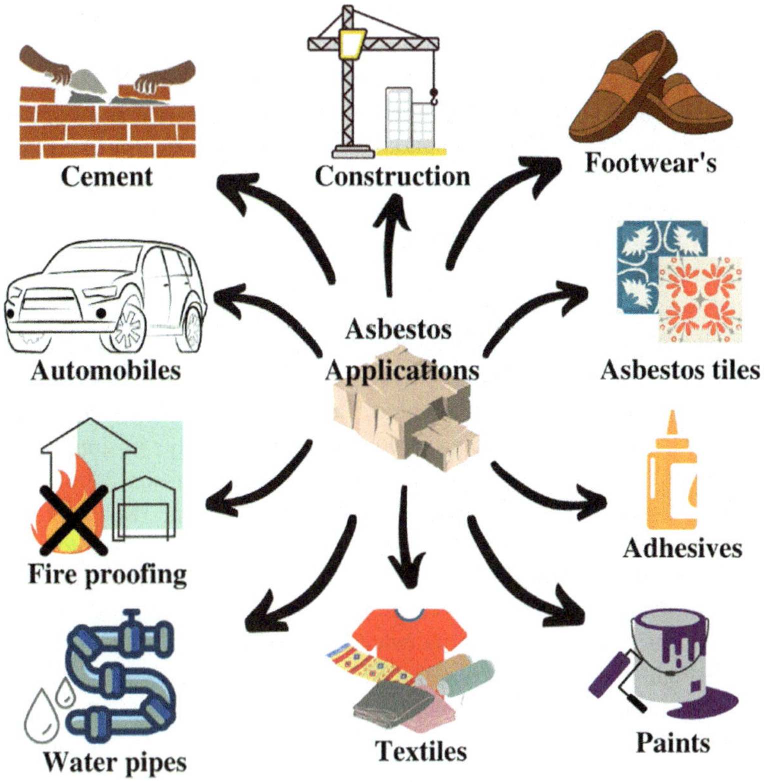

Fig. 14.1 Application of asbestos mineral in different industrial sectors

cardiovascular-related diseases among the occupational workers' (Rong et al. 2015). This mineral not only affects mankind but also environment, by destroying aquatic water bodies, affecting vegetation of agricultural lands, and habitat loss (Kumar et al. 2016). Water bodies located near asbestos mines have fibers and this contaminated water with toxic fibers and heavy metals will be used in agriculture (Kumar and Maiti 2015a) and thus increases toxicity of plants. This fiber consists of Cr and Ni heavy metals which will increase the toxicity of water. These fibers can also contribute to air pollution with release of silicate and high-fluoride content in atmosphere (Subramanian and Madhavan 2005). Considering huge amount of asbestos fibers produced by Asbestos Containing Products (ACP) and Asbestos Containing Waste (ACW), on the 30th Jan 2013 European Parliament quoted "to work with the social partners and other stakeholders at European, national and regional levels to develop and share action plans for asbestos removal and management" (Spasiano and Pirozzi 2017). Initially, the ACW were disposed in controlled landfills, but this would not eliminate the problem of asbestos fiber release. So along with this several scientific findings gave an outcome for detoxification of these wastes, this approach included physical, chemical, and biological remediation of pollutants. Before remediation, the concentration and type of asbestos in contaminated soil is determined and necessary risk assessment could be done for remediation plan (Wroble et al. 2017). Different methods of remediation are available which include physical, chemical, and biological, though, physical and chemical approaches are costly and itself can lead to pollution. These methodologies are responsible for changing the chemical and morphological properties of asbestos fibers, thus resulting in inert and safe compounds. Most of the physical methods include thermal, mechanical, and vitrification (Spasiano and Pirozzi 2017; Paolini et al. 2018). Among these, thermal treatment for asbestos remediation is the most effective method but has relatively high energy consumption. Microwave assisted detoxification of asbestos was done with different heating mechanisms (Yoshikawa et al. 2015) depending upon decreased thickness of chrysotile fibers. Thermal treatments are being successfully developed based on chrysotile and amphibole asbestos undergoing dihydroxylation and recrystallization (Tomassetti et al. 2020). These physical methods are involved in detoxification of asbestos or giving out non-hazardous products which can be further degraded. Alternative to physical methods, there are several chemical methods which require lower temperature and does chemical degradation of asbestos fibers either by using acidic environment (Teir et al. 2007; Turci et al. 2010) reducing agents with high temperatures (Porcu et al. 2005) or using supercritical fluids with hydrothermal conditions (Anastasiadou et al. 2010). Also, mechano-chemical methods resulted in morphological changes of asbestos fibers when chrysotile asbestos fibers were subjected to oxalic acid along with powered ultrasound. Oxalic acid's leaching effect on the structural cations, operating at the same time as strong acoustic cavitation, detoxifies the mineral (Turci et al. 2008). Due to high performance, less toxic environmental production, and cost-effectiveness, biological methods have lately been given greater importance. There are 2 main approaches in biological methods: Bioremediation and Phytoremediation for clearance of asbestos or asbestos generated waste. Microorganisms such as bacteria, lichens, algae are already in use for remediating the toxic heavy metals from contaminated sites; similarly, microorganisms are in use for removal of hazardous asbestos. Weathering of asbestos by siderophore production from fungus and bacteria leads to extraction of iron from fibers and thus reducing asbestos toxicity (Mohanty et al. 2018). Furthermore, great attention is given to phytoremediation in order to reduce the toxicity of asbestos from environment.

14.2 Asbestos Toxicity

Despite the highly toxic nature of asbestos fibers, the processing of asbestos is still continuing. All forms of asbestos which are being processed can cause all of the acute respiratory distress syndromes with no limiting exposure levels. Depending on the fiber dimension they are deposited in the lungs and can be carried to pleural or various other organs and tissues (Jung et al. 2020; Barlow et al. 2017). Due to the biopersistent nature of these nano sized fibers, they internalize and gradually accumulate especially in lungs, lymph nodes (Dodson and Atkinson 2006), pleura (Miles et al. 2008) and larynx (Roh et al. 2016). These nano sized fibers exhibit carcinogenic activity because of the presence of iron at its surface which causes inflammation and scarring of the lungs that restricts lung expansion. The active iron present on the surface binds with the DNA of pulmonary cells and induces the development of hydrogen peroxide from immune cells that induce ROS and trigger cell damage; even a very minimal quantity of iron stimulates radical activity for which the location of iron within the fiber is significant (Pollastri et al. 2015). Progressive fiber dissolution can provide the surface area with bulk iron concentration. For example, Chrysotile is not rich in irons, but the expected rates of fiber dissolution are higher than amosite or crocidolite. Therefore, iron is more usable, despite its low content, and the capacity of chrysotile fibers to produce surface-related HO is forecast to be as high as crocidolite fibers with greater iron richness (Pollastri et al. 2015). The measurement of the release of H_2O_2 and other types of reactive oxygen (ROS) is thus essential and remains subject to review (Balamurugan et al. 2018).

14.2.1 Mechanism of Toxicity

A recent epidemiological study carried out by Dusinska et al. (2004) on the toxic consequences of asbestos on humans has shown that, compared to non-asbestos exposed individuals, asbestos workers had significantly greater chromosome aberrations with oxidized pyrimidine foundation and DNA alkylation damage. Meanwhile, oxidative stress resulted in increased levels of DNA double strand breaks in chrysotile-exposed asbestos workers (Marczynski et al. 1994). Recent research suggests that oxidative damage to DNA can also possibly be due to modifications or oxidation of the base of cytosine (Valinluck and Sowers 2007). The cytotoxic effects of crocidolite asbestos fibers were the result of several events, including: (a) secretion of apoptotic factors due to fiber release of calcium ions; (b) interaction with the complex of mitochondrial cytochrome oxidase; (c) increased mitochondrial generation of reactive oxygen species. Koi and Barrett (1986) reported that due to asbestos there is a high chance of tumor suppression gene loss which accounts for its carcinogenicity. Several mechanisms have been suggested as the possible iron reactions to catalyze DNA oxidation, contributing to cancer susceptibility to asbestos fibers and dust. At genetic level deletion or hypermethylation at the CDKN2A/ARF locus on chromosome 9p21, which carries three significant tumor suppressor genes (i.e., p15, p16INK4A, and p14ARF) is a major event associated with asbestos-induced diffuse malignant mesothelioma (Murthy and Testa 1999). Nymark et al. (2008) discovered that lung carcinoma in asbestos exposed workers is due to the point mutations in p53 tumor suppressor gene and loss of heterozygosity. He also stated that the relationship between lung tumors exposed to asbestos and genomic changes in chromosomes 19p13, 2p16, and 9q33.1 indicated an allelic mismatch (Nymark et al. 2013).

In aquatic macrophytes, duckweed *Lemma gibba* chrysotile asbestos exhibited phytotoxic effects, which showed abnormalities in chlorophyll, carotenoids, free sugar, starch, protein content and inhibitory effects on root length and biomass (Trivedi et al. 2004). The primary characteristics of asbestos to cause ROS mediated oxidative stress have been described in the aquatic macrophyte *Lemma gibba* duckweed. Asbestos exposure has substantially decreased the amount of oxidized glutathione and ascorbate, which are antioxidants in the system (Trivedi et al. 2007).

14.3 Risks with Asbestos: Environmental and Health Risks

It is difficult to understate the impact of asbestos on occupational and ecological health. Exposure to all forms of asbestos can cause severe longstanding issues including malignant mesothelioma (MM), ARD, and lung cancer. Currently, the World Health Organization reports that exposure to asbestos causes more than half the worldwide deaths from occupational cancers (Arsenic, metals, fibers, and dusts 2012).

14.3.1 Environmental Risks

Asbestos production by extracting and processing goods containing asbestos produces waste that may pose a major environmental danger. The manufacturing and smashing of host rocks, resulting in the elimination of small asbestos fibers comprising heavy metals seems to be the most effective method of asbestos exposure. These fibers are long suspended in the air and accumulated into the soil that is leached into the surface and transudate by water into the surface and groundwater. In addition, increased toxic Cr, Ni and other metals contribute to scarce growth of the vegetation in mine sites (Kumar and Maiti 2015b). Another major sources responsible for the introduction of asbestos into environment are the disposal of toxic industrial asbestos wastes, asbestos rock erosion, and wind erosion (Kumar et al. 2016). Trivedi and Ahmad (2013) revealed that toxic fibers and metals are imparted into the vegetation cover by the dissipation of these fibers in the waterways that are the primary sources of drinking water. In addition, these materials possess, along with many other metals, significant quantities of toxic Cr and Ni that maximize the toxic effects of water resources. Epidemiological findings have shown that both positive and negative correlations between asbestos-contaminated drinking water and community asbestos exposure (Berndt and Brice 2008).

14.3.2 Health Risks

Exposure to asbestos and asbestos-related products causes a significant effect on humans which results in chronic disorders and health risks. Because of its curly and pliable form, it is easily inhaled and deposited on the airway, i.e., nose, throat, and lungs (Morgan 1997). Once fine fibers penetrate the nostrils or gastrointestinal tract, they become bound to the lungs, resulting in cancer and other associated disorders (De Guire et al. 2005). Asbestos exposure-related primary diseases include tumors, especially mesothelioma, lung cancer, and are predominantly concerned with five common noncancerous conditions, i.e., chronic pleural or pericarditis; diffuse pleural inflammation; pleural plaques; rounded atelectasis; and asbestosis, all of which have distinct radiological findings and prognoses (Solbes and Harper 2018). In both, a steady decline in lung capacity and progressive radiological results, asbestosis is a recurrent and eventually lethal disease. On the other hand Pease and Kratzke (2017) documented that environmental exposure to asbestos strongly associated with Mesothelioma (cancer that grows on the body in each mesothelial surface) and non-small cell lung cancer (NSCLC). Generally, after the initiation of a symptomatic pleural effusion, mesothelioma will first come to medical attention. This is normally one-sided and contributes to dyspnea. Care needs to be taken not to assume that in a person with asbestos toxicity, mesothelioma is simply a peritoneal effusion. As previously discussed, with a prevalence of 10 percent, benign perennial effusions are normal in these patients. The majority of patients with a history of exposure will be exposed 20–40 years before the occurrence of clinically evident mesothelioma and potentially 5–10 years after the ultimate cancer cell emergence (Wang et al. 2017). Asbestos-related lung cancers were first known to be scar cancers (American Thoracic 2004), which later on changed as more awareness of the fiber risk became apparent in employees with other types of asbestos-related disorder. A study of Quebec employees working in chrysotile asbestos mine found that the concentration of fiber in the lungs of asbestos-dead chrysotile miners and millers was almost twice as high as that of those dying from lung cancer. Both types of lung cancer cells have been linked with exposure to asbestos, which further supports the idea of exposure plays a direct part in their genesis, regardless of smoking (Sir Anthony Newman Taylor 2017).

14.4 Asbestos Cleaning Strategies

14.4.1 Physical Methods of Remediation

Physical methods majorly include thermal treatments depending on serpentine and amphibole asbestos which undergoes dehydroxylation and recrystallization which are being developed for remediation. The method of thermal decomposition has 3 stages: loss of adsorbed water, removal of OH (hydroxyl group), and then crystallization of amorphous material (Paolini et al. 2018). Asbestos fibers are unstable at higher temperatures of about 500–600 °C, leading to changes in physical and chemical properties, e.g., chrysotile [$Mg_3Si_2O_5(OH)_4$], will lose its OH (hydroxyl) group and gets converted to forsterite (Mg_2SiO_4). Further increase in temperature leads to crystallization of forsterite and enstatite ($MgSiO_3$) (Gualtieri et al. 2011; Zaremba et al. 2010). At temperature 100–250 °C, chrysotile dehydrates to lose 1–2% of its weight, and at higher temperatures of over 600 °C, chrysotile fibers undergo dehydroxylation, leading to weight loss of 13% (Spasiano and Pirozzi 2017). This thermal treatment includes various methods such as vitrification, microwave assisted, and oxyhydrogen. In order to produce an inert silica product without fibers, vitrification is the easiest procedure where ACW substances are exposed to temperatures greater than 1000 °C. Electric and methane furnaces are generally used to achieve this temperature (Paolini 2018). One such approach by vitrification process which is done at pre-pilot scale was applied to cement-asbestos pipes consisting of chrysotile and crocidolite, to this progressive heating up to 1600 °C was given. This resulted in melting of fibers and its rapid cooling gave monolithic glass. Absence of crystalline phases within glass was estimated by XRD and SEM (Dellisanti et al. 2009). Plasma treatment of microwave air is an electricity discharge which uses microwave as electricity. The wavelength of the electromagnetic radiation is between infrared lights and greater range radio waves. This higher frequency of electromagnetic waves rapidly decomposes and modifies the fibrous structure of asbestos and transform into inert magnesium oxide, the composition is based on forsterite (Mg_2SiO_4) a major crystalline phase with absence of hazardous minerals after thermal treatment (Falciglia et al. 2018). Treatment of ACW done by initial pretreatment of waste by dark fermentation followed by hydrothermal reaction at different conditions leading to degradation of chrysotile mineral (Spasiano and Pirozzi 2017). One of the studies showed that microwave irradiation on ACW could decompose the asbestos fibers to less toxic magnesium silicate (Leonelli et al. 2006). This process of thermal remediation is cost effective up to ten-fold lower than the cost for disposal of toxic waste and producing non-hazardous and environmentally friendly minerals (Granat et al. 2015). An alternative method of oxyhydrogen treatment, which uses stoichiometric gas mixture of 1:2 (oxygen and hydrogen) generated by electrolysis of water. Due to this, temperature rises up to 570 °C and will release energy of 241.8 kJ per mole of hydrogen, with this method temperature can rise up to 2800 °C. This method allows degradation of asbestos of about 99%. But the only disadvantage is high energy consuming method due to electrolysis of water in order to produce gaseous mixture (Paolini et al. 2018; Min et al. 2008).

14.4.2 Chemical Treatment of Asbestos

Denaturation of asbestos fibers by chemical treatment is another approach to convert toxic fibers to non-hazardous end product. For treatment, chemical additives are added at low melting temperatures which can react with chemical structure of asbestos fibers. The main advantage of this method is it requires less energy cost because decomposition occurs at room temperature, and the major drawback is duration of treatment, cost associated with reagent consumption, and further waste liquid treatment is required (Paolini et al. 2018). It is important to hydrolyze the bond between oxygen and silicon; two main approaches are carried out for degradation of asbestos fibers. One is the use of highly basic pH which will degrade asbestos fibers to give free silanols. Another approach uses hydrofluoric acid to get silicon fluoride (SiF_4) (Paolini et al. 2018). Asbestos structural modification was observed when fibers were treated with 2.5 mol L^{-1} H_2SO_4 (sulfuric acid) for 10–24 h and heating at 100 °C (Nam et al. 2014; Kiyoji 2009). One study reported, complete dissolution of chrysotile fibers when treated with sulfuric acid derived from fluorosulfonic acid (FSO_3H), as it acts on brucite layer of chrysotile asbestos leading to form Mg^{+2} ions and precipitation of $MgSO_4.H_2O$ (kieserite) and amorphous (MgO) (Sugama et al. 1998). Fluoride compounds are also used to break down asbestos by altering physical and chemical structure as fluorides react with silicon in asbestos fibers to destroy them (Nocito 2014). Furthermore by mechano-chemical method by using microwave and chemical treatment, this will convert anisotropic fibers to isotropic fibers and allows asbestos inertization and reusability (Kusiorowski et al. 2015). Also, in situ chemical conversion of chrysotile asbestos to a non-toxic form of asbestos was converted by usage of oxalic acid (polycarboxylic acid), and sulphuric acid aided oxalic treatment can increase the conversion rate (Brown 2004). In another analysis, ACW was dissolved with a solution of phosphoric acid that can degrade asbestos fiber and provide end products of calcium and magnesium phosphate that can be used in agriculture (Pawelczyk et al. 2017). Studies also demonstrated asbestos decomposition by Freon-decomposed acidic gas which presents hydroxyl group. This technique provides several advantages of low working temperature and energy consumption, with short process time (Paolini et al. 2018). Treating asbestos with strong base can degrade silicate components of fibers by hydrolysis of Si–O bond, reaction driven by OH- anion (Paolini et al. 2017), similarly with strong acid which will result in free silanols moieties. Asbestos silicates can also be degraded by reaction of reducing agents and reaction is self-propagating due to oxidation and reduction. It was observed that asbestos fibers degraded by thermochemical process with a mixture of ferric oxide and elementary magnesium (Porcu et al. 2005). These methods for remediating the contaminants is cost effective, less time consuming, and shows large scale applications, but although there is usage of chemicals is a major drawback. Thus, after remediating further waste liquid management is to be done, which eventually increases the cost, consumption of time, and labor. So, the better alternative suggested by recent findings is making use of bioremediation and phytoremediation for decomposition of ACW.

14.4.3 Bioremediation of Asbestos

Currently, there are no good techniques for decontaminating areas polluted with asbestos. However, several researchers have reported that some potential microorganisms are able to tolerate asbestos fibers either by removing them from environment or breaking them down into less toxic forms. This new solution of remediating asbestos-contaminated sites is called bioremediation. Simply stated, bioremediation is a technique that offers the possibility of decontamination or removal of pollutants from the substrate by using microorganism's natural biological activity (i.e., the use of fungi or bacteria for asbestos cleanup) (Pande et al. 2020; Juwarkar et al. 2014). Asbestos is a commonly used raw material in industries for the manufacturing of asbestos-based materials which had driven a rapid increase in the amount of asbestos waste and thus lead to potentially lethal diseases. However, it is far more challenging to remove polluted soil near asbestos factories and mills than to strip asbestos sheets from houses (Kumar et al. 2016). Several experiments have been devoted to the weathering of asbestos fibers in the presence of fungi, lichens and bacteria. To name some, Iron is the foremost factor accountable for the formation of free radicals, which can damage DNA and trigger cancer, elimination or separation of iron will prove to be a remedy (Balamurugan et al. 2018). Many microorganisms require iron to produce energy and retain iron atoms in soil minerals and extract them using chemical links called siderophores. This compound acts on the ferric ions by forming a 1:1 complex which is solubilized and transported from the environment to the cell via cell membrane of microbe and reduced to ferrous form (Mahbubul A. F. Jalal 1991) (Fig. 14.2). Iron-reduced fibers do not produce free radicals that are carcinogenic. In specific, certain fungi emanate siderophores and other chelating compounds that can remove iron ions from asbestos fibers, thereby reducing their exposure (Martino et al. 2003; Daghino et al. 2008). Fungus such as *Verticillium* sp. absorbed 7.3% iron from asbestos fibers during 20 days of its growth in crocidolite suspension and in addition, reduced high-oxidative HO. radicals that cause DNA damage (Daghino et al. 2008). Subsequently, soil bacteria such as Pseudomonas have developed several unique uptake strategies, especially the development of siderophores, small molecules (200–2000 Da) with a high affinity for Iron (Fe^{3+}) produced in iron-limited conditions (Goldberg 2000). It has been reported that the bacteria *Bacillus mucilaginosus* have facilitated biologically induced serpentine rock dissolution. Its metabolism contributes to the production of organic acids and other organic ligands that promote the dissolution of silica and magnesium.

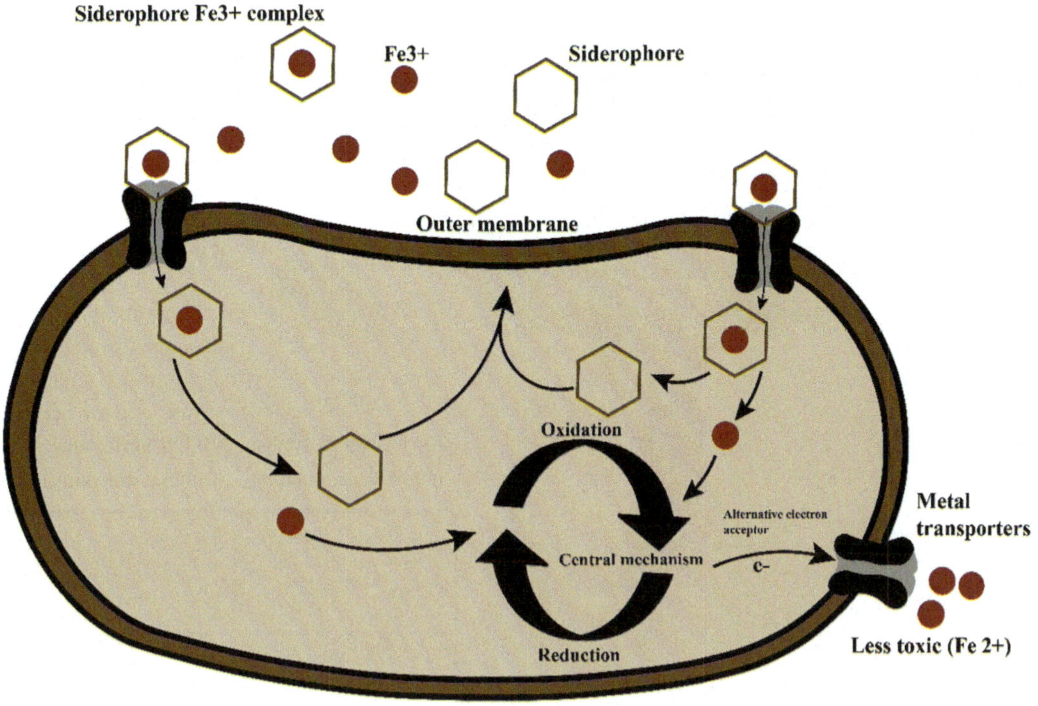

Fig. 14.2 The process of bacteria mediated remediation of ferric ions present on the surface of the asbestos fibers using siderophores

The XRD study showed a substantial decrease in serpentine peak sharpness and width and increasing amorphous content after 20 and 30 days of contact between *Bacillus mucilaginosus* culture and serpentine rock (Yao et al. 2013).

Likewise, the composite organism lichens, derived from algal or cyanobacteria that live among the multi-fungal filaments, was found to have the capacity to grow and form colonizing of sites with protrusion of chrysotile fibers on chrysotile-containing rocks or on asbestos cement roofing (Favero-Longo et al. 2005). The main step of this mineral weathering process is to secrete chelating agents, such as oxalic acid, secreted by lichens during their metabolic activity (e.g., *Acarospora cervina, Candelariella itelline* and *Candelariella*). Precisely, oxalic acid can form a soluble organic acid (glushinskite) that can remove Mg^{2+} ions from chrysotile fibers, which can be easily leached during precipitation (Favero-Longo et al. 2013).

14.4.4 Phytoremediation of Asbestos

Plants mediate to reduce the toxicity and contamination of the polluted areas. Plants such as grasses, herbs, shrubs, or trees in association with microorganisms are known to remediate the contaminated site through degradation of toxins (Gomes 2012; Rajkumar et al. 2013; Cameselle and Gouveia 2019). So, this is the "Green Technology" involving plants having great potential for removal of contaminants from soil, water or air. Plant mediated remediation occurs by various methods: extraction and incorporation of contaminants in plant parts, phytostabilization of contaminants in the subsurface, release of plant enzymes into rhizosphere area of contamination, and microbial mediated degradation in rhizosphere region (Fig. 14.3) (Juwarkar et al. 2010). Along with its great potential, it is very easy technique which does not require any expertise or expensive equipment for remediation (Mandal et al. 2014; Liu et al. 2018). One such

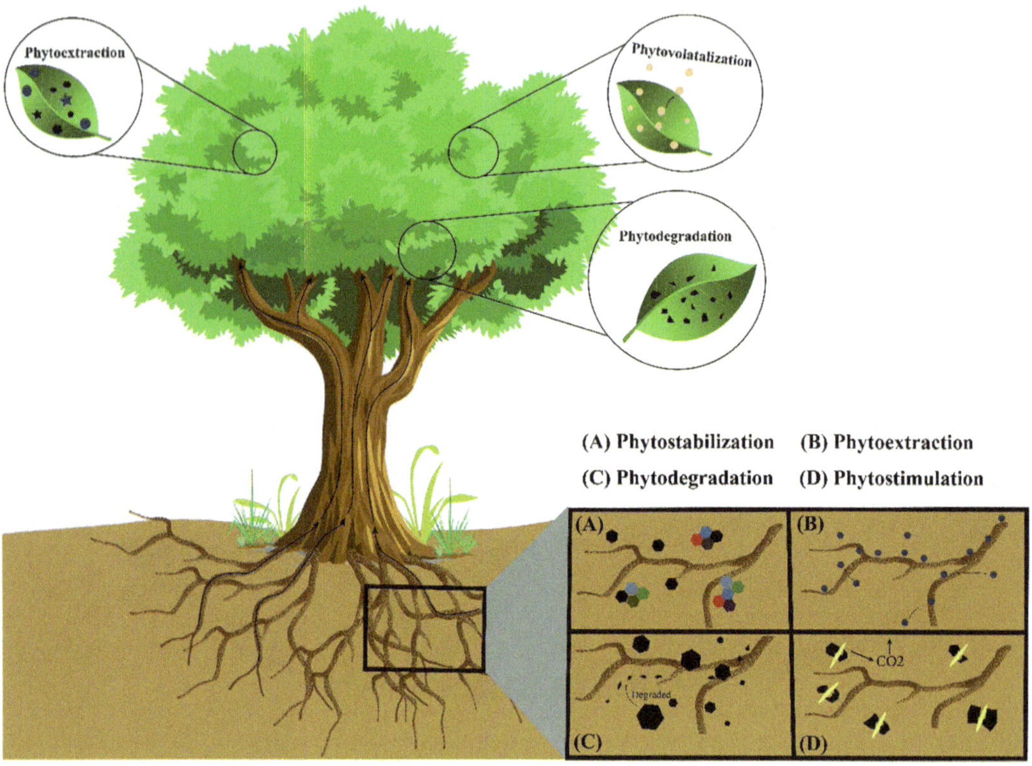

Fig. 14.3 Schematic representation of phytoremediation strategies for asbestos

study done for remediating waste from chromite-asbestos mines having metal and metalloid made use of aromatic grasses, *Cymbopogon citratus* and *Chrysopogon zizanioides* by supplying different manures for faster growth. This results in extraction of toxic metals Cr and Ni from waste, thus reducing toxicity of waste (Kumar and Maiti 2015b). Naturally growing plant species *Minuartia* and *Thymus*, which can revegetate on spoiled mine, having asbestos contamination (Kumar et al. 2016). The areas with natural plant growth can reduce the asbestos dispersion to 50% in comparison to barren areas. Plants are responsible to stabilize topsoil and minimize erosion, which could help limit asbestos exposure via air (Alkorta et al. 2010; Pandey and Bajpai 2019). It was observed that chromite-asbestos mine waste (CMV) having Cr and Ni were accumulated in grasses (*C. dactylon* and *S. nutans*) and legumes (*A. concinna* and *C. ajan*) (Kumar et al. 2015). This shows plants have potential to accumulate metals thus facilitating phytoremediation of CMV. Furthermore, it was also found that there is a release of organic acids from plant rhizospheres or soil microbes which can extract metals and alter charge of fibers, thus reducing the toxicity of contaminants (Daghino et al. 2006; Favero-Longo et al. 2013; Holmes and Lavkulich 2014; Mohanty et al. 2018). Phytostabilization is a cost-effective remediation technique that does not allow asbestos fibers to spread into the atmosphere. This method remediates the soil containing heavy metals, in many of the chromite mine regions. Thus, phytostabilization and phytoremediation could be advantageous technologies for treatment of asbestos-contaminated sites.

14.5 Substitution of Asbestos

During mining and crushing of asbestos mineral rock, there is a lot of waste generated termed ACW consists of asbestos fibers which have potential environmental and health risks. These fibers cause contamination to soil, water and air along with severe health conditions. Due to these effects in 2010, American Public Health Association along with three international health organizations asked for a global ban on usage of asbestos (Karen Selby 2020). Until now 67 countries and territories worldwide have banned the usage of asbestos in order to control the risks and hazards (Kazan-Allen 2018). After terminating the usage of asbestos, a substitute which is eco-friendly, economical, higher availability, and technologically should be applicable at wider range. According to various researches several substitutes of asbestos could be used, which are mentioned in Table 14.1.

14.6 Laws and Regulations for Usage of Asbestos

There are several laws and regulations into amendment for controlling and limiting the usage of asbestos minerals. Whereas the United States has not fully banned asbestos, and the mineral is still found in building materials such as gaskets and roofing materials (Lucarelli 2020). But about 67 countries and territories have banned the usage of asbestos due to its health risk to cause mesothelioma and diverse environmental hazards. Asbestos Laws and Regulations by the U.S. Environmental Protection Agency (EPA) are made to protect the public from asbestos exposure. Different asbestos-related laws by EPA have been made to prevent and reduce asbestos hazards. One such law The Asbestos Hazard Emergency Response Act (AGHERA) made by educational agencies for inspecting school buildings materials, preparing asbestos management plans, and take regarding actions. Furthermore, Asbestos Information Act (Public Law 100–577) provided transparency and identifying the companies making use of asbestos to make related products and report the production to EPA (USEPA 2020a). Now considering the Clean Air Act (CAA) on protecting the environment, "This act protects the obligations of the EPA to safeguard and improve the air quality and stratospheric ozone layer and contains provisions for the EPA to establish national emissions regulations for dangerous air pollutants, along with asbestos (USEPA 2020a). Restrictions on Discontinued Uses of Asbestos Rule strengthens The Agency's ability to review asbestos products that are no longer in market, and products manufactured should be notified to EPA at least 90 days prior to manufacture, import, or processing of asbestos or asbestos-containing products. But the user will be prohibited until EPA conducts thorough analysis and provides certain restrictions on its use (USEPA 2020a). Several Laws for protecting Workers from asbestos exposure are made under Sect. 6 of the Toxic Substance Control Act (TSCA) along with EPA. General laws made for protection of employees cover 4 risks divided into 2 generic groups: occupational diseases, accidents and occupational risks, and non-occupational diseases, accidents, and non-occupational risks (Poyatos 2016). In 1989, there was Partial Ban on manufacturing, processing, importing, and distributing the asbestos consisting products, but in April 2019 there was Final Rule to ensure that asbestos-containing products should no longer exist in market and also cannot return to commerce until agency evaluates and puts restriction and prohibition on its use (USEPA 2020b). Current regulations in India have been made to protect health of workers and also control environmental pollution. The Indian Factories Act (1948) is in amendment for protection of workers involved in the manufacturing, handling, and processing of asbestos. If industry is found with usage of this mineral, it would be listed as hazardous industry under Schedule 1 of the Factory Act. Asbestos is listed as a hazardous waste in the Environment (Protection) Act (1986) under Sects. 6, 8, and 25. Thus, under Environment protection act, in October 1998, the Ministry of Environment and Forest prohibits the import of waste asbestos due to human and environmental health hazards. There were about two to three million active workers suffering from exposure to asbestos and other dangerous fibers. The government has also made regulations for 'No use of asbestos' clause in recently-released development plans (DP) of 12 regions

Table 14.1 Substitutes of asbestos

S. No	Substitutes for asbestos	Characteristics	Uses	Reference
1	Glass wool (Fiber glass)	Resistance to heating and corrosion, high seam intensity	Construction, automobile, and spare parts, furniture industry, boats, etc	Costa and Orriols (2012); Lee et al. (1981)
2	Rock wool (High-temperature insulation wool)	Crafted from alkaline-earth silicates, which are less biopersistent than ceramic refractory fibers	Thermal insulation	Costa and Orriols (2012); Fyles et al. (1999)
3	Stone and slag wool	Heat and sound resistance, high seam strength	Buildings, automobile industries, filtration, hydroponics growth media	Costa and Orriols (2012); Kyoung-Woo and Jeong (2014)
4	Cement-bonded wool fibers	Greater toughness index, low, medium, and high density	Construction industry	Karade (2010); Wolfe (1999)
5	Para-aramid (polyamide fibers)	High strength, durability, and heat resistance	Aviation and sports industry, thermoplastic materials, tiers and rubber products	IARC (2012)
6	Attapulgite (palygorskite)	Light weight & water absorbing capacity	Building and friction material, packing	Forte and Mudd (1975); Washabaugh (1981); Zhang (2009); IARC (2012)
7	Sepiolite	Viscosity improver and sedimentation preventer	Absorbents, environmental deodorants, cosmetics, animal nutrition, detergents, rubber, etc	IARC (2012); Alvarez (1984); Guillon (1990)
8	Wollastonite	Chemically inert, unique cleavage property	Ceramic, plastic, rubber, paint and coating products	IARC (2012)
9	Erionite	Brittle, wool like fibrous, chemically inert, alkaline-earth metals.	Animal feed, wastewater treatment, gas absorbents	Dogan et al. (2008); Irigaray et al. (2007)
10	Banana peel	Strong resistance, stiffness, and specific gravity	Automotive brake pads	Idris et al. (2015)
11	Palm kernel fibers	Increased coefficient of friction, strength properties, stiffness, and specific gravity	Automotive brake pads	Ikpambese et al. (2016); Ibhadode (2008)
12	Bagasse	High friction coefficient	Automotive brake pads	Aigbodion (2010)
13	Palm slag	Strong thermal characteristics, strength of compression	Automotive brake pads	Ruzaidia (2011)
14	Cashew nutshell liquid	Better specific gravity, viscosity, and moisture content	Surface coatings, paints, varnishes, and brake linings and clutch facings in automotive industry	Telascrêa et al. (2014); Chaudhari (2012)
15	Kenaf Fibers	High thermal conductivity, wear resistant	Automotive brake pads	Namessan (2013)
16	Amorphous Silica	Low bulk density, pozzolanic activity, porous structure	Packaging, welding curtains, fire blankets, fire pads, etc	Yu et al. (1998)

(continued)

Table 14.1 (continued)

S. No	Substitutes for asbestos	Characteristics	Uses	Reference
17	Polyurethane Foam	Thermal resistance	Roofing materials, design of movie theatre sets, car upholstery	Wit Witkiewicz and Zieliński (2006)
18	Cellulose Fiber	Durability, fire resistance	Textile, chemical filters, fiber-reinforcement composites	Bendaikha (2015)
19	Flour Fillers	Heat resistant, bending strength, modules of elasticity	Building insulation	Nayak (2016)
20	Thermoset Plastic Flour	Insulating material, Opaque, chemical resistance, self-extinguishing, Rigid and brittle	Electrical insulation, brake linings of vehicles, switch gear, colored housing appliances	Saiter et al. (2012)

(INDIAN-FINAL-NAP-25–4-17). This clause is part of the regulations specified by the Urban Development Department for year 2011–2031 for development of tourism and hospitality services under the regional plans (Vyas 2018). Several other laws for safety and health requirements relating to occupational exposure to asbestos were made in order to protect the nation at environmental and individual levels.

14.7 Conclusion

Anthropogenic activities have widely spread the pollution in environment; one among them is asbestos mineral, which has varied physical structure and chemical properties. Thus, remediation of this pollutant is required in order to protect health of individuals and environment. Several alternatives are available for remediating ACW, but all treatments cannot be applicable at higher scale. Physical and chemical processes, though, are costly and can trigger environmental contamination often. Biological approaches have increasingly gained more recognition because they are cost-efficient, highly effective, and minimally toxic to the environment. These methods include bioremediation and phytoremediation majorly used to clean up the ACW and convert it to less or non-toxic substance which may not impact human health as well as the environment. The microbes and plants have the potential to extract the heavy metals, either through degradation or by accumulating them followed by reduction of metals. Several case studies have also been supportive for biosorption of heavy metals from contaminated sites containing asbestos waste. Thus, this method reduces the toxicity of asbestos fibers. These methods are widely applicable but have several drawbacks as it is time consuming and is not a scalable method to remediate the complete asbestos generated waste.

14.8 Future Prospects

The use of asbestos was increasing after 1800's with outcome of more than thousands of products. Later investigation suggested the drawbacks of asbestos due to several health risks and environmental pollution. And currently, 67 countries and territories have banned the usage of asbestos, thus making a hypothesis to future usage and effects of asbestos for regions that have banned the asbestos mineral usage completely; a) will have remnants of asbestos fibers in environment which has to be remediated and, b) the effects of asbestos exposed people suffering from complications. But eventually incidence rate of asbestosis, mesothelioma, etc. may increase in both cases. For the places that have not yet been banned, there can be an increase in occupational and non-occupational risk for asbestos-related diseases and it can also impact the environmental conditions. Thus, it implies on an individual basis of exposure and intensity of exposure, its effect could not be generalized.

References

Aigbodion VSUA, Hassan SB, Asuke F, Agunsoye JO (2010) Development of asbestos - Free brake pad using bagasse. Tribology in Industry 32(1):12–17

Alkorta I, Becerril JM, Garbisu C (2010) Phytostabilization of metal contaminated soils. Rev Environ Health 25(2):135–146. https://doi.org/10.1515/reveh.2010.25.2.135

Alvarez A (1984) Sepiolite: Properties and Uses. In: Palygorskite — Sepiolite: Occurrences, Genesis and Uses. Developments in Sedimentology, pp 253–287. https://doi.org/10.1016/s0070-4571(08)70044-x

American Thoracic S (2004) Diagnosis and initial management of nonmalignant diseases related to asbestos. Am J Respir Crit Care Med 170(6):691–715. https://doi.org/10.1164/rccm.200310-1436ST

Anastasiadou K, Axiotis D, Gidarakos E (2010) Hydrothermal conversion of chrysotile asbestos using near supercritical conditions. J Hazard Mater 179(1–3):926–932. https://doi.org/10.1016/j.jhazmat.2010.03.094

Balamurugan TST, Mani V, Hsieh C-C, Huang S-T, Peng T-K, Lin H-Y (2018) Real-time tracking and quantification of endogenous hydrogen peroxide production in living cells using graphenated carbon nanotubes supported Prussian blue cubes. Sens Actuators, B Chem 257:220–227. https://doi.org/10.1016/j.snb.2017.10.151

Barlow CA, Grespin M, Best EA (2017) Asbestos fiber length and its relation to disease risk. Inhal Toxicol 29 (12–14):541–554. https://doi.org/10.1080/08958378.2018.1435756

Boudjema HL, Bendaikha H (2015) Studies on the properties of cellulose fibers-reinforced thermoplastic starch composites. J Chem Mater Res 3:21–25

Berndt ME, Brice WC (2008) The origins of public concern with taconite and human health: Reserve mining and the asbestos case. Regul Toxicol Pharmacol 52(1 Suppl):S31–S39. https://doi.org/10.1016/j.yrtph.2007.09.019

Bhattacharya S, John PJ, Ledwani L (2015) Bacterial weathering of Asbestos. SILICON 7(4):419–431. https://doi.org/10.1007/s12633-014-9260-9

Brown P (2004) *In-situ* treatment of asbestos-containing material. US Patent Number US20060106272A1

Cameselle C, Gouveia S (2019) Phytoremediation of mixed contaminated soil enhanced with electric current. J Hazard Mater 361:95–102. https://doi.org/10.1016/j.jhazmat.2018.08.062

CDC (2012) ASBESTOS. https://www.cdc.gov/niosh/topics/asbestos/default.html. Acessed on 20 January 2021

Chaudhari APNJT, Mahale DM, Haldankar PM, Sonawane SP (2012) Physical properties of cashew nut shell liquid. International Journal of Application or Innovation in Engineering & Management 1(2):172–175

Costa R, Orriols R (2012) Man-made mineral fibers and the respiratory tract. Arch Bronconeumol 48(12):460–468. https://doi.org/10.1016/j.arbres.2012.04.006

Daghino S, Martino E, Vurro E, Tomatis M, Girlanda M, Fubini B, Perotto S (2008) Bioweathering of chrysotile by fungi isolated in ophiolitic sites. FEMS Microbiol Lett 285(2):242–249. https://doi.org/10.1111/j.1574-6968.2008.01239.x

Daghino S, Turci F, Tomatis M, Favier A, Perotto S, Douki T, Fubini B (2006) Soil fungi reduce the iron content and the DNA damaging effects of asbestos fibers. Environ Sci Technol 40(18):5793–5798. https://doi.org/10.1021/es060881v

David SR, Jaouen A, Ihiawakrim D, Geoffroy VA (2021) Biodeterioration of asbestos cement by siderophore-producing Pseudomonas. J Hazard Mater 403:123699. https://doi.org/10.1016/j.jhazmat.2020.123699

De Guire L, Labrèche F, Poulin M, Dionne M (2005) The use of chrysotile asbestos in Québec [Internet]. Québec, Canada: Intitut National de Santé Publíque/Government of Québec 11

Dellisanti F, Rossi PL, Valdrè G (2009) Remediation of asbestos containing materials by Joule heating vitrification performed in a pre-pilot apparatus. Int J Miner Process 91(3–4):61–67. https://doi.org/10.1016/j.minpro.2008.12.001

Dodson RF, Atkinson MA (2006) Measurements of asbestos burden in tissues. Ann N Y Acad Sci 1076:281–291. https://doi.org/10.1196/annals.1371.015

Dogan AU, Dogan M, Hoskins JA (2008) Erionite series minerals: mineralogical and carcinogenic properties. Environ Geochem Health 30(4):367–381. https://doi.org/10.1007/s10653-008-9165-x

Dusinska M, Collins A, Kazimirova A, Barancokova M, Harrington V, Volkovova K, Staruchova M, Horska A, Wsolova L, Kocan A, Petrik J, Machata M, Ratcliffe B, Kyrtopoulos S (2004) Genotoxic effects of asbestos in humans. Mutat Res 553(1–2):91–102. https://doi.org/10.1016/j.mrfmmm.2004.06.027

Falciglia PP, Roccaro P, Bonanno L, De Guidi G, Vagliasindi FGA, Romano S (2018) A review on the microwave heating as a sustainable technique for environmental remediation/detoxification applications. Renew Sustain Energy Rev 95:147–170. https://doi.org/10.1016/j.rser.2018.07.031

Favero-Longo SE, Turci F, Fubini B, Castelli D, Piervittori R (2013) Lichen deterioration of asbestos and asbestiform minerals of serpentinite rocks in Western Alps. Int Biodeter Biodegr 84:342–350. https://doi.org/10.1016/j.ibiod.2012.07.018

Favero-Longo SE, Turci F, Tomatis M, Castelli D, Bonfante P, Hochella MF, Piervittori R, Fubini B (2005) Chrysotile asbestos is progressively converted into a non-fibrous amorphous material by the chelating action of lichen metabolites. J Environ Monit 7 (8):764–766. https://doi.org/10.1039/b507569f

Forte WB, Mudd PJ (1975) US Patent No 3, 907, 725. US Patent and Trademark Office, Washington DC

Fyles KM, Shorrock P, Davidson MR (1999) US Patent No. 5, 962, 354. US Patent and Trademark Office, Washington DC

Goldberg JB (2000) *Pseudomonas*: global bacteria. Trends Microbiol 8(2):55–57. https://doi.org/10.1016/s0966-842x(99)01671-6

Gomes HI (2012) Phytoremediation for bioenergy: challenges and opportunities. Environ Technol Rev 1 (1):59–66. https://doi.org/10.1080/09593330.2012.696715

Granat K, Nowak D, Pigiel M, Florczak W, Opyd B (2015) Application of microwave radiation in innovative process of neutralising asbestos-containing wastes. Arch Civ Mech Eng 15(1):188–194. https://doi.org/10.1016/j.acme.2014.05.012

Gualtieri AF, Giacobbe C, Sardisco L, Saraceno M, Gualtieri ML, Lusvardi G, Cavenati C, Zanatto I (2011) Recycling of the product of thermal inertization of cement-asbestos for various industrial applications. Waste Manag 31(1):91–100. https://doi.org/10.1016/j.wasman.2010.07.006

Guillon JSRABD (1990) Use of sepiolite in manufacturing processes of fiber reinforced products containing mica. US Patent Number EP0454222A1

Holmes EP, Lavkulich LM (2014) The effects of naturally occurring acids on the surface properties of chrysotile asbestos. J Environ Sci Health A Tox Hazard Subst Environ Eng 49(12):1445–1452. https://doi.org/10.1080/10934529.2014.928558

IARC Working Group on the Evaluation of Carcinogenic Risks to Humans (2012) Arsenic, metals, fibres, and dusts. IARC Monogr Eval Carcinog Risks Hum 100 (PT C):11

Ibhadode AOA, Dagwa IM (2008) Development of asbestos-free friction lining material from palm kernel shell. J Braz Soc Mech Sci Eng 30(2):166–173. https://doi.org/10.1590/S1678-58782008000200010

Idris UD, Aigbodion VS, Abubakar IJ, Nwoye CI (2015) Eco-friendly asbestos free brake-pad: Using banana peels. J King Saud Univ - Eng Sci 27(2):185–192. https://doi.org/10.1016/j.jksues.2013.06.006

Ikpambese KK, Gundu DT, Tuleun LT (2016) Evaluation of palm kernel fibers (PKFs) for production of asbestos-free automotive brake pads. J King Saud Univ - Eng Sci 28(1):110–118. https://doi.org/10.1016/j.jksues.2014.02.001

Irigaray P, Newby JA, Clapp R, Hardell L, Howard V, Montagnier L, Belpomme D (2007) Lifestyle-related factors and environmental agents causing cancer: an overview. Biomed Pharmacother 61(10):640–658

Jung HS, Park EK, Cha JS, Lee JW, Lee JC, Jang J, Kim S, Oak C, Yates DH, Kim H (2020) Characteristics of asbestos fibers in lung tissue from occupational and environmental asbestos exposure of lung cancer patients in Busan. Korea. Sci Rep 10(1):20359. https://doi.org/10.1038/s41598-020-77291-9

Juwarkar AA, Misra RR, Sharma JK (2014) Recent trends in bioremediation. In: Geomicrobiology and Biogeochemistry. Soil Biology. pp 81–100. https://doi.org/10.1007/978-3-642-41837-2_5

Juwarkar AA, Singh SK, Mudhoo A (2010) A comprehensive overview of elements in bioremediation. Rev Environ Sci Biotechnol 9(3):215–288. https://doi.org/10.1007/s11157-010-9215-6

Karade SR (2010) Cement-bonded composites from lignocellulosic wastes. Constr Build Mater 24 (8):1323–1330. https://doi.org/10.1016/j.conbuildmat.2010.02.003

Karen Selby R (2020) Mesothelioma & Asbestos Worldwide. https://www.asbestos.com/mesothelioma/worldwide/.

Kazan-Allen L (2018) Current Asbestos Bans. International Ban Asbestos Secretariat. Available online: http://ibasecretariat.org/alpha_ban_list.php.

Kilburn KH (2000) Indoor air effects after building renovation and in manufactured homes. Am J Med Sci 320(4):249–254. https://doi.org/10.1016/s0002-9629(15)40835-3

King D (2020) History of Asbestos. https://www.asbestos.com/asbestos/history/.

Kiyoji Y, Toshiya, N. (2009) Method of conducting treatment of non-asbestos treatment of asbestos containing construction material and gypsum composition obtained from the same. Japanese Patent No 2009018228

Koi M, Barrett JC (1986) Loss of tumor-suppressive function during chemically induced neoplastic progression of Syrian hamster embryo cells. PNAS USA 83(16):5992–5996. https://doi.org/10.1073/pnas.83.16.5992

Kumar A, Maiti SK (2015) Assessment of potentially toxic heavy metal contamination in agricultural fields, sediment, and water from an abandoned chromite-asbestos mine waste of Roro hill, Chaibasa. India. Environ Earth Sci 74(3):2617–2633. https://doi.org/10.1007/s12665-015-4282-1

Kumar A, Maiti SK (2015) Effect of organic manures on the growth of *Cymbopogon citratus* and *Chrysopogon zizanioides* for the phytoremediation of chromite-asbestos mine waste: A Pot scale experiment. Int J Phytoremediation 17(1–6):437–447. https://doi.org/10.1080/15226514.2014.910174

Kumar A, Maiti SK, Tripti PMNV, Singh RS (2015) Grasses and legumes facilitate phytoremediation of metalliferous soils in the vicinity of an abandoned chromite–asbestos mine. J Soils and Sediments 17 (5):1358–1368. https://doi.org/10.1007/s11368-015-1323-z

Kumar A, Prasad MNV, Maiti SK, Tripti (2016) Asbestos. In: Environmental Materials and Waste. pp 285–305. https://doi.org/10.1016/b978-0-12-803837-6.00013-5

Kusiorowski R, Zaremba T, Piotrowski J, Podwórny J (2015) Utilisation of cement-asbestos wastes by thermal treatment and the potential possibility use of obtained product for the clinker bricks manufacture. J Mater Sci 50(20):6757–6767. https://doi.org/10.1007/s10853-015-9231-6

Kyoung-Woo KIM, Jeong YS (2014) Experimental study on the comparison of the material properties of glass wool used as building materials. Mater Sci 20(1):103–107. https://doi.org/10.5755/j01.ms.20.1.3714

Lee KP, Barras CE, Griffith FD, Waritz RS, Lapin CA (1981) Comparative pulmonary responses to inhaled inorganic fibers with asbestos and fiberglass. Environ Res 24(1):167–191. https://doi.org/10.1016/0013-9351(81)90143-2

Leonelli C, Veronesi P, Boccaccini DN, Rivasi MR, Barbieri L, Andreola F, Lancellotti I, Rabitti D, Pellacani GC (2006) Microwave thermal inertisation of asbestos containing waste and its recycling in traditional ceramics. J Hazard Mater 135(1–3):149–155. https://doi.org/10.1016/j.jhazmat.2005.11.035

Liu L, Li W, Song W, Guo M (2018) Remediation techniques for heavy metal-contaminated soils: Principles and applicability. Sci Total Environ 633:206–219. https://doi.org/10.1016/j.scitotenv.2018.03.161

Lucarelli J (2020) Asbestos laws and regulations. https://www.mesothelioma.com/lawyer/legislation. Accessed on 20 January 2021

Mahbubul A. F. Jalal DvdH (1991) CRC Handbook of microbial iron chelates. Isolation and spectroscopic identification of fungal siderophores.

Mandal A, Purakayastha TJ, Ramana S, Neenu S, Bhaduri D, Chakraborty K, Manna MC, Rao AS (2014) Status on phytoremediation of heavy metals in India- A Review. Int J Bio-res Stress Manage 5(4). https://doi.org/10.5958/0976-4038.2014.00609.5

Marczynski B, Czuppon AB, Marek W, Reichel G, Baur X (1994) Increased incidence of DNA double-strand breaks and anti-ds DNA antibodies in blood of workers occupationally exposed to asbestos. Hum Exp Toxicol 13(1):3–9. https://doi.org/10.1177/096032719401300102

Martino E, Prandi L, Fenoglio I, Bonfante P, Perotto S, Fubini B (2003) Soil fungal hyphae bind and attack asbestos fibers. Angew Chem Int Ed Engl 42(2):219–222. https://doi.org/10.1002/anie.200390083

Miles SE, Sandrini A, Johnson AR, Yates DH (2008) Clinical consequences of asbestos-related diffuse pleural thickening: A review. J Occup Med Toxicol 3:20–20. https://doi.org/10.1186/1745-6673-3-20

Min SY, Maken S, Park J-W, Gaur A, Hyun JS (2008) Melting treatment of waste asbestos using mixture of hydrogen and oxygen produced from water electrolysis. Korean J Chem Eng 25(2):323. https://doi.org/10.1007/s11814-008-0055-9

Mohanty SK, Gonneau C, Salamatipour A, Pietrofesa RA, Casper B, Christofidou-Solomidou M, Willenbring JK (2018) Siderophore-mediated iron removal from chrysotile: Implications for asbestos toxicity reduction and bioremediation. J Hazard Mater 341:290–296. https://doi.org/10.1016/j.jhazmat.2017.07.033

Morgan A (1997) Acid leaching studies of chrysotile asbestos from mines in the Coalinga region of California and from Quebec and British Columbia. Ann Occup Hyg 41(3):249–268. https://doi.org/10.1016/s0003-4878(97)00001-x

Murray R (1990) Asbestos: A chronology of its origins and health effects. Br J Ind Med 47(6):361–365. https://doi.org/10.1136/oem.47.6.361

Murthy SS, Testa JR (1999) Asbestos, chromosomal deletions, and tumor suppressor gene alterations in human malignant mesothelioma. J Cell Physiol 180(2):150–157. https://doi.org/10.1002/(sici)1097-4652(199908)180:2%3c150::Aid-jcp2%3e3.0.Co;2-h

Nam SN, Jeong S, Lim H (2014) Thermochemical destruction of asbestos-containing roofing slate and the feasibility of using recycled waste sulfuric acid. J Hazard Mater 265:151–157. https://doi.org/10.1016/j.jhazmat.2013.11.004

Namessan NO, Maduako, JN, Iya, S (2013) Comparative study of the effects of treatment techniques on the thermal and frictional properties of kenaf (*Hibiscus canabinus*) fibre reinforced brake pads. Afr J Sci Technol 12(2)

Nayak L (2016) The mineral fibre: Asbestos - Its manufacture, properties, toxic effects and substitutes. Nat Environ Pollut Technol 15(2):477–482

Nocito T (2014) Asbestos: Eliminating the threat. https://www.environmental-expert.com/articles/asbestos-eliminating-the-threat-432164.

Nymark P, Aavikko M, Makila J, Ruosaari S, Hienonen-Kempas T, Wikman H, Salmenkivi K, Pirinen R, Karjalainen A, Vanhala E, Kuosma E, Anttila S, Kettunen E (2013) Accumulation of genomic alterations in 2p16, 9q33.1 and 19p13 in lung tumours of asbestos-exposed patients. Mol Oncol 7(1):29–40. https://doi.org/10.1016/j.molonc.2012.07.006

Nymark P, Wikman H, Hienonen-Kempas T, Anttila S (2008) Molecular and genetic changes in asbestos-related lung cancer. Cancer Lett 265(1):1–15. https://doi.org/10.1016/j.canlet.2008.02.043

Pande V, Pandey SC, Sati D, Pande V, Samant M (2020) Bioremediation: an emerging effective approach towards environment restoration. Environmental Sustainability 3(1):91–103. https://doi.org/10.1007/s42398-020-00099-w

Pandey VC, Bajpai O (2019) Phytoremediation: From theory toward practice. In: Pandey VC, Bauddh K (eds) Phytomanagement of Polluted Sites. Elsevier, pp 1–49. https://doi.org/10.1016/B978-0-12-813912-7.00001-6

Paolini V, Guerriero E, Bacaloni A, Rotatori M, Benedetti P, Mosca S (2017) Simultaneous sampling of vapor and particle-phase carcinogenic polycyclic aromatic hydrocarbons on functionalized glass fiber filters. Aerosol Air Qual Res 16(1):175–183. https://doi.org/10.4209/aaqr.2015.07.0476

Paolini V, Tomassetti L, Segreto M, Borin D, Liotta F, Torre M, Petracchini F (2018) Asbestos treatment technologies. J Mater Cycles Waste Manag 21(2):205–226. https://doi.org/10.1007/s10163-018-0793-7

Pawelczyk A, Bozek F, Grabas K, Checmanowski J (2017) Chemical elimination of the harmful properties of asbestos from military facilities. Waste Manag 61:377–385. https://doi.org/10.1016/j.wasman.2016.11.041

Pease DF, Kratzke RA (2017) Oncolytic viral therapy for mesothelioma. Front Oncol 7:179. https://doi.org/10.3389/fonc.2017.00179

Pollastri S, D'Acapito F, Trapananti A, Colantoni I, Andreozzi GB, Gualtieri AF (2015) The chemical environment of iron in mineral fibres. A combined X-ray absorption and Mossbauer spectroscopic study. J Hazard Mater 298:282–293. https://doi.org/10.1016/j.jhazmat.2015.05.010

Porcu M, Orrù R, Cincotti A, Cao G (2005) Self-propagating reactions for environmental protection: Treatment of wastes containing asbestos. Ind Eng Chem Res 44(1):85–91. https://doi.org/10.1021/ie040058c

Radetzki M (2010) Peak Oil and other threatening peaks—Chimeras without substance. Energy Policy 38(11):6566–6569. https://doi.org/10.1016/j.enpol.2010.07.049

Rajkumar M, Prasad MN, Swaminathan S, Freitas H (2013) Climate change driven plant-metal-microbe interactions. Environ Int 53:74–86. https://doi.org/10.1016/j.envint.2012.12.009

Roh S, Park S, Tae G, Song J (2016) A case of laryngeal cancer induced by exposure to asbestos in a construction site supervisor. Ann Occup Environ Med 28:34. https://doi.org/10.1186/s40557-016-0114-3

Ronald W. Wolfe Ag (1999) Durability and strength of cementbonded wood particle composites made from construction waste. Composites and manufactured products 49 (2)

Rong Y, Luo X, Zhang Z, Cui X, Liu Y, Chen W (2015) Occupational exposure to asbestos and cardiovascular related diseases: A meta-analysis. Prev Med Rep 2:920–926. https://doi.org/10.1016/j.pmedr.2015.10.005

Ruzaidia CMHK, Shamsul JB, Abdullah MMA (2011) Comparative study on thermal, compressive, and wear properties of Palm slag brake pad composite with other fillers. Adv Mat Res 328–330:1636–1641. https://doi.org/10.4028/www.scientific.net/AMR.328-330.1636

Saiter JM, Dobircau L, Leblanc N (2012) Are 100% green composites and green thermoplastics the new materials for the future? Int J Polym Sci 2012:280181. https://doi.org/10.1155/2012/280181

Sir Anthony Newman Taylor PC, Paul Blanc, Anthony Pickering (2017) Occupational Lung Disorders

Solbes E, Harper RW (2018) Biological responses to asbestos inhalation and pathogenesis of asbestos-related benign and malignant disease. J Investig Med 66(4):721–727. https://doi.org/10.1136/jim-2017-000628

Spasiano D, Pirozzi F (2017) Treatments of asbestos containing wastes. J Environ Manage 204(Pt 1):82–91. https://doi.org/10.1016/j.jenvman.2017.08.038

Subramanian V, Madhavan N (2005) Asbestos problem in India. Lung Can 49(1):S9–S12. https://doi.org/10.1016/j.lungcan.2005.03.003

Sugama T, Sabatini R, Petrakis L (1998) Decomposition of chrysotile asbestos by fluorosulfonic acid. Ind Eng Chem Res 37(1):79–88. https://doi.org/10.1021/ie9702744

Teir S, Revitzer H, Eloneva S, Fogelholm C-J, Zevenhoven R (2007) Dissolution of natural serpentinite in mineral and organic acids. Int J Miner Process 83(1–2):36–46. https://doi.org/10.1016/j.minpro.2007.04.001

Telascrêa M, Leão AL, Ferreira MZ, Pupo HFF, Cherian BM, Narine S (2014) Use of a cashew nut shell liquid resin as a potential replacement for phenolic resins in the preparation of panels – A Review. Mol Cryst Liq Cryst 604(1):222–232. https://doi.org/10.1080/15421406.2014.968509

Tomassetti L, Di Giuseppe D, Zoboli A, Paolini V, Torre M, Paris E, Guerriero E, Petracchini F, Gualtieri AF (2020) Emission of fibres and atmospheric pollutants from the thermal treatment of asbestos containing waste (ACW). J Clean Prod 268. https://doi.org/10.1016/j.jclepro.2020.122179

Trivedi AK, Ahmad I (2013) Impact of chrysotile asbestos contaminated soil on foliar nutrient status of plants. Ind J Plant Physiol 18(3):301–307. https://doi.org/10.1007/s40502-013-0036-9

Trivedi AK, Ahmad I, Musthapa MS, Ansari FA (2007) Environmental contamination of chrysotile asbestos and its toxic effects on antioxidative system of *Lemna gibba*. Arch Environ Contam Toxicol 52(3):355–362. https://doi.org/10.1007/s00244-006-0056-9

Trivedi AK, Ahmad I, Musthapa MS, Ansari FA, Rahman Q (2004) Environmental contamination of chrysotile asbestos and its toxic effects on growth and physiological and biochemical parameters of *Lemna gibba*. Arch Environ Contam Toxicol 47(3):281–289. https://doi.org/10.1007/s00244-004-3161-7

Turci F, Colonna M, Tomatis M, Mantegna S, Cravotto G, Fubini B (2010) New detoxification processes for asbestos fibers in the environment. J Toxicol Environ Health A 73(5):368–377. https://doi.org/10.1080/15287390903442678

Turci F, Tomatis M, Mantegna S, Cravotto G, Fubini B (2008) A new approach to the decontamination of asbestos-polluted waters by treatment with oxalic acid under power ultrasound. Ultrason Sonochem 15(4):420–427. https://doi.org/10.1016/j.ultsonch.2007.08.007

USEPA (2020a) Asbestos Laws and Regulations. https://www.epa.gov/asbestos/asbestos-laws-and-regulations#epalaws. Accessed on 20 January 2021

USEPA (2020b) EPA Actions to Protect the Public from Exposure to Asbestos. https://www.epa.gov/asbestos/epa-actions-protect-public-exposure-asbestos. Accessed on 20 January 2021

USGS (2020) Report Shows Global Asbestos Trends. Mesothelioma.com. https://www.mesothelioma.com/blog/usgs-2020-report-shows-global-asbestos-trends/.
Valinluck V, Sowers LC (2007) Endogenous cytosine damage products alter the site selectivity of human DNA maintenance methyltransferase DNMT1. Cancer Res 67(3):946–950. https://doi.org/10.1158/0008-5472.CAN-06-3123
Virta RL (2005) Mineral commodity profiles – asbestos. Circular, Online only, Version 1.0 edn. https://doi.org/10.3133/cir1255KK
Vyas S (2018) New rules to make State free of asbestos. https://www.thehindu.com/news/cities/mumbai/new-rules-to-make-state-free-of-asbestos/article23469360.ece.
Wallis SL, Emmett EA, Hardy R, Casper BB, Blanchon DJ, Testa JR, Menges CW, Gonneau C, Jerolmack DJ, Seiphoori A, Steinhorn G, Berry TA (2020) Challenging global waste management - Bioremediation to Detoxify Asbestos. Front Environ Sci 8. https://doi.org/10.3389/fenvs.2020.00020
Wang X, Wang X, Hodgson L, George SL, Sargent DJ, Foster NR, Ganti AK, Stinchcombe TE, Crawford J, Kratzke R, Adjei AA, Kindler HL, Vokes EE, Pang H (2017) Validation of progression-free survival as a surrogate endpoint for overall survival in malignant mesothelioma: Analysis of cancer and leukemia Group B and North Central Cancer Treatment Group (Alliance) Trials. Oncologist 22(2):189–198. https://doi.org/10.1634/theoncologist.2016-0121
Washabaugh FJ (1981) Asbestos-free friction material incorporating attapulgite clay. US Patent Number US4373037A
WHO (2014) Health impacts of chemicals - asbestos. https://www.who.int/ipcs/assessment/public_health/asbestos/en/.
WHO (2018) Asbestos: elimination of asbestos-related diseases. WHO. https://www.who.int/en/news-room/fact-sheets/detail/asbestos-elimination-of-asbestos-related-diseases.
Witkiewicz W, Zieliński A (2006) Properties of the polyurethane (PU) light foams. Adv Mater Sci 6 (2):35–51
Wroble J, Frederick T, Frame A, Vallero D (2017) Comparison of soil sampling and analytical methods for asbestos at the Sumas Mountain Asbestos Site-Working towards a toolbox for better assessment. PLoS One 12 (7):e0180210. https://doi.org/10.1371/journal.pone.0180210
Yao M, Lian B, Teng HH, Tian Y, Yang X (2013) Serpentine dissolution in the presence of bacteria *Bacillus mucilaginosus*. Geomicrobiol J 30(1):72–80. https://doi.org/10.1080/01490451.2011.653087
Yoshikawa N, Kashimura K, Hashiguchi M, Sato M, Horikoshi S, Mitani T, Shinohara N (2015) Detoxification mechanism of asbestos materials by microwave treatment. J Hazard Mater 284:201–206. https://doi.org/10.1016/j.jhazmat.2014.09.030
Yu DP, Hang QL, Ding Y, Zhang HZ, Bai ZG, Wang JJ, Feng SQ (1998) Amorphous silica nanowires: Intensive blue light emitters. Appl Phys Lett 73(21):3076–3078
Zaremba T, Krząkała A, Piotrowski J, Garczorz D (2010) Study on the thermal decomposition of chrysotile asbestos. J Therm Anal Calorim 101(2):479–485. https://doi.org/10.1007/s10973-010-0819-4
Zhang J, Xie S, Ho YS (2009) Removal of fluoride ions from aqueous solution using modified attapulgite as adsorbent. J Hazard Mater 165(1–3):218–222

Chromium Contamination in Soil and Its Bioremediation: An Overview

15

Avijit Bakshi and Ashis Kumar Panigrahi

Abstract

Heavy metal contamination in the environment has been considered as an important threat to the life in recent days. Chromium contamination is also listed among the potential threat to the human and animals as well as plants. Chromium is a ubiquitous metal having three main oxidation states viz., Cr^{2+}, Cr^{3+} and Cr^{6+}. Among these, divalent form is unstable. Chromium and its particulates are excreted into the environment from different industries like tanneries, textiles, ore mining, printing-photographic houses, dyeing factories, electroplating workshops and medical industries. Hexavalent chromium having carcinogenic potentiality is considered to be the most toxic form because it can readily cross the biomembrane of organisms. Chromium can contaminate soil, groundwater and surface water. To render the contaminated resource reusable, chromium must be removed physically or by using the techniques of bioremediation. Bioremediation has been considered as the future of waste management technologies for sustainable development. The process includes the involvement of plants and microbes that are capable of absorbing, degrading and removing contaminated chromium from the environment. Usually, the process can be practiced both ex situ and in situ taking the advantage of natural homeostasis mechanism of environment. Among these two, in situ practice is cheaper and environment friendly.

Keywords

Bioremediation · Chromium · Contamination · Environment · Microbes

A. Bakshi (✉)
Ecotoxicology, Fisheries and Aquaculture Extension Laboratory, Department of Zoology, University of Kalyani, Kalyani 741235, India

A. K. Panigrahi
University of Burdwan, Burdwan, India

15.1 Introduction

Due to intensification of anthropogenic activity in last century, pollution has been reported to be increased greatly. Negative impact of pollution is further enhanced synergistically when organic and inorganic pollutants coexist in environment. The phenomenon of mixed pollution or co-pollution poses hazardous and impulsive consequences on human and ecosystem health. The interactions between different toxic pollutants exert complex impact on soil biota along with the individual toxicity of each pollutant (Lacalle et al. 2020). Mostly, the presence of both inorganic and/or organic pollutants in soil is

J. A. Malik (ed.), *Advances in Bioremediation and Phytoremediation for Sustainable Soil Management*,
https://doi.org/10.1007/978-3-030-89984-4_15

attributed by more than one type of pollutant. In recent days, inorganic pollution is predominantly fortified by the contamination of heavy metals mainly excreted from the industries (Bakshi 2016). The situations are becoming more hazardous when heavy metals coexist with different organic pollutants. Mixed pollution of soil with Cr(VI) and organic pesticides or fertilizers has been observed recently in different regions of the world which often exceed the maximum allowable concentration of the toxicant (Aparicio et al. 2018a, b; Aparicio et al. 2019).

Chromium in found to be contaminated in air, water and soil. Among the two stable states, Cr (III) is said to be natural in origin, mostly leached out from ores whereas, origin of hexavalent chromium is mostly anthropogenic (Shahid et al. 2017; Bakshi and Panigrahi 2018). Major part of environmental chromium of anthropogenic origin comes from industries like stainless steel manufacturing plants, electroplating, chrome plating, tannery, alloy cast irons, paint production houses, rubber manufacturing and leather industries, wood treatment, dyeing factories, etc. (Bakshi and Panigrahi 2018). Cr(VI) has been estimated to have 1000-fold more mutagenic and cytotoxic impact than Cr(III) (Biedermann and Landolph 1990). Cr(VI) is more soluble than Cr (III) as trivalent form shows greater tendency to precipitate in soil (Zayed and Terry 2003).

Environmental contamination of chromium in recent days has gained consideration as an important hazardous issue of concern because the concentration of Cr is increasing substantially in air, water and soil (Choppala et al. 2016; Bakshi 2016; Bakshi and Panigrahi 2018). The main reason behind the increase is mostly anthropogenic. Chromium has distinct property of having more than one stable oxidation state. Main two species of chromium, i.e., Cr(III) and Cr(VI) vary greatly in their sorption method and availability in soil which enable them to absorb and translocate differentially in the aerial part of the plant. Generally, Cr is accumulated in roots and is poorly translocated in the shoots (Jaison and Muthukumar 2016). However, transfer of chromium from soil to root is highly regulated by several physiological factors (plant type, root type, rate of absorption, root surface area, transpiration rate, etc.) and also by some soil properties (soil texture, pH, organic content, cation exchange properties) (Banks et al. 2006; Shahid et al. 2017). Organisms having chromium hyperaccumulation property with higher uptake and shoot translocation rate have also been reported. Owing to this property, hyperaccumulator plants like *Spartina argentinensis, Amaranthus dubius, Dyera costulata, Pteris vittata,* etc., can be used as bioremediating agent (de Oliveira et al. 2016).

Bioremediation is now becoming a major promising technology to remove the contamination of heavy metals. It is a very challenging technology involving incorporation of Cr hyperaccumulator organisms into a metal or pollutant contaminated environment to up take the pollutant or metal. In the practice of bioremediation, it is very essential to approve that the remediation process should not produce further harm to the environment. Thus, the term Gentle Remediation Options(GRO) is recently used to acknowledge the technology as risk management strategy ensuring net gain (or at least no loss) to the soil which is advantageous than traditional bioremediation processes. GROs include technologies such as bioaugmentation, biostimulation, phytoremediation and vermiremediation. These processes, nowadays, have been considered as effective risk management strategies to decrease the contaminant/pollutant to local acceptors through extraction or in situ stabilization of contaminant (Cundy et al. 2013). These biological techniques of management of soil pollution can be beneficial environmentally and are also proved to be cost effective (Agnello et al. 2016). Objective of this review article is to provide maximum updated and consolidated information on recent researches on soil chromium contamination and its remediation methods.

15.2 Methodology

Attempt has been made to prepare an utmost consolidated scientific review on the research topic. Extensive scrutiny has been done during compilation and consolidation of the available scientific data in order to make it more inclusive and significant for the future researchers. Used data has been collected, from various reputed science journals, different governmental or non-governmental published reports (particularly from national/international agencies) and published doctoral or postdoctoral theses. Precedence has been given only to the reproducible articles that are indexed in science journal database like Web of Science, Copernicus, Scopus, Google Scholar, PubMed, etc. The articles highlighting ambiguous research methodologies are strictly avoided. Key words, for searching the articles, have been judiciously chosen and examined based on systematic scientific approaches. The experimental findings (both laboratory and field) from our own studies have been encompassed thoroughly at various parts of the article specially to improve the essence of the review.

15.3 General Chemistry of Chromium

Chromium is an abundant metal element with the symbol Cr. It is the first element among group six elements in periodic table and also considered as a transitional element. It is hard with steely gray lustrous and also is used as the main additive component of stainless steel (Brandes et al. 1956). It has been considered as the 21st most available element in rocks and has also been estimated as the seventh most abundant element on earth (Katz and Salem 1994). Some primary physical and chemical properties of chromium are listed in Table 15.1. Trivalent and hexavalent chromium both exist in two ionic forms within the normal value of pH and Eh of soil. Two trivalent ionic forms are Cr^{3+} and CrO^{2-}, while $Cr_2O_7{}^{2-}$ and $CrO_4{}^{2-}$ are two available ionic forms in soil. Cr(III) is poorly soluble at or above pH 4.0 and exhibits complete precipitation at pH 5.5 (Mohanty and Patra 2011).

15.4 Chromium in Environment

Chromium is a ubiquitous metal found in soil, water, volcanic dusts and air. The origin may be natural though major emission is done by anthropogenic activities. In India, small towns have an average of 0.02 $\mu g/m^3$ of chromium concentration in air (Bakshi 2016). Chromium is naturally found in waterbodies with concentrations ranging between 1 and 2 mg/l in dissolved form (Bakshi and Panigrahi 2018). Chromium is found to exist naturally in the form of bound chemical to crystalized iron oxides and primary rock-derivative forms (Shanker et al. 2005; Quantin et al. 2008) and is naturally found as chromite ($FeCr_2O_4$) in ultra-mafic rocks (Avudainayagam et al. 2003). The metal is also found in some other ore like vauquelinite ($CuPb_2CrO_4$–PO_4OH), bentorite [$Ca_6(CrAl)_2(SO_4)_3$] tarapacaite (K_2CrO_4) and crocoite ($PbCrO_4$) (Avudainayagam et al. 2003; Babula et al. 2008). Specks of these chemicals undergo weathering and get absorbed in soil particles forming compounds with soil organic matter (Hsu et al. 2015). Igneous and sedimentary rocks contain a very low amount of chromium (5–120 ppm) while mafic (170–200 ppm) and ultra-mafic rocks (1600–3400 ppm) contain comparatively higher amount of chromium within it (Kabata-Pendias 2010). Oxidation states and ore forms are listed in Table 15.2. Generally, four chromium isotopes are available in nature, i.e., Cr-54 (2.4%), Cr-53 (9.6%), Cr-52 (83.8%) and Cr-50 (4.3%), but some other available isotopes of anthropogenic origin are also reported (Eisler 1986).

Natural level of chromium in Earth's crust ranges from 0.1 to 0.3 ppm. However, the concentration may vary with different conditions. Studies have revealed that background concentration or average concentration of chromium is different in various countries (Table 15.3). The amount of chromium rises with the clay

Table 15.1 Physical and chemical properties of chromium

Properties	Value
State at STP	Solid
Color	Silvery white
Atomic number	24
Group	6
Period	4
Block	d-block
Electron per shell	2, 8, 13, 1
Elemental category	Transition metal
Atomic mass	51.996 g/mol
Density	7.19 g cm^{-3} at 20 °C (at melting point liquid state 6.3 g cm^{-3})
Hardness	8.5 Mohs
Melting point	2130.2 K
Boiling point	2963 K
Mass number	51.9961
Isotopes	6
Position	First series of transition metal
Subgroup	VIB
Electronic configuration	(Ar) $3d^5 4s^1$
Electronegativity	1.66 (pauling scale)
Oxidation states	Cr(II), Cr(III), Cr(IV), Cr(V), Cr(VI)
Stable oxidation states	Cr(III), Cr(VI)
Ionic radius	0.061 nm (trivalent); 0.044 nm (hexavalent)
Atomic radius	128 pm
Van der Waals radius	0.127 nm
Electrical conductivity (1 mohm/cm)	77.519
Thermal conductivity	93.9 J/m s K
Heat of fusion	21.0 kJ/mol
Heat of atomization	397 kJ/mol
Heat of vaporization	347 kJ/mol
Molecular heat capacity	23.35 J/mol-K

percentage of the soil (Shahid et al. 2017). Average chromium concentration of most of the countries ranges between of 15 and 100 ppm (Table 15.3). Cr concentration in freshwater ranges from 0.1 to 117 ppm, while in marine water Cr ranges from 0.2 to 50 ppm (Nriagu 1988).

MATC value or maximum allowable toxicant concentration of chromium in humans is estimated at 64 ppm in soil (CCME 2015). The allowable levels of chromium in soil vary in different countries like Poland (150 ppm), Czech Republic (100–200 ppm), Austria (100 ppm), Canada (64 ppm), Serbia (100 ppm), etc. (Ding et al. 2014). Maximum permissible value of hexavalent chromium in drinking water is 0.1 ppm (ATSDR 2012). Nriagu (1988) reported that air samples from urban areas contain 0.015–

Table 15.2 Different oxidation states of chromium and its availability in environment

Chemical state	Oxidation state	Ore/compound form	Stable/unstable	Occurrence
Elemental chromium	0	–	–	Not found naturally
Divalent chromium	Cr^{2+}	Cr_2Si, CrSe, $CrBr_2$, $CrCl_2$, $CrFe_2$	Unstable, readily oxidized to Cr^{3+}	Rarely found, not stable
Trivalent chromium	Cr^{3+}	$FeCr_2O_4$,$CrCl_3$, CrF_3,	Stable	Found naturally as ores mainly ferrochromite
Tetravalent chromium	Cr^{4+}	CrF_4, CrO_2, Sr_2CrO_4, Ba_2CrO_4	Unstable	Not found naturally
Pentavalent chromium	Cr^{5+}	CrF_5	Unstable	Not found naturally
Hexavalent chromium	Cr^{6+}	$CaCrO_4$, $PbCrO_4$, $BaCrO_4$, K_2CrO_4, $K_2Cr_2O_7$, $(NH4)_2CrO_4$,	Stable	Most toxic, rarely found in nature, anthropogenic origin

Table 15.3 Availability of chromium and its concentration in soils

S. No	Cr Conc.	Parameter	Country	References
1	5–3000 ppm	Background Conc	India	Shanker et al. (2005)
2	17–727 ppm 21–313 ppm	Cultivated soil (paddy field) Uncultivated soil	Bangladesh	Bakshi (2016)
3	2–60 ppm	Natural Conc	Turkey	Isıklı et al. (2003)
4	147.28 ppm	Cultivating soil	China	Qu et al. (2015)
5	100 ppm	Average Conc	Caribbean Island	Mandal and Voutchkov (2011)
6	59.5 ppm	Average Conc	Poland	Kabata-Pendias (2010)
7	22 ppm	Average Conc	Sweden	Eriksson (2001)
8	58 ppm	Average Conc	Japan	Takeda et al. (2004)
9	54 ppm	Average Conc	USA	Burt et al. (2003)
10	94.8 ppm	Average Conc	Finland	Salminen et al. (2005)
11	276 ppm	Mine soil	Slovakia	Kulikova et al. (2019)
12	7–65.50 ppm 10.75–102.80 ppm	Silt and clay Dry soil	East Kenya	Mwamburi (2016)

0.03 μg m^{-3} of chromium, whereas the value is very low (0.59×10^{-8} to 1.29×10^{-3} μg m^{-3}) in remote areas.

Chromium contamination in soil is evident in different countries around the world specifically in India, China, Eastern Africa, Eastern Europe and South America (Ericson 2011; Shahid et al. 2017). According to international guideline, maximum chromium emission limit is 50–100 kg/year, but actual estimated discharge is very much higher in air (30 metric tons/year), water and soil.

15.5 Speciation of Chromium in Soil

Atmospheric chromium is mainly found in particulate form, whereas natural gaseous forms of chromium are very scanty (Seigneur and Constantinous 1995). Biogeochemical nature of any metal is not only dependent on total metal content in atmosphere because most of the metal shows speciation forming its different chemical species (Rafiq et al. 2017). Study of chemical speciation of chromium is very important for understanding its risk assessment and bioremediation method (Saifullah et al. 2015). Valence state of chromium varies from 0 to 6, whereas, trivalent and hexavalent chromium are more stable than any other state (Table 15.2). Chromium is a typical metal which shows contrary effects at different oxidation states. Trivalent chromium is slowly mobile in soil and less toxic to the organisms than the hexavalent form. Hexavalent chromium has mutagenic, teratogenic and carcinogenic potential on the living organisms (Prado et al. 2016). It is highly reactive with other metallic ions confirming its toxicity. Both trivalent and hexavalent forms of chromium coexist in nature. The oxidation and/or reduction of chromium species in soil are thermodynamically spontaneous so when chromium is contaminated from any source it can undergo both oxidation and reduction reaction (Dhal et al. 2013; Ding et al. 2016). Conversation of chromium into various species can include not only reduction and oxidation but also hydrolysis, precipitation etc. (Zayed and Terry 2003; Di Palma et al. 2015). Speciation of chromium is regulated by some parameters like metal concentration in soil, availability of competing cations, cation exchange capacity, pH of soil, redox potential (Eh) and also soil microbiological environment (Shahid et al. 2014; Taghipour and Jalali 2016).

15.5.1 Impact of Soil pH on Chromium Speciation

Soil pH is an important factor regulating the speciation of chromium showing negative correlation with metal mobility (Shahid et al. 2017). Desorption of trivalent chromium from solid to solution state is highly observed at low soil pH, whereas Cr(VI) adsorption inhibits at increased pH (Dias-Ferreira et al. 2015). Addition of organic materials and lime to soil affects the sorption of Cr(III) and Cr(VI) (Taghipour and Jalali 2016). It has been reported that Cr(III) sorption increases when lime is added to the soil, whereas sorption of Cr(VI) has been found to decrease slightly at the same time (Bolan and Thiagarajan 2001). Increased pH is attributed to release of hydroxyl ions affecting surface negative ion balance which promotes Cr(III) precipitation and decreased sorption of Cr (VI) in soil. Shahid et al. (2017) have shown that trivalent chromium is predominated in low pH soil, but its availability decreases when pH of soil rises up to 5. Cr(III) transforms into $Cr(OH)_2^+$ species at a pH of below 5.

The toxic hexavalent chromium ($HCrO_4$) is found to be predominant under acidic pH (range 1–6.5), whereas $CaCrO_4$ and CrO_4 have been reported to co-dominate at alkaline conditions (range 8–12). Specifically, CrO_4 has been reported to be predominant at high pH values greater than 12.

15.5.2 Impact of Soil Eh or Redox Potential on Chromium Speciation

Eh or redox potential of soil is determined by measuring its tendency to accept or donate electron. Thus, it depicts the oxidation and reduction potential of soil (Shahid et al. 2017). Low Eh level of soil facilitates reduction reactions increasing soil pH whereas high value of Eh promotes oxidation processes lowering soil pH. This can be achieved by the consumption and production of protons in soil, respectively (Frohne et al. 2015). Otero and Macias (2003) classified three different states of Eh, i.e., anoxic (less than 100 mV), oxic (greater than 350 mV) and suboxic (greater than 100 mV but less than 350 mV).

Chromium speciation is highly regulated by redox potential of the soil as chromium exhibits various oxidation states (Xiao et al. 2015). Reduced soil condition promotes conversion of hexavalent chromium in less toxic trivalent form (Rupp et al. 2010). Neutral to alkaline pH of soil can store Cr(VI) if it contains high oxygen level. Hexavalent form has strong oxidizing property in acidic soil because of its high Eh (+1.38 V) value (Shadreck 2013).

15.5.3 Impact of Soil Organic Content on Chromium Speciation

Organic content of soil plays a significant role in bioavailability, mobility and sorption/desorption of chromium (Shahid et al. 2013). Having complex chemical nature due to the presence of various organic matter of different structure, composition and molecular weight, organic content of soil can act as carrier of chromium enhancing coupling and storage of metal (Quenea et al. 2009). Organic matter mediated reduction of toxic hexavalent chromium into trivalent form is regulated by soil pH and redox potential. High level of organic content facilitates microbial growth maintaining reduced condition of soil altering redox potential (Shahid et al. 2017). Positive ions from organic matter complex and colloidal property of soil facilitate higher retention of toxic Cr(VI) acting as electron shuttle for bio-reduction (Choppala et al. 2016; Shahid et al. 2017). Organic content promotes reduction of Cr(VI) into Cr(III) and also stimulates microbial propagation enhancing rate of bio-reduction (Banks et al. 2006; Ashraf et al. 2017). Amendment of organic manure (seaweed, farm yard manure, black carbon, compost, etc.) in soil can help in reducing the toxicity of hexavalent chromium, thus the technique is very much popular for soil reclamation and remediation (Shahid et al. 2015).

15.5.4 Impact of Soil Microbial Diversity on Chromium Speciation

Microbial diversity of soil is highly related to the mineralization and immobilization of nutrient in soil. Thus, soil microbes can play significant role to determine biogeochemical property of heavy metals in soil–plant system (Ahemad 2015). Several authors have reported that reduction of Cr(VI) to Cr(III) can be done under both aerobic and/or anaerobic environment (Zeng et al. 2016; Qian et al. 2016). Microbe-assisted reduction of hexavalent chromium to Cr(III) is highly dependent on the bacterial strain. The process may use NADPH or $NADH^+$-dependent chromium reductase under aerobic condition or may use hexavalent chromium directly as an electron acceptor under anaerobic environment. Another complex pathway of reduction of chromium has been reported by several researchers. In this process, some inter or intracellular compounds like amino acid, nucleotides, sugars, vitamin, glutathione readily help to reduce Cr(VI) aerobically.

15.6 Biological Importance of Chromium

Chromium has toxic as well as beneficial impact on organisms. Early reports have described that chromium has some stimulatory growth effect on the plants (Smith et al. 1989). Though recent researchers have claimed that though chromium may have some beneficial effects on human health, it has no role in plant metabolism (Shanker et al. 2005).

Hexavalent chromium is toxic for organisms, but trivalent chromium has some importance as a bio-element as it performs some exceptional role in metabolism (Barabasz et al. 1998). Chromium concentration in fetus or new borne is higher than the adult. After birth, the concentration of chromium in different tissues begins to reduce with the age of the individual except the ling tissue. Chromium can be deposited in lung through inhalation of chromium contaminated air. Chromium has some impact on human health particularly the trivalent from. In organisms' body, chromium has a potential impact on the metabolism of insulin. It also plays some significant role in different cellular enzyme reactions. Glucose tolerance factor (GTF) is also regulated by the concentration of trivalent chromium in the body. It helps to up take blood glucose in the tissue in collaboration with insulin. In the absence of optimum amount of chromium, insulin activity is blocked elevating the sugar level in blood (Bielicka et al. 2005). Chromium is also taken as a supplement to reduce body weight and fat, reduce cholesterol and enhance muscle mass. Trivalent chromium has potential impact in the metabolism of humans and animals (Kendrick et al. 1992). Moreover, chromium is an essential minor element which can enhance carbohydrate, lipid and protein metabolism and can also improve insulin sensitivity (Bakshi 2016). Researches have established that chromium supplement like chromium picolinate having anticoagulant action can impede hyperglycemia induced atherosclerosis (Ganguly et al. 2017). Nonetheless, a vast research is also available claiming non-essentiality of chromium in animal body (Di Bona et al. 2011).

15.7 Toxicity of Chromium

Chromium, especially its hexavalent form, is one of the major toxicants of the soil. The maximum amount of contamination is occurred through human activities. It has various adverse impacts on animals, plants and even on the beneficial microorganisms of the soil (Ahemad 2015). Trivalent form of chromium is found naturally in soil. Due to its low solubility property, Cr(III) is found to have a great tendency of absorbing on soil particles. The absorbed metal is used as an essential nutrient by soil organisms like microbes and plants (Bosnir et al. 2013). But hexavalent metallic form is a potential toxic pollutant for the soil organisms especially of anthropogenic origin (Bojanowska 2002). Very small amount of transformation of trivalent chromium to the hexavalent form occurs in natural way of oxidation within ultra-mafic- and serpentinite-derived sediment/ soil (Oze et al. 2007).

15.7.1 Toxicity in Plants

Chromium is very much resistant to corrosive environmental or chemical agents. This characteristic feature makes it very useful as a protective coating agent. Being highly used in electroplating industries, resistant alloy manufacturing industries, cement manufacturing industries, electronic gadgets, aircrafts and also in nuclear reactor vessels, chromium is eliminated into the soil directly or indirectly (Ahemad 2015). Plants can accumulate both Cr(III) and Cr (VI) from the soil; however, the detailed mechanism of uptake is still unclear (Shahid et al. 2017) as the metal does not have any essential role in plant metabolism (Oliveira 2012). The capability of chromium uptake by plants is depended upon the plant type and Cr species (Shukla et al. 2007). Trivalent chromium is basically up taken through passive mechanism and does not require energy (Shanker et al. 2005). Hexavalent chromium is highly toxic for the plant as it, as an oxyanion/dichromate ion form, can easily traverse biomembranes by diffusion. The entry is mostly regulated by non-specific anion channels that are present in the cell membrane for transporting sulfate and phosphate ions (Nickens et al. 2010; de Oliveira et al. 2016). The degree of toxicity is further enhanced after the reduction which produce free radicals and reactive oxygen speciesROS) within the cell (Alam and Ahmed 2013). The sequential metabolic reduction of hexavalent chromium forms stable trivalent state through several intermediate unstable valence states and free radicals within the cell (Salnokow and Zhitkovich 2008). Trivalent chromium has a great affinity to cellular proteins and nucleic acids of the cell. Thus, it causes DNA damage ensuring the inhibition of DNA replication and RNA transcription process (Nickens et al. 2010). Chromosomal aberrations like polyploidy and mitotic cell cycle arrest are very common cellar damages due to the toxicity of chromium. Other cellular alterations like changes in enzymatic activity (especially, invertase, amylase, catalase, RNAse, Fe-reductase and peroxidase), decrease of nutrient uptake and water potential, disorganization of chloroplast, inhibition of electron transport process, etc., are reported to be observed due to chromium toxicity (Ramírez-Díaz et al. 2008). Several plant physiological processes are also affected by chromium intoxication. Reduced photosynthesis rate, delayed seed germination and abnormality in transpiration process are also documented as the result of chromium-induced toxicity (Ahemad 2015). Phytotoxicity studies also confirm consequences like stunted growth, chlorosis, falling of younger leaves, senescence reduced crop yield, reduction in dry weight due to chromium contamination (Misra et al. 1994, 2004).

15.7.2 Toxicity to Animals

In animals, chromium can enter into the body through various routes and in both oxidative conditions, trivalent and/or hexavalent. Increased absorption of chromium in either state can lead to renal failure, liver malfunctioning and even death in organisms (Lippmann 2000). Acute exposure of hexavalent chromium can lead to serious damage in liver and kidneys (Avudainayagam et al. 2003). Hexavalent chromate ion is said to be iso-structural with phosphate and sulfate ion and likewise can penetrate the cell membrane of gastrointestinal cells (Costa 1997). After entry through GI tract, hexavalent chromium can easily arrive in different tissues and can establish some non-cancerous symptoms in respiratory system, GI tract, immune system, kidney and liver cells in rat or mice (Costa 1997; Avudainayagam et al. 2003). In acute exposure to sub-lethal concentrations of chromium, freshwater fishes show different symptoms like erratic swimming, loss of body balance, hyperactivity and increased mucus secretion (Bakshi and Panigrahi 2018). In chronic exposure of hexavalent chromium, fishes show decreased level of glycogen, total lipid and total protein in different tissues like liver, muscle and gill (Bakshi 2016). Disintegration of gill lamellae, necrosis in gill epithelia and hepatocytes, glomerular disorganization, etc., are also observed in freshwater fishes during chronic exposure to sub-lethal concentrations (Velma et al. 2009; Bakshi 2016). Chromium is well documented as potential carcinogen and genotoxic agent as it can break DNA phosphate backbone and cross-linking proteins in mammalian cells (Lippmann 2000). Reduction of hexavalent chromium in trivalent form produces oxygen radicals and other cytoplasmic reducing agents in the cell which are capable of producing serious damage at cellular level like DNA damage, abnormal metabolic paths, etc. (Avudainayagam et al. 2003).

15.7.3 Toxicity to Humans

Chromium gets accumulated in human body mainly via food chain contamination (Ahmad et al. 2016). In humans, chromium is connected with various pathologies, including carcinogenic alterations. Hexavalent chromium-induced epigenetic modifications, genomic instability and multistage carcinogenesis are three possible pathways of cellular damage. The degree of damage is highly associated with different external factors (viz. doses, pH etc.) and internal factors (viz. enzymatic polymorphism, action of carrier proteins, efficiency of DNA repair mechanisms, endogenous reducing cycle, etc.) though variability in chromium-induced alteration is also observed due to individual genetic polymorphism (Pavesi and Moreira 2020). Compounds of hexavalent chromium are highly oxidizing agents mostly having irritating and corrosive nature. It can easily transform into trivalent state after entering into the cell traversing plasma membrane. Reduction of hexavalent chromium can be toxic if it takes place near the nucleus of the target cell (Dayan and Paine 2001). Though the reduction reactions can be considered as a process of detoxification method of Cr toxicity extracellularly (or at distant place from the nucleus), Cr^{6+} can also be reduced by a number of chemicals intracellularly, viz. hydrogen peroxide, ascorbic acid, glutathione reductase (GSH), etc. During the reduction by GSH, some reactive intermediates can be formed which are responsible for various cellular damages like degeneration of DNA, cellular fat and proteins or disruption of cellular integrity (De Mattia et al. 2004). Respiratory problems like asthma, chronic bronchitis, pharyngitis, chronic rhinitis, bronchial congestion, hyperemia, polyps in upper bronchial tract, ulceration of nasal mucosa and nasal septal perforation are evident in man during prolonged exposure to hexavalent chromium (Dayan and Paine 2001). Allergic signs on skin and other symptoms like dryness, erythema, scaling, papules, swelling, etc., have also been reported (MacKie 1981; Adams 1983). Pulmonary carcinogenic property of chromium is reported to be associated with prolonged inhalation of less soluble or soluble hexavalent form (Luippold et al. 2003). Cr(VI) toxicity also affects renal system of humans particularly damaging glomerulus, proximal convoluted tubule and other parts of renal tubule. Elevated concentration of ß2-microglobulin in urine indicates renal tubular damage after chronic chromium exposure (Lindberg and Vesterberg 1983).

15.8 Bioremediation of Chromium

Contamination of heavy metals in soil is mostly done by anthropogenic activities. Normally, the contaminated soil is used for landfills though it is certainly not an eco-friendly solution for the problem. Apart from this, use of heavy metal contaminated soil in landfills proved to be not only expensive but also harmful for the environment. In last three decades, scientific uses of weed plants and/or microbes to eliminate heavy metals from soil or water have been proved to be low-cost alternative and also eco-friendly. Bioremediation is a process which uses living organisms, particularly microbes or plants to eliminate toxic heavy metals from the environment taking the advantage of natural heavy metal absorption potential of experimental organism that already exists within it (Kumar et al. 1995; Kiling 1997). Recent researches of bioremediation of chromium ion include uses of many prokaryote and eukaryote microbes and plants. The process can be practiced directly in contaminated areas, i.e.,in situ or at the exterior site i.e.,ex situ. Among these two processes, the former one is cheaper and eco-friendlier.

Depending upon the type of organism used in bioremediation, the process can be broadly classified into two categories, *i.e.,* microbial remediation and phytoremediation.

15.8.1 Microbial Remediation of Chromium

15.8.1.1 Use of Bacteria and Algae in Chromium Remediation

Microbial remediation relies on the pervasiveness and versatility of microorganisms that are capable of responding to a broad spectrum of adverse environments that are capable of transforming a number of pollutants into nutrients and converting them into non-toxic or less toxic compounds. Many prokaryotes and eukaryote microbes are used successfully in this technology to achieve eco-friendly removal of heavy metals mainly from industrial discharges. Many researches and trials have been done to find out potential microbial bio-accumulator of soil chromium. The microbes are known to produce and to release metal chelating chemicals to the environment to increase the absorption of chromium. These organisms are used in various ways to achieve the specific goal.

Biosorption

Biosorption involves group of processes recruiting living or dead biomass to remove heavy metals or other pollutants from solution (Kisielowska et al. 2010). In this process, surface adsorption of metal from the soil gathers at the cell surfaces of the microbe, linking with extracellular polymers (Tarekegn et al. 2020) (Table 15.4).

The practical application of biosorption for the control of soil chromium contamination chiefly uses the reversibility processes. During the desorption process of metals linked by microbes, applications of weak mineral acid solutions (like 0.1 M HCl) or various chelating agents (like 10 mM EDTA) are found to be

Table 15.4 Biosorption by microbes and their efficiency

Type	Microbes	Metal ion Conc. (mg/l)	Sorption efficiency (%)	Remarks
Bacteria	*Acinetobacter* sp	16	87	Bhattacharya et al. (2014)
	Sporosarcina saromensis	50	82.5	Ran et al. (2016)
	Bacillus cereus	1500	81	Nayak et al. (2018)
	Bacillus circulans MN1	1500	96	Chaturvedi (2011)
	Bacillus cereus	1	78	Singh et al. (2013a, b)
	Bacillus subtilis	0.57	99.6	Kim et al. (2015)
	Bacillus sp b	50–37.06	47	Kumar et al. (2011)
	Staphylococcus	4.108	45	Kumar et al. (2011)
	Pseudomonas aeruginosa (PCP 2)	6.4	72	Kumar et al. (2011)
	Pseudomonas aeruginosa (P)	570–2	99.6	Benazir et al. (2010)
	Immobilized B. subtilis (B bead)	570–2	99.3	Benazir et al. (2010)
Algae	*Spirogyra* sp.	5	98.23	Mane and Bhosle (2012)
	Spirulina sp.	5	98.23	Mane & Bhosle (2012)

beneficial (Sklodowska 2000). In the pH of range 5–7, metal ions, like trivalent chromium are strongly linked to the amount of microbial biomass (Tarekegn et al. 2020).

Bioaccumulation

When absorption rate of any chemical by any organism is higher than the elimination rate, the chemical or contaminant is retained within the organisms' body with a phenomenon of bioaccumulation. Being a toxicokinetic process, bioaccumulation affects sensitivity of any organism toward certain contaminant. It is economically beneficial only when average chromium concentration is high in the environment, but it is different from biosorption with respect to metal removal from cell and recovery related to cellular structural transformation (Mohanty and Patra 2011; Bakshi 2016). Thus, the possibility of practical application in several cycles is quite impossible (Sklodowska 2000).

Biotransformation

Microbial transformation of chromium includes the chemical processes like reduction, oxidation, methylation and demethylation. Isolated gram negative bacteria from tannery effluent have shown significant capacity of reduction of hexavalent chromium into trivalent form facilitating biotransformation (Kisielowska et al. 2010). The sites of these reactions are mainly vacuoles and cell surfaces, whereas the reactions may take place in extracellular spaces (Sklodowska 2000).

Bioprecipitation

Bioprecipitation or biocrystallization of chromium can be done by using microbial enzymatic activity but has not proved to be cost and time effective. Theoretically, bioprecipitation of chromium may occur on the surface of the cell or inside the cell by direct enzymatic activity or by galactosis of secondary metabolites (Sklodowska 2000; Tarekegn et al. 2020).

Bioaugmentation

In bioaugmentation microbes are added to biotransform or biodegrade contamination. Added microbe may be those that are already present in the contaminated site or may be a complete new species. Microbes that are already present in the contaminated site helping in accumulation of contaminant are added in more number to facilitate bioaugmentation (Mohanty and Patra 2011). Presence of nutrients in the site is very essential for the propagation of added microbes from other origin (Quagraine et al. 2005). The technique for removal of chromium from contaminated soil is not practiced in situ mainly due to its efficiency with respect to cost and time, but it is carrying a huge opportunity of research in future. Augmentation of native or indigenous microbes is not accepted as advantageous process of treatment. Concept of inoculation of indigenous microbes in combination with microbes of other origin with a greater efficiency of chromium removal is getting popularized in recent days of researches. Acclimatization of foreign microbes and formulation of proper ratio with indigenous microbes have kept the key of success in bioaugmentation process (Mohanty and Patra 2011).

Biostimulation

Biostimulation is a process in which nutrients, oxygen or other electron donors and electron acceptors are enriched to mitigate heavy metal contamination problem enhancing microbial activity in soil (Leung 2004). During biostimulation, viable native contaminant-degrading microbe population get their basic nutrient as well as microbiota from its original site (Mohanty and Patra 2011). Successful biostimulation process requires amendment to achieve the correct environment for degradation of contaminated chromium in soil below the permissible limit (Quagraine et al. 2005).

15.8.1.2 Use of Fungi in Chromium Remediation

Organisms like yeast and some other filamentous fungi offer another option of treatment of chromium contaminated soil. Rhizoferrin, a siderophore in *Mucorales,* exhibits increased rate of chromium uptake (Pillichshammer et al. 1995). Chemical-treated mycelia of selected fungi like *Mucor mucedo, Rhizomucor miehei,* etc., have proved to be an efficient chromium binder (Wales and Sagar 1990). *Candida tropicalis, C. utilis, Penicillium chrysogenum, Saccharomyces cerevisiae,* etc., have excellent property of biosorption of chromium (Volesky and Holan 1995; Benazir et al. 2010). Chromate-resistant species of *Aspergillus* spp and *Candida* spp isolated from chromium contaminated soil have shown bio-reduction property (Table 15.5) of hexavalent chromium (Paknikar and Bhide 1993; Ramirez et al. 2000; Congeevaram et al. 2007).

Phytoremediation

As stated earlier, chromium is available mostly in hexavalent and trivalent form in the environment. Hexavalent chromium is absorbed in roots mostly by active transport, whereas trivalent form is up taken by osmosis (Barros et al. 2006). Plasma membrane of root cells comes in contact with the metal and regulates the entry through channels of essential ions (Liu et al. 2011; Hayat et al. 2012). Several studies confirm that trivalent form of the chromium competes with ionic forms of iron (Fe), sulfur (S) and phosphorus (P) for binding with membrane-bound carrier proteins (Cervantes et al. 2001; Shanker et al. 2005; Liu et al. 2011). López-Bucio et al. (2014) experimentally proved that chromate inhibits sulfate absorption in cells when applied for a short time duration. After entry through roots, chromium ions start to translocate to shoots very slowly as it is preferred to be retained in the roots (Sundaramoorthy et al. 2010; Singh et al. 2013). Within the root tissues, it inhibits cell proliferation restricting the growth of root. Shorter length of root leads to decrease in nutrient and water supply which results in stunted growth of shoots (Shanker et al. 2005). Phytoremediation is a process with low or no harmful impact on environment thus is also termed as green remediation. The phytoremediation methods include several processes such as phytoextraction, phytovolatilization, rhizofiltration, phytodetoxification and phytostabilization.

Table 15.5 Role and efficiency of fungi in bioremediation

Type	Microbes	Metal ion Conc. (mg/l)	Sorption efficiency (%)	Remarks
Fungi	*Aspergillus* sp.	100	92	Congeevaram et al. (2007)
	Saccharomyces cerevisiae (Y)	570–25	95	Benazir et al. (2010)

Phytoextraction

Phytoextraction is the major process which represents the phytoremediation process. It involves the plants which can accumulate heavy metals from water, sediments and soils. It is the best procedure to eliminate the toxic metals from the soil without hampering the soil structure and fertility. Phytoextraction, also termed as phytoaccumulation, is best practiced where contaminants are at low concentration and very close to the upper surface of the soil (Brooks et al. 1998). The selected plants can accumulate the toxic material into their biomass when introduced to the contaminated sites. It concentrates and precipitates the toxic materials into the plant body. Two major processes are evident for phytoextraction: (a) induced phytoextraction and (b) continuous phytoextraction.

Induced phytoextraction is basically termed as chelate-assisted phytoextraction where artificial chelates have been added to increase the uptake and mobility of chromium in soil. Continuous phytoextraction is a natural process depending on the ability of plant to remediate. In this case of phytoextraction, plant growth repetitions are crucially controlled. The basic phenomenon of phytoextraction of chromium involves mobilization of ions by chelation or reduction in soil which is then taken up and sequestrated in roots. After sequestration in roots, chromium is transported to xylem facilitating shoot trafficking and redistribution to different parts of the plant through phloem (Mohanty and Patra 2011).

Hyperaccumulator plant species have been effectively used in this process. The plants exhibit the tendency to accumulate the toxic metal in the tissues. The removed metals can be recycled to produce bio-ore through phytomining process (Ghosh and Singh 2005; Mohanty and Patra 2011). Plant growth rate, resistance to diseases, element selectivity and method of harvesting should be considered as significant parameters of selecting hyperaccumulator plants for phytoextraction (Mohanty and Patra 2011). Plants such as bush morning glory (*Ipomoea carnea*), pricklyburr or dhutura (*Datura innoxia*), tall reed (*Phragmites karka*), Chaakvad or chakunda (*Cassia tora*), *Lantana camara*, brown mustard (*Brassica juncea*), field mustard (*Brassica campestris*), southern cut grass (*Leersia hexandra*), field bindweed (*Convolvulus arvensis*), Krishna Siris (*Albizia amara*), Bermuda grass (*Cynodon dactylon*) *and* camphorwood (*Pluchea indica*) are reported to have potential chromium bioaccumulation property (Torresdey et al. 2004; Shanker et al. 2005; Ghosh and Singh 2005; Sampanpanish et al. 2006; Zhuang et al.2007). Thus, these can be used as experimental hyperaccumulators. *Typha angustifolia* has exhibited high tolerance to chromium and can be effectively used as hyperaccumulator (Dong et al. 2007). Induced phytoextraction is achieved by the help of production of oligopeptides like metallothioneins and phytochelatins. These proteins bind with chromium forming stable complexes which facilitate tissue translocation of the metal. Addition of EDTA, a synthetic chelating agent, to the soil can enhance chromium uptake from soil (Huang et al. 1997).

Phytoextraction and plant-assisted remediation processes are most convenient way of chromium removal from contaminated soil. But it is highly effective only when contaminant exists within 1 m from soil surface. Chemical chelators used in phytoextraction may be toxic for the plants. These chemicals may increase the uptake of chromium but may exert some potential threat to plant growth (Mohanty and Patra 2011).

Phytovolatilization

Phytovolatilization, another type of phytobioremediation, involves up taking the contaminants from soil, transforms them into a volatile form and further elimination of the metal toxicant through transpiration into the atmosphere. During phytovolatilization, the contaminants can traverse the biomembrane of the root cells and finally enter into the leaves. From the leaves, volatile chromium can be transpired out. Phytovolatilization has been practiced successfully to remove mercury from soil (Henry 2000). Theoretically, the process can be done in case of chromium pollution. The disadvantage of phytovolatilization is the volatile gas which is transpired out by the plant and may be harmful for the environment.

Rhizofiltration

Rhizofiltration is a process which can remove heavy metals mainly from flowing water or aqueous waste streams through the plants with extensive root system. The process is beneficial in controlling soil chromium pollution if it is used on the aqueous wastewater before polluting the soil. Hyperaccumulator plants, both from terrestrial and aquatic origin, are used to up take the contaminated chromium from the wastewater. Rhizofiltration can be practiced to partially treat agricultural runoff, mine drainage and industrial discharges to remove chromium from the environment. Soils of the riverbed can be contaminated with chromium if aquatic chromium pollution takes place in the river. Use of water hyacinth (*Eichornia crassipes*) to remove chromium from riverine system is evident in different rivers of India (Bakshi 2016). Several terrestrial plants like sunflower, tobacco, rye, Indian mustard, spinach and some member of Brassicaceae family have been identified to have the potential of rhizofiltration (Mohanty and Patra 2011). Rhizofiltration is advantageous and can be easily practiced because it can be used in both in situ and ex situ conditions. The practice is also beneficial with non-hyperaccumulator plant species.

Phytodetoxification

This is an in situ method of phytoremediation which involves detoxification of contaminant heavy metal through phyto-chelation, oxidation and reduction reactions. Metal chelators like DTPA, EDTA, maleic acid, salicylic acid, carboxylic acid glycine, etc., are used in chromium remediation through chelation (Henry 2000). Some vegetable plants like cucumber and pumpkin are used for remediation as successful phytodetoxifying agents. Several other plant species especially crops, wetland plants and algae have the potential to reduce Cr(VI) to Cr (III) through the formation of intermediate oxidation form of chromium, *i.e.,* Cr (V).

Phytostabilization

Phytostabilization is a process in which plants help in transforming toxic metals in comparatively low toxic forms. Firstly, root exudates and root excreted organic acids form complexes of chromium in soil then up take the complexes and precipitates in root (Mohanty and Patra 2011). Later, the complexes get adhere to the cell wall and vacuoles of the root cell. Due to low xylem loading of the complexes, very minimum amount of metal is translocated in shoots. Basically, low contaminant mobility and bioavailability in soil are ensured by the process of phytostabilization. It can happen through sorption, precipitation, complex formation and reduction (Salt et al. 1995). This in turn prevents the soil erosion and formation of hazardous leachates to restrict the toxic chromium into limited site of contamination.

15.9 Future Scope of Research

Chromium toxicity and its bioremediation are getting much attention by recent researches. The review chapter, highlighting the consequences of chromium contamination in soil and bioremediation methods, has shown that there are very few instances of potent chromium absorbing bioremediation agents. More research is needed to find out further microbes or plants which can eliminate chromium from contaminated sites. More investigations should be done to understand the speciation of chromium in case of long-term contamination sites as laboratory-based data cannot interpret the natural way of speciation. Results obtained from field study have been found to differ from the laboratory data (Avudainayagam et al. 2003). The future aim of the research should be specific goal oriented.

A detailed research is needed to clearly understand the role of inorganic and organic amendments in chromium speciation. The different processes of Cr adsorption–desorption, soil–plant transfer, toxicity pathways and more beneficial detoxification should be studied more extensively. Role of different transporter proteins and their responsible genes for chromium (Cr-III and Cr-VI) uptake and speciation in plant tissues should be investigated thoroughly. Additional research should be carried to understand the reason behind low translocation of chromium from root to shoot. Reason behind the enhanced sequestration of chromium in roots is still understudied. More efforts should be given to find out eco-friendly hyperaccumulator organisms. Use of organic ligands to minimize the impact of soil pollution is widely explored, but the detailed mechanism of different organic ligands like phytochelatins, glutathione, methionine, vitamins, proteins and amino acids etc., to detoxify the soil is still unexplored. Chromium exhibits hermetic property in plant, but the exact toxic values of chromium in soil are not established till now which can be investigated in future. Lesser data regarding cost–benefit ratio is available for chromium bioremediation. Future researchers can evaluate the case-specific cost–benefit ratio for different bioremediation methods.

15.10 Conclusion

Being ubiquitous, chromium can be found to be present in air, water and soil. It is an important environmental toxicant with carcinogenic and mutagenic effects. The main contributors of chromium pollution are large-scale and small-scale industries such as chrome plating, electroplating industries, alloy cast iron industries, photo printing industries, stainless steel industries, rubber factory, leather factory and dyeing factories. The stress imposed by chromium contamination affects plants in different ways, including physiological and metabolic stress. Contamination of chromium in animals and humans can happen directly or through food chain posing adverse impact on GI tract, heart and pulmonary system, immune system, etc. Attempts have been made to mitigate the problem of chromium contamination in soil by the help of bioremediation methods. Phytoremediation, biostimulation, bioaugmentation, microbial remediation, vermiremediation etc., are used to eliminate chromium from soil. These processes are more advantageous and eco-friendly than traditional ways of dumping and/or shifting of contaminated soil or landfills. Thus, to practice eco-friendly approaches of remediation, more research needs to be done on phytovolatilization, phytoextraction, phytodetoxification, phytostabilization, rhizofiltration etc. Discovering new native hyperaccumulator species to mitigate the problem of chromium contamination in soil should get more concern to investigate. Furthermore, introduction of new, more effective, affordable, eco-friendly cost–benefit remediation processes in Cr-contaminated area should be taken as challenges to investigate.

Acknowledgements Authors are grateful to the authorities of Department of Zoology, University of Kalyani, Kalyani, Nadia, West Bengal, India, for their support during the research.

References

Adams RM (1983) Occupational skin disease. Grune & Stratton, New York, p 92

Agnello AC, Bagard M, Van Hullebusch ED, Esposito G, Huguenot D (2016) Comparative bioremediation of heavy metals and petroleum hydrocarbons cocontaminated soil by natural attenuation, phytoremediation, bioaugmentation and bioaugmentation-assisted phytoremediation. Sci Total Environ 563: 693–703

Ahemad M (2015) Enhancing phytoremediation of chromium-stressed soils through plant-growth-promoting bacteria. J Genet Eng Biotechnol 13:51–58. https://doi.org/10.1016/j.jgeb.2015.02.001

Ahmad I, Javed M, Asghar HN, Shahid M (2016) Differential effects of plant growth-promoting rhizobacteria on maize growth and cadmium uptake. J Plant Growth Regul 35:303–315

Alam MZ, Ahmad S (2013) Multi-metal biosorption and bioaccumulation by *Exiguobacterium* sp. ZM-2. Ann Microbiol 63:1137–1146. https://doi.org/10.1007/s13213-012-0571-z

Aparicio JD, Garcia-Velasco N, Urionabarrenetxea E, Soto M, Álvarez A, Polti MA (2019) Evaluation of the effectiveness of a bioremediation process in experimental soils polluted with chromium and lindane. Ecotoxicol Environ Saf 181:255–263

Aparicio JD, Raimondo EE, Gil RA, Benimeli CS, Polti MA (2018) Actinobacteria consortium as an efficient biotechnological tool for mixed polluted soil reclamation: experimental factorial design for bioremediation process optimization. J Hazard Mater 342:408–417

Aparicio JD, Saez JM, Raimondo EE, Benimeli CS, Polti MA (2018) Comparative study of single and mixed cultures of actinobacteria for the bioremediation of co-contaminated matrices. J Environ Chem Eng 6:2310–2318

Ashraf A, Bibi I, Niazi NK, Ok YS, Murtaza G, Shahid M, Kunhikrishnan A, Mahmood T (2017) Chromium(VI) sorption efficiency of acid-activated banana peel over organo-montmorillonite in aqueous solutions. Int J Phytoremediat 19(7):605–6013. https://doi.org/10.1080/15226514.2016.1256372

ATSDR (2012) Toxicol profile of chromium. Agency for toxic substances and drug registry, Atlanta, GA. US Department of Health and Human Services

Avudainayagam S, Megharaj M, Owens G, Kookana RS, Chittleborough D, Naidu R (2003) Chemistry of chromium in soils with emphasis on tannery waste sites. In: Ware GW (ed) Reviews of environmental contamination and toxicology, vol 178. Springer, New York, pp 53–91. https://doi.org/10.1007/0-387-21728-2_3

Babula P, Adam V, Opatrilova R, Zehnalek J, Havel L, Kizek R (2008) Uncommon heavy metals, metalloids and their plant toxicity: a review. Environ Chem Lett 6:189–213

Bakshi A, Panigrahi AK (2018) A comprehensive review on chromium induced alterations in fresh water fishes. Toxicol Rep 5:440–447. https://doi.org/10.1016/j.toxrep.2018.03.007

Bakshi A (2016) Analysis of anthropogenic disturbances and impact of pollution on fish fauna of River Churni with special reference to chromium pollution. PhD Thesis, University of Kalyani, Shodhganga. http://hdl.handle.net/10603/241694

Banks M, Schwab A, Henderson C (2006) Leaching and reduction of chromium in soil as affected by soil organic content and plants. Chemosphere 62:255–264

Barabasz W, Chmiel M, Galus A, Paśmionka I (1998) Ecotoxicology of chromium (In Polish). Chem Inż Ekol 5(8–9):665

Barros MASD, Zola AS, Arroyo PA, Tavares CRG, Sousa-Aguiar EF (2006) Chromium uptake from tricomponent solution in zeolite fixed bed. Adsorption 12(4):239–248

Benazir JF, Suganthi R, Rajvel D, Pooja MP, Mathithumilan B (2010) Bioremediation of chromium in tannery effluent by microbial consortia. Afr J Biotech 9(21):3140–3143

Bhattacharya A, Gupta A, Kaur A, Malik D (2014) Efficacy of Acinetobacter sp. B9 for simultaneous removal of phenol and hexavalent chromium from co-contaminated system. Appl Microbiol Biotechnol 98 (23):9829–9841. https://doi.org/10.1007/s00253-014-5910-5

Biedermann KA, Landolph JR (1990) Role of valence state and solubility of chromium compounds on induction of cytotoxicity, mutagenesis, and anchorage independence in diploid human fibroblasts. Cancer Res 50:7835–7842

Bielicka A, Bojanowska I, Wis′niewski A (2005) Two faces of chromium—pollutant and bioelement. Pol J Environ Stud 14(1):5–10

Bojanowska I (2002) Recovery of chromium from sludge formed after neutralization of chromic wastewater. Polish J Environ Stud 11(2):117

Bolan NS, Thiagarajan S (2001) Retention and plant availability of chromium in soils as affected by lime and organic matter amendments. Soil Res 39:1091–1103

Bosnir J, Puntaric D, Cvetkovic Z, Pollak L, Barusic L, Klaric I, Miskulin M, Puntaric I, Puntaric E, Milosevic M (2013) Effects of magnesium, chromium, iron and zinc from food supplements on selected aquatic organisms. Coll Antropol 37:965–971

Brandes EA, Greenaway HT, Stone HEN (1956) Ductility in chromium. Nature 178(4533):587–587

Brooks RR, Chambers MF, Nicks LJ, Robinson BH (1998) Phytomining. Trends Plant Sci 1:359–362

Burt R, Wilson M, Mays M, Lee C (2003) Major and trace elements of selected pedons in the USA. J Environ Qual 32:2109–2121

CCME (2015) Canadian soil quality guidelines for the protection of environmental and human health. Canada Council of Ministers of the Environment, Winnipeg, Canada

Cervantes C, Campos-García J, Devars S, Gutiérrez-Corona F, Loza-Tavera H, Torres-Guzmán JC, Moreno-Sánchez R (2001) Interactions of chromium with microorganisms and plants. FEMS Microbiol Rev 25:335–347

Chaturvedi MK (2011) Studies on chromate removal by chromium-resistant Bacillus sp. isolated from tannery effluent. J Environ Prot 2(1):76. https://doi.org/10.4236/jep.2011.21008

Choppala G, Kunhikrishnan A, Seshadri B, Park JH, Bush R, Bolan N (2016) Comparative sorption of chromium species as influenced by pH, surface charge and organic matter content in contaminated soils. J Geochem Explor 184:255–260. https://doi.org/10.1016/j.gexplo.2016.07.012

Congeevaram S, Dhanarani S, Park J, Dexilin M, Thamaraiselvi K (2007) Biosorption of chromium and nickel by heavy metal resistant fungal and bacterial isolates. J Hazard Mater 146(1–2):270–277. https://doi.org/10.1016/j.jhazmat

Costa M (1997) Toxicity and carcinogenicity of Cr(VI) in animal models and humans. Crit Rev Toxicol 27 (5):431–442. https://doi.org/10.3109/10408449709078442

Cundy AB, Bardos RP, Church A, Puschenreiter M, Friesl-Hanl W, Müller I, Neu S, Mench M, Witters N, Vangronsveld J (2013) Developing principles of sustainability and stakeholder engagement for "gentle" remediation approaches: the European context. J Environ Manag 129:283–291

Dayan AD, Paine AJ (2001) Mechanisms of chromium toxicity, carcinogenicity and allergenicity: review of the literature from 1985 to 2000. Hum Exp Toxicol 20 (9):439–451

De Mattia G, Bravi MC, Laurenti O, De Luca O, Palmeri A, Sabatucci A, Mendico G, Ghiselli A (2004) Impairment of cell and plasma redox state in subjects professionally exposed to chromium. Am J Ind Med 46(2):120–125

de Oliveira LM, Gress J, De J, Rathinasabapathi B, Marchi G, Chen Y, Ma LQ (2016) Sulfate and chromate increased each other's uptake and translocation in As-hyperaccumulator *Pteris vittata*. Chemosphere 147:36–43

Dhal B, Thatoi H, Das N, Pandey B (2013) Chemical and microbial remediation of hexavalent chromium from contaminated soil and mining/metallurgical solid waste: a review. J Hazard Mater 250:272–291

Di Bona KR, Love S, Rhodes NR, McAdory D, Sinha SH, Kern N, Kent J, Strickland J, Wilson A, Beaird J, Ramage J, Rasco JF, Vincent JB (2011) Chromium is not an essential trace element for mammals: effects of a "'low-chromium"' diet. J Biol Inorg Chem 16:381–390

Di Palma L, Gueye M, Petrucci E (2015) Hexavalent chromium reduction in contaminated soil: a comparison between ferrous sulphate and nanoscale zerovalent iron. J Hazard Mater 281:70–76

Dias-Ferreira C, Kirkelund GM, Ottosen LM (2015) Ammonium citrate as enhancement for electrodialytic soil remediation and investigation of soil solution during the process. Chemosphere 119:889–895

Ding C, Li X, Zhang T, Ma Y, Wang X (2014) Phytotoxicity and accumulation of chromium in carrot plants and the derivation of soil thresholds for Chinese soils. Ecotoxicol Environ Saf 108:179–186

Ding W, Stewart DI, Humphreys PN, Rout SP, Burke IT (2016) Role of an organic carbon-rich soil and Fe (III) reduction in reducing the toxicity and environmental mobility of chromium (VI) at a COPR disposal site. Sci Total Environ 541:1191–1199

Dong J, Wu F, Huang R, Zang G (2007) A chromium tolerant plant growing in Cr-contaminated land. Int J Phytoremediat 9:167–179

Eisler R (1986) Chromium hazards to fish, wildlife, and invertebrates: a synoptic review. US Fish and Wildlife Service Biological Report 85(1.6), p 60

Ericson B (2011) Common global pollution issues: Blacksmith Institute's experience. In: Presentation to the 10th meeting of the international committee on contaminated Land, Washington, DC

Eriksson J (2001) Concentrations of 61 trace elements in sewage sludge, farmyard manure, mineral fertiliser, precipitation and in oil and crops. Swedish Environmental Protection Agency, Stockholm

Frohne T, Diaz-Bone RA, Du Laing G, Rinklebe J (2015) Impact of systematic change of redox potential on the leaching of Ba, Cr, Sr, and V from a riverine soil into water. J Soils Sed 15:623–633

Ganguly R, Sahu S, Ohanyan V, Haney R, Chavez RJ, Shah S, Yalamanchili S, Raman P (2017) Oral chromium picolinate impedes hyperglycemia-induced atherosclerosis and inhibits proatherogenic protein TSP-1 expression in STZ-induced type 1 diabetic ApoE$^{-/-}$ mice. Sci Rep 7:45279. https://doi.org/10.1038/srep45279

Ghosh M, Singh SP (2005) A review on phytoremediation of heavy metals and utilization of its by-products. Appl Ecol Environ Res 3(1):1–18

Hayat S, Khalique G, Irfan M, Wani AS, Tripathi BN, Ahmad A (2012) Physiological changes induced by chromium stress in plants: an overview. Protoplasma 249:599–611. https://doi.org/10.1007/s00709-011-0331-0

Henry JR (2000) An overview of phytoremediation of lead and mercury. National Network of Environmental Management Studies (NNEMS) Report, pp 1–31

Hsu LC, Liu YT, Tzou YM (2015) Comparison of the spectroscopic speciation and chemical fractionation of chromium in contaminated paddy soils. J Hazard Mater 296:230–238

Huang JW, Chen J, Berti WR, Cunningham SD (1997) Phytoremediation of lead contaminated soils-role of synthetic chelates in lead phytoextraction. Environ Sci Tech 31:800–806

Isıklı B, Demir T, Ürer S, Berber A, Akar T, Kalyoncu C (2003) Effects of chromium exposure from a cement factory. Environ Res 91(2):113118

Jaison S, Muthukumar T (2016) Chromium accumulation in medicinal plants growing naturally on tannery

contaminated and non-contaminated soils. Biol Trace Elem Res 175(1):223–235
Kabata-Pendias A (2010) Trace elements in soils and plants. CRC Press, Boca Raton, Florida
Katz SA, Salem H (1994) The biological and environmental chemistry of chromium. VCH Publishers Inc., New York, NY, p 214
Kendrick MJ, May MT, Plishka MJ, Robinson KD (1992) Metals in biological systems. Ellis Horwood Limited, Chichester, p 183
Kiling J (1997) Phytoremediation of organics moving rapidly into field trials. Environ Sci Tech 31(3):129A-129A
Kim IH, Choi JH, Joo JO, Kim YK, Choi JW, Oh BK (2015) Development of a microbe-zeolite carrier for the effective elimination of heavy metals from seawater. J Microbiol Biotechnol 25(9):1542–1546. https://doi.org/10.4014/jmb.1504.04067
Kisielowska E, Hołda A, Niedoba T (2010) Removal of heavy metals from coal medium with application of biotechnological methods. Górnictwo I Geoinzynieria 34:93–104
Kulikova T, Hiller E, Jurkovič Ľ, Filová L, Šottník P, Lacina P (2019) Total mercury, chromium, nickel and other trace chemical element contents in soils at an old cinnabar mine site (Merník, Slovakia): anthropogenic versus natural sources of soil contamination. Environ Monit Assess 191(5):263. https://doi.org/10.1007/s10661-019-7391-6
Kumar P, Dushenkov V, Motto H, Raskin I (1995) Phytoextraction: the use of plants to remove heavy metals from soils. Environ Sci Technol 29:1232–1238
Kumar R, Bhatia D, Singh R, Rani S, Bishnoi NR (2011) Sorption of heavy metals from electroplating effluent using immobilized biomass *Trichoderma viride* in a continuous packed-bed column. Int Biodeterior Biodegradation 65(8):1133–1139. https://doi.org/10.1016/j.ibiod.2011.09.003
Lacalle RG, Aparicio JD, Artetxe U, Urionabarrenetxea E, Polti MA, Soto M, Garbisu C, Becerril JM (2020) Gentle remediation options for soil with mixed chromium (VI) and lindane pollution: biostimulation, bioaugmentation, phytoremediation and vermiremediation. Heliyon 6(8):e04550. https://doi.org/10.1016/j.heliyon.2020.e04550
Leung M (2004) Bioremediation: techniques for cleaning up a mess. Biotech J 2:18–22
Lindberg E, Vesterberg O (1983) Urinary excretion of proteins in chromeplaters, exchromeplaters and referents. Scand J Work Environ Health 9(6):505–510
Lippmann M (2000) Environmental toxicants: human exposures and their health effects, 2nd edn. Wiley-Interscience, New York, pp 173–191
Liu J, Chang-Qun D, Xue-Hong Z, Yi-Nian Z, Hu C (2011) Characteristics of chromium (III) uptake in hyperaccumulator *Leersia hexandra* Swartz. Environ Exp Bot 74:122–126
López-Bucio J, Hernández-Madrigal F, Cervantes C (2014) Phosphate relieves chromium toxicity in *Arabidopsis thaliana* plants by interfering with chromate uptake. Biometals 27:363–370. https://doi.org/10.1007/s10534-014-9718-7
Luippold R, Mundt K, Austin R, Liebig E, Panko J, Crump C, Crump K, Proctor D (2003) Lung cancer mortality among chromate production workers. Occup Environ Med 60(6):451–457
MacKie RM (1981) Clinical dermatology. Oxford University Press, New York, Toronto
Mandal A, Voutchkov M (2011) Heavy metals in soils around the cement factory in Rockfort, Kingston Jamaica. Int J Geosci 2(1):48
Mane PC, Bhosle AB (2012) Bioremoval of some metals by living Algae *Spirogyra* sp. and *Spirullina* sp. from aqueous solution. Int J Environ Res 6(2):571–576
Misra AK, Pattnaik R, Thatoi HN, Padhi GS (1994) Study on growth and N_2 fixation ability of some leguminous plant species for reclamation of mine spoilt areas of Eastern Ghats of Orissa. Final Technical Report submitted to Ministry of Environment and Forests, Govt of India, New Delhi, India
Misra AK, Thatoi HN, Dutta B, Pattnaik MM, Padhi GS (2004) Stabilisation and restoration of ecosystem in iron and chromite mine waste areas of Eastern Ghats of Orissa through application of microbial technology. Final Technical Report submitted to Ministry of Environment and Forests, Govt of India, New Delhi, India
Mohanty M, Patra HK (2011) Attenuation of chromium toxicity by bioremediation technology. In: Whitacre DM (ed) Reviews of environmental contamination and toxicology, vol 210. Springer, New York, pp 1–34
Mwamburi J (2016) Chromium distribution and spatial variations in the finer sediment grain size fraction and unfractioned surficial sediments on Nyanza gulf, of Lake Victoria (East Africa). J Waste Manage 2016, Article ID 7528263:15. https://doi.org/10.1155/2016/7528263
Nayak AK, Panda SS, Basu A, Dhal NK (2018) Enhancement of toxic Cr (VI), Fe, and other heavy metals phytoremediation by the synergistic combination of native *Bacillus cereus* strain and *Vetiveria zizanioides* L. Int J Phytorem 20(7):682–691. https://doi.org/10.1080/15226514.2017.1413332
Nickens KP, Patierno SR, Ceryak S (2010) Chromium genotoxicity: a double-edged sword. Chem Biol Interact 188(2):276–288. https://doi.org/10.1016/j.cbi.2010.04.018
Nriagu JO (1988) Production and uses of chromium. Adv Environ Sci Technol 20:81–103
Oliveira H (2012) Chromium as an environmental pollutant: insights on induced plant toxicity. J Bot 2012:375843
Otero XL, Macias F (2003) Spatial variation in pyritization of trace metals in saltmarsh soils. Biogeochemistry 62:59–86
Oze C, Bird DK, Fendorf S (2007) Genesis of hexavalent chromium from natural sources in soil and groundwater. Proc Natl Acad Sci 104(16):6544–6549. https://doi.org/10.1073/pnas.0701085104

Paknikar KM, Bhide JV (1993) Aerobic reduction and biosorption of chromium by a chromate resistant *Aspergillus* spp. In: Torma AE, Apel ML, Brierley CL (eds) Biohydrometallurgical technologies. The Minerals, Metals and Materials Society, Warrendale, PA, pp 237–244

Pavesi T, Moreira JC (2020) Mechanisms and individuality in chromium toxicity in humans. J Appl Toxicol 40(9):1183–1197

Pillichshammer M, Pumpel T, Poder R, Eller K, Klima A, Schinner F (1995) Biosorption of chromium to fungi. Biometal 8(2):117–121

Prado C, Ponce SC, Pagano E, Prado FE, Rosa M (2016) Differential physiological responses of two Salvinia species to hexavalent chromium at a glance. Aquat Toxicol 175:213–221

Qian J, Wei L, Liu R, Jiang F, Hao X, Chen GH (2016) An exploratory study on the pathways of Cr (VI) reduction in sulfate-reducing up-flow anaerobic sludge bed (UASB) reactor. Sci Rep 6:23694

Qu M, Li W, Zhang C, Huang B, Zhao Y (2015) Assessing the pollution risk of soil chromium based on loading capacity of paddy soil at a regional scale. Sci Rep 5:18451

Quagraine EK, Peterson HG, Headley JV (2005) In situ bioremediation of naphthenic acids contaminated tailing pond waters in the Athabasca oil sands region—demonstrated field studies and plausible options: a review. J Environ Sci Heal 40:685–722

Quantin C, Ettler V, Garnier J, Šebek O (2008) Sources and extractibility of chromium and nickel in soil profiles developed on Czech serpentinites. Comptes Rendus Geosci 340:872–882

Quenea K, Lamy I, Winterton P, Bermond A, Dumat C (2009) Interactions between metals and soil organic matter in various particle size fractions of soil contaminated with waste water. Geoderma 149:217–223

Rafiq M, Shahid M, Abbas G, Shamshad S, Khalid S, Niazi NK, Dumat C (2017) Comparative effect of calcium and EDTA on arsenic uptake and physiological attributes of *Pisum sativum*. Int J Phytoremediat 19(7):662–669. https://doi.org/10.1080/15226514.2016.1278426

Ramirez R, Calvo MC, Avila-Rodriguez M, Gutierrez-Corona JF (2000) Chromate resistance and reduction in a yeast strain isolated from industrial waste discharges. In: Raynal JA, Nucklos JR, Reyes P, Ward M (eds) Environmental engineering and health sciences, Section 4: environmental engineering application. Water Resources Publications, LCC, Englewood, CO, pp 437–445

Ramírez-Díaz MI, Díaz-Pérez C, Vargas E, Riveros-Rosas II, Campos García J, Cervantes C (2008) Mechanisms of bacterial resistance to chromium compounds. Biometals 21(3):321–332. https://doi.org/10.1007/s10534-007-9121-8

Ran Z, Bi W, Cai QT, Li XX, Min LIU, Dong HU, Chun FAN (2016) Bioremediation of hexavalent chromium pollution by *Sporosarcina saromensis* M52 isolated from offshore sediments in Xiamen, China. Biomed Environ Sci 29(2):127–136. https://doi.org/10.3967/bes2015.014

Rupp H, Rinklebe J, Bolze S, Meissner R (2010) A scale-dependent approach to study pollution control processes in wetland soils using three different techniques. Ecol Eng 36:1439–1447

Saifullah SM, Zia-Ur-Rehman M, Sabir M, Ahmad HR (2015) Phytoremediation of Pb-contaminated soils using synthetic chelates. In: Hakeem K, Sabir M, Ozturk M, Mermut A (eds) Soil remediation and plants. Academic Press, San Diego, pp 397–414

Salminen R, Batista MJ, Bidovec M, Demetriades A, De Vivo B, De Vos W, Duris M, Gilucis A, Gregorauskiene V, Halamic J, Heitzmann P (2005) Geochemical atlas of Europe. Part 1—background information, methodology and maps. Geological Survey of Finland. Otamedia Oy, Espoo, p 525

Salnokow K, Zhitkovich A (2008) Genetic and epigenetic mechanisms in metal carcinogenesis and cocarcinogenesis: nickel, arsenic, and chromium. Chem Res Toxicol 21(1):28–44. https://doi.org/10.1021/tx700198a

Salt DE, Smith RD, Raskin I (1995) Phytoremediation. Annu Rev Plant Physiol Plant Mol Biol 49:643–668

Sampanpanish P, Pongsapich W, Khaodhiar S, Khan E (2006) Chromium removal from soil by phytoremediation with weed plant species in Thailand. Water Air Soil Pollut Focus 6:191–206

Seigneur C, Constantinou E (1995) Chemical kinetic mechanism for atmospheric chromium. Environ Sci Technol 29(1):222–231

Shadreck I (2013) Chromium, an essential nutrient and pollutant: a review. Afr J Pure Appl Chem 7:310–317

Shahid M, Austruy A, Echevarria G, Arshad M, Sanaullah M, Aslam M, Nadeem M, Nasim W, Dumat C (2014) EDTA-enhanced phytoremediation of heavy metals: a review. Soil Sediment Contam Int J 23:389–416

Shahid M, Dumat C, Pourrut B, Abbas G, Shahid N, Pinelli E (2015) Role of metal speciation in lead-induced oxidative stress to *Vicia faba* roots. Russ J Plant Physiol 62:448–454

Shahid M, Shamshad S, Rafiq M, Khalid S, Bibi I, Niazi NK, Dumat C, Rashid MI (2017) Chromium speciation, bioavailability, uptake, toxicity and detoxification in soil-plant system: a review. Chemosphere 178:513–533. https://doi.org/10.1016/j.chemosphere.2017.03.074

Shahid M, Xiong T, Castrec-Rouelle M, Leveque T, Dumat C (2013) Water extraction kinetics of metals, arsenic and dissolved organic carbon from industrial contaminated poplar leaves. J Environ Sci 25:2451–2459

Shanker AK, Cervantes C, Loza-Tavera H, Avudainayagam S (2005) Chromium toxicity in plants. Environ Int 31:739–753

Shukla O, Dubey S, Rai U (2007) Preferential accumulation of cadmium and chromium: toxicity in *Bacopa monnieri* L. under mixed metal treatments. Bull Environ Contam Toxicol 78:252–257

Singh HP, Mahajan P, Kaur S, Batish DR, Kohli RK (2013) Chromium toxicity and tolerance in plants. Environ Chem Lett 11:229–254. https://doi.org/10.1007/s10311-013-0407-5

Singh N, Verma T, Gaur R (2013) Detoxification of hexavalent chromium by an indigenous facultative anaerobic *Bacillus cereus* strain isolated from tannery effluent. Afr J Biotech 12(10):1091–1103

Sklodowska A (2000) Biological methods of heavy metal leaching—biohydrometallurgy. Adv Microbiol 39 (1):73–89

Smith S, Peterson PJ, Kwan KHM (1989) Chromium accumulation, transport and toxicity in plants. Toxicol Environ Chem 24(4):241–251. https://doi.org/10.1080/02772248909357496

Sundaramoorthy P, Chidambaram A, Ganesh KS, Unnikannan P, Baskaran L (2010) Chromium stress in paddy: (i) nutrient status of paddy under chromium stress; (ii) phytoremediation of chromium by aquatic and terrestrial weeds. Comptes Rendus Biol 333:597–607

Taghipour M, Jalali M (2016) Influence of organic acids on kinetic release of chromium in soil contaminated with leather factory waste in the presence of some adsorbents. Chemosphere 155:395–404

Takeda A, Kimura K, Yamasaki S (2004) Analysis of 57 elements in Japanese soils, with special reference to soil group and agricultural use. Geoderma 119:291–307

Tarekegn MM, Salilih FZ, Ishetu AI (2020) Microbes used as a tool for bioremediation of heavy metal from the environment. Cogent Food Agric 6:1783174. https://doi.org/10.1080/23311932.2020.1783174

Torresdey JLG, Videa JRP, Montes M, Rosa G, Diaz CB (2004) Bioaccumulation of cadmium, chromium and copper by Convolvulus arvensis: Impact on plant growth and uptake of nutritional elements. Bioresour Technol 92:229–235

Velma V, Vutukuru SS, Tchounwou PB (2009) Ecotoxicology of hexavalent chromium in fresh water fish: a critical review. Rev Environ Health 24(2):129–145

Volesky B, Holan ZR (1995) Biosorption of heavy metals. Biotechnol Prog 11:235–250

Wales DS, Sagar BF (1990) Recovery of metal ions by microfungal filters. J Chem Technol Biotechnol 49:345–355

Xiao W, Ye X, Yang X, Li T, Zhao S, Zhang Q (2015) Effects of alternating wetting and drying versus continuous flooding on chromium fate in paddy soils. Ecotoxicol Environ Saf 113:439–445

Zayed AM, Terry N (2003) Chromium in the environment: factors affecting biological remediation. Plant Soil 249:139–156

Zeng J, Gou M, Tang YQ, Li GY, Sun ZY, Kida K (2016) Effective bioleaching of chromium in tannery sludge with an enriched sulfur-oxidizing bacterial community. Bioresour Technol 218:859–866

Zhuang P, Yang QW, Wang HB, Shu WS (2007) Phytoextraction of heavy metals by eight plant species in the field. Water Air Soil Pollut 84:235–242

Heavy Metal Detection in Soil and Its Treatment (Bioremediation) with Nanomaterials

16

Jyoti Sharma, Amit Chattree, Shabnam Dan, and Mohd. Imran

Abstract

The industrial effluents, oil spillage on land and water, improper discard of materials containing the heavy metals and their compounds lead to heavy metal pollution. The metals are taken up by plants and animals through their absorbent property causing the entrapment of heavy metals in the food web. The heavy metals prove to be highly toxic and cause several disorders in plants and animals including the genetic disorders. The detection and their treatment are a necessity in order to develop a clean and sustainable environment for the survival. A number of techniques have been developed for the detection of heavy metals at their precisely low concentrations of ppm and ppb. The techniques used for the detection are atomic absorption spectrometry (AAS), flame photometry, inductively coupled plasma mass spectrometry (ICP-MS), ion chromatography (IC) systems. However, the treatment of soil samples for heavy metal removal has extensive scope of research and development and one such aspect is using nanomaterials. Nowadays, nanoparticles are encompassed in the removal of toxic heavy metals from soil, which opens up broad scope for research and progress. Even nanoparticles are adsorbed on biomaterial and form bio-nanocomposites which are used in treatment of pollutants from wastewater. The adsorbent capacity of nanocomposites has shown great impact in the treatment of soil pollutants. Development of chemical-based magnetic nanoparticles has shown excellent efficiency in the removal of cadmium (Cd), chromium (Cr), copper (Cu), iron (Fe), etc. The application of zerovalent iron oxide nanoparticles assists in immobilization of heavy metals, i.e., in-situ remediation. Apart from removal of metal toxicity, nanomaterial-based optical and electrochemical sensors are also being developed for heavy metal detection. The advantage of using nanomaterials for the detection and bioremediation of heavy metals is their precision, accuracy and robustness. Fluorescence resonance energy transfer (FRET)-based sensors have also provided a new avenue for sensitive and quantitative detection of specific heavy metals that pose environmental risks. The

J. Sharma (✉)
School of Basic and Applied Sciences, Shobhit Institute of Engineering and Technology (Deemed to-be-University), Meerut, Uttar Pradesh 250110, India
e-mail: jyoti2@shobhituniversity.ac.in

A. Chattree · S. Dan
Department of Chemistry, Sam Higginbottom University of Agriculture, Technology and Sciences, Prayagraj (Allahabad), UP 211007, India

Mohd.Imran
Department of Chemistry, Chandigarh University, Mohali, Punjab 140413, India

J. A. Malik (ed.), *Advances in Bioremediation and Phytoremediation for Sustainable Soil Management*,
https://doi.org/10.1007/978-3-030-89984-4_16

wide range of application and their efficiency make these nanomaterials a great contributor in sustainable development.

16.1 Introduction

Biosphere is the source of life that allows the existence of living beings on earth. Recent human activities have introduced household sewage and toxic industrial wastes into the surroundings causing pollution of the environment. Furthermore, the use of insecticides in agriculture has severely damaged the groundwater and soil quality. The industrial effluents, oil spillage and partial discard of materials, carrying heavy metals and their derivatives result in the degradation of the water and soil standards. These heavy metals are directly or indirectly absorbed by plants, animals, etc., causing their biomagnification in the food web (Alvarenga et al. 2012). Their toxic effects lead to several disorders in plants and animals, also causing various genetic disorders in the consumers. Hence, the survival of future generation depends on the recovery of soil and water sources from heavy metal pollutants, with the help of advanced technologies. A number of techniques have been developed for the detection of heavy metals at really low concentrations (ppm/ppb), viz. atomic absorption spectrometry (AAS), flame photometry, inductively coupled plasma mass spectrometry (ICP-MS), ion chromatography (IC) systems, etc. (Hoang et al. 2013; Li et al. 2013; Yuan et al. 2013; Kaur et al. 2015a, b).

However, one more aspect of remediation of heavy metals from soil can be accomplished by using nanomaterials that have elaborated the scope of research and development. Earlier, natural or synthetic adsorbents were used for the same but later on it was observed that they are not efficient enough to treat the quality of soil and water from heavy metal deterioration. So, there was always a necessity for the alternative material that can be used as a powerful technique for wastewater and soil treatment; considering every aspect in mind, nanocomposites have been found suitable for this purpose. Therefore, the larger surface area of nanoparticles and the advantageous characteristic of nanocomposite, i.e., adsorption, show a great impact in the removal of soil pollutants.

Additionally, development of chemical-based magnetic nanoparticles has been observed as the excellent efficient sources for the formation of bio-composites and thus is utilized as a potential way of eliminating the heavy metals like cadmium (Cd), chromium (Cr), copper (Cu), iron (Fe), etc. (Lin et al. 2017; Ang et al. 2010). Consequently, the application of zerovalent iron oxide nanoparticles (USEPA et al. 1997, 2004) assists in the immobilization of heavy metals, i.e., in-situ remediation (Cui et al. 2013). Also, the rare property of nanomaterials like precision, accuracy, robustness, etc., allowed the researchers to develop different nanomaterial-based sensors (Borah et al. 2015; Kaur et al. 2015a, b) of optical and electrochemical nature that not only permits the removal of metal toxicity but also aids in the heavy metal detection. Their contribution is applied to develop fluorescence resonance energy transfer (FRET)-based sensors that show a wide application in the environmental technology domain. The wide range of application and effectiveness has made nanotechnology and nanomaterials a great contributor in sustainable development. In order to promote the broad applications of nanocomposites for the treatment of soil, extensive research is recommended.

16.2 Heavy Metal Pollution and Detection Techniques

A few decades earlier, rapid progress of humans by succeeding the industrialization pathway led to various environment-related issues, and contamination of soil from heavy metals is one of them. The human-induced wastes from industries, sewage sludge, etc., are responsible for all types of pollutant moieties released into the surroundings as contaminants, free radicals, greenhouse gases, heavy metals, etc. The direct accession of heavy metals into the soil usually takes place from industrial effluents of mine activities, sewage wastes, coal oxidation, petrochemical spillage, etc. (Wuana and Okieimen 2011). These heavy metals directly or indirectly are consumed by plants and animals and thus induced into their bodies to cause various diseases. The efflux of these heavy metals, like mercury, lead, arsenic, etc., has hazardous toxic effects on the ecosystem and also continues as the potent cause of several chronic and genetic diseases (Ang et al. 2010). In comparison with water molecules, the toxic heavy metals are five times larger as observed in several ions, elements, etc. (Ugulu 2015). As nature has a solution to cope up with these contaminants by utilizing microorganisms, so most of the organic pollutants cannot be a threat except for the heavy metal pollutants. Therefore, their discharge into the surroundings resulted in their increment and deviated the biosphere. Even an extremely low-level diffusion of heavy metals can emit toxicity, which deeply permeates into the soil. The food-producing crops have a severe damage in the form of food quality and safety from their absorption and thus cause reduction in the production.

So, the researchers and scientists felt the need of removing heavy metals from soil which led to the development of several techniques with an aim of protecting the humanity and the environment. Although the challenge of removing the heavy metals is not an easy job and the developed advanced techniques are not basically efficient, due to the limitation of influencing the soil standards characteristically, still a number of remedial techniques have been investigated and developed with a desire of controlling the toxicity of heavy metals and preventing the soil health from pollutants (Lajayer et al. 2017). Accordingly, there are several techniques developed for the detection of heavy metals preferably as atomic absorption spectrometry (AAS), flame photometry, inductively coupled plasma mass spectrometry (ICP-MS), ion chromatography (IC) systems, etc. (Hoang et al. 2013; Li et al. 2013; Yuan et al. 2013; Kaur et al. 2015a, b). These techniques are suitable for detecting the heavy metal concentration at very low concentration on the parts per million (ppm) or parts per billion (ppb) scale.

16.3 Nanoparticles and Their Phenomenal Properties

The term 'nanotechnology' was first introduced by Nobel laureate Richard P. Feynman in the year 1959 during his infamous lecture "There's Plenty of Room at the Bottom" (Feynman 1960); since then, a tremendous growth has been observed in the form of fabricating various types of nano-based materials at nano-size scale. The stereo-structure of nanoparticles ranged below 100 nm, like particulate matter (Laurent et al. 2010). The shape of these nanoparticles can be zero dimension, one dimension, two dimensions or three dimensions (Tiwari et al. 2012). It was found by researchers that due to their physicochemical properties (like optical properties) these materials seek a great importance. With the variation in the size and shape, these nanoparticles of the same metals exhibit different colors and respective characteristics, which can be of great importance in bioimaging (Dreaden et al. 2012). The molecularity of these particles are complex as they contain 3 layers i.e. (a) surface layer, (b) shell layer, and (c) core (Shin et al. 2016). Nanoparticles exhibit many unique physicochemical properties like larger surface area, greater mechanical strength, higher chemical reactivity and optical activity. The optical and electronic properties of nanoparticles are directly proportional to each other (Eustis and El-Sayed 2006). With respect to their magnetic properties, researchers show great importance from an electric point of view as in biomedicine, magnetic liquid, MRI and environmental treatment. It was reported that these nanoparticles illustrate a great performance when their particle size ranges between 10 and 20 nm (Reiss and Hütten 2005). The magnetic properties can be influenced because of the random electronic distribution of nanoparticles, and they usually confide on the processing synthetic procedure and methodology (Qi et al. 2016). Additionally, with the involvement of nanoparticles, the mechanical properties show a major enhancement as noted by researchers in nano-fabrication, nano-manufacturing, surface engineering, etc.

Moreover, in order to measure the mechanical properties, the desired parameters involve elastic modulus, hardness, stress and strain, etc., along with other accessory parameters such as surface coating and coagulation (Guo et al. 2014). Their application has been seen in the delivery of drugs (Lee et al. 2011), as chemical and biological sensors (Barrak et al. 2019), gas sensing, etc. (Mansha et al. 2016; Rawal and Kaur 2013; Ullah et al. 2017).

16.4 Nanomaterials: A Remedy

Earlier, the production of a rare material revealing advanced physio-chemical characteristics was not an easy process due to insufficient knowledge of incorporating nano-sized particles with any bulk materials (Grimsdale and Müllen 2005; Aljerf and Nadra 2018). But with the production of nanomaterials (like carbon nanomaterials), they have grasped the opportunity to enhance their application properties (like magnetic properties) to another level (Liu et al. 2000; Dai et al. 1996; Chatterjee et al. 2003; Krishna et al. 1997; Shiraishi et al. 2004; Sharon et al. 1997; Fang 2010). Thereafter, a rapid rise in the production of magnetic nanoparticles has been reported due to the understanding of magnetic relaxation theory given by Tartaj et al. (2005) on ferro- and ferri-magnetic materials. The implementation of magnetic carbon nanomaterials has been observed as catalyst, supercapacitor, biosensor, etc. (Kondo and Fukuda 1997; Reetz et al. 1998; Varlan et al. 1995; Rossi et al. 2014; Reddy and Yun 2016; Creamer et al. 2016). Usually, these nanoparticles with magnetism properties can be applied in medical sciences as in MRI, cancer screening, diagnosis of hyperthermia, etc. (Elsherbini 2011). As a supercapacitor, due to the porous nature of carbon magnetic nanoparticles such as graphene, it has been reported that it exhibits efficient performance in the capacitor (Zhu et al. 2013; Noked et al. 2012; Kiyohara et al. 2014), and thus, recently lithium ion batteries started to gain all the attention for its better enactment because of increase in electrical conductivity and capacitance power as a result of amalgamation of polymer and magnetic nanoparticles. Therefore, due to its potentiality it is not incorrect to say that the next generation will likely to produce more magnetic carbon nanomaterial, and in the continuation of producing such goods, the methods utilized will be arc discharge, chemical vapor deposition, laser pyrolysis and metallic reduction–pyrolysis (Du et al. 2005; Liu et al. 2003; Suh and Suslick 2005; Guo et al. 2007; Rakhi et al. 2010; Gai et al. 2017). But due to their higher cost and complex nature, it will be little difficult to deal with them practically.

The advancement of nanotechnology leads to produce various nanoparticles, like gold nanorods, that show a tremendous implementation in the biomolecule spotting (Liu and Tan 2018; Wang et al. 2017, 2016; Zhang et al. 2017), photo-thermal therapy (Zhu et al. 2014; Monem et al. 2014) and detection of heavy metals. The techniques that involved the usage of gold nanorods exhibit better selectivity and sensitivity, and thus emit the power of detecting the heavy metals' presence, even at low percentage within a short period of time. This will ultimately help in the guidance and monitoring their dosage concentration in the environment. A technique called surface plasmon resonance when accumulated with gold nano-rods carries out an important role in the detection of lead (Pb), a heavy metal, in the bio-related samples (Lan and Lin 2014; Lee and Huang 2011). In a similar way, the presence of copper (heavy) metal in its derivative compound with gold nano-rods can be observed via measuring the intensity of scattered light (Jing et al. 2014). Although the detection power is quite not high, due to the high sensitivity and selectivity of gold nano-rods, it can be used for detecting lead metal in soil (Zhang et al. 2014).

The assembly of hydroxyapatite and nanoparticles resulted in the formation of nanohydroxyapatite (NHA) that has been reported as a controller of increasing the soil acidity and decreasing cadmium availability which is because of its novel chemical properties and larger surface area caused by small-sized particles (Cui et al. 2013; He et al. 2013).

16.5 Why Nanomaterials?

The reason behind the vast application of nano-based material relies on their large surface area, physicochemical properties, cost-effective production, eco-friendliness, biodegradability, etc. Due to the conjunction of nanoparticles and biomaterial, the porosity of biomaterial can be enhanced to a higher level that leads to adaptation of another unique characteristic, i.e., adsorption. These materials are non-toxic and can be easily degradable with the help of microbial organisms. Additionally, the nano-based material, either synthetic or bio-composite, does not cause any harm to the environment. Despite its cost-effective manner of synthesis, it is still a challenging work to incorporate the nano-sized particles with the biomaterials.

16.6 Bio-nanocomposites

Due to the entanglement about biosphere and renewable issues, this era has alleged remarkable growth in the augmentation of bio-composites (La Mantia and Morreale 2011; Satyanarayana et al. 2009). The nature of these bio-composites is compatible with the environment, does not cause any sabotage to the surroundings and thus is eco-friendly and expressed to be a suitable alternative of synthetic-based bio-composites (Laner et al. 2012; Gurunathan et al. 2015). Bio-composite term is also admitted by the name eco-composite or green composite (Hughes et al. 2002). Moreover, in this decade the terminology bio-nanocomposite is getting the attention of worldwide researchers that is associated with balanced blending of natural polymer with an inorganic moiety, measured on nanometer scale (Pande and Sanklecha 2017; Bhatia and Kurian 2008). Bio-nanocomposites exhibit some special properties like antimicrobial, thermal, biological, mechanical and barrier (Ahmed et al. 2018).

In dispersion through numerous composite materials, natural fibers are augmented as the primary source of developing the bio-composites. These sources not only are cost effective but also reflect several functional–structural properties (Shanks et al. 2004). Regardless of the fact, scientists are focussing on those materials which are compatible with the environment in respect of utilization, production and evacuation (Gurunathan et al. 2015). Other than advantageous characteristics of natural fiber stationed composites, additional stereo-configurational features of lignocellulosic fiber play a vital role in the preparation of composites and illustrate interfacial adhesive property between fiber matrices (La Mantia and Morreale 2011; Satyanarayana et al. 2009; Zini and Scandola 2011; Jawaid and Abdul Khalil 2011; John and Thomas 2008; Dittenber and GangaRao 2012; Bledzki and Gassan 1999). Within the last few years, experts studied different natural fibers and their specific features in order to depict their potency for the reclamation of synthetic fiber like carbonated fiber, glass, etc. (Bansal et al. 2017; Shahzad 2012; Gurunathan et al. 2015). Materials on nanometer scale depict some rare but beneficial essence that shows high impact by modifying permeability, thermal stability, mechanical property, packaging applications, etc. (Pande and Sanklecha 2017).

16.7 Removal of Heavy Metals Using Nanomaterials

Nano-electrolysis degrades pollutants such as chlortetracycline (CTC). The studies have shown that pure CTC compounds were removed up to an extent of 96%. One more USP of the material is that the intermediate compounds produced were non-toxic such as carbon dioxide, water and ammonium ions (Liu et al. 2000).

Another type of material developed has been biogenic manganese oxide (BioMnOx), which works as bio-magnets. Bio-palladium nanocrystals have proven to be efficient in removal of heavy metals from soil and water. Bio-magnetic nanoparticles are developed using iron oxides, iron and aluminum alloys. The removal has been efficient not only to heavy metals but also to halogenated compounds and recalcitrant pollutants (Kumari et al. 2019).

Gold nano-rods have shown exquisite performance in removal of lead from agricultural soils. The application of humic acid along with gold nano-rods leads to spike in lead mobility. The sensitivity of gold nano-rods increased with the combination of iron and manganese. The process has proved to be rapid and convenient (Liu et al. 2020).

Dyes as pollutant have been a great issue for soil and water. Fabricated nanomaterial graphene oxide and tourmaline oxide in a ranging ratio of 1:1 to 1:10 were developed using hydrothermal methods. Dyes such as methylene blue and heavy metals such as mercury were removed using the nanomaterial. The variation in the ratio affects the removal capacity of the nanomaterial. The maximum capacity observed was 0.5 g of methylene blue per gram of water and around 0.3 g of mercury per gram of water. They have shown excellent stability and removal capability retention even after repeated cycles of application. They may prove to be a milestone in the field of wastewater treatment (Lin et al. 2019). They further may prove to be effective in application to the heavy metal removal from agricultural hydrated soil.

Nano-flowers were designed by polyol method (mediated using ethylene glycol). Iron oxide nanoparticles fabricated nano-flowers proved to be a great asset for cadmium, chromium and lead ions from wastewater and hydrated soil. Adsorption experiments revealed the adsorbent dosages and pH conditioning of the nanomaterial. The maximum absorption capacity for cadmium and chromium ions was observed to be 0.017 and 0.02 g per gram of water, while it was maximum for lead ions at 0.03 g per gram of water. The versatility of the material proves the iron nanoparticles-based flowers an efficient and sustainable alternative to conventional adsorbents (Tsedenbal et al. 2020).

Besides the above-discussed nanomaterials, the following nanomaterials are also found to be effective in the heavy metal remediation of soil:

16.7.1 Carbon-Based Nanomaterials

Nanomaterials based on carbon and its compounds are of potential use in industries. They have several cutting-edge properties such as ease of chemical and physical alteration and availability of large surface area. Some nanomaterials also have the capability to eliminate contaminants such as heavy metals, organic and inorganic compounds (Krivorotov et al. 2004). For the treatment of heavy metals, generally the following types of nanomaterials are used.

16.7.1.1 Carbon Nanotubes

These are categorized into single-walled carbon nano-tubes (SWCNTs) and multi-walled carbon nano-tubes (MWCNTs). They have the length of less than 1000 nm. However, there diameter is restricted to only 1–3 nm (Gupta et al. 2016). These are applied in the treatment of heavy metal-contaminated water and soil. CNTs exhibit these properties due to their large surface area and high adsorption capacity.

16.7.1.2 Graphene-Based Nanomaterials

Another class of nano-materials which are specifically used for waste water treatment and contaminated soil treatment. These are applicable widespread in the maintenance of sustainable environment due to their properties such as its mechanical strength, and thermal and electrical conductivity (Novoselov et al. 2012). Apart from graphene, its oxides and reduced oxides are also having bio-remedial properties.

16.7.2 Silica-Based Nanomaterials

Silica-based nanomaterials are also potential candidates for bio-remediation; their usage is expanded by the surface modification of amino ($-NH_2$) and thiol ($-SH$) groups. These modifications lead to development of silica as a material for bio-composites (Mahmood et al. 2018). Kotsyuda and co-workers synthesized silica nanospheres which were biofunctionalized by 3-aminopropyl and phenyl groups (Kotsyuda et al.

2017). This enhanced the application of the composite as copper and cationic thiazine dye eliminator. Apart from this, the material showed enhanced antimicrobial activity as well. Jha and co-workers reported about the modern nanotechnological approaches for wastewater treatment (Jha et al. 2020).

16.7.3 Zerovalent Metal-Based Nanomaterials

It is well known that silver (Ag) has significant anti-bacterial and cleansing properties which are well exploited in wastewater treatment. However, zerovalent Ag nanoparticles have shown enhanced antimicrobial activity. Similarly, zerovalent zinc has been reported to have exquisite dioxin degrading capacity (Srinivasan et al. 2017).

16.7.4 Metal Oxide-Based Nanomaterials

The properties that are well exploited in the wastewater and contaminated soil treatment are these possess high grade of selectivity toward heavy metals. They are categorized under various categories of oxides of aluminum, iron, zirconium, titanium, etc. (Yang et al. 2019).

16.7.5 Nanocomposites

These are classified majorly as hybrid nanocomposites. These are made up of one or more than one nanomaterial and a carrier material. These are developed for therapeutic purposes along with bio-remediation. The materials also prove to be rigorous and versatile with only one type of composite being applicable in both environmental and therapeutic applications. Few of them are discussed below:

16.7.5.1 Inorganic-Supported Nanocomposites

Materials such as activated carbon (AC), CNTs and materials such as zeolite, bentonite etc. AC is simplest type of adsorber used for water treatment. It is mostly used in water purifiers and filters (Tounsadi et al. 2016). ACs have exquisite adsorbing properties for heavy metals such as chromium (Cr), lead (Pb) and cadmium (Cd).

16.7.5.2 Organic Polymer-Supported Nanocomposites

Polymeric nanocomposites such as polyvinyl alcohol (PVA) possess many exceptional properties which are many a times enhanced or modified as a novel property to the polymer by the addition of nanomaterials to it. Prakash and researchers developed a polymeric bio-composite of PVA using electrospinning method for the treatment of water and other biological fluids, specifically for the removal of bacteria and lipopolysaccharides (LPS), which are also known as bacterial endotoxins (Prakash et al. 2020). Many other polymeric materials such as polystyrene (PS) and polyaniline (PAN) are also used to fabricate nano-bio-composites for treatment of heavy metals (Rajakumar et al. 2014).

16.7.5.3 Magnetic Nanocomposites

Magnetic composites offer easy separation capability. They comprise mainly magnetic iron and its oxides. There are three common approaches for their fabrication: (a) surface modification of iron/iron oxide nanoparticles by amino and thiol groups, (b) encapsulation of materials such as humic acid, polyethylineamine and manganese oxide and (c) coating of iron or its oxides with porous materials such as graphene oxides and CNTs. However, the integration process and the fabrication often get tedious and expansive restricting its applicability.

16.8 Conclusion and Futuristic Trends

This chapter showed the recent progress of nanoparticles, bio-nanocomposites, properties, types, methods of synthesis, modifications, potent natural wastes and its applications. In order to fulfill the biopolymer demands in the market, the production rate should be higher to a certain limit. Regardless of the fact of growing demands of biopolymers, they still lack some special properties like barrier, mechanical, thermal properties; therefore, it is necessary to modify the biopolymer nature by incorporating nanoparticles such as nano-clays, nanocellulose, and carbon nanotubes, to make the processing smoother. Bio-nanocomposites possess biodegradable and biocompatible nature; hence, they are applicable for medical applications like tissue engineering, bone replacement, drug delivery, etc., with a better heat resistive ability as well as mechanical properties. Food packaging can be highly influential by using bio-nanocomposite film because it enhances the shelf life of goods and has protective ability. Additionally, the major advantage is that they are worthwhile from the economic point of view and are also eco-friendly. Also, bio-nanocomposites express adsorptive nature; therefore, they can be used to eliminate micropollutants like dyes, heavy metallic ions, etc., from wastewater and contaminated soil. There are many hurdles that obstruct the race of commercialization worldwide, which includes poor conjugation property between matrix and fiber, bio-fiber orientation, uniform nanoparticle dispersion, etc., that demands comprehensive research. Within the last few years, tremendous growth has been reported and significant research leads to prolonged breakthrough in different fields, mainly in medicine and electronics. In order to promote the widespread applications of bio-nanocomposites, extensive research is recommended to directionalize the objective for heavy metal treatment of soil and wastewater.

References

Ahmed S, Ikram S, Kanchi S, Bisetty K (2018) Bionanocomposite materials: concept, applications, and recent advancements. Biocomposites. Pan Stanford Publishing Pte Ltd., Singapore, pp 219–236

Aljerf L, Nadra R (2019) Developed greener method based on MW implementation in manufacturing CNFs. Int J Nanomanuf 15(3):269–289

Alvarenga P, Palma P, de Varennes A, Cunha-Queda AC (2012) A contribution towards the risk assessment of soils from the Sao Domingos Mine (Portugal): chemical, microbial and ecotoxicological indicators. Environ Pollut 16(1):50–56

Ang LH, Tang LK, Ho WM, Hui TF, Theseira GW (2010) Phytoremediation of Cd and Pb by four tropical timber species grown on an ex-tin mine in Peninsular Malaysia. Int J Environ Ecol Eng 4(2): 70–74

Bansal G, Singh VK, Gope PC, Gupta T (2017) Application and properties of chicken feather fiber (CFF) a livestock waste in composite material development. J Graphic Era Univ 5(1):16–24

Barrak H, Saied T, Chevallier P, Laroche G, M'nif A, Hamzaoui AH, (2019) Synthesis, characterization, and functionalization of ZnO nanoparticles by N-(trimethoxysilylpropyl) ethylenediamine triacetic acid (TMSEDTA): Investigation of the interactions between Phloroglucinol and ZnO@ TMSEDTA. Arab J Chem 12(8):4340–4347

Bhatia SK, Kurian JV (2008) Biological characterization of Sorona polymer from corn-derived 1, 3-propanediol. Biotech Lett 30(4):619–623

Bledzki AK, Gassan J (1999) Composites reinforced with cellulose based fibers. Prog Polym Sci 24(2):221–274

Borah SB, Bora T, Baruah S, Dutta J (2015) Heavy metal ion sensing in water using surface plasmon resonance of metallic nanostructures. Groundw Sustain Dev 1(1–2):1–11

Chatterjee AK, Maheshwar S, Rangan B (2003) Alkaline fuel cell: carbon nanobeads coated with metal catalyst over porous ceramic for hydrogen electrode. J Power Sources 117(1–2):39–44

Creamer AE, Gao B, Wang S (2016) Carbon dioxide capture using various metal oxyhydroxide–biochar composites. Chem Eng J 283:826–832

Cui HB, Zhou J, Zhao QG, Si Y, Mao JD, Fang GD, Liang JN (2013) Fractions of Cu, Cd, and enzyme activities in a contaminated soil as affected by applications of micro- and nanohydroxyapatite. J Soils Sediment 13:742–752

Dai H, Hafner JH, Rinzler AG, Colbert DT, Smalley RE (1996) Nanotubes as nanoprobes in scanning probe microscopy. Nature 384(6605):147–150

Dittenber DB, GangaRao HVS (2012) Critical review of recent publications on use of natural composites in infrastructure. Compos Part A Appl Sci Manuf 43 (8):1419–1429

Dreaden EC, Alkilany AM, Huang X, Murphy CJ, El-Sayed MA (2012) The golden age: gold nanoparticles for biomedicine. Chem Soc Rev 41(7):2740–2779

Du J, Liu Z, Li Z, Han B, Sun Z, Huang Y (2005) Carbon nanoflowers synthesized by a reduction–pyrolysis–catalysis route, Mater Lett Elsevier 59(4:456–458

Elsherbini AA (2011) Magnetic nanoparticle-induced hyperthermia treatment under magnetic resonance imaging. Magn Reson Imaging 29(2):272–280

Eustis S, El-Sayed MA (2006) Why gold nanoparticles are more precious than pretty gold: noble metal surface plasmon resonance and its enhancement of the

radiative and nonradiative properties of nanocrystals of different shapes. Chem Soc Rev 35(3):209–217
Fang Y (2010) A low-concentration hydrothermal synthesis of biocompatible ordered mesoporous carbon nanospheres with tunable and uniform size. Angew Chem Int Ed 122(43):8159–8163
Feynman RP (1960) There's plenty of room at the bottom. Am J Phys 74(9):22–36
Gai C, Zhang F, Lang Q, Liu T, Peng N, Liu Z (2017) Facile one-pot synthesis of iron nanoparticles immobilized into the porous hydrochar for catalytic decomposition of phenol. Appl Catal B 204:566–576
Grimsdale AC, Müllen K (2005) The chemistry of organic nanomaterials. Angew Chem Int Ed 44(35):5592–5629
Guo D, Xie G, Luo J (2014) Mechanical properties of nanoparticles: basics and applications. J Phys D Appl Phys 47(1):13001
Guo Z, Park S, Wei S, Pereira T, Moldavan M, Karki AB, Young DP, Hahn HT (2007) Flexible high-loading particle-reinforced polyurethane magnetic nanocomposite fabrication through particle-surface-initiated polymerization. Nanotechnology 18(33):335–704
Gupta VK, Moradi O, Tyagi I, Agarwal S, Sadegh H, Shahryari-Ghoshekandi R, Makhlouf ASH, Goodarzi M, Garshasbi A (2016) Study on the removal of heavy metal ions from industry waste by carbon nanotubes: Effect of the surface modification: a review. Crit Rev Environ Sci Technol 46(2):93–118
Gurunathan T, Mohanty S, Nayak SK (2015) A review of the recent developments in biocomposites based on natural fibres and their application perspectives. Compos A Appl Sci Manuf 77:1–25
He M, Shi H, Zhao XY (2013) Immobilization of Pb and Cd in contaminated soil using nano-crystallite hydroxyapatite. Procedia Environ Sci 18:657–665
Hoang CV, Oyama M, Saito O, Aono M, Nagao T (2013). Monitoring the presence of ionic mercury in environmental water by plasmon-enhanced infrared spectroscopy. Nat Sci Rep 3:Article no 1175
Hughes M, Chen GZ, Shaffer MS, Fray DJ, Windle AH (2002) Electrochemical capacitance of a nanoporous composite of carbon nanotubes and polypyrrole. Chem Mater 14(4):1610–1613
Jawaid M, Abdul Khalil HPS (2011) Cellulosic/synthetic fibre reinforced polymer hybrid composites: a review. Carbohydrate Polymer 89(1):1–18
Jha SK, Singh HR, Prakash P, Fatimah F (2020) Nanobiotechnological interventions for the removal of toxic pollutants. In: Removal of toxic pollutants through microbiological and tertiary treatment, Elsevier, vol 1, pp 335–347
Jing C, Shi L, Liu X, Long YT (2014) A single gold nanorod as a plasmon resonance energy transfer based nanosensor for high-sensitivity Cu (II) detection. Analyst 139(24):6435–6439
John MJ, Thomas S (2008) Biofibres and biocomposites. Carbohydrate. Polymer 71(3):343–364
Kaur B, Srivastava R, Satpati B (2015) Ultratrace detection of toxic heavy metal ions found in water bodies using hydroxyapatite supported nanocrystalline ZSM-5 modified electrodes. New J Chem 39(7):5137–5149
Kaur R, Mandal K, Kumar R, Singh B (2015) Analytical method for determination of fipronil and its metabolites in vegetables using the QuEChERS method and gas chromatography/mass spectrometry. J AOAC Int 98(2):464–471
Kiyohara K, Shioyama H, Asaka K (2014) Thermodynamics of nano-porous carbon materials as adsorbents and electrochemical double-layer capacitor electrodes —Implications from computer simulation studies. Carbon 76:469–470
Kondo A, Fukuda H (1997) Preparation of thermo-sensitive magnetic hydrogel microspheres and application to enzyme immobilization. J Ferment Bioeng 84(4):337–341
Kotsyuda SS, Tomina VV, Zub YL, Furtat IM, Melnyk IV (2017) Bifunctional silica nanospheres with 3-aminopropyl and phenyl groups: Synthesis approach and prospects of their applications. Appl Surf Sci 420:782–791
Krishna KM, Soga T, Mukhopadhyay K, Sharon M, Umeno M (1997) Photovoltaic solar cell from camphoric carbon A natural source. Sol Energy Mater Sol Cells 48(1–4):25–33
Krivorotov IN, Emley NC, Garcia AG, Sankey JC, Kiselev SI, Ralph DC, Buhrman RA (2004) Temperature dependence of spin-transfer-induced switching of nanomagnets. Phys Rev J Achieve 130(5):1677–1686
Kumari P, Alam M, Siddiqi WA (2019) Usage of nanoparticles as adsorbents for waste water treatment: an emerging trend. Sustain Mater Technol 22:e00128
La Mantia FP, Morreale M (2011) Green composites: A brief review. Compos A Appl Sci Manuf 42(6):579–588
Lajayer BA, Ghorbanpour M, Nikabadi S (2017) Heavy metals in contaminated environment: destiny of secondary metabolite biosynthesis, oxidative status and phytoextraction in medicinal plants. Ecotoxicol Environ Saf 145:377–390
Lan YJ, Lin YW (2014).A non-aggregation colorimetric method for trace lead (II) ions based on the leaching of gold nanorods. Anal Methods 6(18):7234–7242
Laner D, Crest M, Scharff H, Morris JW, Barlaz MA (2012) A review of approaches for the long-term management of municipal solid waste landfills. Waste Manage 32(3):498–512
Laurent S, Forge D, Port M, Roch A, Robic C, Vander Elst L, Muller RN (2010) Magnetic iron oxide nanoparticles: synthesis, stabilization, vectorization, physicochemical characterizations, and biological applications. Chem Rev 108(6):2064–2110
Lee JE, Lee N, Kim T, Kim J, Hyeon T (2011) Multifunctional mesoporous silica nanocomposite

nanoparticles for theranostic applications. Acc Chem Res 44(10):893–902
Lee YF, Huang CC (2011) Colorimetric assay of lead ions in biological samples using a nanogold-based membrane. ACS Appl Mater Interfaces 3(7):2747–2754
Li M, Gou H, Al-Ogaidi I, Wu N (2013) Nanostructured sensors for detection of heavy metals: a review. ACS Sustain Chem Eng 1(7):713–723
Lin L, Zhang J, Su H, Li J, Sun L, Wang Z, Liu Z (2019) Towards Super-Clean Graphene. Nat Commun 10 (1):1–7
Lin WC, Li Z, Burns MA (2017) A drinking water sensor for lead and other heavy metals. Anal Chem 89 (17):8748–8756
Liu C, Cheng HM, Cong HT, Li F, Su G, Zhou BL, Dresselhaus MS (2000) Synthesis of macroscopically long ropes of well-aligned single-walled carbon nanotubes. Adv Mater 12(16):1190–1192
Liu J, Shao M, Chen X, Yu W, Liu X, Qian Y (2003) Large-scale synthesis of carbon nanotubes by an ethanol thermal reduction process. J Am Chem Soc 125(27):8088–8089
Liu S, Ni L, Ma F, Fu H (2020) Analysis of intracellular copper ions content and distribution in algae by gold nanorods method. Chemosphere 259:127512
Liu S, Tan Y (2018) Detection of microalgae superoxide dismutase (SOD) using GNRs-based resonance light scattering system. Integr Biol 10(3):159–165
Mahmood H, Vanzetti L, Bersani M, Pegoretti A (2018) Mechanical properties and strain monitoring of glass-epoxy composites with graphene-coated fibers. Compos A Appl Sci Manuf 107:112–123
Mansha M, Qurashi A, Ullah N, Bakare FO, Khan I, Yamani ZH (2016) Synthesis of In2O3/graphene heterostructure and their hydrogen gas sensing properties. Ceramic Int 42(9):11490–11495
Monem AS, Elbialy N, Mohamed N (2014) Mesoporous silica coated gold nanorods loaded doxorubicin for combined chemo–photothermal therapy. Int J Pharm 470(1–2):1–7
Noked M, Okashy S, Zimrin T, Aurbach D (2012) Composite carbon nanotube/carbon electrodes for electrical double-layer super capacitors. Angew Chem 124(7):1600–1603
Novoselov KS, Fal'Ko VI, Colombo L, Gellert PR, Schwab MG, Kim K (2012) A roadmap for graphene. Nature 490:192–200
Pande VV, Sanklecha VM (2017) Bionanocomposite: A Review. Austin J Nanomed Nanotechnol 5(1):1045
Prakash P, Singh HR, Jha SK (2020) Preparation, characterization and application of curcumin based polymeric bio-composite for efficient removal of endotoxins and bacterial cells from therapeutic preparations. J Biomater Sci Polym Ed 1–18
Qi M, Zhang K, Li S, Wu J, Pham-Huy C, Diao X, Xiao D, He H (2016) Superparamagnetic Fe3O4 nanoparticles: synthesis by a solvothermal process and functionalization for a magnetic targeted curcumin delivery system. New J Chem 40(5):4480–4491
Rajakumar K, Dinesh KS, Sivanesan S, Sai RL (2014) Effective removal of heavy metal ions using Mn_2O_3 doped polyaniline nanocomposite. J Nanosci Nanotechnol 14(4):2937–2946
Rakhi RB, Lim X, Gao X, Wang Y, Wee ATS, Sethupathi K, Ramaprabhu S, Sow CH (2010) Electron field emission from magnetic nanomaterial encapsulated multi-walled carbon nanotubes. Appl Phys A 98:195
Rawal I, Kaur A (2013) Synthesis of mesoporous polypyrrole nanowires/nanoparticles for ammonia gas sensing application. Sens Actuators, A 203:92–102
Reddy DHK, Yun YS (2016) Spinel ferrite magnetic adsorbents: alternative future materials for water purification? Coord Chem Rev 315:90–111
Reetz MT, Zonta A, Vijayakrishnan V, Schimossek K (1998) Entrapment of lipases in hydrophobic magnetite-containing solgel materials: magnetic separation of heterogeneous biocatalysts. J Molecular Catalyst A: Chem 134(1–3):251–258
Reiss G, Hütten A (2005) Applications beyond data storage. Nat Mater 4(10):725–726
Rossi LM, Costa NJ, Silva FP, Wojcieszak R (2014) Magnetic nanomaterials in catalysis: advanced catalysts for magnetic separation and beyond. Green Chem 16(6):2906–2933
Satyanarayana KG, Arizaga GGC, Wypych F (2009) Biodegradable composites based on lignocellulosic fibers – an overview. Prog Polym Sci 34(9):982–1021
Shahzad A (2012) Hemp fiber and its composites – a review. J Compos Mater 46(8):973–986
Shanks RA, Hodzic A, Wong S (2004) Thermoplastic biopolyester natural fiber composites. J Appl Polym Sci 91(4):2114–2121
Sharon M, Mukhopadhyay I, Mukhopadhyay K (1997) A photoelectrochemical solar cell from camphoric p-carbon semiconductor. Sol Energy Mater Sol Cells 45 (1):35–41
Shin WK, Cho J, Kannan AG, Lee YS, Kim DW (2016) Cross-linked composite gel polymer electrolyte using mesoporous methacrylate-functionalized SiO_2 nanoparticles for lithium-ion polymer batteries. Sci Rep 6:26332
Shiraishi M, Takenobu T, Kataura H, Ata M (2004) Hydrogen adsorption and desorption in carbon nanotube systems and its mechanisms. Appl Phys A 78 (7):947–953
Srinivasan R, Maity A, Singh KK, Ghosh PK, Kumar S, Srivastava MK, Kumari B (2017) Influence of copper oxide and zinc oxide nano-particles on growth of fodder cowpea and soil microbiological properties. Range Manage Agrofor 38(2):208–214
Suh WH, Suslick KS (2005) Magnetic and porous nanospheres from ultrasonic spray pyrolysis. J Am Chem Soc 127(34):12007–12010
Tartaj P, Morales MP, Gonzalez-Carreno T, Veintemillas-Verdaguer S, Serna CJ (2005) Advances in magnetic nanoparticles for biotechnology applications. J Magn Magn Mater 290:28–34

Tiwari JN, Tiwari RN, Kim KS (2012) Zero-dimensional, one dimensional, two-dimensional and three-dimensional nanostructured materials for advanced electrochemical energy devices. Progress Mater Sci 57 (4):724–803

Tounsadi H, Khalidi A, Abdennouri M, Barka N (2016) Activated carbon from Diplotaxis Harra biomass: Optimization of preparation conditions and heavy metal removal. J Taiwan Inst Chem Eng 59:348–358

Tsedenbal B, Lee JE, Kim MS, Huh SH, Koo BH (2020) Flower-Like α-Fe2O3 for Removal of Heavy Metals from Wastewater. J Nanosci Nanotechnol 20 (11):6753–6759

Ugulu I (2015) Determination of heavy metal accumulation in plant samples by spectrometric techniques in Turkey. Appl Spectrosc Rev 50(2):113–151

Ullah H, Khan I, Yamani ZH, Qurashi A (2017) Sonochemicaldriven ultrafast facile synthesis of SnO2 nanoparticles: growth mechanism structural electrical and hydrogen gas sensing properties. Ultrason Sonochem 34:484–490

USEPA (2004) Contaminated Sediment. http://www.epa.gov/OST/cs/aboutes/index.html. Accessed on 21 January 2021

USEPA (United States Environmental Protection Agency) (1997) The incidence and severity of sediment contamination in surface water of the United States. Volume 1, National Sediment Quality Survey, EPA 823-R-97-006; United States Environmental Protection Agency

Varlan AR, Suls J, Jacobs P, Sansen W (1995) A new technique of enzyme entrapment for planar biosensors. Biosens Bioelectron 10(8):xv–xix

Wang L, Li W, Wu B, Li Z, Pan D, Wu M (2017) Room-temperature synthesis of graphene quantum dots via electron-beam irradiation and their application in cell imaging. Chem Eng J 309:374–380

Wang L, Li W, Wu B, Li Z, Wang S, Liu Y, Wu M (2016) Facile synthesis of fluorescent graphene quantum dots from coffee grounds for bioimaging and sensing. Chem Eng J 300:75–82

Wuana RA, Okieimen FE (2011) Heavy metals in contaminated soils: a review of sources, chemistry, risks and best available strategies for remediation. Int Sch Res Not 2011

Yang J, Hou B, Wang J, Tian B, Bi J, Wang N, Li X, Huang X (2019) Nanomaterials for the removal of heavy metals from wastewater. Nanomaterials (basel) 9(3):424

Yuan X, Setyawati MI, Tan AS, Ong CN, Leong DT, Xie J (2013) Highly luminescent silver nanoclusters with tunable emissions: cyclic reduction–decomposition synthesis and antimicrobial properties. NPG Asia Materials 5(2):e39–e39

Zhang Y, Jiang T, Tang L (2017) Near-infrared photoluminescence biosensing platform with gold nanorods-over-gallium arsenide nanohorn array. Biosens Bioelectron 97:278–284

Zhang Z, Chen Z, Qu C, Chen L (2014) Highly sensitive visual detection of copper ions based on the shape-dependent LSPR spectroscopy of gold nanorods. Langmuir 30(12):3625–3630

Zhu J, Chen M, Qu H, Luo Z, Wu S, Colorado HA, Guo Z (2013) Magnetic field induced capacitance enhancement in graphene and magnetic graphene nanocomposites. Energy Environ Sci 6(1):194–204

Zhu Y, Ramasamy M, Yi DK (2014) Antibacterial activity of ordered gold nanorod arrays. ACS Appl Mater Interfaces 6(17):15078–15085

Zini E, Scandola M (2011) Green composites: an overview. Polym Compos 32(12):1905–1915

Microplastics and Synthetic Polymers in Agricultural Soils: Biodegradation, Analytical Methods and Their Impact on Environment

17

M. Mizanur Rahman, Zahurul Haque, Nazmul Huda, Md. Amdadul Huq, Mamoona Rauf, Md. Maharub Hossain Fahim, and Muhammad Arif

Abstract

Microplastics (MPs) and synthetic polymers (SPs) are emerging pollutants/contaminants worldwide. Due to the significance of soil environment and the demand from scientific communities for increased soil research, it is expected that related studies will rise steeply in the years to come. This present analysis aims to provide an overview of existing information about contamination in soil ecosystems by MPs and SPs. We precisely summarize the types, source, functional analytical methods, exposure routes, contamination of MPs in soils. We also carefully explain the influence of MPs on soil physicochemical properties, plants, and soil biota and determine what we are capable of learning from available data. The chapter critically assesses the efficient MP biodegradation strategies, showing the role of microorganisms and enzymes in the processes with influencing factors of biodegradation. The chapter also outlines the problems of MPs pollution, which would be an emphasis on source management and cleanup.

Keywords

Agricultural soil · Biodegradation · Contamination · Environmental pollution · Microplastics (MPs) · Plastic biodegradability · Synthetic polymers (SPs)

17.1 Introduction

Microplastics (MPs) and synthetic polymers (SPs) are soundless hazards which are measured as a substantial problem in varied environments (Sarker et al. 2020). MPs are heterogeneously diverse plastics (>5 mm in diameter) comprising of plastic granules, fibers, and fragments, which are evolving contaminants (Guo et al. 2020; Cózar et al. 2014). Plastic manufacture has enlarged from 1.5 to 335 million tons in 2016 worldwide (Sarker et al. 2020; PlasticsEurope 2018). The present stages of plastic production, disposal/usage pattern, demographic data, and small recovery rate are the foundations of rising accumulation of plastic waste (Guo et al. 2020).

M. Mizanur Rahman (✉) · Z. Haque · N. Huda · Md.M. H. Fahim
Department of Biotechnology and Genetic Engineering, Faculty of Biological Science, Islamic University, Kushtia 7003, Bangladesh

Md.A. Huq
Department of Food and Nutrition, College of Biotechnology and Natural Resource, Chung-Ang University, Gyeonggi-do, Anseong-si 17546, Republic of Korea

M. Rauf
Department of Botany, Abdul Wali Khan University, Mardan, Pakistan

M. Arif
Department of Biotechnology, Abdul Wali Khan University, Mardan, Pakistan

J. A. Malik (ed.), *Advances in Bioremediation and Phytoremediation for Sustainable Soil Management*,
https://doi.org/10.1007/978-3-030-89984-4_17

Although less than 5% plastics are recyclable materials and domesticated, nonetheless large amounts (4.8–12.7 million tons) of plastic waste are coming into the ocean (Sutherland et al. 2010; Guo et al. 2020; Jambeck et al. 2015). Subsequently, productions of carbon dioxide have increased due to plastics production worldwide (PlasticsEurope 2018).

In recent years, plastics have have been regarded as the severely dangerous pollutants in the environment (Zhang et al. 2015; Sarker et al. 2020). Soil is observed to be the main transportation route for MPs (Zhang et al. 2015; Horton et al. 2017). Disposition of MPs has not been studied in the soil as an important research area with biotic toxicological assay (Hurley and Nizzetto 2018; Zhang et al. 2015). In mulching purposes, polypropylene (PP) and low-density polyethylene (LDPE) are applied as main sources of MPs in agricultural soil (Sarker et al. 2020; Blasing and Amelung 2018). Additional sources comprise composting with littering, sewage, suburban runoff, and sludge wastewater irrigation (Corradini et al. 2019). Their toxic effects have been recognized in marine animals due to occurrence of numerous plastic debris in oceans and coastal watercourses (Hossain et al. 2020, 2019).

Recently, it was demonstrated that the toxicity of MPs transfer from agriculture field to the food chain and thus to humans. Consequently, MPs could be treated as the upcoming hazards to sustainable agricultural and food safety. This chapter focuses on overview of existing information about contamination in soil ecosystems by MPs and SPs with the types, source, exposure routes, contamination level and fate of the MPs in soils.

17.2 Biodegradation and Plastic Biodegradability

17.2.1 Biodegradation

Biodegradation is a biological process which involves the degradation and assimilation of polymers into their simpler and nontoxic forms such as carbon dioxide (CO_2), methane, water, and biomass using living microorganisms (Kumar and Maiti 2016; Raaman et al. 2012). Polymers are converted biochemically by reducing their molecular mass, mechanical strength and the external properties (John and Salim 2020). *Pseudomonas flourescens*, *Pseudomonas aeroginosa*, and *Penicillium simplicissimum* are the most common biodegrading strains (Ahmed et al. 2018; Raziyafathima et al. 2016).

Biological degradation of high molecular weight polymers is primarily regulated by two steps that may occur in soil, water, or human beings (Eskander and Saleh 2017; Tokiwa et al. 2009). The first step is known as fermentation/fragmentation, where macromolecular chain with high molecular weight is converted into oligomers (Fig. 17.1). The second step is known as a mineralization in which oligomers and monomers are mineralized into water, carbon dioxide (CO_2), methane, and biomass by using microorganisms (Agarwal 2020). Biodegradation is affected by many environmental factors, including the accessibility of light, pH, oxygen, temperature, moisture, microorganisms, and type of enzyme and concentration of enzyme. Under different conditions, the identical polymer display diverse rates of degradation (Agarwal 2020).

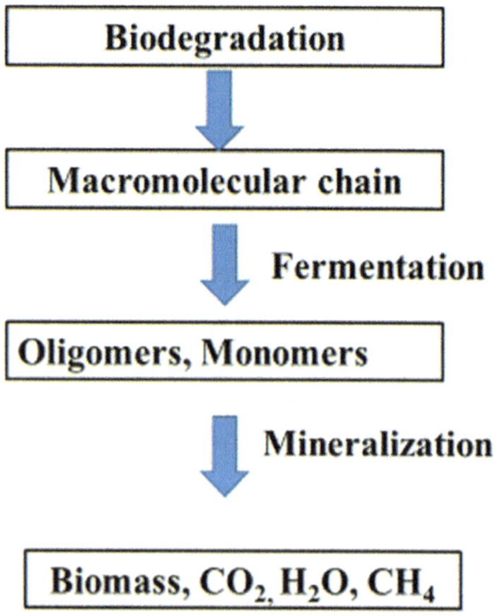

Fig. 17.1 Process of biodegradation

17.2.2 Plastic Biodegradability

The word "biodegradation' applies to materials which are either decomposed or mineralized into ultimate products of carbon dioxide and water when exposed to a particular microbial environment. Polymers that are degraded in this way are referred as biodegradable polymers which, instead of raw materials, are very much dependent on the chemical composition of polymers (Kijchavengkul and Auras 2008). Plastics are degraded by various mechanisms such as chemical, photo, thermal, and biological degradation. Plastic degradation is a physical or chemical alteration in polymers caused by environmental causes, for example, heat, light, humidity, and biological and chemical activity (Tokiwa et al. 2009).

Bioplastics (BPs) are made of bio-based and biodegradable plastics. However, in our view, BPs consist of biodegradable (such as fossil-based plastics) or bio-based plastics (biomass produced plastics or renewable sources) (Tokiwa et al. 2009). BPs are classified as "plastics' where the carbon (100%) is extracted from renewable sources, for example maize starch and soybean cellulose and protein in agricultural and forestry sectors (Alshehrei 2017).

The biodegradability of plastics can be defined as a preliminary point for biological processes or as the interruption of plastic polymers or monomers (Annemette 2019). Without leaving detectable toxic traces, biodegradable BPs are completely degraded by microorganisms (Jain et al. 2010). The utmost popular forms of biodegradable polyester plastics include polylactic or polylactate acid (PLA), poly propiolactone (PPL), poly 3-hydroxybutyrate (PHB), polycaprolactone (PCL), poly 4-hydroxybutyrate (P4HB), polyethylene succinate (PES), poly ester carbonate (PEC) (PHBV), poly butylene succinate (PBS), and poly 3-hydroxybutyrate-co-3-hydroxyvalerate plastics (Northcott and Pantos 2018).

17.3 MPs and SPs

MPs obtained from non-biodegradable polymers can create a potential threat to health and the environment (Annemette 2019). MPs particle size in ranging from 100 nm to 5 mm (ng et). The incorporation of MPs into the environment can be thriven by laundering cosmetic beads and textile, fabrics or indirectly by breaking up of bigger plastic parts (mechanical degradation). Because of their small size, many species at almost all food chains, especially in marine ecosystems, including zooplankton, coral, fish, birds, and marine mammals, consume microplastics easily. It was noted that seabirds (99%) had swallowed MPs, and by 2050, more than 600 aquatic animals (nearly 15%) which are predicted to be affected by MPs ingestion or predicament in MPs marine litter (UNEP 2016). In addition to causing direct landscape issues, the pervasive presence of plastic waste and MPs poses possible environmental threats to living species, including humans (Shen et al. 2019; Diepens and Koelmans 2018; Miranda and Carvalho-Souza 2016; Fossi et al. 2012).

SPs are described as polymers that are manufactured artificially. They are also recognized as man-made polymers. Polyethylene (PE), polyamides (nylon), polystyrene (PS), poly vinyl chloride (PVC), teflon, epoxy, synthetic rubber, and some others are several examples of SPs (Verma 2004). In a regulated environment, SPs are mainly derived from petroleum oil and consist of carbon–carbon bonding. Using synthetic polymers, millions of daily applications are made. The groups of thermoplastics, thermosets, elastomers, and synthetic fibers fall into these applications (Shrivastava 2018). SPs have a number of appearances, for example, in the manufacturing of corrective lenses, some transparent polymers may be formed into specific shapes. To be distorted from one form to another, the polymer rubber used in tires must be flexible enough (Ouellette and Rawn 2015).

17.3.1 Precise Classification of MPs and SPs

Plastics are widely classified as natural, semi-synthetic or synthetic depending on their source of origin. Natural polymers are classified as materials generally found in nature or derived from animals

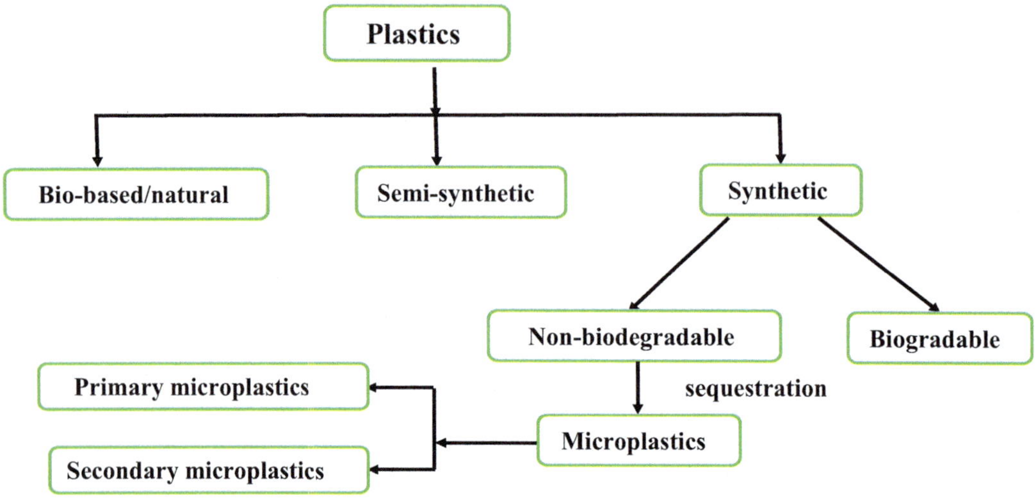

Fig. 17.2 Classification of MPs and SPs

and plants. Proteins and nucleic acids that exist in the human body, cellulose, natural rubber, silk, and wool are some examples of natural polymers. Semisynthetics polymers are the polymers produced by chemical modification of the natural polymers. Vulcanized rubber, cellulose acetate, and rayon are among the more popular ones (Shrivastava 2018). Again, SPs are classified into two groups, biodegradable and non-biodegradable polymers. Then non-biodegradable polymers sequestrate into MPs (Fig. 17.2).

Moreover, MPs are classified based on source as primary and secondary MPs (Duis and Coors 2016; Thompson 2015; Cole et al. 2011). Primary MPs are firmly formulated for uses, including medicine vectors, cosmetic abrasives, and applications of automotive and aerospace (Auta et al. 2017). Secondary MPs are derived from large plastics, in which they are increasingly broken into small sections by various, dynamic physical factors such as temperature, UV light, waves, and wind (Rocha-Santos and Duarte 2015).

17.3.2 Emission Sources of MPs in Soils

MPs enter the soil via mainly two sources such as direct and indirect source. In cultivation, plastic mulch products, greenhouse products and soil conditioners are direct sources. Indirect sources involve the use of wastewater, littering and biological substances (Duis and Coors 2016). In addition, MPs penetrate soil from numerous channels, including landfill sites, soil alteration, land application of sewage sludge, drainage, irrigation, compost and organic fertilizers, remnants of agricultural mulching film, tire wear and tear, and atmospheric deposition, etc. (Guo et al. 2020). Because of population size, resources, existence, and effectiveness of waste management activities, MPs emissions per capita differ significantly across countries (Ziajahromi et al. 2016; Nizzetto et al. 2016). In Europe, 63,000 to 430,000 tons per year of MPs were found in agroecosystems by biosolids alone, while 44,000 to 300,000 tons per year of the MPs were in North America (Nizzetto et al. 2016).

17.4 Exposure Routes of MPs and SPs in Soil

The prevalence of plastics in the world, whether as MPs debris or as MPs have been broadly recognized as a global issue (Gionfra 2018). Plastics are recalcitrant polymers released to the atmosphere by uncontrolled usage leading to accumulation and increased water and soil

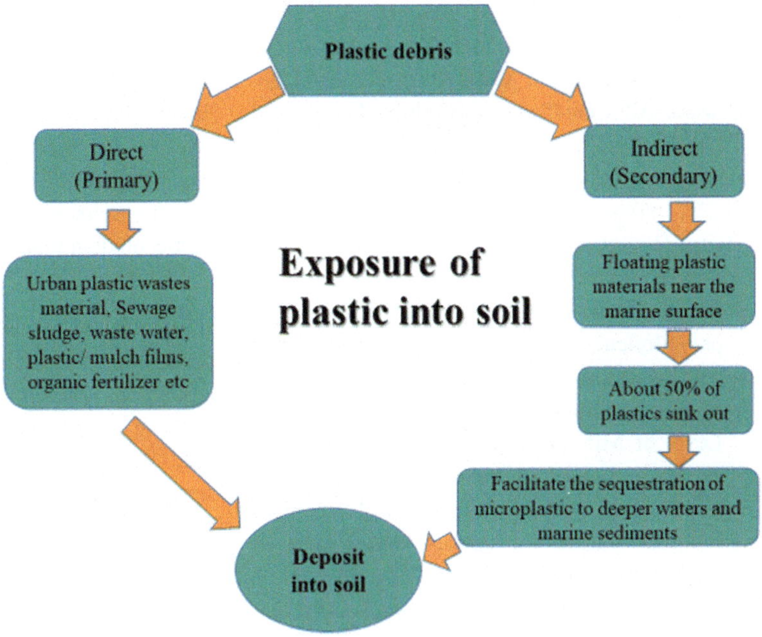

Fig. 17.3 Multiple exposure routes of MPs and SPs in soil

contamination. It consists of approximately 80% of the litter leads to accumulate in agricultural land, waste disposal, and water bodies. Therefore, plastics have a wide variety of uses extending from agricultural, commercial, and domestic applications. An example of common application in the agriculture sector involves polyethylene soil mulching (Iram et al. 2019). Since practices of recycling or otherwise handling plastic wastes have not been preserved, remaining plastic wastes have deposited in the environment (Awasthi 2020). Most of this waste is dumped near water bodies, in urban drainage systems, in which it flows into rivers and ends up in oceans in different types, such as MPs (particles of 5 mm in small fragments of large pieces of plastic) (Hale et al. 2020). Global production of plastics has reached alarming proportions; plastics were manufactured 322 million tons globally in 2015. In 2015, plastic waste produced 6.300 million tons, 9% of which was recycled, 12% burned and remaining 79% sent to spilled or to landfill sites (Gionfra 2018). A study estimated that a large amount (about 110 000–730 000 tons per year) of MPs were emitted to cultivated fields in North America and Europe (Awasthi 2020).

Argo plastics can leak via wind or river transport in the marine environment. Plastic and MPs contamination can be seen in the oceans more significantly. Furthermore, over 80% of plastic contained on the ground has been made, consumed, and removed from marine environments (de Souza Machado et al. 2018). The practices of urban wastewater treatment plants as irrigations on agricultural land are commonly used method and an important source of primary MP pollution of soil (Nizzetto et al. 2016). Nutrient combinations N, P and K are encapsulated in a nutrient tablet, a polymer coating. It is also a significant cause of contamination of MPs. The nutritional pill does not decay after the introduction of the nutrients. MPs can be released into the environment in two ways: direct or primary and indirect or secondary (Fig. 17.3). In primary way MPs are released into the environment from domestic goods, such as microbeads, direct depleting and inadequate wastewater treatment, e.g., losses through the waste collection, industrial spills or discharges from landfill places (Lechner and Ramler 2015). Secondary MP pollution causes, on the other hand, include deliberate statement (illegal dumping), untreated waste or accidental wounding (such as fishing gear) (Boucher and Friot 2017). During municipal waste collection, sorting, transport and waste disposal, MPs are released. Additional plastic products being used for agricultural purposes, which also reflect possible sources of MP contamination in soil, are bottles, wrapping and netting (Horton et al. 2017). The plastic mulching is used to cover plant, seedlings, and shoots by using plastic films on crops. Plastic mulches are usually consisting of polyethylene and it is not easily dissolved in the soil, which is connected to MPs residue deposition (Steinmetz et al. 2016).

Level of MPs in the oceans has been extensively studied. Agricultural overflow from drainage and farmland can result in involvement of agricultural plastics or sewage-sludge derived fibers and microbeads.

In the above circumstances, it can be said that MPs persistence within the soil is greatly related to the direct exposure of MP sources. For reducing the exposure of MP, recycling efficacy should be increased, and public awareness should be raised. By using biodegradable plastic, it could be helpful for reducing the presence of MP for long duration.

17.5 Biodegradation of MPs and SPs

The recycling process is currently growing, but since more additives are used in their processing, the recycling rates are very low in maximum plastic materials (Song et al. 1998). Compared to other waste management technology, biodegradation is consistent (microbial mineralization) (Schink et al. 1992). Biodegradation using microorganisms offers a simple method of cleaning such plastic residues. Microbial enzymes are used to manage pollutants and help to establish an ecosystem that is environmentally friendly (Pathak and Navneet 2017). Breakdown of macromolecular chains by microbes is termed as biodegradation (Agarwal 2020). By the biological activity upon a material that causes any physical or chemical alteration is known as biodegradation (Alshehrei et al. 2017). Hydrolysis is the most critical form of enzymatic polymer cleavage reaction (Artham and Doble 2008; Schink et al. 1992). Some microorganisms have shown the ability for biodegradation of plastic content (Table 17.1).

A few steps have been taken in the process of plastic biodegradation (Fig. 17.4) and could be defined by particular terms:

Biodeterioration describes the results of the physical and chemical deterioration of microbial populations and other decomposing organisms resulting in a gradual deterioration of the plastics with changes in their physical, mechanical, and chemical properties (Lucas et al. 2008; Iram et al. 2019).

Biofragmentation is the enzymatic activity which cleaves polymeric plastics by ectozymes or free radicals secreted by micro-organisms into oligomers, dimers or monomers (Lucas et al. 2008). The use of different enzymes released via the microorganisms, including lipase, proteinase K., hydrogenase etc. are involved in plastic biodegradation (Ghosh et al. 2013).

Microbial assimilation period resembles to the breakdown in previous stages of the low-molecular organisms, which have contributed to substantial gas evolution and mineralization (Harrison et al. 2018). Microbial cell membrane receptors recognize and activate certain dispersed molecules through the membrane to reach the cells. Increased bio-transformations of non-realizable plastic fragments by a cell membrane receptor lead to the generation of products that can easily spread into the cell (Lucas et al. 2008). Most cases can measure the stage rate by calculating the evolution of the gas or by growing the biomass of the selected microorganism, if carried out in a bioreactor (Harrison et al. 2018).

Plastics are being degraded very slowly. At first physical factors such as pH, temperature, and UV initiate this process (Devi et al. 2016). Biodegradability is also influenced by the chemical composition and source of the polymer (Muthu 2014). Microbes with different cleavage bond and enzyme activities achieve the process of

Table 17.1 Selected characteristics of the main commercial synthetic polymers along with microorganisms and enzymes involved in these polymers' biodegradation

Type of polymers	Name of plastics		Recycle ID code	Chemical formula	Structure	R group	Plastic Density ($g.cm^{-3}$)	T_m (°C)	Crystallinity (%)	Lifespan (year)	Uses of plastics	Microorganisms	Sources of microbes	Enzymes
Synthetic polymers (Microplastics)	Polyethylene (PE)	Low-density Polyethylene (LDPE)	4 LDPE	$(C_2H_4)_n$	Homo-polymer	Hydrogen	0.917–0.965	140–143	50	10–600	Squeezable Bottles, Frozen food bags, flexible container lids	*Rhodococcus rhodochorus*, *R. ruber* C208, *Staphylococcus epidermis*, *Brevibacillus borstelensis* 707, *Bacillus sp.*, *Pseudomonas sp.*, *Aspergillus sp.*, *Penicillium funiculosum*, *Gliocladium virens*, *Chaetomium globosum*, *Pullularia pullulans*, *Fusarium sp.* AF4, *Tremetes versicolor*	Sewage treatment plants and waste management landfills, Dumping soil, Garden soil	Laccase, Alkane hydroxylase
		High-density Polyethylene (HDPE)	2 HDPE	$(C_2H_4)_n$	Homo-polymer	Hydrogen	0.917–0.965	140–143	50	10–600	Water, milk, and juice bottles, retail bags, and trash	*Amycolatopsis sp.*, *Bacillus sp.*, *Rhodococcus rhodochorus*, *Listeria sp.*, *Micrococcus sp.*, *Vibrio sp.*, *Rhodococcus rhodochorus*, *Arthrobacter sp.*, *Pseudomonas sp.*,, *Trametes versicolor*, *Phanerochaete chrysosporium* ME-446*Aspergillus sp.*	Garden soil, Mangrove soil	Manganese peroxidase

(continued)

Table 17.1 (continued)

Type of polymers	Name of plastics	Recycle ID code	Chemical formula	Structure	R group	Plastic Density (g.cm^{-3})	T_m (°C)	Crystallinity (%)	Lifespan (year)	Uses of plastics	Microorganisms	Sources of microbes	Enzymes
	Polyvinyl chloride (PVC)	3 PVC	$(C_2H_3Cl)_n$	Homo-polymer	Chorine	1.16–1.58	115–245	0	50–100$_+$	Curtains, automobiles, bottles, raincoats, shoes soles, agriculture, electricity pipes, garden hoses	*Pseudomonas sp.*, *Ochrobactrum* TD, *Aspergillus sp.*, *Phanerochaete chrysosporium*, *Lentinus tigrinus*, *Thermomonospora fusca*, *Streptomyces sp.*, *Poliporusversicolor*, *Phanerochaete chrysosporium* ME 446, *Pleurotus sp.*, *Bacillus cereus*, *Acanthopleurobacter pedis*	Not reported	Not reported
	Polypropylene (PP)	5 PP	$(C_3H_6)_n$	Homo-polymer	Methyl	0.90–0.91	165	50	10–600	Bottle cups, straws, car seats, batteries, bumpers, syringes	Not reported	Not reported	Not reported
	Polyethylene terephthalate (PET) / Polyester	01 PET	$(C_{10}H_8O_4)_n$	Homo-polymer	Carboxyl and hydroxyl	1.37–1.45	280	0–50	450	meat package, carbonated soft drink bottle, clothing, food package, textile fibers	*Pseudomonas fluorescens*, *P. chlororaphis*, *P. putida*, *P. aeruginosa*, *P. protegens* BC2 12, *Ochrobactrum sp.*, *Ideonella sakaiensis* 201-F6	Waste sites and dumping situations	Lipase, Poly-urethanase, Esterase, Protease, Hydrolases, Lipase, Cutinase

(continued)

Table 17.1 (continued)

Type of polymers	Name of plastics	Recycle ID code	Chemical formula	Structure	R group	Plastic Density ($g.cm^{-3}$)	T_m (°C)	Crystallinity (%)	Lifespan (year)	Uses of plastics	Microorganisms	Sources of microbes	Enzymes
	Polyurethane (PU)	07 O	$(R-(N=C=O)_n$	Hetero-polymer	Isocyanate and polyol	1.20	400			Automotive, sponges, life jacket, clothing	*Aureobasidium pullulans*, *Rhodococcus equi.*, *Pseudomonas sp.*, *Corynebacterium sp.*, *Bacillus sp.*, *Arthrobacter globiformis*, *Aspergillus terreus*, *Curvularia senegalensis*, *Chaetomium globosum*, *Actinetobacter calcoaceticus*, *Actinetobacter gerneri* P7, *Fusarium solani*, *Rhizopus delemar*, *Comamonas acidovorans* TB-35, *Trichoderma sp.*, *Pestalotiopsis microspore*, *Cladosporium sp.*,	Soil samples	Aryl acylamidase, Ureases, Esterases, Proteases, Lipase
	Polystyrene (PS)	6 PS	$(C_8H_8)_n$	Homo-polymer	Phenyl	1.04–1.1	240	0	50–80	Disposable cups, food packaging materials, laboratory ware, electronic device	*Actinomycete* sp., *Tenebrio molitor* (mealworm), *Exiguobacterium sp.* YT2, *Zophobas morio* (superworm), *Enterobacter sp.*, *Citrobacter sedlakii*, *Alcaligenes sp.*, *Brevundimonas diminuta*, *Pseudomonas putida* AJ, *Bacillus sp.*	Soil samples, Gut of mealworm, Rural market setting	Alkane hydroxylase

(continued)

Table 17.1 (continued)

Type of polymers	Name of plastics	Recycle ID code	Chemical formula	Structure	R group	Plastic Density ($g.cm^{-3}$)	T_m (°C)	Crystallinity (%)	Lifespan (year)	Uses of plastics	Microorganisms	Sources of microbes	Enzymes
	Polycarbonate (PC)	07 O		Homo-polymer	Carbonate		52–150			Safety visor, lens in glasses, baby bottles, roofs	*Roseateles depolymerans* 61A, *Amycolatopsis sp.* HT-6, *Candida cylindracea*, *Chromobacterium viscosus*, *Pseudomonas sp.*	Not reported	Cholesterol esterase, Lipase, Lipoprotein lipase
	Polyamide (PA) / Nylon (NY)	07 O		Homo-polymer	Amide	1.13–1.35	190–276			Shoes, clothing, rainwear	*Agromyces sp.*, *Tremetes versicolor*, *Flavobacterium sp.*, *Pseudomonas sp.* NK87, White-rot fungus IZU-154	Soil samples, Compost and activated sludge	Manganese peroxidase, Nylon hydrolase, Laccase
	Polytetrafluoro-ethylene (PTFE)	07 O								Chemicals, electronics, kitchens utensils	Not reported	Not reported	Not reported
	Polymethyl-acrylate (PMA)	07 O	$(C_4H_6O_2)_n$			1.17–1.20					*Cyanobacteria*	Not reported	Not reported
References	Pathak and Navneet (2017); Devi et al. (2016); Siracusa (2019); Wu et al. (2016)	Devi et al. (2016); Wu et al. (2016)	Tokiwa et al. (2009); Wu et al. (2016)	Pathak and Navneet (2017)	Pathak and Navneet (2017)	Zhu et al. (2019); Glaser (2019; Horton et al. (2018)	Pathak and Navneet (2017); Tokiwa et al. (2009)	Glaser (2019); Ojeda (2013)	Glaser (2019); Ojeda (2013)	Siracusa (2019); Devi et al. (2016); Alshehrei (2017); Tokiwa et al. (2009)	Tokiwa et al. (2009); Glaser (2019); Devi et al. (2016); Iram et al. (2019); Fesseha et al. (2019) Pathak and Navneet (2017); Northcott and Pantos (2018); Wu et al. (2016)	John and Salim (2020); Tokiwa et al. (2009); Glaser (2019);	Iram et al., (2019); Fesseha et al., (2019); Wu et al., (2016); Tokiwa et al., (2009); Glaser (2019)

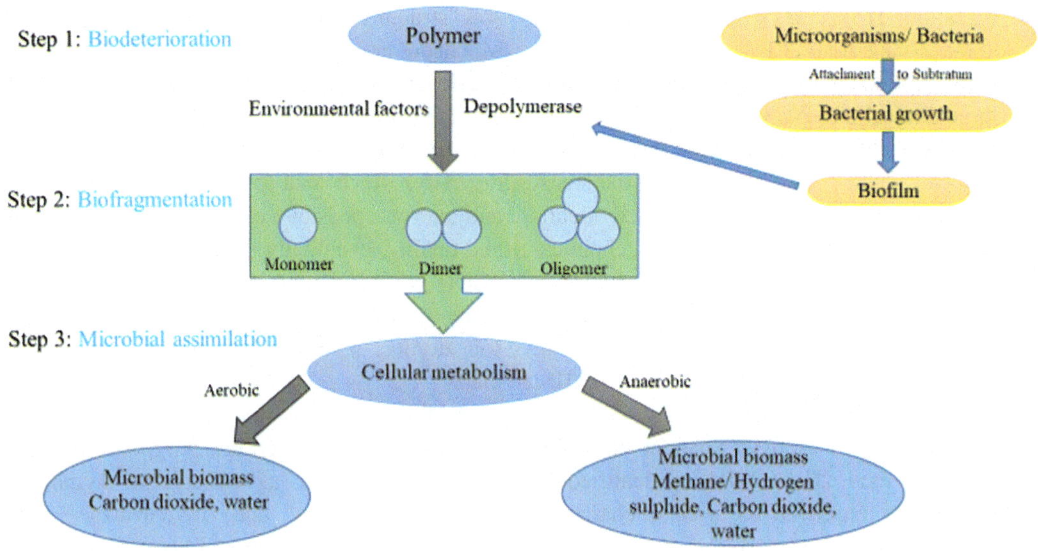

Fig. 17.4 The process of plastic biodegradation

degradation. Two types of enzymes are involved as extracellular and intracellular depolymerases (Dey et al. 2012). In the aerobic degradation process, oxygen is the necessary terminal electron acceptor. In addition, the production of CO_2 and H_2O occurs in aerobic conditions during plastic degradation due to cellular biomass of microorganisms (Glaser 2019). The aerobic system is more effective as compared to anaerobic conditions. The anaerobic method generates low energy due to the absence of O_2 when considering the energy output (Gottschalk 2012). The difference between anaerobic and aerobic degradation is very significant because the anaerobic conditions have been found to promote slower biodegradation kinetics (Glaser 2019).

C(plastic) + $O_2 \rightarrow H_2O + CO_2$ + C(residue) + C(biomass) (Alshehrei 2017).

C(plastic) $\rightarrow CH_4 + H_2O + CO_2$ + C(residue) + C(biomass) (Alshehrei 2017).

Alshehrei (2017) reported that biodegradation of polymers involves some steps as following:

1. The microorganism attachment to the surface of the polymer.
2. Development of the microorganism.
3. Ultimate degradation of the polymer.

The relation of the microorganism to the polymer's surface helps to produce biofilm (Pathak and Navneet 2017) (Fig. 17.2). For the degradation of the natural environment, biofilm formation is important (Sivan et al. 2006).

17.6 Analytical Methods of MPs and SPs in Soils

Analytical methods of MPs in soils comprise four steps- (a) extraction, (b) cleanup, (c) identification, and (d) quantification.

(a) Density fractionation methods during extraction procedure are extensively applied to abstract MPs from the soil complex matrix, since the density values of the frequently (0.8 to 1.4 gcm^{-3}) detected MPs are smaller than soil particles (2.6–2.7 gcm^{-3}) (Hidalgo-Ruz et al. 2012; Claessens et al. 2013; Hidalgo-Ruz et al. 2012; Scheurer and Bigalke 2018; Li et al. 2019). Pressurized fluid extraction (PFE) methods for extraction has several benefits including full automation, high efficiency, and low cost (Fuller and Gautam 2016; Li et al. 2019).

(b) Cleanup is a procedure which is used to eliminate SOM and/or other organic accessories from MPs. Currently applied cleanup procedures include peroxide digestion (H_2O_2), alkaline digestion (NaOH), and acid digestion (HNO_3, H_2SO_4) (Brady and Weil 2000; Zhang et al. 2018; Enders et al. 2017).

(c) Identification of MPs is generally based on the chemical and physical properties of isolated elements in combination with the subsequent extraction and cleanup steps. Consequently, the generally applied identification methods comprise of chemical identification and physical identification (such as mass spectrometry and spectral analysis) (Nor and Obbard 2014; Peng et al. 2017; Shim et al. 2017; Wang et al. 2019; Eriksen et al. 2013; Blasing and Amelung 2018; Paul et al. 2019; Shan et al. 2018; Corradini et al. 2019). MPs are identified by naked eyes based on the precise characteristics (shape, color, or surface texture). Dissecting or stereoscopic microscopy with image software are extensively applied for the smaller (i.e., < 1 mm) MPs in soils (Zhang and Liu 2018; Liu et al. 2018). Since visual sorting exhibits error rates of 20–70%, it is considered to be questionable (Eriksen et al. 2013).

(d) Quantification of MPs in soils includes weighing, counting, instrumental calculation, and mathematical analysis. Counting is the utmost applied quantitative method among them. Unfortunately, counting is a massive assignment (Li et al. 2019), yet weighing is more appropriate for soil samples with high MPs concentrations (Zhang et al. 2018). Mathematical analysis roughly calculates the mass of MPs in the soil. Additionally, some studies quantified MPs concentration in soils by using an instrument (e.g. vis–NIR, TGA-MS) (David et al. 2018; Li et al. 2019; Corradini et al. 2019).

17.7 Effect of MPs on the Soil Properties, Soil Biota and Plants

The presence of MPs could alter soil physicochemical properties such as water holding capacity, bulk density, nutrition contents, and soil structure (Rillig et al. 2019; de Souza Machado et al. 2018; Wan et al. 2018). Soil nature could influence the movement of MPs, and MPs alter the soil properties as soil function and structure as well as microbial composition/diversity, which lead to animal and plant values and current possible concerns for food safety and quality, eventually threatening human health (Table 17.2) (Rillig et al. 2019). MPs can increase water evaporation which may lead to soil drying, with possible negative values for plant

Additionally, fluctuations in the overall structure of soil affect the progression of soil accumulation such as affect root symbionts, including N-fixers and mycorrhiza, which change the microbial community composition in soil (de Souza Machado et al. 2018). Interestingly, plant activities comprehensively depend on the soil biota and their composition/diversity (Rilling et al. 2019). However, considering plants are a main constituent in terrestrial ecosystems and the occurrence of MPs, additional research should be comprised with the various types of plastic particles, soil conditions, and plant species, due to systematically assess the potential associations of MPs contamination to the agricultural soil (Wang et al. 2019). The collective effects of MPs and their related contaminants on the soil microorganisms are very few studied.

17.8 Techniques for Determining the Biodegradability of Polymers

Various techniques could be used in combination for determining the biodegradability of polymers (Raddadi and Fava 2019). These methods include visual observations, molecular weight measurement, physical property evaluation, chemical element analysis, gas formation study, radiolabeling, etc. These techniques are summarized in Table 17.3. The assessment of observable changes in plastics designate degradation which comprises of the formation of holes or cracks, de-fragmentation, roughening of the

Table 17.2 Effect of MPs on soil properties, soil biota and plants

Microplastic effects	Effects on the soil properties	Effects on the soil biota	Effects on the plants
	– Decline soil bulk densities which decrease infiltration resistance for better soil aeration and plant roots – Increase water evaporation – Affect the process of soil aggregation – Effects on soil fertility and nutrients – Increase the concentration of nitrogen, phosphorus and dissolved organic carbon (DOC) in soil – Play role in toxic concentrating chemicals such as heavy metals and hydrophobic organic contaminants on their surface – Stimulate the transport activities of chemicals – Increase the flexibility of organic contaminants in soil.	– Mycorrhiza and N-fixers affect the root symbiosis – Decrease soil enzyme activities (fluorescein diacetate hydrolysis and dehydrogenase), microbial biomass, and functional diversity with increasing concentrations of MPs residue – Influence in mortality – Reduction in growth rate – Increased Zinc exposure to earthworm –Reproduction inhibition – Gut damages – Decrease in body weight – Reproduction inhibition – Modifications in expression of genes	– Reduce the root and shoot biomass – Adversarial effects on wheat reproductive and vegetative growth – Modifications in leaf and root characters and biomass
References	Rilling et al. (2019); Guo et al. (2020); de Souza Machado et al. (2018); Wang et al. (2019)	Wang et al. (2019); Rilling et al. (2019); de Souza Machado et al. (2019); Ng et al. (2018)	Wang et al. (2019); Rilling et al. (2019); Li et al. (2019)

surface, and fluctuations in color or establishment of biofilms on the surface (Ikada 1999). Highly sophisticated observations could be needed to obtain the degradation mechanism information by using transmission optical microscopy, SEM or atomic force microscopy (Alshehrei 2017). Physical properties can be examined by using various methods, as for example: density and viscosity by IIT-GPC, morphology by SEM, amorphous and crystalline region by X-ray diffraction, and melting and glass transition temperature by TG analysis (John and Salim 2020).

FTIR is used to determine the disappearance or formation of functional groups (Arutchelvi et al. 2008; John and Salim 2020). TLC, GCMS and NMR are used to determine the molecular distribution and weight of the degraded intermediates or products (Arutchelvi et al. 2008; John and Salim 2020). CO_2 evolution can be determined by Gas Chromatography (Hoffmann et al. 1997; Raddadi and Fava 2019). Radiolabeling technique is used as substrate for the development of microbial growth with carbon isotope ^{14}C for labeling the carbon in the polymer (Alshehrei 2017). Overall analytical methods

Table 17.3 Techniques for determining the biodegradability of polymers

Methods	Analytical approach	Comments	References
Visual observations	SEM, TEM, AFM	– Applied to designate degradation include the establishment of cracks or holes, roughening of the surface, changes in development or color, and de-fragmentation of biofilms on the surface – Used as indication of any microbial attack by visual changes of parameter	Alshehrei (2017); Ikada (1999)
Molecular weight measurement	TLC, GC, NMR, GC–MS	– Used to evaluate the change of polymer molar mass – Observed the distribution and molecular weight of the degraded intermediates or products	Arutchelvi et al. (2008); John and Salim (2020)
Physical properties evaluation	SEM, HT-GPC, X-ray diffraction, Thermogravimetric (TG) analysis	– Used to measure density, contact angle, melting temperature (T_m), viscosity, glass transition temperature (T_g), amorphous regions, and changes in the crystalline	Witt et al. (2008); John and Salim (2020)
Chemical element analysis	FTIR	– Used to analysis the disappearance or establishment of functional groups	Arutchelvi et al. (2008); John and Salim (2020)
Mechanical features query	Dynamic Mechanical Analysis	– Used to analysis elastic modulus, tensile strength, and elongation at break	Harrison et al. (2018); John and Salim (2020)
Gas formation (carbon dioxide and/or methane) study	GC, Titration with barium hydroxide	– Gives direct information on the polymer to metabolic product and the bioconversion of the carbon backbone	John and Salim (2020); Hoffmann et al. (1997); Raddadi and Fava (2019)
Radiolabeling	Not reported	– Applied as substrate for the development of microbial growth with carbon isotope ^{14}C for labeling the carbon in the polymer – The mineralization is distinguished by the measurement of radioactive gas ($^{14}CO_2$, $^{14}CH_4$) – Limited application due to the cost and difficulties of preparing the radioactive polymer – Need to specific measurement for the disposal and management of the radiolabeled samples	Raddadi and Fava (2019)
Metabolic activity estimation	Protein analysis, ATP assays, and FDA analysis	– Applied to screen microorganisms which may degrade a certain polymer	John and Salim (2020); Arutchelvi et al. (2008);
Other analytical techniques reported recently	RIfS	– Valuable method for the assessment dissimilarity to the physical thickness of biodegradable polymer	Raddadi and Fava (2019)

(continued)

Table 17.3 (continued)

Methods	Analytical approach	Comments	References
		– Applied for the observing enzymatic biodegradation of PCL – Not used in the circumstance of polymers/plastics.	
	EA/IRMS	– Applied for the assessment of carbon stable isotopes ($\delta^{13}C$) – It could be reflected the biodegradation of plastic material by increase of $\delta^{13}C$	

* FTIR: Fourier-transform infrared spectroscopy, GC: gas chromatography, NMR: nuclear magnetic resonance spectroscopy, GC–MS: gas chromatography with mass spectrometry, HT-GPC: high temperature gel permeation chromatography, FDA: fluorescein diacetate analysis, TLC: thin layer chromatography, SEM: scanning electron microscope, AFM: atomic force microscopy, TEM: transmission electron microscopy, RIfS: Reflectometric interference spectroscopy, EA/IRMS: Elemental analyzer/isotope ratio mass spectrometry

are generally applied for assessing the polymer biodegradation/conventional plastics with other techniques such as RIfS and EA/IRMS. On the contrary, EA/IRMS is a technique based on the assessment of carbon stable isotopes ($\delta^{13}C$) that reflect the biodegradation of plastic material by increase of $\delta^{13}C$ values (Raddadi and Fava 2019).

17.9 Factors Affecting Biodegradation of Plastics

Biodegradations of plastic are affected by numerous factors that comprise of microorganism's type, nature of pretreatment, and polymer characteristics. The characteristics of polymer include its mobility, molecular weight, crystallinity, substituent present, and functional groups existing in its structure. Table 17.4 summarizes the various factors that directly affect the biodegradation of MPs. Biodegradations of polymers are affected by two main factors, namely characteristic features of polymer and exposure condition. Exposure circumstances are further classified as biotic and abiotic factors. Microorganisms can enhance the degradation of MPs, which is pronounced implication to combat MP pollution (Devi et al. 2016). For example, *Zalerion maritimum* reveals high removal productivities of MPs; but *Nia vibrissa* showed lower biodegradation productivities under the similar circumstances (Shen et al. 2019).

Abiotic factors such as pH, moisture, and temperature affect the hydrolysis reaction rates through degradation (Iram et al. 2019). The high moisture content and temperature increase in microbial activity and hydrolysis reaction rates (Devi et al. 2016). The kinetics of polymer degradation rely on several environmental factors such as humid air, dry air, a landfill, soil, freshwater, sewage, a marine environment, or a composting environment (Fesseha et al. 2019). Configuration plasticity plays a significant role in polymer biodegradation (Iram et al. 2019). The high plasticity of polymer has high accessible for microbes. Nevertheless, the copolymer biodegradability depends on the comonomer types (Devi et al. 2016). Among the factors affecting biodegradation of plastics, the presence microbial species and plastic properties play a crucial role in MPs biodegradation.

17.10 Strategies to Resolve the Question of MPs

Strategies to resolve the problem of MPs pollution could be focused on the cleanup, source of remediation, and control. Questions of concern are pointed below-

(1) Plastic products should be banned to eliminate the main source of MPs.
(2) Applicability of biodegradable materials. The highest eco-friendly and creative method is to practice biodegradable plastics. Both

Table 17.4 List of several factors affecting biodegradation of plastics

Factors		Remarks	References
Biotic	Microbial species	Presence of suitable microbial species can initiate the biodegradation process	Shen et al. (2019)
	Extracellular enzymes	Different microorganisms are produced extracellular enzymes which may have active sites and may able to biodegrade polymers	Devi et al. (2016); Shen et al. (2019)
	Initial biomass	Initial biomass is one of the key players of plastic biodegradation	Shen et al. (2019)
	Biosurfactants	The biodegradation process is enriched by the accumulation of biosurfactants.	Iram et al. (2019)
Abiotic	Temperature	Changes the temperature increase/decrease the microbial activity and hydrolysis reaction rates	Devi et al. (2016); Iram et al. (2019)
	Moisture	Hydrolytic movement of microorganisms is enlarged with changed of moisture content	Iram et al. (2019)
	Oxygen	Sufficient amount of oxygen should be present in usable form	Kumar et al. (1982)
	pH	pH affects the rate of degradation and alters microbial growth rate	Iram et al. (2019)
	UV radiation	The ultraviolet (UV) radiation acts a significant role in initiating weathering such as mechanical stress with cracking and stiffening	Devi et al. (2016); Glaser (2019)
	Nutrients	Even if the polymer acts as the source of sole carbon, but other vital elements are needed for microbial usage	Kumar et al. (1982)
	Infrared radiation	Near infrared and visible radiation may contribute to the weathering procedure of biodegradation	Glaser (2019)
	Additives, impurities and intermediate products	Biodegradation processes are exposed to inhibit by a variety of agents such as impurities, additives, and intermediate products which can prevent or retard degradation	Kumar et al. (1982); Devi et al. (2016)
Plastic properties	Shape	Polymers are easy to degrade by enzyme in large surface area	Iram et al. (2019)
	Molecular weight	Biodegradability decreases as the molecular weight increases	Iram et al. (2019); Devi et al. (2016)
	Density	Plastics having lower density degrade faster than higher	Fesseha et al. (2019)
	Functional groups	The availability of functional groups increases hydrophobicity	Fesseha et al. (2019)
	Hydrophobicity	Hydrophilic degradation is quicker as compared hydrophobic	Shen et al. (2019)
	Molecular chain branching / Structural complexity	Biodegradation is inhibited by molecular chain branching	Kumar et al. (1982); Fesseha et al. (2019)
	Molecular bonds	Occurrence of simply breakdown bonds as like amide or ester bonds (ester > ether > amide > urethane)	Fesseha et al. (2019); Alshehrei (2017)
	Crystallinity	Polymer crystallinity can play a strong role. An amorphous region of polymer plastic degrades faster than crystalline	Devi et al. (2016); Fesseha et al. (2019)
	Blend	Molecular compositions of plastic material affect biodegradation	Devi et al. (2016)
	tacticity	The stereochemical arrangement of polymers has dramatic effects on the physical properties of the polymer	Devi et al. (2016)
	Comonomers	Accumulation of comonomer into polymer structure improved the abnormality of the polymer chain	Devi et al. (2016)
	Physical form	Nature and physical structure of the polymer (e.g. powder, pellets, films, or fibers)	Fesseha et al. (2019)
	Melting point	Enzyme proficiently degrades at low melting point. But high melting point, polymers are less degraded	Iram et al. (2019)
	Hardness / flexibility	Soft polymers degrade faster than hard ones	Fesseha et al. (2019)

fossil-based and bioplastics can be proficiently degraded. Microbe development and active enzymes can degrade plastics with high value compounds.

(3) Improved reuses recycle and recovery of plastics. Biodegradable/biocompatible plastics as like poly-hydroxyalkanoates (PHA), polylactatide (PLA), and others are commercially accessible which may substitute traditional plastics.

(1) Upgraded separation proficiency at wastewater treatment plant (WWTP). The ability of current WWTP should be promoted to eliminate MPs skillfully and to avoid MPs from the incoming surface, for example, ocean, river, and so on.

(2) Development of bioremediation and cleanup skills. Besides, worldwide collaborations are required to clean up the plastic remains from the ocean, which decrease the main source of ocean MPs. Forthcoming study should be required to develop the approaches for in situ biodegradation of MPs by improving natural attenuation, by adding of microorganisms or by using native microflora.

17.11 Knowledge Gaps and Future Research Challenges

Based on this review, the understanding of MPs in agricultural soil is progressing, but there is a notable deficiency of the appropriate information. Despite progress in the identification, measurement, and isolation of MPs in agricultural soil, there are still numerous scientific difficulties existing. Here, we highlighted some key knowledge gaps that are essential to be followed;

(1) The characteristics of MP pollution in agricultural soil, sustaining mechanisms of toxicity and their possible ecological effects should be broadly studied in the future.

(2) Very few researchers have studied the MPs exposure and their effects on reproductive and vegetative growth of a few plants, whereas more than 200,000 plant species are present worldwide.

(3) There is a need to study how plants can accumulate MPs from the soil.

(4) More scientific studies are needed regarding the effects of the MPs on human health.

(5) Additionally, it should produce high-value compounds, synthetic biology to generate microorganisms from plastic waste by improving circular use of plastics.

(6) Future research should be focused on monomers and oligomers formed from MPs.

17.12 Conclusion

MPs are tiny, heterogeneously mixed plastics that are ubiquitous in arable soils, entering soil environments through sewage irrigation, agricultural mulching films, landfills, and other outlets. Some factors, such as soil characteristics and soil biota, affect the horizontal and vertical movement of MPs in the agricultural soil, and MPs modify the soil structure when they are mixed into soil aggregates. MPs are also capable of interacting with other factors such as impacting soil function and health, and have higher adsorption potential for harmful pollutants, exacerbating soil contamination and increasing antagonistic effects on microorganisms and human health. Additionally, MPs are readily consumed by soil organisms due to their minor size and pass through the food chain; the absorption of MPs cause both physiological and mechanical destruction. MPs also have possible effects on the plant growth where MPs can transport and accumulate in plants. Here, we suggest many areas of the soil MPs for future study, and possible remediation steps are immediately required to moderate the hazard factors by MP contamination. Bioremediation of MP-polluted soil is a promising and environmentally sustainable measure. The application of biodegradable plastics, genetically modified organisms, and changes in industrial degradation facilities should be encouraged to ensure environmental protection and sustainability.

References

Agarwal S (2020) Biodegradable polymers: present opportunities and challenges in providing a microplastic-free environment. Macromol Chem Phys 221(6):2000017. https://doi.org/10.1002/macp.202000017

Ahmed T, Shahid M, Azeem F, Rasul I, Shah AA, Noman M, Hameed A, Manzoor N, Manzoor I, Muhammad S (2018) Biodegradation of plastics: current scenario and future prospects for environmental safety. Environ Sci Pollut Res Int 25(8):7287–7298. https://doi.org/10.1007/s11356-018-1234-9

Alshehrei F (2017) Biodegradation of synthetic and natural plastic by microorganisms. Appl Environ Microbiol 5(1):8–19. https://doi.org/10.12691/jaem-5-1-2

Annemette KMP (2019) A review of standards for biodegradable plastics. Industrial Biotechnology Innovation Centre IBioIC, pp 28

Artham T, Doble M (2008) Biodegradation of aliphatic and aromatic polycarbonates. Macromol Biosci 8:14–24

Arutchelvi J, Sudhakar M, Arkatkar A, Doble M, Bhaduri S, Uppara PV (2008) Biodegradation of polyethylene and polypropylene. Ind J Biotechnol 7 (1):9–22

Auta HS, Emenike CU, Fauziah SH (2017) Distribution and importance of microplastics in the marine environment: a review of the sources, fate, effects, and potential solutions. Environ Int 102:165–176. https://doi.org/10.1016/j.envint.2017.02.013

Awasthi AK, Tan Q, Li J (2020) Biotechnological potential for microplastic waste. Trends Biotechnol 38(11):1196–1199

Blasing M, Amelung W (2018) Plastics in soil: analytical methods and possible sources. Sci Total Environ 612:422–435. https://doi.org/10.1016/j.scitotenv.2017.08.086

Boucher J, Friot D (2017) Primary microplastics in the oceans: a global evaluation of sources. IUCN, Gland, Switzerland

Brady NC, Weil RR (2000) Nature and properties of soils. Macmillan Publishing Company, New York

Claessens M, Cauwenberghe VL, Vandegehuchte MB, Janssen CR (2013) New techniques for the detection of microplastics in sediments and field collected organisms. Mar Pollut Bull 70:227–233

Cole M, Lindeque P, Halsband C, Galloway TS (2011) Microplastics as contaminants in the marine environment: a review. Mar Pollut Bull 62:2588–2597. https://doi.org/10.1016/j.marpolbul.2011.09.025

Corradini F, Bartholomeus H, Lwanga EH, Gertsen H, Geissen V (2019) Predicting soil microplastic concentration using vis-NIR spectroscopy. Sci Total Environ 650:922–932

Cózar Z, Echevarría F, González-Gordillo JI, Irigoien X, Ubeda B, HernándezLeón S, Palma ÁT, Navarro S, García-de-Lomas J, Ruiz A, Fernández-de-Puelles ML, Duarte C (2014) Plastic debris in the open ocean. P Natl Acad Sci USA 111:10239–10244. https://doi.org/10.1073/pnas.1314705111

David J, Steinmetz Z, Kucerik J, Schaumann GE (2018) Quantitative analysis of poly (ethylene terephthalate) microplastics in soil via thermogravimetry-mass spectrometry. Anal Chem 90:8793–8799

de Souza Machado AA, Lau CW, Kloas W, Bergmann J, Bachelier JB, Faltin E (2019) Microplastics can change soil properties and affect plant performance. Environ Sci Technol 53:6044–6052

de Souza Machado AA, Lau CW, Till J, Kloas W, Lehmann A, Becker R (2018) Impacts of microplastics on the soil biophysical environment. Environ Sci Technol 52:9656–9665

Devi RS, Kannan VR, Natarajan K (2016) The role of microbes in plastic degradation. Environ Waste Manage 341-370. https://doi.org/10.1201/b19243-13

Dey U, Mondal NK, Das K, Dutta S (2012) An approach to polymer degradation through microbes. IOSR J Pharm 2:385–388

Diepens NJ, Koelmans AA (2018) Accumulation of plastic debris and associated contaminants in aquatic food webs. Environ Sci Technol 52:8510–8520. https://doi.org/10.1016/j.coesh.2017.10.006

Duis K, Coors A (2016) Microplastics in the aquatic and terrestrial environment: sources (with a specific focus on personal care products), fate and effects. Environ Sci Eur 28:1–25. https://doi.org/10.1186/s12302-015-0069-y

Enders K, Lenz R, Beer S, Stedmon CA (2017) Extraction of microplastic from biota: recommended acidic digestion destroys common plastic polymers. Ices J Mar Sci 74:326–331

Eriksen M, Mason S, Wilson S, Box C, Zellers A, Edwards W (2013) Microplastic pollution in the surface waters of the Laurentian Great Lakes. Mar Pollut Bull 77:177–182

Eskander S, Saleh HEIDM (2017) Environmental science and engineering. In: Kumar P, Gurjar BR (eds) Biodegradation and bioremediation, vol 8. Studium Press, New Delhi, India

Fesseha H, Abebe F (2019) Degradation of plastic materials using microorganisms: a review. Public Health Open J 4(2):57–63. https://doi.org/10.17140/PHOJ-4-136

Fossi MC, Panti C, Guerranti C, Coppola D, Giannetti M, Marsili L (2012) Are baleen whales exposed to the threat of microplastics? A case study of the Mediterranean fin whale (*Balaenoptera physalus*). Mar Pollut Bull 64:2374–2379

Fuller S, Gautam A (2016) A procedure for measuring microplastics using pressurized fluid extraction. Environ Sci Technol 50:5774–5780

Ghosh SK, Pal S, Ray S (2013) Study of microbes having potentiality for biodegradation of plastics. Environ Sci Pollut Res 20(7):4339–4355

Gionfra S (2018) Plastic pollution in soil. Institute for European environmental policy (IEEP). https://ieep.eu/news/isqaper-exploring-plastic-pollution-in-soil. Accessed 15 October 2020

Glaser JA (2019) Biological degradation of polymers in the environment. In: Gomiero A (ed) Plastics in the environment, IntechOpen. https://doi.org/10.5772/intechopen.85124

Gottschalk G (2012) Bacterial metabolism. Springer-Verlag, New York, p 281

Guo JJ, Huang XP, Xiang L, Wang YZ, Li YW, Li H, Cai QY, Mo CH, Wong MH (2020) Source, migration and toxicology of microplastics in soil. Environ Int 137:105263. https://doi.org/10.1016/j.envint.2019.105263

Hale RC, Seeley ME, Guardia LMJ, Mai L, Zeng EY (2020) A global perspective on microplastics. J Geophys Res 125(1):e2018JC014719. https://doi.org/10.1029/2018JC014719

Harrison JP, Boardman C, O'Callaghan K, Delort AM, Song J (2018) Biodegradability standards for carrier bags and plastic films in aquatic environments: a critical review. R Soc Open Sci 5:171792.https://doi.org/10.1098/rsos.171792

Hidalgo-Ruz V, Gutow L, Thompson RC, Thiel M (2012) Microplastics in the marine environment: a review of the methods used for identification and quantification. Environ Sci Technol 46:3060–3075

Hoffmann J, Řezníčèková I, Vaňòková S, Kupec J (1997) Manometric determination of biological degradability of substances poorly soluble in aqueous environments. Int Biodeterior Biodegrad 39:327–332

Horton AA, Dixon SJ (2018) Microplastics: an introduction to environmental transport processes. WIREs Water 5:e1268. https://doi.org/10.1002/wat2.1268

Horton AA, Walton A, Spurgeon DJ, Lahive E, Svendsen C (2017) Microplastics in freshwater and terrestrial environments: evaluating the current understanding to identify the knowledge gaps and future research priorities. Sci Total Environ 586:127–141. https://doi.org/10.1016/j.scitotenv.2017.01.190

Hossain MS, Rahman MS, Uddin MN, Sharifuzzaman SM, Chowdhury SR, Sarker S, Chowdhury MSN (2020) Microplastic contamination in Penaeid shrimp from the Northern Bay of Bengal. Chemosphere 238:124688. https://doi.org/10.1016/j.chemosphere.2019.124688

Hossain MS, Sobhan F, Uddin MN, Sharifuzzaman SM, Chowdhury SR, Sarker S, Chowdhury MS (2019) Microplastics in fishes from the Northern Bay of Bengal. Sci Total Environ 690:821–830. https://doi.org/10.1016/j.scitotenv.2019.07.065

Hurley RR, Nizzetto L (2018) Fate and occurrence of micro(nano)plastics in soils: knowledge gaps and possible risks. Curr Opin Environ Sci Health 1:6–11. https://doi.org/10.1016/j.coesh.2017.10.006

Ikada E (1999) Electron microscope observation of biodegradation of polymers. J Environ Polym Degrad 7:197–201

Iram D, Riaz RA, Iqbal RK (2019) Usage of potential micro-organisms for degradation of plastics. Open J Environ Biol 4(1):007–0015. https://doi.org/10.17352/ojeb.000010

Jain R, Kosta S, Tiwari A (2010) Polyhydroxyalkanoates: a way to sustainable development of bioplastics. Chron Young Sci 1:10–15

Jambeck JR, Geyer R, Wilcox C, Siegler TR, Perryman M, Andrady A, Narayan R, Law KL (2015) Plastic waste inputs from land into the ocean. Science 347:768–771. https://doi.org/10.1126/science.1260352

John N, Salim R (2020) Biodegradation of plastics by microbes-a review. AIP Conf Proc 2263:020003. https://doi.org/10.1063/5.0016828

Kijchavengkul T, Auras R (2008) Perspective: compostability of polymers. Polym Int 57(6):793–804

Kumar GS, Kalpagam V, Nandi US (1982) Biodegradable polymers: prospects, problems, and progress. J Macromol Sci C 22(2):225–260. https://doi.org/10.1080/07366578208081064

Kumar S, Maiti P (2016) Controlled biodegradation of polymers using nanoparticles and its application. RSC Adv 6:67449–67480

Lechner A, Ramler D (2015) The discharge of certain amounts of industrial microplastic from a production plant into the River Danube is permitted by the Austrian legislation. Environ Pollut 200:159–160

Li J, Song Y, Cai Y (2019) Focus topics on microplastics in soil: analytical methods, occurrence, transport, and ecological risks. Environ Pollut 257:113570.https://doi.org/10.1016/j.envpol.2019.113570

Liu MT, Lu SB, Song Y, Lei LL, Hu J, Lv W (2018) Microplastic and mesoplastic pollution in farmland soils in suburbs of Shanghai, China. Environ Pollut 242:855–862

Lucas N, Bieniame C, Belloy C, Queneudec M, Silvestre F, Nava-Saucedo JE (2008) Polymer biodegradation: mechanisms and estimation techniques. Chemosphere 73:429–442

Miranda DDA, Carvalho-Souza GFD (2016) Are we eating plastic-ingesting fish? Mar Pollut Bull 103:109–114

Muthu SS (2014) Roadmap to sustainable textiles and clothing: environmental and social aspects of textiles and clothing supply chain. Springer NY, USA

Ng EL, Lwanga EH, Eldridge SM, Johnston P, Hu HW, Geissen V (2018) An overview of microplastic and nanoplastic pollution in agroecosystems. Sci Total Environ 627:1377–1388

Nizzetto L, Futter M, Langaas S (2016) Are agricultural soils dumps for microplastics of urban origin? Environ Sci Technol 50:10777–10779

Nor NHM, Obbard JP (2014) Microplastics in Singapore's coastal mangrove ecosystems. Mar Pollut Bull 79:278–283

Northcott G, Pantos O (2018) Biodegradation and environmental impact of oxo-degradable and polyhydroxyalkanoate and polylactic acid biodegradable plastics. Northcott Research Consultants Limited (NRC Ltd) and Institute of Environmental Science and Research. https://www.pce.parliament.nz/media/196536/oxo-and-biodegradable-plastics-report-northcott-and-pantos.pdf Accessed 15 October 2020

Ojeda T (2013) Polymers and the environment. In: Yılmaz F (ed) Polymer science. Intech Open, London, pp 1–34. https://doi.org/10.5772/51057

Ouellette JR, Rawn JD (2015) Synthetic polymers. In: Ouellette JR, Rawn JD (eds) Principles of organic chemistry. Elsevier, pp 397–419. https://doi.org/10.1016/B978-0-12-802444-7.00015-X

Pathak VM, Navneet, (2017) Review on the current status of polymer degradation: a microbial approach. Bioresour Bioprocess 4:15

Paul A, Wander L, Becker R, Goedecke C, Braun U (2019) High-throughput NIR spectroscopic (NIRS) detection of microplastics in soil. Environ Sci Pollut Res Int 26:7364–7374

Peng JP, Wang JD, Cai LQ (2017) Current understanding of microplastics in the environment: occurrence, fate, risks, and what we should do. Integr Environ Asses 13:476–482

PlasticsEurope (2018) Plastics—the facts 2018: an analysis of European plastics production. Demand and waste data. pp 60. https://www.plasticseurope.org/application/files/6315/4510/9658/Plastics_the_facts_2018_AF_web.pdf

Raaman N, Rajitha N, Jayshree A, Jegadeesh R (2012) Biodegradation of plastic by Aspergillus spp. isolated from polythene polluted sites around Chennai. J Acad Ind Res 1:313–316

Raddadi N, Fava F (2019) Biodegradation of oil-based plastics in the environment: existing knowledge and needs of research and innovation. Sci Total Environ 679:148–158

Raziyafathima M, Praseetha PK, Rimal IRS (2016) Microbial degradation of plastic waste: a review. J Pharm Chem Biol Sci 4(2):231–242

Rillig MC, Lehmann A, de Souza Machado AA, Yang G (2019) Microplastic effects on plants. New Phytol 223 (3):1066–1070

Rocha-Santos T, Duarte AC (2015) A critical overview of the analytical approaches to the occurrence, the fate and the behavior of microplastics in the environment. Trend Anal Chem 65:47–53. https://doi.org/10.1016/j.trac.2014.10.011

Sarker A, Deepo DM, Nandi R (2020) A review of microplastics pollution in the soil and terrestrial ecosystems: a global and Bangladesh perspective. Sci Total Environ 733:139296.https://doi.org/10.1016/j.scitotenv.2020.139296

Scheurer M, Bigalke M (2018) Microplastics in swiss floodplain soils. Environ Sci Technol 52:3591–3598

Schink B, Janssen PH, Frings J (1992) Microbial degradation of natural and of new synthetic polymers. FEMS Microbiol Rev 103(2/4):311–316

Shan J, Zhao J, Liu L, Zhang Y, Wang X, Wu F (2018) A novel way to rapidly monitor microplastics in soil by hyperspectral imaging technology and chemometrics. Environ Pollut 238:121–129

Shen M, Zhu Y, Zhang Y, Zeng G, Wen X, Yi H (2019) Micro(nano)plastics: unignorable vectors for organisms. Mar Pollut Bull 139:328–331

Shim WJ, Hong SH, Eo SE (2017) Identification methods in microplastic analysis: a review. Anal Methods 9:1384–1391

Shrivastava A (2018) Introduction to plastics engineering, In: Shrivastava A (ed) Plastics design library, introduction to plastics engineering, William Andrew Publishing, pp 1–16. https://doi.org/10.1016/B978-0-323-39500-7.00001-0

Siracusa V (2019) Microbial degradation of synthetic biopolymers waste. Polymers (Basel) 11(6). https://doi.org/10.3390/polym11061066

Sivan A, Szanto M, Pavlov V (2006) Biofilm development of the polyethylene degrading bacterium *Rhodococcus ruber*. Appl Microbiol Biotechnol 72 (2):346–352

Song JJ, Yoon SC, Yu SM, Lenz RW (1998) Differential scanning calorimetric study of poly (3-hydroxyoctanoate) inclusions in bacterial cells. Int J Biol Macromol 23:165–173

Steinmetz Z, Wollmann C, Schaefer M, Buchmann C, David J, Troger J, Muñoz K, Fror O, Schaumann G (2016) Plastic mulching in agriculture. trading short-term agronomic benefits for long-term soil degradation? Sci Total Environ 15(550):690–705. https://doi.org/10.1016/j.scitotenv.2016.01.153

Sutherland WJ, Clout M, Isabelle M, Daszak CP (2010) A horizon scan of global conservation issues for 2010. Trends Ecol Evol 25(1):1–7. https://doi.org/10.1016/j.tree.2009.10.003

Thompson (2015) Microplastics in the marine environment: sources, consequences and solutions. In: Melanie B, Lars G, Michael K (eds) Marine anthropogenic litter. Springer International Publishing, pp 185–200. https://doi.org/10.1007/978-3-319-16510-3_7

Tokiwa Y, Calabia BP, Ugwu CU, Aiba S (2009) Biodegradability of plastics. Int J Mol Sci 10:3722–3742

UNEP (2016) Marine plastic debris and microplastics—global lessons and research to inspire action and guide policy change. https://wedocs.unep.org/handle/20.500.11822/7720 Accessed 15 October 2020

Verma NK, Khanna SK, Kapila B (2004) Comprehensive chemistry XII. Laxmi Publications, New Delhi, India

Wan Y, Wu C, Xue Q, Hui X (2018) Effects of plastic contamination on water evaporation and desiccation cracking in soil. Sci Total Environ 654:576–582

Wang W, Ge J, Yu X, Li H (2019) Environmental fate and impacts of microplastics in soil ecosystems: Progress and perspective. Sci Total Environ 708:134841.https://doi.org/10.1016/j.scitotenv.2019.134841

Witt U, Einig T, Yamamoto M, Kleeberg I, Deckwer WD, Müller RJ (2008) Biodegradation of aliphatic–aromatic copolyesters: evaluation of the final biodegradability and ecotoxicological impact of degradation intermediates. Chemosphere 44:289–299

Wu W, Yang J, Criddle CS (2016) Microplastics pollution and reduction strategies. Front Environ Sci Eng 11:6. https://doi.org/10.1007/s11783-017-0897-7

Zhang GS, Hu XB, Zhang XX, Li J (2015) Effects of plastic mulch and crop rotation on soil physical properties in rain-fed vegetable production in the mid-Yunnan plateau, China. Soil till Res 145:111–117. https://doi.org/10.1016/j.still.2014.09.010

Zhang GS, Liu YF (2018) The distribution of microplastics in soil aggregate fractions in southwestern China. Sci Total Environ 642:12–20

Zhang S, Yang X, Gertsen H, Peters P, Salanki T, Geissen V (2018) A simple method for the extraction and identification of light density microplastics from soil. Sci Total Environ 616:1056–1065

Zhu F, Zhu C, Wang C, Gu C (2019) Occurrence and ecological impacts of microplastics in soil systems: a review. Bull Environ Contam Toxicol 102(6):741–749. https://doi.org/10.1007/s00128-019-02623-z

Ziajahromi S, Neale PA, Leusch DDL (2016) Wastewater treatment plant effluent as a source of microplastics: review of the fate, chemical interactions and potential risks to aquatic organisms. Water Sci Technol 74:2253–2269

18 Bioremediation of Tannery Effluent Contaminated Soil: A Green Approach

Tamil Selvan Silambarasan, Manickam Dakshinamoorthi Balakumaran, Subramaniyam Suresh, Velramar Balasubramanian, Muthusamy Sanjivkumar, Balasundaram Sendilkumar, and Ramamurthy Dhandapani

Abstract

Water contamination is an important concern for our daily life and the earth's environment, especially terrestrial and marine systems. The largest toxic wastes are released from tannery industries. A greater concentration of heavy metals is found in tannery effluents such as chromium, lead, cadmium, nickel, copper and zinc. However, the cost and performance of effluent treatment are high. Therefore, the effluent is directly released by the tannery industries to water reservoirs and agricultural areas. Algal technology is a sustainable method to remove heavy metals and other contaminants released from the tannery industries. Algal and plant-based effluent treatment is more effective and highly economical. The microalgae are used to remove hazardous chemicals and heavy metals from effluent contaminated soil. This chapter provides the complete information about the advantages of using microalgae and plant-based phytoremediation for the decontamination of effluents released from tannery industries.

Keywords

Microalgae · Cyanobacteria · Bioadsorption · Plant · Heavy metals · Phytoremediation · Tannery

18.1 Introduction

Nature has given the earth four spheres: biosphere, lithosphere, hydrosphere and atmosphere. These all domains are essential for sustaining and preserving a better environment. Unnatural amounts of chemical contaminants have been generated by rapid industrialization and

T. S. Silambarasan (✉) · B. Sendilkumar
Department of Microbiology, School of Allied Health Sciences, Vinayaka Missions Research Foundation (Deemed to be University), Salem, Tamil Nadu 636308, India

M. D. Balakumaran
Department of Biotechnology, Dwaraka Doss Goverdhan Doss Vaishnav College, Chennai, Tamil Nadu 600106, India

S. Suresh
Department of Biotechnology, College of Science and Humanities, SRM Institute of Science and Technology, Ramapuram Campus, Chennai, Tamil Nadu 600089, India

V. Balasubramanian
Department of Biotechnology, Amity Institute of Biotechnology (AIB), Amity University, Raipur, Chhattisgarh 493225, India

M. Sanjivkumar
Department of Microbiology, K.R. College of Arts and Science, K.R. Nagar, Kovilpatti, Tamil Nadu 628503, India

T. S. Silambarasan · R. Dhandapani
Fermentation Technology Laboratory, Department of Microbiology, School of Biosciences, Periyar University, Salem, Tamil Nadu 636011, India

J. A. Malik (ed.), *Advances in Bioremediation and Phytoremediation for Sustainable Soil Management*,
https://doi.org/10.1007/978-3-030-89984-4_18

demographic movements from rural areas to urban centres. Consequently, all the spheres became contaminated due to human activities, population explosion and industrial development (Tamil Selvan et al. 2020; Baroni et al. 2007). There seem to be two essential routes of toxic metal penetration throughout the ecosystem: natural sources—that include volcanic explosions, deforestation, sea salt and anthropogenic sources—include mining and processing, steel manufacturing, paint and colouring industrial sectors, and leather-based tanneries. Many toxic metals and other substances are being discharged into the ecosystem through these sources. Metal ions are becoming the main health hazards if exceeding a permissible limit. Even though the metal concentrations do not fairly cross certain levels, biomagnification occurs, and related chronic poisoning is recognized as immeasurable inside the biological systems. There are many different types of toxic metals which include arsenic (As), chromium (Cr), cadmium (Cd), mercury (Hg), cobalt (Co), gold (Au), lead (Pb), palladium (Pd), platinum (Pt), nickel (Ni), ruthenium (Ru), silver (Ag), zinc (Zn), etc., present in the tannery waste effluents. Industrial development could be correlated to a large degree of environmental pollution. Primary pollutants from leather manufacturing industries are considered to contain main toxic metals (Shokrzadeh and Saeedi Saravi 2009; Kermani et al. 2010).

Heavy metals are one of the harmful materials achieving dangerously high levels. Pollution by toxic metals is one of the world's most prevalent environmental issues. Various companies deliver and release waste containing toxic metals such as the mines of polymetallic materials, metal wrapping up, energy and resource manufacturing, fertilizer and pesticide manufacturers and technologies, metalworking, alloy and iron, electroplating, electroosmosis, leather working, photography, electronic component and metal fabrications. Metals, as a side effect, have become a resource and contribute to significant industrial pollution that directly affects human health and the environment (Joshi 2017; Lakherwal 2014).

A waste product is described as a material that partially results from living organisms throughout the ecosystem, which has a harmful ecological impact. The waste product includes a wide variety of substances, ranging from a vast quantity of resources which depreciate the habitats by incorporating harmful compounds which may be neurotoxic or excitotoxic compounds (Alkorta et al. 2004). Toxins may be categorized on a larger scale as either exotoxins, emitted by organisms, or endotoxins, released mostly when bacteria are lysed. Heavy metal refers to any metallic organic compound which is significantly noxious and volatile or hazardous at small quantities. They are mainly the variants of mercury, nickel, cadmium, arsenic, chromium and lead. Heavy metals cannot be either destroyed or deteriorated and reach human bodies to a limited extent through the food chain, consuming air and water. Other heavy metals like copper, selenium and zinc are critical for the sustenance of the metabolic rate and can cause toxicity at large concentrations, even though as a significant agent. Increasing anthropological activities introduces increasingly toxic metal ions to a material environment that disrupts the biodiversity. Metals of increased concentrations have a possible toxic influence on total plant growth and metabolism. Small and heavy steel ion effluents from textile, leather, tannery, electroplating, colouring, dyes, paints and other metalworking and processing industries produce significant quantities of harmful metal ions as wastes. But on the other side, tannery contaminants are categorized among all toxic wastes as the largest pollutants (Ali et al. 2013).

Nearly 90% of chromium-based tanning agents are used by the leather industry to process raw skins/hides into leather. Metals such as aluminium, titanium, copper and zirconium are commonly used for multiple end uses, apart from chromium. Therefore, during tanning procedures, waste produces higher heavy metal concentrations. Several techniques have also been established to extract the heavy metals from the effluent for traditional effluent treatment before treating them. Phytoremediation has become one of the eco-friendly methods used to isolate soil and wastewater from heavy metals. It is widely used for the remediation of unfertilized fields into fertile farm lands. In addition, metal-absorbed plants are being used in the leather industry for various uses, such as tanning and preservative agents. Likewise, algae and fungi are being employed to remove heavy metals from tannery residue and could be used as auxiliary additives for metal polysaccharides throughout post-tanning phases.

One of the fast-growing conventional industries with high environmental risks is the leather industry. The processing of leather deals with the transfer of putrescible material to non-putrescible material, such as skin/hide. In order to achieve the final special characteristics of leather, a number of chemical processes and physical procedures are carried out throughout the process. Due to its special leather attributes, chrome tanning is commonly practised worldwide. However, because of its strong cancer-causing aspect, Cr(VI) generation in the waste stream poses a significant warning to geotechnical engineering (carcinogenic). Vegetable tanning is done in addition to chrome due to its non-toxic virtue; it is seen as an alternative to chrome tanning. Vegetable-tanned leathers do not, however, complement the perfect chrome-tanned leathers. Many other tanning products are taken in with their benefits and costs, such as phosphonium, oxazolidine and silica-based tannings. Researchers are also focusing on the production of natural plant-derived products and tanning microbes such as dialdehyde starch, dialdehyde cellulose, dialdehyde alginic acid, dialdehyde pectin and fungal scleraldehyde (*Schizophyllum commune*).

The tannery industries produce vast quantities of wastewater and loops that contain high organic matter and extremely harmful materials at significantly higher levels than natural environment. In recent years, landfills of polluted water or toxic waste tannery have become a usual place contributing to contaminated water and soil. It includes high concentrations of organic toxicants, which would be affecting plants and quickly absorbed by the soil. Correspondingly, high amounts of heavy metals could almost persistently accumulate in the soils and can pose a diagnosable disorder for crop growth and human health (Saranya and Shanthakumar 2019). Tannery sediment is characterized by excellent alkaline composition consisting of residue leather slurry incorporating heavy metal, high oxygen absorption, a high salt and chemical solid matrix, etc.

The leather processing involves significant and unintended industrial area growth; the living ecosystem has indeed been affected. As during dyeing processes, animal skins are treated with various additives and processed with hexavalent chromium or tannins, salts and leather-making colours to extract the fat, skin, hair and other undesirable substances. Untreated wastewater pollutants arising from this process are often contaminated with significant amounts of metals and other toxic pollutants. The specific chromium sulphate is used as tanning agent which generates Cr- concentrated wastewater. While Cr frequently occurs in its trivalent form in the leather industry, Cr^{6+} is typically converted to more hazardous hexavalent form through its disposal. Consumption of Cr^{6+} is related to cancer, renal injury, allergies and asthma in humans (Aliabadi et al. 2006). Cadmium (Cd) and lead (Pb) are commonly used for the tannery industry dyes with bright colours. Cadmium sulphide is being used as a yellow colour pigment or red pigment in combination with cadmium selenide. Accumulation of Cd in tissues is associated with cardiovascular, renal, neurological, muscle and sympathetic and parasympathetic nervous disorders. Several other toxic and harmful chemicals have been used in a variety of different leather processes including ammonium salt, sulphide, chlorobenzene, formic acid, sodium hydrochloride, sodium hydroxide and sulphuric acid. Recovery of metals through tanning is significantly impacted through the processing of toxic pollutants, capital investment and operating expenditure and is incompatible after treatment enactment. Consequently, in several less developed nations, pollutants are removed in tannery wastewater, particularly among those with stringent environmental standards. For the last seven decades, contaminants from this sector have been drained directly into the soil and groundwater ecosystem bodies (Alvarez-Ayuso et al. 2007).

Significant solid wastes are released from the tanning industries like leather particles and animal fat, skin, fur, etc. However, a substantial quantity of waste is also deposited into surrounding residential areas throughout the discharge of a leather processing wastes, causing extreme episodes of pollution during the rainy season. The release through the marine ecosystem of tannery effluents and related toxicants has been of significant public health and environmental concern. Untreated wastewater pollution affects nearly two million civilians, whereas several parameters, viz., the texture, soil chemical properties and ecological provenance, affect metal mobility; Cr typically had higher mobility, while Cd, Ni and Pb are having low mobility. They can cause soil pollution and contaminate agriculture and drinking water, thus adversely affecting water and crops' quality. Therefore, Cr is more mobile in the soil and more likely to enter groundwater and pollute the ecosystem. Conferring to the significant health and environmental issues faced by the tannery industry, groundwater and soil seem to be keep suffering through ecological pollutants from tannery wastes (Chowdhury et al. 2015; Chan et al. 2014). A brief overview on the different biological processes used in the handling of leather wastes and tannery sludge is discussed in this chapter.

18.2 Tannery Industrial Process for Leather Production

The leather processing involves the treatment of several chemicals for the execution of the hide hollowing process, termed as bleaching. The chemicals are discharged through three major tannery wastewater groups:

1. Chromium-free waste effluents, viz., household-filling, deliming, bathing, prey waters and contentious machinery, are containing high pH iron sulphate.
2. Higher concentration chromium waste effluents, viz., sammying, re-tanning and tanyard tanning.
3. Low concentration chromium waste effluents, viz., soaking, fat-liquoring and dyeing.

These waste effluents are directly discharged to the ecosystem, thereby contaminating the natural composition of varied environments (Tamil Selvan et al. 2020). The overall leather manufacturing process involves treatment, sweetening, liming, dehairing, deliming, race baiting, sorting, degrading and tanning. All these processes are carried by the use of chemicals, viz., chromium sulphite, sodium sulphide, calcium salts, sodium bicarbonate, acids, hydrogen peroxide, fat, ammonium salts, liquor and other substances which release organic chlorinated phenols, organic Cr(VI) and other toxic chemicals directly into the surroundings (Chowdhury et al. 2015; Chan et al. 2014).

Fluoridated phenols and chromium are appearing to be directly related to tanning waste. The fluoridated phenols such as 3,5, dichlorophenol are incredibly toxic and affect the cellular organizations of plants and animals including humans. Extensive exposure to such compounds leads to olfactory paralysis, intense lung and eye inflammation, lung necrosis and loss of consciousness in humans. Another toxic disposal from the tanning plants is chromium-contaminated sediment generated and combined with several other wastes and toxic substances as a by-product of waste management. As a raw resource at various levels, leather processing involves significant quantities of chemicals and generates harmful chemicals into the atmosphere.

18.3 Sources of Soil Contamination from Tannery Effluents

Soil is essential both as development immutability and for improving and enhancing the sustenance and life of animals and plants. In numerous cases, its endurance and configuration are influenced by the composition of certain pollutants occurring from the dumping of untreated wastewater, particularly of the spill from tannery, including inorganic type toxins. Leather factories are the primary sources of heavy metal pollution for a local agricultural soil when used for cultivation. Continued drainage sediment may cause accumulation of toxic metals including Cd, Zn, Cr, Ni, Pb and Mn in the surface soils (Bai and Abraham 2001). Later, they may be released into water and soil situations accessible to plants, due to decreased soil retention capability. Soil contamination and toxicity by the accumulation of heavy metals cause harmful effects on plants and threats to human health. Cr(VI) is one of the toxic contaminants released from the tannery effluents which is neurotoxic, carcinogenic, highly poisonous and harmful in higher exposures, mostly persists throughout the ecosystem and is less capable of disintegration (Bahafid et al. 2013; Bai and Abraham 2003).

Tannery factories seem to be the main predominant pollutants in the agricultural soil community as the effluents are used for irrigation purposes. Sustained drainage raw sewage usage could induce the concentration of heavy toxic metals like Cd, Zn, Cr, Ni, Pb and Mn and liberation into the soil. Plants and micro-organisms uptake these toxic metals leading to decrease in soil metal ion storing ability. The pollution by cytotoxic and other noxious elements in soil reflects a harmful impact not only on plants but also for humans. The chromium concentration gets very substantial in soils as the agricultural land is irrigated by pollutants abundant in tannery effluents, including a large chromium sulphur composition (Arshad et al. 2008; Ali et al. 2013).

18.4 Effects of Chromium from Tannery Effluents

Chromium (Cr) is a significant element excreted as waste from leather processing industries used throughout chrome tanning leather. The tanning industries that generally employ fundamental chromium sulphate for the tanning phase have been the significant factor in generating strong Cr flow to a natural ecosystem. A significant proportion of the chemicals used for the manufacturing of leather are not directly consumed but are released to the atmosphere. The wastewater derived from the tannery industry includes different Cr [Cr(VI) and Cr(III)] species which are interchangeable depending on pH, organic material, oxidation reaction material and decreasing wastewater prediction performance (Apte et al. 2005).

18.4.1 Effects on Ecosystem

Industrial tannery waste is a huge pollution issue and influences the water sources by adulterating them and reducing their consistency (Visconti et al. 2020). Tannery waste sludge also contains toxic chemicals that are the critical issue of stress as they prove to be discordant with the microbial degradation process. This mainly affects depletion of oxygen concentration which would be harmful to aquatic lives and facilitates the anaerobic process leading to noxious fume emission (Azimi et al. 2016; Barakat 2011; Vangronsveld et al. 2009).

18.4.2 Effects on Plant Growth

Wastewater discharged in agricultural land impacts soil fertility; the toxic substances are biomagnified within the food web at various trophic stages. Tannery sewage induces phytotoxic effect with a high concentration of heavy metals inducing stress in plants such as salinity stress that affect different metabolic processes to reduce plant reproductive development and consequently affect its respiratory and photosynthetic process. Moreover, the accumulation depends on plant species, the components, biocompatibility, pH, electrostatic interaction, biological oxygen demand, temperature and root migration. Tannery wastewater influences maize, soya bean and wheat seed germination. The wide range of salts including chlorides and sulphates in tannery wastewater can hinder adequate cultivation progress in maize as they prevent seed germination and seedling growth, with even low adulteration of pollutants due to excess chromium, dissolved solids, chlorides, sulphides, high BOD and COD. More than 80% of wastewater compositions appeared harmful for growing plants in both vegetation and propagation phases. Cr(VI) affects plant enzyme amylase (that also plays a vital role in germinating seeds via substitute hydrolysis of starch and transpiration rate) and significantly reduces plants' growing period. Chromium throughout groundwater is typically considered to have the chronic toxicity. Tannery sludge consists of hair, sorting things, fragments, hide/skin trim, leather trimmings, dust buffing, leather finishing residues, general plant waste, affecting the meristematic process and reducing germination in plants (Roychowdhury et al. 2019; Pogrzeba et al. 2017).

18.4.3 Effects on Health of Humans and Animals

Chromium is however an important metal which contributes to animal and human gluconeogenesis, but the Cr(VI) form is highly toxic, mutagenic and carcinogenic. Also, Cr(VI) is highly mobile in most environments, primarily because of its solubility and its tremendous accessibility, affecting the ecosystem. Cr(VI) is very accessible to living organisms through multiplexing mechanisms such as ingestion, respiration and absorption. Chronic exposure to tannery employees between five months to 14 years would be a significant risk factor in developing genomic disruption disorders because of the misinformation on harmful tannery-related symptoms, such as dizziness, headache, inflammation of the eyes, skin or lungs, allergic reactions, liver toxicity, kidney, or nervous systems or failure triggered by oxygen deprivation, or long-term diseases, such as occupational asthma, ulcers and bronchitis, are temporarily impaired (Tumolo et al. 2020; Rambabu et al. 2020).

18.5 Methods to Remove Chromium

Cr is engrossed to surface soil and groundwater by pollution from effluent, mainly from leather processing factories, etc. Biosorption/bioaccumulation and enzyme treatment are the methods through which microbes associate with heavy metal substances and pledge their degradation and restoration. Microbial apparatuses such as fungi, bacteria and algae have increasingly been shown to be adsorbent substances and remove toxic metals. The polluted metal community responds to the hazardous accumulation of heavy metals and becomes responsive to metals (Elahi et al. 2020).

18.5.1 Phytoremediation Mechanism of Chromium

The specific mechanism of metabolism takes places in biomass-making layers of cell walls. The bioadsorption technique has been a vital aspect of the whole physicochemical pathway of biosorption. Atomic associations or physiochemical bioadsorption could be used in their adsorption system. Ionic binding affinities even contribute significantly throughout the metal bioadsorption in microalgal cell walls (carboxyles, amines, hydroxyles, phosphates and sulphuryl groups). Adsorption is a fast process, although the precipitation is rapid and based on energy. The nature of metals within the microalgal cells could increase the quality of the microalgae accumulation, detoxification and/or efflux (Caporale et al. 2018; Capuana 2011).

Heavy metals could penetrate the cells to provide a physiological impact on cell development. The evolutionary multiplication typically regulates them through the algal cytoplasmic membrane. In comparison with the passive chemical gradient pathway for the adsorption of heavy metals, the second category of absorption mechanism is extremely specific, unstable, isoform and depending on ATP (Barbosa et al. 2015). Both the processes are triggered during particular situations and the need for specialized mechanisms. Because the cell wall contacts metal ions, a complex is formed which acts as a prerequisite for all the microbes to take up metals. When the membrane is straightened, metal is brought into the microalgal cell via cytoplasmic membrane and is again transported into chloroplasts. The interaction between heavy metals like Cd^{2+} and Co^{2+} with Fe^{2+}, Ni^{2+}, Zn^{2+} and Mg^{2+} with biological compounds inhibits the role of both and contributes to oxidative cell pressure (Capuana 2011) (Figs. 18.1 and 18.2).

18.5.2 Micro-organisms for the Reduction of Chromium

Because of their versatile abilities, the role of micro-organisms is gaining significance in various applications. Researchers have concentrated on recognizing the significance and requirements of the use of possibly the best micro-organisms in the various fields of science. A number of micro-organisms are being found to be capable of eliminating Cr(VI) emissions. The micro-organisms which maintain similar characteristics and show improvement for Cr pollution reduction or removal are isolated from polluted and unpolluted Cr(VI) environments.

18.5.2.1 Chromium Reduction by Algae

Algae play a significant role in recovering the environment from different toxins to its original form. Further, the phycoremediation strategy is modelled by utilizing algae that reduce and transform harmless contaminants from the atmosphere. The utilization of micro- or macroalgae in the absorption or bioaugmentation of pollutants, including hazardous waste minerals, is regarded as phycoremediation. Algae are extraordinarily adaptable and therefore can emerge in different climates heterotrophically, autotrophically or mixotrophically (Ozyigit et al. 2020). Algae serve a crucial responsibility for reducing the metal concentration in reservoirs

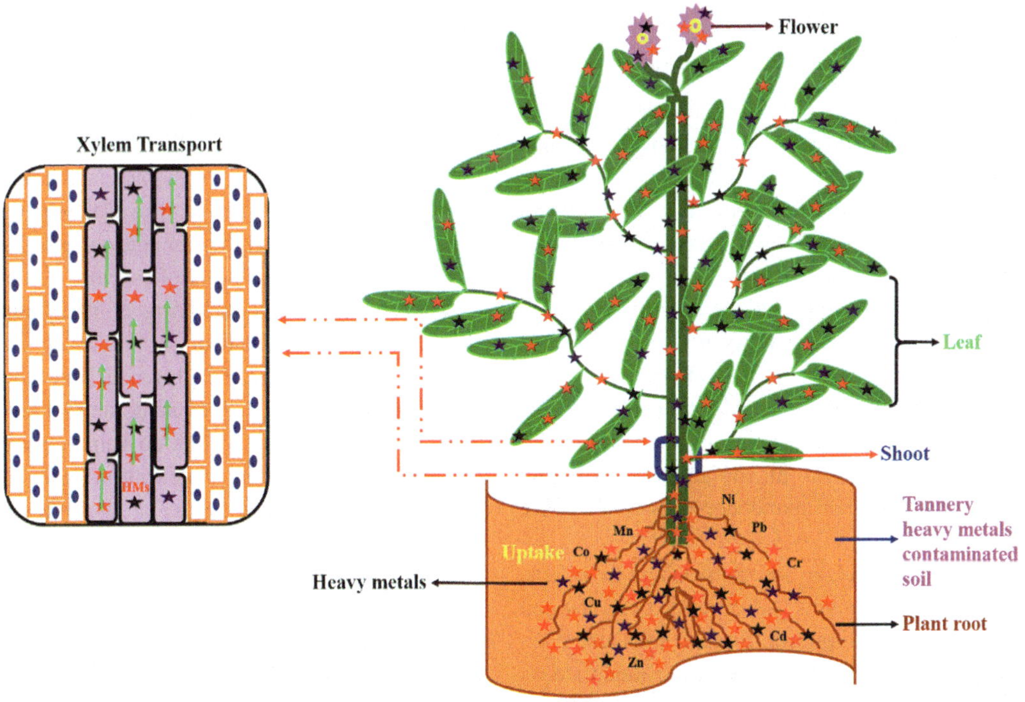

Fig. 18.1 Schematic representation of mechanism of phytoremediation of heavy metal ions by plants

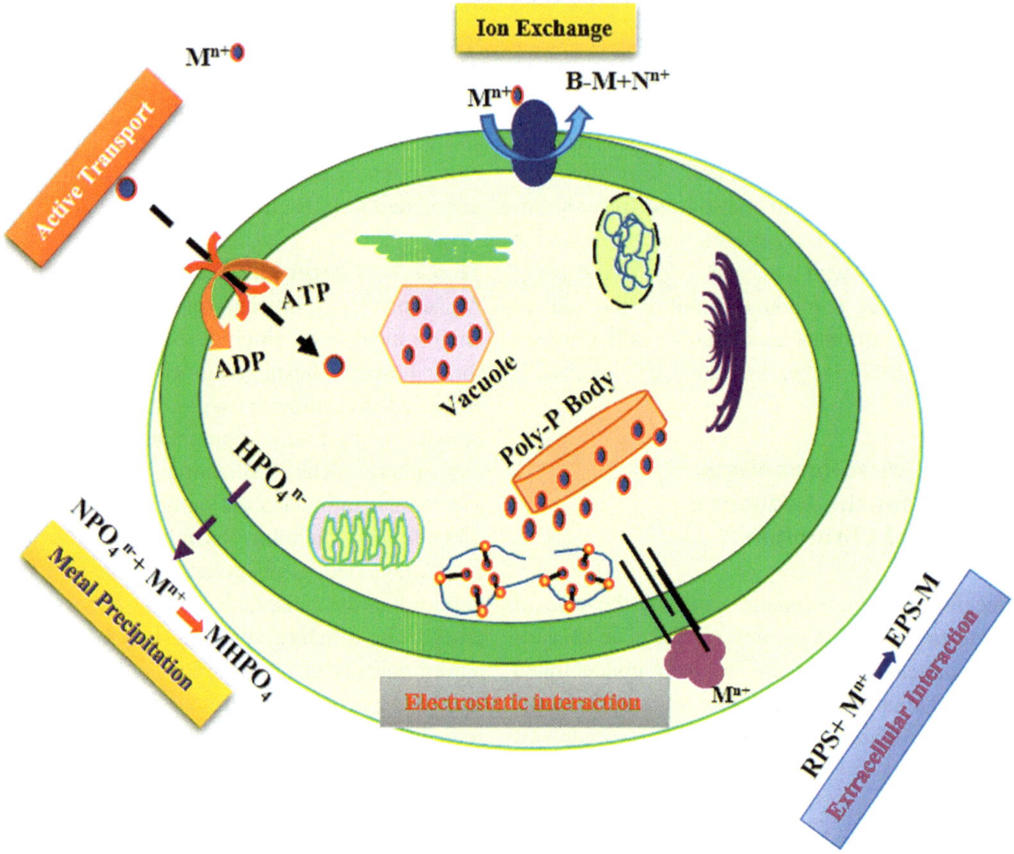

Fig. 18.2 Schematic representation of various mechanisms of heavy metal removal by microalgae

and coastlines throughout ecological systems. They can deteriorate or absorb dangerous heavy and toxic chemicals from the atmosphere, including polyphenols, hydrocarbons, herbicides and carbon tetrachloride, due to increased exposure levels than lake environmental soil. Mixotrophic contaminant-degrading algae are exceptional vehicles for mitigation and absorption of oils and lubricants. Algae absorb CO_2 and generate O_2 by photosynthesis process and enhance BOD amount of contaminated soil and productive nutrient abundance (Lin et al. 2020; Rai et al. 2020).

Active microalgae absorb metals in two steps. The first phase entails "adsorption" upon the cell membrane (adsorption). Similar ions are then transferred rapidly to a membrane identified as biosorption unit. The second stage is cell-dependent which includes the absorption or intracellular penetration of toxic substances. Different metals, such as Cu, Pb, Cd, Hg, Co, Mg, Zn, Ti and Ni, are stored and execute key responsibilities in algal polyphosphate bodies, i.e. conservation and detoxifying of metal ions (Ubando et al. 2020; DalCorso et al. 2019; Das et al. 2016). As they have played a significant role in the sequestration of contaminants in algal cells, they are also essential combinations of multipurpose polymeric materials. Algae are recognized for efficient soils or groundwater removal of organic matter by absorption and then processed as biomass. Over the period, biomass can break down to release nitrogen back into the earth or atmosphere (N_2O) that can be recovered or lost (Agarwal et al. 2020). Algae play a significant role in pH correction, slurry reduction and total dissolved solids reduction, while various methods or steps are required for alternative treatments. The algae or cyanobacteria are susceptible to sunlight, alkalinity, humidity, nutrient components, chemical and physical modifications, and hence cyanobacteria have largely been used for financial and low-keeping remediation technologies at hazardous and contaminated sites. Cyanobacteria are used effectively for the remediation of milk liquid waste by transferring water and nutrients into biomass. Further work remains required to apply cyanobacteria to phytoremediation of sewage, whether in natural aquatic environments or in polluted situations (Leong et al. 2020) (Table 18.1).

Additional forms of microalgae like *Oscillatoria salina, Aphanocapsa* sp., *Synechococcus* sp. and *Plectonema terebrans* are known to decontaminate the marine-polluted habitats and have effectively been shown to biodegrade oil spills in different world regions. Several algae forms like *Stigeoclonium tenue, Anabaena inaequalis, Chlorella* sp., *Westiellopsis prolifica* and *Synechococcus* sp. are metal tolerant and thus oftenly used to remove hazardous contaminants. The Zn and Cd accumulation was documented with microalga *Scenedesmus obliquus*, and increased uptake with increased phosphoric concentration was also demonstrated in the media, where it was found that Se accumulation was inhibited. *Oedogonium rivulare* and *Cladophora glomerata* also accumulate metals, viz., Co, Cu, Cd and Pb. A filamentous freshwater alga, *Spirogyra hatillensis*, exhibited continuous Cr, Mn, Fe and Ni absorption from the contaminated sites.

Phycoremediation shows a benefit over other chemical processes, since it is simple and inexpensive technique to extract algal mass from the treated effluent (Chojnack et al. 2005; Doshi et al. 2007; Monteiro et al. 2009). Algal cells are capable to remove Cr(VI) from the adulterated settings. Its first step is to bond the Cr(VI) electrons to a algal species/surface of cells. This process continues efficiently which is independent of the metabolism of the cells. The second stage of intracellular metal absorption occurs from combined proliferation and phytoremediation activities of a surface. This process needs metabolic energy from the cells and is much slow (Sen and Ghosh Dastidar 2010). *Chlorella miniata* (Han et al. 2007) and *Cladophora albida,* green algae (Deng et al. 2009), are known to degrade Cr(VI) from the tannery effluent and biosorbed/bioreduced to Cr (III) on the microalgae biomass. The various forms of microbial substance algae have several benefits, like economic regeneration, the potential recovery for metals, lower organic waste generation, high-quality algal product in dissolved wastes and a large area proportion (Gupta et al. 2009).

The phototropic and autotrophic algae require little growth nutrients produced by large biomass associated with the biosorbents of other microbes. These adsorbents are used for multitude removal of contaminants. Algae biomass can be used for bioremediation of heavy metal-contaminated sludge through adsorbent or penetration through membranes. Phycoremediation includes use of different forms of algae and cyanobacteria to absorb or remove toxic substances and for revival of heavy metals. Algae contain different surface chemistry substituents, including carboxyl group, hydroxyl group, phosphate and amide group, which function as metal chelating sites. In order to remove cadmium (Cd^{2+}), copper (Cu^{2+}) and lead (Pb^{2+}) compounds which have been separated through alkaline phase in varying environments, algal dried biomass from *Chlorella vulgaris* is known to have the greatest outcomes. The findings showed that *Chlorella vulgaris* proliferation is highly effective for the removal of cadmium (Cd^{2+}), copper (Cu^{2+}) and lead (Pb^{2+}) heavy metals from mixed tannery sludge of 50 mg dm^{-3} of each metal ion at 95.5, 97.7 and 99.4%, respectively.

Algal biomass removes the contamination and heavy metal loads, viz., Cu-73.2–98%, Cr-81.2–96%, Zn-65–98% and Pb-75–98% during their growth cycle. The *Scenedesmus* sp. microalgal biomass has removed 100% of highest concentrations of Cr-12.8 mgg^{-1} tannery effluent sludge dumped soil (Malik, 2004). The efficacy of chromium, copper and lead reduction by the

Table 18.1 Phytoremediation of heavy metals by different algae

Algae	Heavy metals	Removal efficiency%	References
Spirulina sp.	Cr (150 mg kg^{-1}), Cd (248 mg kg^{-1}) and Pb (221 mg kg^{-1})	96.54	Doshi et al. (2007). Chojnacka et al. (2005)
Scenedesmus obliquus, Scenedesmus abundans	Zn (180 mg kg^{-1}) and Cd (106 mg kg^{-1})	98.61	Monteiro et al. (2009)
Chaetoceros calcitrans	Cr (205 mg kg^{-1}), Cd (105 mg kg^{-1}) and Zn (287 mg kg^{-1})	96.53	Sjahrul (2012)
Chlorella vulgaris	Ag (98 mg kg^{-1}), Cd (103 mg kg^{-1}), Cu (116 mg kg^{-1}), Zn (218 mg kg^{-1}), Pb (83 mg kg^{-1}), Ra (52.5 mg kg^{-1}), Ni (210 mg kg^{-1}), U (25 mg kg^{-1}) and Fe (148 mg kg^{-1})	83.25	Aksu and Deonmez (2006), Vogel et al. (2010)
Chlorella sorokiniana	Cd (223 mg kg^{-1}), Cu (109 mg kg^{-1}), Cr (163 mg kg^{-1}), Pb (106 mg kg^{-1}) and Fe (129 mg kg^{-1})	91.6	Akhtar et al. (2008)
Anabaena inaequalis, Anabaena spiroides, Asterionella formosa	Cr (189 mg kg^{-1}), Pb (206 mg kg^{-1}) and Cd (127 mg kg^{-1})	97.26	Doshi et al. (2007), Tien et al. (2005)
Cladophora glomerata	Cd (90 mg kg^{-1}), Cu (105 mg kg^{-1}), Cr (123 mg kg^{-1}), Pb (93 mg kg^{-1}), Fe (143 mg kg^{-1}), Ni (189 mg kg^{-1}), Zn (193 mg kg^{-1}), Mn (184 mg kg^{-1}), Cs (23 mg kg^{-1}) and Sr (19 mg kg^{-1})	90.95	Sargin et al. (2016), Yalçın et al. (2012)
Nostoc sp.	Cr (168 mg kg^{-1}), Cu (185 mg kg^{-1}), Hg (53 mgkg^{-1}), Cd (163 mg kg^{-1}), Pb (182 mg kg^{-1}),	86.23	Abd El-Hameed et al. (2018)
Tetraselmis chuii	Ni (98 mg kg^{-1}), Cr (105 mg kg^{-1}), Zn (143 mg kg^{-1}), Mn (113 mg kg^{-1}), Fe (121 mg kg^{-1}), Pb (83 mg kg^{-1}), Cu (134 mg kg^{-1}), Co (102 mg kg^{-1}) and Cd (121 mg kg^{-1})	92.35	Sjahrul (2012)
Sargassum sp.	Cr (101 mg kg^{-1}), U (15 mg kg^{-1}), Pb (93 mg kg^{-1}), Ni (135 mg kg^{-1}), Cd (126 mg kg^{-1}), Cu (101 mg kg^{-1}) and Zn (89 mg kg^{-1})	90.63	Ungureanu et al. (2015)
Spirogyra sp.	Cr (124 mg kg^{-1}), Ni (184 mg kg^{-1}), Mn (162 mg kg^{-1}) and Fe (102 mg kg^{-1})	95.52	Gupta and Rastogi (2008a, b), Yaqub et al. (2012)
Microcystis sp.	Ni (163 mg kg^{-1}), Cr (106 mg kg^{-1}), Zn (123 mg kg^{-1}), Fe (141 mg kg^{-1}) and Pb (83 mg kg^{-1})	94.54	Tien et al. (2005)

examined organism improved significantly throughout the soil remediation process. The algae treatment of *Scenedesmus* spp., *Scenedesmus acutus* and *Selenastrum* sp. is known to remove effectively 97.5% of Cr(IV), 73.2–98.1% of Cu, 87.5–90.8% of Zn and 75.3–99% Pb from tannery effluent polluted area. Bioaccumulation prospective of blue-green and green algae has been shown to increase the removal of toxic heavy metals naturally in tannery-contaminated soil sites (Dwivedi et al. 2010; Shashirekha et al. 2008). Microalgae could be used in metal-contaminated soils because of their potential to tolerant metals (Malik 2004) and the small cell size. Microalgae have a large surface-to-volume ratio available for immediate relationship with the natural ecosystem; cations could effectively associate with the mechanically abundant cell membrane categories.

18.6 Chromium and Other Heavy Metal Reduction by Plants

The phytoremediation method uses plants for the remediation of various pollutants. This is achieved by phytoextraction, phytodegradation, phytovolatilization, phytostabilization and phytostimulation. It is an economical methodology that enables plants and soil microbial communities to mitigate environmental pollutants (Das and Osborne 2018; Greipsson 2011). Phytoremediation is a bioremediation method that entails the removal, conversion, transformation, stabilization and/or destruction of soil and groundwater pollutants from different kinds of plants (Willey 2007; TechTree 2019). In phytostabilization, pollutants are stabilized with in roots or even in the rhizobia and are maintained in the soil in a non-toxic form. In phytoextraction, plants capture and accumulate toxins inside the plant system which become utilizable biomass within the aboveground parts. In phytodegradation, biodegradation of pollutants requires the enzymes released from roots and the contaminants are metabolized in tissues. In phytovolatilization, pollutants are picked up from the plant roots, converted directly as gases and gradually released into the environment (Table 18.2).

The plant roots facilitate the degradation of toxins in complex microbial soil ecosystems and bioadsorbents and thereby remove Cr from tannery soil and wastewater. Terrestrial plants such as *Cyperus alternifolius*, *Jasminum* sp., *Borassus aethiopum* sp., *Typha domingensis, Hyptis-suaveolens, Typha* sp., *Parawaldeckia, Eichhorniacraassipes, Chrysopogon zizanioides, Cyprus papyrus*, *Polygonum coccineum*, *Trichoderma, Brachiaria mutica* and *Cyperus rotundus* have been reported as efficient plants to remove Cr from the tannery effluents (Vankar and Bajpai 2008; Dotro et al. 2009; Santiago and Santhamani 2010).

18.6.1 Various Plant Species Used for the Biodegradation of Tannery Effluents

The ornamental plants could help to treat heavy metal-polluted soils with tannery effluent and are typically grown under stressful conditions. Aerial parts of the plant metabolize tannery effluent pollutants absorbed from soil by the roots (Liu et al. 2009). Some ornamental plants like *Pavonia lasiopetala, Dendranthema morifolium, Dracaena sanderiana* and *Plumbago auriculata* are known to potentially adsorb Cr, Hg, Ni, As, Cd, Pb and Zn by 7.8, 9.3, 7.6, 5.1, 8.9, 2.9 and 6.8%, respectively (Luo and Chen 2016). *P. lasiopetala* and *Pl. auriculata* plant species have evolved with specialized salt glands to metabolize excessive toxic ions on the surface of the leaf to minimize their toxicity. Increased concentrations of heavy metals could decrease soil fertility, reduce the yield, lead to poor farming products, pollute groundwater and subsurface, and damage human and ecosystem health via the food web (Tadesse and Seyoum 2015; Sivakumar et al. 2016; Chakrabarty et al. 2017; Bekele 2018).

Chromium metal-contaminated soil was analysed at 10 mg Cr per kilogram of soil to quantify the phytoremediator activities of marigold (*Tagetes patula*) and also the ornamental arum (*Syngonium* sp.). The marigold and mature ornamental plants were grown, and at the seedling stage, the highest concentration of Cr was 1987.12, 5.45 and 3.50 mg/kg in root, shoot and leaf of marigold, respectively. From these measurements with both the plants, it could be inferred that these have the hyperaccumulative Cr characteristics and hence can be used as potential phytoremediators. However, in removal of Cr from soil, the ornamental arum would be a

Table 18.2 Phytoremediation of heavy metals by different plant species

Plant species	Heavy metals' utilized level	References
Quamoclit pennata L., *Antirrhinum majus L., Celosia Cristata L. Var. pyramidalis .Nerium oleander L Chrysanthemum indicum L Calendula officinalis L*	Pb (5000 mg kg^{-1}) and Cd (80 mg kg^{-1})	Cui et al. (2013), Trigueros et al. (2012). Mani et al. (2015), Tabrizi et al. (2015)
Calendula officinalis L	Cu (400 mg kg^{-1})	Goswami and Das (2016)
Panicum maximum Jacq., Tagetes erecta L., Salvia splendens Sellow ex J.A Cosmos sulphureus Cav, Gladiolus grandiflorus L., Tagetes erecta L., Helianthus annus L., Schultes, Tagetes eracta L., Chrysanthemum indicum L., Helianthus annuus L Alyssum montanum L. and Daphne jasminea Sm Chlorophytum comosum (Thunb.) Jacques Iris lactea Pall. var. chinensis (Fisch.), Iris hexagona Walter. and Iris tectorum	Cd (0–400 mg kg^{-1})	Han et al. (2007), Lal et al. (2008), Liu et al. (2009), Bosiacki (2008), Wiszniewska et al. (2017), Wang et al. (2012), Rungruang et al. (2011), Han et al. (2015)
Celosia cristata L., Lonicera japonica Thunb, Helianthus annuus L., Zantedeschia aethiopica (L.) Spreng and *Tagetes patula L.,*	Cr (400 mgkg^{-1}), Fe (275 mg kg^{-1}), Cu (450 mg kg^{-1}), Mn (600 mg kg^{-1}), Pb (350 mg kg^{-1}), Zn (370 mg kg^{-1})	Chatterjee et al. (2012), Casierra-Posada et al. (2014)
Erica andevalensis Cabezudo & Rivera and Erica australis L	Al (170 mg kg^{-1}), As (240 mg kg^{-1}), Fe (240 mg kg^{-1}), Mn (325 mg kg^{-1})	Pérez-López et al. (2014)
Mirabilis jalapa L Impatiens balsamin L., Iris pseudacorus L and Tagetes erecta L	Cr 0–102.5 mgkg^{-1}	Miao and Yan (2013), Caldelas et al. (2012a, b)
Helianthus annuus L. and Hydrangea paniculata Zantedeschia aethiopica (L.) Spreng, Canna indica L., Carex hirta L., Miscanthus sinensis Andersson and Phragmites australis Cav	Pb and Cu (10,000 mg kg^{-1})	Forte and Mutiti (2017), Macci et al. (2015)
Buddleja asiatica Lour. and Buddleja paniculata Wall Iris pseudacorus L	Pb (1000 mg kg^{-1}) and Fe (500 mg kg^{-1})	Waranusantigul et al. (2008), Zhong et al. (2010)
Nymphaea spontanea Landon	Cr (VI) (100 mg kg^{-1}) and Cu (150 mg kg^{-1})	Choo et al. (2006)
Talinum triangulare L	Cd, Cu, Pb and Ni (250 mg kg^{-1})	Rajkumar et al. (2009)
Calendula alata Rech. fil	Pb and Cs (500 mg kg^{-1})	Borghei et al. (2011)

greater phytoremediator than marigold. The germination of seeds of an alfalfa plant is significantly supressed by 60 ppm of Cr and Cd and 80 ppm of Ni and Cu. The development of a root and shoot was induced by 50 ppm Cu, Cr, Zn and Ni. Alfalfa plants have not yet been capable of growing and developing in soil that contains Cd and Cr at 35 ppm and Cu and Ni at 45 ppm concentration (Gerhardt et al. 2017; Koźmińska et al. 2018).

The two ornamental plants, *Zinnia* and marigold, are the better candidates to reduce lead concentration from tannery effluent polluted soils. The higher concentration of lead in the soil (26.9 g/kg) accumulates in the plants across most treatments, thus concluding the better potentiality of *Zinnia elegans* and marigold. *Georgina,* native ornamental plant (Dahlia), has been used to remediate toxic heavy metal at 10–50 mg/kg of nickel and 15–65 mg/kg of lead in tannery effluent contaminated soil (Sunita 2012). The development and efficacy of leguminous plants in N-soil depend on a symbiotic relationship with N-fixing microbes. Symbiotic microbes such as *Mesorhizobium metallidurans* and a *Rhizobium* genus that have been detected in Zn-contaminated soil have been regarded as the candidate metal-tolerant in tannery effluent polluted soils. The Cd- and Cu-resistant strains of *Bradyrhizobium, Rhizobium* sp.*, Mesorhizobium* sp. and *Lotus corniculatus* are known to reduce the concentration of heavy metals, viz., Ni, Co and Cr from tannery-contaminated soil. Rapidly growing herbaceous plants, like *Arundo donax* L. and *Miscanthus sinensis* L., grow on tannery polluted soils and build a strong, green barrier within a couple of years.

18.6.1.1 Metallophytes

Plants which can grow and tolerate high levels of heavy metal-contaminated soil are called metallophytes. Metallophytes are predominantly from Brassicaceae plant family and are classified into three groups: excluders, indicators and accumulators (Bothe 2011). Metal excluders absorb toxic metals from the environment and facilitate through root branches, however limiting the transportation to the aboveground parts. Metal indicators absorb pollutants from polluted soil to their upper components (Sheoran et al. 2011; Malik and Biswas 2012). Hyperaccumulation of heavy metals acts as a defence action in the first place in infectious plant species. More than 400 species of plants were recognized as hyperaccumulators with slower growth and less development of biomass (Kramer 2018). For phytoremediation and phytomining of toxic metals and precious metals, the field utilization of natural hyperaccumulators can be performed simultaneously. Other effective heavy metal hyperaccumulators include *Azolla pinnata, Arabispaniculata* (Cd-0.7–1.2 g/kg), *Ipomoea alpine, Eleocharis acicularis* (Cu-12.3–20.2 g/kg), *Euphorbia cheiradenia* (Pb-1.14 g/kg), *Alyssum corsicum, Alyssum markgrafii* (Ni-18.1–19.3 g/kg), *Thlaapi caerulescens, Potentilla griffithii* (Zn-19.4–9.6 g/kg), *Pteris vittata, Pteris ryukyuensis* (As-2.3–3.6 g/kg), *Phragmites australis, Pteris vittata* (Cr-4.8–20.6 g/kg) *and Schima superba* (Mn-62.3 g/kg) (Chehregani and Malayeri 2007; Calheiros et al. 2008; Zeng et al. 2009; Bani et al. 2010; Kalve et al. 2011; Sakakibara et al. 2011).

The *R. communis, A. indica, D. strictus, M. azadirachta* and *C. sebestena* fruit plants accumulate and translocate toxic metal chromium (Cr) in the plant upper parts, viz., stem and leaves. The concentration of Cr accumulated in the range 4.22–34.44 μg g^{-1} followed by *A. indica* > *R. communis* > *D. strictus* > *C. sebestena* > *M. azadirachta* (Pulford et al. 2001; Vyslouzilova et al. 2003; Hammer et al. 2003; Keller et al. 2003; Rotkittikhun et al. 2006; Fischerova et al. 2006). The heavy metal uptake is estimated by proliferating metal concentration, dry biomass and uptake ratio as Cr > Pb > Ni > Cd in *Brassica juncea, Palmarosa* and *Trishna* sp. The chromium uptake ranges were 4.88–18.33 kg ha^{-1} by *Brassica juncea* crop. *Trishna* showed a higher uptake of chromium from 25.50 to 54.78 kg ha^{-1} from tannery sludge-contaminated soil. The nickel uptake range of *Brassica juncea* crop was 1.73 to 3.62, *Palmarosa* crop 1.02–2.31 and *Trishna* crop 1.23–2.2 kg ha^{-1}. The maximum Pb uptake by *Trishna* crop was 1.4–5.22, *Brassica juncea* crop 0.05–0.26 and *Palmarosa* 0.02–0.15 kgha^{-1}. The Cd uptake range from *Trishna* crop was 1.32–1.59, *Palmarosa* crop 1.06–1.45 and *Brassica juncea* crop 0.96 to 1.25 kg ha^{-1} (Gupta and Sinha 2006, 2007).

The economically important medicinal plants *F. bengalensis, C. coronarium, D. sissoo and T.*

erecta are grown in tannery effluent contaminated soil. The plants accumulate and uptake the heavy metals, viz., Zn, Cr, Mn and Cu. These metals are essential for the plant growth and uptake heavy metal concentration from Zn-344.18 μg g^{-1}, Mn-131.78 μg g^{-1}, Cu-62.36 μg g^{-1} and Cr-55.63 μg g^{-1} followed by *D. sissoo* > *F. bengalensis* > *C. coronarium* > *T. erecta*. The trees *A. indica, A. procera, D. regia, D. sissoo, D. strictus, E. camaldulensis, J. curcas, P. dulce, S. cumini* and *T. arjuna* are known to accumulate and utilize the chromium toxic heavy metal from tannery sludge dumping land areas. The maximum Cr accumulation range was noted as 11.97–29.53 μg g^{-1} for *A. indica, and* the lowest accumulation of Cr range 7.89–11.32 μg g^{-1} was shown by *S. cumini* plant. In plants, the chromium metal accumulates in the leaves, but the difference was most apparent between chromium (Cr) agglomeration in young plants and fully mature plants (Rotkittikhun et al. 2006; Fischerova et al. 2006; Vyslouzilova et al. 2003; Keller et al. 2003; Hammer et al. 2003; Pulford et al. 2001; Gupta and Sinha 2007; Gupta and Sinha 2006).

The flowering plants, *V. zizanoides, C. winterianus, C. citrates* grasses, *C. coronarium and T. erecta,* exhibited healthy growth in tannery-wastewater-contaminated soils. The plants accumulate Cr in their roots and aerial parts. The grasses uptake the concentration of Cr from 0.39 to 14.88 mg kg^{-1}, and flowering plants range from 0.19 to 11.85 mg kg^{-1} (Chen et al. 2004; Chiu et al. 2005; Chintakovid et al. 2008). Similarly, the roots of some grasses uncovered additional accumulation of chromium (Cr) than flowering plants. This could be due to stringy roots in the grasses consuming large surface area, which accumulated extra heavy metals than tap roots (flowering plants). The woody plants grown abundantly on tannery sludge dump area are known to show higher range of chromium removal from the contaminated soil in 3 to 12 months of phytoremediation (Gupta and Sinha 2007; Chintakovid et al. 2008). The woody plants, *P. alba, E. tereticornis, T. arjuna, D. strictus and P. juliflora*, hyperaccumulate and remove the heavy metals, viz., Cr-70.22, Ni-59.21, Cd-58.4, Fe-49.75, Mn-30.95, Zn-22.80, Cu-20.46, Pb-14.05% (Giachetti and Sebastiani 2006; Gupta and Sinha 2007).

18.7 Conclusion

The threat of heavy metal contamination is a major concern because it affects the quality of the environment. Owing to its large chemical content, which includes salinity, original products, sulphide, chromium, chloride, sodium, tannery effluent is a significant environmental risk. Agriculture, human beings and animals are harmed by these components found in the waste. Therefore, it is very important to eliminate these high-strength, hazardous contaminants from wastewater. In contrast to other physiochemical approaches, biological treatment of wastewater is more favourable and cost-effective.

As a way of cleaning Cr, not just from tannery waste but also to clean other heavy metals from many industrial effluents, bioremediation provides a true and realistic solution. Heavy metal hyperaccumulators are now the most suitable method for phytoremediation of tannery waste contaminants. Algal technique is a safe way of extracting heavy metals emitted from the tannery industry and other pollutants. Effluent treatment based on algae and plants is more reliable and highly economical.

Acknowledgements The authors are thankful to Fermentation Technology Laboratory, Department of Microbiology (DST-FIST Supported Department), School of Biosciences, Periyar University, Salem, Tamil Nadu, India, for providing laboratory facilities and Tamil Nadu State Council for Science and Technology, Tamil Nadu, India, for providing the financial support (TNSCST/S&T Project/VR/ES/01/2016–2017).

References

Abd El-Hameed MM, Abuarab ME, Mottaleb SA, El-Bahbohy RM, Bakeer GA (2018) Comparative studies on growth and Pb (II) removal from aqueous solution by Nostoc muscorum and Anabaena variabilis. Ecotoxicol Environ Saf 165:637–644

Agarwal P, Giri BS, Rani R (2020) Unravelling the role of rhizospheric plant-microbe synergy in phytoremediation: a genomic perspective. Curr Genom 21(5):334–342

Akhtar N, Iqbal M, Zafar SI, Iqbal J (2008) Biosorption characteristics of unicellular green alga Chlorella sorokiniana immobilized in loofa sponge for removal of Cr (III). J Environ Sci 20:231–239

Aksu Z, D€onmez G (2006) Binary biosorption of cadmium (II) and nickel (II) onto dried Chlorella vulgaris: co-ion effect on mono-component isotherm parameters. Process Biochem 41(4):860–868

Ali H, Khan E, Sajad MA (2013) Phytoremediation of heavy metals—Concepts and applications. Chemosphere 91:869–881

Aliabadi M, Morshedzadeh K, Soheyli H (2006) Removal of hexavalent chromium from aqueous solution by Lignocellulosic solid wastes. Int J Envi Sci Techn 3:321–325

Alkorta I, Hernández-Allica J, Becerril JM, Amezaga I, Albizu I, Garbisu C (2004) Recent findings on the phytoremediation of soils contaminated with environmentally toxic heavy metals and metalloids such as zinc, cadmium, lead, and arsenic. Rev Environ Health 3:71–90

Alvarez-Ayuso E, Garcia-Sanchez A, Querol X (2007) Adsorption of Cr(VI) from synthetic solutions and electroplating wastewaters on amorphous aluminium oxide. J Hazard Mater 142:191–198

Apte AD, Verma S, Tare V, Bose P (2005) Oxidation of Cr(III) in tannery sludge to Cr(VI): field observations and theoretical assessment. J Hazard Mater 121:215–222

Arshad M, Silvestre J, Pinelli E, Kallerhoff J, Kaemmerer M, Tarigo A, Shahid M, Guiresse M, Pradere P, Dumat C (2008) A field study of lead phytoextraction by various scented Pelargonium cultivars. Chemosphere 71:2187–2192

Azimi A, Azari A, Rezakazemi M, Ansarpour M (2016) Removal of heavy metals from industrial wastewaters: a review. Chem Biochem Eng Revi 4:37–59

Bahafid W, Joutey NT, Sayel H, Ghachtouli NEL (2013) Mechanism of hexavalent chromium detoxification using Cyberlindnera fabianii yeast isolated from contaminated site in Fez (Morocco). J Mater Environ Sci 4(6):840–847

Bai RS, Abraham TE (2001) Biosorption of chromium (VI) from aqueous solution by Rhizopus nigricans. Bioresour Technol 79:73–81

Bai RS, Abraham TE (2003) Studies on Chromium (VI) adsorption-desorption using immobilized fungal biomass. Bioresour Technol 87:17–26

Bani A, Pavlova D, Echevarria G, Mullaj A, Reeves RD, Morel JL, Sulce S (2010) Nickel hyperaccumulation by the species of Alyssum and Thlaspi (Brassicaceae) from the ultramafic soils of the Balkans. Bot Serb 34:3–14

Barakat M (2011) New trends in removing heavy metals from industrial wastewater. Arab J Chem 4:361–377

Barbosa B, Boléo S, Sidella S, Costa J, Duarte MP, Mendes B, Cosentino SL, Fernando AL (2015) Phytoremediation of heavy metal-contaminated soils using the perennial energy crops Miscanthus spp. and Arundo donax L. Bio Energy Res 8:1500–1511

Baroni L, Cenci L, Tettamanti M, Berati M (2007) Evaluating the environmental impact of various dietary patterns combined with different food production systems. Eur J Clin Nut 61:279

Bekele M (2018) Phytoremediation of tannery waste water using horizontal sub surface flow constructed wetland. College of Biological and Chemical Engineering, Madison

Borghei M, Arjmandi R, Moogouei R (2011) Potential of Calendula alata for phytoremediation of stable cesium and lead from solutions. Environ Monit Assess 18:63–68

Bosiacki M (2008) Accumulation of cadmium in selected species of ornamental plants. Acta Sci Pol Hortorum Cultus 7:21–31

Bothe H (2011) Plants in heavy metal soils. In: Sherameti I, Varma A (eds) Detoxification of heavy metals. Springer, Berlin, Heidelberg, Germany, pp 35–57

Caldelas C, Araus J, Febrero A, Bort J (2012) Accumulation and toxic effects of chromium and zinc in Iris pseudacorus L. Acta Physiol Plant 34:1217–1228

Caldelas C, Bort J, Febrero A (2012b) Ultrastructure and subcellular distribution 637 of Cr in Iris pseudacorus L. Using TEM and Xray microanalysis. Cell Biol Toxicol 28(1):57–68

Calheiros CSC, Rangel AOSS, Castro PML (2008) The effects of tannery wastewater on the development of different plant species and chromium accumulation in Phragmites australis. Arch Environ Contam Toxicol 55:404–414

Caporale AG, Adamo P, Capozzi F, Langella G, Terribile F, Vingiani S (2018) Monitoring metal pollution in soils using portable-XRF and conventional laboratory-based techniques: evaluation of the performance and limitations according to metal properties and sources. Sci Total Environ 643:516–526

Capuana M (2011) Heavy metals and woody plants-biotechnologies for phytoremediation. iFor Biogeosci For 4(1):7–15

Casierra-Posada F, Blanke MM, Guerrero-Guío JC (2014) Iron tolerance in Calla Lilies (Zantedeschia aethiopica). Gesunde Pflanz 66:63–68

Chakrabarty T, Afrin R, Mia Y, Hossen Z (2017) Phytoremediation of chromium and some chemical parameters from tannery effluent by using water. Res Agric Livest Fish 4(3):151–156

Chan A, Salsali H, McBean E (2014) Heavy metal removal (copper and zinc) in secondary effluent from wastewater treatment plants by microalgae. ACS Sustainable Chem Eng 2(2):130–137

Chatterjee S, Singh L, Chattopadhyay B, Datta S, Mukhopadhyay S (2012) A study on the waste metal remediation using floriculture at East Calcutta Wetlands, a Ramsar site in India. Environ Monit Assess 184:5139–5150

Chehregani A, Malayeri BE (2007) Removal of heavy metals by native accumulator plants. Int J Agri Biol 9:462–465

Chen Y, Shen Z, Li X (2004) The use of vetiver grass (Vetiveria zizanioides) in the phytoremediation of soils contaminated with heavy metals. Appl Geochem 19(10):1553–1565

Chintakovid W, Visoottiviseth P, Khokiattiwong S, Lauengsuchonkul S (2008) Potential of the hybrid marigolds for arsenic phytoremediation and income generation of remediators in Ron Phibun District. Thailand. Chemosphere 70(8):1532–1537

Chiu KK, Ye ZH, Wong MH (2005) Enhanced uptake of As, Zn, and Cu by Vetiveria zizanioides and Zea mays using chelating agents. Chemosphere 60(10):1365–1375

Chojnacka K, Chojnacki A, Górecka H (2005) Biosorption of Cr^{3+}, Cd^{2+} and Cu^{2+} ions by blue-green algae Spirulina sp.: kinetics, equilibrium and the mechanism of the process. Chemosphere 59:75–84

Choo T, Lee C, Low K, Hishamuddin O (2006) Accumulation of chromium (VI) from aqueous solutions using water lilies (Nymphaea spontanea). Chemosphere 62:961–967

Chowdhury M, Mostafa MG, Biswas TK, Mandal A, Saha AK (2015) Characterization of the effluents from leather processing industries. Environ Process 2 (1):173–187

Cui S, Zhang T, Zhao S, Li P, Zhou Q, ZhangQ HanQ (2013) Evaluation of three ornamental plants for phytoremediation of Pb-contamined soil. Int J Phytoremed 15:299–306

DalCorso G, Fasani E, Manara A, Visioli G, Furini A (2019) Heavy metal pollutions: state of the art and innovation in phytoremediation. Int J Mol Sci 20:3412

Das N, Bhattacharya S, Maiti MK (2016) Enhanced cadmium accumulation and tolerance in transgenic tobacco overexpressing rice metal tolerance protein gene OsMTP1 is promising for phytoremediation. Plant Physiol Biochem 105:297–309

Das A, Osborne JW (2018) Bioremediation of heavy metals. In: Nanotechnology, food security and water treatment. Springer, Berlin

Doshi H, Ray A, Kothari IL (2007) Bioremediation potential of live and dead Spirulina: spectroscopic, kinetics and SEM studies. Biotechnol Bioeng 96 (6):1051–1063

Dotro G, Palazolo P, Larsen D (2009) Chromium fate in constructed wetlands treating tannery wastewaters. Water Environ Res 81(6):617–625

Dwivedi D, Srivastava S, Mishra S, Kumar A, Tripathi RD, Rai UN, Dave R, Tripathi P, Chakraborty D, Trivedi PK (2010) Characterization of native microalgal strains for their chromium bioaccumulation potential: phytoplankton response in polluted habitats. J Hazard Mater 173:95–101

Elahi A, Arooj I, Bukhari DA, Rehman A (2020) Successive use of microorganisms to remove chromium from wastewater. Appl Microbiol Biotechnol 14:1–5

Fischerová Z, Tlustoš P, Száková J, Šichorová K (2006) A comparison of phytoremediation capability of selected plant species for given trace elements. Environ Pollut 144:93–100

Forte J, Mutiti S (2017) Phytoremediation potential of Helianthus annuus L. and Hydrangea paniculata Siebold. In: Copper and Lead- Contaminated Soil. Water Air Soil Pollut 228 (2):77

Gerhardt KE, Gerwing PD, Greenberg BM (2017) Opinion: Taking phytoremediation from proven technology to accepted practice. Plant Sci 256:170–185

Giachetti G, Sebastiani L (2006) Metal accumulation in poplar plant grown with industrial wastes. Chemosphere 64(3):446–454

Goswami S, Das S (2016) Copper phytoremediation potential of Calandula officinalis L. and the role of antioxidant enzymes in metal tolerance. Ecotox Environ Safe 126:211–218

Greipsson S (2011) Phytoremediation. Nat Educ Knowl 3 (10):7

Gupta AK, Sinha S (2006) Chemical fractionation and heavy metal accumulation in the plant of Sesamum indicum (L.) var. T55 grown on soil amended with tannery sludge: Selection of single extractants. Chemosphere 64(1):161–73

Gupta AK, Sinha S (2007) Assessment of single extraction methods for the prediction of bioavailability of metals to Brassica juncea L. Czern. (var. Vaibhav) grown on tannery waste contaminated soil. J Hazard Mater 149(1):144–50

Gupta V, Rastogi A (2008) Biosorption of lead from aqueous solutions by green algae Spirogyra species: kinetics and equilibrium studies. J Hazard Mater 152:407–414

Gupta VK, Rastogi A (2008) Equilibrium and kinetic modeling of cadmium(II) biosorption by nonliving algal biomass Oedogonium sp. from aqueous phase. J Hazard Mater 153:759–766

Hammer D, Kayser A, Keller C (2003) Phytoextraction of Cd and Zn with Salix viminalis in field trials. Soil Use Manag 19:187–192

Han YL, Yuan HY, Huang SZ, Guo Z, Xia B, Gu J (2007) Cadmium tolerance and accumulation by two species of Iris. Ecotoxicology 16:557–563

Han Y, Chen G, Chen Y, Shen Z (2015) Cadmium toxicity and alleviating effects of exogenous salicylic acid in Iris hexagona. Bull Environ Contam Toxicol 95:796–802

Joshi NC (2017) Heavy metals, conventional methods for heavy metal removal, biosorption and the development of low cost adsorbent. Eur J Phar Med Res 4:388–393

Kalve S, Sarangi BK, Pandey RA, Chakrabarti T (2011) Arsenic and chromium hyperaccumulation by an ecotype of Pteris vittata-prospective for phytoextraction from contaminated water and soil. Curr Sci 100:888–894

Keller C, Hammer D, Kayser A, Richner W, Brodbeck M, Sennhauser M (2003) Root development and heavy metal phytoextraction efficiency: comparison of different plant species in the field. Pl Soil 249:67–81

Kermani AJN, Ghasemi MF, Khosravan A, Farahmand A, Shakibaie MR (2010) Cadmium bioremediation by metal-resistant mutated bacteria isolated from active sludge of industrial effluent. Ira J Envir Health Sci Eng 7(4):279–286

Koźmińska A, Wiszniewska A, Hanus-Fajerska E, Muszyńska E (2018) Recent strategies of increasing metal tolerance and phytoremediation potential using genetic transformation of plants. Plant Biotechnol Rep 12:1–14

Kramer U (2018) The plants that suck up metal. Ger Res 40(3):18–23

Lakherwal D (2014) Adsorption of heavy metals: a review. Int J Env Res Devel 4:41–48

Lal K, Minhas P, Chaturvedi R, Yadav R (2008) Extraction of cadmium and tolerance of three annual cut flowers on Cd-contaminated soils. Bioresour Technol 99:1006–1011

Leong YK, Chang JS (2020) Bioremediation of heavy metals using microalgae: recent advances and mechanisms. Bioresour Technol 303:122886

Lin Z, Li J, Luan Y, Dai W (2020) Application of algae for heavy metal adsorption: a 20-year meta-analysis. Ecotoxicol Environ Saf 190:110089

Liu Z, He X, Chen W, Yuan F, Yan K, Tao D (2009) Accumulation and tolerance characteristics of cadmium in a potential hyperaccumulator—Lonicera japonica Thunb. J Hazard Mater 169:170–175

Luo J, Qi S, Peng L, Wang J (2016) Phytoremediation efficiency of Cd by Eucalyptus globulus transplanted from polluted and unpolluted sites. Int J Phytoremed 18:308–314

Macci C, Peruzzi E, Doni S, Iannelli R, Masciandaro G (2015) Ornamental plants for micropollutant removal in wetland systems. Environ Sci Pollut Res 22:2406–2415

Malik A (2004) Metal bioremediation through growing cells. Environ Inter 30:261–278

Malik N, Biswas AK (2012) Role of higher plants in remediation of metal contaminated sites. Sci Rev Chem Commun 2:141–146

Mani D, Kumar C, Patel N, Sivakumar D (2015) Enhanced clean-up of lead-contaminated alluvial soil through Chrysanthemum indicum L. Int J Environ Sci Technol 12:1211–1222

Miao Q, Yan J (2013) Comparison of three ornamental plants for phytoextraction potential of chromium removal from tannery sludge. J Mater Cycles Waste Manage 15:98–105

Monteiro CM, Castro PML, Malcata FX (2009) Use of the microalga Scenedesmus obliquus to remove cadmium cations from aqueous solutions. World J Microbiol Biotechnol 25:1573–1578

Pérez-López R, Márquez-García B, Abreu MM, Nieto JM, Córdoba F (2014) Erica andevalensis and Erica australis growing in the same extreme environments: phytostabilization potential of mining areas. Geoderma 230:194–203

Pogrzeba M, Rusinowski S, Krzyzak J (2017) Macroelements and heavy metals content in Panicum virgatum cultivated on contaminated soil under different fertilization. Int J Agric for 63:69–76

Pulford ID, Watson C, McGregor SD (2001) Uptake of chromium by trees: prospects for phytoremediation. Environ Geo Chem Health 23:307–311

Rai PK, Kim KH, Lee SS, Lee JH (2020) Molecular mechanisms in phytoremediation of environmental contaminants and prospects of engineered transgenic plants/microbes. Sci Total Environ 705:135858

Rajkumar K, Sivakumar S, Senthilkumar P, Prabha D, Subbhuraam C, Song Y (2009) Effects of selected heavymetals (Pb, Cu, Ni, and Cd) in the aquatic medium on the restoration potential and accumulation in the stem cuttings of the terrestrial plant, Talinum triangulare Linn. Ecotoxicology 18:952–960

Rambabu K, Thanigaivelan A, Bharath G, Sivarajasekar N, Banat F, Show PL (2020) Biosorption potential of Phoenix dactylifera coir wastes for toxic hexavalent chromium sequestration. Chemosphere 28:128809

Rotkittikhun P, Kruatrachue M, Chaiyarat R, Ngernsansaruay C, Pokethitiyook P, Paijitprapaporn A, Baker AJ (2006) Uptake and accumulation of lead by plants from the Bo Ngam lead mine area in Thailand. Environ Pollut 144(2):681–688

Roychowdhury R, Roy M, Zaman S, Mitra A (2019) Phytoremediation potential of castor oil plant (Ricinus communis) grown on fly ash amended soil towards lead bioaccumulation. Innov Food Sci Emerg Technol 6:156–160

Rungruang N, Babel S, Parkpian P (2011) Screening of potential hyperaccumulator for cadmium from contaminated soil. Desalin Water Treat 32:19–26

Sakakibara M, Ohmori Y, Ha NTH, Sano S, Sera K (2011) Phytoremediation of heavy metal contaminated water and sediment by Eleocharis acicularis. Clean: Soil, Air, Water 39:735–741

Santiago M, Santhamani S (2010) Remediation of chromium contaminated soils: potential for phyto and bioremediation. In: 19th world congress of soil science, soil solutions for a changing world, pp 211–214

Saranya D, Shanthakumar S (2019) Green microalgae for combined sewage and tannery effluent treatment: performance and lipid accumulation potential. J Environ Manage 241:167–178

Sargın I, Arslan G, Kaya M (2016) Efficiency of chitosan–algal biomass composite microbeads at heavy metal removal. React Funct Polym 98:38–47

Shashirekha V, Sridharan MR, Mahadeswara S (2008) Biosorption of trivalent chromium by free and immobilized blue green algae: kinetics and equilibrium studies. J Environ Sci Health, Part a: Toxic/hazard Sub Environ Eng 43:390–401

Sheoran V, Sheoran A, Poonia P (2011) Role of hyperaccumulators in phytoextraction of metals from contaminated mining sites: a review. Crit Rev Environ Sci Technol 41:168–214

Shokrzadeh M, Saeedi Saravi SS (2009) The study of heavy metals (zinc, lead, cadmium, and chromium) in

water sampled from Gorgan coast (Iran). Toxicolog Environ Chem 91(3):405–407
Sivakumar P, Kanagappan M, Das SSM (2016) Phytoremediation of tannery waste polluted soil using Hyptis suaveolens (Lamiaceae). Int J Pure Appl Biosci 4 (1):265–272
Sjahrul M, Arifin D (2012) Phytoremediation of Cd^{2+} by Marine Phytoplanktons, Tetraselmis chuii and Chaetoceros calcitrans. Int J Chem 4(1):69–74
Tabrizi L, Mohammadi S, Delshad M, Motesharezadeh B (2015) Effect of arbuscular mycorrhizal fungi on yield and phytoremediation performance of pot marigold (Calendula officinalis L.) under heavy metals stress. Int J Phytoremed 17:1244–1252
Tadesse AT, Seyoum LA (2015) Evaluation of selected wetland plants for removal of chromium from tannery wastewater in constructed wetlands. Ethiopia. Afr J Environ Sci Technol 9(5):420–427
Tamil Selvan S, Velramar B, Ramamurthy D, Balasundaram S, Sivamani K (2020) Pilot scale wastewater treatment, CO_2 sequestration and lipid production using microalga, Neochloris aquatica RDS02. Int J Phytorem 22(14):1462–1479
Tien CJ, Sigee DC, White KN (2005) Copper adsorption kinetics of cultured algal cells and freshwater phytoplankton with emphasis on cell surface characteristics. J Appl Phycol 17:379–389
Trigueros D, Mingorance M, Oliva SR (2012) Evaluation of the ability of Nerium oleander L. to remediate Pb-contaminated soils. J Geochem Explor 114:126–133
Tumolo M, Ancona V, De Paola D, Losacco D, Campanale C, Massarelli C, Uricchio VF (2020) Chromium pollution in European water, sources, health risk, and remediation strategies: an overview. Int J Environ Res Public Health 17(15):5438
Ubando AT, Africa AD, Maniquiz-Redillas MC, Culaba AB, Chen WH, Chang JS (2020) Microalgal biosorption of heavy metals: a comprehensive bibliometric review. J Hazard Mater 402:123431
Ungureanu G, Santos S, Boaventura R, Botelho C (2015) Biosorption of antimony by brown algae S. muticum and A. nodosum. Environ Eng Manag J 14:455–463
Vangronsveld J, Herzig R, Weyens N, Boulet J, Adriaensen K, Ruttens A, Thewys T, Vassilev A, Meers E, Nehnevajova E (2009) Phytoremediation of contaminated soils and groundwater: lessons from the field. Environ Sci Pollut Res 16:765–794
Vankar PS, Bajpai D (2008) Phyto-remediation of chrome-VI of tannery effluent by Trichoderma species. Desalination 222(1–3):255–262
Visconti D, Álvarez-Robles MJ, Fiorentino N, Fagnano M, Clemente R (2020) Use of Brassica juncea and Dactylis glomerata for the phytostabilization of mine soils amended with compost or biochar. Chemosphere 260:127661
Vogel M, Gunther A, Rossberg A, Li B, Bernhard G, Raff J (2010) Biosorption of U(VI) by the green algae Chlorella vulgaris in dependence of pH value and cell activity. Sci Total Environ 409(2):384–395
Vyslouzilova M, Tlustos P, Száková J, Pavlíková D (2003) As, Cd, Pb and Zn uptake by Salix spp. clones grown in soils enriched by high loads of these elements. Plant Soil Environ 49:191–196
Wang Y, Yan A, Dai J, Wang N, Wu D (2012) Accumulation and tolerance characteristics of cadmium in Chlorophytum comosum L.: a popular ornamental plant and potential Cd hyperaccumulator. Environ Monit Assess 184:929–937
Waranusantigul P, KruatrachueM PP, Auesukaree C (2008) Evaluation of Pb phytoremediation potential in Buddleja asiatica Lour. and B. paniculata. Water Air Soil Pollut 193:79–90
Wiszniewska A, Hanus-Fajerska E, Muszyńska E, Smoleń S (2017) Comparative assessment of response to cadmium in heavy metaltolerant shrubs cultured in vitro. Water Air Soil Pollut 228:304
Yalçın S, Sezer S, Apak R (2012) Characterization and lead(II), cadmium(II), nickel(II) biosorption of dried marine brown macro algae Cystoseira barbata. Environ Sci Pollut Res 19:3118–3125
Yaqub A, Mughal M, Adnan A, Khan W, Anjum K (2012) Biosorption of hexavalent chromium by Spirogyra sp.: equilibrium, kinetics and thermodynamics. J Anim Plant Sci 22(2):408–415
Zeng X, Ma LQ, Qiu R, Tang Y (2009) Responses of non-protein thiols to Cd exposure in Cd hyperaccumulator *Arabis paniculata* Franch. Environ Exp Bot 66:242–248
Zhong S, Shi J, Xu J (2010) Influence of iron plaque on accumulation of lead by yellow flag (Iris pseudacorus L.) grown in artificial Pb contaminated soil. J Soils Sediments 10:964–970

19 Production of Safer Vegetables from Heavy Metals Contaminated Soils: The Current Situation, Concerns Associated with Human Health and Novel Management Strategies

Hafiz Muhammad Tauqeer, Veysel Turan, and Muhammad Iqbal

Abstract

Vegetables play a chief part in the human diet and provide the essential nutrients and vitamins necessary to perform numerous essential physiological functions in the human body. Unfortunately, the consumption of vegetables laden with heavy metals (HMs) is among the most imperative issues of recent years because of their toxic impacts on human health. The toxic HMs accumulated in vegetables after their release into the ecosystem through diverse natural and human-centered activities. The prolonged use of synthetic agrochemicals, irrigation of agricultural lands with untreated municipal and industrial effluents, inappropriate dumping of solid waste, and various other industrial activities are the main causative factors of HMs accumulation in productive soils. The mobility of HMs in the soil and their accumulation in vegetables is remarkably influenced by several soil and plant factors that control their bioavailability. Reduction in growth, biomass, yield and poor nutritional quality are the key symptoms of HMs toxicity after their absorption by the vegetables. Health risks to humans via the consumption of HMs contaminated vegetables have been investigated through different risk assessment equations. Interestingly, different novel remediation techniques such as phytoremediation, immobilization, water management strategies, and applications of microbial inocula could be practiced for safer vegetable production for human consumption from HMs polluted soils.

Keywords

Bioaccumulation · Food chain · Heavy metals · Human health index · Remediation · Vegetables

19.1 Introduction

Rapid industrialization and urban sprawls have significantly increased problems associated with food security, sustainable agriculture, and safe food production (Rai 2018; Toth et al. 2016;

The original version of this chapter was revised: For detailed information, please see Correction. The correction to this chapter is available at https://doi.org/10.1007/978-3-030-89984-4_28

H. M. Tauqeer
Department of Environmental Science, Hafiz Hayat Campus, University of Gujrat, Gujrat, Pakistan

V. Turan
Department of Soil Science and Plant Nutrition, Faculty of Agriculture, Bingöl University, Bingöl, Turkey

M. Iqbal (✉)
Department of Environmental Sciences and Engineering, Government College University, Faisalabad, Pakistan
e-mail: iqbal.farhad@gmx.at

J. A. Malik (ed.), *Advances in Bioremediation and Phytoremediation for Sustainable Soil Management*,
https://doi.org/10.1007/978-3-030-89984-4_19

Saumel et al. 2012; Clarke 2011). Among different problems, soil pollution with heavy metals (HMs) such as Cd, Pb, Cr, As, Ni, and Hg are becoming a serious environmental concern in recent years (Kumar et al. 2019; Gupta et al. 2018; Oves et al. 2012).

Mainly, anthropogenic activities such as rapid industrialization, aerosols production through the combustion of fossil fuels, mining processes, aerial deposition from smelters, applications of agrochemicals like herbicides or metallo-pesticides, phosphate fertilizers which release diverse HMs such as Cr, Hg, Cd, and Ni in agricultural soils, irrigation with untreated industrial or municipal wastewater, improper handling and dismantling of hazardous waste, additions of livestock manures as well as sewage sludge have significantly accelerated soil contamination with HMs (El-Kady and Abdel-Wahhab 2018; Gall et al. 2015; Woldetsadik et al. 2017; Kihampa et al. 2011; Luo et al. 2009; Chary et al. 2008). The toxic effects of HMs appeared on soil (micro)organisms which ultimately damage soil quality, and its fertility consequently affects safe food production after their deposition in the soil (Gadd 2010).

Vegetables are the most vital part of the human diet and are widely consumed due to the provision of essential nutrients such as carbohydrates, proteins, antioxidants, vitamins, dietary fibers, and essential minerals. Unfortunately, vegetables produced from HMs contaminated soils situated near industrial sources have higher concentrations of HMs in them than others (Slavin and Lloyd 2012). The accumulation and biotoxic effects of HMs are entirely influenced by their concentrations, source of contamination, chemical fraction and speciation, mode of deposition, the accumulation capacity of vegetables, soil, and other environmental factors (Yadav et al. 2018; Lente et al. 2014). Vegetables accumulate HMs either by absorption through their roots or by aerial deposition. Heavy metals are taken up by the vegetables and absorbed in the apoplast of roots which subsequently encourage aerial transport. It was reported that tubers and leafy vegetables accumulate higher concentrations of HMs because roots and leaves of herbaceous plants retain very high concentrations compared to fruits and stems (Singh et al. 2015; Agrawal et al. 2007). Hereafter, this loading of HMs in vegetables and their edible parts from contaminated soils becomes a grave concern owing to the risk of metal toxicity in animals and humans. Humans may experience reduced intellectual abilities in children, dementia in adults, dysfunctions of central nervous system, renal and gastrointestinal failure, insomnia, visionary loss and osteoporosis upon accelerated exposure to HMs (Rai et al. 2018; Emamverdian et al. 2015; Gall et al. 2015; Jan et al. 2011; Gadd 2010). Different risk assessment models are being used to evaluate potential hazards from the exposures to these HMs (Kamunda et al. 2016; Zhou et al. 2016).

Thus, there is a dire need to remediate such HMs affected soils that can pose serious threats to human health. Several remediation techniques have been adopted to reduce HMs accumulation in vegetables. These strategies include phytomanagement (Radziemska et al. 2020), immobilization (Xu et al. 2019; Wang et al. 2014), water management strategies, cropping patterns (De Juan et al. 1996), and applications of different microbial inocula (Edelstein and Ben-Hur 2018). Apart from this, laws have been enforced in many countries to control the release of HMs from different industries. Hence, this chapter aims to highlight HMs toxicity, their accumulation and transfer in vegetables, and associated health risks by consuming the HMs polluted foodstuff.

19.2 Soil Pollution with HMs

Major sources of soil pollution with HMs are categorized as natural and anthropogenic activities. Among natural phenomena, geological rock formation is the most important natural source of HMs discharge in the environment (Gupta et al. 2019). Generally, large quantities of Mn, Co, Cr, Ni, Cu, Zn, Cd, Sn, Pb, and Hg are released by geological processes. Similarly, some igneous rocks such as hornblende, augite, and olivine also share considerable amounts of Ni, Co, Zn, and

Cu in the soils. Moreover, increased levels of different HMs were observed among the categories of sedimentary rocks in the order of shale > limestone > sandstone (Nagajyoti et al. 2010). The volcanic eruption is also contributing its share in releasing Zn, Al, Mn, Ni, Cu, Hg, and Pb, along with some hazardous and toxic gases (Nagajyoti et al. 2010).

Industrial sources of HMs pollution include smelting, mining, transport of ores, metal recycling, and finishing activities. Estimatedly, ore mining is the major source of the release of different HMs in the environment (Yang et al. 2018; Duruibe et al. 2007). Runoff from mine wastes and weathering of metallic materials also contribute to the contamination of water bodies and surrounding lands due to leaching (Li et al. 2015; Pandey et al. 2016). The long-term use of industrial and municipal wastewater considerably increased HMs accumulation in agricultural soils (Turan et al. 2018). Numerous scientists reported the considerable concentrations of different HMs in arable soils followed by in vegetables (Ratul et al. 2018; Chabukdhara et al. 2016; Prashar and Prasad 2013). For example, higher concentrations of HMs were found in tomatoes when irrigated by sewage water (Alghobar and Suresha 2017).

Similarly, the applications of industrial effluents released from electroplating and Pb-acid batteries could cause the contamination of soil with Ni and Pb (Shahbaz et al. 2018; Khan et al. 2020). The atmospheric deposition also results in the precipitation of HMs on soil or nearby vegetation, thus increasing soil pollution with HMs. High-temperature processes, e.g., casting and smelting are involved in releasing different HMs in vapors and particulate forms. These vapors chemically react with water vapors present in the air and produce aerosols. Later, these aerosols are dispersed by the wind (commonly known as a dry deposition) or deposited by rainfall (wet deposition) causing contamination of water and soil (Chen et al. 2014). Energy production units, for example, coal-burning power plants, nuclear power stations, and petroleum combustion also emit different toxic HMs (Liao et al. 2016; Chen et al. 2014).

19.3 Factors Influencing the Mobility and HMs Accumulation in Vegetables

Several soil factors controlled the mobility and accumulation of HMs in vegetables from agricultural soils. The pH values of agricultural soils, an important factor, play a pivotal part in controlling the solubility of HMs. For instance, mobility of HMs increased at acidic pH whereas decreased at alkaline pH (Sheoran et al. 2016). This is because of the adsorption of HMs onto the surfaces of negatively charge soil constituents such as organic matter, the mineral-based clays such as silicates and others as well as the (hydro) oxides of Mn, Al, and Fe. Similarly, the anion exchange capacity (AEC) increases at acidic pH owing to an increase in overall net positive charge which enhanced the bioavailability of HMs and vice versa (Bhargava et al. 2012). Additionally, the presence of organic components in the soil also restricts the solubility of HMs due to the occurrence of more active binding sites and the abundance of ionic and polar functional groups like amino, phenol and carboxyl groups. These functional groups are released from the breakdown of fulvic and humic acids which are soluble at all pH levels. Inner sphere complexation, adsorption, and ion exchange are the key mechanisms involved in retaining HMs by organic matter (Evans 1989). The bioavailability of HMs in agricultural soils was also increased due to a rise in temperature owing to the rapid breakdown of organic matter (Silveira et al. 2003). For instance, rise in temperature significantly increased Zn and Cd transfer from the soil to different parts of plants (Cornu et al. 2016). Likewise, the soil texture also affects the uptake and bioaccumulation of HMs in vegetables. The highest bioavailability of HMs was observed in sand and loam followed by fine-textured and clay loam soils due to the abundant fine pores in fine-textured soils compared to coarse-textured soils (Sheoran et al. 2010). The lowest bioavailability of HMs was observed in soils having higher CEC values such as clay due to their much high adsorption potential (Bhargava et al. 2012).

19.3.1 Factors Associated with Vegetables

The accumulation of HMs in different vegetables varied among them owing to different morphological, physiological, and anatomical traits of plants (Yadav et al. 2018). Branch density, leaf inclination angle, stomata size and density, leaf area, the structure and shape of plant canopy are other factors that favor HMs accumulation in vegetables from aerial deposition (Shahid et al. 2017). Likewise, the transpiration rate also controls HMs uptake and their accumulation in vegetables. Initially, HMs are absorbed by the root apoplast and later ascend with transpiration channels via xylem tissues. Later, HMs were transported to aerial parts of vegetables and subsequently accumulated under the influence of transpiration. Plants that have high and flourishing transpiration rates accumulate higher quantities of HMs. Thus, leafy vegetables store much larger amounts of HMs than non-leafy vegetables owing to their higher transpiration and translocations rates (Hao et al. 2019). Likewise, the transport of HMs from roots to stem followed by fruit during translocation and transpiration processes is longer in non-leafy vegetables which may be attributed to their much lower accumulation (Khan et al. 2009).

19.4 Accumulation of HMs in Vegetables

The accumulation of HMs in vegetables depends upon several plants (vegetable type) and soil factors (bioavailability). Generally, leafy vegetables are good accumulators of HMs as compared to fruits. For example, spinach and lettuce are more efficient in accumulating Cd, when compared with French beans and peas (Alexander et al. 2006).

Much lower Cd uptake was observed in leafy vegetables compared to solanaceous, roots, alliums, melon, and legumes (Yang et al. 2010). The accumulation of different HMs in the vegetable of six different categories (legume, stalk, melon, solanaceous, root, and leafy vegetables) was investigated grown on HMs contaminated agricultural land. Results suggested that leafy vegetables significantly accumulated the higher concentrations of HMs with the least accumulation in melon vegetables. The Pb, As, and Cd concentrations were found above the threshold levels of food contaminants set by the China National Standard (Zhou et al. 2016). Likewise, the accumulation of Cd, Ni, Cr, As, Pb, and Hg were evaluated in different vegetables and the results suggested that *Chicorium endive* and *Coriandrum sativum* L. accumulated Pb and As respectively, while, *Spinacia oleracea* L as well as *Ipomea aquatica,* Forssk and *Phaseolus vulgaris* L. accumulated Cr, Cd, Hg, and Ni, respectively (Anarado et al. 2019; Kumar et al. 2014). The concentrations of Pb, Ni, Cr, and Cd in *Abelmoschus esculentus* were estimated collected from HMs contaminated soil irrigated with wastewater. *Abelmoschus esculentus* remarkably accumulated the concentrations of these HMs above their recommended values (Balkhair and Ashraf 2016). Leafy vegetables such as spinach, cabbage, parsley, and lettuce were also able to store the higher concentrations of Pb in contrast to stem (garlic and white radish) and fruit vegetables (cucumber, pumpkin, capsicum, green beans, and eggplant). However, average values of As, Cr, Se, and Zn in vegetables were higher than their standard values (Cao et al. 2014). Likewise, concentrations of numerous HMs were also assessed in radish, tomato, lady finger, cauliflower, brinjal, spinach, and cabbage (Chauhan and Chauhan 2014). Reportedly, much higher transport of different HMs in roots, stems, and leaves were observed in onion, lettuce, cabbage, and spinach. All reported values were higher than their standard values set by FAO and the WHO/EU combined limits (Akan et al. 2013).

19.5 Toxic Effects of HMs on Vegetables After Their Accumulation

Different plants show variable toxic symptoms on exposure to higher concentrations of HMs. Biomass reduction, growth inhibition, alterations in photosynthesis pigments, restricted water uptake are the usual key indicators of HMs toxicity in plants (Edelstein and Ben-Hur 2018; Sridhar et al. 2011). Numerous studies revealed that HMs stress in plants alters their spectral reflectance, which could cause different biochemical and physiological disorders in them and thus influence nutrients uptake by the vegetables (Sridhar et al. 2017, 2011). Interface with key nucleic acids, (de) activation of essential enzymes, disturbance in electron transport pathways and membrane injury are the known HMs toxicity in plants at the cellular level (Chen et al. 2003). For instance, the higher Cd uptake in lettuce caused a significant reduction of shoot biomass owing to Cd-induced chromosomal aberration (Monteiro et al. 2009; Seregin and Kozhevnikova 2006). Furthermore, alterations in protein synthesis, photosynthetic pigments, and respiration rates significantly reduced morphological traits of leaves of different plants grown on HMs contaminated soils (Chaves et al. 2011). Similarly, the excessive uptake and accumulation of HMs in vegetables resulted in the overproduction of oxygen-based non-radical species such as hydrogen peroxide (H_2O_2), organic hydroperoxide (ROOH), and singlet oxygen as well as oxygen-based free radicals such as peroxyl ($RO_2^{\bullet}$), alkoxyl ($RO^{\bullet}$), hydroxyl (OH•) and superoxide anion radicals ($O_2^{\bullet-}$) (Shahid et al. 2014; Circu and Aw 2010).

19.6 Human Health After the Exposure to HMs Through the Intake of Contaminated Vegetables

The substantial accumulation of HMs in vegetables is of serious concern due to damaging human health even in much lower concentrations (Manzoor et al. 2018). Toxic HMs entered into the food chain via soil-plant-humans and soil-plant-animal-humans pathways, which caused detrimental effects in humans after exposure (Edelstein and Ben-Hur 2018; McLaughlin et al. 2000). Nevertheless, the biotoxic effects of HMs entirely depend upon the total and bioavailable concentrations, speciation, time, and dose of exposure (Manzoor et al. 2018). The ingestion of HMs contaminated vegetables resulted in the depletion of certain crucial nutrients in humans which further caused malnutrition disabilities, growth retardation, neurological and immunological disorders, renal failure, reduced intellectual abilities as well as gastrointestinal and other types of cancer (Türkdogan et al. 2003; Iyengar and Nair 2000). Chronic or acute Pb poisoning damages the gastrointestinal tract and the central nervous system in children (Markowitz 2000). Likewise, appetite loss, abdominal pain, hallucinations, headache, fatigue, arthritis, hypertension, and kidney failure are the symptoms of acute Pb exposure (Khan et al. 2020; Jaishankar et al. 2014). Long-lasting contact with Pb caused congenital disabilities, autism, and damage to brain tissues, dyslexia, hyperactivity, muscular weakness, a significant reduction in weight, psychosis, and even could lead to death (Martin and Griswold 2009). Abnormal heartbeat, leukocytes, vomiting, nausea, damage to blood vessels, reduction of erythrocytes as well as pricking feelings in different body parts, while cancer, hypertension, cardiovascular failure, diabetes mellitus, skin itching, neurological, peripheral, and pulmonary disorders are the common symptoms of acute and chronic As poisoning in humans (Smith et al. 2002). Likewise, the negative impacts of HMs in pregnant women and on the growth of the fetus have been substantially available in the literature. For instance, exposure to HMs affects the ovary resulting in damage to the female reproductive system and disturbing the hormonal production and their discharge mechanisms (Silberstein et al. 2006). Exposure to Pb during pregnancy caused its accumulation in the blood which resulted in premature birth, weight loss in neonates, stillbirths, and hypertension, and even spontaneous abortions (Grant et al. 2013).

19.7 Prediction of Health Risks Associated with Contaminated Vegetables Through Different Models

19.7.1 Risk Evaluation Theory

The risk evaluation process is adopted to determine the health effects caused by HMs in humans after exposure to them. The risk assessment approach mainly contains (i) hazard determination, (ii) exposure estimation, (iii) toxicity assessment (dose-response), and (iv) risk classification. Hazard determination mainly aims to examine the presence, amount, and spatial dispersion of HMs in an ecosystem in a given time (Chen et al. 2015; Huang et al. 2014; Shakoor et al. 2017). In recent findings, many researchers identified the presence of HMs in the ecosystem owing to natural or anthropogenic events recognized as a possible hazard for the community. Different risk assessment models are being used to evaluate potential hazards from these HMs after the acute and chronic exposures (Kamunda et al. 2016; Zhou et al. 2016).

19.7.2 Estimating the Daily HMs Intake

Different methods have been used to estimate health risk assessment based on Provisional Tolerable Daily Intake (PTDI) by consuming HMs enriched vegetables (Chary et al. 2008). The expression for the estimation of daily HMs intake is as follows

$$\mathrm{DIM} = C_{\mathrm{metal}} \times C_{\mathrm{factor}} / B_{\mathrm{averageweight}}$$

In the above expression C_{metal}, C_{factor}, $D_{\mathrm{food\ intake}}$ and $B_{\mathrm{average\ weight}}$ represent HMs concentration in vegetable (mg kg^{-1}), conversion factor, daily intake of HMs enriched vegetables, and average body weight, respectively. The values of DIM were higher for vegetable samples collected from wastewater irrigation zone in contrast to vegetables irrigated with groundwater (Mahmood and Malik 2014).

19.7.3 Hazard Quotients

The hazard quotient index has been previously used to estimate the human health risks associated with HMs intake after consuming vegetables. It is the ratio between the estimated and the standard doses (RD). If the ratio value is less than 1 represents no risk to humans from exposure to toxic HMs. If the values of HQ are equal or greater than 1, it shows a high risk to populations. The expression of HQ is given below

$$\mathrm{HQ} = \left[W_{\mathrm{plant}}\right] \times \left[\mathrm{Metal}_{\mathrm{plant}}\right] / R_f D \times B$$

In the above equation, W_{plant} is the dry weight of HMs in the consumable parts of vegetables (mg d^{-1}), M_{plant} represents the amount of HMs in vegetables (mg kg^{-1}), R_fD expressed standard of reference dose of a HM for food (mg d^{-1}), and B expressed the average body weight (kg).

19.7.4 Health Risk Index

The health risk index calculates the relationship between daily HM intake and standard dose. The mathematical expression of HRI is as follows

$$\mathrm{HRI} = \mathrm{DIM} / R_f D$$

It is assumed that the population is at higher risk if HRI values are found higher than 1 in them. Results of HRI revealed that the consumption of HMs contaminated vegetables poses a serious health risk to humans. It was mainly due to irrigation with wastewater having very higher HMs concentrations (Mahmood and Malik 2014).

19.7.5 Carcinogenic Risk

The populations consuming HMs contaminated vegetables may experience cancer risk, which is estimated by the following expression.

$$\mathrm{CR} = \mathrm{CDI} \times \mathrm{SF}$$

Cancer risk is 10–100 times higher in children exposed to Ni and Cr by consuming contaminated foodstuff. Likewise, As also possess serious potential carcinogenic risk in children when exceeded from its tolerable level (Cao et al. 2014).

19.8 Management of HMs Contaminated Soils for Safer Vegetable Production

This section covers different management strategies that remove, render or reduce the uptake of higher concentrations of HMs by the vegetables from the soil environment.

19.8.1 Phytoremediation

Phytoremediation is a "green solution" technique that involve plants to partially or eliminate HMs from the environment (Ali et al. 2013). It can also be used with other remediation methods such as immobilization and other primitive methods as the final step in the remediation process (Radziemska et al. 2019, 2020). Phytoremediation has several advantages such as being cost-effective, high acceptance rate by the community, no harm to the environment, controlling HMs from the root zones of trees, minimal risk of secondary pollution as well as the potential to eliminate multiple HMs from a single site (Tauqeer et al. 2019). Poor plant establishment, growth inhibition because of HMs toxicity, prior knowledge about the site and environmental conditions, required large time, increased solubility and transport of HMs which further enhanced the risk of secondary pollution are the disadvantages of phytoremediation (Tauqeer et al. 2019).

19.8.2 Immobilization

In recent years, the in-situ immobilization remediation method has gained the attention of scientists worldwide owing to its vast applicability, easy availability of raw materials as well as lower labor and energy requirements (Zhai et al. 2018). Numerous organic and inorganic amendments have been known to reduce HMs uptake by vegetables grown on HMs polluted soils (Arshad et al. 2016; Kumar and Chopra 2014). These amendments not only reduced HMs uptake by the vegetables but also improved soil conditions that further supported plant establishment and maintain their nutritional quality (Xu et al. 2019). Likewise, iron and silicon-rich material significantly increased the growth of *B. Chinensis* by reducing As and Cd uptake compared to alkaline clay and synthetic zeolite (Yao et al. 2017). Phosphorus (P) is also a key component of vegetables development in the agricultural system. Phosphorus applications also significantly control HMs uptake by forming a stable metal complex, increasing soil pH and CEC (Yin et al. 2016).

Organic materials have also been considered to be effective additives in reducing HMs bioavailability in agricultural soils (Shan et al. 2016). Compost, pig manure, and wheat straw had noticeably restricted Cd transport to the roots and aerial parts of radish. During the experiment, it was observed that pig manure was the most efficient amendment in reducing Cd uptake compared to wheat straw (Shan et al. 2016). Similarly, in a field experiment, poultry, swine, and cattle manure were added to the Cd polluted soil during a four-year vegetable production period. It was noticed that these amendments had significantly decreased Cd concentrations and its uptake by spinach (Sato et al. 2010). Likewise, biochar, "a substance produced from organic residues such as agricultural wastes, plant, and animal wastes" under the limited supply of oxygen, has recently gained the attention of scientists worldwide due to its vast applications as fertilizer and potential amendment in immobilizing numerous environmental contaminants (Awad et al. 2017; Woldetsadik et al. 2016; Wang et al. 2015). Biochar applications have significantly increased the growth of turnips (*Brassica rapa* L.) by lowering HMs uptake. It was observed that peanut shell-derived biochar was efficient in decreasing HMs uptake by turnips in contrast to soybean, sewage sludge, and rice straw amendments (Khan et al. 2015). Furthermore, paper-mill sludge biochar had also considerably reduced Zn and Cd uptake, while improving the yield of lettuce (Kim et al. 2015). Similarly, biochar applications also reduced HMs concentrations in garlic (Song et al. 2014), Jack bean (Puga et al. 2015) and pepper (Xu et al. 2016).

19.8.3 Water Management Strategies

Constant and prolonged water applications also influence the HMs accumulation in soils and vegetables. Irrigation of contaminated agricultural lands with water significantly increased HMs uptake by vegetables at their critical growth (Tack et al. 2017). However, continuous and long-term field monitoring is required to explore this fact. Likewise, irrigation of arable lands with fresh and surface waters as well as municipal and industrial wastewaters influence HMs accumulation in vegetables (Asgari and Cornelis 2015; Qureshi et al. 2016). Additionally, modes of water use such as surface, drip, and other irrigation practices may also reduce HMs accumulation in soil profile and vegetables grown on them. Reportedly, the use of subsurface pressure-compensating drip irrigation method was able to reduce HMs accumulation in the soil profile and cauliflower curds (Singh et al. 2020).

19.8.4 Soil Applications of Different Microbial Inocula

Soil-microbe-plant interaction plays a key role owing to its potential in improving the growth, yield, nutritional quality, and restricting HMs accumulation in plants. This interaction not only increased microbial mediated HMs tolerance in plants but also improved the overall traits of plants (Tiwari and Lata 2018).

This possibly could be due to precipitation, absorption, and accumulation of HMs in the cell walls of microbes, conversion of HMs into less toxic form through oxidation-reduction reactions, exclusion of HMs from their cell as well as encapsulation (Tiwari and Lata 2018 and references therein). Likewise, the applications of arbuscular mycorrhizal fungi (AMF) in arable lands polluted with HMs have been extensively revealed (Riaz et al. 2020; Chang et al. 2018). Arbuscular mycorrhizal fungi are unique and diverse microorganisms directly associated with the host plant and soil, increasing the minerals and water acquisition and their uptake by the plants which ensure plant establishment under HMs stress (Khan et al. 2020). The presence of AMF in HMs contaminated soils encourage the plant growth through developing root system, by improving the growth and surface area of root hair which increased nutrient acquisition under HMs stress (Pavithra and Yapa 2018).

19.9 Conclusion and Way Forward

Vegetables are the key component of the human diet and provide essential mineral nutrients to maintain numerous physiological functions. Also, they are a good accumulator of HMs without showing any toxic symptoms and pose a severe risk to human health after exposure by consuming HMs contaminated vegetables. Thus, there is a need to take effective remedial measures to control HMs accumulation in vegetables grown on contaminated soils. Applications of different novel remediation techniques such as phytoremediation, water management strategies and utilization of microbial inocula control HMs accumulation in vegetables. It is further suggested that more lab-scale and field studies are required to understand different mechanisms occurring on molecular levels that affect the nutritional components of vegetables produced from HMs contaminated soils.

References

Agrawal SB, Singh A, Sharma RK, Agrawal M (2007) Bioaccumulation of heavy metals in vegetables: a threat to human health. Terr Aquat Environ Toxicol 1:13–23

Akan JC, Kolo BG, Yikala BS, Ogugbuaja VO (2013) Determination of some heavy metals in vegetable samples from Biu local government area, Borno State, North Eastern Nigeria. Int J Environ Monit Anal 1:40–46

Alexander PD, Alloway BJ, Dourado AM (2006) Genotypic variations in the accumulation of Cd, Cu, Pb and Zn exhibited by six commonly grown vegetables. Environ Pollut 144:736–745

Alghobar MA, Suresha S (2017) Evaluation of metal accumulation in soil and tomatoes irrigated with sewage water from Mysore city, Karnataka, India. J Saudi Soc Agric Sci 16:49–59

Ali H, Khan E, Sajad MA (2013) Phytoremediation of heavy metals—concepts and applications. Chemosphere 91:869–881

Anarado CE, Anarado CJO, Okeke MO, Ezeh CE, Umedum NL, Okafor PC (2019) Leafy vegetables as potential pathways to heavy metal hazards. J Agric Chem Environ 8:23

Arshad M, Ali S, Noman A, Ali Q, Rizwan M, Farid M, Irshad MK (2016) Phosphorus amendment decreased cadmium (Cd) uptake and ameliorates chlorophyll contents, gas exchange attributes, antioxidants, and mineral nutrients in wheat (Triticum aestivum L.) under Cd stress. Arch Agron Soil Sci 62:533–546

Asgari K, Cornelis WM (2015) Heavy metal accumulation in soils and grains, and health risks associated with use of treated municipal wastewater in subsurface drip irrigation. Environ Monit Assess 187:410

Awad GE, Wehaidy HR, Abd El Aty AA, Hassan ME (2017) A novel alginate–CMC gel beads for efficient covalent inulinase immobilization. Colloid Polym Sci 295:495–506

Balkhair KS, Ashraf MA (2016) Field accumulation risks of heavy metals in soil and vegetable crop irrigated with sewage water in western region of Saudi Arabia. Saudi J Biol Sci 23:32–44

Bhargava A, Carmona FF, Bhargava M, Srivastava S (2012) Approaches for enhanced phytoextraction of heavy metals. J Environ Manage 105:103–120

Cao S, Duan X, Zhao X, Ma J, Dong T, Huang N, Wei F (2014) Health risks from the exposure of children to As, Se, Pb and other heavy metals near the largest coking plant in China. Sci Total Environ 472:1001–1009

Chabukdhara M, Munjal A, Nema AK, Gupta SK, Kaushal RK (2016) Heavy metal contamination in vegetables grown around peri-urban and urban-industrial clusters in Ghaziabad, India. Hum Ecol Risk Assess Int J 22(3):736–752

Chang Q, Diao FW, Wang QF, Pan L, Dang ZH, Guo W (2018) Effects of arbuscular mycorrhizal symbiosis on growth, nutrient and metal uptake by maize seedlings (Zea mays L.) grown in soils spiked with Lanthanum and Cadmium. Environ Pollut 241:607–615

Chary NS, Kamala CT, Raj DSS (2008) Assessing risk of heavy metals from consuming food grown on sewage irrigated soils and food chain transfer. Ecotox Environ Safe 69(3):513–524

Chauhan G, Chauhan UK (2014) Human health risk assessment of heavy metals via dietary intake of vegetables grown in wastewater irrigated area of Rewa, India. Int J Sci Res Publ 4(9):1–9

Chaves LHG, Estrela MA, de Souza RS (2011) Effect on plant growth and heavy metal accumulation by sunflower. J Phytol 3(12)

Chen YX, Lin Q, Luo YM, He YF, Zhen SJ, Yu YL, Wong MH (2003) The role of citric acid on the phytoremediation of heavy metal contaminated soil. Chemosphere 50(6):807–811

Chen Y, Wu P, Shao Y, Ying Y (2014) Health risk assessment of heavy metals in vegetables grown around battery production area. Sci Agric 71:126–132

Chen M, Xu P, Zeng G, Yang C, Huang D, Zhang J (2015) Bioremediation of soils contaminated with polycyclic aromatic hydrocarbons, petroleum, pesticides, chlorophenols and heavy metals by composting: applications, microbes and future research needs. Biotechnol Adv 33(6):745–755

Circu ML, Aw TY (2010) Reactive oxygen species, cellular redox systems, and apoptosis. Free Radical Biol Med 48(6):749–762

Clarke BO (2011) Review of emerging organic contaminants in biosolids and assessment of Turan

Cornu JY, Bakoto R, Bonnard O, Bussiere S, Coriou C, Sirguey C, Nguyen C (2016) Cadmium uptake and partitioning during the vegetative growth of sunflower exposed to low Cd^{2+} concentrations in hydroponics. Plant Soil 404(1–2):263–275

De Juan JA, Tarjuelo JM, Valiente M, Garcia P (1996) Model for optimal cropping patterns within the farm based on crop water production functions and irrigation uniformity I: Development of a decision model. Agric Water Manag 31(1–2):115–143

Duruibe JO, Ogwuegbu MOC, Egwurugwu JN (2007) Heavy metal pollution and human biotoxic effects. Int J Phys Sci 2(5):112–118

Edelstein M, Ben-Hur M (2018) Heavy metals and metalloids: Sources, risks and strategies to reduce their accumulation in horticultural crops. Sci Hortic 234:431–444

El-Kady AA, Abdel-Wahhab MA (2018) Occurrence of trace metals in foodstuffs and their health impact. Trends Food Sci Technol 75:36–45

Emamverdian A, Ding Y, Mokhberdoran F, Xie Y (2015) Heavy metal stress and some mechanisms of plant defense response. Sci World J 2015:756–772

Evans RM, Palmiter RD, Brinster RL (1989) US Patent No. 4,870,009. US Patent and Trademark Office, Washington, DC

Gadd GM (2010) Metals, minerals and microbes: geo microbiology and bioremediation. Microbiology 156 (3):609–643

Gall JE, Boyd RS, Rajakaruna N (2015) Transfer of heavy metals through terrestrial food webs: a review. Environ Monit Assess 187:201–214

Grant K, Goldizen FC, Sly PD, Brune MN, Neira M, van den Berg M, Norman RE (2013) Health consequences of exposure to e-waste: a systematic review. Lancet Glob Health 1(6):350–361

Gupta N, Yadav KK, Kumar V, Kumar S, Chadd RP, Kumar A (2018) Trace elements in soil-vegetables interface: translocation, bioaccumulation, toxicity and amelioration—a review. Sci Total Environ 651:2927–2942

Gupta N, Gedam VV, Moghe C, Labhasetwar P (2019) Comparative assessment of batch and column leaching studies for heavy metals release from coal fly ash bricks and clay bricks. Environ Technol Innov 16:1004–1046

Hao J, Wei Z, Wei D, Mohamed TA, Yu H, Xie X, Zhao Y (2019) Roles of adding biochar and montmorillonite alone on reducing the bioavailability of heavy metals during chicken manure composting. Biores Technol 294:122–199

Huang Z, Pan XD, Wu PG, Han JL, Chen Q (2014) Heavy metals in vegetables and the health risk to population in Zhejiang China. Food Control 36 (1):248–252

Iyengar GV, Nair PP (2000) Global outlook on nutrition and the environment: meeting the challenges of the next millennium. Sci Total Environ 249(1–3):331–346

Jaishankar M, Tseten T, Anbalagan N, Mathew BB, Beeregowda KN (2014) Toxicity, mechanism and health effects of some heavy metals. Interdiscip Toxicol 7(2):60–72

Jan AT, Ali A, Haq Q (2011) Glutathione as an antioxidant in inorganic mercury induced nephrotoxicity. J Postgrad Med 57:72–77

Kamunda C, Mathuthu M, Madhuku M (2016) Health risk assessment of heavy metals in soils from Witwatersrand gold mining basin, South Africa. Int J Environ Res Public Health 13:663

Khan S, Farooq R, Shahbaz S, Khan MA, Sadique M (2009) Health risk assessment of heavy metals for population via consumption of vegetables. World Appl Sci J 6(12):1602–1606

Khan S, Waqas M, Ding F, Shamshad I, Arp HPH, Li G (2015) The influence of various biochars on the bioaccessibility and bioaccumulation of PAHs and potentially toxic elements to turnips (*Brassica rapa* L.). J Hazard Mater 300:243–253

Khan MA, Ramzani PMA, Zubair M, Rasool B, Khan MK, Ahmed A, Iqbal M (2020) Associative effects of lignin-derived biochar and arbuscular mycorrhizal fungi applied to soil polluted from Pb-acid batteries effluents on barley grain safety. Sci Total Environ 710:136294

Kihampa C, Mwegoha WJS, Shemdoe RS (2011) Heavy metal concentrations in vegetables grown in the vicinity of the closed dumpsite. Int J Environ Sci 2:889–895

Kim HS, Kim KR, Kim HJ, Yoon JH, Yang JE, Ok YS, Kim KH (2015) Effect of biochar on heavy metal immobilization and uptake by lettuce (Lactuca sativa L.) in agricultural soil. Environ Earth Sci 74:1249–1259

Kumar V, Chopra AK (2014) Ferti-irrigation effect of paper mill effluent on agronomical practices of Phaseolus vulgaris (L.) in two seasons. Commun Soil Sci Plan 45:2151–2170

Kumar V, Thakur RK, Kumar P (2019) Assessment of heavy metals uptake by cauliflower (Brassica oleracea var. botrytis) grown in integrated industrial effluent irrigated soils: a prediction modeling study. Sci Hortic 257:108–132

Lente I, Ofosu-Anim J, Brimah AK, Atiemo S (2014) Heavymetal pollution of vegetable crops irrigated with wastewater in Accra, Ghana. West Afr J App Ecol 22:41–58

Li N, Kang Y, Pan W, Zeng L, Zhang Q, Luo J (2015) Concentration and transportation of heavy metals in vegetables and risk assessment of human exposure to bioaccessible heavy metals in soil near a waste-incinerator site, South China. Sci Total Environ 521:144–151

Liao J, Wen Z, Ru X, Chen J, Wu H, Wei C (2016) Distribution and migration of heavy metals in soil and crops affected by acid mine drainage: Public health implications in Guangdong Province, China. Ecotox Environ Safe 124:460–469

Luo L, Ma Y, Zhang S, Wei D, Zhu YG (2009) An inventory of trace element inputs to agricultural soils in China. J Environ Manag 90:2524–2530

Mahmood A, Malik RN (2014) Human health risk assessment of heavy metals via consumption of contaminated vegetables collected from different irrigation sources in Lahore, Pakistan. Arab J Chem 7:91–99

Manzoor J, Sharma M, Wani KA (2018) Heavy metals in vegetables and their impact on the nutrient quality of vegetables: a review. J Plant Nutr 41:1744–1763

Markowitz M (2000) Lead poisoning: a disease for the next millennium. Curr Probl Pediatr 30:62–70

Martin S, Griswold W (2009) Human health effects of heavy metals. Environ Sci Technol Briefs Citizens 15:1–6

McLaughlin MJ, Zarcinas BA, Stevens DP, Cook N (2000) Soil testing for heavy metals. Commun Soil Sci Plan 31:1661–1700

Monteiro MS, Santos C, Soares AMVM, Mann RM (2009) Assessment of biomarkers of cadmium stress in lettuce. Ecotox Environ Safe 72(3):811–818

Nagajyoti PC, Lee KD, Sreekanth TVM (2010) Heavy metals, occurrence and toxicity for plants: a review. Environ Chem Lett 8(3):199–216

Oves M, Khan MS, Zaidi A, Ahmad E (2012) Soil contamination, nutritive value, and human health risk assessment of heavy metals: an overview. Toxicol Heavy Metals Leg Biorem 23:1–27

Pandey B, Suthar S, Singh V (2016) Accumulation and health risk of heavy metals in sugarcane irrigated with industrial effluent in some rural areas of Uttarakhand, India. Process Saf Environ Prot 102:655–666

Parashar P, Prasad FM (2013) Study of heavy metal accumulation in sewage irrigated vegetables in different regions of Agra District India. Open J Soil Sci 3 (1):1–8

Pavithra D, Yapa N (2018) Arbuscular mycorrhizal fungi inoculation enhances drought stress tolerance of plants. Groundw Sustain Dev 7:490–494

Puga AP, Abreu CA, Melo LCA, Paz-Ferreiro J, Beesley L (2015) Cadmium, lead, and zinc mobility and plant uptake in a mine soil amended with sugarcane straw biochar. Environ Sci Pollut Res 22:17606–17614

Qureshi AS, Hussain MI, Ismail S, Khan QM (2016) Evaluating heavy metal accumulation and potential health risks in vegetables irrigated with treated wastewater. Chemosphere 163:54–61

Radziemska M, Wyszkowski M, Bęś A, Mazur Z, Jeznach J, Brtnický M (2019) The applicability of compost, zeolite and calcium oxide in assisted remediation of acidic soil contaminated with Cr (III) and Cr (VI). Environ Sci Pollut Res 26:21351–21362

Radziemska M, Bęś A, Gusiatin ZM, Cerdà A, Jeznach J, Mazur Z, Brtnický M (2020) Assisted phytostabilization of soil from a former military area with mineral amendments. Ecotoxicol Environ Saf 188:109–124

Rai PK, Kumar V, Lee SS, Naddem R, Ok YS, Kim KH, Tsang DSW (2018) Nanoparticle plant interaction: implications in energy, the environment, and agriculture. Environ Int 119:1–19

Ratul AK, Hassan M, Uddin MK, Sultana MS, Akbor MA, Ahsan MA (2018) Potential health risk of heavy metals accumulation in vegetables irrigated with polluted river water. Int Food Res J 25:44–57

Riaz M, Kamran M, Fang Y, Wang Q, Cao H, Yang G, Wang X (2020). Arbuscular mycorrhizal fungi-induced mitigation of heavy metal phytotoxicity in metal contaminated soils: a critical review. J Hazard Mater 123919

Sato A, Takeda H, Oyanagi W, Nishihara E, Murakami M (2010) Reduction of cadmium uptake in spinach (Spinacia oleracea L.) by soil amendment with animal waste compost. J Hazard Mater 181(1–3):298–304

Saumel I, Kotsyuk I, Hölscher M, Lenkereit C, Weber F, Kowarik I (2012) How healthy is urban horticulture in high traffic areas? trace metal concentrations in vegetable crops from plantings within inner city neighbourhoods in Berlin, Germany. Environ Pollut 165:124–132

Seregin I, Kozhevnikova AD (2006) Physiological role of nickel and its toxic effects on higher plants. Russ J Plant Physiol 53:257–277

Shahbaz AK, Iqbal M, Jabbar A, Hussain S, Ibrahim M (2018) Assessment of nickel bioavailability through chemical extractants and red clover (Trifolium pratense L.) in an amended soil: related changes in various parameters of red clover. Ecotox Environ Safe 149:116–127

Shahid M, Austruy A, Echevarria G, Arshad M, Sanaullah M, Aslam M, Dumat C (2014) EDTA-enhanced phytoremediation of heavy metals: a review. Soil Sediment Contam: an Int J 23:389–416

Shahid M, Dumat C, Khalid S, Schreck E, Xiong T, Niazi NK (2017) Foliar heavy metal uptake, toxicity and detoxification in plants: a comparison of foliar and root metal uptake. J Hazard Mater 325:36–58

Shakoor MB, Nawaz R, Hussain F, Raza M, Ali S, Rizwan M, Ahmad S (2017) Human health implications, risk assessment and remediation of As-contaminated water: a critical review. Sci Total Environ 601:756–769

Shan H, Su S, Liu R, Li S (2016) Cadmium availability and uptake by radish (Raphanus sativus) grown in soils applied with wheat straw or composted pig manure. Environ Sci Pollut Res 23:15208–15217

Sheoran V, Sheoran AS, Poonia P (2010) Soil reclamation of abandoned mine land by revegetation: a review. Int J Soil Sediment Water 3:13–25

Sheoran V, Sheoran AS, Poonia P (2016) Factors affecting phytoextraction: a review. Pedosphere 26:148–166

Silberstein T, Saphier O, Paz-Tal O, Trimarchi JR, Gonzalez L, Keefe DL (2006) Lead concentrates in ovarian follicle compromises pregnancy. J Trace Elem Med Biol 20:205–207

Silveira MLA, Alleoni LRF, Guilherme LRG (2003) Biosolids and heavy metals in soils. Sci Agric 60:793–806

Singh S, Parihar P, Singh R, Singh VP, Prasad SM (2015) Heavy metal tolerance in plants: role of transcriptomics, proteomics, metabolomics, and ionomics. Front Plant Sci 6:11–43

Singh D, Patel N, Patra S, Singh N, Roy T, Caucci S, Hettiarachchi H (2020). Efficacy of drip irrigation in controlling heavy-metal accumulation in soil and crop. J Environ Eng Sci :1–13

Slavin JL, Lloyd B (2012) Health benefits of fruits and vegetables. Adv Nutr 3:506–516

Smith AH, Lopipero PA, Bates MN, Steinmaus CM (2002) Arsenic epidemiology and drinking water standards. Science 296(5576):2145–2146

Song XD, Xue XY, Chen DZ, He PJ, Dai XH (2014) Application of biochar from sewage sludge to plant cultivation: Influence of pyrolysis temperature and biochar-to-soil ratio on yield and heavy metal accumulation. Chemosphere 109:213–220

Sridhar BM, Vincent RK, Roberts SJ, Czajkowski K (2011) Remote sensing of soybean stress as an indicator of chemical concentration of biosolid amended surface soils. Int J Appl Earth Obs 13 (4):676–681

Sridhar SGD, Sakthivel AM, Sangunathan U, Balasubramanian M, Jenefer S, Rafik MM, Kanagaraj G (2017) Heavy metal concentration in groundwater from besant nagar to sathankuppam, south Chennai, Tamil nadu, India. Appl Water Sci 7:4651–4662

Tack FM (2017) Watering regime influences Cd concentrations in cultivated spinach. J Environ Manage 186:201–206

Tauqeer HM, Hussain S, Abbas F, Iqbal M (2019) The potential of an energy crop Conocarpus erectus for lead phytoextraction and phytostabilization of chromium, nickel, and cadmium: an excellent option for the management of multi-metal contaminated soils. Ecotoxicol Environ Saf 173:273–284

Tiwari S, Lata C (2018) Heavy metal stress, signaling, and tolerance due to plant-associated microbes: an overview. Front Plant Sci 9:452

Toth G, Hermann T, Da Silva MR, Montanarella L (2016) Heavy metals in agricultural soils of the European Union with implications for food safety. Environ Int 88:299–330

Turan V, Khan SA, Iqbal M, Ramzani PMA, Fatima M (2018) Promoting the productivity and quality of brinjal aligned with heavy metals immobilization in a wastewater irrigated heavy metal polluted soil with biochar and chitosan. Ecotox Environ Safe 161:409–419

Türkdoğan MK, Kilicel F, Kara K, Tuncer I, Uygan I (2003) Heavy metals in soil, vegetables and fruits in the endemic upper gastrointestinal cancer region of Turkey. Environ Toxicol Pharmacol 13:175–179

Wang T, Sun H, Mao H, Zhang Y, Wang C, Zhang Z, Sun L (2014) The immobilization of heavy metals in soil by bioaugmentation of a UV-mutant Bacillus subtilis 38 assisted by NovoGro biostimulation and changes of soil microbial community. J Hazard Mater 278:483–490

Wang ZJ, Ghasimi S, Landfester K, Zhang KA (2015) Photocatalytic suzuki coupling reaction using conjugated microporous polymer with immobilized palladium nanoparticles under visible light. Chem Mater 27:1921–1924

Woldetsadik D, Drechsel P, Keraita B, Marschner B, Itanna F, Gebrekidan H (2016) Effects of biochar and alkaline amendments on cadmium immobilization, selected nutrient and cadmium concentrations of lettuce (Lactuca sativa) in two contrasting soils. Springerplus 5:397–409

Woldetsadik D, Drechsel P, Keraita B, Itanna F, Gebrekidan H (2017) Heavy metal accumulation and health risk assessment in wastewater-irrigated urban vegetable farming sites of Addis Ababa, Ethiopia. Int J Food Contam 4:9–27

Xu G, Zhang Y, Sun J, Shao H (2016) Negative interactive effects between biochar and phosphorus fertilization on phosphorus availability and plant yield in saline sodic soil. Sci Total Environ 568:910–915

Xu C, Qi J, Yang W, Chen Y, Yang C, He Y, Lin A (2019) Immobilization of heavy metals in vegetable-growing soils using nano zero-valent iron modified attapulgite clay. Sci Total Environ 686:476–483

Yadav KK, Gupta N, Kumar A, Reece LM, Singh N, Rezania S, Khan SA (2018) Mechanistic understanding and holistic approach of phytoremediation: a review on application and future prospects. Ecol Eng 120:274–298

Yang J, Guo H, Ma Y, Wang L, Wei D, Hua L (2010) Genotypic variations in the accumulation of Cd exhibited by different vegetables. J Environ Sci 22:1246–1252

Yang Q, Li Z, Lu X, Duan Q, Huang L, Bi J (2018) A review of soil heavy metal pollution from industrial and agricultural regions in China: pollution and risk assessment. Sci Total Environ 642:690–700

Yao Y, Sun Q, Wang C, Wang PF, Ding SM (2017) Evaluation of organic amendment on the effect of cadmium bioavailability in contaminated soils using the DGT technique and traditional methods. Environ Sci Pollut Res 24:7959–7968

Yin H, Tan N, Liu C, Wang J, Liang X, Qu M, Liu F (2016) The associations of heavy metals with crystalline iron oxides in the polluted soils around the mining areas in Guangdong Province, China. Chemosphere 161:181–189

Zhai X, Li Z, Huang B, Luo N, Huang M, Zhang Q, Zeng G (2018) Remediation of multiple heavy metal-contaminated soil through the combination of soil washing and in situ immobilization. Sci Total Environ 635:92–99

Zhou H, Yang WT, Zhou X, Liu L, Gu JF, Wang WL, Zou JL, Tian T, Peng PQ, Liao BH (2016) Accumulation of heavy metals in vegetable species planted in contaminated soils and the health risk assessment. Int J Environ Res Public Health 13:289–304

Importance of Vermicomposting and Vermiremediation Technology in the Current Era

20

Jackson Durairaj Selvan Christyraj,
Melinda Grace Rossan Mathews,
Ravichandran Subramaniam,
Beryl Vedha Yesudhason,
Karthikeyan Subbiahanadar Chelladurai,
and Johnson Retnaraj Samuel Selvan Christyraj

Abstract

Since prehistoric times the technique of composting has been used by farmers to recycle wastes into useful products that are able to improve plant growth. With industrial growth and population explosion, wastes generated have been significantly improved in previous years. Earthworms have played an important role in the organic waste mitigation by colonizing the wastes and eliminating it by intake, digestion, and assimilation. Today, vermicomposting biotechnology is established all over the globe for its environmental sustainability qualities. Degradation of waste and vermicomposting has proven to be economically and environmentally desired technology over traditional microbial degradation and composting technologies. Vermiremediation can accelerate the process of micro and phytoremediation in combination. Even though it may have many advantages over conventional remediation, certain limitations such as extreme soil conditions, survival rate, feed concentration, and climatic conditions can affect the process.

Keywords

Vermiaccumulation · Vermicomposting · Vermiculture · Vermidegradation · Vermiremediation · Vermitechnology · Vermitransformation

20.1 Introduction

20.1.1 Composting Technology

Significant amounts of organic wastes are derived from the agricultural systems that can be a global threat in the form of soil pollution. The risk of waste accumulation can therefore be mitigated by turning these wastes into something beneficial. Clay tablets from 2300 BC were found with written references to composting, but their use and defined protocols were scientifically studied and published by the beginning of the twentieth century (Fitzpatrick et al.. 2005). Composting can be defined as a technique by which organic wastes are processed into beneficial materials which is a result of biological activities under controlled conditions. Aim of the composting is to maintain the biodegradable organic matter (BOM) level in wastes to minimize the unpleasant odors, combat weed seeds, and pathogens, and finally produce standardized

J. D. S. Christyraj · M. G. R. Mathews · R. Subramaniam · B. V. Yesudhason · K. S. Chelladurai · J. R. S. S. Christyraj (✉)
Regeneration and Stem Cell Biology Lab, Centre for Molecular and Nanomedical Sciences, International Research Centre, Sathyabama Institute of Science and Technology (Deemed to be University), Chennai, Tamil Nadu 600119, India
e-mail: johnnbt@sathyabama.ac.in

J. A. Malik (ed.), *Advances in Bioremediation and Phytoremediation for Sustainable Soil Management*,
https://doi.org/10.1007/978-3-030-89984-4_20

organic fertilizers suitable for various applications in agriculture (Haga 1999).

20.1.2 Vermicomposting and Its Significance

Activities of earthworm such as movement (water infiltration), burrowing (aeration), feeding (waste reduction) and casting (nutrient and microbial biomass enrichment) in the soil have been widely studied for its application in sustainable agriculture, thus they are profoundly known as nature's ploughmen and farmer's friends. Vermicomposting (Latin *vermis* – worm) is a simple mesophilic biological process of composting in which the synergistic effects of earthworms and microorganisms result in the conversion of organic waste into useful vermicompost under favorable conditions. Important parameters such as bedding, feed, moisture content, temperature, pH, salinity and aeration accelerates the composting. Epigeic worms like *Eisenia fetida, Perionyx excavatus, Eudrilus eugeniae* are the most commercially preferred type for the process of composting. Vermicomposting plays an important role in the degradation of biodegradable household and municipal solid wastes, maintenance of soil homeostasis, and production of low-cost biofertilizers.

20.1.3 Concept of Vermiremediation

Vermiremediation is a technology that deals with earthworms to convert organic wastes into valuable compost. Vermiremediation is defined as removing the non-recyclable chemicals from the soil using earthworms (Rodriguez-Campos et al. 2014). It is a cost-effective method that can eliminate agricultural wastes, kitchen debris, garbage, sewage matters with the help of earthworms. Various studies reported that hazardous sources like Polycyclic aromatic hydrocarbons (PAHs), polychlorinated biphenyls (PCBs), heavy metals and pesticides have been successfully resolved by vermiremidiation technology (Rodriguez-Campos et al. 2014; Rorat et al. 2017). Several studies revealed that earthworms naturally biodegrade or bioaccumulate wide range of toxicants like PAHs (Pattnaik and Reddy 2011). The natural ancillary of earthworms with biotic and abiotic factors and ensuing growth of microorganisms assists the vermiremediation process. Vermiremediation technology focuses on the removal of chemical pollutants along with reducing the salinity of soil and enhances the physical, chemical, biological qualities of the soil.

20.2 Vermicomposting

20.2.1 Composition

Nutrient composition of vermicompost can highly differ based on the organic substrate provided. It comprises nitrogen, carbon, phosphorous, calcium, sodium, potassium, magnesium, copper, zinc, manganese, and iron (Nagavallemma et al. 2004). Other than the micro and macronutrients, soluble potassium, nitrates, phosphorous, and calcium were also abundantly present in the vermicompost. Even though they have good nutrient quantities, to compete with the inorganic synthetic fertilizers, rock phosphate and fly ash was used to increase the phosphorous content in the form of phosphate through vermicomposting (Table 20.1).

20.2.2 Vermicultures

Earthworm has inhabited the globe for over 20 million years, feeding on organic matter such as leaf litters and animal excreta at the surface of the soil. Earthworms are natural bioreactors and they belong to the phylum *Annelida* and class *Clitellata*. Vermiculture can be defined as the rearing of earthworms in biodegradable organic matter under controlled conditions. Almost 4400 species of earthworms were widely distributed globally and were classified into three types based on their habitat and behavior. While every species of earthworms has the ability to degrade organic wastes, only few species (8% of total species)

Table 20.1 Composition, nutrient form, quantity, and functions of vermicompost (Garg and Gupta 2009; Ronen 2016)

Nutrient family	Available nutrients	Form	Quantity	Functional roles in plant
Macronutrient	Organic carbon (C)	Carbon dioxide	9.5–17.98%	Structural development
	Nitrogen (N)	Nitrate and ammonia	0.5–1.50%	Formation of protein
	Phosphorous (P)	Phosphate	0.1–0.30%	Synthesis of protein and proper metabolism
	Potassium (K)	Potassium ion	0.15–0.56%	Enzymatic activity and water absorption
	Sulphur (S)	Sulphate	128–548 mg kg^{-1}	Main constituent in amino acids, coenzymes and vitamins
Micronutrient	Calcium (Ca) and Magnesium (Mg)	Calcium and magnesium ion	22.67–47.60 meq/100 g	Root permeability, enzymatic activity, and proper metabolism
	Copper (Cu)	Copper ion	2–9.50 mg kg^{-1}	Proper metabolism
	Iron (Fe)	Ferrous and ferric	2–9.30 mg kg^{-1}	Plays a crucial role in chlorophyll formation and electron transfer
	Zinc (Zn)	Zinc ion	5.70–11.50 mg kg^{-1}	Enzyme activation

Table 20.2 Different vermiculture specifications

Category	Characteristics	Species	Optimum temperature (in Celsius)	Average body weight (in Grams)
Epigeic (upon the earth)	Surface dwelling worms	*Eisenia fetida*	18–25	0.5
		Eudrilus eugeniae	20–25	1.0
		Perionyx excavatus	25–30	1.0
Aneic (out of the earth)	Deep burrowing worms makes vertical burrows	*Lampito mauriti*	18–30	1.0
		Drawida willsi	20–25	0.1

have been successful for commercial usage (Table 20.2). On the basis of the ability to naturally populate in the waste, rates of consumption, digestion, and aggregation of organic matters, capacity to endure environmental stresses, increased reproductive rates and production of cocoons, brief hatching periods, accelerated growth and development rate in adults are ideal for the vermicomposting (Karmegam and Daniel 2009). Epigeic and aneic worms are used for vermicompost production whereas endogeic worms are unfit. *Eisenia fetida* is generally used across the world, while *Eudrilus eugeniae* is used prominently in tropical countries. Polyculture of *Eisenia fetida*, *Eudrilus eugeniae* and *Perionyx excavatus* is known to be employed for better production of vermicompost.

20.2.3 Steps Involved

During the process of vermicomposting, earthworms sustain the aerobic conditions in waste, consume and assimilate the solids, transform it into aggregates and excrete the partially stable product (Singh et al. 2011). The key attributes of earthworms in vermicomposting are breaking of organic wastes and integrating with soil particles for enhanced microbial activity and for the blending of humid material into the soil. They are known to eliminate about 40% - 60% volume of organic wastes and consume organic matter equivalent to their body weight in a day.

Vermicomposting can be differentiated into three phases (Garg and Gupta 2009):

Precomposting phase—Before being given to earthworms as feed, the organic substrate is partially digested for about 15 days after the collection and separation of wastes from inorganic substances. In this step, easily degradable compounds are decomposed and the possible toxic substances that may be hazardous to earthworms are removed. This phase is otherwise known as thermophilic phase.

Mesophilic phase—At this stage, earthworms voraciously feed on the substrates, grind the wastes in their gizzard and excrete them out as cast. This results in the blending of degraded wastes into soil and the surface area is increased for the enhancement of microbial activity.

Maturation and stabilization phase—During the phase, mineralization and humification occur in which humic acids and fulvic acids are synthesized. Cast is converted into valuable vermicompost by microbial activity. This phase elongates based on the source of organic waste and the efficacy of the worm to decompose the substrates.

The process of vermicomposting is widely classified as gut and cast associated processes. In gut associated process (direct), surface area of the substrate and enzymatic activities are increased, pathogenic organisms and heavy metals are reduced. In cast associated process (indirect), production of organic acids (citric acid, formic acid, acetic acid, and oxalic acid), humification and mineralization occurs to produce stable vermicompost (Mupambwa and Mnkeni 2018).

20.2.4 Types of Vermicomposting Systems

Vermicomposting systems (Table 20.3) differ based on the species, requirements, and the application of the vermiculture. Most commonly windrows (continuous flow, stacking, batching, and wedge system), tanks, bins, pits, beds (low-cost floor, gantry-fed, raised gantry-fed, and dorset-wedge style beds), and cement rings are employed (Edwards et al. 2010).

20.2.5 Factors Affecting the Vermicomposting Systems

20.2.5.1 Temperature

Temperature is the primary factor that affects the ability of earthworms to breakdown organic matter by influencing their behavior, metabolism and reproduction. Optimum temperature is maintained between the range of 25–37 °C. Temperature rising beyond the optimum increases the anaerobic microbial activity that results in the decreased oxygen content, thus providing unfavorable conditions for the worms, and significant loss of nitrogen is also witnessed. At higher temperatures, the activity of the worm ceases. Temperature dropping below the optimum results in the survival of pathogens. *Eisenia fetida*, recorded by different reports, can withstand temperature gradients as high as 45 °C and as low as 5 °C.

20.2.5.2 pH

pH of the soil and the feed given can greatly alter the process of vermicomposting and it should be maintained within the range of 4.5–9.0. At the initial stages, the pH is found in the alkaline range and at the final stage of vermicomposting, it passes progressively to the acidic range. pH fluctuations at the end of the process are due to the presence of organic acids and carbon

Table 20.3 Different vermicomposting systems

Vermicomposting system	Type of the system	Description	Disadvantages
Windrows	Open composting system	Substrate mixed with the soil and placed in narrow piles	Poor oxygenation, excess moisture and prone to predators
Tanks	In-container composting system	Laid above the ground and built with bricks	High cost and maintenance
Bins	Closed composting system	Large containers stacked in racks	Labor intensive
Pits	Open composting system	Burrowed below the land	Easily get water clogged
Beds	Open composting system	Laid on the ground with feed stock at the surface up to 1.5 feet	Large area requirement, relatively slow and labor intensive
Cement rings	Open composting system	Laid above the ground in the measurement of 90 × 30 cm	Labor intensive

accumulation. Even though most of the species prefer neutral pH (7.0), few of them likely to grow in slightly acidic environments (4.5 pH) such as *Megascolex*; and few of them such as *Eisenia fetida* grows in slightly alkaline substrates (7.0–8.0 pH). Feeds such as peat moss can induce acidification, which can be adjusted by the addition of calcium carbonate.

20.2.5.3 Moisture

Moisture content of the soil plays a pivotal role in the activity of earthworms. Around 70% - 90% of moisture should be maintained for the survival of the worms. At optimum level, the population and activity are known to be exponential. This can be maintained by regular sprinkling of water in the substrate. With the presence of excessive moisture content, anaerobic condition arises and growth of pathogens will increase along with production of foul smell. During the dry conditions, the sexual maturation is delayed in the worms and they may enter into the hibernation phase called diapause in which they coil themselves into a moisture encased burrow (Diapause chambers). By losing their total body water content, some species, such as *Lumbricus terrestris* can tolerate dry conditions.

20.2.5.4 Feed

Source and quantity of the food given to the worms govern their growth and survival rate. Each species has its own feed preferences and each feed stocks has their own pros and cons (Table 20.4). Feeds commonly given are animal manures, leaf litters, saw dust, fresh peels, hays, pre-composted food wastes, sea weeds, poultry droppings, and animal wastes. Animal manures and wastes give good nutrition to the vermiculture but most of them need to be pre-composted. Food wastes and fresh peels are easily digestible and highly variable with their nutrient content based on their source (Munroe 2007).

Feeds play a role in maintaining other optimal conditions such as pH, moisture and aeration. It is necessary to avoid greasy and fatty foods. These types of feed will obstruct the burrows, resulting in a decrease in the content of oxygen. During the metabolization of the feed, pH and moisture content will also be affected. The feed should contain less than 0.5% of the salt content

Table 20.4 Types of feed, their pros, and cons

Feed	Pros	Cons
Animal manures	High nutrient content	Requisite of precomposting
Sea weeds	Presence of both macronutrient and micronutrient	High amounts of salt content
Poultry droppings	Nitrogen rich	High protein levels
Agro-based wastes	Good bedding material	Less moisture content
Vegetable wastes	Nitrogen rich	High moisture content
Municipal solid wastes	High micronutrient content	Presence of pathogen
Industrial sludge	High nutrient content	Presence of and heavy metals

(Kumar et al. 2008). When aquatic weeds are used as an organic substrate, their salinity should be minimized by repeated rinsing before they are introduced into the vermicomposting systems.

20.2.5.5 Density

To ensure higher reproductive rate and proper gut associated composting process, densities of the earthworms in the vermicomposting systems should be monitored and maintained at 1.60 kg worms per square meter. It highly depends on the feed, soil texture and other optimal factors. Earthworms copulate frequently and at the higher density, mating and cocoon production is decreased significantly due to substrate inadequacy.

20.2.5.6 Carbon and Nitrogen Ratio

Since the main substrate used in vermicomposting is of organic origin, carbon and nitrogen (C:N) ratio plays a significant role in the process. This parameter should be monitored and optimized for effective bioconversion of the organic wastes by earthworms. It plays a vital role in the growth, maturation and composting efficiency of the worms. The optimum ratio is reported to be in the range of 25–30 parts C per unit N. Carbon acts as the energy source and nitrogen is involved in protein synthesis. If the carbon levels are higher, it leads to the acidification of the substrate, and low levels of nitrogen result in the disruption of the growth leading to death eventually. The index of waste maturation is indicated when the C:N ratio is lowered below 20 from the initial substrate ratio of 25. Nitrogen content increases by the release of the excretory material into the substrate. By the effect of mineralization and degradation of organics, carbon content decreases which reflects in the low C:N ratio content.

20.2.5.7 Growth and Reproduction Rate

The acceleration of the process of vermicomposting depends heavily on the population which in turn, has scaled in terms of growth and reproduction rate. Certain metals such as copper, lead and chromium can diffuse into the cocoon and result in damages to the embryo.

20.2.6 Vermicast, Vermiwash, Vermicomposting Leachate and Vermicompost Tea

Vermicast is widely used as soil conditioner and biofertilizer. They have increased nitrogen (5x), magnesium (2x), potash (7x) and calcium (1½ x) content present in the topsoil. Many studies have proven that the addition of vermicast to the soil improves the plant nutrient content, germination, growth, and productivity. This was achieved by increasing the ability of soil to retain the porosity, moisture and nutrients, pest repulsion, disease resistance against several pathogens, and promotes beneficial microorganism's growth.

Vermiwash is a liquid collection of mucus, excreta from earthworms, microbial biomass and the micronutrients flushing from the vermicomposting system (Quaik and Ibrahim 2013). For the production of vermiwash, cow dung was preferred over other organic substrates such as leaf litter, kitchen and animal waste. They are reported to enhance the carotenoid content, nutrient concentration (N, P, K, Mg and Ca), Indole acetic acid (IAA) content and chlorophyll levels.

Vermicomposting leachate can be defined as the leachate from the excess water flushed out to maintain the optimum moisture levels. It is otherwise known as worm bed leachate and consists of assimilated minerals and nutrients. Substrates such as cow dung, sheep manure, vegetable waste, and green forage were used (Quaik and Ibrahim 2013). Application of the vermicomposting leachate has resulted in increased plant growth, seed germination, fruit yield and nutrient composition.

Vermicompost teas are the aqueous extracts of vermicompost through aerated and non-aerated processes. Their quality and efficacy were uplifted by adding grains, kelps, sugars, humic acids, fish emulsion and sea weed extracts (Radovich et al. 2011). They have shown enhanced plant growth, yield, nutrient quality and nitrogen content through root and leaf surface application. In the aerated process, steps involved are pumping and maintenance of oxygen into the vermicompost containing chamber. At the non-aerated process, the chamber should be undisturbed for a week, containing vermicompost and known quantity of water (Quaik and Ibrahim 2013).

20.2.7 Applications

20.2.7.1 Biofertilizers

Fertilizer act as a catalyst in providing nutrition for the growth and yield of the plants. A biofertilizer is an organic product containing specific microorganisms in concentrated form, which are derived from plant roots or the soil of the root zone (Mishra and Dash 2014). They promote growth by increasing the availability of the primary nutrients when applied to the soil. They have shown great potential as plant supplements and are eco-friendly. From the reported studies (Tadayyon et al. 2018 and Roychowdhury et al. 2017), the use of vermicompost as biofertilizer has increased the traits of plants such as crop growth rate (CGR), pod length, pod width, grain yield, chlorophyll a and b, essential oil percentage, and essential oil yield were evaluated. The results revealed that the use of vermicompost biofertilizer increased the measured quantitative traits. One of the key benefits of using vermicompost in plant growth is its ability to release exchangeable nutrients slowly into the soil (Saranraj and Stella 2012). Other than the use of vermicompost in agriculture, compost worms and their cocoons were directly inoculated into the agroecosystems.

20.2.7.2 Biogas Production

Biogas is a mixture of gases generated in anaerobic conditions by the degradation of organic matter, composed mainly of methane, carbon dioxide, nitrogen and oxygen. It is often used by farmers in agro-based countries to generate biogas digesters, and billions of people around the globe have gained numerous benefits from this cheaper and eco-friendly technology (Ali et al. 2015). Addition of vermicompost was studied to induce the biogas production of methanogenic microorganisms because of its nutrient content and wide array of microbial biomass. The study conducted by Zhang and Yang (2007), has proved that the vermicompost consists of anaerobic microorganisms that can produce methane through anaerobic fermentation.

20.2.7.3 Industrial Waste Treatment

Biotransformation and bioremediation of industrial sludges from paper and pulp industries, palm oil mill, sugar industry, food industry, milk processing industry, distillery industry, tannery industry, and textile industry is performed efficiently using vermicomposting process (Lee et al. 2018 and Rupani et al. 2010). Instead of opting for landfilling, these sludges can be recycled and reused for its nutrient content (Mn, Cu, Zn, N, and P). Studies have shown that 1 g of earthworm can ingest and assimilate 4 g of sludges within 5 days.

Paper and pulp mill sludges were vermicomposted using *Lumbricus terrestris*, *Eisenia fetida*,

Eisenia anderi and *Perionyx excavatus*. In sugar industries, by-products such as bagasse, fermentation yeast sludge, pressmud sludge and cane trash can be composted using *Eudrilus eugeniae* and *Eisenia fetida* with organic substrate such as manures and dungs. Biomass residues, by-products, and effluents of the palm oil mills were composted instead of dumping in landfills using *Eisenia fetida* were studied by Sabrina et al. (2009) and the resultant compost was used as the organic manure.

20.2.7.4 Solid Waste Management

Due to rapid urbanization, large quantities of solid wastes are generated. These wastes are managed by dumping in the landfills that result in not only the contamination of soil and water but also the emission of greenhouse gases into the environment. Domestic wastes such as garbage, glass, paper, plastics and textiles can be composted using vermisystems. A study conducted by Suthar (2009), postulates that *Eisenia fetida* has effectively bio-transformed vegetable market solid wastes into biofertilizers.

20.2.7.5 Terrestrial Weed Management

Weeds can be identified as the alien invasive plants that grew in the uninvited land with no beneficial qualities. They rapidly germinate, occupy a larger portion of the land and suppress the native vegetation growth. So, weeds are basically a nuisance to the ecosystem. There are several terrestrial weed management techniques; one among them is usage of vermicomposting technology. Weeds such as *Argemone Mexicana*, *Parthenium hysterophorus*, *Lantana camara*, *Colocasia esculenta*, *Hydrilla Verticillata*, *Ageratum conyzoides*, *Saccharum spontaneum* and *Galinsoga pur-viflora* has caused distress globally (Saha et al. 2018). These weeds can be eliminated by the vermicultures through mechanisms such as destruction of seeds by grinding in the gizzard and stimulation of seed dormancy through enzymatic degradation. Studies have shown that the weeds can be vermicomposted into a quality end product using *Eisenia fetida* (Yadav and Garg 2011 and Devi and Khwairakpam 2020) and *Eudrilus eugeniae* (Malins and Gunselmen 1994) with cow dung and animal manures. They have shown high amounts of nutrient content and extensive microflora.

20.2.7.6 Biological Inactivation of Pathogens and Parasites in Organic Wastes

Pathogens such as bacteria, fungi and parasites inhabit in the organic wastes which can enter into the medium of air, water and soil during the dumping of wastes into large piles or landfills. This contamination can be susceptible to the human, animal, and plant systems and can cause serious health issues. So, the complete sanitation of organic wastes is a mandatory step during waste management. Even though vermicomposting is a mesophilic process, pathogens can be eliminated through the activities of worms and vermistabilization.

Commonly seen pathogens in the organic wastes are fecal coliforms, *Streptococcus*, *Enterococcus*, *Salmonella*, *Shigella*, Helminths and enteric viruses from human biosolids, animal manures, animal slurry, septic sludge, agricultural wastes and effluents. Pathogens were reduced by worms through grinding, enzymatic digestion, coelomic fluid secretion of antibiotic agents and helping in proliferation of beneficial microorganisms such as *Pseudomonas*, *Paenibacillus*, *Mucor*, *Azoarcus*, *Acaligenes*, *Burkholderia*, *Acidobacterium* and *Spiroplasm* (Swati and Hait 2018).

20.3 Vermiremediation

20.3.1 Process and Mechanisms Involved

Organic pollutants and toxic metals are the major contaminants of soil. The vermiremediation has three processes such as pre-vermirediation (decomposed organic waste composted with cow dung 1:1) ratio (Biruntha et al. 2019), vermiremediation and post-vermiredimediation. In general, vermiremediation has three phases such as accumulation (vermi accumulation), extraction (vermi extraction) and transformation (vermidegradation). The process is influenced by the factors (biotic and abiotic) as well as the life cycle of the earthworm. One of the vermiremideation mechanism phase is drilodegradation (degradation of organic pollutants by microbes—vermi drilo degradation). After the process of transformation, the metabolites are produced by microbes (Shi et al. 2020).

Earthworms can bioaccumulate and biotransform the contaminants naturally and reduce pollution. Two possible mechanisms include internal and extrinsic mechanisms in vermiremediation. The part of internal mechanism is trophic soil microorganisms, regulating the enzyme activity and metabolism, increase phytoextraction. Extrinsic mechanisms are mobility, speciation of contaminants and physical activity of earthworms. In addition to the biotransformation and biodegradation of heavy metals, earthworms bioaccumulate multiple heavy metals in their bodies (Zeb et al. 2020).

20.3.1.1 Nutritional and Dermal Uptake

Earthworms ingest chemicals from soil pollutants and porewater, by their skin (dermal) uptake or by oral intake. Dissolved organic compounds from porewater or from weakly associated droplets on the surface of particulate matter are absorbed by the body wall, subsequently transported in the body of earthworms. Oral nutritional elements are ingested, digested, and used for earthworm's life cycle. In earthworm digestion, enzymes are produced by anterior intestine and nutrients are absorbed by the posterior intestine. Microorganisms play a vital role in the production of food for earthworms. The fungi and protozoa are the major sources of food rather than bacteria and algae (minor and moderate diet) (Edwards et al. 1988). As a result, organic compounds in the soil are translocated to earthworm tissues (Shi et al. 2014).

20.3.1.2 Vermiaccumulation

Vermiaccumulation is the process by which earthworms ingest the chemicals and eventually accumulate the contaminants in their bodies and decrease the level of pollutants in soil. Earthworms absorb the contaminants by way of epidermal or nutritional uptake. Vermiaccumulation process is equated with physiological characters such as lipid content of earthworm tissue, number of contaminants in the soil pollutants, physiochemical properties of aqueous solubility concentration. Earthworms can persist in heavy metal contaminated (Cd, Cu, Pb, Zn) soils and accumulate the metals. Based on the essential and non-essential elements, earthworm uptake patterns are varied. Stürzenbaum (2013) reported that *Lampito mauritii, Allolobophora rosea, Eisenia fetida, Nicodrilus caliginosus* species of earthworms accumulate metals by persuading metallothionein.

20.3.1.3 Biotransformation and Vermitransformation

Biotransformation process is the one that can modify chemical pollutants (xenobiotics) into minerals with the help of earthworms and microorganisms. Oxidation, conjugation, enzymatic transformation (Cytochrome P450 dependent monooxygenases) plays a vital role in the degradation of chemical pollutants into non-toxic minerals. Cysteine-rich metallothioneins proteins bind to the divalent cations (Cu^{2+}, Cd^{2+}, Hg^{2+}) counteract toxicity by binding with metal ions. (Dzul-Caamal et al. 2020).

Biotransformation process can be categorized into phase I and phase II steps. Phase I reactions modify hydrophobic chemicals to more reactive products via hydrolysis and oxidation. Phase II is conjugation reactions that add polar groups to the phase I products. Thus, phase II metabolites are even more water-soluble. So, it can be easily excreted. Enzymes like CYP450 exhibit a wide range of substrate specificity that contributes more to biotransformation process (Zeb et al. 2020).

Vermitransformation is a process of ingesting the organic pollutants by earthworms and degraded with the help of enzymes such as (CYP450, peroxidase), or with gut microbes, or by both. In another words, vermitransformation is a quick conversion of slowly degradable wastes to cost-effective organic fertilizer by the action of earthworms and the associated microorganisms. Vermitransformation is one of the pathways of vermiremediation which mainly focuses on chemical pollutants, yet the process is not fully understood (Zeb et al. 2020).

Vermitransformation is an important ecotoxicological process for which some mysteries should be unraveled; (i) the potential of earthworm in vermitransformation of organic contaminants is unknown; (ii) whether ingested organic pollutants are digested by earthworms or by gut microbes. (iii) The overall contribution in vermiremediation process. A recent study revealed the importance of green manure alteration and precomposting for an effective vermitransformation of industrial waste (Karmegam et al. 2021).

20.3.1.4 Biodegradation and Vermidegradation

Biodegradation refers to the degradation of complex chemical compounds by the action of microorganisms under the typical environmental condition. Biodegradation takes a single or sequence of biological reactions for converting the complex substrate into a simple molecule. The role of earthworms in biodegradation is to stimulate microbial activity and induce the contact of the microbial population with organic chemical compounds in the soil. The earthworm gut is a drilosphere that is helpful in mineralization and decomposition of soil organic compounds (Brown et al. 2000). Earthworm gut epithelium secretes digestive enzymes such as lipases, proteases, phosphatases, glucosidases, cellulases with the combined action of intestinal mucus which in combination play vital role in biodegradation process (Sanchez et al. 2020).

Earthworms play a crucial part in the degradation of organic debris by digesting the substrates, increasing the aeration, and emphasize the soil for microorganisms. Vermidegradation is managed through interactions of microorganisms and earthworms on organic matters. Aira and Aira (2011) reported that earthworms are the intermediates for increasing the exposure of microorganisms on the contaminants in the soil.

Earthworms enhance PCB degradation by augmentation of PCB degrading bacteria (*Rhodococcus* sp. and *Ralstonia eutrophus*) into the soil, and also providing suitable growth conditions for PCB degrading bacteria (Brown et al. 2000).

20.3.2 Remediation of Organic Pollutants

Earthworms contribute to the organic pollutant remediation by modifying the degrading microorganisms, nourish with carbon and nitrogen content, and increasing the aeration in the polluted soil (Hoeffner et al. 2019). Polychlorinated biphenyls (PCB), Polybrominated diphenyl ethers (PBDEs), polycyclic aromatic hydrocarbon (PAHs), pesticides and insecticides, crude oil, petroleum hydrocarbons (TPH) are the organic chemicals that pollute the soil.

Earthworms stimulate hydrocarbon degradation (Table 20.5) by increasing fungal and bacterial population, enhancing aeration, discharging the degradable carbon, grazing, and the formation of cluster of microbes (Sinha et al. 2008).

Tharakan (2006) reported that *Eisenia fetida* inoculated with sludge and sterile soil in vermicomposting bioreactors, 55–66% of PCB is reduced in the soil and PCB levels are increased in earthworm biomass. Mass balance analysis shows that most of the PCB content is bioaccumulated in earthworms.

20.3.3 Remediation of Heavy Metals

Earthworms have the potential to remediate heavy metals in the polluted soil by bioaccumulating in their body (Swati and Hait 2017). The previous data supported that increased copper and cadmium in the soil increases the metal content in the earthworm body. Earthworms can absorb or retain heavy metals with the help of gut tissue and chloragogen tissue in their bodies (Sivakumar et al. 2003). The heavy metal remediation and bioaccumulation rate is also

Table 20.5 Vermiremediation efficiency of organic pollutants

Earthworm species	Organic pollutant	Effect of pollutant	Concentration / kg	Efficiency (%)	References
Eisenia fetida	Diesel	Depleation of hydrocarbons	1014 μL	43–52	(Fernández et al. 2011)
Hyperiodrilus africanus	Crude oil	The earthworms Biodegrated toluene, xylene, benzene,	5 mL	38–90	(Ekperusi and Aigbodion 2015)
Eisenia fetida	Anthracene	Earthworms increased degradation by enhancing microbial activity	500 mg	93	(Coutiño-González et al. 2010)
Eisenia Andrei	PAH	PAHs and heavy metals accumulated in earthworm body	3965.86 μg	85	(Rorat et al. 2017)
Eisenia fetida, Dendrobena veneta, E. Andrei, Microorganisms, Fungi	Crude oil	Earthworms, N-fixing bacteria, photosynthetic bacteria, fungi enhanced the removal efficiency of hydrocarbons	100 g	99	(Chachina et al. 2015)

dependent on physicochemical properties of soil, exposure time, and earthworm species. Davies (2003) reported that earthworms' uptake rate of lead (Pb) is faster with high Pb-treatment than low Pb-treatment. The external environment also influences the Pb uptake by *Eisenia fetida*.

Earthworms accumulate heavy metals based on the metal species. Heavy metals like cadmium, mercury, zinc is highly bioaccumulated by the earthworms (Richardson et al. 2015). Remediation of Cd, Pb, Zi, Cu, Mn by the earthworm species *Perionyx excavates, Eudrilus eugeniae, Eisenia fetida* in urban wastes process, shows that *Eudrilus eugeniae* absorbed higher metal concentration than other species. Comparatively Cd accumulation in earthworm tissue is more than any other metals (Zn, Pb, Cu, Mn) (Pattnaik and Reddy, 2011).

20.3.4 Fly Ash Remediation

Fly ash is the lightest type of coal ash. It is produced when the coal combusts during power generation. Fly ash causes severe environmental pollution due to the presence of heavy metals, hydrocarbons, PAHs and PCBs. Vermicomposting helps in reducing the pollutants in fly ash and is suitable for fly ash management.

Wang (2013) reported that fly ash and phosphoric rock is stabilized after 60 days of vermiremediation process. The earthworms bioaccumulate the heavy metals in their tissues. The pH is decreased in the intermediate process and neutralized at the end by earthworms. Metals Pb, Cu, Cd, As are highly available in the extractable metal form.

Usmani and Kumar (2017) investigated with epigeic and epi-endogeic earthworm species for fly ash metal remediation and evaluated the changes in earthworm number, biomass and nutrient content while fly ash remediation process. Cow dung and fly ash mixture in 1:3 ratio were used. *Eisnia fetida, Eudrilus eugeniae, Lumbricus rubellus* reduced 58.82% of Cr, 71.94% of Ni, 51.67% of Cu respectively. Comparatively *E. eugeniae* produced maximum metallothionein in fly ash remediation.

20.3.5 Advantages and Limitations

Vermiremediation has diverse advantages than other remediation technologies. The major advantages are eco-friendly nature, more efficient when compared with physiochemical remediation and phytoremediation technologies. Vermiremediation improves soil fertility by increasing microbial growth, improve nutrients, organic matter and biotic factors. Earthworms contribute to remediation of heavy metals by inhibiting the binding and affinity of chemical compounds with soil particles and enhance the metal compound availability.

Only 10–15% of chemical pollutants are ingested and digested and utilized by earthworm, rest of the matters are excreted with mucus layer called vermicasting. It is enhanced with NKP, and soil microbes (nitrogen fixers, mycorrhizal fungus). The earthworm gut treats the organic matter with the process called 'humification'. So, one-fourth of organic matter is converted as humus, which acts as a slow-release fertilizer in the soil. The vermiremediation and vermicomposting have greater economic and environmental significance in the polluted land for not only rectifying them but also for improving soil quality.

Apart from the remediation of organic pollutants and heavy metal contaminations, vermiremediation has few limitations. This technique only works in slightly contaminated soils because of the restricted survival of earthworms. Earthworms keep away from unnatural soil conditions, such as higher salt content, pH, heavy metals, crude oils, organic contaminants, which might reduce the surviving activity of earthworms (Rodriguez-Campos et al. 2014). Climatic conditions such as extreme heat or cold greatly decrease the activity of earthworms. 8–57% of water content must be present for burrowing and surviving in the soil (Richardson et al. 2009). Improper disposal of contaminants bioaccumulated in earthworms can affect the food chain system.

20.4 Conclusion and Future Direction

Vermicomposting and vermiremediation are eco-friendly, fast-growing remediation techniques. Numerous studies and research works have been investigated in the past decades. Based on the reports, many aspects of vermicomposting and vermiremediation have been growing for their realization and importance. Vermicomposting technology is known to be very effective when its critical factors are monitored and maintained properly. Earthworms have wider applications in sustainable agriculture and waste management technologies. Vermiaccumulation, vermidegradation, vermitransformation are the important procedures that enhance the vermicomposting process.

More fundamental research should be conducted for understanding the vermitransformation and vermidegradation process in earthworms. The contribution, mechanisms of different stages, capacity in vermiremediation should be clarified. Enhancement strategies like adding surfactants, soil amendments, agronomic practices and enhanced biomass are the potential ways to increase vermiremediation. After the vermiremediation, the effect of contaminants on earthworms, inoculation, colonization, collection of earthworms from polluted soil, safe post-harvest disposal should be investigated seriously.

References

Aira M, Domínguez J (2011) Earthworm effects without earthworms: inoculation of raw organic matter with worm-worked substrates alters microbial community functioning. PLoS one 6(1):e16354

Ali U, Sajid N, Khalid A, Riaz L, Rabbani MM, Syed JH, Malik RN (2015) A review on vermicomposting of organic wastes. Environ Prog Sustain Energy 34 (4):1050–1062

Biruntha M, Mariappan P, Selvi BK, Paul JAJ, Karmegam N (2019) Vermiremediation of urban and agricultural biomass residues for nutrient recovery and vermifertilizer production. Waste Biomass Valorization 1–15

Brown GG, Barois I, Lavelle P (2000) Regulation of soil organic matter dynamics and microbial activityin the drilosphere and the role of interactionswith other edaphic functional domains. Eur J Soil Biol 36(3–4):177–198

Chachina S, Voronkova N, Baklanova O (2015) Biological remediation of the engine lubricant oil-contaminated soil with three kinds of earthworms, *Eisenia fetida*, *Eisenia andrei*, *Dendrobena veneta*, and a mixture of microorganisms. Procedia Eng 113:113–123

Coutiño-González E, Hernández-Carlos B, Gutiérrez-Ortiz R, Dendooven L (2010) The earthworm Eisenia fetida accelerates the removal of anthracene and 9, 10-anthraquinone, the most abundant degradation product, in soil. Int Biodeterior Biodegradation 64 (6):525–529

Davies NA, Hodson ME, Black S (2003) The influence of time on lead toxicity and bioaccumulation determined by the OECD earthworm toxicity test. Environ Pollut 121(1):55–61

Devi C, Khwairakpam M (2020) Feasibility of vermicomposting for the management of terrestrial weed Ageratum conyzoides using earthworm species *Eisenia fetida*. Environ Technol Innov 18:100696

Dzul-Caamal R, Vega-López A, Osten JR (2020) Distribution of heavy metals in crop soils from an agricultural region of the Yucatan Peninsula and biochemical changes in earthworm Eisenia foetida exposed experimentally. Environ Monit Assess 192:1–15

Edwards CA, Arancon NQ, Sherman RL (eds) (2010) Vermiculture technology: earthworms, organic wastes, and environmental management. CRC Press, Boca Raton, Florida, pp 623

Edwards CA, Fletcher K (1988) Interactions between earthworms and microorganisms in organic-matter breakdown. Agric Ecosyst Environ 24(1–3):235–247

Ekperusi OA, Aigbodion IF (2015) Bioremediation of heavy metals and petroleum hydrocarbons in diesel contaminated soil with the earthworm: Eudrilus eugeniae. Springerplus 4(1):1–13

Fernández MD, Pro J, Alonso C, Aragonese P, Tarazona JV (2011) Terrestrial microcosms in a feasibility study on the remediation of diesel-contaminated soils. Ecotoxicol Environ Saf 74(8):2133–2140

Fitzpatrick GE, Worden EC, Vendrame WA (2005) Historical development of composting technology during the 20th century. HortTechnology 15(1):48–51

Garg V, Gupta R (2009) Vermicomposting of agro-industrial processing waste. In: Biotechnology for agro-industrial residues utilisation. Springer, pp 431–456

Haga K (1999) Development of composting technology in animal waste treatment-review. Asian-Australas J Anim Sci 12(4):604–606

Hoeffner K, Monard C, Cluzeau D, Santonja M (2019) Response of temperate anecic earthworm individual biomass to species interactions. Appl Soil Ecol 144:8–11. https://doi.org/10.1016/j.apsoil.2019.06.017

Karmegam N, Daniel T (2009) Investigating efficiency of Lampito mauritii (Kinberg) and Perionyx ceylanensis Michaelsen for vermicomposting of different types of organic substrates. Environment 29(3):287–300. https://doi.org/10.1007/s10669-008-9195-z

Karmegam N, Jayakumar M, Govarthanan M, Kumar P, Ravindran B, Biruntha M (2021) Precomposting and green manure amendment for effective vermitransformation of hazardous coir industrial waste into enriched vermicompost. Bioresour Technol 319:124136. https://doi.org/10.1016/j.biortech.2020.124136

Lee LH, Wu TY, Shak KPY, Lim SL, Ng KY, Nguyen MN, Teoh WH (2018) Sustainable approach to biotransform industrial sludge into organic fertilizer via vermicomposting: a mini review. J Chem Technol Biotechnol 93(4):925–935

Malins DC, Gunselman SJ (1994) Fourier-transform infrared spectroscopy and gas chromatography-mass spectrometry reveal a remarkable degree of structural damage in the DNA of wild fish exposed to toxic chemicals. Proc Natl Acad Sci USA 91(26):13038–13041

Mishra P, Dash D (2014) Rejuvenation of biofertilizer for sustainable agriculture and economic development. Consilience 11:41–61

Munroe G (2007) Manual of on-farm vermicomposting and vermiculture. Organic Agricul Centre of Canada 39:40

Mupambwa HA, Mnkeni PNS (2018) Optimizing the vermicomposting of organic wastes amended with inorganic materials for production of nutrient-rich organic fertilizers: a review. Environ Sci Pollut Res Int 25(11):10577–10595

Nagavallemma K, Wani S, Lacroix S, Padmaja V, Vineela C, Rao MB, Sahrawat K (2004) Vermicomposting: recycling wastes into valuable organic fertilizer. Global Theme on Agroecosystems Report no. 8

Pattnaik S, Reddy MV (2011) Heavy metals remediation from urban wastes using three species of earthworm (*Eudrilus eugeniae*, *Eisenia fetida* and *Perionyx excavatus*). J Environ Chem Ecotoxicol 3(14):345–356

Quaik S, Ibrahim MH (2013) A review on potential of vermicomposting derived liquids in agricultural use. Int J Sci Res 3(3):1–6

Radovich T, Pant A, Hue N, Sugano J, Arancon N (2011) Promoting plant growth with compost teas. The Food Provider, pp 3

Richardson JB, Görres JH, Jackson BP, Friedland AJ (2015) Trace metals and metalloids in forest soils and exotic earthworms in northern New England, USA. Soil Biol Biochem 85:190–198. https://doi.org/10.1016/j.soilbio.2015.03.001

Richardson DR, Snyder BA, Hendrix PF (2009) Soil moisture and temperature: tolerances and optima for a non-native earthworm species, *Amynthas agrestis* (Oligochaeta: Opisthopora: Megascolecidae). Southeast Nat 8(2):325–334

Rodriguez-Campos J, Dendooven L, Alvarez-Bernal D, Contreras-Ramos SM (2014) Potential of earthworms to accelerate removal of organic contaminants from soil: a review. Appl Soil Ecol 79:10–25. https://doi.org/10.1016/j.apsoil.2014.02.010

Ronen E (2016) Micro-elements in agriculture. Practical Hydroponics Greenhouses 164:35

Rorat A, Wloka D, Grobelak A, Grosser A, Sosnecka A, Milczarek M, Jelonek P, Vandenbulcke F, Kacprzak M (2017) Vermiremediation of polycyclic aromatic hydrocarbons and heavy metals in sewage sludge composting process. J Environ Manage 187:347–353. https://doi.org/10.1016/j.jenvman.2016.10.062

Roychowdhury D, Mondal S, Banerjee S (2017) The effect of biofertilizers and the effect of vermicompost on the cultivation and productivity of maize—a review. Adv Crop Sci Technol 05:1–4. https://doi.org/10.4172/2329-8863.1000261

Rupani PF, Singh RP, Ibrahim MH, Esa N (2010) Review of current palm oil mill effluent (POME) treatment methods. World Appl Sci J 11(1):70–81

Sabrina DT, Hanafi MM, Mahmud TMM, Azwady AAN (2009) Vermicomposting of oil palm empty fruit bunch and its potential in supplying of nutrients for crop growth. Compost Sci Util 17(1):61–67. https://doi.org/10.1080/1065657X.2009.10702401

Saha B, Devi C, Khwairakpam M, Kalamdhad AS (2018) Vermicomposting and anaerobic digestion—viable alternative options for terrestrial weed management—a review. Biotechnol Rep 17:70–76. https://doi.org/10.1016/j.btre.2017.11.005

Sanchez-Hernandez JC, Capowiez Y, Ro KS (2020) Potential use of earthworms to enhance decaying of biodegradable plastics. ACS Sustain Chem Eng 8(11): 4292–4316. https://doi.org/10.1021/acssuschemeng.9b05450

Saranraj P, Stella D (2012) Vermicomposting and its importance in improvement of soil nutrients and agricultural crops. J Nat Sci Res 1(1):14–23

Shi Z, Liu J, Tang Z, Zhao Y, Wang C (2020) Vermiremediation of organically contaminated soils: concepts, current status, and future perspectives. Appl Soil Ecol 147:103377. https://doi.org/10.1016/j.apsoil.2019.103377

Shi Z-M, Xu L, Hu F (2014) A hierarchic method for studying the distribution of phenanthrene in eisenia fetida. Pedosphere 24(6):743–752. https://doi.org/10.1016/S1002-0160(14)60061-8

Singh RP, Embrandiri A, Ibrahim MH, Esa N (2011) Management of biomass residues generated from palm oil mill: vermicomposting a sustainable option. Resources, Resour Conserv Recyc 55(4):423–434. https://doi.org/10.1016/j.resconrec.2010.11.005

Sinha RK, Bharambe G, Ryan D (2008) Converting wasteland into wonderland by earthworms—a low-cost nature's technology for soil remediation: a case study of vermiremediation of PAHs contaminated soil. Envir 28(4):466–475

Sivakumar S, Kavitha K, Sivaraj R, Prabha D, Subburam V (2003) Effect of cadmium and mercury on the survival morphology and burrowing behaviour of the earthworm Lambito Mauritii (Kinberg). Indian J Environ Prot 23:792–799

Stürzenbaum SR, Höckner M, Panneerselvam A, Levitt J, Bouillard J-S, Taniguchi S, Dailey L-A, Ahmad Khanbeigi R, Rosca EV, Thanou M, Suhling K, Zayats AV, Green M (2013) Biosynthesis of luminescent quantum dots in an earthworm. Nat Nanotechnol 8(1):57–60. https://doi.org/10.1038/nnano.2012.232

Suthar S (2009) Vermicomposting of vegetable-market solid waste using Eisenia fetida: impact of bulking material on earthworm growth and decomposition rate. Ecol Eng 35(5):914–920. https://doi.org/10.1016/j.ecoleng.2008.12.019

Swati A, Hait S (2017) Fate and bioavailability of heavy metals during vermicomposting of various organic wastes—a review. Process Saf Environ Prot 109:30–45. https://doi.org/10.1016/j.psep.2017.03.031

Swati A, Hait S (2018) A comprehensive review of the fate of pathogens during vermicomposting of organic wastes. J Environ Qual 47(1):16–29. https://doi.org/10.2134/jeq2017.07.0265

Tadayyon A, Naeimi MM, Pessarakli M (2018) Effects of vermicompost and vermiwash biofertilizers on fenugreek (Trigonella foenum) plant. Commun Soil Sci Plant Anal 49(19):2396–2405. https://doi.org/10.1080/00103624.2018.1510950

Tharakan J, Tomlinson D, Addagada A, Shafagati A (2006) Biotransformation of PCBs in contaminated sludge: potential for novel biological technologies. Eng Life Sci 6(1):43–50. https://doi.org/10.1002/elsc.200620117

Usmani Z, Kumar V, Mritunjay SK (2017) Vermicomposting of coal fly ash using epigeic and epi-endogeic earthworm species: nutrient dynamics and metal remediation. RSC Adv 7(9):4876–4890. https://doi.org/10.1039/C6RA27329G

Vinod Kumar G, Renuka G, Anoop Y (2008) Potential of Vermicomposting Technology in Solid Waste Management. In: Pandey A, Soccol CR, Larroche C (eds) Current developments in solid-state fermentation. Springer, New York, NY, pp 468–511

Wang L, Zheng Z, Zhang Y, Chao J, Gao Y, Luo X, Zhang J (2013) Biostabilization enhancement of heavy metals during the vermiremediation of sewage sludge with passivant. J Hazard Mater 244–245:1–9. https://doi.org/10.1016/j.jhazmat.2012.11.036

Yadav A, Garg VK (2011) Vermicomposting—an effective tool for the management of invasive weed Parthenium hysterophorus. Bioresour Technol 102(10):5891–5895. https://doi.org/10.1016/j.biortech.2011.02.062

Zeb A, Li S, Wu J, Lian J, Liu W, Sun Y (2020) Insights into the mechanisms underlying the remediation potential of earthworms in contaminated soil: a critical review of research progress and prospects. Sci Total Environ 740:140145. https://doi.org/10.1016/j.scitotenv.2020.140145

Zhang B, Yang S (2007) Enhancement of methane production by using vermicompost as additive. In: 2007 ASAE annual meeting, american society of agricultural and biological engineers, pp 1

21 Biological Indicators of Soil Health and Biomonitoring

Ujjwal Ranjan Dahiya, Jhumki Das, and Subia Bano

Abstract

Soil contamination is a significant problem today, with the potential of impacting the terrestrial environment and food production by reducing crop yields. Even though soil organisms (including the microbiome) are an integral part of overall soil productivity and health, the impact of pollution on these organisms can be insufficiently estimated using the traditional approaches like chemical analysis and setting up thresholds. An alternative approach of measuring soil organism's response to contaminant exposure and then using them as biomarker for identifying level of pollutants and their adverse effect is gaining popularity. Invertebrates present in soil remain directly in contact with pore-water and soil and can serve as pollution bioindicators, when used in combination with contaminants and the response of biomarker in their presence. This chapter targets to summarize and present the literature available on use of biomarkers for assessing soil pollution, evaluate the effect of remediation trials and the role of soil organisms in the agroecosystems. A brief introduction to soil ecosystem and diversity of life forms available in soil is provided in the beginning of chapter, followed by discussing the key pollutants and available methods for the assessment of soil quality. The chapter includes a brief discussion on key organisms residing in soil ecosystem and describes the concepts of sentinel's species, bioindicators and biomarkers. A summarized literature review on biomarkers usage for soil quality assessment with the help of soil organisms is also discussed in the chapter. Also, characteristics and classification of soil biomarkers are discussed for better understanding. Selection of relevant organisms, limitation of biomarkers, and the use of multiple biomarkers are also touched upon to comprehensively summarize the advantage of organism-based soil quality assessment.

Keywords

Bioindicators · Biomarkers · Bioremediation · Earthworms · Microbiome · Sentinel species · Soil pollution

U. R. Dahiya
CSIR—Institute of Genomics and Integrative Biology, New Delhi 110021, India

J. Das
Department of Pediatrics, PGIMER Chandigarh, Chandigarh 160011, India

S. Bano (✉)
Elvesys Microfluidics Innovation Centre, 75011 Paris, France

21.1 Introduction

Soil is the productive layer of terrestrial ecosystem, supporting the growth of vegetation, microbiome and large number of organisms of

J. A. Malik (ed.), *Advances in Bioremediation and Phytoremediation for Sustainable Soil Management*,
https://doi.org/10.1007/978-3-030-89984-4_21

higher order and thus is a critical factor in the ecological balance (Adhikari and Hartemink 2016). Anthropogenic activities such as industrialization, continuous cultivation on same land, use of chemical fertilizers apart from indirect contamination, and accumulation of harmful compounds in soil can significantly influence the flora, fauna, and ultimately human health (Lavelle et al. 2006; Power 2010). Over the past few years, continuously increasing usage of hazardous chemicals and toxins has resulted into changes in structure of soil and thus impacting the fertility and ecosystem dwelling on it. These contaminants after reaching to soil can accumulate, infiltrate, immobilized or transformed into other unwanted compounds (Abbasi et al. 2014). Owing to the critical importance held by the health of soil in supporting the terrestrial life and food security for humans, recently multiple environmental agencies along with the scientific community are giving heeds to the assessment and monitoring of soil pollution (Cachada et al. 2017). Conventional methods of pollution and contamination estimation are based on chemical testing, and marking threshold value has limited relevance in the case of soil pollution estimation. As recognized by many soil researchers, instead of the estimating the amount of pollutants, measuring the bioavailability of these pollutants makes more sense in the case of soil. This is because few characteristics of soil such as pH, amount of organic matter, and composition can affect the bioavailability of the pollutants and tolerance level of inhabiting organisms to them (Bradham et al. 2006; Spurgeon et al. 2006). New comprehensive approaches integrating chemical analysis along with measurement of multiple prognostic biomarkers are coming up for estimation of soil degradation and pollution (Criel et al. 2008).

21.2 Conventional Approaches for Measuring Soil Pollution

Soil pollution has been increased tremendously over the last few decades and has become a global issue. A wide range of contaminants from different sources can pollute the soil ecosystem. Soil pollution poses a critical threat to the residing organism and future of man being; thus, soil pollution assessment is an essential step for protecting the ecosystem (Teng et al. 2014). Several techniques are used to monitor the soil contamination, such as chemical, geophysical, and biological techniques.

For the measurement of particular inorganic, organic, or radioactive pollutants in the soil, analytical methods are used. In all these methods, the contaminated soil sample is collected from site and is taken to testing laboratories for estimation of pollutants after multiple processing steps (Xu et al. 2009). Analysis through analytical techniques such as gas chromatography, mass spectrometry, atomic adsorption spectrophotometry, fluorescence spectroscopy and nuclear magnetic resonance (NMR) is followed by extraction (Reed and Martens 1996; Kögel-Knabner 1997).

Geophysical techniques are used to examine changes in the physical properties of soil and the contaminants to address large areas of soil pollution. Geophysical approaches do not require disturbance to soil ecosystem; however, the method is not effective in comprehensive identification of contaminants (Hirsch et al. 1982). Biological techniques use species to track or forecast changes in soil contaminant concentrations over time as measures of soil pollution or by-products of contaminant's biodegradation processes(Harmsen 2007; Bastida et al. 2008). For example, earthworms are widely used as a biomarker is in soil pollution assessment due to their sensitivity to toxic chemicals and other unique biological advantages.

The traditional physicochemical approaches for assessment of soil pollution depends upon the concentration of pollutants in soil; however, these are inadequate and does not represent toxic effects on the life forms (Andrews et al. 2004). Thus, there is a need for development of newer techniques in soil contamination monitoring which can bridge the gap between the sub organism and the behavior these organism shows in response to stress stimuli and thus can represent both practical application and theoretical significance.

21.3 Soil Pollution and Its Threat for Biodiversity and Food Security

Gradual accumulation of persistent hazardous compounds, radioactive substances, salts, chemicals, dangerous disease-causing microbes in the soil is termed as soil pollution and can endanger growth and survival of dependent flora and fauna. Towards reduction and management of soil quality, measurement of contamination level is critical and often require deployment of indirect methods (Adhikari and Hartemink 2016). Soil contamination can occur though several processes, such as: discharge into the soil of toxic waste, percolation into the soil of contaminated water, sewage from a dump, rupture of underground storage tanks, excess use of pesticides, herbicides or fertilizer, filtration of solid waste. Soil pollution also results from atmospheric deposition from smelting, incomplete combustion of many substances, transportation and radionuclide deposition from atmospheric weapons testing and nuclear accidents (Anastopoulos and Kyzas 2015). Soil pollution causing hazardous chemicals are petroleum hydrocarbons, heavy metals (cadmium lead copper zinc arsenic), benzene ethyl-benzene, xylene, toluene, pesticides, and solvents. Pharmaceuticals, hormones and toxins, endocrine disruptors, and biological pollutants, such as bacteria and viruses are also leading to soil pollution (Schwab 2020). Soil pollution affects agricultural practices severely by reducing nitrogen fixation and soil fertility due to loss of soil and nutrients thereby reducing crop yield. This also leads to increased erodibility and deposition of silt in tanks and reservoirs (Zahran 1999). All the major ecosystem services rendered by soil are severely affected by contamination of soil. Food security is endangered as crop yields are reduced significantly along with contamination of agriproducts with hazardous chemicals and pesticides. Many pollutants are transferred from the soil to the surface and ground water, including nutrients such as nitrogen and phosphorus, causing direct human health concerns due to the ingestion of contaminated water and environmental damage by eutrophication. Pollutants also have adverse effects on the local microbiome of the soil and other inhabiting life forms, thus endangering the ecosystem as whole (Briški and Vuković Domanovac 2019).

21.4 Potentially Toxic Elements and Pollutants of Soil Ecosystem

Increased industrialization caused production of tons of waste which includes hazardous compounds. These wastes accumulated every day in the environment without any previous treatment (Chen 2007). Soil is the major and cheaper substitutes for the disposal of these hazardous compounds, which is responsible for contamination. The effect of these contaminants on invertebrates and plants are discussed below;

Sewage sludge The Sewage sludges generation (generated in treatment stations) and their final disposal is the greatest challenge of current century. Sewage sludge from urban and semi-urban township is often recycled or reused for agricultural use (Singh and Agrawal 2007). However, this sewage sludge composition is rich in chemicals (metals, heavy metals and organic chemical compounds) and biological compounds (pathogens/microorganism). Whenever, these contaminants come in contact with human being, flora and fauna may cause severe contamination and disease. Thus, disposal of these contaminates must be done carefully to avoid the contamination risks to the environment and living being (Werther and Ogada 1999).

Dioxins Dioxins are persistent organic pollutants. The chemical names of dioxins are 7-dibenzo-p-dioxins (PCDDs), 10 polychlorinated dibenzofurans (PCDFs) and 12 polychlorinated biphenyls (PCBs). Dioxins are produced through combustion of municipal or industrial waste and are released to the environment. They are released in the environment as a byproduct and generated through the combustion of industrial and municipal waste (Stephens et al. 1995). Dioxins caused the real threat in Seveso, Italy. The disposal of dioxins is a concerning issue with respect to environmental contamination. Once dioxins enter in the body, they are absorbed by the fat tissue, because of their chemical stability. In this way they persist longer in the body (Alcock and Jones 1996).

Agrochemicals Currently, fertilizers, pesticides and insecticide are being in frequent used due to population growth, demand and supply, and to increase the agriculture production and plan safety. Agrochemicals are identified as one of the main chemical pollutants and its byproduct cause contamination to the environment and the ecosystem (Sánchez-Bayo and Tennekes 2015). However, the impact of these agrochemicals on the invertebrates and its consequences on the tropical chain is still not very clear (Mantecca et al. 2006). However, these agrochemicals may impact the invertebrate's survival, reproduction, behavior, biomass alteration and tissue and cellular lesions (Tanaka 2003; Francis et al. 2014; Khanna and Gupta 2018).

MetalsMany industrial and biologically important metals are classified in the category of heavy metals or trace metals. Heavy metal pollution has emerged due to anthropogenic activity. The most toxic heavy metals are As, Cd, Hg, Pb, Ti and U which are introduced into environment due to mining, pesticides, fertilizers, contaminated irrigation water and industrial contribution (Appenroth 2010). These metals are long lasting in the soil up to thousands of years (Masindi and Muedi 2018). They can interact with the soil microorganism through plants and transfer to the food chain through the plants and through water contaminations and impact the wildlife and human health (Park et al. 2011).

Polycyclic aromatic hydrocarbons (PAHs) PAHs are the composition of 100 different compounds that are originated by combustion, incomplete burning of oil, gas, coal, garbage, combustion of organic matter and petroleum refining (Bispo et al. 1999). They enter in the environment through the water, volcanic eruption and anthropogenic activities. Significant amounts of PAHs get released and accumulate into soil, because of the complexity of their chemical structure, low solubility, longer degradation time and long persistent timing in the soil which enhances the probability of exposure to the humans and animals (Haritash and Kaushik 2009). Through the environment exposure they transferred into invertebrates by assimilation of plant and soil materials contaminants or by cuticle (Cerniglia 1992).

Vinasse Vinasse is the product of the alcohol production, which is composed of water (97%), an organic matter and mineral elements. The distilling industries are impacting the growth of this residue, which is traditionally discharged in open areas or next to water bodies and impacting water, soil and air pollution. The vinasses can impact the changes in the population of soil microorganism by altering their nitrification, denitrification, and fixation ability (Rajagopal et al. 2014). As fertilizer or soil conditioner, the recycled volume of treated vinasses becomes an option of great interest (Jiang et al. 2012). However, the impact of vinasses on the soil and water pollution are not well explored (Prado et al. 2016).

Pharmaceuticals, hormones and toxins, endocrine disruptors, and biological pollutants, such as bacteria and viruses, also lead to soil pollution. Soil pollution affects agricultural practices severely by reducing the nitrogen fixation and soil fertility due to loss of soil and nutrients thereby reducing the crop yield. This also leads to increased erodibility and deposition of silt in tanks and reservoirs. Soil pollution severely degrades the major ecosystem services provided by soil (Lavelle et al. 2006). Due to soil contamination, food safety is reduced by both reducing crop yields due to toxic contaminant levels and causing crops generated from contaminated soils to be unhealthy for animals and humans to eat. Many pollutants, including nutrients such as nitrogen and phosphorus, are transferred from the soil to surface and ground water, causing direct human health problems through eutrophication due to contaminated water intake and environmental damage.

Pollutants also have detrimental effects on soil microorganisms and larger soil organisms, thus impacting the ecology of the soil and the services rendered by the affected organisms (Wagg et al. 2014).

21.5 Soil Ecosystem and Diversity of Its Resident Organisms

Soil is the central organizing element in terrestrial ecosystems, with a multitude of ecological and geochemical functions. It is a complex dynamic system composed of biotic and abiotic components representing the primary habitat and port of biological activity and diversity that supports several ecosystem services. The factors like local climate, topography, material, biota, and time affect soil formation and its characteristics which influence all its functions and services (Palm et al. 2014).

The soil ecosystem and its overall impact rely on the functions of the underlying natural processes in the soil and the composition of the soil ecosystem (soil biotic and abiotic components and the interactions within and between them). Soil organisms either spend part of their life cycle in soil or complete their entire life cycle within this ecosystem (or its surfaces having decaying logs and litter) (Altieri 1999). However, despite the huge efforts made by soil ecologists in the last few decades to describe and understand soil communities, the true extent of soil biodiversity remains relatively unknown. More than 1000 species can be accommodated by moist tropical soil within an area of 1-m square. Millions of bacteria and thousands of fungal hyphae can be housed in only one gram of soil. Multiple process is undertaken by inhabiting organisms toward maintenance of soil health and fertility (Wall et al. 2015). They control a large proportion of the transformation of organic matter and of the flow of carbon and nutrients in terrestrial ecosystems (Pavao-Zuckerman 2018).

Based on the body width, soil organisms can be grouped as: mesofauna, microfauna, macrofauna, and large animals. Plants and lichens having the capability of photosynthesis also forms the integral part of soil ecosystem (Lehmann et al. 2017). Microorganisms (fungi, bacteria, archea, cyanobacteria, yeast, actinomycetes, and myxomycetes) present in soil forms represent the majority of diversity and plays active role in decomposition of organic material added to soil. Organic matter is converted by microorganisms into plant nutrients that are assimilated by plants. In symbiotic relationships with plants, these species also play an important role in enhancing nutrient absorption (e.g., aiding N fixation) and/or controlling plant hormones (Schimel and Schaeffer 2012). Soil microfauna include organisms like and mites, nematodes, protozoa, and rotifers. Protozoa regulate multiple processes in soil, viz. cycling of nutrients, maintenance of population size of bacteria and fungi, crucial rhizosphere dispersal (Foster 1988). When microfauna comes in contact with roots, either by feeding on those roots or by modifying plant defenses or hormones, showing their deleterious effects on plants, organisms like Acari, Collembola, Protura, Diplura Tardigrada, and Enchytraeidae are part of mesofauna present in soil. These species have minimal burrowing potential and live-in soil pores, feeding on organic matter, microflora, microfauna, and other invertebrates in general. They live in close contact with the soil's air and water and are therefore very dependent on soil aeration and humidity. These species contribute to the cycling of nutrients, function as food for other soil organisms, suppression of pests and diseases, and participate in the distribution of soil biota. Soil macrofauna (>2 mm) includes macroarthropods (e.g., insects, isopods, spiders,) along with soft-bodied organisms (e.g., annelids, gastropods). These are mainly responsible for litter fractionation and predation on other soil-dwelling organisms and are often called "ecosystem engineers.' These species contribute to various soil functions, such as decomposition and cycling of nutrients, suppression of pests and diseases, penetration of water (e.g., burrowing behaviors) and the regulation of other biota as predators (Maraun et al. 2001; Bradford et al.

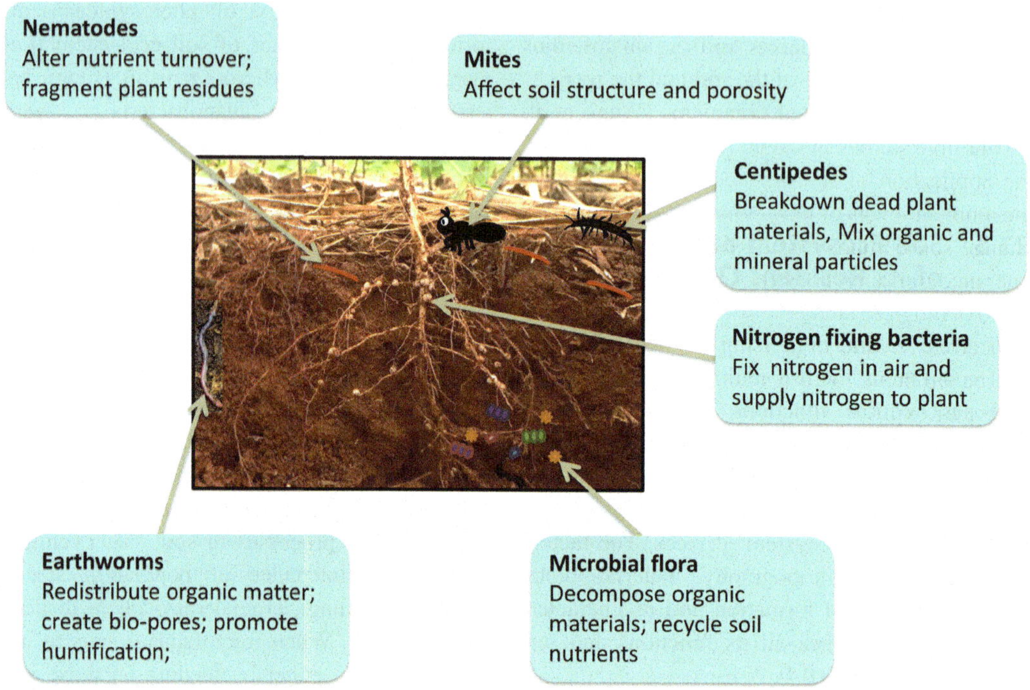

Fig. 21.1 Soil ecosystem and the key functional organisms involved in its maintenance

Table 21.1 Types of soil biota and their role in ecosystems

Type of biomass	Examples	Main function
Microflora (20–200 μm)	Fungi, bacteria, actinomycetes	Breakdown organic matter; immobilize and mineralize nutrients
Microfauna (100 μm)	Mites, nematodes, protozoa and rotifers	Maintenance of population of fungi and bacteria
Mesofauna (100–2 mm)	Acarina, Collembola, enchytraeids	Regulate fungal and microfaunal populations; changes nutrient turnover; plant residues fragmentation, create biopores; promote humification
Macrofauna (>2 mm)	Centipedes, millipedes, earthworms	Mix organic and mineral particles; redistribute organic matter and microorganisms; create biopores; promote humification; produce fecal pellets

2002). The detail information about soil biota and its role in ecosystem is described in Fig. 21.1 and Table 21.1.

21.5.1 Sentinel Species

As the name suggests "sentinel species' represent those organisms in a particular ecosystem which can serve as biological risk monitor, as they can accumulate toxic elements or pollutant in their tissue mass without significant adverse effect (Donnelly et al. 2013). Typically, the use of sentinel species in ecotoxicology is to estimate the amount bioavailable toxic compound for any pollutant; however, these organisms can also be used in conjugation of chemical analysis method to improve sensitivity or to identify and summarize a pollution indication. Invertebrates present in soil resides are in direct contact of the groundwater and soil and thus can be considered as excellent sentinel species for the assessment of various biomarkers to evaluate soil contamination (Kammenga et al. 2000).

Multiple studies in the literature have assessed the application of sentinel species in estimating the ambient levels of new compounds (which can change soil composition), thus giving an idea of tolerance for any contaminant. Researchers in ecotoxicological studies also emphasizes on the subjectivity of the term "bioavailability,' which can be different for different species. The time course of exposure along with the spatial aspect for which sentinel integrates the pollution signal should be given dues consideration for better estimation of bioavailability. In the view of the biological and environmental determinants for assimilation of toxic compounds, calibration of sentinel is required against the source concentration. When the aim is not determining bioavailability to resident population of soil ecosystem, transplanted sentinel species (matched in terms of physiological state, sex and age) can also be used to determine pollution level (Galloway et al. 2004).

21.5.2 Terrestrial Invertebrates

Terrestrial soil mesofauna arthropods such as Isopoda, Collembola, and Diplopoda are among the most suitable species for determining the effects of the presence of harmful substances present in the soil due to their direct interaction with pollutants present in the soil (Hopkin 1990; Gräff et al. 1997). Annelids, in special the Oligochaeta is one of the most important representatives of the edaphic macrofauna that are frequently used in toxicity tests. Through their movement and ingestion of polluted soil or leaf litter, these invertebrates come into contact with a wide range of contaminants found in this compartment (Ribera and Saint-Denis 1999; Dominguez and Edwards 2010). Earthworms are widely used as a biomarker in soil pollution assessment due to their sensitivity to toxic chemicals and other unique biological advantages. Collembola can be chosen a bioindicators, as they are present in all ecosystems, abundantly and can be easily collected in sufficient number to allow statistical analyses. Moreover, they have a short life cycle, making them respond quickly to some type of stress applied in the ecosystem (Xu et al. 2009). Isopoda is of the largest orders of crustaceans with approximately 10,000 thousand described species, mostly marine (Ruggiero et al. 2015). These have already been used in toxicity tests of soil, and the main parameters of evaluation were abundance of individuals (Faulkner and Lochmiller 2000), survival (Stanek et al. 2006), and reproduction rates (Niemeyer et al. 2009). The key toxic agents measured using Isopoda are metals, since these invertebrates bioaccumulate these materials.

Diplopods are colonizers of various layers of soil; these animals can be greatly influenced by the deposition of organic compounds and metals complex substances in the soil. They are frequently used in the recycling of nutrients, aeration, and fertilization of soil (Da Silva Souza et al. 2014). Studies available in the literature on diplopods use as bioindicators of the soil are related to metals, and the effect of organic pollutants and complex mixtures on these invertebrates is relatively less known (Souza and Fontanetti 2011). It is shown through the histological, histochemical as well as the ultrastructural analysis that such substances are toxic to diplopods, since different tissular and cellular alterations were observed in the midgut and perivisceral fat body of these invertebrates (Galloway et al. 2004).

21.5.3 Higher Plants

Plants can offer important information about the cytotoxic, genotoxic and mutagenic potential of substances, even when exposed for short term. Plants have shown satisfactory results in studies with complex mixtures, indicating that these are sensitive enough to detect the adverse effects of environmental samples (Majer et al. 2002). There are some advantages of employing higher plants for use as bioindicator such as: (a) they have low-cost cultivation and easy maintenance, as compared to mammals (Parmar et al. 2016); (b) no filtration or dilution of contaminant takes place before its interaction to plants (Steinkellner et al. 1999); (c) plants being eukaryotes with complex structure and organization resemble the animal system more closely than microbes; (d) much simpler and agile approaches can be used while estimation of contaminants using plants; (e) the assays can be carried out under a broad range of environmental conditions, temperature and pH; (f) higher plants can regenerate more quickly; (g) plants in combination of microbial assays can be used to evaluate the promutagens or mutagenic metabolites; and (h) plants are showing higher genotoxicity in response to carcinogenic agents (Williams and Burdock 2009).

In the higher plants, onion (*A. cepa*) is mostly utilized for determination of the cellular toxicity, toxicity at gene level and effects of mutagenic elements which exists in the soil. Its characteristic cellular kinetics shows positive influence on the roots (Arya and Mukherjee 2014). In this way, the abnormalities and mitotic activity in the cell cycle of the meristematic cells roots can be easily investigated (Grant 1994). The genus Tradescantia has been utilized to study the genetic alteration for the detection of mutations accelerated by different substances which are exist in the water, soil and air and analyzed by micronuclei in the mother cell of pollen grain (Trad-MCN). *Vicia faba* extensively used in cytological studies and physiological experimentations (Khadra et al. 2012).

21.5.4 Special Case of Earthworms: Their Role in Bioremediation and Putative Application as Biomarker

The microorganism present in soil strata plays an active role in bioremediation, if soil is contaminated by organic pollutant. Among other soil organism, earthworms have crucial role in pesticide degradation. The earthworm plays a critical role in soil structure and function and helpful to eradicate pesticides from the soil. Earthworms influence the soil biological and physiological properties, due to this they considered as important biomonitoring candidate for soil pollutant. Their borrowing and feeding characteristics represent a driving force for attracting the population of soil microhabitats. They also inducing the physical changes in soil by increasing the soil porosity and aggregates, which impact the soil aeration, water permeation, and plants root development (Nahmani et al. 2007).

Thy also influence the biological properties by influencing the plant productivity and stimulate the growth and development of root and shoot (Blouin et al. 2013). To prove this hypothesis, Xiao and his colleagues conducted a metanalysis, which confirm that under the influence of earthworm, growth of plant has increased by 20%. This might be possible because of the impact of earthworm which influences soil texture, nutrient mineralization, and microbial communities (Brown 1995). Although the impact of earthworm on plants and plant biodiversity is still not very clear (Edwards 2004). Additionally, change in plant performance directly affect the herbivores, in that case, earthworms may circuitously modify the aboveground herbivore populations (Scheu 2003). Earthworms also contribute in the proliferation and dispersion of microbes by promoting high levels of microbial activity and biomass. This might be due to higher soil enzyme activities in earthworm-rich soil (Lavelle et al. 2006). The contribution of earthworms toward the role of carbon sequestration of soil and biological catalysts for greenhouse gas emissions is a topic of debate. Lubbers and his group (2013) have done metanalysis to investigate the involvement of earthworm in increasing the soil CO_2 and N_2O emissions (33% and 42%, respectively). However, this environmental effect was significant for CO_2 emissions in the short term; a reduction in CO_2 emissions was observed in the longer term (>200 days) (Lubbers et al. 2013). This research was collaborated with other group, who concluded that earthworms showed less impact on carbon mineralization and higher impact on carbon stabilization. Earthworms are conventionally categorized into three epigeic, endogeic, and anecic functional groups (Bertrand et al. 2015).

21.5.4.1 Epigeic Earthworms

These are present on the soil surface and rarely burrow into the soil. Generally, they are present in the forest soil and living in the litter layer. This group of earthworms generally feeds organic matters accumulated in the soil surface such as dead and decaying plant roots and leaf litter. They form temporary burrows.

21.5.4.2 Endogeic Earthworms

These are soil-dwellers. They are getting nutrients from soil and organic matters. These earthworms burrow intensely in the top most (10–15 cm) soil surface. Although in search of food, sometimes they come on the surface. They form temporary horizontal burrows (Bertrand et al. 2015).

21.5.4.3 Anecic Earthworms

This species includes large earthworms that form the permanent burrows 3 m deep below the soil surface. The species belongs to the ecological group which generally obtains the food from dead organic matter that tow into their dugout. They also swallow some soil minerals and even collect green leaves and grasses that also drug into the burrows. Anecic earthworms usually deposit the organic residues mixed with feces at the entrance of their burrows which referred as middens (Brown 1995). These deposits are rich in organic matter decomposes and faunal diversity, which is considered to be hotspots. In summary, earthworm casts, burrow walls, and the middens are active hotspot rich in the microenvironments that helps in harmful organic matters and pesticide immobilization and degradation. There are three main processes which have great impact on the environmental fate of pesticides in the troposphere compartments. First, the organic matter deposition in these microsites enhances the binding affinity of hydrophobic pesticides with the organic ligands, which contribute in biodegradation and increasing their persistence. Second, earthworm activity may stimulate the growth and proliferation of indigenous soil microorganisms which might help in pesticides degradation. And latter, these microorganisms can collect the pesticide residues through absorption by their skin and gastrointestinal epithelium (Araneda et al. 2016).

21.6 The Concept of Bioindicators

Any biological response or organisms present in the soil, which shows typical symptoms or measurable attributes when exposed certain pollutants is called as bioindicators of soil ecosystem. Bioindicator concept is a qualitative way of determining or indicating the presence of toxins or contaminants (Gerhardt 2002). Chemical, behavioral, or physiological changes are often exhibited by the bioindicator organisms (or group of organisms) to deliver information regarding changes in the local environment or presence of contaminants. Thus, bioindicators are observed and studied in terms of their composition, biochemical metabolism, population structure, morphology, and cell structure (McGeogh 1998). These indicators can not only suggest direct contamination of soil environment but also reveal the indirect biotic effects on the species and ecosystem. The use of bioindicator organisms is also justified by the fact that studying the species itself is the best way to ascertain the status of harmful effect or pollution on itself and ecosystem. Selection of a suitable bioindicator is an important choice in successful evaluation of ecotoxicology (Van Gestel and Van Brummelen 1996). Diverse organisms residing in soil and belonging to wider range of life forms can be used as bioindicators, and this includes microbial community, soil microinvertebrates (mites, spirngtail, nematodes etc.) (Tang et al. 2002), macroinvertebrates (snails, spider, insect, and earthworms), and the local fauna (Catling 2005). Among these diverse organisms, earthworm (especially Eisenia and Lumbricus species) in particular has received a lot attention by the researcher as it is known to bioaccumulate toxic compounds (Pérès et al. 2011). Some earthworm species like allolobophora chlorotica (endogeic species) is known to be geophagus (soil eating) and makes horizontal burrows in soil strata.

An ideal bioindicator should be key species of the ecosystem, sensitive, uniform distribution, longevity, easy for sampling and should have its genome sequence and annotate. The chosen organism for bioindicator should have a known biology with established knowledge about physiology, anatomy, metabolic pathways, and relative role in the ecosystem (Chou et al. 2003). More robust approach of biomonitoring with the help of bioindicators includes use of more than one species belonging to different ecological habitats with the soil ecosystem to be assessed.

Both passive and active monitoring of pollutants in soil can be undertaken with the help of biomonitors, which involve studying specific properties of the organism to gather information on nearby chemical and physical attributes. Biomonitors are often called bioaccumulative markers (Dórea 2008). There are many types of bioindicators, depending on the chosen organism and their application. The bioindicators on the basis of their respective application can be categorized as:

- Ecological indicators are organisms which show sensitivity toward habitat fragmentation, pollution, and stress on local ecosystem.
- Biodiversity indicators generally correspond to the species taxa, where richness of species can give an estimate of overall biodiversity prevalent in the ecosystem or community.
- Those species or group of organisms which shows predictable changes in response to changes in ecological factors are known as environmental indicator (Burger and Gochfeld 2001).

21.7 Biomarkers for Assessment of Soil Pollution

Biomarker in assessment of soil pollution represents the usage of biological parameters (Bioindicators) and their measurement for identification of negative effects of pollutants which exist in the soil and can serve as early warning for environment risk (Galloway et al. 2004). Studying biomarkers enables the researchers to observe biologicals influence of the contaminants and how any xenobiotics affects the fauna. A comprehensive definition for biomarkers was proposed by Depledge et al. in 1993 and is the most widely accepted one. The authors described biomarkers as "adaptive biological responses to the stressors, which manifests as cellular, biochemical, histological, behavioral or physiological variations" (Depledge and Fossi 1994). Another group of researcher mentions "biomarkers are molecular, cellular, biochemical, or physiological change in cells, tissue, bodily fluids, or organs of the organisms, when it is exposed to xenobiotic" (Lam and Gray 2003). The measurement of biomarkers for assessing the toxic effects is sensitive and shows the very beginning of responses toward disturbances caused by contaminants or other environmental factors (McCarthy and Shugart 2018). Introduction of a foreign contaminant or pollutant to soil ecosystem can affect the life forms at various levels, starting from the very molecular levels to species, community, population or orders (Sánchez-Bayo and Tennekes 2015).

On exposure to foreign substances and pollutants, diverse and multiple pathways can become operative or altered in the organisms residing in soil ecosystem. Such metabolic changes often result in structural change, functional change or can influence the inherent physiological mechanisms in the organisms. The lower taxonomical levels are first influenced by pollutants due to the hierarchical architecture of ecosystems. Therefore, activation of signal transduction cascade pathways, transcriptional reprogramming, changes in enzyme activity, post-translational modifications of proteins, and changes in other molecular effectors are among the few early characteristics caused by exposure to contaminants (Knackmuss 1996). Gradually the overall exposure level, degree of abuse, and toxicity of the contaminants will manifest in form abnormalities or change in cells, organs, and higher forms of life. However, this should also be understood that organisms abused by toxic pollutants and foreign chemicals will tend to develop adaptive responses toward maintaining the homeostasis or begin the equilibrium at an altered state or level (Alexander 2000). These responses for maintenance of homeostasis and pristine condition will tend to restore the balance by detoxification of the environment and can serve as biomarkers for effective monitoring of soil conditions (Hyne and Maher 2003).

Multiple reports over the past few years suggests use of several biomarkers in testing method for particular toxicants, many of which showed promising results both with and without use of biomonitors. The efficiency of such system depends on creation of sophisticated multiple target system for detection of environmental hazards with the advantage of being rapid and economical simultaneously. Just like ecology deals at the extent of populace dynamics and community structure, ecotoxicology at the level of species and organisms (survival), biomarker study works at the suborganism level and can provide the opportunity for early intervention. The use of biomarkers will help in chalking out remedial plans with clear priorities for fast action against environmental restoration.

21.7.1 Biomarker Selection for Effective Assessment of Soil Pollution

The selection of biomarkers in a biomonitoring program is a critical point. In addition to this, multibiomarker approach provides an innovative way for measuring several parameters at a time and can be useful for identification of specific pollutants and also useful for monitoring the effect of these pollutants on health status (He and Yu 2010). There are three stages where these biomarkers can be used for biomonitoring programs (McCarthy and Shugart 2018):

1 Identification of unknown chemical contaminant.
2 Discovering the accurate pollutants and identifying their contamination strength.
3 Collecting the information related to risk assessment, to evaluate the long-term negative impact on the community and population level. It is useful to identify biomarkers of susceptibility and biomarker of mutagenicity (micronucleus).

21.7.2 Classification of Biomarkers

There are multiple ways in which biomarkers used for assessment of soil can be grouped and categorized. On the basis of determinant parameter, generally biomarkers can be categorized into different group such as biomarkers of exposure, effect, and susceptibility (Chambers et al. 2002). Biomarkers of exposure deal with the organisms which are already in contact with the pollutant and show an early sign of exposure to micropollutants. They can be useful to give a quantitative and qualitative estimation of contact to several compounds. Though any changes in biomarker of exposure may not be directly related to create any antagonistic effects on the organism and the population, it could be helpful to give the information about the population health status that is in the transition phase from homeostasis to disease sates or recovering from illness. On the other hand, biomarkers of effect provide the information about toxicant's mechanism of action which are showing correlation with the intensity of biomarker modification to the intensity of adverse effects (Chambers et al. 2002). Finally, biomarkers of susceptibility give information regarding the possibility of developing a pathological stress syndrome (acquired or intrinsic) upon exposure to contaminants (Gastaldi et al. 2007). All three types of biomarkers have their own impact for soil assessment. Biomarkers of exposure can be used for the chemical analyses which are short-lived and provide more relevant indication of exposure (Hagger et al. 2006). In contrast, biomarkers of effect can be used for qualitative measurement for hazard identification which gives the information about their occurrence and provides a mechanism responsible for it. Based on the biological characteristics of the biomarker, they can be grouped in many categories such as morphological biomarkers, molecular biomarkers (enzyme, protein and other biomolecules), behavioral biomarkers, DNA biomarkers, histological, cytological and Omics biomarkers (Fig. 21.2).

21.7.2.1 Morphological Biomarkers

Morphological biomarkers are used for the damage detection at cellular and tissue level. These morphological variations may provide qualitative indications of a functional variation to the surrounding environment (Van der Oost et al. 2003). Mostly, histology and ultrastructure are used for the recognition of morphological change in variety of organs (Johnson et al. 1993), as gills and liver. The investigation on gills of fish and bivalves are frequently used because any variation in gill and liver may lead to the impairment of gaseous exchange, ionic balance, metabolites excretion, and several other functions (Cruz et al. 2015). The comprehend shape and size of the whole body provides knowledge about impact of potential pollutants. Although this is very insensitive method which can change disease condition and nutritional level, it can be used for prescreening of biomarker to get the information about biomarker of exposure and effects and energy reserves (Linde-Arias et al. 2008). The cheap and rapid screening time of this biomarker makes it more assessable to evaluate the initial effects of toxic chemicals in fish (Van der Oost et al. 2003).

21.7.2.2 Molecular Biomarkers

Biochemical, proteins, and enzymes fall under the category of molecular biomarkers which can be helpful to assess the changes induced or inhibited by the presence of pollutants (McCarthy and Shugart 2018). These molecular markers provide qualitative information about the damages caused by certain pollutants and provide the initial information about the adverse impact of the exposure, before onset of visible damages. Some examples of molecular markers are metallothioneins, heat shock proteins (HSP), esterases, and antioxidants enzymes.

Metal concentration is one of the main factors which can be used for the assessment of soil ecotoxicology. Metal exposure enhanced reactive oxygen species (ROS) production, which creates oxidative damage and cause adverse impact (Barreiros et al. 2006). During oxidative damage in the cells, there are enzymatic and non-enzymatic antioxidant reactions, which minimizes the injury (de Freitas et al. 2008). In this way, antioxidant enzymatic activity is helpful to reduce the environmental stress. Glutathione -S-transferase, catalase, superoxide dismutase, and

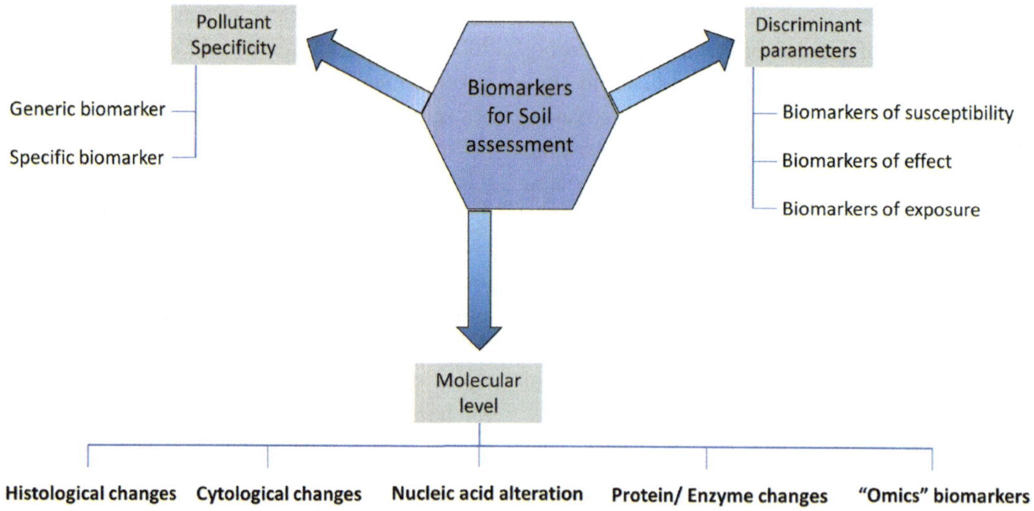

Fig. 21.2 Classification of biomarker based on different attributes

glutathione reductase are the main enzymes which are found involved in the antioxidant reactions (Mishra et al. 2006). Glutathione enzyme plays major impact on the defense of cells against ROS and xenobiotics. The glutathione peroxidase enzyme neutralizes toxic chemicals in the cells by converting the oxidized form to the reduced form. The associations between heavy metal intoxication, free radical foraging, and glutathione metabolism are crucial for cell survival (Łaszczyca et al. 2004; Aigerim et al. 2015).

Metallothioneins are cysteine-rich universal cytosolic proteins. It is a specific biomarker of heavy metals exposure such as Cu and Zn, Cd, and Hg. High binding affinity of this protein helps in reducing the toxic effects of heavy metal. The metallothioneins biomarker is mostly employed for the evaluation and monitoring of environmental pollution (Kammenga et al. 2000; Sánchez-Bayo and Tennekes 2015). The heat shock proteins are present in cytoplasm which plays a major role in the protein folding and maturation. Heat shock protein has their prime function of protecting the cells from physical–chemical stress and denaturation (Feder and Hofmann 1999). It is identified that heat shock proteins have very high affinity with toxic materials such as heavy metals, polycyclic aromatics hydrocarbons, polychlorinated biphenyls and pesticides in various soil microorganisms (Köhler et al. 1992; Eckwert et al. 1997).

Esterase is another class of enzymes which is highly influenced by organophosphoric (OP) and biological pesticides like carbammates. Acetylcholinesterase and carboxylesterase are the two major esterases enzymes which are used for soil pollution monitoring (van den Brink et al. 2011). The activity of Acetylcholine esterases can be simply evaluated for soil pesticides pollution in variety of terrestrial invertebrates like isopods, snails, and earthworms (Pohanka et al. 2011; Calisi et al. 2013). Acetylcholinesterase is the key cholinesterase present in earthworms (Rault et al. 2007). Its functionality has been recognized and biochemically categorized in limited species of earthworm. The pre-clitellar part of the animal shows highest AChE activity which is mainly associated with the functionality of enzymes in the dorsal brain situated close to the prostomium.

21.7.2.3 Alteration in DNA: Genotoxicity Biomarkers

Among the various pollutants and toxic compounds, some can induce carcinogenic and mutagenic damages to the nucleic acid of residing organisms. Such chemicals are chemically dioxins, polycyclic aromatic carbons, acrylamide, and other agents (Shugart 2000). Multiple mode of DNA damage including breakage of double bond, sever fragmentation of nucleosome or chromosome and crosslinking are the common modes of DNA damage induced by these foreign chemicals. The induced structural damage can be quantitatively measured and use as a biomarker (McCarthy and Shugart 2018). The release of chemical and physical substances which impact the genotoxic level in the terrestrial ecosystems which increase the frequency of mutations and influence population size as a result it can cause species extinction and affect ecosystem (Majer et al. 2002). In this way, it is very important to conduct a numerous tests to assess the probability of genotoxicity in soil samples. Coelomocytes are also useful for ecotoxicological investigation and application for the assessment of the genotoxicity effect of pollutants on earthworms. Many mutagenic and carcinogenic environmental pollutants, such as benzopyrenes and other carcinogenic aromatic hydrocarbons in the soil can react with the DNA and affect the DNA and chromosomal structure and integrity (Lourenço et al. 2012). In this way, it is worth to investigate this biomarker quantitatively. The comet assay (or single cell gel electrophoresis) and the micronucleus test are mostly used to evaluate the DNA damage, oxidative DNA damage. These methods are extensively used as genotoxic biomarkers due to its sensitivity and specificity. The micronucleus test is generally used for the assessment of in situ genotoxic pollution (Smaka-Kincl et al. 1996; Çavaş and Ergene-Gözükara 2005). This technique is very sensitive and used to measure the chromosomal break and chromosomal damage collected throughout the lifetime of the cell in vertebrate and invertebrates. The comet assay is also an effective genotoxic biomarker which is used for the detection of DNA damage in the earthworms coelomocytes exposed to genotoxic compounds, both in vitro and in vivo (Reinecke and Reinecke 2004; Casabé et al. 2007). Comet assay is a significant tool for the evaluation of contamination because of its sensitivity and easy handling.

21.7.2.4 Histological and Cytological Biomarkers

Among other biological characteristics used as biomarkers, cytology and histology are prominently critical branches for understanding the compensatory mechanism along with the detrimental effect of the contaminants in soil. Cells being the structural and functional unit of life, it is also the place toxic pollutant added to the soil ecosystem will be stored, metabolized, and detoxified with the help of various metabolic pathways (Sanchez-Hernandez 2006). Both structural and functional alteration can be observed in the cells or group of cells, when the organism is exposed to toxic compounds and chemical stressors. Some of these measurable changes are signal transduction, membrane potential changes, membrane structure, cellular arrangements, lysosomal membrane, and cytoskeletal changes (Kammenga et al. 2000). Cell-based biomarker for assessment of soil pollution has been reported in multiple reports in diverse organism like earthworms (Lionetto et al. 2011), isopods (Lemos et al. 2010), collembolans (Augulyte et al. 2008), and nematodes (Sochová et al. 2006). Accumulation of lipids in tissue is considered a biomarker for contamination by organic pollutants while lipofuscin accumulation indicates general stress (Hodson et al. 2008). Immune system of soil invertebrates consists of circulating cells known as coelomocytes and haemocytes and provide inherent defense against external environment agents. These circulating immune cells can also serve as a first barrier against external chemical and toxin shocks and thus can be utilized as biomarkers. Literature suggests rounding of these cells, enlargement and loss of pseudopods on exposure to pollutants (Leomanni et al. 2015). Destabilization of lysosomal membrane is another important cytological biomarker, which indicates toward general stress in the environment and has been commonly studied (Svendsen et al. 1996; Lionetto et al. 2011). Both organic and inorganic toxic element accumulate in lysosomes by making the membrane of the organelle leaky, and this will eventually cause leakage of degrading enzymes in cytoplasm and death of cells (Regoli 1992).

21.7.2.5 Behavioral Biomarkers

For the assessment of ecological risk or for the indication of already present environment toxins, behavioral studies are promising tool. In a generalized way, it represents the response of a resident species in terms of its habits, physiology, and interaction to other species, when it is exposed to some external pollutant or stress (Depledge and Fossi 1994; Filser et al. 2008). Most of the alterations in the behavior of an organism upon its exposure to toxic soil are indirect manifest of underlying biological damages to sensory organs or nervous system. Owing to the ease of recording behavioral changes and thus data collection, large number of soil toxicity assessment studies based on this method is reported over the past few years (Capowiez et al. 2010; Žižek and Zidar 2013). One of the most common behavioral changes after exposure to chemical contaminants is avoidance (most commonly employed), in which the organism tends to show a defense mechanism for its protection (Matos-Moreira et al. 2011; Martínez Morcillo et al. 2013). An ISO 2008 certification is in-fact rendered to avoidance behavioral biomarker test. Other behavioral aspects being studied include tactile sensitivity, predatory behavior, chemoreception, aggressiveness, burrow formation and alteration in locomotion capabilities. Another advantage of behavior study is that of being non-invasive, allowing recording of data without harming the organism. However, the lack of standardized protocol for this approach is one of the practical limitations, apart from biases due other factors (not related to soil contamination).

21.7.2.6 Omics Biomarkers

New generation biological methods which allow comprehensive assessment and study for any change are particular biomolecule or pathways such as transcriptomics (mRNA), genomics (DNA), metabolomics (regulatory pathways) that can be utilized as biomarkers. Such biomarker is supposed to be much sensitive and can enable detection to uncharacterized contaminants in the soil ecosystem. For example, *Caenorhabditis elegans* (nematode) is soil organisms often studies in environmental toxicology, as it is well-characterized organism with its genome sequenced and annotated (Brulle et al. 2010). Other soil organisms like earthworm and collembola have also been well-characterized recently for their use in eco-toxicological studies (Pirooznia et al. 2007). The effects and toxicity of Dechlorane plus (DDC-CO) (Polychlorinated flame retardant) on *E. foetida* were assessed using Illumina RNAseq method, which confirmed increased oxidative stress, DNA repair inhibition and neuronal damage even at acute toxicity level of this compound (Zhang et al. 2014)y. Recent studies are reporting a combinatorial use of metabolomics and transcriptomics for estimating alterations in energetic metabolic pathways (oxidative phosphorylation) (Bundy et al. 2008). Such findings clearly indicate that gene expression and additional methods to omics are now able to be incorporated into complex procedures for ecological risk assessment and that scientific and interpretative bottlenecks have been resolved.

21.8 Biomarkers for Assessing Soil Pollution, Future Directions, and Limitations

Conventional chemical analysis and ecotoxicological methods are limited and expensive for the ever-increasing and diverse pollutants of soil ecosystem (Arshad and Martin 2002). Biomarker as discussed above provides an all-natural and integrated exposure response to soil ecosystem degradation with the help of chosen bioindicator species. Biomarker-based assessment of soil provides a comprehensive effect of the toxicant and its bioavailability to soil organisms (Santorufo et al. 2012). Bioindicator species-based soil assessment is however complicated by the presence of more than one pollutant having diverse chemical nature which can lead to synergistic response and biased estimation. Soil ecosystem represents a complex matrix which has the capacity to bind the pollutants and influence its bioavailability in a variable manner (Bradl 2004). Thus for using an organism for assessing contamination in wild, its adaptation to the presence of pollutant must be known (Peakall 1994).

A good correlation can be established between the degree of toxicity or contamination in the soil environment and the alteration biological response (biomarker), which can be used to formulate a regression formula to estimate the level of pollution. For a correct interpretation of the estimations, time kinetics of biomarker (in response to particular contaminant) along with dose dependence dynamics should be considered (Hagger et al. 2006). A measurement bias of pollutant level is often expected when the levels in the environment is too high, this is why a prior standardization with the help of biomarker with the help of general indices of organism's health is required (Dagnino et al. 2008).

A measurement calibration with the help of known responses (e.g., temperature, reproductive cycle, etc.) along with the maximum and minimum recorded values is also necessary to make sure unbiased estimation of the degree of contamination. Seasonal variation along with climate can have an impact on bioindicator species and thus can limit the biomarker-based soil assessment to certain time of the year.

Syntrophic associations can be formed by microbial assemblies, contributing to the breakdown of recalcitrant polymers, including plastics, and a broad range of other organic pollutants. Degradation can also not be accomplished individually, and the degradation of these compounds is crucial to these dynamic group interactions. It is very interesting and significance to comprehend how these groups function cooperatively interact and behave in bioremediated environments, if the ecosystem is natural or man-made (e.g., rivers and lakes) (e.g., sewage systems and landfills). These environmental microbiomes can interact with, for example, protozoans, microarthropods, and nematodes at different trophic levels and with a number of higher species, including well-known interactions with key players in ecosystems such as earthworms in the soil. Similar interactions can occur both in the planktonic and benthic phases in marine and freshwater environments; indeed, studies are already investigating the microbiomes of microplastics and their role in the spread of pathogens, as well as the concomitant degradation of polymers. Microbial species interact with their biotic and abiotic ecosystems at the molecular level, becoming integral parts of geochemical processes (e.g., nitrogen, carbon, sulfur, and water) and also helping to establish the infrastructure that keeps the ecosystem in place, such as the exoproteins produced in soils by the bacterial community holding the soil particles together. The determination of their resistance and responses to contaminants and climatic impacts is a major challenge in studying such microbial assemblages. The complex mixture or organic and inorganic contaminants that can strongly bind to the environmental matrix are also contaminated environments. Much like the human gut microbiome is vital for the breakdown of plant polymers and important for helping to meet our nutritional needs, so the earth's microbiome is crucial for helping to recycle nutrients from natural and anthropogenic sources for a healthier climate.

21.9 Conclusion

The continuous release of toxic pollutants and harmful chemical because of industrialization and increase in agricultural outputs is straining terrestrial ecosystem by contaminating life supporting layer of earth crust (soil). We discussed the soil ecosystems, its pollutants, and approaches undertaken for the assessment of pollution. The chapter discusses the diversity of inhabiting life forms in the soil and their prospective role as biomarker for assessing pollution level of soil, while emphasizing on key species and sentinel organism. Bioindicators and different biomarkers used for assessing soil pollution was discussed along with comprehensive literature review on the same. Biomarkers for soil pollution assessment, its characteristics, and classification are concisely elaborated for lucid understanding. The technicalities involved in choosing a suitable organism and bioindicator are explained for successful environmental monitoring. In addition, the combined use of anatomy, biochemistry, genotoxicity and the assessment of injured sentinel species are exciting as they provide information about the impact of pollutants on exposed organisms and better consistency gained in the study. Finally, the relevant conditions and limitations for use for biomarkers-based assessment of soil pollution is discussed along with the necessity to take relevant controls. We assume that it would be of great benefit to the future work of scientists in ecotoxicology to consider the value of the integrated usage of multiple techniques in evaluating the substrates toxicity. In addition, the use of a multibiomarker method would make it possible to classify the major contaminant groups suggested in the soil contamination scenario. We can perceive the use of biomarkers as an area of research full of potential and opportunities for biomonitoring or bioremediation program. Equally the research community and the government bodies are employed for environmental conservation; their use is a suggested instrument.

References

Abbasi A, Sajid A, Haq N, Rahman S, Misbah ZT, Sanober G, Ashraf M, Kazi AG (2014) Agricultural pollution: an emerging issue. improvement of crops in the era of climatic changes, vol 1. Springer, New York, NY, pp 247–387

Adhikari K, Hartemink AE (2016) Linking soils to ecosystem services—a global review. Geoderma 262:101–111

Aigerim S, Valentina K, Bayan Y, Karlygash Y, Nelli R, Berikbay K, Anar T, Raushan G (2015) Assessment of changes in biochemical indices of free radical oxidation in the semen of men living in the Aral Sea region. Free Radic Biol Med 86:S2129. https://doi.org/10.1016/j.freeradbiomed.2015.07.081

Alcock RE, Jones KC (1996) Dioxins in the environment: a review of trend data. Environ Sci Technol 29:346–355

Alexander M (2000) Aging, bioavailability, and overestimation of risk from environmental pollutants. Environ Sci Technol 20:4259–4255

Altieri MA (1999) The ecological role of biodiversity in agroecosystems. Agric Ecosyst Environ 74:19–31. https://doi.org/10.1016/S0167-8809(99)00028-6

Anastopoulos I, Kyzas GZ (2015) Composts as biosorbents for decontamination of various pollutants: a review. Water Air Soil Pollut 226:2345–2349. https://doi.org/10.1007/s11270-015-2345-2

Andrews SS, Karlen DL, Cambardella CA (2004) The soil management assessment framework: a quantitative soil quality evaluation method. Soil Sci Soc Am J 68(6):1945–1962. https://doi.org/10.2136/sssaj2004.1945

Appenroth K-J (2010) Soil heavy metals: definition of "heavy metals" and their role in biological systems. Springer, Berlin, p 642

Araneda AD, Undurraga P, Lopez D, Saez K, Barra R (2016) Use of earthworms as a pesticide exposure indicator in soils under conventional and organic management. Chil J Agric Res 76:356–362. https://doi.org/10.4067/S0718-58392016000300014

Arshad MA, Martin S (2002) Identifying critical limits for soil quality indicators in agro-ecosystems. Agr Ecosyst Environ 88(2):153–160

Arya SK, Mukherjee A (2014) Sensitivity of *Allium cepa* and *Vicia faba* towards cadmium toxicity. J Soil Sci Plant Nutr 14:447–458. https://doi.org/10.4067/S0718-95162014005000035

Augulyte L, Kliaugaite D, Racys V, Jankunaite D, Zaliauskiene A, Andersson PL, Bergqvist PA (2008) Chemical and ecotoxicological assessment of selected biologically activated sorbents for treating wastewater polluted with petroleum products with special emphasis on polycyclic aromatic hydrocarbons. Water Air Soil Pollut 195(1):243–256. https://doi.org/10.1007/s11270-008-9743-7

Barreiros AL, David JM, David JP (2006) Oxidative stress: relations between the formation of reactive species and the organism's defense. Quim Nova 29 (1):113–123

Bastida F, Zsolnay A, Hernández T, García C (2008) Past, present and future of soil quality indices: A biological perspective. Geoderma 147:159–171

Bertrand M, Barot S, Blouin M, Whalen J, Oliveira T, Roger-Estrade J (2015) Earthworm services for cropping systems: A review. Agron Sustain Dev 35:553–567

Bispo A, Jourdain MJ, Jauzein M (1999) Toxicity and genotoxicity of industrial soils polluted by polycyclic aromatic hydrocarbons (PAHs). Organic Geochemistry 30(8):947:952

Blouin M, Hodson ME, Delgado EA, Baker G, Brussard L, Butt K, Dai J, Dendooven L, Peres G, Tondoh J, Cluzeau D, Brun J (2013) A review of earthworm impact on soil function and ecosystem services. Eur J Soil Sci 6(2):161–182

Bradford MA, Tordoff GM, Eggers T, Jones TH, Newington JE (2002) Microbiota, fauna, and mesh size interactions in litter decomposition. Oikos 99:317–323. https://doi.org/10.1034/j.1600-0706.2002.990212.x

Bradham KD, Dayton EA, Basta NT, Schroder J, Payton M, Lanno RP (2006) Effect of soil properties on lead bioavailability and toxicity to earthworms. Environ Toxicol Chem 25:769–775

Bradl HB (2004) Adsorption of heavy metal ions on soils and soils constituents. J Colloid Interface Sci 227 (1):1–18. https://doi.org/10.1016/j.jcis.2004.04.005

Briški F, Vuković Domanovac M (2019) Environmental Microbiology. Phys Sci Rev 2(11):1–22. https://doi.org/10.1515/psr-2016-0118

Brown GG (1995) How do earthworms affect microfloral and faunal community diversity? Plant Soil 170:209–231. https://doi.org/10.1007/BF02183068

Brulle F, Morgan AJ, Cocquerelle C, Vandenbulcke F (2010) Transcriptomic underpinning of toxicant-mediated physiological function alterations in three terrestrial invertebrate taxa: a review. Enviro. Pollut 158:2793–2808

Bundy JG, Sidhu JK, Rana F, Spurgeon DJ, Svendsen C, Wren JF, Sturzenbaum SR, Morgan AJ, Kille P (2008) Systems toxicology: approach identifies coordinated metabolic responses to copper in a terrestrial non-model invertebrate, the earthworm *Lumbricus rubellus*. BMC Biol 6:2–21. https://doi.org/10.1186/1741-7007-6-25

Burger J, Gochfeld M (2001) On developing bioindicators for human and ecological health. Environ Monit Assess 66:23–46. https://doi.org/10.1023/A:1026476030728

Cachada A, Rocha-Santos TAP, Duarte A (2017) Soil and pollution: an introduction to the main issues. Soil pollution. Academic Press, UK, pp 1–28

Calisi A, Zaccarelli N, Lionetto MG, Schettino T (2013) Integrated biomarker analysis in the earthworm Lumbricus terrestris: application to the monitoring of soil heavy metal pollution. Chemosphere 90(11):2637–2644. https://doi.org/10.1016/j.chemosphere.2012.11.040

Capowiez Y, Dittbrenner N, Rault M, Triebskorn R, Hedde M, Mazzia C (2010) Earthworm cast production as a new behavioural biomarker for toxicity testing. Environ Pollut 158:388–393. https://doi.org/10.1016/j.envpol.2009.09.003

Casabé N, Piola L, Fuchs J, Oneto ML, Pomaparoto L, Basak S, Gimenez R, Massaro R, Papa J, Kesten E (2007) Ecotoxicological assessment of the effects of glyphosate and chlorpyrifos in an Argentine soya field. J Soils Sediments 7:232–239. https://doi.org/10.1065/jss2007.04.224

Catling PM (2005) A potential for the use of dragonfly (Odonata) diversity as a bioindicator of the efficiency of sewage lagoons. Can Field-Naturalist 119:233–236. https://doi.org/10.22621/cfn.v119i2.111

Çavaş T, Ergene-Gözükara S (2005) Micronucleus test in fish cells: a bioassay for in situ monitoring of genotoxic pollution in the marine environment. Environ Mol Mutagen 46(1):64–70. https://doi.org/10.1002/em.20130

Cerniglia CE (1992) Biodegradation of polycyclic aromatic hydrocarbons. Biodegradation 3:351–368. https://doi.org/10.1007/BF00129093

Chambers JE, Boone JS, Carr RL, Chamber HW, Straus DL (2002) Biomarkers as predictors in health and ecological risk assessment. Hum Ecol Risk Assess 8(1):165–176. https://doi.org/10.1080/20028091056809

Chen J (2007) Rapid urbanization in China: a real challenge to soil protection and food security. Catena 69:(1)1–15. https://doi.org/10.1016/j.catena.2006.04.019

Chou CL, Paon LA, Moffatt JD, King T (2003) Selection of bioindicators for monitoring marine environmental quality in the Bay of Fundy. Atlantic Canada. Mar Pollut Bull 46(6):756–762. https://doi.org/10.1016/S0025-326X(03)00045-6

Criel P, Lock K, Van Eeckhout H, Oorts K, Smolder E, Janssen C (2008) Influence of soil properties on copper toxicity for two soil invertebrates. Environ Toxicol Chem 27(8):1748–1755. https://doi.org/10.1897/07-545.1

da Cruz AL, Prado TM, da Maciel LA, Couto RD (2015) Environmental effects on the gills and blood of *Oreochromis niloticus* exposed to rivers of Bahia, Brazil. Ecotoxicol Environ Saf 111:23–31.https://doi.org/10.1016/j.ecoenv.2014.09.022

Da Silva ST, Christofoletti CA, Bozzatto V, Fontanetti CS (2014) The use of diplopods in soil ecotoxicology—a review. Ecotoxicol Environ Saf 67:190–205

Dagnino A, Sforzini S, Dondero F, Fenoglio S, Bona E, Jena J, Viarengo A (2008) A Weight-of-Evidence approach for the integration of environmental "Triad" data to assess ecological risk and biological vulnerability. Integr Environ Assess Manag 4(3):314–326. https://doi.org/10.1897/IEAM_2007-067.1

da Souza T, Fontanetti CS (2011) Morphological biomarkers in the Rhinocricus padbergi midgut exposed to contaminated soil. Ecotoxicol Environ Saf 74(1):10–18https://doi.org/10.1016/j.ecoenv.2010.09.009

de Freitas DR, S Jr VI, Masuda A (2008). Expression and enzymatic activity of glutathione s-transferase in tissues of Boophilus microplus females. Revista brasileira de parasitologia veterinaria= Brazilian journal of veterinary parasitology: Orgao Oficial do Colegio Brasileiro de Parasitologia Veterinaria 17(2):99–104. https://doi.org/10.1590/s1984-29612008000200007

Depledge MH, Fossi MC (1994) The role of biomarkers in environmental assessment (2). Invertebrates Ecotoxicol 3:161–172. https://doi.org/10.1007/BF00117081

Dominguez J, Edwards CA (2010) Biology and ecology of earthworm species used for vermicomposting. Vermiculture Technology: Earthworms, Organic Wastes, and Environmental Management. CRC Press, USA, pp 25–37

Donnelly RK, Harper GL, Morgan AJ, Orozco-Terwengel P, Pinto-Juma GA, Bruford MW (2013) Nuclear DNA recapitulates the cryptic mitochondrial lineages of *Lumbricus rubellus* and suggests the existence of cryptic species in an ecotoxological soil sentinel. Biol J Linn Soc 110:780–795. https://doi.org/10.1111/bij.12171

Dórea JG (2008) Persistent, bioaccumulative and toxic substances in fish: human health considerations. Sci Total Environ 1:93–114. https://doi.org/10.1016/j.scitotenv.2008.06.017

Do Prado EAF, Vitorino ACT, Marchi G, Muniz DHF, Souza TA (2016) Vinasse dynamics on soil solution under sugarcane crop: inorganic and organic ion analysis. Water Air Soil Pollut 227:145https://doi.org/10.1007/s11270-016-2846-7

Eckwert H, Alberti G, Köhler HR (1997) The induction of stress proteins (hsp) in *Oniscus asellus* (Isopoda) as a molecular marker of multiple heavy metal exposure: I. Principles and Toxicological Assessment. Ecotoxicology 801:127–162. https://doi.org/10.1023/A:1018682928839

Edwards CA (2004) Earthworm ecology, 2nd edn. CRC Press, USA, p 456

Faulkner BC, Lochmiller RL (2000) Increased abundance of terrestrial isopod populations in terrestrial ecosystems contaminated with petrochemical wastes. Arch Environ Contam Toxicol 39:86–90. https://doi.org/10.1007/s002440010083

Feder ME, Hofmann GE (1999) Heat-shock proteins, molecular chaperones, and the stress response: Evolutionary and ecological physiology. Annu Rev Physiol 61:243–282

Filser J, Koehler H, Ruf A et al (2008) Ecological theory meets soil ecotoxicology: challenge and chance. Basic Appl Ecol 9:346–355. https://doi.org/10.1016/j.baae.2007.08.010

Foster RC (1988) Microenvironments of soil microorganisms. Biol Fertil Soils 6:189–203. https://doi.org/10.1007/BF00260816

Francis MR, Raja L, Inbarani E, Regi H, Nicolas J, Paul N, Thomas R, Earnest P, Kiran R, War S, Alex R, Balraj V, Mohan V (2014) Perceptions of farmers' and farmworkers' wives on the use and hazards of agrochemicals in rural vellore. New Solut 23:625–642. https://doi.org/10.2190/NS.23.4.f

Galloway TS, Brown RJ, Browne MA, Dissanayake A, Lowe D, Jones MB, Depledge MH (2004) Ecosystem management bioindicators: the ECOMAN project—a multi-biomarker approach to ecosystem management. Mar Environ Res 58(2–5):233–237

Gastaldi L, Ranzato E, Caprì F, Hankard P, Peres G, Canesi L, Viarego A, Pons G (2007) Application of a biomarker battery for the evaluation of the sublethal effects of pollutants in the earthworm Eisenia andrei. Comp Biochem Physiol C Toxicol Pharmacol 146:398–405. https://doi.org/10.1016/j.cbpc.2007.04.014

Gerhardt A (2002) Bioindicator species and their use in biomonitoring. UNESCO, Ed Encycl life Support Syst Oxford, UNESCO, EOLSS, UK, p 50

Gräff S, Berkus M, Alberti G, Kohler HR (1997) Metal accumulation strategies in saprophagous and phytophagous soil invertebrates: a quantitative comparison. Biometals 10:45–53. https://doi.org/10.1023/A:1018366703974

Grant WF (1994) The present status of higher plant bioassays for the detection of environmental mutagens. Mutat Res Regul Pap 310(2):175–185. https://doi.org/10.1016/0027-5107(94)90112-0

Hagger JA, Jones MB, Leonard DR, Owen R, Galloway TS (2006) Biomarkers and integrated environmental risk assessment: are there more questions than answers? Integr Environ Assess Manag 2(4):312–329

Haritash AK, Kaushik CP (2009) Biodegradation aspects of polycyclic aromatic hydrocarbons (PAHs): a review. J Hazard Mater 169(1–3):1–15. https://doi.org/10.1016/j.jhazmat.2009.03.137

Harmsen J (2007) Measuring bioavailability: from a scientific approach to standard methods. J Environ Qual 36(5):1420–1428. https://doi.org/10.2134/jeq2006.0492

He Z, Yu W (2010) Stable feature selection for biomarker discovery. Comput Biol Chem 34(4):215–225

Hirsch RM, Slack JR, Smith RA (1982) Techniques of trend analysis for monthly water quality data. Water Resour Res 18(1):107–121. https://doi.org/10.1029/WR018i001p00107

Hodson L, Skeaff CM, Fielding BA (2008) Fatty acid composition of adipose tissue and blood in humans and its use as a biomarker of dietary intake. Prog Lipid Res 47(5):348–380

Hopkin SP (1990) Critical concentrations, pathways of detoxification and cellular ecotoxicology of metals in terrestrial arthropods. Funct Ecol 4(3):321–327. https://doi.org/10.2307/2389593

Hyne RV, Maher WA (2003) Invertebrate biomarkers: Links to toxicosis that predict population decline. Ecotoxicol. Environ Saf 54(3):366–74. https://doi.org/10.1016/s0147-6513(02)00119-7

Jiang ZP, Li YR, Wei GP, Liao Q, Su TM, Meng YC, Zhang HY, Lu CY (2012) Effect of long-term vinasse application on physico-chemical properties of sugarcane field soils. Sugar Tech 4(4):412–417. https://doi.org/10.1007/s12355-012-0174-9

Johnson LL, Stehr CM, Olson OP, Myers MP, Pierce SM, Wlgren CA, McCalm BB, Varanasi U (1993) Chemical contaminants and hepatic lesions in winter flounder (*Pleuronectes americanus*) from the northeast coast of the United States. Environ Sci Technol 27 (13):2759–2771. https://doi.org/10.1021/es00049a015

Kammenga JE, Dallinger R, Donker MH, Kohler HR, Simonsen V, Triebskorn R, Weeks JM (2000) Biomarkers in terrestrial invertebrates for ecotoxicological soil risk assessment. Rev Environ Contam Toxicol 164:93–147

Khadra A, Pinelli E, Lacroix MZ, Bousquet-Melou A, Hamdi H, Merlina G, Guiresse M, Hafidi M (2012) Assessment of the genotoxicity of quinolone and fluoroquinolones contaminated soil with the *Vicia faba* micronucleus test. Ecotoxicol Environ Saf 76 (2):187–192. https://doi.org/10.1016/j.ecoenv.2011.10.012

Khanna R, Gupta S (2018) Agrochemicals as a potential cause of ground water pollution : a review. Int J Chem Stud 6(3):985–990

Knackmuss HJ (1996) Basic knowlegde and perpectives of bioelimination of xenobiotic compounds. J Biotechnol 51(3):287–295

Kögel-Knabner I (1997) 13C and 15N NMR spectroscopy as a tool in soil organic matter studies. Geoderma 80 (3–4):243–270. https://doi.org/10.1016/S0016-7061(97)00055-4

Köhler HR, Triebskorn R, Stöcker W, Kloetzel PM, Albrti GM (1992) The 70 kD heat shock protein (hsp 70) in soil invertebrates: a possible tool for monitoring environmental toxicants. Arch Environ Contam Toxicol 22(3):334–338. https://doi.org/10.1007/BF00212095

Lam PKS, Gray JS (2003) The use of biomarkers in environmental monitoring programmes. Mar Pollut Bull 46(2):182–186. https://doi.org/10.1016/S0025-326X(02)00449-6

Łaszczyca P, Augustyniak M, Babczyńska A, Bednarska K, Kafel A, Migula P, Wilczek G, Witas I (2004) Profiles of enzymatic activity in earthworms from zinc, lead and cadmium polluted areas near Olkusz (Poland). Environ Int 30:901–910. https://doi.org/10.1016/j.envint.2004.02.006

Lavelle P, Decaëns T, Aubert M, Barot S, Blouin M, Bureau F, Margerie P, Mora P, Rossi JP (2006) Soil invertebrates and ecosystem services. Eur J Soil Biol 42:S3–S15. https://doi.org/10.1016/j.ejsobi.2006.10.002

Lehmann A, Zheng W, Rillig MC (2017) Soil biota contributions to soil aggregation. Nat Ecol Evol 1:1828–1835. https://doi.org/10.1038/s41559-017-0344-y

Lemos MFL, Van Gestel CAM, Soares AMVM (2010) Developmental toxicity of endocrine disrupters bisphenol a and vinclozolin in a terrestrial isopod. Arch Environ Contam Toxicol 59(2):274–281. https://doi.org/10.1007/s00244-010-9474-9

Leomanni A, Schettino T, Calisi A, Gorbi S, Mezzelani M, Regoli F, Lionetti MG (2015) Antioxidant and oxidative stress related responses in the Mediterranean land snail *Cantareus apertus* exposed to the carbamate pesticide Carbaryl. Comp Biochem Physiol Part C Toxicol Pharmacol 168:20–27. https://doi.org/10.1016/j.cbpc.2014.11.003

Linde-Arias AR, Inácio AF, Novo LA, Alburquerque C, Moreira JC (2008) Multibiomarker approach in fish to assess the impact of pollution in a large Brazilian river Paraiba Do Sul. Environ Pollut 156(3):974–979. https://doi.org/10.1016/j.envpol.2008.05.006

Lionetto MG, Caricato R, Calisi A, Schettino T (2011) Acetylcholinesterase inhibition as a relevant biomarker in environmental biomonitoring: New insights and perspectives. Ecotoxicol Around Globe 13:321213

Lourenço J, Pereira R, Silva A, Carvalho F, Oliveira J, Malta M, Paiva A, Goncalves F, Mendao S (2012) Evaluation of the sensitivity of genotoxicity and cytotoxicity endpoints in earthworms exposed in situ to uranium mining wastes. Ecotoxicol Environ Saf 75:46–54. https://doi.org/10.1016/j.ecoenv.2011.08.024
Lubbers IM, van Groenigen KJ, Fonte SJ, Six J, Brussaard L, van Groenigen JW (2013) Greenhouse-gas emissions from soils increased by earthworms. Nat Climate Change 3(3):187–194. https://doi.org/10.1038/nclimate1692
Majer BJ, Tscherko D, Paschke A, Wennrich R, Kundi M, Kandeler E, Knasmuller S (2002) Effects of heavy metal contamination of soils on micronucleus induction in Tradescantia and on microbial enzyme activities: A comparative investigation. Mutat Res Genet Toxicol Environ Mutagen 515(1–2):111–124. https://doi.org/10.1016/S1383-5718(02)00004-9
Mantecca P, Vailati G, Bacchetta R (2006) Histological changes and Micronucleous induction in the Zebra mussel Dreissena polymorpha after Paraquat exposure. Histol Histopathol 21(8):829–840
Maraun M, Alphei J, Beste P, Bonkowski M, Buryn R, Migge S, Peter M, Schaefer M, Scheu S (2001) Indirect effects of carbon and nutrient amendments on the soil meso- and microfauna of a beechwood. Biol Fertil Soils 34:222–229. https://doi.org/10.1007/s003740100403
Martínez Morcillo S, Yela JL, Capowiez Y, Mazzia C, Rault M, Sanchez-Hernandez JC (2013) Avoidance behaviour response and esterase inhibition in the earthworm, Lumbricus terrestris, after exposure to chlorpyrifos. Ecotoxicology 22(4):597–607. https://doi.org/10.1007/s10646-013-1051-3
Masindi V, Muedi KL (2018) Environmental Contamination by Heavy Metals. In: Saleh HEDM, Aglan RF (eds) Heavy Metals. IntechOpen. https://doi.org/10.5772/intechopen.76082
Matos-Moreira M, Niemeyer JC, Sousa JP, Cunha M, Carral E (2011) Behavioral avoidance tests to evaluate effects of cattle slurry and dairy sludge application to soil1. Rev Bras Ciência Do Solo 35(4):1471–1477. https://doi.org/10.1590/s0100-06832011000400041
McCarthy JF, Shugart LR (2018) Biomarkers of environmental contamination. CRC Press, Boca Raton, p 467
McGeogh MA (1998) The selection, testing and application of terrestrial insects as bioindicators. Biol Rev 73 (2):181–201. https://doi.org/10.1111/j.1469-185X.1997.tb00029.x
Mishra S, Srivastava S, Tripathi RD, Seth CS, Gupta DK (2006) Lead detoxification by coontail (Ceratophyllum demersum L.) involves induction of phytochelatins and antioxidant system in response to its accumulation. Chemosphere 65(6):1027–1039. https://doi.org/10.1016/j.chemosphere.2006.03.033
Nahmani J, Hodson ME, Black S (2007) A review of studies performed to assess metal uptake by earthworms. Environ Pollut 45(2):402–424
Niemeyer JC, Santos VC, Araújo PB, da Silva EM (2009) Reproduction of *Cubaris murina* (Crustacea: Isopoda) under laboratory conditions and its use in ecotoxicity tests. Brazilian J Biol 69(1):137–142. https://doi.org/10.1590/s1519-69842009000100017
Palm C, Blanco-Canqui H, DeClerck F, Gatere L (2014) Conservation agriculture and ecosystem services: an overview. Agric Ecosyst Environ 187(1):87–105. https://doi.org/10.1016/j.agee.2013.10.010
Park JH, Choppala GK, Bolan NS, Chung JW, Chuasavathi T (2011) Biochar reduces the bioavailability and phytotoxicity of heavy metals. Plant Soil 348:439. https://doi.org/10.1007/s11104-011-0948-y
Parmar TK, Rawtani D, Agrawal YK (2016) Bioindicators: the natural indicator of environmental pollution. Front Life Sci 9(2):110–118. https://doi.org/10.1080/21553769.2016.1162753
Pavao-Zuckerman MA (2018) Encyclopedia of Ecology, 1st edn. Elsevier, USA, pp 126–265
Peakall DB (1994) The role of biomarkers in environmental assessment (1) Introduction. Ecotoxicology 3 (3):157–160. https://doi.org/10.1007/BF00117080
Pérès G, Vandenbulcke F, Guernion M et al (2011) Earthworm indicators as tools for soil monitoring, characterization and risk assessment. An example from the national Bioindicator programme (France). Pedobiologia (jena) 54:S77–S87. https://doi.org/10.1016/j.pedobi.2011.09.015
Pirooznia M, Gong P, Guan X et al (2007) Cloning, analysis and functional annotation of expressed sequence tags from the Earthworm *Eisenia fetida*. BMC Bioinformatics 8:S7
Pohanka M, Hrabinova M, Kuca K, Simonato JP (2011) Assessment of acetylcholinesterase activity using indoxylacetate and comparison with the standard Ellman's method. Int J Mol Sci 12(4):2631–2640. https://doi.org/10.3390/ijms12042631
Power AG (2010) Ecosystem services and agriculture: Tradeoffs and synergies. Philos Trans R Soc B Biol Sci 365:2959–2971
Rajagopal V, Paramjit SM, Suresh KP, Singh Y, Rao N, Nirmale A (2014) Significance of vinasses waste management in agriculture and environmental quality—review. African J Agric Res 9(38):2862–2873. https://doi.org/10.5897/ajar2014.8819
Rault M, Mazzia C, Capowiez Y (2007) Tissue distribution and characterization of cholinesterase activity in six earthworm species. Comp Biochem Physiol B Biochem Mol Biol 147(2):340–346. https://doi.org/10.1016/j.cbpb.2007.01.022
Reed S, Martens D (1996) Methods of soil analysis (Part 3): chemical Methods. SSSA and ASA: Madison, WI, p 703–722
Regoli F (1992) Lysosomal responses as a sensitive stress index in biomonitoring heavy metal pollution. Mar Ecol Prog Ser 84(1):63–69. https://doi.org/10.3354/meps084063
Reinecke SA, Reinecke AJ (2004) The comet assay as biomarker of heavy metal genotoxicity in earthworms. Arch Environ Contam Toxicol 46(2):208–215. https://doi.org/10.1007/s00244-003-2253-0

Ribera D, Saint-Denis M (1999) The worm *Eisenia fetida* : Interests and perspectives in terrestrial ecotoxicology. Bull La Soc Zool Fr 124(4):411–420

Ruggiero MA, Gordon DP, Orrell TM, Bailly N, Bourgoin T, Brusca RC, Cavalier-Smith T, Guiry MD, Kirk PM (2015) A higher level classification of all living organisms. PLoS One 10(4):e0119248. https://doi.org/10.1371/journal.pone.0119248

Sánchez-Bayo F, Tennekes HA (2015) Environmental risk assessment of agrochemicals—a critical appraisal of current approaches. In: Toxicity and hazard of agrochemicals. IntechOpen. https://doi.org/10.5772/60739

Sanchez-Hernandez JC (2006) Earthworm biomarkers in ecological risk assessment. Rev Environ Contam Toxicol 188:85–126

Santorufo L, Van Gestel CAM, Rocco A, Maisto G (2012) Soil invertebrates as bioindicators of urban soil quality. Environ Pollut 161:57–63. https://doi.org/10.1016/j.envpol.2011.09.042

Scheu S (2003) Effects of earthworms on plant growth: patterns and perspectives. Pedobiologia 47:846–856

Schimel JP, Schaeffer SM (2012) Microbial control over carbon cycling in soil. Front Microbiol 3:348

Schwab AP (2020) Managing global resources and universal processes: pollutants. Taylor & Francis publications, UK, p 12

Shugart LR (2000) DNA damage as a biomarker of exposure. Ecotoxicology 9:329–340

Singh RP, Agrawal M (2007) Effects of sewage sludge amendment on heavy metal accumulation and consequent responses of Beta vulgaris plants. Chemosphere 67(11):2229–2240. https://doi.org/10.1016/j.chemosphere.2006.12.019

Smaka-Kincl V, Stegnar P, Lovka M, Toman MJ (1996) The evaluation of waste, surface and ground water quality using the Allium test procedure. Mutat Res Genet Toxicol 368(3–4):171–179. https://doi.org/10.1016/S0165-1218(96)90059-2

Sochová I, Hofman J, Holoubek I (2006) Using nematodes in soil ecotoxicology. Environ Int 32(3):374–383. https://doi.org/10.1016/j.envint.2005.08.031

Spurgeon DJ, Lofts S, Hankard PK et al (2006) Effect of pH on metal speciation and resulting metal uptake and toxicity for earthworms. Environ Toxicol Chem 25 (3):788–796

Stanek K, Drobne D, Trebše P (2006) Linkage of biomarkers along levels of biological complexity in juvenile and adult diazinon fed terrestrial isopod (Porcellio scaber, Isopoda, Crustacea). Chemosphere 64(10):1745–1752. https://doi.org/10.1016/j.chemosphere.2005.12.070

Steinkellner H, Kassie F, Knasmüller S (1999) Tradescantia-micronucleus assay for the assessment of the clastogenicity of Austrian water. Mutat Res 426 (2):113–116

Stephens RD, Petreas MX, Hayward DG (1995) Biotransfer and bioaccumulation of dioxins and furans from soil: chickens as a model for foraging animals. Sci Total Environ 175(3):253–273. https://doi.org/10.1016/0048-9697(95)04925-8

Svendsen C, Meharg AA, Freestone P, Weeks JM (1996) Use of an earthworm lysosomal biomarker for the ecological assessment of pollution from an industrial plastics fire. Appl Soil Ecol 3(2):99–107. https://doi.org/10.1016/0929-1393(95)00085-2

Tanaka Y (2003) Ecological risk assessment of pollutant chemicals: extinction risk based on population-level effects. Chemosphere 53(4):421–425. https://doi.org/10.1016/S0045-6535(03)00016-X

Tang T, Cai Q, Liu J (2002) River ecosystem health and its assessment. Chinese J Appl Ecol 13(9):1191–1194

Teng Y, Wu J, Lu S, Wang Y, Jiao X, Song L (2014) Soil and soil environmental quality monitoring in China: a review. Environ Int 69:177–199

van den Brink PJ, Mann RM, Hanson ML (2011) Ecological impacts of organic chemicals on freshwater ecosystems. Ecol Impacts Toxic Chem (Open Access) 27:138–164

Van der Oost R, Beyer J, Vermeulen NPE (2003) Fish bioaccumulation and biomarkers in environmental risk assessment: a review. Environ Toxicol Pharmacol 13 (2):57–149

Van Gestel CAM, Van Brummelen TC (1996) Incorporation of the biomarker concept in ecotoxicology calls for a redefinition of terms. Ecotoxicology 5(4):217–225

Wagg C, Bender SF, Widmer F, Van Der Heijden MGA (2014) Soil biodiversity and soil community composition determine ecosystem multifunctionality. Proc Natl Acad Sci 111(14):5266–5270. https://doi.org/10.1073/pnas.1320054111

Wall DH, Nielsen UN, Six J (2015) Soil biodiversity and human health. Nature 528(7580):69–76

Werther J, Ogada T (1999) Sewage sludge combustion. Prog Energy Combust Sci 25:55–116

Williams LD, Burdock GA (2009) Genotoxicity studies on a high-purity rebaudioside a preparation. Food Chem Toxicol 47(8):1831–1836. https://doi.org/10.1016/j.fct.2009.04.046

Xu GL, Schleppi P, Li MH, Fu SL (2009) Negative responses of Collembola in a forest soil (Alptal, Switzerland) under experimentally increased N deposition. Environ Pollut 157(7):2030–2036. https://doi.org/10.1016/j.envpol.2009.02.026

Zahran HH (1999) Rhizobium-Legume symbiosis and nitrogen fixation under severe conditions and in an arid climate. Microbiol Mol Biol Rev 63(4):968–989. https://doi.org/10.1128/mmbr.63.4.968-989.1999

Zhang L, Ji F, Li M, Cui Y, Wu B (2014) Short-term effects of Dechlorane Plus on the earthworm *Eisenia fetida* determined by a systems biology approach. J Hazard Mater 273:239–246. https://doi.org/10.1016/j.jhazmat.2014.03.018

Žižek S, Zidar P (2013) Toxicity of the ionophore antibiotic lasalocid to soil-dwelling invertebrates: avoidance tests in comparison to classic sublethal tests. Chemosphere 92(5):570–575. https://doi.org/10.1016/j.chemosphere.2013.04.007

22 Molecular Tools for Monitoring and Validating Bioremediation

Aurobinda Das and Suresh Kumar Panda

Abstract

Bioremediation may be defined as the mechanism by which a biological agent (e.g., microbes, fungi, plants, and enzymes) is used to minimise soil, groundwater, and air contaminant mass and toxicity. Bioremediation aims at reducing contamination and targeting the amounts of pollutants in water and/or soil by either biodegradation (of organic matter) or biotransformation (of metals). The monitoring tests may provide necessary details on the bioremediation process and usually include the use of analytical techniques (microbiological, biochemical or molecular) to determine, under certain situations, the status and effectiveness of bioremediation. By considering several micro-organisms with diverse genomes, expressed transcripts and proteins, bioremediation has become able to take advantage of approaches powered by genomics to observe, map and evaluate its path. Owing to the lack of sensitive methods of identification, the classification of microbial species poses restrictions. Conventional microbiological approaches, such as isolation of pure colonies and further study of their physiological and biochemical properties, are not always well suited for studies of microbial communities and their behaviour. In particular, more than 99% of microbial population in soils could not be segregated due to a lack of awareness of their physiological needs. Molecular techniques were thus created to compensate for the disadvantages found in conventional methods of culture. Molecular techniques based on 16S rRNA, RTPCR, and real time PCR polymerase chain reaction (PCR), microarray, fingerprinting, and metaproteomics have become standard instruments for detecting and quantifying microorganisms previously added to soils. Similarly, other less traditional approaches such as FISH (fluorescent in situ hybridization), DGGE (denaturing gradient gel electrophoresis), TGGE (temperature gradient gel electrophoresis) and metagenomics had been studied and applied to research and better understand thec microbial degradation of polycyclic aromatic hydrocarbons (PAHs). The new genomic and metagenomic methodologies are intended to improve the exploration of new catabolic practises and to provide, from a sustainable development perspective, statistical and reliable reporting for the maintenance and clean-up of contaminated areas and wastes.

A. Das (✉)
Department of Zoology, Centurion University of Technology and Management, Bhubaneswar, Odisha, India

S. K. Panda
Department of Ecology and Environmental Sciences, Pondicherry University, Puducherry, India

J. A. Malik (ed.), *Advances in Bioremediation and Phytoremediation for Sustainable Soil Management*,
https://doi.org/10.1007/978-3-030-89984-4_22

Keywords

Bioremediation · DGGE · Metagenomics · Microarray · Molecular tools · Monitoring · PCR

22.1 Introduction

Due to the global industrialization, there is huge problem of pollution and it is becoming a serious problem for the world. This occurs due to the extensive use of chemicals, pesticides, solvents, heavy metals which tend to cause all types of pollution i.e. soil, water and air. Due to the ever increasing population and tremendous pollution in environment, remediation of soil and contaminated sites is strongly needed.

Bioremediation can be defined as, it is the mechanism where organisms specifically microorganisms like bacteria, fungi, microalgae are used to degrade or detoxify the contaminants or can say, process by which they minimise the soil, ground water and air contaminants mass and toxicity (Megharaj et al. 2011). Nowadays many scientists, microbiologists and biotechnologists have been attracted towards the bioremediation for the environmental control as it applies the biological agents i.e. micro-organisms like bacteria, yeast, fungi, and algae for the treatment of contaminated soils or water**.** One of the components of bioremediation is the use of plants to remove the contaminants from environment and this is known as phytoremediation (Mishra and Mohan 2017; Folch et al. 2013; Kim et al. 2014; Roy et al. 2015).

In traditional method, remediation is not a sustainable solution as contaminants only moves from one place to another like dig and dump, hence it causes the risk at the time of transportation, handling. Bioremediation is the cost effective method and environmentally safe method to decontaminate the soil, water or air. The main principle behind it is, that microorganisms like algae, fungi, bacteria uses the carbon, nitrogen and oxygen as a nutrients from the contaminants and thereby degrading the toxic products (Megharaj et al. 2011). Bioremediation is also a challenging and an integrated approach which involves the principles from microbiology, engineering, chemistry, ecology and geology. The main advantage of the bioremediation is that it minimises transportation cost as is done on site directly and do not disturb the other settings. There are various techniques which have been invented and developed by researchers and scientist for bioremediation, but there is no single bioremediation technique which serves as a 'silver bullet' to restore the polluted environment (Azubuike et al. 2016; Verma and Jaiswal 2016; Megharaj et al. 2011). Due to the all types of pollutions, these mixed types of contaminants are becoming the main causes of the different kinds of degenerative diseases. To make environment clean and safe from the contaminants for human habitation and food consumption, a low budget solutions appears to be a very difficult task and hence decontamination of the environment is challenging. In developed countries, it has been more progressed as compare to India. New molecular techniques can provide the best opportunities to initiate the required microbial cultures and will assist the development of different methods for decontamination and describing the bacterial diversity along with its functionality during the bioremediation (Shekhar et al. 2020; Frutos et al. 2010; Smith et al. 2015).

22.2 High-Throughput Techniques for Characterisation of Contaminated Sites

For the study of microbial species, there are different types of methods applied which are culture independent molecular techniques. These are consistent with high-performance configurations such as fingerprinting approaches, PCR, microarrays, metagenomics, metaproteomics, metatranscriptomics, and metabolomics in real time. In order to explain the complexity, function, diversity, and structure of the microbial species or microbial population, there are different strategies developed on the basis of analysis of nucleic acids of microbial species (Fig. 22.1) (DeLong 2005; Paerl and Steppe 2003).

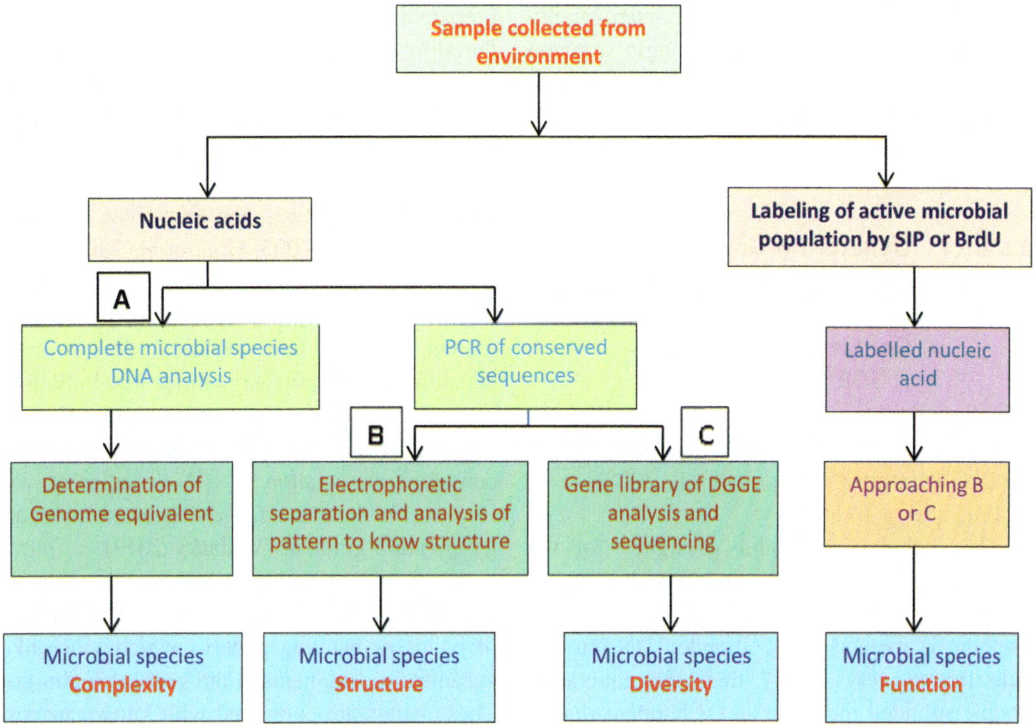

Fig. 22.1 Analysis of total nucleic acids of microbial community to explain the complexity, structure, diversity and function of the environmental microbial species

Description of Fig. 22.1 In the figure the representation of complexity, structure, function and diversity is assessed from the environmental microbial population for which nucleic acid is extracted from the microbial species and subsequent assays are done. Different steps and molecular approaches involved are DNA and RNA analysis, labelling of microbial species with the help of stable isotope probing (SIP) or Bromodeoxyuridine (BrdU), labelling of nucleic acids; all these can be correlated with information of microbial species (DeLong 2005; Yannopoulos et al. 2015).

22.2.1 Fingerprinting Technique

In soil microbial ecology, molecular techniques were implemented around 15 years ago and enabled the study of ecology of microbial species that cannot be cultivated in synthetic media and yet constitute the largest amount of soil micro-organisms. There are two different types of molecular finger printing methods. (1) Denaturing gradient gel electrophoresis i.e. DGGE and (2) Terminal restriction fragment length polymorphism i.e. TRFLP (Huang and Ye 2020; Tang et al. 2009), also known as genetic finger printing method. These are the most popular molecular finger printing methods used in the microbial ecology. These methods are used in single or in combination for both the sequencing and cloning. To determine whether all microbial species have recovered or responded to remedial action, microbial fingerprinting methods are used. Microbial finger printing methods provide overall profile of the micro-organism (Karpouzas and Singh 2010; Rittmann et al. 2006). Genetic fingerprinting technique provides specific pattern of the complete profile of micro-organisms and is depending upon the separation of amplicons after polymerase chain reaction or functional genes using universal or specific primers. These finger printing techniques are high-throughput design which includes TRFLP, length heterogeneity analysis by PCR i.e. LH-PCR, Single stranded conformation polymorphism (SSCP), Denaturing gradient gel electrophoresis (DGGE) and ribosomal intergenic spacer analysis (RISA) (Stenuit et al. 2009).

22.2.1.1 Denaturing Gradient Gel Electrophoresis (DGGE)/ Temperature Gradient Gel Electrophoresis (TGGE)

DGGE method was first described by Fischer and Lerman for mutation analysis or detection of gene polymorphism (Siqueira et al. 2005).

Principle of DGGE

This technique is mainly designed for the detection of single point mutations that means single nucleotide polymorphism. In this electrophoresis method single stranded DNA molecule moves very slowly than the equivalent double stranded molecule and it happens due to the more interaction of nucleotides in gel matrix which are not bonded from the single strand DNA. Opposite to this, in double stranded DNA, due to hydrogen bonding in nucleotides, they move fast and easily through gel matrix. In DGGE method, Polyacrylamide gel is used which act as or having a role of increasing gradient of chemical denaturant. The denaturants employed are usually urea and formamide, where DNA molecules pass by electrophoresis. When DNA passes through the gel matrix, every molecule of it get denaturated at particular concentration of denaturant based on its % GC (Gas Chromatography) content and also proper arrangement of bases in sequence (McAuliffe et al. 2005; Hout et al. 2006; Strathdee and Free 2013).

Melting temperature is correlated with the % content of G-C bonds in the sequence. If GC bonds are higher in number, higher is the melting temperature T_m. In different regions of DNA molecules, different T_m can be observed and these regions are called as melting domains which are determined by the DNA sequence within that domain. Melting domains denature at different points in the gel. The melting domains prepare the single stranded branches which cause the slow movement of the fragment within the gel matrix. This branching pattern is done by the GC clamp. It is the region of the GC content where the PCR primers cause the prevention of the two strands form dissociating thereby allowing the analysis for any DNA sequence (McAuliffe et al. 2005; Hout et al. 2006; Maitra 2018). Both DGGE and TGGE are same techniques and rely on the same principle where a gradient temperature replaces the chemical denaturant in the polyacryl amide gel. In both the methods GC-lamps of 30–50 nucleotides is attached to the one end which is essential for the complete dissociation of DNA fragment with increase in denaturing agent concentration and temperature gradient (Viglasky 2013).

If mixed population of samples is taken for the analysis, the migration is shown in the form of bands or banding patterns which looks likes varying size fragments in the gel electrophoresis. These bands are compared with known markers whose sequencing is known and it can be used for the identification of the species present in the sample to be taken (McAuliffe et al. 2005; Hout et al. 2006). Instead of using gradient temperature, chemical denaturant is also used which is called as Temperature gradient gel Electrophoresis (TGGE) (Strathdee and Free 2013; Mohan et al. 2011).

Denaturing gradient gel ingredients various reagents are used as gradient gel reactants like, acrylamide. Two concentrations are used for denaturant, 0% denaturant and 80% denaturant. Urea and formamide are used as chemical denaturants. These solutions should be made in the fume hood in glass containers and should be covered in aluminium foil and stored at lower temperature. Apart from the denaturing agent, there are also different reagents used in the technique of DGGE like, gel loading and running solutions, silver staining solutions (Strathdee and Free 2013).

Applications of DGGE Method

This technique is widely used in the microbial ecology. Specifically it is used in the degradation of oil. By the use of micro sensor detectors, microbial diversity of the contaminant could be monitored. Oil pollution is becoming the serious problem nowadays. It can be a waste from the industrial area, from chemical reactions or from the waste material. So it is very much important to degrade the oil based waste materials which otherwise could harm the marine life too. Most of the biodegradation studies have been performed on the oil-degrading stains or in the laboratory scale under simulated conditions. But, it is not possible to copy or duplication of the natural condition in the laboratory and which may lead to other bacterial community growth under in situ conditions (Abed and Grötzschel 2005).

Applications in Oil Biodegradation Studies

DGGE method is useful for evaluating the bacterial diversity in oil polluted sites, population dynamics, bioremediation studies and to monitor the enrichment of culture. It is also beneficial for the characterization of petroleum degrading consortia and to estimate the active members from the community. Other applications include detecting the microheterogeniety in rRNA encoding genes, screening of clone library, determination of PCR and comparison of different DNA extraction protocols (Eyers et al. 2006; Abed and Grötzschel 2005).

Limitations of DGGE Method

Efficiency of Extracting DNA DNA is isolated from microbial cells, and the lytic mechanism does not generally have the same potential or sensitivity to release DNA. These differences are due to the different cell wall structure of the micro-organisms. Hence, the challenging task is to remove DNA from all organisms with the same quality.

Differential Amplification of 16S rDNA It is caused by re-annealing of the template DNA strands. The amplification can be missing in the PCR product.

No Reliable Identification If fragments resolved by DGGE method would be more than 500 bp, the reliable identification is not possible; since sequencing DGGE bands have to be compared which are very short fragments.

Applicable to High Concentrations Only High concentrations are expressed on gel, but on DGGE banding patterns, all micro-organisms delivered in habitat may not appear. It can be overcome by DGGE in combination with the rDNA based technique which might show the good specificity and allow the detection of the samples at lower concentration also (Siqueira et al. 2005).

22.2.1.2 Terminal Restriction Fragment Length Polymorphism (TRFLP)

This method has been used to study the microbial community organisation in anaerobic digestion. It is the most powerful tool amongst all the methods and is able to generate the information about the genomic structure and gene expression (De Vrieze et al. 2018; Stenuit et al. 2009). This approach is a methodology that is independent of culture and has high resolution, hence is more detailed than other techniques dependent on cultivation. Now a day this technique is successfully applied to composition and diversity analysis of soil microbial species under the different environmental conditions (Nithya and Pandian 2012; Sheng et al. 2012).

Principle of TRFLP

It is based on the position of the restriction site which is closer to the labelled end of an amplified gene sequence. This technique is used for profiling of microbial communities which uses the 5' fluorescently labelled primer during the PCR reaction thereby the PCR products get digested with the restriction enzymes, known as terminal restricted fragments TRFs which are then separated on an automated DNA sequence. In this analysis, the fragments which are labelled terminally are detected only. Hence the method is named as Terminal restriction fragment length polymorphism.

Applications of TRFLP

It is used in the measurement of the composition of microbial species in compost systems and also applied in the waste water treatment failure by monitoring the microbial communities. This method is used in the dynamics in full scale treatment of industrial waste water and is functional in the microbial structure analysis in bioremediation of anthracene (Louati et al. 2013; Bharagava et al. 2019). The method is useful for the understanding of biogeographically pattern of soil bacterial communities and investigation of biotic and abiotic factors (Bharagava et al. 2019).

22.2.1.3 Length Heterogeneity Analysis by PCR (LH-PCR)

LH-PCR method is similar to TRFLP method. The difference between two methods is that TRFLP method identifies the PCR fragments length variation based on restriction site variability, whereas LH-PCR analysis distinguishes different organisms which is based on natural variations in the length of 16S ribosomal DNA sequences. This method is limited in microbial diversity. It is only used for the aquatic environment (Ritchie et al. 2000). Length heterogeneity occurs due to the mutation and natural length of polymorphism within genes of microorganisms. To distinguish from the microbial communities, unique profile is generated by targeting the hypervariable region of 16S RNA gene. This method is applicable in an atypical environmental media (Dash and Das 2018).

22.2.1.4 Fluorescence in Situ Hybridization (FISH)

This technique is rapid and sensitive which allows identification and determination of environmental micro-organisms without making their cultures. In this method, there is hybridization of the microbial cells with DNA probe which is already labelled with fluorescent molecule and it allows the microscopic detection. FISH is very important tool to count, monitor and examine the changes which are present in the microbial community in the environmental premises. FISH probes are designed on the basis of ribosomal DNA sequence but it can be also designed for the functional genes for the detection and expression of PAH-catabolism (polycyclic aromatic hydrocarbons) and genes like naphthalene dioxygenase and Toluene dioxygenase (Galvão et al. 2005; Bakermans and Madsen 2002).

Applications

Simazine is an herbicide and its degradation depends on the abiotic and biotic process. Rousseaux et al. (2001) and Strong et al. (2002) have identified that several micro-organisms are capable of using the s-triazine herbicide as a nitrogen and carbon source for growth. Culture independent approaches are more suitable for the prediction of natural attenuation potential of contaminated soil. FISH technology is the best approach to detect the metabolic potential of soil bacterial population. For this purpose, specific oligonucleotide probe for the detection of gene by using optimised FISH technology and its protocols on four different types of soils, with different Simazine exposures have been designed (Martín et al. 2008).

FISH method is mainly used for the detection of the 16S and 23 rDNA genes. Special precautions must to be taken while using the FISH technique directly on soil or on sediment samples (Quintero and Zafra 2016; Amann et al. 1995). Cytogenetics is used within this technique of comparative genomic hybridization for the detection of quantitative differences in the chromosomes of their patients which could study whole chromosomes on microscopic scale (Parra and Windle 1993; Claire 2008).

Limitations

FISH technology is having the low sensitivity and low availability of sequence specific probes which is a drawback of this method. Further, the radioisotopes labelling requires long exposure of time and the spatial resolution is also low. Another limitation of FISH technique is that probe consumption and hybridisation time is more (Huber et al. 2018).

22.2.1.5 Single Stranded Conformation Polymorphism (SSCP)

This method shows the difference in DNA sequences which is based on the electrophoresis method by which DNA molecules get separated. This method was initially used for the detection of novel polymorphism and point mutations (Dash and Das 2018). After some cases, it was observed that DNA molecules are available in the form of a secondary folded structure called as heteroduplex, which relies on the encoding of DNA and enables DNA to travel on the gel. But major limitation of this method is, at the time of DNA sequencing, other factors also contribute to the secondary conformation of DNA, which makes the complications of analysis (Svozil et al. 2008). Instead of this constraint, in atypical contexts, this approach has a wide scope for its use in group and demographic diversity. Schwieger and Tebbe (1998) analysed the difference between cultivated pure soil culture microorganisms and non-cultivated rhizosphere microbial communities by using the SSCP technique.

22.2.1.6 Ribosomal Intergenic Spacer Analysis (RISA)

There are spacer regions present in the small and large ribosomal subunit coding genes. Ribosomal intergenic spacer analysis technique targets these intergenic spaces and the spacer regions has been reported to contain the nucleotide sequence and length heterogeneity from the microbial species (Ciesielski et al. 2013; Rittmann et al. 2006). In the targeted atypical climate, RISA profiles of most microbial communities are present and can be created to a significant degree (Hong et al. 2010; Kovacs et al. 2010). RISA technique was successfully found to be a useful tool in the monitoring of the variation in methanogenicarchaeal diversity for about 6 months in an atypical environment like digester treatment in biomass (Ciesielski et al. 2013).

22.2.2 Real-Time PCR

Vast microbial diversity and its characterization and quantification in an atypical environment is a challenging task. In this condition, quantitative real time PCR is the best option and promising tool for the analysis of microbial diversity. This technique is based on the identification of DNA molecules in real time bound to fluorescent molecules whose strength increases when some parameter or functional gene markers are amplified. Furthermore, the procedure is paired with a reverse transcriptase reaction that can be used to determine the extent of gene expression. In this method different mathematical models can be used for the calculation of relative expression of targeted genes from the environmental samples for comparison to corresponding housekeeping genes (Pfaffl 2001). Foti et al. (2007) described that several genes can be targeted by using their primer sets in the RT-PCR (Real-Time Polymerase Chain Reaction). This can be applicable for the determination of ammonia oxidiser, methane oxidiser and sulphate reducers. Some metabolic pathways like photosynthesis, nitrogen acquisition, and carbon fixation have been estimated from the open ocean microbial species by RT-PCR method. It is the independent method of expression level of corresponding genes (Frias-Lopez et al. 2008).

22.2.3 DNA Microarrays

Nucleic acid microarrays or simply DNA microarrays are the group of technologies where specific DNA sequences used which are either deposited or synthesised in 2 dimension or sometime in 3 dimensions array on the surface. It is generated in such a way that the substrate is covalently or noncovalently connected to the DNA molecule. The DNA array is used to monitor the solution of the labelled nucleic acid mixtures and afterwards tie these targets together. This is used to determine the concentration of nucleic acid species in solution. This method had been used specially for providing the high-throughput of microbial communities from the environmental samples (Bumgarner 2013; Golyshin et al. 2003; Kube et al. 2005). The significant advantage of Microarray is its rapid evaluation and replication. However, cross-hybridization is the major limitation for this method specifically when dealing with environmental samples (Rastogi and Sani 2011; Bharagava et al. 2019).

Disadvantages

This method is not useful for identification and detection of novel prokaryotic taxa. If genus does not have a corresponding probe on the microarray, it could be completely ignored (Bharagava et al. 2019).

Applications

It is classified into two main classes depending on probes used (1) 16S rRNA gene microarray, (2) functional gene array. Microarray methods are not based on sequencing depth to provide detailed insight into the microbial community. They provide the rigorous annotations for the various genes present on the chip. Approaches based on microarray are also a helpful compliment to an additional line of proof and sequencing based approach. Molecular methods are culturally based because these microbial cultures are less precise, time consuming and therefore not successful in discovering the uncultivable culture for their classification and identification. The use of traditional molecular technique is used to address the limitations of Metagenomic approaches that advance our awareness of the full scale characterization of the composition, structure and behaviour of the microbial community during bioremediation at a polluted site (Bharagava et al. 2019; Deng et al. 2016; Zafra et al. 2014).

22.2.4 Metagenomics

Metagenomics is also known as ecogenomics, environmental genomics or community genomics. It is the study of metagenome and metagenome can be defined as genetic material which is present in the environmental sample. Thus, samples are retrieved directly from the setting. It offers the different genomic approaches for the environmental characterization of the microbial population and the unfolding of genomes from uncultured microbes and thus shows the variety of genes important to taxonomy and phylogenetics, certain catabolic genes and the whole activity (Uhlik et al. 2013). In simple language, Metagenomic analysis involves the isolation of DNA form the environmental samples, cloning the DNA on a vector, then transferring of this clones in to the host bacteria which is followed by the screening of the resultant. These clones are screened with help of markers which are termed as Anchors e.g. 16S rRNA and recA (Handelsman 2004; Peng et al. 2015).

22.2.4.1 Concept of Metagenomics

It was first introduced by Handelsman et al. (1998) but Pace et al. (1985) studied it earlier. In metagenomic study they first carried out the analysis of environmental microbial species. Initially it was considered as only the screening of environmental microbial communities but is now regarded as the function based screening (Handelsman et al. 1998). In addition, another approach to Metagenomics, i.e. sequence-based scanning, has been improved by high throughput NGS (Next generation sequencing) technologies. Metagenomic studies have been conducted with the use of high throughput microarrays in addition to sequence base screening. These methods have been used in the analysis of microbial communities and to monitor the environmental biochemical process (Handelsman et al 1998).

22.2.4.2 Metatranscriptomics

It addresses the extraction work and the mRNA examinations. This includes details on the regulatory and expression profiles of the microbial complex populations in the environmental sample. Metatranscriptomics gives the snap shot of gene expression from the given sample and at given time under specific conditions by selecting the total mRNA (Aguiar-Pulido et al. 2016). Limitations of this method can be overcome by direct cDNA sequencing method which employs the NGS technologies and provides the affordable access to metatranscriptome and allow the profiling of whole genome expression of microbial communities. By using this method, ammonia oxidizing archaea in soil ecosystem were studied and recently, Shi et al. (2009) showed the contribution of small RNAs in many environmental processes (Bharagava et al. 2019). It is the subset of Metagenomics and provides the valuable information about whole gene expression. Metagenomic studies rely mainly on genomic contents and identification of microbes present in the community. Metatranscriptomics is relying on utilisation of mRNA separated from environmental samples and it is the suitable approach for eukaryotic gene pool analysis and serves as the bioinformatics pipeline for analysis of information obtained from Metatranscriptomics analysis (Mukherjee and Reddy 2020).

22.2.4.3 Metaproteomics

This approach is also called as community proteomics, environmental proteomics or community proteogenomics. It is the study of all protein samples collected from environmental sources. Ram et al. (2005) has conducted a more comprehensive metaproteomics study, where he analysed the gene expressions, metabolic functions of natural acid mine drainage microbial biofilm and some important key activities. Near about 2000 different kinds of proteins from the five different micro-organisms were identified. It was very difficult task for the detection and identification of all the proteins from the different micro-organisms which were including the uneven species distribution and the broad range of protein expression levels within micro-organisms. This approach has the great potential to connect microbial communities' genetic diversity and behaviours with their impact on the environment (Simon and Daniel 2011).

Approach for Metaproteomics

Its investigation is depending upon the bottom up and top down approach. Bottom up approach needs a digested protein which should be in gel or gel free. Its identification is done by LC–MS (Liquid chromatography-Mass Spectroscopy). The top down approach do not require the digested proteins, where separated proteins are directly detected by the LC–MS. One more advanced target approach is for specific peptides for selective reactive monitoring (Abiraami et al. 2020).

Applications

Biomass degradation By using Metagenomics and Metaproteomics, Aylward et al. (2012) studied the biomass degradation on the leaf cutter ant named Atta. This is actually done by fungus grown on ant nest which secrets the exoenzymes for degradation of the lignocellulosic polymers. Metaproteomics reflects an important role in the study of such degradation pathways.

Soil restoration Use of large amount of chemicals makes the soils deteriorated and ultimately decreases the plant productivity. Soil restoration has become the essential parameter. The researchers have concluded that the organic amendments are impacted directly on the ecosystem (Bastida et al. 2012).

Bioremediation Guazzaroni et al. (2013) studied Metagenomics and Metaproteomics for the microbial diversity where naphthalene was exposed to PAH-polluted soil. Naphthalene degradation was analysed by the researchers successfully. Taylor and Williams (2010) studied the effect of toluene on the bacterial soil by using metaproteomes and showed that ABC transporters exposed the toxic substances out from the bacterial cell wall. Bastida et al. (2012) determined the compost assisted bioremediation in the petroleum contaminated soil and detected 0.55% of proteins form them.

22.2.4.4 Metabolomics

Metabolome word was used very long ago for the low molecular mass compounds i.e. metabolites which are generated and modified by microorganisms. Metabolomics has entered in the biology field to fasten the functional analysis of genes. This approach has many applications in various fields like human and animal nutrition, obesity studies, enzyme discovery, and cancer therapy and also in bioremediation (Villas-Bôas and Bruheim 2007; Tang et al. 2009). It is the analysis of all metabolites of the small molecules which are released from the micro-organisms into the immediate environment. The metabolome is known to be an indicator that is useful for environmental protection and very useful for pathway research, drug development, and pharmacogenomics applications. The key goal of the metabolomics is to enhance the understanding of role of microbiome in the transition of nutrients and contaminants and other abiotic factors which can have the effect on the homeostasis of the host environment (Aguiar-Pulido et al. 2016). During the communication of the bacteria, signalling process is involved like quorum sensing which is related to the gene expression responses for changing the cell population density. By using the combination of chromatographic techniques, identification and quantification of metabolites are carried out e.g. liquid chromatography and gas chromatography. Nuclear magnetic resonance and mass spectroscopy are the methods used for identification and quantification of metabolites (Aguiar-Pulido et al. 2016; Bharagava et al. 2019).

22.2.4.5 Fluxomics

It is useful for the determination of the rates of metabolic reactions within a biological entity. In this method a set of metabolic flux and the fluxome is the representative of the phenotype so, it captures the metabolism in its functional interaction with the environment and the genome. It is depending on the information received from the metabolites which are very away from the genes and proteins; this is the main advantage of this method over genomics and proteomics (Bharagava et al. 2019).

22.2.4.6 Next Generation Sequencing (NGS) Technologies

These are the most powerful gene sequencing technologies which are useful for providing greater insightforecologyofmicroorganismsmediatedwith different processes. These processes include, waste water treatment, detoxification and degradation of environmental pollutants and destruction of pathogens. NGS technologies have various kinds of applications which include sequencing of whole genome from single genome and shot gun metagenome sequencing. There are different modern sequencing technologies which include illumina sequencing, Roche 454 sequencing, and solid sequencing. These methods have brought the revolution in the field of environmental genomics and microbial ecology (Mohan et al. 2011; Kim et al. 2013).

22.2.4.7 Workflow for Metagenomics

The approaches for the Metagenomics are usually applied in two different ways, (1) Targeted Metagenomics and (2) Shotgun Metagenomics. In the first method of targeted Metagenomics, diversity of single gene is probed for the identification of full complement sequence of a specific gene from the environment and it is usually employed for the investigation of both the phylogenetic diversity and relative abundance of a specific gene from environmental sample (Techtmann and Hazen 2016; Bharagava et al. 2019). The approach of this method is regularly used for the investigation of the diversity of small molecules and subunits of rRNA sequences in the environmental samples. Many microbiologists and microbial ecologists applied this technique routinely for the small subunits of rRNA sequencing for understanding the taxonomy and diversity of an environment. It is also used to examine the effect of modifying the composition of microbial population on environmental pollutants.

The environmental samples are generated in the targeted metagenomics, DNA is extracted and the gene of interest is amplified by PCR using primers uniquely engineered to amplify the greatest variety of gene sequences. After amplification of genes, sequencing is to be done with the help of NGS which tends to form the thousands of small rRNA reads per sample and which can probe hundreds of samples simultaneously (Parada et al. 2016; Klindworth et al. 2013; Bharagava et al. 2019).

In shotgun metagenomics, microbial community from the environment is probed through the genomic sequencing. In this technology, DNA is extracted from the environmental samples and fragmented for the preparation of sequencing library which are then further sequenced for the determination of the total genomic content of the sample. It is very powerful technique by which functional potential of the microbial community can be identified however it is limited for depth of sequencing. It is having the good coverage of the entire genomic content of every organism form the community. In addition to this, the analysis of Metagenomic sequencing data is very complicated and is involving the accurately annotating diverse gene sequence, from which many of them have no homologs in the current sequence databases (Delmont et al. 2012; Bharagava et al. 2019). Many computational approaches have also been applied for the Metagenomic sequencing to have a full understanding of the functional capacity of individual organisms within the ecosystem (Delmont et al. 2012; Bharagava et al. 2019).

The aim of many studies is to link a functional gene using a phylogenetic anchor to a taxonomic classification. Metagenomic sequencing is difficult unless adequate sequencing depth is reached and the reads can be assembled correctly into the sufficiently long contigs. Many computational methods have been implemented to solve these issues. To unravel the Metagenomic sequence into complete genome, a full understanding of functional sequences into complete genome is assembled (Imelfort et al. 2014; Sharon and Banfield 2013; Bharagava et al. 2019).

22.2.4.8 Applications of Metagenomics in Bioremediation

In an ecological matrix, it is very useful method and plays a very important role to understand the microbial community composition and dynamics. It is an innovative tool for understanding the microbial degradation and detoxification of the polluted sites of organic and inorganic contaminants. It is very useful for identifying possible microbial degraders that are responsible for the degradation and detoxification of a certain form of pollutants (Zwolinski 2007). For the determination of the gene pool of enzymes involved in the degradation of the anthropogenic pollutant, some environmental micro-organisms related to bioremediation are very beneficial (Galvão et al. 2005). Nucleic acids are specifically isolated from environmental samples in cultivation-dependent experiments, and are technically regarded as representatives of whole microbial species genomes found in the ecosystem or the environment (Desai et al. 2010). These methods have been applied in the novel gene families and the micro-organisms which are involved in the bioremediation of xenobiotic. In current era, DNA microarray method has been used for the microbial ecological research and also used in determination of the microbial communities, efficacy of various types of bioremediation processes. Martin et al. (2006) investigated the screening of metagenomic libraries for identification of genes which are involved in bioremediation (Martin et al. 2006). Kim et al. (2013) studied the catabolic pathways of *Pseudomonas putida* KT 2440 by using the combined proteomic approach which was depending upon the MS and cleavable isotope coded affinity tag analysis. The field of metagenomics is expanding globally for the exploration of the environmental issues and also the research has been increased on the cellular metabolites within the cellular structure of micro-organisms, known to be a metabolomics. Various types of inventions have been done recently on applied microbial metabolome analysis for the study of biodegradation of anthropogenic pollutants (Kim et al. 2013; Bharagava et al. 2019). Jennings et al. (2009) studied the transcriptomics analysis of the cis-dichloroethene on the *Polaromonas* species strain for the identification of the genes regulated by the cDCE (cis-dichloroethene) using DNA microarray. Investigation based on the proteomics are useful for the determination of changes in the composition and abundance of proteins, and are also used for the identification of key proteins involved in the physiological responses of microbial communities exposed to anthropogenic pollutants (Jennings et al. 2009). Keum et al. (2008) studied the comparative metabolite analysis of *Sinorhizobium* species at the time of degradation of phenanthrene. Ye and Zang (2011) studied the pathogenic bacteria in sewage treatment plant by using metagenomic approach. Ma et al. (2015), Hu et al. (2012) and Wang et al. (2012) studied the characterization of structure and composition of microbial species in waste water treatment plant. Wang et al. (2012) analysed the metagenomic profiling of antibiotic resistance of genes and mobile genetic elements in tannery waste water treatment plant.

22.3 Applications of Molecular Tools in the Contaminated Sites for Characterisation of Microbial Community

It is very essential to use the appropriate monitoring tools for the real time analysis and depends upon the physicochemical situation of the site which undergoes reclamation by dynamical characteristics of the microbial species involved in the degradation activities. Molecular tools as already described are helpful for monitoring and quantifying the biotransformation of pollutants. These techniques possess the high throughput characteristics. Among all techniques, fingerprinting approaches like DGGE, RISA and TRFLP are applied for increasing success to characterize the use of bioremediation projects (Scow and Hicks 2005; Hong et al. 2010).

With the help of combination of DGGE and RST (Ribosomal Sequence Tag) method, the habitat specific array design and the validation of bulk of the probes is done. There is a strong bonding between total HCH (Hexachlorocyclihexane) concentration and the probe signals which correspond to unknown *Proteobacteria.* Gao et al. (2007) studied the development of microarray technology with its potential for detection of catabolic genes. In this study they combined the whole microbial community RNA amplification with community genome microarrays for determination of the structure and function of microbial community in low biomass ground water samples which are contaminated with hydrocarbons, organic solvents, nitrates and uranium. Neufeld et al. (2006) discovered a Ribosomal Sequence Tag (RST) array for characterization and comparison of different types of Hexachlorocyclihexane (HCH) contaminated soils. These types of isomers are included in the group of various organic pollutants. For bioremediation of such soils, RST array was developed to make a target for abundant PCR-amplified phylotypes in the soil samples. Egorova et al. (2017) also studied and developed the bioremediation technique by using the *Rhodococcus wratislaviensis* strain which was isolated from organochlorine contaminated soil. This strain was bioaugmented in chronically HCH contaminated soil. The results showed that due to bioaugmentation HCH degradation is prominent. Zhang et al. (2020) studied the bioremediation of oil contaminated soil using agricultural waste via microbial consortium, where agriculture waste like wheat bran and swine waste water were used for the bioremediation. The investigation was successful for reducing the pollution of agriculture wastes, exploring the novel model for bioremediation of oil-contaminated soil. The researchers reproduced the different mechanisms in vitro and successfully build up the pesticide degrading genes. LinA was produced from the metagenome of lindane free soil and the homologous sequence of LinA was detected. This method was industrialised by Boubakri et al. (2006) and named as Metagenomic DNA shuffling process. The main target was to assess the potential of environmental metagenome to furnish the appropriate genes.

22.4 Conclusion

In the removal and detoxification of organic and inorganic contaminants, microbes play a very significant role which is helpful for the environment and for recycling of the biochemical minerals in the ecosystem. Hence the thorough knowledge of the microorganisms and the methods for the bioremediation in the contaminated environment is very necessary which would help to get and understanding of the actual mechanism of bioremediation. Knowledge regarding searching of the key enzymes required for catabolic gene is also very much necessary. Hence, for better understanding of metagenomics different methods of finger printing have been applied for studying the microbial communities and their structure, function, composition, diversity and the contaminated matrix. This chapter mainly provides an overview of the bioremediation concept and different molecular tools used for the monitoring and validating the bioremediation process. Molecular biology techniques are the useful tools for determining the bioremediation. These studies are becoming so comprehensive in the recent years which aimed at better understanding of the molecular mechanisms which are involved in the metabolism of hydrocarbons, and the interaction between soil microorganisms. Due to the direct access of the environmental microbes independently in the ecosystem or environment, molecular tools have achieved a great succession in the field of microbial ecology and ecosystem yet its true practical potential is nowadays emerging in the field of bioremediation. In recent years, many researchers have studied different microbial species and the metagenomic technologies are expected to boost greater discovery of new updated catabolic activities and provide time to time updates for the management regarding the reclamation of contaminated sites.

References

Abed RM, Grötzschel S (2005) Applications of denaturing gradient gel electrophoresis (DGGE) and microsensor techniques in oil biodegradation studies. In: Developments in earth and environmental sciences, vol 3, Elsevier. pp 75–91

Abiraami TV, Singh S, Nain L (2020) Soil metaproteomics as a tool for monitoring functional microbial communities: promises and challenges. Rev Environ Sci Bio/technol 1:1–30

Aguiar-Pulido V, Huang W, Suarez-Ulloa V, Cickovski T, Mathee K, Narasimhan G (2016) Metagenomics, metatranscriptomics, and metabolomics approaches for microbiome analysis: supplementary issue: bioinformatics methods and applications for big metagenomics data. Evol Bioinform 12:EBO-S36436

Amann RI, Ludwig W, Schleifer KH (1995) Phylogenetic identification and in situ detection of individual microbial cells without cultivation. Microbiol Rev 59 (1):143–169

Aylward FO, Burnum KE, Scott JJ, Suen G, Tringe SG, Adams SM, Barry KW, Nicora CD, Piehowski PD, Purvine SO, Starrett GJ (2012) Metagenomic and metaproteomic insights into bacterial communities in leaf-cutter ant fungus gardens. ISME J 6(9):1688–1701

Azubuike CC, Chikere CB, Okpokwasili GC (2016) Bioremediation techniques–classification based on site of application: principles, advantages, limitations and prospects. World J Micro Biotech 32(11):180

Bakermans C, Madsen EL (2002) Detection in coal tar waste-contaminated groundwater of mRNA transcripts related to naphthalene dioxygenase by fluorescent in situ hybridization with tyramide signal amplification. J Microbio Methods 50(1):75–84

Bastida F, Algora C, Hernández T, García C (2012) Feasibility of a cell separation-proteomic based method for soils with different edaphic properties and microbial biomass. Soil Biol Biochem 45:136–138

Bharagava RN, Purchase D, Saxena G, Mulla SI (2019) Applications of metagenomics in microbial bioremediation of pollutants: from genomics to environmental cleanup. In: Microbial diversity in the genomic era. Academic press, pp 459–477

Boubakri H, Beuf M, Simonet P, Vogel TM (2006) Development of metagenomic DNA shuffling for the construction of a xenobiotic gene. Gene 375:87–94

Bumgarner R (2013) Overview of DNA microarrays: types, applications, and their future. Curr Protoc Mol Biol 101(1):22–31

Ciesielski S, Bułkowska K, Dabrowska D, Kaczmarczyk D, Kowal P, Możejko J (2013) Ribosomal intergenic spacer analysis as a tool for monitoring methanogenic archaea changes in an anaerobic digester. Curr Microbiol 67(2):240–248

Claire OC (2008) Fluorescence in situ hybridization (FISH), cytogenetics can now go "FISH-ing" for chromosomal abnormalities, which are deletions an duplications that can cause disease. How exactly does FISH work. Nature Education. Chromosom Cytogenet 1(1):1–8

Dash HR, Das S (2018) Molecular methods for studying microorganisms from atypical environments. J Microbiol Methods 45(1):89–122

De Vrieze J, Ijaz UZ, Saunders AM, Theuerl S (2018) Terminal restriction fragment length polymorphism is an "old school" reliable technique for swift microbial community screening in anaerobic digestion. Sci Rep 8(1):1–12

Delmont TO, Simonet P, Vogel TM (2012) Describing microbial communities and performing global comparisons in the 'omic era. ISME J 6(9):1625–1628

DeLong EF (2005) Microbial community genomics in the ocean. Nat Rev Microbio 3(6):459–469

Deng F, Liao C, Yang C, Guo C, Ma L, Dang Z (2016) A new approach for pyrene bioremediation using bacteria immobilized in layer-by-layer assembled microcapsules: dynamics of soil bacterial community. RSC Adv 6(25):20654–20663

Desai C, Pathak H, Madamwar D (2010) Advances in molecular and "-omics" technologies to gauge microbial communities and bioremediation at xenobiotic/anthropogen contaminated sites. Bioresour Technol 101(6):1558–1569

Egorova DO, Buzmakov SA, Nazarova EA, Andreev DN, Demakov VA, Plotnikova EG (2017) Bioremediation of hexachlorocyclohexane-contaminated soil by the new Rhodococcuswratislaviensis strain Ch628. Water Air Soil Pollut 228(5):183

Eyers L, Smoot JC, Smoot LM, Bugli C, Urakawa H, McMurry Z, Siripong S, El Fantroussi S, Lambert P, Agathos SN, Stahl DA (2006) Discrimination of shifts in a soil microbial community associated with TNT-contamination using a functional ANOVA of 16S rRNA hybridized to oligonucleotide microarrays. Envir Sci Tech 40(19):5867–5873

Folch A, Vilaplana M, Amado L, Vicent T, Caminal G (2013) Fungal permeable reactive barrier to remediate groundwater in an artificial aquifer. J Hazard Mater 262:554–560

Foti M, Sorokin DY, Lomans B, Mussman M, Zacharova EE, Pimenov NV, Kuenen JG, Muyzer G (2007) Diversity, activity, and abundance of sulfate-reducing bacteria in saline and hypersaline soda lakes. Appl Environ Microbiol 73(7):2093–2100

Frias-Lopez J, Shi Y, Tyson GW, Coleman ML, Schuster SC, Chisholm SW, DeLong EF (2008) Microbial

community gene expression in ocean surface waters. Proc Natl Acad Sci 105(10):3805–3810
Frutos FJG, Escolano O, García S, Babín M, Fernández MD (2010) Bioventing remediation and ecotoxicity evaluation of phenanthrene-contaminated soil. J Hazard Mater 183(1–3):806–813
Galvão TC, Mohn WW, de Lorenzo V (2005) Exploring the microbial biodegradation and biotransformation gene pool. Trends Biotechn 23(10):497–506
Gao H, Yang ZK, Gentry TJ, Wu L, Schadt CW, Zhou J (2007) Microarray-based analysis of microbial community RNAs by whole-community RNA amplification. Appl Envirn Micro 73(2):563–571
Golyshin PN, Dos Santos VAM, Kaiser O, Ferrer M, Sabirova YS, Lünsdorf H, Timmis KN (2003) Genome sequence completed of *Alcanivorax borkumensis*, a hydrocarbon-degrading bacterium that plays a global role in oil removal from marine systems. J Biotech 106(2–3):215–220
Guazzaroni ME, Herbst FA, Lores I, Tamames J, Peláez AI, Lopez-Cortés N, Alcaide M, Del Pozo MV, Vieites JM, Von Bergen M, Gallego JL (2013) Metaproteogenomic insights beyond bacterial response to naphthalene exposure and biostimulation. ISME J 7(1):122–136
Handelsman J (2004) Metagenomics: application of genomics to uncultured microorganisms. MMBR 68 (4):669–685
Handelsman J, Rondon MR, Brady SF, Clardy J, Goodman RM (1998) Molecular biological access to the chemistry of unknown soil microbes: a new frontier for natural products. Chem Biolog 5(10):R245–R249
Hong JW, Fomina M, Gadd GM (2010) F-RISA fungal clones as potential bioindicators of organic and metal contamination in soil. J Appl Microbiol 109(2):415–430
Hout AHVD, Ouweland AMVD, Luijt RBVD, Gille HJ, Bodmer D, Brüggenwirth H, Mulder IM, Vlies PVD, Elfferich P, Huisman MT, Berge AMT (2006) A DGGE system for comprehensive mutation screening of BRCA1 and BRCA2: application in a Dutch cancer clinic setting. Hum Mutat 27(7):654–666
Hu T, Zheng Y, Zhang Y, Li G, Qiu W, Yu J, Cui Q, Wang Y, Zhang C, Zhou X, Feng Z (2012) Identification of a novel Getah virus by Virus-Discovery-cDNA random amplified polymorphic DNA (RAPD). BMC Microbial 12(1):305
Huang H, Ye L (2020) Biological technologies for cHRPs and risk control. In: High-risk pollutants in wastewater. Elsevier, pp 209–236
Huber D, von Voithenberg LV, Kaigala GV (2018) Fluorescence in situ hybridization (FISH): history, limitations and what to expect from micro-scale FISH? Micro Nano Eng 1:15–24
Imelfort M, Parks D, Woodcroft BJ, Dennis P, Hugenholtz P, Tyson GW (2014) GroopM: an automated tool for the recovery of population genomes from related metagenomes. Peer J 2:e603
Jennings LK, Chartrand MM, Lacrampe-Couloume G, Lollar BS, Spain JC, Gossett JM (2009) Proteomic and transcriptomic analyses reveal genes upregulated by cis-dichloroethene in Polaromonas sp. strain JS666. Appl Environ Microbiol 75(11):3733–44
Karpouzas DG, Singh BK (2010) Application of fingerprinting molecular methods in bioremediation studies. In: Bioremediation. Humana Press, pp 69–88
Keum YS, Seo JS, Li QX, Kim JH (2008) Comparative metabolomic analysis of Sinorhizobium sp. C4 during the degradation of phenanthrene. Appl Microbiol Biotechnol 80(5):863–72
Kim M, Lee KH, Yoon SW, Kim BS, Chun J, Yi H (2013) Analytical tools and databases for metagenomics in the next-generation sequencing era. Genomics Inform 11(3):102
Kim S, Krajmalnik-Brown R, Kim JO, Chung J (2014) Remediation of petroleum hydrocarbon-contaminated sites by DNA diagnosis-based bioslurping technology. Sci Total Environ 497:250–259
Klindworth A, Pruesse E, Schweer T, PepliesJ, Quast C, Horn M, Glöckner FO (2013) Evaluation of general 16S ribosomal RNA gene PCR primers for classical and next-generation sequencing-based diversity studies. Nucleic Acids Res 41(1):e1
Kovacs A, Yacoby K, Gophna U (2010) A systematic assessment of automated ribosomal intergenic spacer analysis (ARISA) as a tool for estimating bacterial richness. Res Microbiol 161(3):192–197
Kube M, Beck A, Zinder SH, Kuhl H, Reinhardt R, Adrian L (2005) Genome sequence of the chlorinated compound–respiring bacterium Dehalococcoides species strain CBDB1. Nat Biotech 23 (10):1269–1273
Louati H, Said OB, Got P, Soltani A, Mahmoudi E, Cravo-Laureau C, Duran R, Aissa P, Pringault O (2013) Microbial community responses to bioremediation treatments for the mitigation of low-dose anthracene in marine coastal sediments of Bizerte lagoon (Tunisia). ESPR 20(1):300–310
Ma Q, Qu Y, Shen W, Zhang Z, Wang J, Liu Z, Li D, Li H, Zhou J (2015) Bacterial community compositions of coking wastewater treatment plants in steel industry revealed by Illumina high-throughput sequencing. Bioresour Technol 179:436–443
Maitra S (2018) In situ bioremediation—an overview. Res J Life Sci Bioinform Pharmacet Chem Sci 576–598
Martin HG, Ivanova N, Kunin V, Warnecke F, Barry KW, McHardy AC, Yeates C, He S, Salamov AA, Szeto E, Dalin E (2006) Metagenomic analysis of two enhanced biological phosphorus removal (EBPR) sludge communities. Nat Biotechnol 24(10):1263–1269
Martín M, Gibello A, Lobo C, Nande M, Garbi C, Fajardo C, Barra-Caracciolo A, Grenni P, Martínez-Iñigo MJ (2008) Application of fluorescence in situ hybridization technique to detect simazine-degrading bacteria in soil samples. Chemosphere 71(4):703–710
McAuliffe L, Ellis RJ, Lawes JR, Ayling RD, Nicholas RA (2005) 16S rDNA PCR and denaturing gradient gel electrophoresis; a single generic test for

detecting and differentiating Mycoplasma species. Sci Total Environ 54(8):731–739

Megharaj M, Ramakrishnan B, Venkateswarlu K, Sethunathan N, Naidu R (2011) Bioremediation approaches for organic pollutants: a critical perspective. Environ Int 37(8):1362–1375

Mishra M, Mohan D (2017) Bioremediation of contaminated soils: an overview. Adaptive soil management: from theory to practices. Springer, Singapore, pp 323–337

Mohan SV, Mohanakrishna G, Srikanth S (2011) Biohydrogen production from industrial effluents. In: Biofuels. Academic press, pp 499–524

Mukherjee A, Reddy MS (2020) Metatranscriptomics: an approach for retrieving novel eukaryotic genes from polluted and related environments. Biotech 10(2):71

Neufeld JD, Mohn WW, De Lorenzo V (2006) Composition of microbial communities in hexachlorocyclohexane (HCH) contaminated soils from Spain revealed with a habitat-specific microarray. Environ Micro 8 (1):126–140

Nithya C, Pandian SK (2012) Evaluation of bacterial diversity in Palk Bay sediments using terminal-restriction fragment length polymorphisms (T-RFLP). Appl Biochem Biotechnol 167(6):1763–1777

Paerl HW, Steppe TF (2003) Scaling up: the next challenge in environmental microbiology. Environ Micro 5(11):1025–1038

Parada AE, Needham DM, Fuhrman JA (2016) Every base matters: assessing small subunit rRNA primers for marine microbiomes with mock communities, time series and global field samples. Environ Microbiol 18 (5):1403–1414

Parra I, Windle B (1993) High resolution visual mapping of stretched DNA by fluorescent hybridization. Nat Genet 5(1):17–21

Peng M, Zi X, Wang Q (2015) Bacterial community diversity of oil-contaminated soils assessed by high throughput sequencing of 16S rRNA genes. Int J Environ Res Public Health 12(10):12002–12015

Pfaffl MW (2001) A new mathematical model for relative quantification in real-time RT–PCR. Nucleic Acids Res 29(9):e45–e45

Quintero WV, Zafra G (2016) Use of molecular tools to monitor microbial communities during the bioremediation of Polycyclic Aromatic Hydrocarbon-contaminated soils. J Indust Pollut Contr 32(2)

Ram RJ, VerBerkmoes NC, Thelen MP, Tyson GW, Baker BJ, Blake RC, Shah M, Hettich RL, Banfield JF (2005) Community proteomics of a natural microbial biofilm. Science 308:1915–1920

Rastogi G, Sani RK (2011) Molecular techniques to assess microbial community structure, function, and dynamics in the environment. Microbes and microbial technology. Springer, New York, NY, pp 29–57

Ritchie NJ, Schutter ME, Dick RP, Myrold DD (2000) Use of length heterogeneity PCR and fatty acid methyl ester profiles to characterize microbial communities in soil. AEM 66(4):1668–1675

Rittmann BE, Hausner M, Loffler F, Love NG, Muyzer G, Okabe S, Oerther DB, Peccia J, Raskin L, Wagner M (2006) A vista for microbial ecology and environ biotech 1096–1103

Rousseaux S, Hartmann A, Soulas G (2001) Isolation and characterisation of new Gram-negative and Gram-positive atrazine degrading bacteria from different French soils. FEMS Microbiol Ecol 36(2–3):211–222

Roy M, Giri AK, Dutta S, Mukherjee P (2015) Integrated phytobial remediation for sustainable management of arsenic in soil and water. Environ Inter 75:180–198

Schwieger F, Tebbe CC (1998) A new approach to utilize PCR–single-strand-conformation polymorphism for 16S rRNA gene-based microbial community analysis. Appl Environ Microbiol 64(12):4870–4876

Scow KM, Hicks KA (2005) Natural attenuation and enhanced bioremediation of organic contaminants in ground water. Curr Opin Biotech 16(3):246–253

Sharon I, Banfield JF (2013) Genomes from metagenomics. Science 342(6162):1057–1058

Shekhar SK, Godheja J, Modi DR (2020) Molecular technologies for assessment of bioremediation and characterization of microbial communities at pollutant-contaminated sites. Bioremediation of industrial waste for environmental safety. Springer, Singapore, pp 437–474

Sheng L, Aijia Z, Rongyu W, Ting C, Mingming Z, Zhaonian Y, Wenxiong L (2012) Effects of sugarcane ratooning cultivation on the alteration of bacterial communities in the rhizosphere soil. Sugar Tech 14 (3):275–283

Shi Y, Tyson GW, DeLong EF (2009) Metatranscriptomics reveals unique microbial small RNAs in the ocean's water column. Nature 459(7244):266–269

Simon C, Daniel R (2011) Metagenomic analyses: past and future trends. Appl Environ Microbiol 77 (4):1153–1161

Siqueira JF Jr, Rôças IN, Rosado AS (2005) Application of denaturing gradient gel electrophoresis (DGGE) to the analysis of endodontic infections. J Endod 31 (11):775–782

Smith E, Thavamani P, Ramadass K, Naidu R, Srivastava P, Megharaj M (2015) Remediation trials for hydrocarbon-contaminated soils in arid environments: evaluation of bioslurry and biopiling techniques. Int Biodeterior Biodegradation 101:56–65

Stenuit B, Eyers L, Rozenberg R, Habib-Jiwan JL, Matthijs S, Cornelis P, Agathos SN (2009) Denitration of 2, 4, 6-trinitrotoluene in aqueous solutions using small-molecular-weight catalyst (s) secreted by Pseudomonas aeruginosa ESA-5. Environ Sci Technol 43 (6):2011–2017

Strathdee F, Free A (2013) Denaturing gradient gel electrophoresis (DGGE). DNA electrophoresis. Humana Press, Totowa, NJ, pp 145–157

Strong LC, Rosendahl C, Johnson G, Sadowsky MJ, Wackett LP (2002) *Arthrobacter aurescens* TC1 metabolizes diverse s-triazine ring compounds. Appl Environ Microbiol 68(12):5973–5980

Svozil D, Kalina J, Omelka M, Schneider B (2008) DNA conformations and their sequence preferences. Nucleic Acids Res 36(11):3690–3706

Tang YJ, Martin HG, Dehal PS, Deutschbauer A, Llora X, Meadows A, Arkin A, Keasling JD (2009) Metabolic flux analysis of *Shewanella* spp. reveals evolutionary robustness in central carbon metabolism. Biotechnol Bioeng 102(4):1161–9

Taylor EB, Williams MA (2010) Microbial protein in soil: influence of extraction method and C amendment on extraction and recovery. Microb Ecol 29(2):390–399

Techtmann SM, Hazen TC (2016) Metagenomic applications in environmental monitoring and bioremediation. J Ind Microbiol Biot 43(10):1345–1354

Uhlik O, Leewis MC, Strejcek M, Musilova L, Mackova M, Leigh MB, Macek T (2013) Stable isotope probing in the metagenomics era: a bridge towards improved bioremediation. Biotechnol Adv 31(2):154–165

Verma JP, Jaiswal DK (2016) Book review: advances in biodegradation and bioremediation of industrial waste. Front Microbiol 6:1555

Viglasky V (2013) Polyacrylamide temperature gradient gel electrophoresis. DNA electrophoresis. Humana Press, Totowa, NJ, pp 159–171

Villas-Bôas SG, Bruheim P (2007) The potential of metabolomics tools in bioremediation studies. Omics J Integr Biol 11(3):305–313

Wang Y, Leung HC, Yiu SM, Chin FY (2012) MetaCluster 5.0: a two-round binning approach for metagenomic data for low-abundance species in a noisy sample. Bioinformatics 28(18):i356–i362

Yannopoulos SI, Lyberatos G, Theodossiou N, Li W, Valipour M, Tamburrino A, Angelakis AN (2015) Evolution of water lifting devices (pumps) over the centuries worldwide. Water 7(9):5031–5060

Ye L, Zhang T (2011) Pathogenic bacteria in sewage treatment plants as revealed by 454 pyrosequencing. Environ Sci Technol 45(17):7173–7179

Zafra G, Absalón AE, Cortés-Espinosa DV (2014) Isolation and selection of a highly tolerant microbial consortium with potential for PAH biodegradation from heavy crude oil contaminated soils. Water Air Soil Pollut 225:1826

Zhang C, Wu D, Ren H (2020) Bioremediation of oil contaminated soil using agricultural wastes via microbial consortium. Sci Rep 10(1):1–8

Zwolinski MD (2007) DNA sequencing: strategies for soil microbiology. Soil Sci Soc Am J 71(2):592–600

23 Bioindication and Biomarker Responses of Earthworms: A Tool for Soil Pollution Assessment

Varun Dhiman and Deepak Pant

Abstract

Hazardous pollutants from anthropogenic activities are continually delivered into various natural spheres including the terrestrial biological system which is a highly influenced ecological sphere confronting the genuine contamination problem. Synthetics like pesticides, insecticides, herbicides, vinasse, polyaromatic hydrocarbons, heavy metals, agrochemicals, dioxins, and toxic sewage are among the potentially harmful pollutants that alter the physicochemical characteristics of the soil by chemical interactions with the soil environment and its dwelling biota, hence upsetting the typical functioning. Accordingly, these pollutants must be checked and monitored to revamp the health of the soil and henceforth utilization of earthworms gives an alternative yet stunning, novel, and biological monitoring tool to evaluate the hazardous impacts of the pollutants through its biomarkers response and bioindication abilities. Earthworms end up being profoundly viable in monitoring the soil pollutants. This chapter significantly reviews the importance of earthworms in pollutants biomonitoring in special reference to the soil ecosystem.

Graphical Abstract

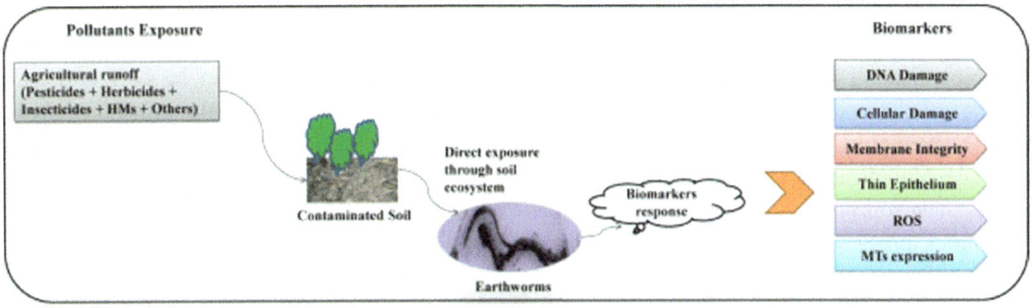

Keywords

Bioindication · Biomarkers · Earthworms · Monitoring · Pollutants · Response · Soil ecosystem

V. Dhiman (✉) · D. Pant
Department of Environmental Sciences, Central University of Himachal Pradesh, Dharamshala 176215, India

J. A. Malik (ed.), *Advances in Bioremediation and Phytoremediation for Sustainable Soil Management*,
https://doi.org/10.1007/978-3-030-89984-4_23

23.1 Introduction

The soil ecosystem is actively involved in the regulation of biogeochemical cycles, disposal of waste, retention of carbon, water filtration, and temperature regulation. All these services are maintained by the inherent communities of the soils intrinsic to life and other fundamental activities on the earth (Pérès et al. 2011). But, in the present-day scenario, the excessive use of pesticides, insecticides, and broad-spectrum chemicals causes negative environmental consequences at physiological and chemical levels in the soil ecosystem (Lionetto et al. 2019). Besides, urban waste, toxic sludge, atmospheric deposition, and industrialization enhance soil pollution (Calisi et al. 2014). The researchers in the most recent decade's centers around the physio-chemical characterization of soil health; however, there is a prerequisite for exceptionally capable and proficient tools to detect the real-time impact of the pollutants as conventional methodologies are not all that compelling (Bünemann et al. 2018; Yang et al. 2020a). Nonetheless, various studies indicate the utility of soil biota as the early warning indicators of pollutants through their biomarkers response and bioindication abilities (Burger 2006; Parmar et al. 2016). Cortet et al. (1999) in their critical review discuss the relevancy of nematodes, mites, isopods, mollusks, and earthworms as exceptionally valuable life forms for contamination bioindication (Cortet et al. 1999).

The growing concerns of soil health account for the developing interest in the improvement of new-age bioindicators and early warning techniques. The soil pollution assessment analysis is a complex process (Ashraf et al. 2014). Subsequently, the use of earthworms as a bioindicator model provides a unique, novel, eco-friendly, cost–benefit, and convenient approach for soil pollution assessment. Earthworms are engaged with the pedogenesis and customarily utilized as agents to indicate the soil fertility, land use impact, and organic matter breakdown (Calisi et al. 2011, 2014). They are straightforwardly confronting the toxic impacts of the soil pollutants through their permeable and highly sensitive skin for the pollutants. Additionally, they ingest the defiled soil particles and accordingly impact the pollutants availability (Wallwork 1983; Jager et al. 2003; Vijver et al. 2003). Because of their higher relevance in standard toxicity testing protocols, earthworms discover their utilization as soil contamination bioindicators. Their mechanism of response and biomarkers generation toward the stress produced by the toxic soil pollutants can provide more extensive information in accessing the level of soil health. Therefore, we, here in this chapter, have focused to explore the bioindication and biomarkers response of earthworms in pollution assessment of the soil ecosystem.

23.2 Biological System and Pollution Biomarkers

The pollution assessment and measuring their toxic impacts in different environmental spheres is a highly difficult process (Ashraf et al. 2014). For doing such complex estimations, nature furnished us with extraordinary sentinel living beings that can distinguish the continuous changes that occurred in the environment. The sentinel species can biosense the extent of pollution by making specific changes in the form of biomarkers response. These biomarkers are expressed as "alterations' in the body of sentinel species. These changes can be best utilized to express the toxicological impacts of a particular pollutant. Therefore, it is used as an early warning indicator. When the biological sentinel species got exposed to a particular or variety of pollutants, adverse and toxic effects have been seen at the molecular and cellular levels. These changes are represented by the molecular and cellular biomarkers in the bioindicator species. These biomarkers viably give the necessary information on the bioavailability and the adverse effects of the contaminants on the environment as these biomarkers can assist us with understanding the biochemical processes of absorption, transportation, and biotransformation of pollutants in the sentinel organisms as well as the environment. Below are the various categories of biomarkers that are useful in measuring the level of pollution and toxicity of particular pollutant existed in the environment.

23.2.1 Exposure

These are the class-specific biomarkers as they are expressed in the body of sentinel species in response to a specific class of pollutants (Scott-Fordsmand and Weeks 2000). Genetic alterations, circulating antibodies, DNA and protein adducts, altered proteins, metallothioneins levels, altered cholinesterase activity, ethoxyresorufin-O-demethylase activity, and altered gene expression are some of the main exposure biomarkers that reflect their expression in response to pollutants exposure in the animal's body.

23.2.2 Histological

These are the biomarkers with a defined cellular origin (Kilty et al. 2007). Elevated troponin levels, altered alanine aminotransferase, transaminase activity, thinning of epithelium lining, altered lysosomal-cytoplasm ratio, and basophil-digestive cell ratio are some of the histological biomarkers that define the morphological damage in the organism's body when exposed to different pollutants (Reddy 2012).

23.2.3 Stress

As the name indicated, these biomarkers are expressed in the animal's body in response to the physiological stress instigated by the toxic impacts of the pollutants (Etteieb et al. 2019). The generation of heat shock proteins in response to tackling temperature variations, acute phase proteins, cortisol, cytokines, alpha-amylase, reactive oxygen species level, MDA levels in the serum and plasma, altered GHS, SOD, thioredoxin reductase, and glutathione peroxidase activities are some of the well-known stress biomarkers in the animal's body (Colacevich et al. 2011; Ali and Naaz 2013).

23.2.4 Genotoxicity

A few pollutants, for example, PAHs, naphthalene, and phenanthrene are notable for their genotoxic potential when exposed to living organisms. These agents cause DNA damage and consequently promote mutations in the organism's body (Hirano and Tamae 2010). These pollutants cause DNA alterations through the phenomenon of oxidative respiration and altered metabolic reactions. The damage to genetic material persists in the form of chromosomal abnormality, distorted sister chromatids, abnormal DNA-DNA crosslinks, and DNA–protein binding. To monitor the genotoxic damage on the exposure of toxic pollutants, there are several genotoxic biomarkers with the help of which we can assess the toxic potential of a particular toxin. For example, increased micronucleus formation, chromosomal aberrations, comet formation, and toxicogenomic signatures are some of the known biomarkers which serve as good genotoxic biomarkers for toxicity assessment (Vasseur and Bonnard 2014; Muangphra et al. 2015).

23.3 Effects of Soil Pollutants on Earthworms

Earthworms are perceived as suitable candidates for the biomonitoring purpose of soil pollutants (Hirano and Tamae 2011). Different investigations have been done to signify the role of earthworms as bioindicators of soil pollution (Haeba et al. 2013). Scientists examined the effect of natural and depleted uranium on the earthworms and noticed genetic and cytotoxic alterations in their body tissues (Giovanetti et al. 2010). It has been discovered that the earthworms have incredible bioaccumulation potential, thus, proved to be helpful for heavy metals biomonitoring (Usmani and Kumar 2015). The study by Natal-de-Luz et. al. observed the centrality of earthworms in the ecological risk assessment of mixed chemical compounds (Natal-da-Luz et al. 2011). Likewise, Qiu et al. utilize *Aporrectodea calignosa* (gray worm) for toxicity assessment of binary mixture of zinc and cadmium (Bart et al. 2018). One of the study reported the bioaccumulation of mixture form of nickel and chlorpyrifos in the body tissues of the earthworm, thus, describes the assimilation pathway of these pollutants in the *lumbricid* earthworms body (Lister et al. 2011). Different studies that evaluated the effects of a variety of soil pollutants on different earthworm species have been presented in Table 23.1.

23.4 Pollutant-Induced Biomarker Responses in Earthworm

As stated above, earthworms are highly sensitive toward pollutants exposure and proved helpful in their biomonitoring. When exposed to pollutants, the body of earthworms reacts to them by expressing cellular, behavioral, morphological, genetic, and biochemical biomarkers (Fig. 23.1). Different pollutants induce a different kind of biomarkers response. Different research reports signify the expression of a variety of biomarkers responses concerning a specific pollutant. These are discussed below.

23.4.1 Methiocarb

The insecticide methiocarb is a carbonic acid derived organic ester that is synthesized from the condensation of 3, 5-dimethyl-4 (methylsulfonyl) phenol with methyl carbamic acid (Fig. 23.2) Researchers investigated the biomarkers response of the *Lumbricus terrestris* toward the methiocarb exposure.

The study was performed under controlled experimental conditions at a temperature of 18 ± 1 °C with a 16:8 h photoperiod ratio of light and dark The exposure of the insecticide was given at different time intervals of 0, 7, and 14 days. The study involves the measurement of altered lysosomal permeability, MTs expression, and granulocyte morphogenetic analysis. With these analyses, other parameters like growth, reproduction, and survival capacity will also be taken into consideration. The results of the study concluded that the used model species of earthworms was very sensitive toward the methiocarb exposure and different biomarkers of effect such as enlarged granulocytes, and destabilized lysosomal membrane was observed to be the potential biomarkers that are helpful in biomonitoring of this specific insecticide (Calisi et al. 2011).

23.4.2 Imidacloprid

Imidacloprid influences the soil health and local soil life forms by enhancing the pollution levels in the terrestrial environment (Knoepp et al. 2012). The native earthworm species *Eisenia fetida* was exceptionally influenced when exposed to Imidacloprid. The risk evaluation of this particular insecticide was evaluated by researchers, and observed genotoxic effects on *Eisenia fetida.* The DNA damage and sperm deformity were observed to be the relevant genotoxic and physiological biomarkers expressed in this particular earthworm species in response to Imidacloprid exposure in the terrestrial ecosystem (Zang et al. 2000).

23.4.3 Pesticides

Certain studies conducted mutual toxicity testing of regularly utilized pesticides. Aldicarb, chlorfluazuron, cypermethrin, metalaxyl, and atrazine are some of the commonly used pesticides that are significantly important in causing soil pollution (Mosleh et al. 2003; Miglani and Bisht 2019). Experimental studies showed the environmental consequences of these pesticides and correlate the expression of the different biomarkers in the earthworm's body with the toxic impacts of these chemicals. For instance, researchers in a study observed the deleterious effects of these pesticides on the earthworm, *Aporrectodea calignosa,* and observed that the soluble protein in the earthworm's body was

Table 23.1 List of investigations on the adverse effects of common soil pollutants in different earthworm species after pollutants exposure

Earthworm Species	Soil Pollutants	Effects	References
Eisenia fetida	1,2,4-trichlorobenzene	• Alterations observed in the ultrastructure of skin and cuticle • Low mucus production and finally disappears	Wu et al. (2012)
	Cadmium and Lead	• Weight loss • Delayed sexual maturity	Urionabarrenetxea et al. (2020)
	Tetraethyl Lead and Lead Oxide	• Inflexible metameric segmentation • Rupturing of skin and cuticle • Coelomic fluid extrusion is observed	Rao et al. (2003)
	Benomyl	• Regeneration of posterior segment is influenced • Teratogenic effects • Groove anomalies • Development of two tails at the posterior end	Zoran et al. (1986), Drewes et al. (1987), Sorour and Larink (2001)
	Carbamates	• Development of tumors and swelling in the body	Yadav et al. (2017)
	Propoxur, Methidathion, Triazophos, Endosulfan, Carbofuran	• Swelling, bursting, and bleeding of the sores have been observed	Dureja et al. (1999), Dureja and Tanwar (2012)
	Integrated toxic effects of Cd, Cu, Pb, and Zn	• Higher mortality rate • Altered sexual activities	Spurgeon and Hopkin (1996)
	Pentachlorophenol	• Affect cocoon production • Infertile cocoons	Van Gestel et al. (1989), Landrum et al. (2006)
	PCBs	• Damaged genetic material • Influence the activity of CAT, POD, and SOD • Altered carbohydrate metabolism • Disrupted osmotic function	Åslund et al. (2011), Duan et al. (2017)
	2, 2′, 4, 4′-tetrabromodiphenyl ether (BDE-47) clothianidin,	• SOD gene transcripts upregulation • Suppressed catalase activity	Xu et al. (2015)
	Imidacloprid, thiacloprid, nitenpyram, and, acetamiprid,	• The altered activity of catalase enzyme • Lower fecundity rate	Wang et al. (2015)
Lampito mauritii	Phosphamidon	• Hyperactivity in the body	Bharathi and Rao (1986), Dureja and Tanwar (2012)
	Monocrotophos and Dichlorvos	• Inhibited and altered AChE activity • Damaged intestinal villi • Degenerated nucleus	Datta et al. (2016), Samal et al. (2019), Kavitha et al. (2020)

(continued)

Table 23.1 (continued)

Earthworm Species	Soil Pollutants	Effects	References
		• Weight loss • Blood sinuses congestion	
Lumbricus terrestris	Benomyl	•Hindered AChE activity • Impairment in locomotion •Mitosis inhibition	Byrde and Richmond (1976), Stringer and Wright (1976), Subaraja and Vanisree (2015)
Lumbricus rubellus and Lumbriculus variegatus	C60 Fullerene Nanoparticles	• Damaged musculature, epidermis, and cuticular part	Van der Ploeg et al. (2011,,
Aporrectodea rosea	Cadmium and Lead	• Inhibition of total antioxidant capacity	Sinkakarimi et al. (2020)
Pontoscolex corethrurus	benzo(a)pyrene	• Loss of weight • Low survival rate	Hernández-Castellanos et al. (2013)
Eisenia andrei	Oil contaminated soil	• Higher mortality rate is observed	Hentati et al. (2013)
Drawida willsi	Carbofuran and malathion	• Lowering acetylcholine esterase activity	Panda and Sahu (2004)
Allolobophora chlorotica	Carbendazim	• Disrupted functioning of giant nerve fibers • Altered burrowing behavior	Ellis et al. (2010)
Perionyx excavatus	Chlorpyrifos and carbofuran	• Highly toxic • Death of earthworms	De Silva and van Gestel (2009)
Enchytraeus crypticus	Nylon microplastics debris	• Significant reduction in the reproduction activity of earthworms	(Lahive et al. 2019)
Aporrectodea tuberculata	Copper and Zinc	• Decreased cytochrome CYP1A and GST activities	Lukkari et al. (2004)
Aporrectodea caliginosa	Pentachlorophenol, copper, and cadmium	• DNA and lysosomal damage are observed	Klobučar et al. (2011)
Aporrectodea rosea and Aporrectodea trapezoides	Cadmium and lead nitrate	• DNA damage • Lipid peroxidation • Decrease in total antioxidant capacity	Sinkakarimi et al. (2020)
Octolasion cyaneum	Glyphosate	• Glutathione S-transferase activity observed to be declined	Salvio et al. (2016)

decreased from the initial proteins levels. Besides this observation, the GPT, AcP, and GOT enzyme activities are at their low. These biochemical changes are considered as important biomarkers response of this snail species which were found to be suitable in determining the environmental toxicity (Mosleh et al. 2003).

23.4.4 Polystyrene Microplastics

Microplastics are one of the major and emerging soil pollutants that are known for their serious environmental consequences (Kumar et al. 2020). They are ubiquitous and non-biodegradable (Smith et al. 2018; Mammo

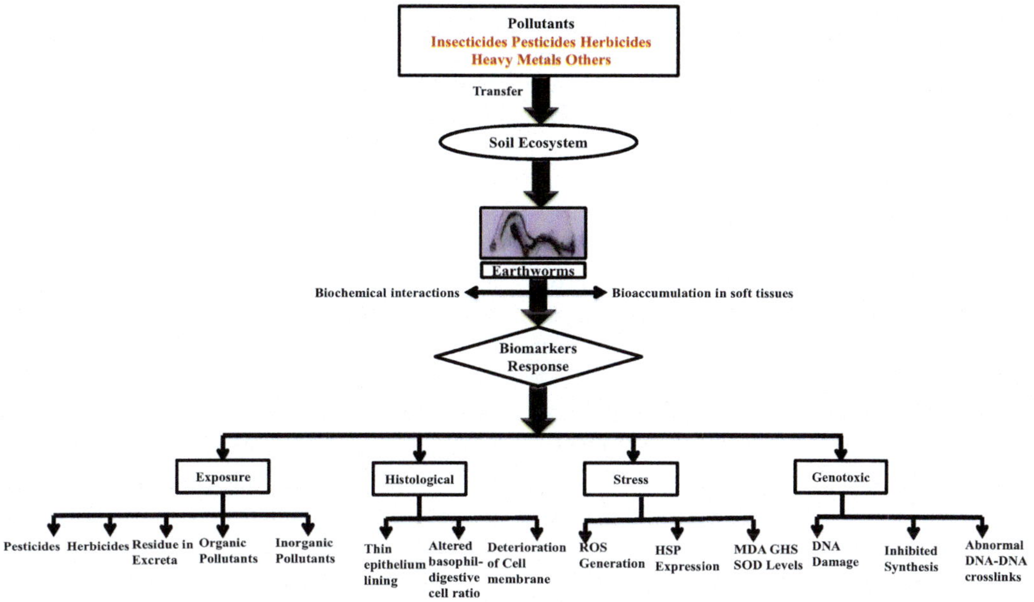

Fig. 23.1 Diagrammatic representation of the different biomarkers response of earthworms to common soil pollutants

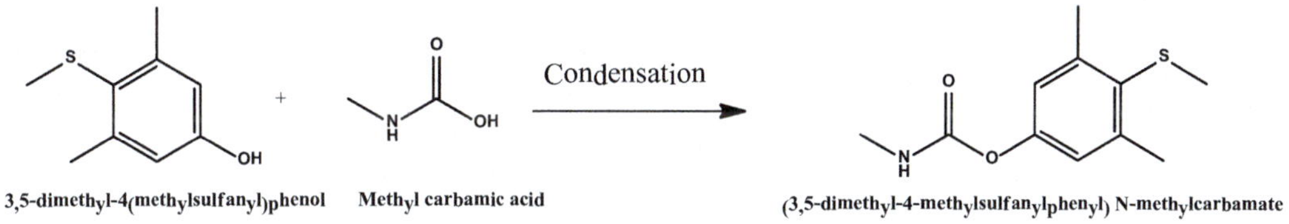

Fig. 23.2 Chemical synthesis of methiocarb. Adapted after: Calisi et al. 2011, Gupta 2011

et al. 2020). Studies on the toxic effects of these polystyrene-based microplastics are one of the enthusiastic areas of research. These polystyrene microplastics (PsMPs) bioaccumulates in the soft and delicate tissues of the soil creatures through the natural way of food chain interactions and causes adverse metabolic functioning (Wang et al. 2019). A recent study shows the toxic consequences of PsMPs on the earthworm species *Eisenia fetida.* Their exposure to *Eisenia fetida* initiates the expression of biomarkers response in the form of DNA damage and oxidative stress. Consequently, the study indicates the histopathological alterations in the intestinal wall of earthworms (Jiang et al. 2020).

23.4.5 Antibiotics

These are widely used biologically active molecules that interact with the soil ecosystem in their pure form (Manyi-Loh et al. 2018; Cycoń et al. 2019). They enter the terrestrial environment through medical waste dumping, domestic sludge, and human excretion (Larsson 2014; Kraemer et al. 2019). In a recent study, researchers explored the environmental effects on the soil ecosystem and the native earthworm species. The study involves the use of different exposure concentrations of ciprofloxacin to the earthworm *Eisenia fetida,* and it was observed that a concentration of 1–2 g/kg of ciprofloxacin exposure causes deformity in DNA while the other biomarkers such as antioxidant enzymatic activity, mRNA expression, HSP 70, MTs, etc. were upregulated (Yang et al. 2020b).

23.4.6 Thifluzamide

The extensive use of fungicides imposes serious environmental concerns (Mahmood et al. 2016). Apart from target organisms, their toxic nature also influences the non-target species present in the soil (Gill and Garg 2014). The fungicide Thifluzamide is one of the commonly used fungicides which are chemically characterized as amide (Yang et al. 2016; Yao et al. 2020). This fungicide disturbs the SDH metabolism in the organisms (Yang et al. 2017). A recent study evaluates the biomarkers response of *Eisenia fetida* concerning the stress induced by the Thifluzamide with different concentrations ranges from 0 to 10 mg/kg. It has been observed that this particular fungicide induces DNA damage, ROS generation, inhibited activities of GST, CAT, POS, and SOD enzymes in the body of *Eisenia fetida* (Yao et al. 2020).

23.4.7 Neonicotinoid Insecticides and Heavy Metals

The extensive use of neonicotinoid insecticides (For example, dinotefuran, thiamethoxam) and heavy metals (e.g., cadmium, zinc, copper) addition in the soil ecosystem causes serious environmental pollution across the globe (Goulson 2013). Recent studies analyzes the mutual toxic impacts of neonicotinoid insecticides and heavy metals on *Eisenia fetida.* This earthworm species is proved to be a very sensitive bioindicator species against the impact of the mutual pollutants. The development of ROS, cellular and DNA damage, deformed midgut cell lining, and disturbed MDA activity are some of the known biomarkers response of *Eisenia fetida* toward the mutual toxic impact of these neonicotinoid insecticides and heavy metals (Yan et al. 2020).

23.4.8 Sunfentrazone

Modern agricultural practices use some specific herbicides. The Sunfentrazone is one of the herbicides that have a wide range of applicability in modern agriculture (Gehrke et al. 2020). Studies reported the toxic effects of this herbicide on some of the aquatic organisms while amphibians also develop abnormalities toward its toxicity (Giesy et al. 2000; Graymore et al. 2001; Mann et al. 2009). Recently, researchers analyze its toxic potential by using *Eisenia fetida* as a model organism for its environmental biomonitoring. Different concentrations ranging from 0.2 to 5.0 mg/kg of the Sunfentrazone have been prepared and the earthworms are exposed to this herbicide in the soil under set laboratory protocols. During the study, the researchers observed the generation of reactive oxygen species in the earthworm's body which was one of the well-established biomarkers of this species toward soil pollutants exposure. Various other biomarkers such as GST, catalase, SOD, guaiacol peroxidase altered activities, and DNA damage are the prominent biomarkers that are highly useful for Sunfentrazone biomonitoring and its associated environmental impacts on soil health (Li et al. 2020). Table 23.2 represents the common, trade name, IUPAC nomenclature, molecular formula, and chemical structures of several soil pollutants.

23.5 Conclusion

It was observed that numerous earthworm species engaged with the biomonitoring and early warning of the soil pollutants. In recent years, the study of earthworm biomarkers proved their utility in contamination detection in terrestrial environments. The DNA damage, anomalous enzymatic functioning, heat shock proteins expression, MTs expression, and so forth are observed to be the prominent biomarkers that help in providing a scientific understanding of earthworm's biomarkers response toward soil pollutants exposure. This article proved to be beneficial for the development and promotion of

Table 23.2 List of common, trade, IUPAC nomenclature, molecular formula, and chemical structures of several soil pollutants (Adapted after: El-Gendy et al. 2020)

Soil Pollutants Class	Pollutants	Trade Name	Action	IUPAC Nomenclature	Molecular Formula	Chemical Structures
Carbamate Insecticide	Methiocarb	Metacil	Activity against insects, rodents, snails, and birds	4-(Dimethylamino)-3-methyl phenyl Nmethylcarbamate	$C_{11}H_{15}NO_2S$	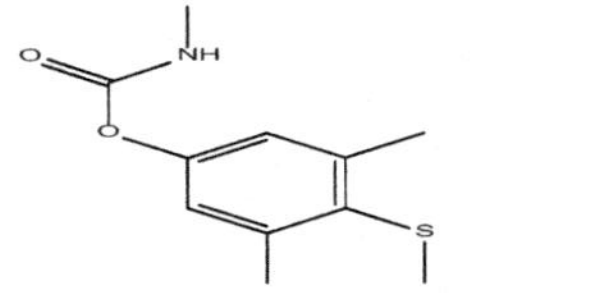
Systemic Insecticide	Imidacloprid	Premise 75, Confidor Provado Admire	Activity against cane beetles, locusts, stink bugs, termites, fleas, aphids, etc.	*N*-{1-[(6-Chloro-3-pyridyl)methyl]-4,5-dihydroimidazol-2-yl} nitramide	$C_9H_{10}ClN_5O_2$	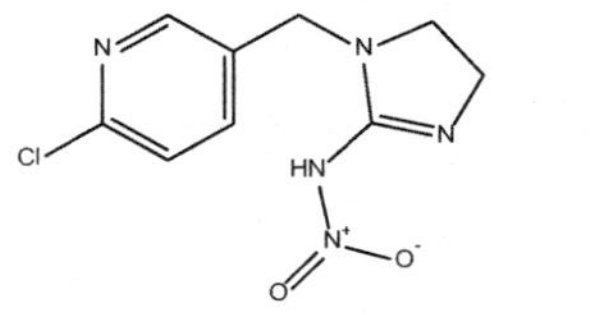
Carbamate insecticide	Aldicarb	Temik, ENT 27093, OMS 771, and UC 21149	Activity against Lygus, cane beetles, locusts, spider mites, stink bugs, termites, fleas, aphids, flea hoppers, etc.	2-Methyl-2-(methylthio) propanal *O*-(*N*-methyl carbamoyl)oxime	$C_7H_{14}N_2O_2S$	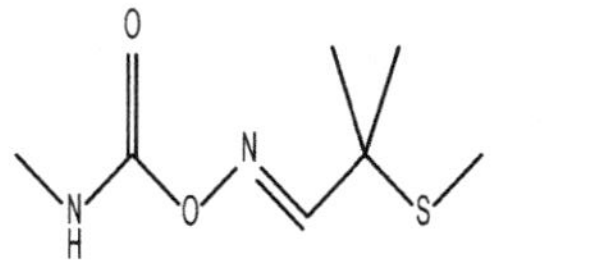
Organochlorine, benzoylurea, Organofluorine insecticide	Chlorfluazuron	Atabron	Inhibit the synthesis of chitin in insects Helps in control of insects such as *Lepidoptera*	*N*-[[3,5-dichloro-4-[3-chloro-5-(trifluoromethyl) pyridin-2-yl] oxyphenyl]carbamoyl]-2,6-difluorobenzamide	$C_{20}H_9Cl_3F_5N_3O_3$	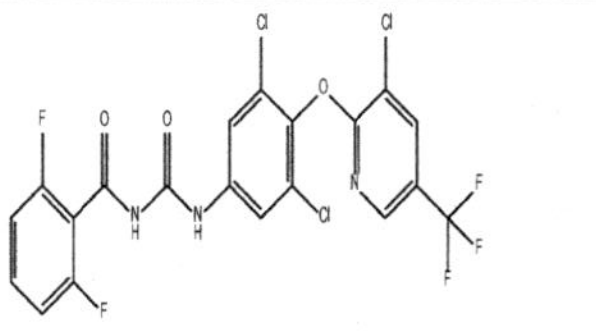
Synthetic pyrethroid pesticide	Cypermethrin	Auzar 25 EC	Kills house termites, cockroaches, etc.	[Cyano-(3-phenoxyphenyl)methyl]3-(2,2-dichloroethenyl)-2,2-dimethylcyclopropane-1-carboxylate	$C_{22}H_{19}Cl_2NO_3$	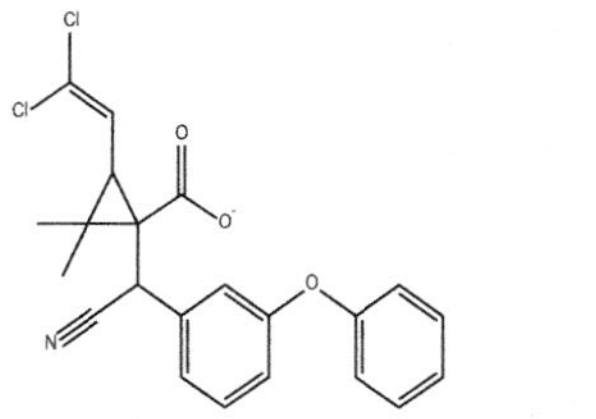

(continued)

Table 23.2 (continued)

Soil Pollutants Class	Pollutants	Trade Name	Action	IUPAC Nomenclature	Molecular Formula	Chemical Structures
Systemic, phenylamide fungicide	Metalaxyl	Mefenoxam	Control on *Phytophthora infestans*	methyl 2-[(2,6-dimethylphenyl)(methoxyacetyl)amino] propanoate	$C_{15}H_{21}NO_4$	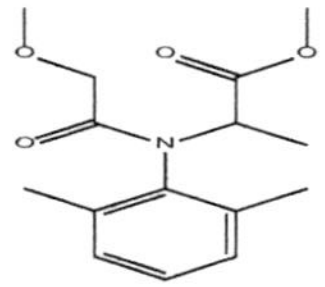
Triazines	Atrazine	Solaro	Inhibit the normal functioning of photosynthesis in the broadleaf weeds by altering the mechanism of photosynthetic electron transport	1-Chloro-3-ethylamino-5-isopropylamino-2,4,6-triazine	$C_8H_{14}ClN_5$	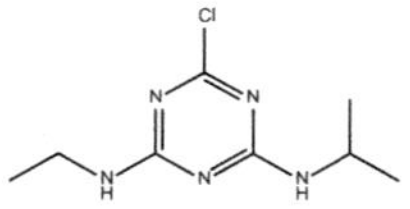
Antibiotic	ciprofloxacin	Cipro, Cipro XR, and ProQuin XR	Activity against bacterial infections	1-cyclopropyl-6-fluoro-4-oxo-7-piperazin-1-ylquinoline-3-carboxylic acid	$C_{17}H_{18}FN_3O_3$	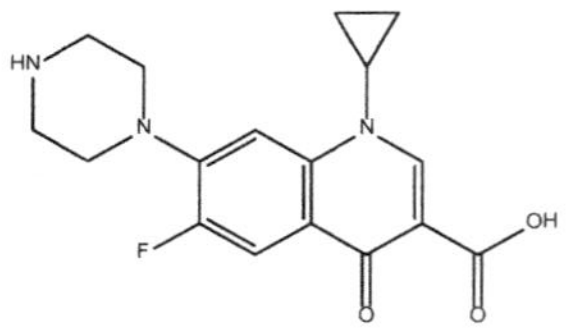
Aromatic fungicide	Thifluzamide	Pulsor	Control *Rhizoctonia*	*N*-[2,6-dibromo-4-(trifluoromethoxy) phenyl]-2-methyl-4-(trifluoromethyl)-1,3-thiazole-5-carboxamide	$C_{13}H_6Br_2F_6N_2O_2S$	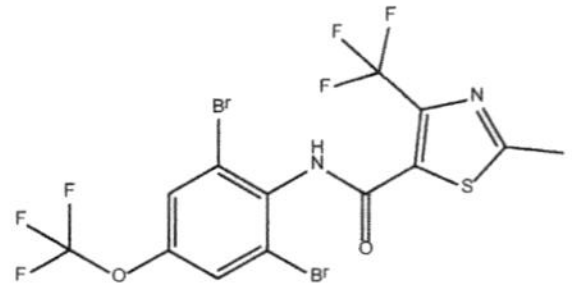
Neonicotinoid	*Dinotefuran*	Oshin 20 SG	Control insects like mole cricket, sawflies, lace bugs, thrips, etc.	2-methyl-1-nitro-3-[(tetrahydro-3-furanyl) methyl] guanidine	$C_7H_{14}N_4O_3$	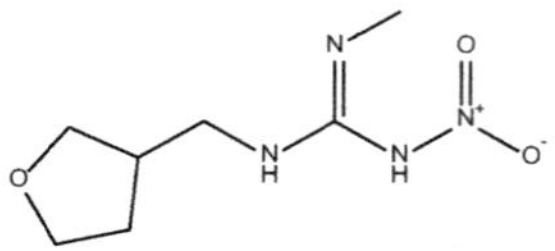

(continued)

Table 23.2 (continued)

Soil Pollutants Class	Pollutants	Trade Name	Action	IUPAC Nomenclature	Molecular Formula	Chemical Structures
Neonicotinoid	Thiamethoxam	Evident 25% WG	Control leaf and soil-dwelling pests.	3-[(2-Chloro-1,3-thiazol-5-yl)methyl]-5-methyl-*N*-nitro-1,3,5-oxadiazinan-4-imine	$C_8H_{10}ClN_5O_3S$	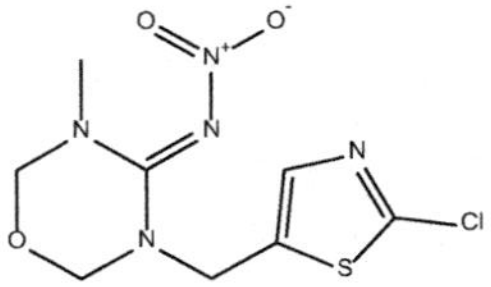
Herbicide	24 Sulfentrazone	Acetochlor	Control weeds, sedges, etc.	*N*-{2,4-Dichloro-5-[4-(difluoromethyl)-3-methyl-5-oxo-4,5-dihydro-1*H*-1,2,4-triazol-1-yl]phenyl} methanesulfonamide	$C_{11}H_{10}Cl_2F_2N_4O_3S$	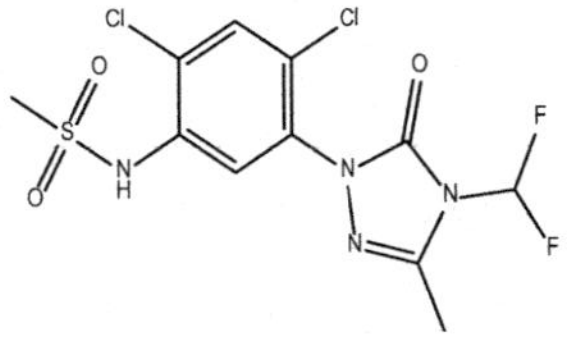

the earthworm-based biosensing approach for soil pollution assessment.

Acknowledgements The authors are grateful to the Central University of Himachal Pradesh and the Central University of Haryana for providing the necessary facilities for the writing of this chapter.

References

Ali A, Naaz I (2013) Earthworm biomarkers: The new tools of environmental impact assessment. Biosci Biotech Res Comm 6:163–169

Ashraf MA, Maah MJ, Yusoff I (2014) Soil contamination, risk assessment and remediation. Environ Risk Assess Soil Contam 1:3–56

Åslund MLW, Simpson AJ, Simpson MJ (2011) 1 H NMR metabolomics of earthworm responses to polychlorinated biphenyl (PCB) exposure in soil. Ecotoxicology 20:836–846

Bart S, Amossé J, Lowe CN et al (2018) Aporrectodea caliginosa, a relevant earthworm species for a posteriori pesticide risk assessment: current knowledge and recommendations for culture and experimental design. Environ Sci Pollut Res 25:33867–33881

Bharathi C, Rao BS (1986) Toxic effects of two organophosphate insecticides, monocrotophos and dichlorvos to common earthworm Lampito mauritii (Kinberg). Water Air Soil Pollut 28:127–130

Bünemann EK, Bongiorno G, Bai Z et al (2018) Soil quality–a critical review. Soil Biol Biochem 120:105–125

Burger J (2006) Bioindicators: a review of their use in the environmental literature 1970–2005. Environ Bioindic 1:136–144

Byrde RJW, Richmond DV (1976) A review of the selectivity of fungicides in agriculture and horticulture. Pestic Sci 7:372–378

Calisi A, Lionetto MG, De Lorenzis E, et al (2014) Metallothionein induction in the coelomic fluid of the earthworm Lumbricus terrestris following heavy metal exposure: a short report. Biomed Res Int

Calisi A, Lionetto MG, Schettino T (2011) Biomarker response in the earthworm Lumbricus terrestris exposed to chemical pollutants. Sci Total Environ 409:4456–4464

Colacevich A, Sierra MJ, Borghini F et al (2011) Oxidative stress in earthworms short-and long-term exposed to highly Hg-contaminated soils. J Hazard Mater 194:135–143

Cortet J, Gomot-De Vauflery A, Poinsot-Balaguer N et al (1999) The use of invertebrate soil fauna in monitoring pollutant effects. Eur J Soil Biol 35:115–134

Cycoń M, Mrozik A, Piotrowska-Seget Z (2019) Antibiotics in the soil environment—degradation and their impact on microbial activity and diversity. Front Microbiol 10:338

Datta S, Singh J, Singh S, Singh J (2016) Earthworms, pesticides and sustainable agriculture: a review. Environ Sci Pollut Res 23:8227–8243

De Silva PMCS, van Gestel CAM (2009) Comparative sensitivity of Eisenia andrei and Perionyx excavatus in earthworm avoidance tests using two soil types in the tropics. Chemosphere 77:1609–1613

Drewes CD, Zoran MJ, Callahan CA (1987) Sublethal neurotoxic effects of the fungicide benomyl on earthworms (Eisenia fetida). Pestic Sci 19:197–208

Duan X, Fu X, Song J et al (2017) Physiological and molecular responses of the earthworm Eisenia fetida to polychlorinated biphenyl contamination in soil. Environ Sci Pollut Res 24:18096–18105

Dureja P, Patra D, Johnson S, Tomar SS (1999) Effect of agrochemicals on earthworms. Toxicol Environ Chem 71:397–404

Dureja P, Tanwar RS (2012) Pesticide residues in soil invertebrates. Pestic Eval Environ Pollution; Hamir, SR, Leo, MLN, Eds 337–359

El-Gendy KS, Gad AF, Radwan MA (2020) Physiological and behavioral responses of land molluscs as biomarkers for pollution impact assessment: a review. Environ Res 110558

Ellis SR, Hodson ME, Wege P (2010) The soil-dwelling earthworm Allolobophora chlorotica modifies its burrowing behaviour in response to carbendazim applications. Ecotoxicol Environ Saf 73:1424–1428

Etteieb S, Tarhouni J, Isoda H (2019) Cellular stress response biomarkers for toxicity potential assessment of treated wastewater complex mixtures. Water Environ J 33:4–13

Gehrke VR, Camargo ER, Avila LA (2020) Sulfentrazone: environmental dynamics and selectivity. Planta Daninha 38

Giesy JP, Dobson S, Solomon KR (2000) Ecotoxicological risk assessment for Roundup® herbicide. In: Reviews of environmental contamination and toxicology. Springer, pp 35–120

Gill HK, Garg H (2014) Pesticide: environmental impacts and management strategies. Pestic Asp 8:187

Giovanetti A, Fesenko S, Cozzella ML et al (2010) Bioaccumulation and biological effects in the earthworm Eisenia fetida exposed to natural and depleted uranium. J Environ Radioact 101:509–516

Goulson D (2013) An overview of the environmental risks posed by neonicotinoid insecticides. J Appl Ecol 50:977–987

Graymore M, Stagnitti F, Allinson G (2001) Impacts of atrazine in aquatic ecosystems. Environ Int 26:483–495

Gupta RC (2011) Toxicology of organophosphate and carbamate compounds. Academic Press

Haeba M, Kuta J, Arhouma ZK, Elwerfalli HMA (2013) Earthworm as bioindicator of soil pollution around Benghazi City, Libya. J Env Anal Toxicol 3

Hentati O, Lachhab R, Ayadi M, Ksibi M (2013) Toxicity assessment for petroleum-contaminated soil using terrestrial invertebrates and plant bioassays. Environ Monit Assess 185:2989–2998

Hernández-Castellanos B, Ortíz-Ceballos A, Martínez-Hernández S et al (2013) Removal of benzo (a) pyrene from soil using an endogeic earthworm Pontoscolex corethrurus (). Appl Soil Ecol 70:62–69
Hirano T, Tamae K (2010) Heavy metal-induced oxidative DNA damage in earthworms: a review. Appl Environ Soil Sci
Hirano T, Tamae K (2011) Earthworms and soil pollutants. Sensors 11:11157–11167
Jager T, Fleuren RHLJ, Hogendoorn EA, De Korte G (2003) Elucidating the routes of exposure for organic chemicals in the earthworm, Eisenia andrei (Oligochaeta). Environ Sci Technol 37:3399–3404
Jiang X, Chang Y, Zhang T et al (2020) Toxicological effects of polystyrene microplastics on earthworm (Eisenia fetida). Environ Pollut 259:113896
Kavitha V, Anandhan R, Alharbi NS et al (2020) Impact of pesticide monocrotophos on microbial populations and histology of intestine in the Indian earthworm Lampito mauritii (Kinberg). Microb Pathog 139:103893
Kilty CG, Keenan J, Shaw M (2007) Histologically defined biomarkers in toxicology. Expert Opin Drug Saf 6:207–215
Klobučar GIV, Štambuk A, Šrut M et al (2011) Aporrectodea caliginosa, a suitable earthworm species for field based genotoxicity assessment? Environ Pollut 159:841–849
Knoepp JD, Vose JM, Michael JL, Reynolds BC (2012) Imidacloprid movement in soils and impacts on soil microarthropods in southern Appalachian eastern hemlock stands. J Environ Qual 41:469–478
Kraemer SA, Ramachandran A, Perron GG (2019) Antibiotic pollution in the environment: from microbial ecology to public policy. Microorganisms 7:180
Kumar M, Xiong X, He M et al (2020) Microplastics as pollutants in agricultural soils. Environ Pollut 265:14980
Lahive E, Walton A, Horton AA et al (2019) Microplastic particles reduce reproduction in the terrestrial worm Enchytraeus crypticus in a soil exposure. Environ Pollut 255:113174
Landrum M, Canas JE, Coimbatore G et al (2006) Effects of perchlorate on earthworm (Eisenia fetida) survival and reproductive success. Sci Total Environ 363:237–244
Larsson DGJ (2014) Antibiotics in the environment. Ups J Med Sci 119:108–112
Li M, Ma X, Saleem M et al (2020) Biochemical response, histopathological change and DNA damage in earthworm (Eisenia fetida) exposed to sulfentrazone herbicide. Ecol Indic 115:106465
Lionetto MG, Caricato R, Giordano ME (2019) Pollution biomarkers in environmental and human biomonitoring. Open Biomark J 9
Lister LJ, Svendsen C, Wright J et al (2011) Modelling the joint effects of a metal and a pesticide on reproduction and toxicokinetics in Lumbricid earthworms. Environ Int 37:663–670
Lukkari T, Taavitsainen M, Soimasuo M et al (2004) Biomarker responses of the earthworm Aporrectodea tuberculata to copper and zinc exposure: differences between populations with and without earlier metal exposure. Environ Pollut 129:377–386
Mahmood I, Imadi SR, Shazadi K, et al (2016) Effects of pesticides on environment. In: Plant, soil and microbes. Springer, pp 253–269
Mammo FK, Amoah ID, Gani KM, et al (2020) Microplastics in the environment: interactions with microbes and chemical contaminants. Sci Total Environ 140518
Mann RM, Hyne RV, Choung CB, Wilson SP (2009) Amphibians and agricultural chemicals: review of the risks in a complex environment. Environ Pollut 157:2903–2927
Manyi-Loh C, Mamphweli S, Meyer E, Okoh A (2018) Antibiotic use in agriculture and its consequential resistance in environmental sources: potential public health implications. Molecules 23:795
Miglani R, Bisht SS (2019) World of earthworms with pesticides and insecticides. Interdiscip Toxicol 12:71–82
Mosleh YY, Ismail SMM, Ahmed MT, Ahmed YM (2003) Comparative toxicity and biochemical responses of certain pesticides to the mature earthworm Aporrectodea caliginosa under laboratory conditions. Environ Toxicol Int J 18:338–346
Muangphra P, Sengsai S, Gooneratne R (2015) Earthworm biomarker responses on exposure to commercial cypermethrin. Environ Toxicol 30:597–606
Natal-da-Luz T, Ojeda G, Costa M et al (2011) Short-term changes of metal availability in soil. II: The influence of earthworm activity. Appl Soil Ecol 49:178–186
Panda S, Sahu SK (2004) Recovery of acetylcholine esterase activity of Drawida willsi (Oligochaeta) following application of three pesticides to soil. Chemosphere 55:283–290
Parmar TK, Rawtani D, Agrawal YK (2016) Bioindicators: the natural indicator of environmental pollution. Front Life Sci 9:110–118
Pérès G, Vandenbulcke F, Guernion M et al (2011) Earthworm indicators as tools for soil monitoring, characterization and risk assessment. An example from the national Bioindicator programme (France). Pedobiologia (jena) 54:S77–S87
Rao JV, Kavitha P, Rao AP (2003) Comparative toxicity of tetra ethyl lead and lead oxide to earthworms, Eisenia fetida (Savigny). Environ Res 92:271–276
Reddy PB (2012) Histopathogical studies as potential and direct biomarkers of pollution. Trends Life Sci 1:27–31
Salvio C, Menone ML, Rafael S et al (2016) Survival, reproduction, avoidance behavior and oxidative stress biomarkers in the earthworm Octolasion cyaneum exposed to glyphosate. Bull Environ Contam Toxicol 96:314–319
Samal S, Mishra CSK, Sahoo S (2019) Setal-epidermal, muscular and enzymatic anomalies induced by certain

agrochemicals in the earthworm Eudrilus eugeniae (Kinberg). Environ Sci Pollut Res 26:8039–8049
Scott-Fordsmand JJ, Weeks JM (2000) Biomarkers in earthworms. In: Reviews of environmental contamination and toxicology. Springer, pp 117–159
Sinkakarimi MH, Solgi E, Colagar AH (2020) Interspecific differences in toxicological response and subcellular partitioning of cadmium and lead in three earthworm species. Chemosphere 238:124595
Smith M, Love DC, Rochman CM, Neff RA (2018) Microplastics in seafood and the implications for human health. Curr Environ Heal Reports 5:375–386
Sorour J, Larink O (2001) Toxic effects of benomyl on the ultrastructure during spermatogenesis of the earthworm Eisenia fetida. Ecotoxicol Environ Saf 50:180–188
Spurgeon DJ, Hopkin SP (1996) Effects of Metal-Contaminated Soils on the Growth, Sexual Development, and Early Cocoon Production of the EarthwormEisenia fetida, with Particular Reference to Zinc. Ecotoxicol Environ Saf 35:86–95
Stringer A, Wright MA (1976) The toxicity of benomyl and some related 2-substituted benzimidazoles to the earthworm Lumbricus terrestris. Pestic Sci 7:459–464
Subaraja M, Vanisree AJ (2015) Cerebral ganglionic variations and movement behaviors of Lumbricus terrestris on exposure to neurotoxin. Ann Neurosci 22:199
Urionabarrenetxea E, Garcia-Velasco N, Marigómez I, Soto M (2020) Effects of elevated temperatures and cadmium exposure on stress biomarkers at different biological complexity levels in Eisenia fetida earthworms. Comp Biochem Physiol Part C Toxicol Pharmacol 231:108735
Usmani Z, Kumar V (2015) Earthvvorms:"The Unheralded Soldiers" standing steadfast against". Res J Environ Sci 9:48–65
Van der Ploeg MJC, Baveco JM, Van der Hout A et al (2011) Effects of C60 nanoparticle exposure on earthworms (Lumbricus rubellus) and implications for population dynamics. Environ Pollut 159:198–203
Van Gestel CAM, Van Dis WA, Van Breemen EM, Sparenburg PM (1989) Development of a standardized reproduction toxicity test with the earthworm species Eisenia fetida andrei using copper, pentachlorophenol, and 2, 4-dichloroaniline. Ecotoxicol Environ Saf 18:305–312
Vasseur P, Bonnard M (2014) Ecogenotoxicology in earthworms: a review. Curr Zool 60:255–272
Vijver MG, Vink JPM, Miermans CJH, van Gestel CAM (2003) Oral sealing using glue: a new method to distinguish between intestinal and dermal uptake of metals in earthworms. Soil Biol Biochem 35:125–132
Wallwork JA (1983) Annelids: the first coelomates. Edward Arnold, London
Wang J, Coffin S, Sun C et al (2019) Negligible effects of microplastics on animal fitness and HOC bioaccumulation in earthworm Eisenia fetida in soil. Environ Pollut 249:776–784
Wang K, Pang S, Mu X et al (2015) Biological response of earthworm, Eisenia fetida, to five neonicotinoid insecticides. Chemosphere 132:120–126
Wu S, Zhang H, Hu Y et al (2012) Effects of 1, 2, 4-trichlorobenzene on the enzyme activities and ultrastructure of earthworm Eisenia fetida. Ecotoxicol Environ Saf 76:175–181
Xu X, Shi Y, Lu Y et al (2015) Growth inhibition and altered gene transcript levels in earthworms (Eisenia fetida) exposed to 2, 2′, 4, 4′-tetrabromodiphenyl ether. Arch Environ Contam Toxicol 69:1–7
Yadav J, Singh D, Yadav J, Kumar D (2017) Organophosphates and carbamates as inhibitors of acetylcholinesterase in Eisenia fetida. Pollut Res 36:277–281
Yan X, Wang J, Zhu L et al (2020) Oxidative stress, growth inhibition, and DNA damage in earthworms induced by the combined pollution of typical neonicotinoid insecticides and heavy metals. Sci Total Environ 754:141873
Yang T, Siddique KHM, Liu K (2020a) Cropping systems in agriculture and their impact on soil health-A review. Glob Ecol Conserv 23:e01118
Yang X, Li Y, Wang X (2020b) Effects of ciprofloxacin exposure on the earthworm Eisenia fetida. Environ Pollut 262:114287
Yang Y, Liu W, Li D et al (2017) Altered glycometabolism in zebrafish exposed to thifluzamide. Chemosphere 183:89–96
Yang Y, Liu W, Mu X et al (2016) Biological response of zebrafish embryos after short-term exposure to thifluzamide. Sci Rep 6:38485
Yao X, Zhang F, Qiao Z et al (2020) Toxicity of thifluzamide in earthworm (Eisenia fetida). Ecotoxicol Environ Saf 188:109880
Zang Y, Zhong Y, Luo Y, Kong ZM (2000) Genotoxicity of two novel pesticides for the earthworm, Eisenia fetida. Environ Pollut 108:271–278
Zoran MJ, Heppner TJ, Drewes CD (1986) Teratogenic effects of the fungicide benomyl on posterior segmental regeneration in the earthworm, Eisenia fetida. Pestic Sci 17:641–652

24

Electrokinetic-Assisted Bioremediation and Phytoremediation for the Treatment of Polluted Soil

Tarun Kumar Kumawat, Vishnu Sharma, Varsha Kumawat, Manish Biyani, Nalinee Kumari, Rashi Garg, and Nirat Kandwani

Abstract

The contamination of soil from heavy metals (HMs), petroleum hydrocarbons (PHCs), and pesticides have become a serious environmental problem in the current world. The pollution has resulted from anthropogenic activities, rapid industrialization, and urbanization. Pesticides are used extensively in farming activities to meet the increasing demand for food and feed. The pollutants change the physicochemical and microbiological characteristics of soil and have mutagenic, carcinogenic, immunotoxic, and teratogenic effects on human health. There is an urgent necessity for sustainable and eco-friendly remediation technologies for the elimination of contaminants from soil. Electrokinetic-assisted remediation (EKR) is an opportune technology for complete remediation of polluted soil including fine-grained soils, which are typically difficult to clean-up using traditional bioremediation and phytoremediation approaches because of several drawbacks. Electrokinetic-Assisted Bioremediation (EKBR) and Electrokinetic-Assisted Phytoremediation (EKPR) are novel and effective technologies for soil remediation which decontaminate heavy metal, remove PHCs and pesticides from polluted soils. This chapter emphasizes electrokinetic-assisted remediation, current development, process, field applications, advantages, disadvantages, and further prospects.

Keywords

Bioremediation · Electrokinetic-assisted remediation · Phytoremediation · Pollutants · Sustainable

T. K. Kumawat (✉) · V. Sharma · R. Garg · N. Kandwani
Department of Biotechnology, Biyani Girls College, Jaipur, Rajasthan 302039, India

V. Kumawat
Naturilk Organic and Dairy Foods Pvt. Ltd, Jaipur, Rajasthan 302012, India

M. Biyani
Department of Bioscience and Biotechnology, Japan Advanced Institute of Science and Technology, Ishikawa, Japan

N. Kumari
Department of Zoology, DPG Degree College, Gurugram, Haryana 122001, India

24.1 Introduction

Soil and sediment pollution is a geo-environmental problem that negatively affects the environment (Amundson et al. 2015; Gomiero 2016; Xu et al. 2019). Anthropogenic activities are causatives of these problems that negatively affect the geo-environment (Gill et al. 2014). Numerous anthropogenic roots of pollutants from farming, mining, smelting,

J. A. Malik (ed.), *Advances in Bioremediation and Phytoremediation for Sustainable Soil Management*,
https://doi.org/10.1007/978-3-030-89984-4_24

electroplating, and other industrial movements are in continuation all over the world that are producing unusual depositions of unwanted quantities of pollutants including petroleum hydrocarbons, polynuclear aromatic, solvents, pesticides, and toxic heavy metals in soil (Moosavi and Seghatoleslami 2013; Jamari et al. 2014; Kanianska 2016; Tuomisto et al. 2017; Ekta and Modi 2018). Several remediation strategies such as bioremediation (Sturman et al. 1995; Gill et al. 2014; Azubuike et al. 2016) and phytoremediation (Arthur et al. 2005; Mosa et al. 2016; Feng et al. 2017; Lajayer et al. 2019) have been applied over the years to mitigate soil contamination with differing degrees of effectiveness. Plant species hold the potential to eliminate/degrade or metabolize a broad range of contaminants via phytoextraction, phytorestoration, phytostabilization, phytodegradation, phytovolatilization, or rhizofiltration (Etim 2012; Martin et al. 2014; Sasse et al. 2018; Rai et al. 2020; Dhaliwal et al. 2020).

Natural attenuation (NA) treatment, biostimulation, and site-specific bio-augmentation have resulted in very low removal/degradation of soil pollutants (Crognale et al. 2020). In recent years, innovations for the remediation of environmental pollutants from soil have gained substantial attention. Amongst them, Electrokinetic Remediation (EKR) is sustainable technology to remove heavy metals, salts, radioactive elements, and organic pollutant from fine-grained and low-permeability soil due to their environmental compatibility, and cost-effectiveness (Klouche et al. 2020a; Pham and Sillanpaa 2020). EKR is an in situ process, so for decontamination, there is no need for soil excavation (De Battisti and Ferro 2007).

Several enhanced electrokinetic remediation technologies have been applied so far, to increase the efficacy of pollutant removal from soil such as Chelating Agent-Enhanced Remediation (Yang et al. 2020), Biosurfactant-Enhanced Electrokinetic Remediation (Tang et al. 2020), Bio-electrokinetic (BEK) Remediation (Sarankumar et al. 2020), Permeable Reactive Barrier (PRB) (Zhao et al. 2016; Yao et al. 2020), Microbial Fuel Cell (MFC)-Enhanced Remediation (Gustave et al. 2020). The current studies on electrokinetic remediation mainly focus on electrokinetic remediation of inorganic and organic pollutants from soil. This chapter emphasizes soil pollutants, electrokinetic-assisted remediation, current development, process, energy consumption, and field applications.

24.2 Soil Pollutants and Pollution

Soil contamination with inorganic substances, including radioactive elements heavy metals, and salts, and organic pollutants, poses threats to human and environment, which in recent years have attracted widespread attention (Sorengard et al. 2020; Wen et al. 2021). The expansion of urbanization and industrial activity has exacerbated significant environmental issues, such as soil contamination, over the last decade (Gnanasundar and Akshai 2020).

24.2.1 Inorganic Contaminants

Soil polluted with inorganic contaminants including radioactive elements, heavy metals, and salts due to certain imbalances and unstoppable anthropogenic processes, such as industrialization, urbanization, and incorrect farming practices pose threats to human health and ecological climate, which in recent years have attracted widespread attention (Singh et al. 2020). The pollution of heavy metals in soil is one of the serious problems and has a huge impact on the environment (Dhaliwal et al. 2020). Usually, heavy metals are found as cations or as retained on soil particles with organic or inorganic bonds. These are responsible for many widespread poisoning activities (Wuana and Okieimen 2011; Tchounwou et al. 2012; Jaishankar et al. 2014; Mao et al. 2016; Palansooriya et al. 2020)."Heavy metals" are a group of elements with an atomic mass of >5 g/cm^3, or >5 times than water (Rajindiran et al. 2015). Lead and arsenic are the soil's major environmental contaminants, so the removal of this metal from the soil is essential in the context of ecological safety

(Selvi et al. 2019; Ait Ahmed 2020). Arsenic contamination in soil is a major problem nowadays and poisoning the human body through crops and vegetables (Shrivastava et al. 2015). With the enhancement of accumulation, heavy metals cause atherosclerosis, melanoma, Alzheimer's disease, Parkinson's disease, etc. (Bakulski et al. 2020). Over the past two decades, substantial research by scientists and experts has concentrated on discovering new ways to eliminate soil pollutants (Cercato and De Donno 2020).

Across the globe, radioactive element contamination of soil and sediments by anthropogenic activities is a major concern. The radioactive substance and waste were produced during the operation of nuclear reactors, uranium mining and milling, nuclear weapons program, nuclear weapons testing, fuel manufacturing units, fuel reprocessing plants, research laboratories working on radionuclides, radioisotope in medicine and industry, accidents and disasters. Huge quantities are produced by coal-fired power plants, which also contained radionuclides elements (Hu et al. 2010; Sharma et al. 2014). Radionuclide-contaminated soils, particularly ^{137}Cs. ^{238}U, ^{239}Pu, and ^{90}Sr, pose a long-term radiation threat to the health of human through exposure via the food chain and other routes (Zhu and Shaw 2000). The primary path of internal radionuclide ingestion in humans is the consumption of food goods tainted with radionuclides (Shaw and Bell 1994).

24.2.2 Organic Contaminants

Soil is a complex environment that supports human activities and ecosystems across a large variety of functions (Upcraft and Guo 2020). The natural ecosystem and public health have been negatively impacted by organic pollutants of soil (Ojuederie and Babalola 2017). The major organic pollutants are Polybrominated Biphenyls (PBBs), Polycyclic Aromatic Hydrocarbons (PAHs), Polychlorinated Biphenyls (PCBs), Polychlorinated dibenzofurans (PCDFs), Polychlorinated dibenzodioxins (PCDDs), organophosphorus and carbamate insecticides (Pesticides), herbicides, fertilizers, and other agriculture product and organic fuels (Gasoline and Diesel).

Organic contaminants contain many insecticides and herbicides that have been used in farming and weed and pest management to fulfill the growing demand for food and feed (Schell et al. 2012; Boudh and Singh 2019; Rajendran et al. 2021). Humans can be affected to pesticides via inhaling soil particles, ingesting soil, and dermal touch (Li 2018). Polychlorinated biphenyls (PCBs) are organic pollutants with hydrophobic properties that inhibit the metabolic process (Burca and Watson 2014). Among the current environmental issues, soil contamination of petroleum (Total Petroleum Hydrocarbon; TPH) is one of the most severe soil pollution problems. The occurrence of petroleum hydrocarbons pollutants in the soil causes major environmental effects and poses a significant risk to humans (Khan et al. 2018a). Petroleum hydrocarbons and their derivatives adversely affect both the environment and human health (Varjani and Upasani 2017; Huang et al. 2019). Dioxin {Polychlorinated dibenzodioxins (PCDDs) and Polychlorinated dibenzofurans (PCDFs)} is an environmental pollutant that is a byproduct of the processes of paper bleaching, herbicide/pesticide production, and incineration of solid/hospital waste (Kimbel et al. 2019; Tu et al. 2021).

24.3 Need for Remediation of Soil Pollutants

Soil is an essential environmental factor that constitutes the ecosystem for the life and growth of human beings (Zhao et al. 2016). The inorganic and organic pollutants in the soil and sediment became very serious worldwide. Such polluted areas are increasing day by day in various countries. There are more than 20 million hectares of land globally polluted by heavy metal (loid)s (Liu et al. 2018). Many of these substances are exceedingly persistent and accumulate beyond acceptable levels in the soil (Ahmad et al. 2017). The pollutants acidify and contaminated the soil and threatening the production of crops, food quality, environmental safety, and public health as well as sustainable expansion (Song et al. 2017). Biological, chemical, physical, and combined processes for remediation have been implemented in recent years to address the problems of contamination of soil and sediments (Khan et al. 2018b). In the majority of situations, the purpose of soil remediation activities is to reduce toxins to levels that are acceptable for usage and to ensure that we're using our land without environmental hazards (Acar and Alshawabkeh 1993).

The best approach to remediation of soil pollutants is the prevention of soil pollution. The soil remediation strategy selected for the polluted soil according to nature, potential hazard, soil characteristics, time, laboratory studies, and feasibility (Lombi and Hamon 2005; Daghan and Ozturk 2015). The remediation of the pollutant from the soil is fundamental for the sustainable development and protection of ecosystems and biodiversity (Stojic et al. 2018). Substantial courtesy has been given to suitable technology for the remediation of harmful contaminants from the land. Among them, electrokinetic (EK) remediation is highlighted because of its versatility and amenability (Andrade and dos Santos 2020). Electrokinetic phenomena (electrosmosis, electrophoresis, electrolysis) in which continuous electricity is produced for the elimination of inorganic and organic contaminants in the polluted soil (Llorente et al. 2014).

24.4 Electrokinetic Assisted Remediation (EKR)

Bioremediation and phytoremediation have been extensively used to improve soils, though it can face some limitations like the term of contaminants, time to be taking in processing (excavation or removal) of contaminants, availability of hyperaccumulator plants, etc. (Mosavat et al. 2012; Couto et al. 2015; Jamil et al. 2015). Electrokinetic (EK) remediation is a new technology for physicochemical remediation which relies on the application of a direct current of low intensity to boost contaminant mobilisation. Since the early 1800s, the concept of electrokinetic remediation has been hypothesized in the context of Electroosmosis. Electrokinetics (EK) uses a low electrical current put in the soil between an anode and a cathode (Fig. 24.1). It was conducted first by the F.F. Reuss in the year 1809 (Reuss 1809; Wall 2010; Biscombe 2017). Electrokinetic remediation is broadly applied to exclude metals, radionuclides, polar inorganic pollutants from soil (Lacatusu et al. 2013). The applied electric potential for EKR is greater than 1 V/cm and the power supply is over than 1 mA/cm^2 (Reddy et al. 2006; Yoo et al. 2015; Li et al. 2019).

In electrokinetic remediation process, the current passes between the electrodes into the soil, which causes several physical and chemical impressions like electrolysis, electromigration, electroosmosis, electrophoresis, electro-oxidation, pH fluctuations, water hydrolysis, etc. (Isosaari et al. 2007; Streche et al. 2018; Head et al. 2020). Numerous studies have reported that electrokinetic remediation is feasible to decontaminate complex toxic contaminants with low power consumption (Cong et al. 2005; Szpyrkowicz et al. 2007; Zheng et al. 2007; Truu et al. 2015; Acosta-Santoyo et al. 2017; Popescu et al. 2017; Meshalkin et al. 2020; Ajiboye et al. 2021).

24.4.1 Electrokinetic Assisted Bioremediation (EKBR)

Electrokinetic bioremediation is an effective technique that can dramatically increase the delivery of nutrients to natural microorganisms and thereby have a substantial opportunity to clean contaminated soils, such as fine-grained soils, which are usually difficult to clean up using conventional methods (Alshawabkeh 2009; Tahmasbian and Sinegani 2016; Karaca et al. 2019; Zhou et al. 2020). The combination of electrokinetic technology and bioremediation allows the absorption of toxins in the form of ions that are also bacterial activity inhibitors. Thus, it allows full remediation of the polluted

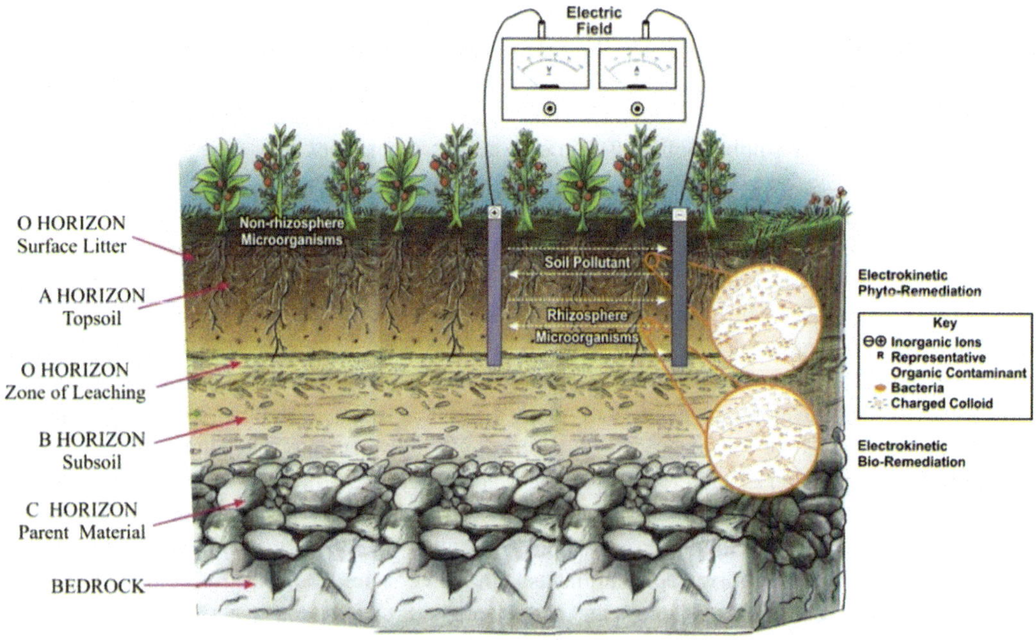

Fig. 24.1 Process of electrokinetic assisted bioremediation and phytoremediation

soil (Chilingar et al. 1997; Gill et al. 2014). The electric field is used in such a remediation procedure to improve the rate of degradation by extending the electrokinetics associated with the transfer of nutrients and adding new bacteria in the absence of indigenous microorganisms (Luo et al. 2005; Dzionek et al. 2016; Ottosen et al. 2019).

Here, a redox reaction involves the electrodes in the presence of bacteria and creates hydrogen ions and oxygen gas at the anode side and hydroxyl ions and hydrogen gas at the cathodes. Hydrogen ions from an acidic presence pass into the cathode via the influence of three processes, namely electroosmotic movement, diffusion, and electromigration. This approach lowers the soil's pH, producing an acidic environment. While hydroxide ions form a fundamental character and move by electromigration and diffusion toward the anode. In electrokinetic bioremediation, the pH of the soil also plays an important role in completing the process. However, bacterial survival and optimal degradation performance are influenced directly by pH (Hassan et al. 2018; Gidudu and Chirwa 2020a).

The oxygen ions can be transferred within the soil and can start an anaerobic biodegradation process because of the high porosity of the silty and sandy soils. While electrical flow often increases the temperature of the polluted soil at a high degree, it has an antagonistic effect on the microorganisms' survival (Virkutyte et al. 2002; Hassan et al. 2016). The cost of electrical power needed for electrokinetic is a big part of the total cost of the electrokinetic-remediation process, according to the literature. Energy expenditure, thus, raises the expense of the bioremediation process and results in the limitation of wide-ranging electrokinetic bioremediation applications (Li et al. 2017; Mao et al. 2019).

24.4.2 Electrokinetic Assisted Phytoremediation (EKPR)

There is an alliance of phytoextractor plants and an electrokinetics system to circumvent the restrictions of conventional phytoremediation for the elimination of both inorganic and organic contaminants from soils. It is termed as "Electrokinetic Assisted Phytoremediation". In this process, a low-voltage electric field (DC) is applied across polluted soil in surrounding area of rising plants to move soluble pollutants out of the soil (Fig. 24.2) (Acar and Alshawabkeh 1993; Virkutyte et al. 2002; O'Connor et al. 2003; Lageman et al. 2005; Sanchez et al. 2020; Siyar et al. 2020). However, little is known about the influence of Electrokinetic-assisted phytoremediation on the biological and physiological properties of soil (Cang et al. 2012). To prevent any harm to the developing plants and soil microflora, the voltage of the electric field and the chemical composition of the electrode shave must be carefully chosen. The electrokinetic assisted phytoremediation technology is capable of remediating soil with mixed contaminants under the proper conditions (Cameselle and Gouveia 2018).

During the electric transient time, hydrogen ions are revealed to accumulate around the anode electrode through water electrolysis. The hydrogen ions lower the pH of the soil around the anode and form an acid front, while the hydroxyl ions raise the pH that produces a base front in the vicinity of the cathode. As an outcome, pollutants are spread around the anode electrode in the acid state and ions transported from the anode to the cathode electrode (Thangavel and Subbhuraam 2004; Dermont et al. 2008). Three electrochemical processes also happen and assist to mobilize soluble pollutants (electroosmosis, electromigration, and electrophoresis) (Kim et al. 2002; Cameselle et al. 2013; Lima et al. 2017).

Amidst, electroosmosis occurs from electrolytic cell's anode to cathode for soil moisture or groundwater. In electromigration, ions and ion compounds are transported to opposite charge electrode. While in electrophoresis, charged particles, or colloid contaminants are embedded in a free state of an electric field and are transported out of the surface (Yeung 2006; Saeedi et al. 2013; Punia and Singh 2018; Ramadan et al. 2018; Klouche et al. 2020b). Usually, the active functioning of phytoremediation–electrokinetic coupled technology depends on the type of current supply, voltage parameters,

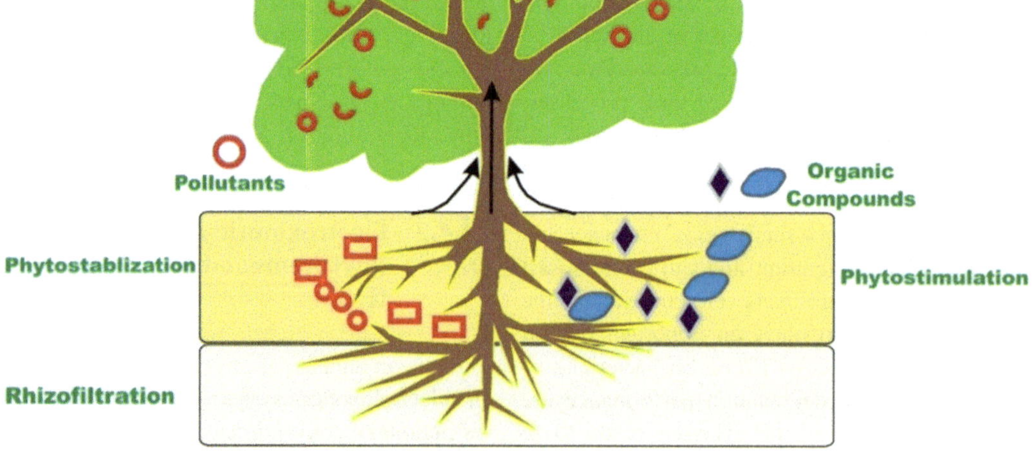

Fig. 24.2 Mechanism of enhanced electrokinetic phytoremediation

pattern of voltage use, pH of the soil, and the addition of promoting factors (Mao et al. 2016). Overall, in this hybrid technology, the plants take positions to eliminate or degrade the contaminants, whereas the electrical flows enhance the plant activity by increasing the bioavailability of pollutants (Hassan et al. 2018) (Table 24.1).

24.5 Source of Energy for Electrokinetic Remediation

In electrokinetic remediation of polluted soil, the electric field is indispensable (Wang et al. 2020). Scientists and researchers evaluated the various energy sources required for the electrokinetic remediation of soil impurities (Vocciante et al. 2016). Usually, an external current source (AC or DC) is actively employed in electrokinetics remediation. Either AC or DC systems with anode and cathode are installed inside the earth. Each electrode (anode and cathode) has a reservoir for the refilling of an ideal electrolytic solution (Kim and Han 2020).

In recent years, some innovations and advances for electrokinetic remediation have been made in the field of energy supplies to achieve better efficiency of pollutant removal from soil and cost-effectiveness. The electrokinetic remediation or process of heavy metals, pesticides, and other organic pollutants was studied on the following power/energy supply systems:

(a) Normal power supply
(b) Solar power supply and,
(c) Microbial fuel cells.

The remediation process can also be triggered by the non stabilised electric current produced by solar panels (Hassan et al. 2018). The solar-based

Table 24.1 Electrokinetic remediation technology for soil pollutant elimination

Electrokinetic technology	Soil pollutant	Observation	References
EKBR	Polycyclic Aromatic Hydrocarbon (PAH)	Removal of PAH (80%) using electrokinetic and *Sphingomonas* sp. L138 and *Mycobacterium frederiksbergense* LB501TG	Wick et al. (2004)
EKBR	Pentadecane	Removal of pentadecane (77.6%) at 0.63 mA/cm^2 after 14 days	Kim et al. (2005)
EKBR	Mercury	Removal of Mercury (78%) by *Lysinibacillus fusiformis* and electrokinetic technique (7 days; 50 V/m)	Azhar et al. (2016)
EKBR	Phenanthrene	Removal of Phenanthrene 65.1% at the anode and 49.9% at cathode using Phe-degrading *Sphingomonas* sp. GY2B	Lin et al. (2016)
EKPR	Zn, Pb, Cu, and Cd	Elimination of heavy metals from polluted soil using Potato plants	Aboughalma et al. (2008)
EKPR	Cd, Cu, Pb, and Zn	Removal of heavy metals using *Brassica juncea* after 40 days	Cang et al. (2011)
EKPR	Cd, Cu, Zn, and Pb	Metal uptake observed from *Brassica napus* and *Nicotiana tabacum* in electrical fields (AC and DC)	Bi et al. (2011)
EKPR	Heavy metals and PAHs	Removal of heavy metal and PAHs using Ryegrass (*Lolium perenne* L.) in AC electric fields	Acosta-Santoyo et al. (2017)
EKPR	Atrazine	Removal of atrazine using Ryegrass (*Lolium perenne* L.) with 0.6 V cm^{-1} DC electric field after 19 days	Sanchez et al. (2020)
EKPR	n-Hexadecane	Removal of n-Hexadecane using Ryegrass after 40 days	Wu et al. (2020b)

*EKPR: Electrokinetic-assisted phytoremediation; EKBR: Electrokinetic-assisted bioremediation

system's power consumption is only 50–55% of the DC-powered system (Jeon et al. 2015). Nowadays, it is also important to investigate the applicability of alternative electricity generation and the application of adequate and inexpensive power outputs for energy fields. In most cases, to produce the electric field in the soil for the mobilization and elimination of the pollutants, direct current (DC) is applied over the electrodes. Electrokinetic remediation driven by DC may lead to an eventual expenditure of electrical energy. Solar energy has the ability for electrokinetic remediation, which transforms sunlight into electricity, to avoid the downside of DC-driven systems. Many scientific papers have been released in the last 20 years on electrokinetic remediation of toxic inorganic pollutants powered through Microbial Fuel Cells (MFC). MFC is an inexpensive, environmentally sustainable, and revolutionary bio-electrochemical technology that transforms the chemical energy of waste matter into electrical energy using extracellular-respiring microbes (Logan and Regan 2006; Wu et al. 2020a). The electrokinetic removal of zinc (Zn) and cadmium (Cd) from polluted paddy soil with MFCs were studied (Chen et al. 2015). Zn (12 mg) and cadmium (0.7 mg) were substantially removed from the contaminated soil after 78 days. The efficacy of Cd and Pb removal using Microbial fuel cells (MFCs) was investigated by Habibul et al. (2016) and found that Cd and Pb in the soil mitigated from anode to cathode.

24.6 Electrokinetic Removal of Inorganic Pollutants

Toxic heavy metals and diverse forms of nutrients and salts contain inorganic contaminants that usually arise in the form of dissolved anions and cations (Goldscheider 2010). Heavy metals and metalloids are among the inorganic pollutants of primary concern due to high toxicity at low concentrations. In soil, heavy metals may be either bound to solid phases or readily used for absorption by organisms (Kumar et al. 2016). Electrokinetic remediation (EKR) has emerged as an optimistic and effective method that can be used to eradicate organic and inorganic pollutants from contaminated land (Kim et al. 2011). Among the prominent technologies developed so far for reclamation of heavy metal polluted soil. EKR has become an effective process, especially in soils with low hydraulic conductivity (Cameselle and Gouveia 2018; Beyrami et al. 2020).

Jeon et al. (2015) found that 32 and 27% of arsenic was eliminated from the soil of a former refinery plant located in Janghang, Chungnam, Republic of Korea, by the normal power supply (Direct Current; DC) and solar power supply, respectively. Cr(VI) was removed at 99.8% in 30 min from the soil of China by using photovoltaic solar panels and a DC-DC converter for electrokinetic remediation (Zhang et al. 2015). Hassan et al. (2015) worked on Two Anode Technique (TAT) using solar cells for remediation (electrokinetic) of Copper polluted soil and observed that 75% Cu was eliminated. The highest removal of Cu (92%) was observed near the anode (Table 24.2).

There are significant threats to the environment from the deposition of lead (Pb) in sediments from anthropogenic activities (Mao et al. 2019). Hussein and Alatabe (2019) researched solar energy for electro-kinetic remediation of Baghdad, Iraq's lead (Pb) contaminated soil, and reported that 90.7, 63.3, and 42.8% of lead elimination were accomplished for sandy, sandy loam, and silty loam soils, respectively. Shu et al. (2019) reported that the removal efficiency of manganese (94.74%) and ammonia nitrogen (88.20%) using Pulse Electric field (PE) were higher than Direct Current (DC).

It has become increasingly important to remediate radionuclide-contaminated soils. Traditional methods for remediation of radioactive elements are expensive and less suitable for large-area contamination (Yan et al. 2021). With the application of physicochemical procedures such as soil cleaning, soil flushing, and soil reclamation of polluted soil with radionuclides can be achieved; however, due to the long treatment period and the associated high costs, they do not succeed (Annamalai et al. 2014; Cameselle and Gouveia 2019). Electrokinetic

Table 24.2 Electrokinetic removal of heavy metals from soil

Heavy metal	Soil sample	Solution	Time duration	Removal efficiency (%)	References
Copper	Red Soil	Lactic acid + NaOH	900 h	81%	Zhou et al. (2004)
Arsenic	Arsenic Contaminated	0.1 M $MgSO_4$ 0.1 M HNO_3	28 days	68%	Baek et al. (2009)
Copper	Kaolin	$NaNO_3$, Citric acid-Sodium citrate buffer	04 days	96.60%	Zhao et al. (2016)
Cadmium Copper Nickle Lead Zinc	Kaolinite clay	Citric acid + Calcium chloride	72 h	98.19% 95.24% 98.95% 86.21% 99.01%	Yuan et al. (2016)
Chromium (Cr^{6+})	Industrial soil	The citric acid (CA) and Polyaspartic acid (PASP)	07 days	94.27% with CA and 93.26% with PASP	Fu et al. (2017)
Lead	Saline	Citric acid and EDTA	168 h	31.5%	Ait Ahmed (2020)

remediation of soil is an emerging decontamination technology for radionuclides (Ugaz et al. 1994). Valdovinos et al. (2016) reported electrokinetic remediation of radionuclide-contaminated Phaeozem soil and observed that 61.0% of ^{99m}Tc and 71.8% of ^{24}Na were removed after 04 h. Purkis et al. (2020) observed high remediation efficiencies radionuclides (80 + % for ^{137}Cs and 50+ % for ^{90}Sr) by electrokinetic remediation (Table 24.3).

One of the most dangerous environmental challenges is salty soils, which retain massive and unsustainable quantities of noxious salt pollutants, thereby harming the ecosystem and, human health (Bessaim et al. 2020). Annamalai et al. (2014) studied electrokinetic removal of trace metals, dyes and inorganic salts from polluted agricultural soil with textile effluent and found 84% (Cl^-) and 68% (SO_4^2) removal efficiency.

24.7 Electrokinetic Removal of Organic Pollutants

Soil pollution by toxic persistent organic pollutants (POPs) such as organochlorinated pesticides, halohydrocarbons, polycyclic aromatic hydrocarbons (PAHs) and polybrominated diphenyl ethers poses a major environmental threat (Manz et al. 2001; Ren et al. 2018). Chemical contaminants are released into the atmosphere because of increased industrialization and processing practices. Hydrophobic organic contaminants (HOCs) are lethal and cannot be eliminated by normal attenuation. (Alcantara et al. 2010). Pham et al. (2009) examined ultrasonic enhanced electrokinetics (EK-US) and electrokinetics alone (EK) experiments to remove {fluoranthene (FLU), phenanthrene (PHE) and hexachlorobenzene (HCB)} persistent organic contaminants (POPs) from kaolin and found that PHE and FLU were easily extracted from EK-US compared to HCB.

To remediate petroleum-contaminated soil, Gidudu and Chirwa (2020b) used a DC-driven electrokinetic reactor with biosurfactant as demulsification. Ni et als. (2018) studied the removal of {dichloro-diphenyl-trichloroethane (DDT) and hexachloro-cyclohexane soprocide (HCH)} organochlorine pesticides from the soil and, found that Enhanced EK-Fenton treatment was better than EK-Fenton-coupled technologies (EF) and Individual Electrokinetic (IE). Souza et al. (2016) investigated the elimination of 2,4-

Table 24.3 Electrokinetic elimination of radioactive elements from soil

Radioactive element	Solution	Time duration	Removal efficiency (%)	Current/Energy	References
^{85}Sr (4892 Bq/kg) U (1027 mg/kg)	CH_3COOH (0.4 M)	5 days	89.5% 80.5%	100 mA	Kim et al. (2003)
Co^{2+} and Cs^{+}	Acetic Acid (0.01 M)	15 days	95.2% 84.2%	20 −30 mA	Kim et al. (2008)
^{60}Co (1042.4 Bq/kg) ^{137}Cs (1185.6 Bq/kg)	Nitric Acid (0.01 M)	20 days	99.7% 64.9%	15 mA/cm^2	Kim et al. (2010)
Pu(−)	Citric acid (0.04 M)	60 days	About 0.4 m^3, or 1/6 starting material remediation (1.7 Bq/g)	33 kWh/m^3	Agnew et al. (2011)
Uranium(VI) Red Soil	Citric acid, Ferric chloride	120 h	61.55 ± 0.41%	0.2559 kW	Xiao et al. (2020)

Dichlorophenoxyacetic acid by Electrokinetic Soil Flushing powered with DC and Photovoltaic (PV) solar panels. After 15 days, elimination of 2,4-D reaches 90.2% by DC power and 73.6% PV solar (Table 24.4).

24.8 Electrokinetic Removal of Co-contamination

The mixtures of inorganic and organic contaminants (Co-contamination or Mix contamination) are found commonly in the environment (Alcantara et al. 2012). The carcinogenic and mutagenic capability of co-existed inorganic and organic contaminants affects human health and, habitats (Mohamadi et al. 2019). The simultaneous elimination of co-contaminants using conventional practices e.g. phytoremediation and bioremediation are often problematic from the soil. These co-contaminants exhibit different characteristics, composition, and properties but synergistic impacts (Maturi et al. 2008; Saberi et al. 2018). Electrokinetic remediation (EKR) is the green, sustainable, and eco-friendly technology to ease the elimination of toxic pollutants from mixed contaminated soil (Cang et al. 2012; Chirakkara and Reddy 2013). However, only limited work about Electrokinetic remediation (EKR) of co-contaminants has been performed globally, and to advance the knowledge of many major mechanism-influencing influences, more study is required (Khodadoust et al. 2005; Colacicco et al. 2010; Ammami et al. 2014).

Lu (2020) examined EKR of cadmium-pyrene mixed polluted soil and observed 56.38% pyrene elimination efficiency adjacent to the electrodes due to the combined effect of electrochemical oxidation and bioremediation. Chirakkara et al. (2016) reported the influence of electrokinetic phytoremediation on contaminated soil spiked by organic (phenanthrene and naphthalene) and heavy metals (cadmium, lead and chromium) pollutants and found substantial reduction of contaminants in soil. Reddy et al. (2006) reported the enhanced electrokinetic remediation of PAHs and heavy metals at former Manufactured Gas Plant. Maturi and Reddy (2008) reported the electrokinetic simultaneous remediation of heavy metals and PAHs from low-permeability kaolin soils using cyclodextrins (Table 24.5).

24.9 Conclusion

Soil contamination from inorganic and organic pollutants poses great harm to people and their surroundings. The association of toxic heavy

Table 24.4 Electrokinetic removal of organic pollutants from soil

Organic pollutant	Soil sample	Solution	Time duration	Removal efficiency (%)	References
Chlorobenzene (CB) and trichloroethylene (TCE)	Clayey loam soddy-podzolic soil	Triton X-100, OS-20, ALM-10	45 h and 34 h	Chlorobenzene, (61%) Ttrichloroethylene (85%)	Kolosov et al. (2001)
Naphthalene and 2,4-DNT	Spiked Soil	Carboxymethyl-β-cyclodextrin	14 days	Naphthalene (83%) and 2,4-DNT (89%)	Jiradecha et al. (2006)
2,6-Dichlorophenol	Kaolinite Clay	$CH_3OH + H_3PO_4$ $KH_2PO_4 + H_3PO_4$	110 h	90%	Polcaro et al. (2007)
Benzanthracene Fluoranthene Pyrene	Contaminated Soil	Hexane	7 days	86.56% 89.78% 80.16%	Alcantara et al. (2009)
Hexachlorobenzene Phenanthrene Fluoranthene	Kaolin	Hexane	10 days	63% 84% 90%	Pham et al. (2009)
PAHs (Fluoranthene, Pyrene, and Benzanthracene)	Kaolin clay	1% Tween 80 and 0.1 M Na_2SO_4	23 days	39.06%	Alcantara et al. (2010)
Gasoil	Spiked Soil	0.1 N of Citric acid	15 days	86.7%	Gonzini et al. (2010)
Phenanthrene	Kaolinite	Hydroxypropyl-â-cyclodextrin + Na_2Co_3	6 days	75%	Jeon et al. (2010)
Oxyfluorfen	Field soil	Water	34 days	63%	Risco et al. (2016)
Total Petroleum Hydrocarbons	Rhodamine B Kaolinite	Hydrogen Peroxide	27 days	58.2%	Popescu et al. (2017)

metals, organic pollutants, and pesticides make the circumstance of pollution more complex. In today's world, soil pollutants have become a major problem, and its prevention is thus desperately required to preserve the environment and public health. There is a current interest in discovering technologies for sustainable remediation to remove toxins from the soil. Electrokinetic is a modern effort at enhancing the remediation process and soil decontamination. Electrokinetic-assisted phytoremediation (EKPR) and Electrokinetic-assisted bioremediation (EKBR) are innovative technology to remove heavy metals, total petroleum hydrocarbon content (TPH), pesticides, heavy metals, radioactive elements, and organic pollutant of contaminated soils. The electrical current needed for electrokinetics is Direct Current and Solar powered. Solar energy is a creative power alternative and can be economically viable for electrokinetic enhanced bioremediation and phytoremediation. Electrokinetic bioremediation and, phytoremediation may be an efficient technique for appropriate remediation in-field application.

Table 24.5 Electrokinetic removal of co-contamination from soil

Co-contaminants	Soil sample	Solution	Time duration	Removal efficiency (%)	References
Reactive Black 5 (RB5) and Cr	Kaolinite Clay	K_2SO_4 (0.1 M)	05 days	RB5 (95%)	Ricart et al. (2008)
Lubricant oil and zinc	Railroad soil	0.1 M HNO_3 and 0.1 M $MgSO_4$ + 0.5wt% Tergitol	17 days	Zn (22.1–24.3%) Lubricant oil (45.1–55.0%)	Park et al. (2009)
Lead and Phenanthrene	Kaolin clay Sandy soil	1% Tween 80 and 0.1 M EDTA	30 days	90% 70%	Alcantara et al. (2012)
Kerosene Phenol Metals	Contaminant Clay	Hexane Distilled Water	21 days	Kerosene (49.8%) Phenol (100%) Metals (26.8–92.49%)	Lukman et al. (2013)
Polycyclic Aromatic Hydrocarbons and Metals	The mixture of Kaolinite, Silt, and Sand	Nitric acid (NA)	10–14 days	PAHs (70.3–89.7%); Metals (76.8–99.9%)	Ammami et al. (2014)
Petroleum Diesel and Heavy Metals	Co-contaminated soil	0.10 M KH_2PO_4	21 days	~95% TPH ~50% As ~20% Cu	Lee et al. (2016)
Decabromodiphenyl ether (BDE-209) and Copper	Field soil	Citric acid, Persulfate and methyl-β-cyclodextrin (MCD)	10 days	Cu (92.5%) BDE-209 (85.6%)	Chen et al. (2019)

Acknowledgements We thank the Director, Research & Development, Biyani Group of Colleges, Jaipur for support and encouragement.

References

Aboughalma H, Bi R, Schlaak M (2008) Electrokinetic enhancement on phytoremediation in Zn, Pb, Cu and Cd contaminated soil using potato plants. J Environ Sci Health A 43(8):926–933

Acar YB, Alshawabkeh AN (1993) Principle of electrokinetic remediation. Environ Sci Technol 27 (13):2638–2647

Acosta-Santoyo G, Cameselle C, Bustos E (2017) Electrokinetic-enhanced ryegrass cultures in soils polluted with organic and inorganic compounds. Environ Res 158:118–125

Agnew K, Cundy AB, Hopkinson L, Croudace IW, Warwick PE, Purdie P (2011) Electrokinetic remediation of plutonium-contaminated nuclear site wastes: results from a pilot-scale on-site trial. J Hazard Mater 186(2–3):1405–1414

Ahmad I, Imran M, Hussain M, Hussain S (2017) Remediation of organic and inorganic pollutants from soil: the role of plant-bacteria partnership. In: Anjum NA (ed) Chemical pollution control with microorganisms. Nova Sci Publisher, pp 197–243

Ait Ahmed O (2020) The removal efficiency of lead from contaminated soil: modeling of cations and anions migration during the electrokinetic treatment. J Environ Sci Health A 55(10):1218–1232

Ajiboye TO, Oyewo OA, Onwudiwe DC (2021) Simultaneous removal of organics and heavy metals from industrial wastewater: a review. Chemosphere 262:128379

Alcantara MT, Gomez J, Pazos M, Sanroman MA (2009) PAHs soil decontamination in two steps: desorption and electrochemical treatment. J Hazard Mater 166:462–468

Alcantara MT, Gomez J, Pazos M, Sanroman MA (2010) Electrokinetic remediation of PAH mixtures from kaolin. J Hazard Mater 179(1–3):1156–1160

Alcantara MT, Gomez J, Pazos M, Sanroman MA (2012) Electrokinetic remediation of lead and phenanthrene polluted soils. Geoderma 173:128–133

Alshawabkeh AN (2009) Electrokinetic soil remediation: challenges and opportunities. Sep Sci Technol 44 (10):2171–2187

Ammami MT, Benamar A, Wang H, Bailleul C, Legras M, Le Derf F, Portet-Koltalo F (2014) Simultaneous electrokinetic removal of polycyclic aromatic hydrocarbons and metals from a sediment using mixed enhancing agents. Int J Environ Sci Technol 11(7):1801–1816

Amundson R, Berhe AA, Hopmans JW, Olson C, Sztein AE, Sparks DL (2015) Soil and human security in the 21st century. Science 348(6235):1261071

Andrade DC, dos Santos EV (2020) Combination of electrokinetic remediation with permeable reactive barriers to remove organic compounds from soils. Curr Opin Electrochem 22:136–144

Annamalai S, Santhanam M, Sundaram M, Curras MP (2014) Electrokinetic remediation of inorganic and organic pollutants in textile effluent contaminated agricultural soil. Chemosphere 117:673–678

Arthur EL, Rice PJ, Rice PJ, Anderson TA, Baladi SM, Henderson KLD, Coats JR (2005) Phytoremediation—an overview. CRC Crit Rev Plant Sci 24(2):109–122

Azhar ATS, Nabila ATA, Nurshuhaila MS, Zaidi E, Azim MAM, Farhana SMS (2016) Assessment and comparison of electrokinetic and electrokinetic-bioremediation techniques for mercury contaminated soil. IOP Conf Ser Mater Sci Eng 160(1): 012077

Azubuike CC, Chikere CB, Okpokwasili GC (2016) Bioremediation techniques–classification based on site of application: principles, advantages, limitations and prospects. World J Microbiol Biotechnol 32(11):180

Baek K, Kim DH, Park SW, Ryu BG, Bajargal T, Yang JS (2009) Electrolyte conditioning-enhanced electrokinetic remediation of arsenic-contaminated mine tailing. J Hazard Mater 161:457–462

Bakulski KM, Seo YA, Hickman RC, Brandt D, Vadari HS, Hu H, Park SK (2020) Heavy metals exposure and Alzheimer's disease and related dementias. J Alzheimer's Dis 76(4):1215–1242

Bessaim MM, Karaca O, Missoum H, Bendani K, Laredj N, Bekkouche MS (2020) Effect of imposed electrical gradient on removal of toxic salt contaminants from alkali-saline low permeable soil during electrokinetic remediation. Arab J Geosci 13(14):666

Beyrami H, Neyshabouri MR, Oustan S (2020) Effects of different treatments and time on electrokinetic remediation of Cd, Pb and Zn from a Calcareous Contaminated Soil. Chin J Chem Eng. https://doi.org/10.1016/j.cjche.2020.09.011

Bi R, Schlaak M, Siefert E, Lord R, Connolly H (2011) Influence of electrical fields (AC and DC) on phytoremediation of metal polluted soils with rapeseed (*Brassica napus*) and tobacco (*Nicotiana tabacum*). Chemosphere 83(3):318–326

Biscombe CJC (2017) The discovery of electrokinetic phenomena: setting the record straight. Angew Chemie Int Ed 56(29):8338–8340

Boudh S, Singh JS (2019) Pesticide contamination: environmental problems and remediation strategies. In: Bharagava R, Chowdhary P (eds) Emerging and eco-friendly approaches for waste management. Springer, Singapore

Burca N, Watson RR (2014) Fish oil supplements, contaminants, and excessive doses. In: Watson RR, De Meester F (eds) Omega-3 fatty acids in brain and neurological health. Academic Press, Boston, pp 447–454

Cameselle C, Chirakkara RA, Reddy KR (2013) Electrokinetic-enhanced phytoremediation of soils: status and opportunities. Chemosphere 93(4):626–636

Cameselle C, Gouveia S (2018) Electrokinetic remediation for the removal of organic contaminants in soils. Curr Opin Electrochem 11:41–47

Cameselle C, Gouveia S (2019) Physicochemical methods for the remediation of radionuclide contaminated sites. In: Gupta D, Voronina A (eds) Remediation measures for radioactively contaminated areas. Springer, Cham, pp 31–49

Cang L, Wang QY, Zhou DM, Xu H (2011) Effects of electrokinetic-assisted phytoremediation of a multiple-metal contaminated soil on soil metal bioavailability and uptake by Indian mustard. Sep Purif Technol 79 (2):246–253

Cang L, Zhou DM, Wang QY, Fan GP (2012) Impact of electrokinetic-assisted phytoremediation of heavy metal contaminated soil on its physicochemical properties, enzymatic and microbial activities. Electrochim Acta 86:41–48

Cercato M, De Donno G (2020) Time-lapse monitoring of an electrokinetic soil remediation process through frequency-domain electrical measurements. J Appl Geophys 175:103980

Chen F, Li X, Ma J, Qu J, Yang Y, Zhang S (2019) Remediation of soil co-contaminated with decabromodiphenyl ether (BDE-209) and copper by enhanced electrokinetics-persulfate process. J Hazard Mater 369:448–455

Chen Z, Zhu BK, Jia WF, Liang JH, Sun GX (2015) Can electrokinetic removal of metals from contaminated paddy soils be powered by microbial fuel cells? Environ Technol Innov 3:63–67

Chilingar GV, Loo WW, Khilyuk LF, Katz SA (1997) Electrobioremediation of soils contaminated with hydrocarbons and metals: progress report. Energ Sour 19(2):129–146

Chirakkara R, Reddy KR (2013) Investigation of plant species for phytoremediation of mixed contaminants in soils. In: Proceedings of 106th annual conference and exhibition, air and waste management association, Chicago, IL, USA pp 1–12

Chirakkara RA, Cameselle C, Reddy KR (2016) Assessing the applicability of phytoremediation of soils with mixed organic and heavy metal contaminants. Rev Environ Sci Bio/technol 15(2):299–326

Chirakkara RA, Reddy KR, Cameselle C (2015) Electrokinetic amendment in phytoremediation of mixed contaminated soil. Electrochim Acta 181:179–191

Colacicco A, De Gioannis G, Muntoni A, Pettinao E, Polettini A, Pomi R (2010) Enhanced electrokinetic treatment of marine sediments contaminated by heavy metals and PAHs. Chemosphere 81:46–56

Cong Y, Ye Q, Wu Z (2005) Electrokinetic behavior of chlorinated phenols in soil and their electrochemical degradation. Process Saf Environ 83(2):178–183

Couto N, Guedes P, Ribeiro AB, Zhou DM (2015) Phytoremediation and the electrokinetic process: potential use for the phytoremediation of antimony and arsenic. In: Ansari A, Gill S, Gill R, Lanza G, Newman L (eds) Phytoremediation. Springer, Cham, pp 199–209

Crognale S, Cocarta DM, Streche C, D'Annibale A (2020) Development of laboratory-scale sequential electrokinetic and biological treatment of chronically hydrocarbon-impacted soils. N Biotechnol 58:38–44

Daghan H, Ozturk M (2015) Soil pollution in Turkey and remediation methods. In: Soil remediation and plants: prospects and challenges, pp 287–312

De Battisti A, Ferro S (2007) Electrokinetic remediation. Methods of remediation of soils and ground waters (EREM 2005). Electrochim Acta 52(10):3345–3348

Dermont G, Bergeron M, Mercier G, Richer-Lafleche M (2008) Soil washing for metal removal: a review of physical/chemical technologies and field applications. J Hazard Mater 152:1–31

Dhaliwal SS, Singh J, Taneja PK, Mandal A (2020) Remediation techniques for removal of heavy metals from the soil contaminated through different sources: a review. Environ Sci Pollut Res 27:1319–1333

Dzionek A, Wojcieszynska D, Guzik U (2016) Natural carriers in bioremediation: a review. Electron J Biotechnol 23:28–36

Ekta P, Modi NR (2018) A review of phytoremediation. J Pharmacogn Phytochem 7(4):1485–1489

Etim EE (2012) Phytoremediation and its mechanisms: a review. Int J Environ Bioenerg 2(3):120–136

Feng NX, Yu J, Zhao HM, Cheng YT, Mo CH, Cai QY, Li YW, Li H, Wong MH (2017) Efficient phytoremediation of organic contaminants in soils using plant–endophyte partnerships. Sci Total Environ 583:352–368

Fu R, Wen D, Xia X, Zhang W, Gu Y (2017) Electrokinetic remediation of chromium (Cr)-contaminated soil with citric acid (CA) and polyaspartic acid (PASP) as electrolytes. Chem Eng J 316:601–608

Gidudu B, Chirwa EMN (2020) The combined application of a high voltage, low electrode spacing, and biosurfactants enhances the bio-electrokinetic remediation of petroleum contaminated soil. J Clean Prod 143:332–339

Gidudu B, Chirwa EMN (2020) Application of biosurfactants and pulsating electrode configurations as potential enhancers for electrokinetic remediation of petrochemical contaminated soil. Sustain 12(14):5613

Gill RT, Harbottle MJ, Smith JWN, Thornton SF (2014) Electrokinetic-enhanced bioremediation of organic contaminants: a review of processes and environmental applications. Chemosphere 107:31–42

Gnanasundar VM, Akshai Raj R (2020) Remediation of inorganic contaminants in soil using electrokinetics, phytoremediation techniques. Mater Today Proc. https://doi.org/10.1016/j.matpr.2020.03.038

Goldscheider N (2010) Delineation of spring protection zones. In: Kresic N, Stevanovic Z (eds) Groundwater hydrology of springs. Butterworth-Heinemann, Boston, pp 305–338

Gomiero T (2016) Soil degradation, land scarcity and food security: reviewing a complex challenge. Sustain 8(3):281

Gonzini O, Plaza A, Di Palma L, Lobo MC (2010) Electrokinetic remediation of gasoil contaminated soil enhanced by rhamnolipid. J Appl Electrochem 40:1239–1248

Gustave W, Yuan Z, Liu F, Chen Z (2020) Mechanisms and challenges of microbial fuel cells for soil heavy metal(loid)s remediation. Sci Total Environ 143865

Habibul N, Hu Y, Sheng GP (2016) Microbial fuel cell driving electrokinetic remediation of toxic metal contaminated soils. J Hazard Mater 318:9–14

Hassan I, Mohamedelhassan E, Yanful EK (2015) Solar powered electrokinetic remediation of Cu polluted soil using a novel anode configuration. Electrochim Acta 181:58–67

Hassan I, Mohamedelhassan E, Yanful EK, Yuan ZC (2016) A review article: electrokinetic bioremediation current knowledge and new prospects. Adv Microbiol 06(01):57–72

Hassan I, Mohamedelhassan E, Yanful EK, Yuan Z-C (2018) Enhancement of bioremediation and phytoremediation using electrokinetics. In: Shiomi N (ed) Advances in bioremediation and phytoremediation. IntechOpen, pp 169–189

Head NA, Gerhard JI, Inglis AM, Nunez Garcia A, Chowdhury AIA, Reynolds DA, de Boer CV, Sidebottom A, Austrins LM, Eimers J, O'Carroll DM (2020) Field test of electrokinetically-delivered thermally activated persulfate for remediation of chlorinated solvents in clay. Water Res 183:116061

Hu QH, Weng JQ, Wang JS (2010) Sources of anthropogenic radionuclides in the environment: a review. J Environ Radioact 101(6):426–437

Huang H, Tang J, Niu Z, Giesy JP (2019) Interactions between electrokinetics and rhizoremediation on the remediation of crude oil-contaminated soil. Chemosphere 229:418–425

Hussein AA, Alatabe MJA (2019) Remediation of lead-contaminated soil, using clean energy in combination with electro-kinetic methods. Pollution 5(4):859–869

Isosaari P, Piskonen R, Ojala P, Voipio S, Eilola K, Lehmus E, Itavaara M (2007) Integration of electrokinetics and chemical oxidation for the remediation of creosote-contaminated clay. J Hazard Mater 144(1–2):538–548

Jaishankar M, Tseten T, Anbalagan N, Mathew BB, Beeregowda KN (2014) Toxicity, mechanism and health effects of some heavy metals. Interdiscip Toxicol 7(2):60–72

Jamari S, Embong Z, Bakar I (2014) Elemental composition study of heavy metal (Ni, Cu, Zn) in riverbank soil by electrokinetic-assisted phytoremediation using XRF and SEM/EDX. AIP Conf Proc 584:221–227

Jamil N, Madun A, Ahmad Tajudin SA, Embong Z (2015) An overview of electrokinetic remediation assisted phytoremediation to remediate barren acidic soil. Appl Mech Mater 773–774:1476–1480

Jeon CS, Yang JS, Kim KJ, Baek K (2010) Electrokinetic removal of petroleum hydrocarbon from residual clayey soil following a washing process, vol 38. CLEAN-Soil, Air, Water, pp 189–193

Jeon EK, Ryu SR, Baek K (2015) Application of solar-cells in the electrokinetic remediation of As-contaminated soil. Electrochim Acta 181:160–166

Jiradecha C, Urgun-Demirtas M, Pagilla K (2006) Enhanced electrokinetic dissolution of naphthalene and 2,4-DNT from contaminated soils. J Hazard Mater 136:61–67

Kanianska R (2016) Agriculture and its impact on land-use, environment, and ecosystem services. In: Almusaed A (ed) Landscape ecology—the influences of land use and anthropogenic impacts of landscape creation. IntechOpen, Rijeka

Karaca O, Cameselle C, Bozcu M (2019) Opportunities of electrokinetics for the remediation of mining sites in Biga peninsula, Turkey. Chemosphere 227:606–613

Khan MAI, Biswas B, Smith E, Naidu R, Megharaj M (2018) Toxicity assessment of fresh and weathered petroleum hydrocarbons in contaminated soil—a review. Chemosphere 212:755–767

Khan NT, Jameel N, Khan MJ (2018) A brief overview of contaminated soil remediation methods. Biotechnol Ind J 14(4):171

Khodadoust AP, Reddy KR, Maturi K (2005) Effect of different extraction agents on metal and organic contaminant removal from a field soil. J Hazard Mater 117:15–24

Kim BK, Baek K, Ko SH, Yang JW (2011) Research and field experiences on electrokinetic remediation in South Korea. Sep Purif Technol 79(2):116–123

Kim D, Han J (2020) Remediation of multiply contaminated ground via permeable reactive barrier and electrokinetic using recyclable food scrap Ash (FSA). Appl Sci 10(4):1194

Kim GN, Jung YH, Lee JJ, Moon JK, Jung CH (2008) An analysis of a flushing effect on the electrokinetic-flushing removal of cobalt and cesium from a soil around decommissioning site. Sep Purif Technol 63:116–121

Kim GN, Lee SS, Shon DB, Lee KW, Chung US (2010) Development of pilot-scale electrokinetic remediation technology to remove ^{60}Co and ^{137}Cs from soil. J Ind Eng Chem 16:986–991

Kim KH, Kim SO, Lee CW, Lee MH, Kim KW (2003) Electrokinetic processing for the removal of radionuclides in soils. Sep Purif Technol 38:2137–2163

Kim SJ, Park JY, Lee YJ, Lee JY, Yang JW (2005) Application of a new electrolyte circulation method for the ex situ electrokinetic bioremediation of a laboratory-prepared pentadecane contaminated kaolinite. J Hazard Mater 118(1–3):171–176

Kim SS, Han SJ, Cho YS (2002) Electrokinetic remediation strategy considering ground strate: a review. Geosci J 6(1):57–75

Kimbel HJ, Nilsson EE, Skinner MK (2019) Environmentally induced epigenetic transgenerational inheritance of Ovarian disease. In: Leung PCK, Adashi EY (eds) The ovary, 3rd ed. Academic Press, pp 149–154

Klouche F, Bendani K, Benamar A, Missoum H, Maliki M, Laredj N (2020) Electrokinetic restoration of local saline soil. Mater Today Proc 22(1):64–68

Klouche F, Bendani K, Benamar A, Missoum H, Maliki M, Mesrar (2020b) Contribution to the remediation of saline soils by electrokinetic process: experimental study. In: Reddy KR, Agnihotri AK, Yukselen-Aksoy Y, Dubey BK, Bansal A (eds) Sustainable environmental geotechnics. Lecture notes in civil engineering, vol 89. Springer, Cham, pp 151–160

Kolosov AY, Popov KI, Shabanova NA, Artem'eva AA, Kogut BM, Frid AS, Zel'venskii VY, Urinovich EM (2001) Electrokinetic removal of hydrophobic organic compounds from soil. Russ J Appl Chem 74:631–635

Kumar A, Schreiter IJ, Wefer-Roehl A, Tsechansky L, Schuth C, Graber ER (2016) Production and utilization of biochar from organic wastes for pollutant control on contaminated sites. In: Prasad MNV, Shih K (eds) Environmental materials and waste. Academic Press, pp 91–116

Lacatusu AR, Cocarta D, Lacatusu R (2013) Ex-situ bioremediation efficiency in removing organic and inorganic compounds from artificially and anthropogenic contaminated soils. Carpathian J Earth Environ Sci 8(1):59–70

Lageman R, Clarke RL, Pool W (2005) Electro-reclamation, a versatile soil remediation solution. Eng Geol 77(3–4):191–201

Lajayer AB, Moghadam KN, Maghsoodi MR, Ghorbanpour M, Kariman K (2019) Phytoextraction of heavy metals from contaminated soil, water and atmosphere using ornamental plants: mechanisms and efficiency improvement strategies. Environ Sci Pollut Res 26(9):8468–8484

Lee JY, Kwon TS, Park JY, Choi S, Kim EJ, Lee HU, Lee YC (2016) Electrokinetic (EK) removal of soil co-contaminated with petroleum oils and heavy metals in three-dimensional (3D) small-scale reactor. Process Saf Environ 99:186–193

Li J, Li R, Li J (2017) Current research scenario for microcystins biodegradation—a review on fundamental knowledge, application prospects and challenges. Sci Total Environ 595:615–632

Li J, Zhang J, Larson SL, Ballard JH, Guo K, Arslan Z, Ma Y, Waggoner CA, White JR, Han FX (2019) Electrokinetic-enhanced phytoremediation of uranium-contaminated soil using sunflower and Indian mustard. Int J Phytoremediation 21(12):1197–1204

Li Z (2018) Health risk characterization of maximum legal exposures for persistent organic pollutant (POP) pesticides in residential soil: an analysis. J Environ Manage 205:163–173

Lima AT, Hofmann A, Reynolds D, Ptacek CJ, Van Cappellen P, Ottosen LM, Pamukcu S, Alshawabekh A, O'Carroll DM, Riis C, Cox E, Gent DB, Landis R, Wang J, Chowdhury AIA, Secord EL, Sanchez-Hachair A (2017) Environmental electrokinetics for a sustainable subsurface. Chemosphere 181:122–133

Lin W, Guo C, Zhang H, Liang X, Wei Y, Lu G, Dang Z (2016) Electrokinetic-enhanced remediation of phenanthrene-contaminated soil combined with sphingomonas sp. GY2B and biosurfactant. Appl Biochem Biotechnol 178:1325–1338

Liu L, Li W, Song W, Guo M (2018) Remediation techniques for heavy metal-contaminated soils: principles and applicability. Sci Total Environ 633:206–219

Llorente I, Fajardo S, Bastidas JM (2014) Applications of electrokinetic phenomena in materials science. J Solid State Electrochem 18:293–307

Logan BE, Regan JM (2006) Electricity-producing bacterial communities in microbial fuel cells. Trends Microbiol 14(12):512–518

Lombi E, Hamon RE (2005) Remediation of polluted soils. In: Hillel D (ed) Encyclopedia of soils in the environment. Elsevier, pp 379–385

Lu Q (2020) Insights into the remediation of cadmium-pyrene co-contaminated soil by electrokinetic and the influence factors. Chemosphere 254:126861

Lukman S, Essa MH, Mu'azu ND, Bukhari A (2013) Coupled electrokinetics-adsorption technique for simultaneous removal of heavy metals and organics from saline-sodic soil. Sci World J 2013:346910

Luo Q, Zhang X, Wang H, Qian Y (2005) The use of non-uniform electrokinetics to enhance in situ bioremediation of phenol-contaminated soil. J Hazard Mater 121 (1–3):187–194

Manz M, Wenzel KD, Dietze U, Schüürmann G (2001) Persistent organic pollutants in agricultural soils of central Germany. Sci Total Environ 277:187–198

Mao X, Han FX, Shao X, Guo K, McComb J, Arslan Z, Zhang Z (2016) Electro-kinetic remediation coupled with phytoremediation to remove lead, arsenic and cesium from contaminated paddy soil. Ecotoxicol Environ Saf 125:16–24

Mao X, Shao X, Zhang Z (2019) Pilot-scale electrokinetic remediation of lead polluted field sediments: model designation, numerical simulation, and feasibility evaluation. Environ Sci Eur 31:25

Martin BC, George SJ, Price CA, Ryan MH, Tibbett M (2014) The role of root exuded low molecular weight organic anions in facilitating petroleum hydrocarbon degradation: current knowledge and future directions. Sci Total Environ 472:642–653

Maturi K, Khodadoust AP, Reddy KR (2008) Comparison of extractants for removal of lead, zinc, and phenanthrene from manufactured gas plant field soil. Prac Periodical Hazardous Toxic Radioactive Waste Manag 12(4):230–238

Maturi K, Reddy KR (2008) Cosolvent-enhanced desorption and transport of heavy metals and organic contaminants in soils during electrokinetic remediation. Water Air Soil Pollut 189(1–4):199–211

Meshalkin VP, Shulayev NS, Pryanichnikova VV (2020) Experimental and theoretical engineering of energy-efficient electrochemical process of soil remediation to remove oil contaminants. Dokl Chem 491(2):61–64

Mohamadi S, Saeedi M, Mollahosseini A (2019) Enhanced electrokinetic remediation of mixed contaminants from a high buffering soil by focusing on mobility risk. J Environ Chem Eng 7:103470

Moosavi SG, Seghatoleslami MJ (2013) Advance in agriculture and biology phytoremediation: a review. Adv Agri Biol 1:5–11

Mosa KA, Saadoun I, Kumar K, Helmy M, Dhankher OP (2016) Potential biotechnological strategies for the cleanup of heavy metals and metalloids. Front Plant Sci 7:303

Mosavat N, Oh E, Chai G (2012) A review of electrokinetic treatment technique for improving the engineering characteristics of low permeable problematic soils. Int J GEOMATE 2(2):266–272

Ni M, Tian S, Huang Q, Yang Y (2018) Electrokinetic-Fenton remediation of organochlorine pesticides from historically polluted soil. Environ Sci Pollut Res Int 25 (12):12159–12168

O'Connor CS, Lepp NW, Edwards R, Sunderland G (2003) The combined use of electrokinetic remediation and phytoremediation to decontaminate metal-polluted soils: a laboratory-scale feasibility study. Environ Monit Assess 84(1–2):141–158

Ojuederie OB, Babalola OO (2017) Microbial and plant-assisted bioremediation of heavy metal polluted environments: a review. Int J Environ Res Public Health 14(12):1504

Ottosen LM, Larsen TH, Jensen PE, Kirkelund GM, Kerrn-Jespersen H, Tuxen N, Hyldegaard BH (2019) Electrokinetics applied in remediation of subsurface soil contaminated with chlorinated ethenes—a review. Chemosphere 235:113–125

Palansooriya KN, Shaheen SM, Chen SS, Tsang DCW, Hashimoto Y, Hou D, Bolan NS, Rinklebe J, Ok YS (2020) Soil amendments for immobilization of potentially toxic elements in contaminated soils: a critical review. Environ Int 134:105046

Park SW, Lee JY, Yang JS, Kim KJ, Baek K (2009) Electrokinetic remediation of contaminated soil with waste-lubricant oils and zinc. J Hazard Mater 169:1168–1172

Pham TD, Shrestha RA, Virkutyte J, Sillanpaa M (2009) Combined ultrasonication and electrokinetic remediation for persistent organic removal from contaminated kaolin. Electrochim Acta 54:1403–1407

Pham TD, Sillanpaa M (2020) Ultrasonic and electrokinetic remediation of low permeability soil contaminated with persistent organic pollutants. In:

Sillanpaa M (ed) Advanced water treatment: electrochemical methods. Elsevier, pp 227–310

Polcaro AM, Vacca A, Mascia M, Palmas S (2007) Electrokinetic removal of 2,6-dichlorophenol and diuron from kaolinite and humic acid-clay system. J Hazard Mater 148:505–512

Popescu M, Rosales E, Sandu C, Meijide J, Pazos M, Lazar G, Sanromán MA (2017) Soil flushing and simultaneous degradation of organic pollutants in soils by electrokinetic-Fenton treatment. Process Saf Environ Prot 108:99–107

Punia T, Singh A (2018) Electrossmotic flow and electroosmosis in soil. Int J Chem Stud 6(2):708–711

Purkis JM, Tucknott A, Croudace IW, Warwick PE, Cundy AB (2020) Enhanced electrokinetic remediation of nuclear fission products in organic-rich soils. Appl Geochem 104826

Rai PK, Kim KH, Lee SS, Lee JH (2020) Molecular mechanisms in phytoremediation of environmental contaminants and prospects of engineered transgenic plants/microbes. Sci Total Environ 705:135858

Rajendran K, Pujari L, Ethiraj K (2021) Biodegradation and bioremediation of S-triazine herbicides. In: Gothandam KM, Ranjan S, Dasgupta N, Lichtfouse E (eds) Environmental biotechnology, vol. 3. Environmental chemistry for a sustainable world. Springer, Cham, pp 31–54

Rajindiran S, Dotaniya ML, Coumar MV, Panwar NR, Saha JK (2015) Heavy metal polluted soils in India: status and countermeasures. JNKVV Res J 49(3):320–337

Ramadan BS, Sari GL, Rosmalina RT, Effendi AJ, Hadrah (2018) An overview of electrokinetic soil flushing and its effect on bioremediation of hydrocarbon contaminated soil. J Environ Manage 218:309–321

Reddy KR, Ala PR, Sharma S, Kumar SN (2006) Enhanced electrokinetic remediation of contaminated manufactured gas plant soil. Eng Geol 85:132–146

Ren X, Zeng G, Tang L, Wang J, Wan J, Liu Y, Yu J, Yi H, Ye S, Deng R (2018) Sorption, transport and biodegradation—an insight into bioavailability of persistent organic pollutants in soil. Sci Total Environ 610–611:1154–1163

Reuss FF (1809) Charge-induced flow. Proc Imp Soc Nat Moscow 1809(3):327–344

Ricart MT, Pazos M, Gouveia S, Cameselle C, Sanroman MA (2008) Removal of organic pollutants and heavy metals in soils by electrokinetic remediation. J Environ Sci Health A 43:871–875

Risco C, Rubi-Juarez H, Rodrigo S, Lopez Vizcaíno R, Saez C, Canizares P, Barrera-Diaz C, Navarro V, Rodrigo MA (2016) Removal of oxyfluorfen from spiked soils using electrokinetic fences. Sep Purif Technol 167:55–62

Saberi N, Aghababaei M, Ostovar M, Mehrnahad H (2018) Simultaneous removal of polycyclic aromatic hydrocarbon and heavy metals from an artificial clayey soil by enhanced electrokinetic method. J Environ Manag 217:897–905

Saeedi M, Li LY, Moradi Gharehtapeh A (2013) Effect of alternative electrolytes on enhanced electrokinetic remediation of hexavalent chromium in clayey soil. Int J Environ Res 7(1):39–50

Sanchez V, Lopez-Bellido FJ, Rodrigo MA, Fernandez FJ, Rodriguez L (2020) A mesocosm study of electrokinetic-assisted phytoremediation of atrazine-polluted soils. Sep Purif Technol 233:116044

Sarankumar RK, Selvi A, Murugan K, Rajasekar A (2020) Electrokinetic (EK) and bio-electrokinetic (BEK) remediation of hexavalent chromium in contaminated soil using alkalophilic bio-anolyte. Indian Geotech J 50(3):330–338

Sasse J, Martinoia E, Northen T (2018) Feed your friends: do plant exudates shape the root microbiome? Trends Plant Sci 23:25–41

Schell LM, Knutson KL, Bailey S (2012) Environmental effects on growth. In: Cameron N, Bogin B (eds) Human growth and development, 2nd ed, pp 245–286

Selvi A, Rajasekar A, Theerthagiri J, Ananthaselvam A, Sathishkumar K, Madhavan J, Rahman PKSM (2019) Integrated remediation processes toward heavy metal removal/recovery from various environments—a review. Front Environ Sci 7:66

Sharma S, Singh B, Manchanda VK (2014) Phytoremediation: role of terrestrial plants and aquatic macrophytes in the remediation of radionuclides and heavy metal contaminated soil and water. Environ Sci Pollut R 22(2):946–962

Shaw G, Bell JNB (1994) Plants and radionuclides. In: Farago ME (ed) Plants and chemical elements, biochemistry, uptake, tolerance and toxicity. VCH Publishers, Weinheim (Federal Republic of Germany), pp 179–220

Shrivastava A, Ghosh D, Dash A, Bose S (2015) Arsenic contamination in soil and sediment in india: sources, effects, and remediation. Curr Pollut Reports 1:35–46

Shu J, Sun X, Liu R, Liu Z, Wu H, Chen M, Li B (2019) Enhanced electrokinetic remediation of manganese and ammonia nitrogen from electrolytic manganese residue using pulsed electric field in different enhancement agents. Ecotoxicol Environ Saf 171:523–529

Singh C, Tiwari S, Singh JS (2020) Biochar: a sustainable tool in soil pollutant bioremediation. In: Bharagava R, Saxena G (eds) Bioremediation of industrial waste for environmental safety. Springer, Singapore, pp 475–494

Siyar R, Doulati Ardejani F, Farahbakhsh M, Norouzi P, Yavarzadeh M, Maghsoudy S (2020) Potential of Vetiver grass for the phytoremediation of a real multi-contaminated soil, assisted by electrokinetic. Chemosphere 246:125802

Song B, Zeng G, Gong J, Liang J, Xu P, Liu Z, Zhang Y, Zhang C, Cheng M, Liu Y, Ye S, Yi H, Ren X (2017) Evaluation methods for assessing effectiveness of in situ remediation of soil and sediment contaminated with organic pollutants and heavy metals. Environ Int 105:43–55

Sorengard M, Ahrens L, Alygizakis N, Jensen P, Gago-Ferrero P (2020) Non-target and suspect screening strategies for electrodialytic soil remediation evaluation: assessing changes in the molecular fingerprints and per- and polyfluoroalkyl substances (PFASs). J Environ Chem Eng 8:104437

Souza FL, Saez C, Llanos J, Lanza MRV, Canizares P, Rodrigo MA (2016) Solar-powered electrokinetic remediation for the treatment of soil polluted with the herbicide 2,4-D. Electrochim Acta 190:371–377

Stojic N, Strbac S, Prokic D (2018) Soil pollution and remediation BT—handbook of environmental materials management. In: International S (ed) Hussain CM. Publishing, Cham, pp 1–34

Streche C, Cocarta DM, Istrate IA, Badea AA (2018) Decontamination of Petroleum-Contaminated Soils Using the electrochemical technique: remediation degree and energy consumption. Sci Rep 8(1):3272

Sturman PJ, Stewart PS, Cunningham AB, Bouwer EJ, Wolfram JH (1995) Engineering scale-up of in situ bioremediation processes: a review. J Contam Hydrol 19(3):171–203

Szpyrkowicz L, Radaelli M, Bertini S, Daniele S, Casarin F (2007) Simultaneous removal of metals and organic compounds from a heavily polluted soil. Electrochim Acta 52(10):3386–3392

Tahmasbian I, Safari Sinegani AA (2016) Improving the efficiency of phytoremediation using electrically charged plant and chelating agents. Environ Sci Pollut Res 23(3):2479–2486

Tang J, He J, Tang H, Wang H, Sima W, Liang C, Qiu Z (2020) Heavy metal removal effectiveness, flow direction and speciation variations in the sludge during the biosurfactant-enhanced electrokinetic remediation. Sep Purif Technol 246:116918

Tchounwou PB, Yedjou CG, Patlolla AK, Sutton DJ (2012) Heavy metal toxicity and the environment. Exp Suppl 101:133–164

Thangavel P, Subbhuraam C (2004) Phytoextraction: role of hyperaccumulators in metal contaminated soils. Proc Indian Natl Sci Acad 70:109–130

Truu J, Truu M, Espenberg M, Nolvak H, Juhanson J (2015) Phytoremediation and plant-assisted bioremediation in soil and treatment wetlands: a review. Open Biotechnol J 9(1):85–92

Tu YJ, Premachandra GS, Boyd SA, Sallach JB, Li H, Teppen BJ, Johnston CT (2021) Synthesis and evaluation of Fe_3O_4-impregnated activated carbon for dioxin removal. Chemosphere 263:128263

Tuomisto HL, Scheelbeek PFD, Chalabi Z, Green R, Smith RD, Haines A, Dangour AD (2017) Effects of environmental change on agriculture, nutrition and health: a framework with a focus on fruits and vegetables. Wellcome Open Res 2:21

Ugaz A, Puppala S, Gale RJ, Acar YB (1994) Electrokinetic soil processing complicating features of electrokinetic remediation of soils and slurries: saturation effects and the role of the cathode electrolysis. Chem Eng Commun 129:183–200

Upcraft T, Guo M (2020) Phytoremediation value chains and modeling. In: Hou D (ed) Sustainable remediation of contaminated soil and groundwater. Butterworth-Heinemann, pp 325–366

Valdovinos V, Monroy-Guzmán F, Bustos E (2016) Electrokinetic removal of radionuclides contained in scintillation liquids absorbed in soil type Phaeozem. J Environ Radioact 162–163:80–86

Varjani SJ, Upasani VN (2017) A new look on factors affecting microbial degradation of petroleum hydrocarbon pollutants. Int Biodeterior Biodegrad 120:71–83

Virkutyte J, Sillanpaa M, Latostenmaa P (2002) Electrokinetic soil remediation—critical overview. Sci Total Environ 289(1–3):97–121

Vocciante M, Caretta A, Bua L, Bagatin R, Ferro S (2016) Enhancements in electrokinetic remediation technology: environmental assessment in comparison with other configurations and consolidated solutions. Chem Eng J 289:123–134

Wall S (2010) The history of electrokinetic phenomena. Curr Opin Colloid Interface Sci 15(3):119–124

Wang Y, Li A, Cui C (2020) Remediation of heavy metal-contaminated soils by electrokinetic technology: Mmechanisms and applicability. Chemosphere 265:129071

Wen D, Fu R, Li Q (2021) Removal of inorganic contaminants in soil by electrokinetic remediation technologies: a review. J Hazard Mater 401:123345

Wick LY, Mattle PA, Wattiau P, Harms H (2004) Electrokinetic transport of PAH-degrading bacteria in model aquifers and soil. Environ Sci Technol 38 (17):4596–4602

Wu Q, Jiao S, Ma M, Peng S (2020) Microbial fuel cell system: a promising technology for pollutant removal and environmental remediation. Environ Sci Pollut Res 27(7):6749–6764

Wu Y, Wang S, Cheng F, Guo P, Guo S (2020b) Enhancement of electrokinetic-bioremediation by ryegrass: sustainability of electrokinetic effect and improvement of n-hexadecane degradation. Environ Res 188:109717

Wuana RA, Okieimen FE (2011) Heavy metals in contaminated soils: a review of sources, chemistry, risks and best available strategies for remediation. Int Sch Res Notices 2011:1–20

Xiao J, Zhou S, Chu L, Liu Y, Li J, Zhang J, Tian L (2020) Electrokinetic remediation of uranium(VI)-contaminated red soil using composite electrolyte of citric acid and ferric chloride. Environ Sci Pollut Res Int 27:4478–4488

Xu J, Liu C, Hsu PC, Zhao J, Wu T, Tang J, Liu K, Cui Y (2019) Remediation of heavy metal contaminated soil by asymmetrical alternating current electrochemistry. Nat Commun 10(1):2440

Yan L, Le Q Van, Sonne C, Yang Y, Yang H, Gu H, Ma NL, Lam SS, Peng W (2021) Phytoremediation of radionuclides in soil, sediments and water. J Hazard Mater 407:124771

Yang X, Zhou M, Cang L, Ji Q, Xie J (2020) Enhanced electrokinetic remediation of heavy-metals contaminated soil in presence tetrasodium N, N-bis(carboxymethyl) glutamic acid (GLDA) as chelator. Int J Electrochem Sci 15:696–709

Yao W, Cai Z, Sun S, Romantschuk M, Sinkkonen A, Sun Y, Wang Q (2020) Electrokinetic-enhanced remediation of actual arsenic-contaminated soils with approaching cathode and Fe^0 permeable reactive barrier. J Soils Sediments 20:1526–1533

Yeung AT (2006) Contaminant extractability by electrokinetics. Environ Eng Sci 23(1):202–224

Yoo JC, Yang JS, Jeon EK, Baek K (2015) Enhanced-electrokinetic extraction of heavy metals from dredged harbor sediment. Environ Sci Pollut Res 22(13):9912–9921

Yuan L, Xu X, Li H, Wang N, Guo N, Yu H (2016) Development of novel assisting agents for the electrokinetic remediation of heavy metal-contaminated kaolin. Electrochim Acta 218:140–148

Zhang S, Zhang J, Cheng X, Mei Y, Hu C, Wang M, Li J (2015) Electrokinetic remediation of soil containing Cr(VI) by photovoltaic solar panels and a DC-DC converter. J Chem Technol Biotechnol 90(4):693–700

Zhao S, Fan L, Zhou M, Zhu X, Li X (2016) Remediation of copper contaminated kaolin by electrokinetics coupled with permeable reactive barrier. Procedia Environ Sci 31:274–279

Zheng XJ, Blais JF, Mercier G, Bergeron M, Drogui P (2007) PAH removal from spiked municipal wastewater sewage sludge using biological, chemical and electrochemical treatments. Chemosphere 68(6):1143–1152

Zhou DM, Deng CF, Cang L (2004) Electrokinetic remediation of a Cu contaminated red soil by conditioning catholyte pH with different enhancing chemical reagents. Chemosphere 56(3):265–273

Zhou H, Xu J, Lv S, Liu Z, Liu W (2020) Removal of cadmium in contaminated kaolin by new-style electrokinetic remediation using array electrodes coupled with permeable reactive barrier. Sep Purif Technol 239:116544

Zhu YG, Shaw G (2000) Soil contamination with radionuclides and potential remediation. Chemosphere 41(1–2):121–128

Monitoring Phytoremediation of Metal-Contaminated Soil Using Remote Sensing

25

Bhabani Prasad Mondal, Rabi Narayan Sahoo, Bappa Das, Priya Paul, Arghya Chattopadhyay, and Sonia Devi

Abstract

Phytoremediation is an effective tool which can be employed to revive the degraded or metal-contaminated soils. However, assessment of contamination caused by heavy metal in soil and its monitoring on long-term basis is essential to assess the efficacy of phytoremediation processes. Conventional techniques for monitoring the contaminated sites are noticeably expensive, time intensive, and destructive in nature. Remote sensing (RS) may assist as an efficient alternative technique for detecting metal contamination and monitoring phytoremediation on a long-term basis. The RS data from various sources at various scales such as proximal sensing data (laboratory and field-based spectroradiometric data), airborne data (dronecollected data), and space-borne data (satellite data) are crucial for monitoring the extent of contamination and to detect changes in land use pattern and surface cover of the polluted site over a time period. Most of the RS based techniques use vegetation reflectivity within the red-edge position of the electromagnetic radiation for indirect estimation of contamination level that is associated with heavy metal and organic carbon (hydrocarbon) concentration in soil. In proximal sensing, laboratory- and field-based spectroscopic data are employed to predict the level of contamination through correlating the characteristic reflectance spectra of the spectrally active soil constituents with metals. To determine the efficiency of phytoremediation, monitoring of revegetation or biorecultivation is also necessary using RS data. One of the most promising techniques to monitor revegetation is to calculate various indices related to soil, vegetation, and moisture through interpreting the remote sensing-based data product. The most frequently used vegetation index such as normalized difference vegetation index (NDVI) helps to measure the phytoproductivity of the polluted area. RS based indices are useful to detect metal-induced vegetation stress. However, a few key limitations are there in obtaining satisfactory results using RS based methods such as complexity of spectra, non-availability of unique spectral feature for particular metal, and noisy spectra due to variation in atmospheric conditions. In spite of

B. P. Mondal (✉) · R. N. Sahoo · B. Das · S. Devi
Division of Agricultural Physics, ICAR-Indian Agricultural Research Institute, New Delhi 110012, India

P. Paul
Division of Plant Physiology, ICAR-Indian Agricultural Research Institute, New Delhi 110012, India

A. Chattopadhyay
Department of Soil Science and Agricultural Chemistry, Benaras Hindu University, Varanasi 221005, India

J. A. Malik (ed.), *Advances in Bioremediation and Phytoremediation for Sustainable Soil Management*,
https://doi.org/10.1007/978-3-030-89984-4_25

so many challenges, RS based techniques are considered as non-destructive, time-saving, and cost-effective alternative techniques especially for large phytoremediation areas. Recently both airborne and space-borne hyperspectral RS data are used for continuous and detailed monitoring of the contaminated areas.

Keywords

Biorecultivation · Hyperspectral RS · Metal contamination · NDVI · Phytoremediation · Remote sensing · Satellite data

25.1 Introduction

Soil is considered as a significant sink of heavy metals discharged into the environment because of varied anthropogenetic activities. Soil gets contaminated by the increasing buildup of those heavy metals released from various sources such as atmospheric deposition, industrial emission, coal combustion, disposal of excessive heavy metals, application of high analysis chemical fertilizers and pesticides in agricultural fields, waste water application for irrigation, sewage-sludge dumping, lead based paint industry, and accumulation of by-products of petroleum and gasoline industries. (Wuana and Okieimen 2011). Soil contaminated by heavy metal poses serious problems and threats to overall environment creating potential health risks to plant, animal, and human beings through entering into the food web.

Adequate protection measures should be adopted to restore the degraded soil ecosystem, caused by heavy metal pollution. Phytoremediation, immobilization, and soil washings are three well-known techniques for remediating the contaminated sites. Among these techniques, phytoremediation is an effective scientific technique to curb the significant level of contamination while preserving the safety issues of the ecosystem. The phytoremediation is a noncorrosive technique to alleviate heavy metal toxicities from the contaminated soil through actively growing green vegetation (Rathod et al. 2012; Kaewtubtim et al. 2016). This technique mainly includes two popular methodologies namely phytoextraction and phytostabilization for remediation purposes. Apart from these two techniques, it also includes phytovolatilization, rhizofiltration, and phytodegradation (Newete and Byrne 2016). Phytoextraction denotes to transportation and accumulation of the pollutants in the foliage tissues or in the above-ground biomass, harvested later. Phytostabilization stabilizes the pollutants in soil system at a non-toxic level. In phytovolatilization, the volatile toxic substances are removed through leaf tissues. When the pollutants are removed from the aquatic system by the rhizosphere, it is known as rhizofiltration. The phytodegradation denotes to the enzymatic break down of toxic compounds inside the plant tissues and translate them into non-toxic harmless compounds. Thus, this green technology is very much effective to remediate the polluted soil through various processes. Therefore, the phytoremediation progression requires to be monitored continuously on a long-term basis as it takes much more time say for example years to decade to decontaminate a polluted area (USEPA 2000). Here, monitoring implies continuous periodic and repeated measurements of the extent of pollution in the polluted areas and observing revegetation and biorecultivation processes in that area with time. Thus, the monitoring or measuring techniques should be rapid as well as cost effective.

Most of the conventional monitoring techniques are time-consuming and destructive in nature. Therefore, they are not useful for continuous monitoring purpose. Remote sensing (RS) is regarded as an effective technological tool to meet up above-mentioned requirements for the continuous and long-term monitoring of the phytoremediation progression and simultaneously to monitor revegetation and biorecultivation of the contaminated areas with time (Ermolaev et al. 2019). There are numerous types of RS techniques ranging from ground-based techniques like proximal sensing using spectroradiometer and airborne sensing by drone and other platforms to space-borne sensing through satellites. Remote sensing techniques primarily relies on the constant of reflectance values within the red-edge position of electromagnetic radiation (EMR), reflected from soil and vegetation of the polluted areas (Noomen et al. 2015). Reflectance within the red-edge position helps to detect the quantity of organic compound (hydrocarbon) pollution within the surroundings (Noomen et al. 2015). Comparing with healthy vegetation, stressed plants grown up in polluted sites exhibit a shift within the reflectance values near to the red-edge position toward shorter wavelengths. The most convenient way to monitor revegetation of the contaminated areas is to calculate various RS based vegetation indices like ratio vegetation index (RVI), normalized difference vegetation index (NDVI), soil-adjusted vegetation index (SAVI), perpendicular vegetation index (PVI), etc. These indices help to interpret the RS data products through correlating soil, vegetation, and hydrological conditions. Among them, the extensively used vegetation index is NDVI. The NDVI value 0.5 indicates vegetation and 0.2–0.3 indicates soil. Ermolaev et al. (2019) studied the efficacy of bioremediation processes and the growth rate of vegetation at a phosphogypsum dump using RS based index mainly NDVI. In this research, spatiotemporal variations in NDVI values indicated the continuous growth and recovery of the vegetation (*Agropyron cristatum*), grown on phosphor gypsumcontaminated dump areas. The RS-based images captured either from airborne or space-borne podiums are analyzed through various machine learning algorithms to interpret the various phenomena related to phytoremediation processes.

This chapter will discuss about the drawbacks of traditional approaches and the efficacy of RS based methods under different scenarios for monitoring the phytoremediation process of the soils contaminated with metals.

25.2 Conventional Techniques of Phytoremediation Monitoring

Phytoremediation is an in-situ remediation technique, and it is popularly recognized as green-remediation or agro-remediation or vegetative-remediation technique. There are several advantages, which make it popular over other methods like (i) it can remediate the area, contaminated by more than one pollutant, (ii) economically viable and cheaper than other technologies, (iii) needs not to excavate and transport of the pollutants from the contaminated areas, (iv) green technology, beneficial for environment, and (v) esthetically pleasant. However, this technique is very much dependent on the growing conditions of the plants. Conventional techniques of phytoremediation monitoring are categorized into three classes, viz (i) performance monitoring, (ii) risk monitoring, and (iii) optimization monitoring (US-DOE 2000). In performance monitoring, effectiveness of the phytoremediation method is determined following the standard limits. The soil concentration range and regulatory standard limits of few important heavy metals are given in Table 25.1.

In risk monitoring, the phytostabilization of heavy metals in the rhizospheric zones is studied, and simultaneously it assures that such type of accumulation does not pose any serious threat to human well-being and ecosystem processes. In the last step (optimization step), various agronomic management practices are optimized to promote optimum root growth to phytostabilize the toxic metals in the rhizospheric region. The plant growth parameters, biomass, yield, and stress-induced plant responses can be measured to consider all types of monitoring steps. Various plant-related parameters like plant health status, metal-induced stress, level of metallic compounds in root, shoot and leaf tissues, rooting deepness and density, leaf geometry, and rate of transpiration are to be monitored for assessing the phytoremediation capability of the growing plant species. Various instruments are used to measure the plant parameters like sap flow logger for measuring water use efficiency and infrared thermometer for measuring photosynthetically active radiation (PAR), intercepted on leaf tissues, etc. Dendrometers are used to witness the leaf growth, while the rhizotron methods are applied to quantity rooting deepness and density. Along with this, the studies of soilrelated parameters like extent of spatial variation of soil

Table 25.1 Concentration range of heavy metals and their safe limits

Heavy metals	Concentration range in soil (ppm)	Regulatory standard limit (ppm)
Zn	150–5000	1500
Hg	< 0.01–1800	270
Pb	1–6900	600
Cr	0.1–345	100
Cd	0.05–3950	100

Adapted after: Wuana and Okieimen (2011)

pollutants, heavy metal speciation, etc., are required for monitoring the phytoremediation process (Rathod et al. 2012). Traditional chemical analyses of soil physical and chemical properties are very much time-consuming, expensive, destructive, and also hazardous to environment also. The other conventional ways to monitor contaminated sites are soil sampling, drilling of soil, and geochemical investigation of toxic metals at the polluted sites (Noomen et al. 2015).

These methods are very much unsuitable for monitoring large areas. Therefore, a rapid and inexpensive technique is needed to avoid all such kind of issues or problems, associated with conventional techniques. Definitely, RS is a potential technology in this respect already mentioned. The remotely detected images can be integrated under geographical information system (GIS) environment to generate spatially auto-correlated maps of the phytoremediation areas at both spatiotemporal scales (Nanni and Dematte 2006; Rathod et al. 2012).

25.3 Potential of Remote Sensing for Phytoremediation Monitoring

Remote sensing denotes to a technique which helps to acquire data related to an entity or phenomena without coming in straight physical connection with it under any conditions like from laboratory and field conditions to airborne and space-borne conditions. RS may be deployed to monitor vegetation conditions as the vegetation has characteristic spectral response in EMR region. Generally, plant shows higher reflectance (around 40–50%) in the near-infrared (NIR) region of EMR spectrum, and it is controlled by internal leaf structure or by plant anatomy. In the visible part, the plant exhibits very less amount of reflectance (generally 10%) near visible region as maximum amount of EMR gets absorbed by plant pigments, particularly by chlorophyll pigments. Chlorophyll absorbs EMR near 0.45 and 0.67 μm regions. Higher chlorophyll pigments broaden the absorption maxima (around 660–680 nm) and shift the maximum slope of the reflectance of red-edge (680–760 nm) toward to longer wavelength, known as "red-edge-shift" (Horler et al. 1983). The green color of the visible portion of EMR gets reflected more from the actively growing vegetation, while red color gets reflected more under stressed conditions or under senescence conditions. Thus, the RS has to potential characterize the stressed condition of plant by using EMR spectrum with the help of different kinds of sensors directly or indirectly. The direct assessment includes the target property that directly affects the sensor characteristics, while in indirect measurements, the other related properties influence the sensor measurements. Reflectance spectroscopy ranging from visible-near-infrared (350–1100 nm) to shortwave infrared (SWIR: 1100–2500 nm) and mid infrared (MIR:2500–25,000 nm) is extensively applied to assess the properties of the soil (Stenberg et al. 2010; Nawar and Mouazen 2017; Mondal and Sekhon 2019) and to study plant characteristics (Das et al. 2020).

We will discuss the efficiency of proximal (ground-based) sensing, airborne sensing, and satellite-borne sensing for monitoring metal contamination and phytoremediation through soil and vegetation monitoring.

25.4 Proximal RS for Studying Metal Contamination in Soil

In the recent years, the proximal sensing has gained much popularity in rapid estimation of soil and vegetation properties (Viscarra Rossel et al.). Proximal sensors may be of two types: active and passive. Further, it can be categorized into invasive and non-invasive types also. Invasive type measures the target properties by making direct connection with that target with the help of a contact probe having in-built light source, whereas the non-invasive type measures the object properties overhead the target surface without coming any physical interaction with it. It is better to use proximal sensing (near sensing) rather than far remote sensing like space-based RS for monitoring level of contaminants or leaf characteristics of the hyper accumulator plants as the proximal technique is close enough to collect single and pure reflectance spectrum from the targeted object both under laboratory-based and field-based (in situ) conditions (Xu et al. 2008). Generally, handheld or portable field spec spectroradiometer is applied to measure the characteristics reflectance spectra of both soil and vegetation samples, illuminating the samples with artificial light source under laboratory-condition and solar radiation under field situations around 350–2500 nm region.

Spectroradiometer is usually employed to record VIS–NIR spectra of the various objects. The visible spectra (350–780 nm) are mainly influenced by the occurrence of iron-bearing minerals and iron oxides like goethite, hematite, etc. The NIR spectra are mainly dominated by the weak overtones and vibrations of the important functional groups and mostly absorbed by soil moisture and organic matter (Song and He 2005). Bending and stretching mechanisms are related to the light absorption depending on the constituents of the assessed samples (Knadel et al. 2017). Thus, based on these basic principles, this spectroradiometric approach can be successfully used for direct prediction of concentration of any toxic metal in soil. However, direct prediction of any toxic metal at a concentration level below 1000 ppm is not possible due to absence of distinct spectral response feature in VIS–NIR region at this concentration (Rathod et al. 2012). Some transition metals like nickel (Ni), copper (Cu), and chromium (Cr) can be noticed by this spectroscopy as they exhibit direct spectral relationship at a concentration more than 4000 ppm (Wu et al. 2007).

For indirect prediction, various statistical methods such as principal component regression (PCR), partial least square regression (PLSR), support vector machine regression (SVMR), random forest (RF), and artificial neural network (ANN) are applied for correlating the coefficient values of reflectance of collected spectra with the measured properties and spectrally active constituents of the soil called chromophores to develop spectral model and then the developed spectral model is employed to estimate several soil and plant-related properties. Before applying the multivariate models, several spectral preprocessing techniques like Savitzky–Golay filtering and smoothening, derivative, continuum removal, standard normal variate (SNV), etc., are employed to get the optimum results from the model. Among these techniques, PLSR is the widely applied technique for estimating properties of the soil (Conforti et al. 2015). Several researchers have used this technique for prediction soil fertilityrelated parameters like soil organic carbon (Vasques et al. 2008; Viscarra Rossel and Behrens 2010), soil available nitrogen (Wenjun et al. 2014), available phosphorus (Mondal and Sekhon 2019), and exchangeable potassium (Gras et al. 2014). Malley and Williams (1997) first reported the possibility of spectroscopic techniques to characterize the level of contamination, caused by the presence of toxic metals in soil. Several other types of pollutions such as hydrocarbon pollution has been reported by Chakraborty and Weindrof (2010) and soil salinity by Farifteh et al. (2008). Thus, soil spectroscopy may be employed to estimate the properties related to soil polluted with harmful metals.

Although NIR spectroscopy is mostly deployed as a portion of proximal RS, this technique has some limitations. This technique underestimates the estimation of dust-borne particles of any toxic metals in heavily polluted soil owing to its narrower spectral range. Mid-infrared (MIR) spectroscopy is better technique than NIR due to wider spectral range, and more spectral wavebands are related to the vibrations of fundamental molecular bonds of hydroxyl, amine, amide, carbonyl, etc. MIR spectra are also sensitive to the inorganic soil components like phosphate and carbonate content of the soils (Siebielec et al. 2004). MIR spectroscopy can be efficiently applied to estimate the heavy metals like Zn, Cu, Cd, and Ni with higher R^2 values (more than 0.80). Some typical MIR spectral wavebands (5848, 5917–5988 nm) may vary with varying level of Pb concentration because it makes complex with functional groups (carboxylic groups and phenolic hydroxyl groups, etc.) existing within organic matter (Dupuy and Douay 2001).

To present the previous work associated with the indirect estimation of toxic metals using proximal RS, in particularly with reflectance spectroscopy, a brief overview is tabulated here (Table 25.2).

From the above discussion, it is clear that various multivariate regression models may be applied to estimate the heavy metals indirectly using reflectance-based spectroscopy (proximal sensing). Although, it is an efficient technique, several challenges are associated with this technique. The challenges are described below.

1. It is very difficult to apply this technique for subsurface level heavy metal monitoring and sensing through reflectance spectra.
2. The accuracy of such technique's prediction is less than the reference technique (chemical analysis).
3. Field-based (in situ) prediction is not so much good like laboratory-based prediction due to heterogeneity under field conditions. Soil surface roughness, variation in surface soil moisture condition, vegetative cover, variation in solar radiation intensity, cloud cover, etc., affect the in-situ spectra.
4. Even under laboratory conditions, variations in instruments, variation in light intensity, and variable sample preparation also produce variable results.

Table 25.2 Summary of indirect prediction of heavy metals using reflectance spectroscopy

Heavy metals	Concentration in soils (ppm)	Type of reflectance spectra used	Regression techniques with prediction accuracy	References
Zn	40–1322	Reflectance spectra	PLSR (R^2 = 0.84)	Vohland et al. (2009)
Cu	21.9–252.6	Ratio of 1344/778 nm	SMLR (R^2 = 0.72	Choe et al. (2009)
As	19.3–403.7	1st derivative reflectance spectra	PLSR (R^2 = 0.61)	Ren et al. (2009)
Pb	18–6530	MIR spectra	ANN (R^2 = 0.94)	Siebielec et al. (2004)
Cr	60.8–104.0	First derivative reflectance spectra	PLSR (R^2 = 0.85)	Wu et al. (2007)
Ni	10.6–59.25	Savitzky–Golay (SG) smooth reflectance spectra	MARS (R^2 = 0.91)	Wu et al. (2011)
Cd	0.17–1.57	Reflectance spectra	Regression (R = 0.76)	Xia et al. (2007)

Adapted after: Rathod et al. (2012)

5. Extreme variations in the concentration of heavy metals are necessary for getting better prediction accuracy. Such extreme variations can only be obtained only in highly contaminated sites, not in the agricultural fields.
6. Simultaneous assessment of multi-metal contamination is often very difficult.

25.5 Monitoring Metal Uptake by Plants During Phytoremediation Using RS

For effective phytoremediation, periodic monitoring of the growth rate and survival rate of vegetation grown up on contaminated site is important. Plant reflectance characteristics change during various growth stages, and the reflectance spectra in the VIS–NIR and MIR regions are also influenced by the extent of heavy metal content of canopy. Spectral reflectivity features are mainly governed by the plant pigments, leaf anatomy, biochemical compositions, and the morphological characteristics of the leaf tissues. Extreme amount of heavy metallic compounds adversely affects the growth rate and the metabolic activities of vegetation (Prasad 2004). Broadly, stress caused by toxic heavy metals is mainly responsible for reduction in chlorophyll concentration (Zengin and Munzuroglu 2005), deformation of internal leaf structure (Smith and Blackshaw 2003), which directly or indirectly influence the reflectance comportment of the plant. Buildup of Zn in excessive amount causes reflectance values to decrease near NIR section and causes a blue shift in the red-edge part of the spectrum. (Sridhar et al. 2007). Lead (Pb) accumulation causes a rise within the reflection coefficient values close to NIR region, whereas accumulation of Cd causes the blue shift within the red-edge position and increases coefficient of reflection in the visible part of the spectrum. The handheld spectroradiometer may be applied for measuring the content of toxic heavy metals like Zn, Cu, and Pb in *Phragmites australis* (Liu et al. 2010). Heavy metal concentration conjointly affects the pigment (chlorophyll) content of the leaf, which in turn affect the spectral response pattern. Leaf chlorophyll content is inversely proportional to the content of heavy metallic compounds present in leaf. The 82% variability of leaf chlorophyll content can be explained by taking into consideration of normalized band depth at three particular wavebands such as green (537 nm), red (667 nm), and NIR (747 nm). The numerous RS based studies have therefore reported a good association between the leaf chlorophyll content and the content of heavy metallic compound of the canopy, influencing the reflectance spectra (Shakya et al. 2008). Li et al. (2008) also found significant correlation between leaf chlorophyll content and Cu concentration, while investigating the biogeochemical response of vegetation in a copper (Cu) mining area. Their research experiment finds that with an increment in Cu concentration, reflectance coefficient of canopy increase causing blue shift from 5 to 15 nm, and simultaneously red shift of about 4.55–8.95 nm. However, the increased Cu concentration decreases the depth of chlorophyll absorption.

Hence, it can be inferred that RS is a very good technological tool applied to distinguish the metal-induced stressed vegetation from the actively growing healthy vegetation, and it is similarly potent to predict the content of heavy metallic compounds in the plant tissue. Along with this, it is very much important to discriminate various morphologically alike plant species, involved in phytoremediation.

25.6 Plant Species Discrimination by RS

Traditional methods of distinguishing the morphologically similar plant species are very much difficult, laborious, and time-consuming. Under wetland ecosystem, this task is much more problematic owing to the inaccessibility of the wetland ecosystem (Mabhungu et al. 2019). Up-to-date mapping of wetland species and the discrimination of morphologically similar species are, however, crucial for the efficient supervision of polluted wetland (Davranche et al. 2010). RS

Table 25.3 Aquatic macrophyte plant species involved in removing heavy metal contamination in wetland ecosystem

Plant species	Heavy metals removed	References
Typha capensis	Zn, Mn, Fe, Ni	Van der Merwe et al. (1990)
Phragmites australis	Zn, Cu, Ni, Cr	Bragato et al. (2006)
Bolboschoenus Maritimus	Zn, Cu, Fe, Al	Shuping et al. (2011)
Typha domingensis	Zn, Cu, Mn, Al, As, Cd, Cr, Ni, Pb, Hg	Bonanno (2013)
Cyperus vaginatus	Zn, Cu, Fe, Mn, Ni, Co, Cd, Pb	Aryal et al. (2016)

Adapted after: Mabhungu et al. (2019)

is an efficient substitute for this purpose also. We have listed few species of vegetations grown under wetland situations, involved in removing of hazardous heavy metallic compounds from wetland ecosystem (Table 25.3).

RS has already shown its capability to discriminate against plant species because various vegetation classes and their associated species in the EMR region have their own spectral reflectance pattern due to different biophysical and biochemical properties (Adam et al. 2010). Several scholars are interested in the introduction of RS for distinguishing various species of vegetation around the globe (Dubula et al. 2016). Dubula et al. (2016) have discriminated several invasive plant species at a natural reservoir of Johansenberg in South Africa. Researchers have showed that NIR region of EMR could be able to distinguish various vegetation species of wetland ecosystem of the reservoir. Pu (2009) used advanced data mining algorithms like ANN and linear discriminant analysis (LDA) to identify and differentiate 11 forest plant species using RS based spectrometric analysis by correlating their spectral characteristics with pigments, extent of leaf water content, and other biochemical characteristics of those plant species.

Although RS is a feasible technique for species discrimination, several challenges are also involved in this regard. Optical RS is not much useful for differentiation of vegetation species under wetland situations due to narrower ecotone of those vegetation units (Zomer et al. 2009; Mabhungu et al. 2019). The field-based hyperspectral RS is ideal for discrimination of species under wetland situations, and it is at least better than aerial and satellitebased multispectral RS because hyperspectral sensors comprise of hundreds of fine adjoining bands in a continuous manner, containing better spectral information (Adam et al. 2010). The visible (400–700 nm) and red-edge (700–730 nm) positions are significant for discriminating the species of vegetation (Zomer et al. 2009). Adam et al. (2012) successfully discriminated four vegetation species grown under marshy land conditions such as *Phragmites australis*, *Cyperus papyrus*, *Echinochola pyramidalis,* and *Thelypteris interrupta* using ASD spectroradiometry.

Vegetation stress, caused by polluted drainage water of acid mine soil area, may be noticed by the field-based spectroscopy. Mangrove plant species on terrestrial ecosystem contaminated by heavy metals can also de differentiated by using field spectrometry (Vaiphasa et al. 2005).

Thus, it is obvious that RS has the ability to discriminate the various species of vegetations grown up under both wetland and terrestrial situations on the basis of their spectral response.

25.7 Metal-Induced Stress Monitoring Using RS Derived Vegetation Indices

Vegetation indices are one kind of spectral indices, generally calculated by the combination of several wave bands and the corresponding reflectance values to increase the spectral response characteristics and to eliminate the background effect. Various RS based indices have been studied extensively, mentioned earlier in the manuscript. In this section, several vegetation indices are given in Table 25.4.

Table 25.4 RS derived vegetation indices for studying heavy metal stress in plants

RS derived vegetation indices	Equations	References
Normalized difference vegetation index (NDVI)	$(R_{NIR}-R_{red})/(R_{NIR} + R_{red})$	Liu et al. (2010)
Ratio vegetation index (RVI)	R_{NIR}/R_{red}	Kooistra et al. (2003)
Difference vegetation index (DVI)	$(R_{NIR}-R_{red})$	Kooistra et al. (2003)
Soiladjusted vegetation index (SAVI)	$(R_{NIR}-R_{red})(1 + L)/(R_{NIR} + R_{red} + L)$; where $L = 0.5$	Davidson and Csillag (2001)
Chlorophyll index (CI)	$(R_{750}-R_{705})/(R_{750} + R_{705})$	Penuelas and Filella (1998)
Normalized pigment chlorophyll index (NPCI)	$(R_{680}-R_{430})/(R_{680} + R_{430})$	Penuelas and Filella (1998)

Adapted after: Rathod et al. (2012)

Among them, NDVI and RVI have been successfully employed to discriminate healthy and metal-induced stressed vegetation. Greater values of those above-mentioned indices indicate the presence of higher amount of chlorophyll in canopy and thus indicating healthy and stress-free vegetation status (Schowengerdt 2006). The NDVI and RVI are also useful in detecting stress, caused by the heavy metallic compounds in various plant species like hardwood assemblages (*Betula populifolia*) (Gallagher et al. 2008). Significantly lower NDVI and RVI values were also reported by Dunagan et al. (2007) for field grown mustard, grown under soil polluted by heavy metallic compounds. Ren et al. (2010) found a significant correlation between NDVI values and concentration of heavy metallic element in the paddy leaves, exhibiting correlation coefficient values -0.76, -0.68, and -0.76 for NDVI and Cu, NDVI and Zn, and NDVI and Pb, respectively. However, a group of researchers also stated that NDVI and RVI were not much useful in detecting plant stress caused by either deficient or in toxic level concentration of heavy metallic element particularly Zn. (Schuerger et al. 2003).

Combination of RS based vegetation indices with multivariate regression models like PLSR, RF, etc., is useful for identifying stress in plants. Kooistra et al. (2004) used PLSR model in together with DVI to correlate the spectral coefficient of reflection values of perennial grass species within the spectral range of 400–1350 nm with the heavy metals concentration (Zn, Cu, Ni, Cd, Pb) and obtained satisfactory results with R^2 (coefficient of determination) values starting from 0.50 to 0.73. However, they also reported poor R^2 values, i.e., below 30% in the cross-validation dataset of the spectral model for herbaceous vegetation types like *Rumex acetosa*, *Cirsium arvensis,* etc., which could be accredited to the change in metal sensitivity among plant species, variable plant morphological characteristics, and the time of spectral data measurement. Heavy metal stress also affects the plant metabolism, and it could be better indicated by the RS based indices like NPCI (Panigada et al. 2010). The back propagation neural network model may be employed to estimate the concentration of heavy metallic elements in paddy leaves by correlating the sensitivity of the spectral indices with the pigment (chlorophyll) content of the canopies (Liu et al. 2011).

It is very much obvious that maximum RS derived diagnostics or indices use the reflectance coefficient values very nearer to the red-edge location due to the sensitivity of this spectral region with plant pigments like chlorophyll. However, some researchers mentioned that this region could not detect the stressed condition of plant due to lower chlorophyll content of canopy cells under metal-induced stress condition (Schuerger et al. 2003; Sridhar et al. 2007). NIR region useful for metal detection is mainly affected by the internal leaf architecture or by leaf anatomy. But the presence of excessive amount of heavy metallic compounds also damages the internal cell architecture of plant leaf tissue. Sridhar et al. (2005) reported the impairment of leaf cell architecture due to phytoextraction of Zn and Cd, whereas the reduction of mesophyll cells in several plant species owing to heavy metallic compound toxicity has been reported by Chmielewska and Chwil (2005).

Sridhar et al. (2007) conducted an experiment on phytoextraction of Zn and Cd by *Hordeum vulgare* and As and Cr by *Pteris vittata* using the hyperspectral sensors. According to their research findings, 800–1300 nm spectral region is linked with foliar structural change due to photo-accumulation of heavy metallic compounds as R_{1110}/R_{810} (ratio index) is very much sensitive to toxic metal accumulation by plant species. They have also compared the performance of ratio index with NDVI and concluded that NDVI performed better in correlating As content with both root and shoot tissues, whereas the ratio index proved its superiority in differentiation As-treated crop species with Cr-treated plant species.

Structural parameters of spinach leaf such as mesophyll structure and its mean surface area, palisade parenchyma, ratio of palisade parenchyma to spongy parenchyma of mesophyll, mean leaf air space ratio, etc., and the leaf arsenic (As) concentration is closely related with the variations in the canopy reflectance pattern near the wavebands of 1048, 1080, 1098 nm, etc. The amount of As metal in the spinach leaf can also be predicted through regression analysis using ratio index of 1048 and 1021 nm wavebands (R_{1048}/R_{1021}) and NIR bands (Bandaru et al. 2010). NIR (998, 1448, 1644, and 2148 nm) and SWIR (1174, 1888, 2140, and 2331 nm) can be employed to discriminate two important species of *Pteris* ferns, namely *Pteris multifidi* and *Pteris cretica mayii* because those wavelengths are closely related with the leaf-geochemistry and chemical composition (Slonecker et al. 2009). Better R^2 value can be obtained from the spectral model, derived from first derivate reflectance spectra instead of raw reflectance spectra. Metal accumulation influences the production of lignin and protein, and it is correlated with the removed continuum of band depth near 1730 nm (Gotze et al. 2010).

Thus, the following studies exhibit that RS derived vegetation indices, combination of various spectral indices, and red-edge indices particularly at infrared region are the potential indicators of metal stress detection through monitoring the physiological growth conditions of the plants during the phytoremediation processes of soils, polluted by heavy metallic elements (Rathod et al. 2012).

25.8 Phytoremediation Monitoring Using Airborne RS

Very few research works have been conducted on phytoremediation monitoring using airborne RS because monitoring of natural phenomena requires several images and the price of aerial hyperspectral imageries with fine resolution is much higher (Smith et al. 2004; Noomen et al. 2015). Airborne imaging spectrometer HyMap (Cocks et al. 1998) sensor is employed to collect multitemporal hyperspectral images for observance the bioremediation of hydrocarbons polluted soil (Noomen et al. 2015). The specifications of HyMap sensors are given below in Table 25.5. VIS, NIR, and SWIR wavelength regions of the sensors are much useful in this regard. Red-edge position (REP) index may be applied to evaluate the phytoremediation progression. The REP can be calculated by following the formula given by Guyot and Baret (1988).

$$\mathrm{REP} = 700 + 40 * \left[\frac{R_i - R_{700}}{R_{740} - R_{700}}\right]$$

where R_i is the coefficient of reflection at inflexion point. A change of REP values toward the longer wavebands occurs if the hydrocarbon (especially benzene) pollution is being remediated by the vegetation. Vegetation reflectance values varies with a certain time period and such variations can be obliterated by using statistical

Table 25.5 Typical design specifications of HyMap sensors

Configuration parameters	Specifications
Field of view (FOV)	60° (512 pixels)
Instantaneous FOV (IFOV)	2.5 mr along track
Swath	2.3 km at 5 m IFOV (along track)
Number of channels	100–200
Spectral bandwidths	10–20 nm
Spectral regions	VIS, NIR, SWIR, MWIR, TIR
Spatial resolution	2–10 m
Width of swath	60–70°
Signal: noise	> 500:1
Flying altitude	2000–5000 m above ground level

Adapted after: Cocks et al. (1998)

normalization procedure. If REP moves to shorter wavelength, it indicates a high degree of pollution of the contaminated sites. The movement of REP position toward shorter wavebands for maize plant, grown up on CO_2 polluted soil, and also for the plants survived on hydrocarbon polluted soil has been observed by many researchers (Yang et al. 2000; Noomen and Skidmore 2009).

Change in the REP values over time indicates the degree of remediation; a positive change indicates a better phytoremediation potential, whereas the negative change indicates poor phytoremediation capability of the vegetation. Noomen et al. (2015) have studied the processes of phytoremediation of a hydrocarbon polluted site from the year 2005 to the year of 2008 and showed two airborne images of 2005 and 2008 and how the threshold value changed over time. For the image of 2005, the variance among the pixels were 1 nm, whereas the variance among the pixels of the image of 2008 corresponded to 0.875. The difference between these two numerical values indicates a positive change which means good phytoremediation.

They have chosen total 17 sites for observing phytoremediation potential. Among them, 11 locations are correctly classified as remediation sites, and thus, they have exhibited that the user accuracy and the total accuracy of the remediation areas are 71 and 65%, respectively.

25.9 Phytoremediation Monitoring Using Satellite-Borne or Space-Borne RS

Improved acquisition of satelliteborne data unfolds good opportunities for periodic monitoring of an area of interest or any phenomena. Satelliteborne RS is very much helpful for monitoring large scale areas. Monitoring spatiotemporal changes of surface cover and identifying different land use units are crucial for monitoring the contaminated sites periodically (Schimid et al. 2013). However, the coarse spatial resolution is a constraint for satellite-borne images. Therefore, an integrated method combing field-based proximal sensing, laboratory measurements, and satellite-borne data from different sources is essential for monitoring periodic changes of the land cover of the contaminated areas. Schimid et al. (2013) used such type of integrated RS based methodology for observing surface characteristics and surface cover changes of mercury mining areas of the three municipal areas in Spain.

The integrated approach mainly comprises of four phases (Fig. 25.1), and the methodology has been described briefly.

In first phase, field-based reflectance spectra of the mercury (Hg) mining site have been recorded using a portable field spec spectroradiometer. Along with this field campaign,

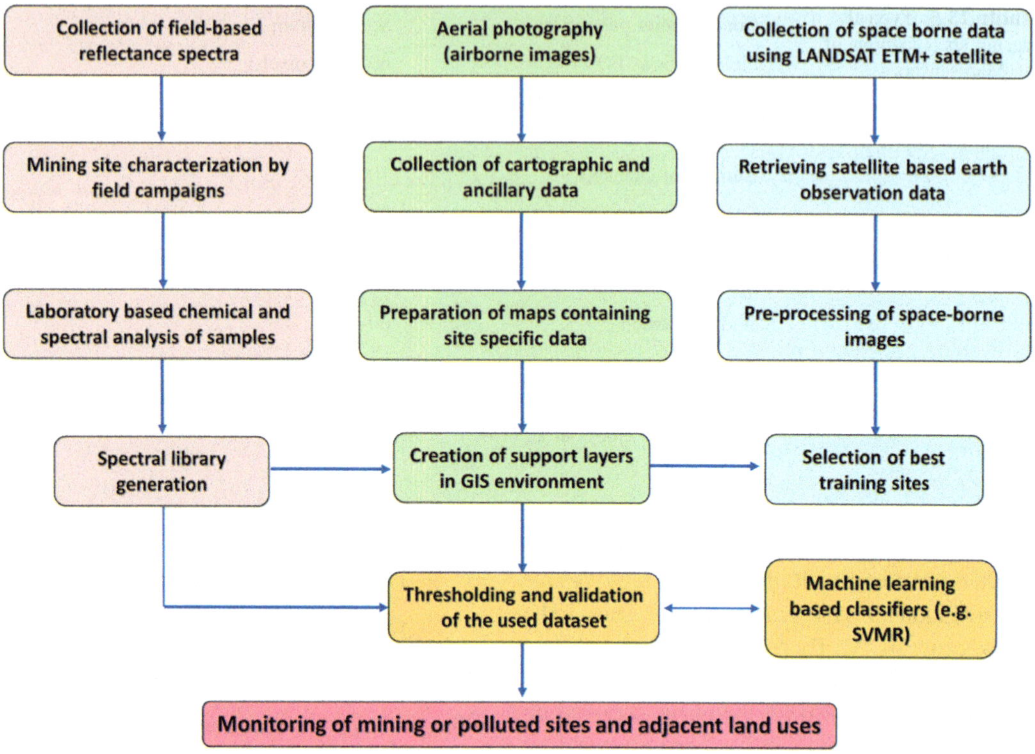

Fig. 25.1 Integrated RS based methodology for monitoring mining areas and land usage and land surface cover changes. Adapted after: Schmid et al. (2013)

laboratory-based all chemical and spectral investigations were carried out following the standard protocols, and later, they were correlated to generate spectral library database (Shepherd and Walsh 2002). In second phase, aerial photography and ancillary data base and the maps of topography, lithology, and vegetation cover have been collected, and a Geographic Information System (GIS) has applied for managing, compiling, and georeferencing the databases to create support layers. Later, Earth Observation satellite data from optical sensors like Landsat ETM + were preprocessed in phase three, i.e., all kinds of atmospheric and radiometric rectifications were performed. In final phase, a machine learning-based supervised classifier, i.e., support vector machine regression (SVMR), has been applied for categorization of all pixels in the raster data into different land usage and surface cover units on the basis of spectral similarity of pixels (Foody and Mathur 2004). Their study recognizes that the RS technology along with field-based and laboratory-based measurements, photogrammetry, and ancillary data are useful to monitor and manage Hg contaminated areas. The study thus demonstrates the possible applications of satellite-based RS for monitoring pollution caused by heavy metallic compounds.

25.10 Future Prospects

Till now, most of the studies specialize in the utilization of proximal sensing for phytoremediation monitoring on the premise of the shift of red-edge position, spectral reflectance factor or features, and numerous RS based vegetation indices. Only few studies are related to aerial and satellitebased hyperspectral RS for monitoring the phytoremediation progression of the metal contaminated soils. Therefore, the research area should be extended to the application of airborne hyperspectral sensors like the Airborne Visible Infrared Imaging Spectrometer-next Generation (AVIRIS-NG), the Compact High Resolution Imaging Spectrometer (CHRIS), etc., and also to the space-borne sensors like Hyperion space-borne hyperspectral imager along with proximal sensors (both imaging and non-imaging types) for better monitoring of the metal-polluted sites and the revegetation during phytoremediation. Nowadays, RS technology is capable to monitor vegetation stress only in severe conditions as a component of phytoremediation. More advanced techniques like thermal infrared sensing, fluorescence spectroscopy, etc., need to be explored to detect early signs of metal-induced abnormalities or stresses in vegetation. Research is also essential to sort out the new techniques for differentiating natural stress from metal-induced stress in the vegetation.

25.11 Conclusion

Undoubtedly, the phytoremediation successfully protects our environment form the ill-effects of the contamination, especially caused by heavy metallic compounds in soil. This chapter has demonstrated the efficiency of various RS based techniques particularly the proximal sensing and also few instances of aerial- and satellite-based RS techniques for monitoring the phytoremediation progression on polluted sites exposed to toxic metals particularly heavy metals. The spectral-based RS techniques (NIR, MIR spectroscopy) are effective in predicting the degree of heavy metal pollution in soil. Correlation of heavy metal concentration with spectrally energetic elements of soil aids in indirect estimation of heavy metals. Along with this, RS is also capable to monitor development rate and survivability of the plants, grown on toxic metal contaminated soils during phytoremediation. RS successfully monitors metal-induced stress in numerous vegetation species to facilitate successful management by measuring a shift near the red-edge position toward shorter wavelength. Moreover, it also assists in discerning the morphologically similar plant species, involved in phytoremediation process. RS derived vegetation indices (NDVI, RVI, etc.) are better indicators of vegetation growth status during phytoremediation. Greater values of such indices indicate higher amount of chlorophyll concentration and better plant health status. Some challenges associated with RS based techniques have also been clearly described here. Despite of so many challenges, RS is a popular technique, deploying to monitor phytoremediation progression of toxic metal-polluted soils. However, since, phytoremediation is still in an evolving stage, multidisciplinary research in collaboration with modern RS based methods is very much essential to make it a commercially feasible green technology throughout the globe.

References

Adam E, Mutanga O, Rugege D (2010) Multispectral and hyperspectral remote sensing for identification and mapping of wetland vegetation: a review. Wetlands Eco Manag 18:281–296. https://doi.org/10.1007/s11273-009-9169-z

Adam EM, Mutanga O, Rugege D, Ismail R (2012) Discriminating the papyrus vegetation (Cyperus papyrus L.) and its co-existent species using random forest and hyperspectral data resampled to HYMAP. Int J Remote Sens 33:552–569. https://doi.org/10.1080/01431161.2010.543182

Aryal R, Nirola R, Beecham S, Kamruzzaman M (2016) Impact of elemental uptake in the root chemistry of wetland plants. Int J Phytoremediation 18:936–942. https://doi.org/10.1080/15226514.2015.1131239

Bandaru V, Hansen DJ, Codling EE, Daughtry CS, White-Hanson S, Green CE (2010) Quantifying arsenic-induced morphological changes in spinach leaves: implications for remote sensing. Int J Remote Sens 31(15):4163–4177. https://doi.org/10.1080/01431161.2010.498453

Bonanno G (2013) Comparative performance of trace element bioaccumulation and biomonitoring in the plant species *Typha domingensis*, *Phragmites australis* (Cav.) Trin. ex Steud. and *Arundo donax* L. Ecotoxicol Environ Saf 97:124–130. https://doiorg/101016/jecoenv201307017

Bragato C, Brix H, Malagoli M (2006) Accumulation of nutrients and heavy metals *in Phragmites australis* (Cav.) Trin. ex Steudel. and *Bolboschoenus maritimus* (L.) *Palla* in a constructed wetland of the Venice lagoon watershed. Environ Pollut 144(3):967–975. https://doiorg/101016/jenvpol200601046

Chakraborty S, Weindorf D (2010) Rapid identification of oil-contaminated soils using visible near infrared diffuse reflectance spectroscopy. J Environ Qual 39:1378–1387. https://doi.org/10.2134/jeq2010.0183

Chmielewska EW, Chwil M (2005) Lead induced histological and ultrastructural changes in the leaves of soybean (*Glycine max*). Soil Sci Plant Nutr 51(2):203–212. https://doi.org/10.1111/j.1747-0765.2005.tb00024.x

Choe E, Kim KW, Bang S, Yoon IH, Lee KY (2009) Qualitative analysis and mapping of heavy metals in an abandoned Au–Ag mine area using NIR spectroscopy. Environ Geol 58(3):477–482. https://doi.org/10.1007/s00254-008-1520-9

Cocks T, Jenssen R, Stewart A, Wilson I, Shields T (1998) The HyMap airborne hyperspectral sensor: the system calibration and performance. In: Schaepman M, Schläpfer D, Itten K (eds) Proceeding of 1st EARSeL workshop on imaging spectroscopy. Zürich, Switzerland, pp 37–42. http://artefacts.ceda.ac.uk/neodc_docs/Hymap_specs.pdf

Conforti M, Castrignanò A, Robustelli G, Scarciglia F, Stelluti M, Buttafuoco G (2015) Laboratory-based Vis–NIR spectroscopy and partial least square regression with spatially correlated errors for predicting spatial variation of soil organic matter content. CATENA 124:60–67. https://doi.org/10.1016/j.catena.2014.09.004

Das B, Manohara KK, Mahajan GR, Sahoo RN (2020) Spectroscopy based novel spectral indices PCA-and PLSR-coupled machine learning models for salinity stress phenotyping of rice. Spectrochim Acta a: Mol Biomol Spectrosc 229:117983. https://doi.org/10.1016/j.saa.2019.117983

Davidson A, Csillag F (2001) The influence of vegetation index and spatial resolution on a two-date remote sensing-derived relation to C4 cpecies coverage. Remote Sens Environ 75:138–151. https://doi.org/10.1016/S0034-4257(00)00162-0

Davranche A, Lefebvre G, Poulin B (2010) Wetland monitoring using classification trees and SPOT-5 seasonal time series. Remote Sens Environ 114 (3):552–562. https://doi.org/10.1016/j.rse.2009.10.009

Dubula B, Tesfamichael SG, Rampedi IT (2016) Assessing the potential of remote sensing to discriminate invasive *Asparagus laricinus* from adjacent land cover types. Cogent Geosci 2:1–17. https://doi.org/101080/23312041201611546 50

Dunagan SC, Gilmore MS, Varekamp JC (2007) Effects of mercury on visible/near-infrared reflectance spectra of mustard spinach plants (Brassica rapa P.). Environ Pollut 148(1):301–311. https://doi.org/10.1016/j.envpol.2006.10.023

Dupuy N, Douay F (2001) Infrared and chemometrics study of the interaction between heavy metals and organic matter in soils. Spectrochim Acta Part A Mol Biomol Spectrosc 57(5):1037–1047. https://doi.org/10.1016/S1386-1425(00)00420-0

Ermolaev NR, Belobrov VP, Yudin SA, Torochkov EL (2019) Monitoring the process of bioremediation and revegetation of a phosphogypsum waste stack by remote sensing. In: IOP Conference series: earth and environmental science, vol 350, no 1. IOP Publishing pp 012057. https://doi.org/10.1088/1755-1315/350/1/012057

Farifteh J, Van der Meer F, Van der Meijde M, Atzberger C (2008) Spectral characteristics of salt-affected soils: A laboratory experiment. Geoderma 145(3–4):196–206. https://doi.org/10.1016/j.geoderma.2008.03.011

Foody GM, Mathur A (2004) A relative evaluation of multiclass image classifica- tion by support vector machines. IEEE Trans Geosci Remote Sens 42:1336–1343. https://doi.org/10.1109/TGRS.2004.827257

Götze C, Jung A, Merbach I, Wennrich R, Gläßer C (2010) Spectrometric analyses in comparison to the physiological condition of heavy metal stressed floodplain vegetation in a standardized experiment. Open Geosci 2(2):132–137. https://doi.org/10.2478/v10085-010-0002-y

Gallagher FJ, Pechmann I, Bogden JD, Grabosky J, Weis P (2008) Soil metal concentrations and productivity of *Betula populifolia* (gray birch) as measured by field spectrometry and incremental annual growth in an abandoned urban Brownfield in New Jersey. Environ Pollut 156(3):699–706. https://doi.org/10.1016/j.envpol.2008.06.013

Gras JP, Barthès BG, Mahaut B, Trupin S (2014) Best practices for obtaining and processing field visible and near infrared (VNIR) spectra of topsoils. Geoderma 214:126–134. https://doi.org/10.1016/j.geoderma.2013.09.021

Guyot G, Baret F (1988) Use of high spectral resolution to follow the state of vegetation cover. In: Spectral signatures of objects in remote sensing, vol 287, pp 279. http://adsabs.harvard.edu/full/1988esasp.287..279g

Horler DNH, Dockray M, Barber J (1983) The red edge of plant leaf reflectance. Int J Remote Sens 4(2):273–288. https://doi.org/10.1080/01431168308948546

Kaewtubtim P, Meeinkuirt W, Seepom S, Pichtel J (2016) Heavy metal phytoremediation potential of plant species in a mangrove ecosystem in Pattani Bay, Thailand. Appl Ecol Environ Res 14(1):367–382. https://doi.org/10.15666/aeer/1401_367382

Knadel M, Gislum R, Hermansen C, Peng Y, de Moldrup P, Jonge LW, Greve MH (2017) Comparing predictive ability of laser-induced breakdown spectroscopy to visible near-infrared spectroscopy for soil property determination. Biosys Eng 156:157–172. https://doi.org/10.1016/j.biosystemseng.2017.01.007

Kooistra L, Leuven R, Wehrens R, Nienhuis PH, Buydens LMC (2003) A comparison of methods to relate grass reflectance to soil metal contamination. Int J Remote Sens 24(24):4995–5010. https://doi.org/10.1080/0143116031000080769

Kooistra L, Salas EAL, Clevers J, Wehrens R, Leuven R, Nienhuis PH, Buydens LMC (2004) Exploring field vegetation reflectance as an indicator of soil contamination in river floodplains. Environ Pollut 127

(2):281–290. https://doi.org/10.1016/S0269-7491(03)00266-5

Li QT, Yang FJ, Zhang B, Zhang X, Zhou GZ (2008) Biogeochemistry responses and spectral characteristics of *Rhus chinensis* Mill under heavy metal contamination stress. J Remote Sens-Beijing 12(2):290. http://en.cnki.com.cn/Article_en/CJFDTOTAL-YGXB200802013.htm

Liu M, Liu X, Li M, Fang M, Chi W (2010) Neural-network model for estimating leaf chlorophyll concentration in rice under stress from heavy metals using four spectral indices. Biosys Eng 106(3):223–233. https://doi.org/10.1016/j.biosystemseng.2009.12.008

Liu M, Liu X, Wu M, Li L, Xiu L (2011) Integrating spectral indices with environmental parameters for estimating heavy metal concentrations in rice using a dynamic fuzzy neural-network model. Comput Geosci 37(10):1642–1652. https://doi.org/10.1016/j.cageo.2011.03.009

Mabhungu L, Adam E, Newete SW (2019) Monitoring of phytoremediating wetland macrophytes using remote sensing: the case of common reed (*Phragmites australis* (cav.) *Trin. Ex steud.*) And the giant reed (*Aarundo donax* L.) A. Appl Eco Environ Res 17 (4):7957–7972. https://doi.org/10.15666/AEER/1704_79577972

Malley DF, Williams PC (1997) Use of near-infrared reflectance spectroscopy in prediction of heavy metals in freshwater sediment by their association with organic matter. Environ Sci Technol 31(12):3461–3467. https://doi.org/10.1021/es970214p

Mondal BP, Sekhon BS (2019) Using diffuse reflectance spectroscopy for assessment of soil phosphorus status of an intensively cropped region. Agric Res J 56:657–661. https://doi.org/10.5958/2395-146X.2019.00102.9

Nanni MR, Dematte JAM (2006) Spectral reflectance methodology in comparison to traditional soil analysis. Soil Sci Soc Am J 70(2):393. https://doi.org/10.2136/sssaj2003.0285

Nawar S, Mouazen AM (2017) Comparison between random forests artificial neural networks and gradient boosted machines methods of on-line Vis-NIR spectroscopy measurements of soil total nitrogen and total carbon. Sensors 17(10):2428. https://doi.org/10.3390/s17102428

Newete SW, Byrne MJ (2016) The capacity of aquatic macrophytes for phytoremediation and their disposal with specific reference to water hyacinth. Environ Sci Pollut Res 23:10630–10643. https://doi.org/101007/s11356-016-6329-6

Noomen M, Hakkarainen A, Van der Meijde M, Van der Werff H (2015) Evaluating the feasibility of multitemporal hyperspectral remote sensing for monitoring bioremediation. Int J Appl Earth Obs Geoinf 34:217–225. https://doi.org/10.1016/j.jag.2014.08.016

Noomen MF, Skidmore AK (2009) The effects of high soil CO_2 concentrations on the first derivative of maize leaf reflectance. Int J Remote Sens 30:481–497. https://doi.org/10.1080/01431160802339431

Panigada C, Colombo R, Meroni M, Rossini M, Cogliati S, Busetto L, Fava F, Picchi V, Migli-Avacca M, Marchesi A (2010) Remote sensing of vegetation status using hyperspectral data. Hyperspectral workshop, Frascati, Italy, ESA SP-683

Penuelas J, Filella I (1998) Visible and near-infrared reflectance techniques for diagnosing plant physiological status. Trends Plant Sci 3(4):151–156. https://doi.org/10.1016/S1360-1385(98)01213-8

Prasad M (2004) Heavy metal stress in plants: from molecules to ecosystems. Springer and Narosa Publishing House, New Delhi, India, p 462. https://ci.nii.ac.jp/ncid/BA73484138?l=en

Pu R (2009) Broadleaf species recognition with in situ hyperspectral data. Int J Remote Sens 30(11):2759–2779. https://doi.org/10.1080/01431160802555820

Rathod PH, Rossiter DG, Noomen MF, Van der Meer FD (2012) Proximal spectral sensing to monitor phytoremediation of metal-contaminated soils. Int J Phytorem 15:405–426. https://doi.org/10.1080/15226514.2012.702805

Ren HY, Zhuang DF, Pan JJ, Shi XZ, Shi RH, Wang HJ (2010) Study on canopy spectral characteristics of paddy polluted by heavy metals. Spectrosc Spectral Anal 30(2):430–434. https://doi.org/10.3964/j.issn.1000-0593(2010)02-0430-05

Ren HY, Zhuang DF, Singh AN, Pan JJ, Qiu DS, Shi RH (2009) Estimation of As and Cu contamination in agricultural soils around a mining area by reflectance spectroscopy: a case study. Pedosphere 19(6):719–726. https://doi.org/10.1016/S1002-0160(09)60167-3

Schmidt T, Rico C, Rodrigues-Rastrero M, Sierra MJ, Diaz-Puente FJ, Pelayo M, Millan R (2013) Monitoring of the mercury mining site Almadèn implementing remote sensing technologies. Environ Res 125:92–102. http://dxdoiorg/101016/jenvres201212014i

Schowengerdt RA (2006) Remote sensing: models and methods for image processing. Elsevier, New York, NY, UK

Schuerger AC, Capelle GA, Di Benedetto JA, Mao C, Thai CN, Evans MD, Richards JT, Blank TA, Stryjewski EC (2003) Comparison of two hyperspectral imaging and two laser-induced fluorescence instruments for the detection of zinc stress and chlorophyll concentration in bahia grass (*Paspalum notatum* Flugge.). Remote Sens Environ 84(4):572–588. https://doi.org/10.1016/S0034-4257(02)00181-5

Shakya K, Chettri MK, Sawidis T (2008) Impact of heavy metals (Copper Zinc and Lead) on the chlorophyll content of some mosses. Arch Environ Contam Toxicol 54:412–421. https://doiorg/101007/s00244-007-9060-y

Shepherd KD, Walsh MG (2002) Development of reflectance spectral libraries for characterization of soil properties. Soil Sci Soc Am J 66:988–998. https://doi.org/10.2136/sssaj2002.9880

Shuping LS, Snyman RG, Odendaal JP, Ndakidemi PA (2011) Accumulation and distribution of metals in *Bolboschoenus maritimus* (Cyperaceae) from a South

African river. Water Air Soil Pollut 216(1–4):319–328. https://doi.org/10.1007/s11270-010-0535-5
Siebielec G, McCarty GW, Stuczynski TI, Reeves JB III (2004) Near-and mid-infrared diffuse reflectance spectroscopy for measuring soil metal content. J Environ Qual 33(6):2056. https://doi.org/10.2134/jeq2004.2056
Slonecker T, Haack B, Price S (2009) Spectroscopic analysis of arsenic uptake in Pteris ferns. Remote Sens 1:644–675. https://doi.org/10.3390/rs1040644
Smith AM, Blackshaw RE (2003) Weed-crop discrimination using remote sensing: a detached leaf experiment. Weed Technol 17(4):811–820. https://doi.org/10.1614/WT02-179
Smith KL, Steven MD, Colls JJ (2004) Use of hyperspectral derivative ratios in the red-edge region to identify plant stress responses to gas leaks. Remote Sens Environ 92:207–217. https://doi.org/10.1016/j.rse.2004.06.002
Song H, He Y (2005) Evaluating soil organic matter with visible spectroscopy. In: IMTC—instrumentation and measurement technology conference. IEEE, Ottawa (Canada), pp 1321–1324. https://doi.org/10.1109/IMTC.2005.1604362
Sridhar BB, Diehl SV, Han FX, Monts DL, Su Y (2005) Anatomical changes due to uptake and accumulation of Zn and Cd in Indian mustard (*Brassica juncea*). Environ Exp Bot 54(2):131–141. https://doi.org/10.1016/j.envexpbot.2004.06.011
Sridhar BB, Han FX, Diehl SV, Monts DL, Su Y (2007) Monitoring the effects of arsenic and chromium accumulation in Chinese brake fern (*Pteris vittata*). Int J Remote Sens 28(5):1055–1067. https://doi.org/10.1080/01431160600868466
Stenberg B, Viscarra Rossel RA, Mouazen A, Wetterlind J (2010) Visible and near infrared spec- troscopy in soil science. In: Donald LS (ed) Advances in agronomy. Academic Press, Burlington (MA), pp 163–215. https://doi.org/10.1016/S0065-2113(10)07005-7
US-DOE (2000) Appendix D: monitoring report proceedings from the workshop on phytoremediation of inorganic contaminants (INEEL/EXT-2000–00207) Idaho Falls (ID): US Department of Energy–Subsurface Contaminants Focus Area Idaho National Engineering and Environmental Laboratory Argonne National Laboratory and Enviro Issues, pp 1–18
USEPA (2000) Introduction to phytoremediation. United State Environment Protection Agency, National Risk Management Research Laboratory Office of Research and Development, pp 56–58
Vaiphasa C, Ongsomwang S, Vaiphasa T, Skidmore AK (2005) Tropical mangrove species discrimination using hyperspectral data: a laboratory study. Estuar Coast Shelf Sci 65:371–379. https://doi.org/101016/jecss200506014
Van der Merwe CG, Schoonbee HJ, Pretorius J (1990) Observations on concentrations of the heavy metals zinc manganese nickel and iron in the water in the sediments and in two aquatic macrophytes *Typha capensis* (Rohrb.) NE Br. and *Arundo donax* L. of a stream affected by goldmine and industrial effluents. Water SA 16(2):119–124. https://www.cabdirect.org/cabdirect/abstract/19911953045
Vasques GM, Grunwald S, Sickman JO (2008) Comparison of multivariate methods for inferential modeling of soil carbon using visible/near-infrared spectra. Geoderma 146:14–25. https://doi.org/10.1016/j.geoderma.2008.04.007
Viscarra Rossel RA, Behrens T (2010) Using data mining to model and interpret soil diffuse reflectance spectra. Geoderma 158(1–2):46–54. https://doi.org/10.1016/j.geoderma.2008.04.007
Viscarra Rossel RA, McBratney A, Minasny B (2010) Proximal soil sensing, 1st edn. Springer Verlag, New York https://doi.org/10.1007/978-90-481-8859-8
Vohland M, Bossung C, Fr HC (2009) A spectroscopic approach to assess trace-heavy metal contents in contaminated floodplain soils via spectrally active soil components. J Plant Nutr Soil Sci 172(2):201–209. https://doi.org/10.1002/jpln.200700087
Wenjun J, Zhou S, Jingyi H, Shuo L (2014) In situ measurement of some soil properties in paddy soil using visible and near-infrared spectroscopy. PLoS ONE 9(8):105708. https://doi.org/10.1371/journal.pone.0105708
Wu Y, Zhang X, Liao Q, Ji J (2011) Can contaminant elements in soils be assessed by remote sensing technology: a case study with simulated data. Soil Sci 176(4):196–205. https://doi.org/10.1097/SS.0b013e3182114717
Wu YZ, Chen J, Ji J, Gong P, Liao Q, Tian Q, Ma HR (2007) A mechanism study of reflectance spectroscopy for investigating heavy metals in soils. Soil Sci Soc Am J 71(3):918–926. https://doi.org/10.2136/sssaj2006.0285
Wuana RA, Okieimen FE (2011) Heavy metals in contaminated soils: a review of sources chemistry risks and best available strategies for remediation. ISRN Ecol. https://doi.org/10.5402/2011/402647
Xia XQ, Mao YQ, Ji JF, Ma HR, Chen J, Liao QL (2007) Reflectance spectroscopy study of Cd contamination in the sediments of the Changjiang River China. Environ Sci Technol 41(10):3449–3454. https://doi.org/10.1021/es0624422
Xu L, Zhou YP, Tang LJ, Wu H-L, Jiang JH, Shen GL, Yu RQ (2008) Ensemble preprocessing of near-infrared (NIR) spectra for multivariate calibration. Anal Chim Acta 616(2):138–143. https://doi.org/10.1016/j.aca.2008.04.031
Yang H, Zhang J, Van der Meer FD, Kroonenberg SB (2000) Imaging spectrometry data correlated to hydrocarbon microseepage. Int J Remote Sens 21:197–202. https://doi.org/10.1080/014311600211091
Zengin FK, Munzuroglu O (2005) Effects of some heavy metals on content of chlorophyll proline and some antioxidant chemicals in bean (*Phaseolus vulgaris* L.) seedlings. Acta Biologica Cracoviensia Series

Botanica 47(2):157–164 bwmeta1.element.agro-article-194d9de5-de29–488e-8a51–3ae04e829833
Zomer RJ, Trabucco A, Ustin SL (2009) Building spectral libraries for wetlands land cover classification and hyperspectral remote sensing. J Environ Manage 90:2170–2177. https://doi.org/101016/jjenvman2007 06028

Application of Artificial Intelligence to Detect and Recover Contaminated Soil: An Overview

26

K. R. Padma and K. R. Don

Abstract

Bioremediation is the chief applicable methodology to control the pollution and contamination of soil. The in-place treatment along with above ground treatment of contaminated soil has created intense scientific growth. The general pollutants of soil include petroleum hydrocarbons, heavy metals, pesticides used in agricultural field which alter the characteristics of soil. Although, microbes are also beneficial to recover the contaminations present in the soil, several artificial intelligence constructed models help in detection of phytotoxicity of soil.

Keywords

Artificial intelligence · Bioremediation · Biostimulants · Heavy metals · Neural network · Soil pollution

K. R. Padma (✉)
Department of Biotechnology, Sri Padmavati Mahila VisvaVidyalayam (Women's) University, Tirupati, AP, India

K. R. Don
Department of Oral Pathology, Saveetha Dental College, Saveetha Institute of Medical and Technical Sciences, Saveetha University, Velappanchavadi, Chennai, Tamil Nadu, India

26.1 Introduction

The soil pollution persists as a serious concern globally. The environmental pollution chiefly involves contamination of soil, water and air. Nevertheless, in order to preclude the contamination of soil, bioremediation is the apt method. Generally, the pollution of soil can threaten the ecosystem and extinguish the food chain. The conventional remediation methods were proclaimed to be expensive treatments earlier but today with the introduction of biological remediation technique, heavy metals, hydrocarbon contaminants which are made up of complex mixtures of aliphatic and aromatic hydrocarbons plus volatile compounds like gasoline and petrol can be degraded effectively. Hence, the involvement of eco-friendly microorganisms was found to be cost-effective and widely utilized method today (Floch et al. 2011; Scelza et al. 2008; Nie et al. 2009; Rimmer et al. 2006; Moreno et al. 2009; Li et al. 2013).

Moreover, the emancipation of several contaminants produces deadly damage to all forms of living beings due to augmented global industrialization (Quintella et al. 2019). The release of several pollutants likes pesticides, heavy metal toxins, oil hydrocarbons cause detrimental impacts on the ecological system. The contamination of soil results in enhanced health effects amongst farmers such as the occurrence of skin cancer due to mutations because of toxins in the

J. A. Malik (ed.), *Advances in Bioremediation and Phytoremediation for Sustainable Soil Management*,
https://doi.org/10.1007/978-3-030-89984-4_26

soil environment (Kuppusamy et al. 2020). The major way of recovery from contaminated environment is bioremediation procedure. This method is a biological process which is been studied intensively as it was found to be eco-friendly and can degrade the contaminants at a cheaper cum faster rate in comparison with the existing methods (Kumar et al. 2018; Soleimani 2014).

Another method of cleaning up the contaminants of soil is the application of biostimulants obtained from animal manures such as pig, poultry and goat which are found to be helpful in removal of pollutants from environment. However, the application of animal manure was found to be as effective as bioremediation technique in removal of contaminants from polluted environment (Ijah et al. 2003; Okolo et al. 2005; Yakubu 2007). Furthermore, there is very few literature or data on the use of animal manure in the biodegradation of petroleum hydrocarbons in a contaminated environment. The total level of petroleum hydrocarbons in soil can be estimated with its physicochemical properties and is nearly ten times in comparison with the ground level. The increased levels in the environment further instigate adversative ecological effects (Florinsky et al. 2002, 2004).

At present, with the advancement of artificial intelligence in all science and technology disciplines, it will be very easy to further develop and apply scientific models. The artificial neural network with either multilayer perception or backpropagation algorithm is helpful to detect the environmental toxicants precisely. Additionally, quantification of the contaminants in the soil is assessed with the support of artificial intelligence technology which estimates the soil parameters by prediction model built with artificial neural networking (Minasny et al. 2002) (Fig. 26.1).

However, the interaction process of soil is difficult to predict with statistical data due to the complex nature of the metal ions. Hence, to understand the geochemistry of the soil, the artificial neural networking plays a key role in successful identification of patterns. Consequently, ANN model approaches help in determining soil parameters along with topographical feature alternations which help in mitigation/ prevention of environmental pollutants. Hence, our article aimed to provide inclusive knowledge

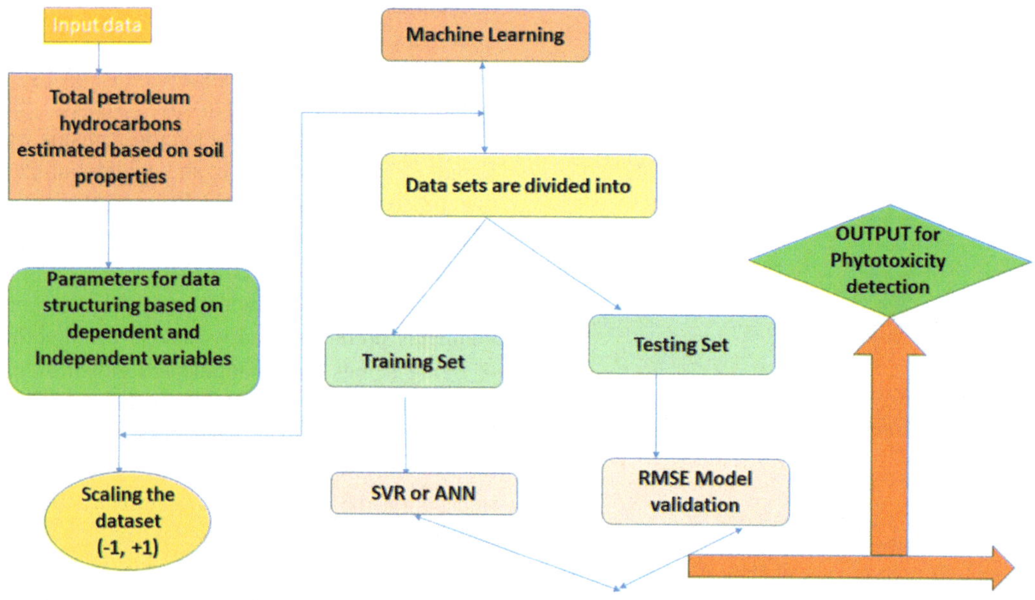

Fig. 26.1 The ANN model to predict the petroleum hydrocarbon contaminants in soil

to readers about the cheapest and latest ANN in bioremediation along with highlighting the bioremediation with microbes, animal/organic waste.

26.2 Industrial Release of Pollutants and Their Toxicity Management with Advancement of ANN Technology

The closer cover-up on the operations of industries such as in manufacture of petroleum-related lamps in stack resulted as main source for contamination of the environmental soil and displayed various harmful ailments and infections around the localities due to release of these residues into soil. Moreover, the chief chemicals that cause threat to human health is polycyclic aromatic hydrocarbons (PAHs), heavy metals such as nickel, chromium, selenium, etc. Upon release of these chemicals into the ecosystem, they cause carcinogenic and mutagenic effects on skin and other parts of the body (Rittler et al. 2007). However, several bio-remedial plans were carried out earlier in order to mitigate the pollutants present in the environment but the current upsurge of ANN modelling has created strong impact as it has the ability to predict the toxicity and provide precise results.

PAHs with more than two bonded benzene rings with carbon and hydrogen atoms in abundance are the major pollutants of the ecosystem. These pollutants are hydrophilic in nature with higher molecular weight and are not soluble in water resulting in high oil affinities. The environmental protection organization of United States has recognized these compounds and classified them into carcinogenic and non-carcinogenic heavy metals (Table 26.1). However, some elements are required for growth of humans but few are harmful and lead to negative effects on human health (Domingo 1994).

Hence, self-organizing maps (SOM) are widely utilized algorithm for collecting data on health plus census (Koua and Kraak 2004; Hatzichristos 2004; Oyana et al. 2005). The SOM algorithm is also employed for soil analysis along with geochemical collection of data (Penn 2005; Fraser 2006; Mele and Crowley 2008; Hosokawa and Hoshi 2001; Ferentinou

Table 26.1 Some PAH compounds regarded as environmental pollutants by EPA (US)

Polycyclic aromatic hydrocarbons (PAHs)	Abbreviations	Molecular weight
Naphthalene	Nap	128.17
Acenaphthylene	Acy	152.19
Acenaphthene	Ace	154.2
Fluorene	Flu	166.2
Phenanthene	Phe	178.2
Anthracene	Ant	178.2
Fluoranthene	Flr	202.3
Pyrene	Pyr	202.3
Chrysene	Chr	228.3
Benzo(a)anthracene	BaA	228.3
Benzo(k)fluoranthene	BkF	252.3
Benzo(b)fluoranthene	BbF	252.3
Benzo(a)pyrene	BaP	252.3
Indeno(123-cd)pyrene	InP	276.3
Dibenz(ah)anthracene	DahA	278.4
Benzo(ghi)perylene	BghiP	276.3

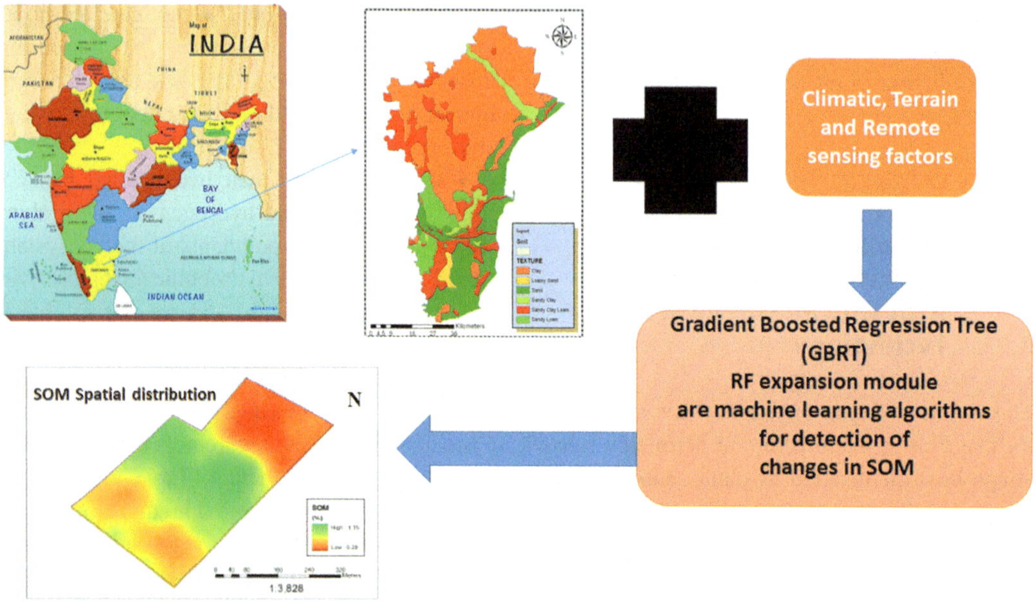

Fig. 26.2 Assessment of SOM spatial distribution employing AI technology

and Sakellariou 2007). The ANN built SOM application is being extensively used as the algorithm for spatial and non-spatial clustering as well as to study the remote sensing too (Wan and Fraser 1993; Wan and Fraser 1994; Ji 2000; Lee and Lathrop 2006; Li and Eastman 2006; Yun and Uchimura 2007; Ehsani and Quiel 2008) (Fig. 26.2).

26.3 Advanced Computing Technology Like ANN

The most advanced computing technology is artificial neural networks (ANNs) which are known as intelligent agents that measure scrutinize information by machine learning process in a much similar manner as human intelligence systems (Fig. 26.3) (Silipo 1999). The alignment of the ANN comprises a medley of different kinds of units regarded as neurons which are coupled together to function as a processing system especially to solve complex problems. Moreover, this device is an advancement in the field of science as it is capable of acquiring prior information. Thus, ANN tool is regarded as most valuable technique for classifying data and patterns recognition (Gurney 1997).

26.4 Self-organizing Mapping Technique (SOM) Application

The primary function of soil organic matter mapping includes incorporation of updated ANN mechanisms based on unsupervised learning particularly with a purpose of decreasing the dimensions in any data set. The discrete map with 2D lattice provides a clear cut idea on the topology (Figs. 26.4 and 26.5). However, the input of the original data creates a grid of neurons and finally the treated one gives the final output space. The result is shown in a two-dimensions lattice with either rectangular/hexagonal grid of neurons (Kohonen 2001). Moreover, the output space can be one-dimensional (Bação et al. 2005) or even three-dimensional (Seiffert and Michaelis 1995; Kim and Cho 2004). The distinguishing topographies of the SOMs include (a) unsupervised learning,

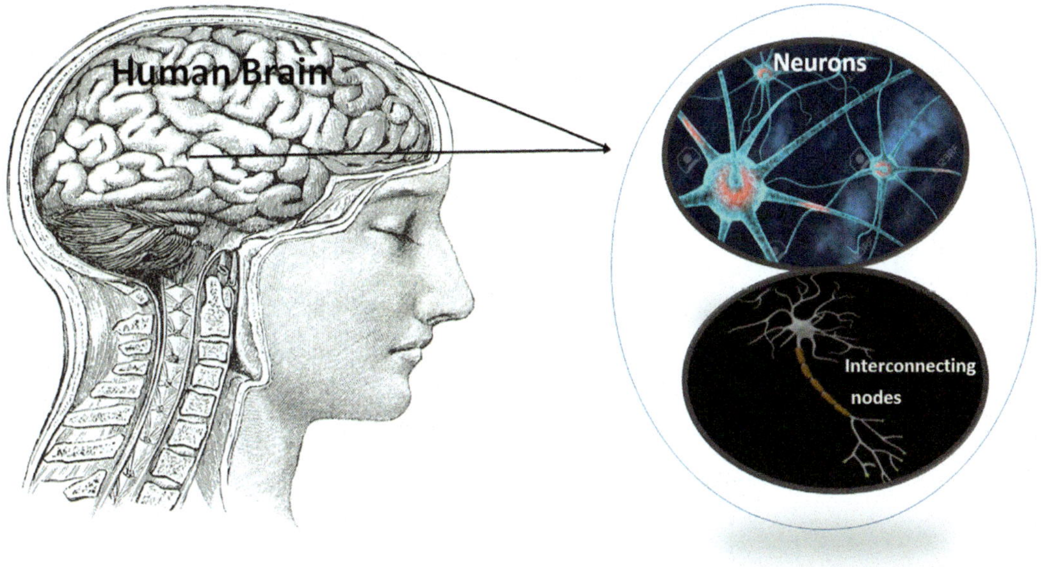

Fig. 26.3 The artificial neural network activities similar to human neurons in brain

Fig. 26.4 One-dimensional and two-dimensional SOM graphical representation

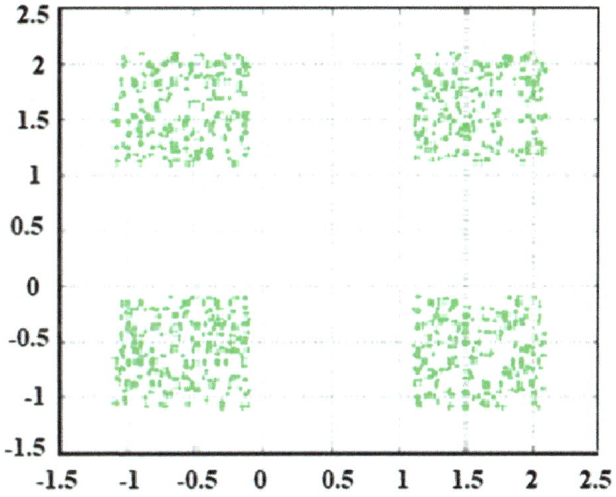

(b) machine learning by processing, (c) major function of ANN is to interpret the input data structure by dimensionality reduction using neural weights, (d) the computer provides training and testing regime and (e) finally, extracts the information and outputs the recognized patterns (Kaski and Kohonen 1996; Kaski, Nikkilä et al. 1998). The final output maps with diverse patterns are the result of topological preservation capacity of SOM.

26.5 Development of ANN with Supervised Learning

The estimation of soil parameters is with MLP network by employing backpropagation learning regime. Although, the most regularly used network is MLP network in engineering setbacks proportional to non-linear mapping (Haykin 1994). The most commonly used device for all

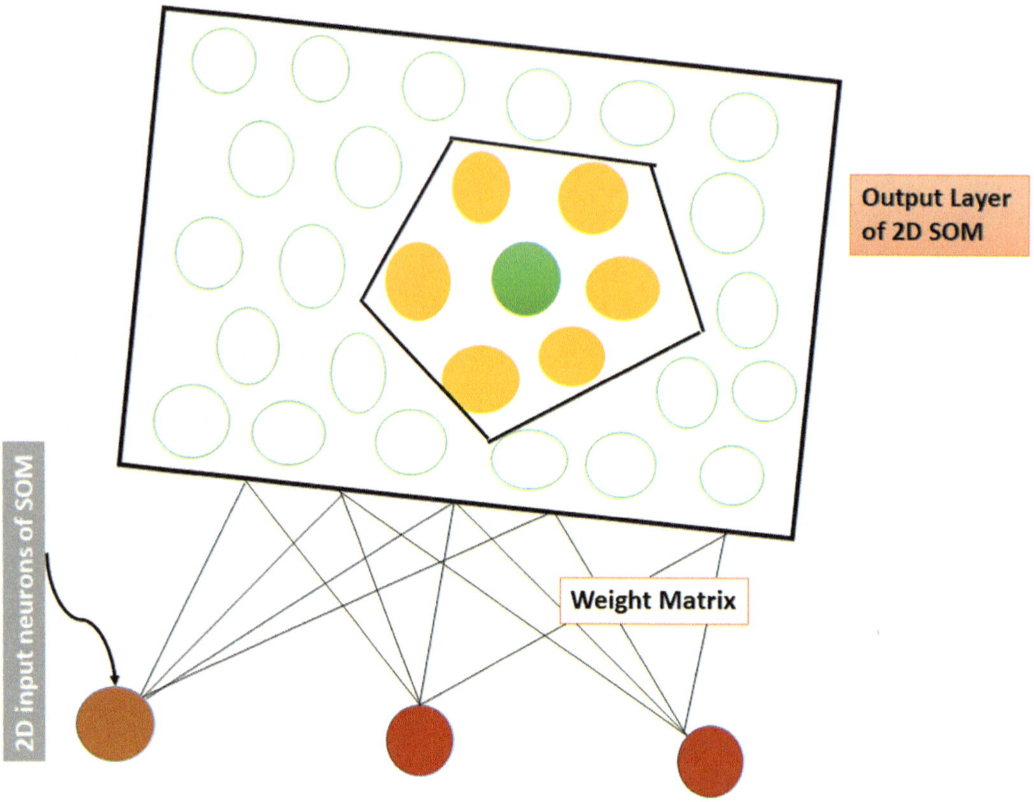

Fig. 26.5 The assessment of 2D SOM input and output layer lattice with diverse patterns

sorts of neural networking is backpropagation method which was developed by Rumelhart et al. (1986). However, this technology utilizes the multilayered feed forward topology based on supervised learning. This BP algorithm employs the gradient descent method with delta learning rule (Rumelhart 1986; Melesse 2005). The BP algorithm is executed with forward pass and backward pass. The output pattern is presented with forward pass through interconnected neuron networking layer. Finally, each layer is compared with next layer until final computed output is reached and the values obtained are calculated with root mean square error (RMSE) method (Degroot 1986). Based on the field data, the artificial neural network is designed.

All the data set is shuffled first and 100% divided accordingly to the 60% for the learning process, for testing 20% sets utilized and for the purpose of verification remaining 20% sets are used so as to avoid bias. However, the modelling of ANN is performed employing MATLAB software package. The estimation of data set is based upon the variables and SOM as the output parameters. The number of input and the hidden layers of neurons are calibrated with parameters such as α, and the number of repetitions by several tests along with trial and errors by usage of Marquardt Levenberg learning rule. However, the tansigmoid function offers best results.

26.6 Bioremediation with ANN Paradigm

The unconditional obligation for promoting ecological improvement of our society with trivial environmental impact is to exterminate the pollutants from the environment. The soil polluted with polycyclic hydrocarbons effectually lead to destruction of the local ecosystems. The acquisition of these pollutants by the aquatic organisms and tissue of plants can result in mutations of offsprings. However, the release of the toxicants such as petroleum due to industrial globalization turns the cultivable terrains into poor soil attributes. Earlier petroleum lights were in demand as they used to enlighten the rural areas (Varjani 2017). Subsequently, the augmented usage of petroleum and its products caused appalling soil along with groundwater contamination (Lim et al. 2016). These petroleum-based hydrocarbons are mentioned as the communal primary energy plus fuel resources globally. The distribution of the petroleum-based products might have resulted in fortuitous emancipation or seepage (Abbasian et al. 2015). The introduction of microorganisms, having the potential to degrade poly aromatic hydrocarbons (PAHs) like bacteria, fungi and microalgae, has proven to be beneficial (Andreolli et al. 2015).

The advancement of artificial intelligence technology with machine learning programme built with neural networking is capable of predicting the behaviour of soil and often measure the responses in forward direction based on the input nodes that lead to directional target outputs of soil parameters. In fact, bioremediation with microbes is also regarded as out-dated technology as with progression of artificial intelligence in all fields of science and technology. The major factor to test the soil parameter is soil organic carbon content. The SOC provides aid to recover the physicochemical properties of soil by enhancing the holding capability of water, nutrients inside the soil and thereby providing durable soil structure (Chivenge et al. 2007; Krauss et al. 1997), soil chemical properties and nutrients holding capacity (Leeper et al. 1993). In order to maintain the terrain, the basic parameter for consideration is SOC which mitigates soil erosion and improves the productivity.

However, the artificial neural network involves artificial intelligence paradigm which is employed to the SOC data and learns the behavioural alterations and further predicts its outcomes. The neural network is trained based on the nature of the application along with strength of the internal data patterns (Mubiru and Banda 2008). The application of the networking depends upon the dynamic relationship and often limited with variable independencies (Fig. 26.6). The archetypal of ANNs can apprehend several types of connexions and can interpret the complex phenomena easily (Sinanoglu 2004).

26.7 Architecture of Artificial Neural Network in Detection and Prediction of Phytotoxicity

Artificial neural networks (ANNs) are biologically inspired by the human brain and work similar to the neurons of the human brain. There is a vide expansion of latest research in the progression of new computational models or ANN architecture for solving complex problems by pattern recognition (Huang 2009). However, the neural networks practice machine learning methods by adjusting and controlling the internal parameters. These ANN neural networks are supple mathematical structures that are efficient in ascertaining the complex non-linear relationships amongst input and output data sets.

The basic application of the ANN is to train the input data sets for successful classification and prediction of the soil organic matter (Zhang and McGrath 2004). The upsurging potential benefit of this process is its cost-effective nature with greater credential in resolving the complex problems with much precision. The programme employed by ANN architecture includes the multilayer perception (MLP) design or backpropagation for resolution of complex problems. The ANN constructed with MLP network has three layers of neurons beginning with input neurons, hidden neurons and finally output

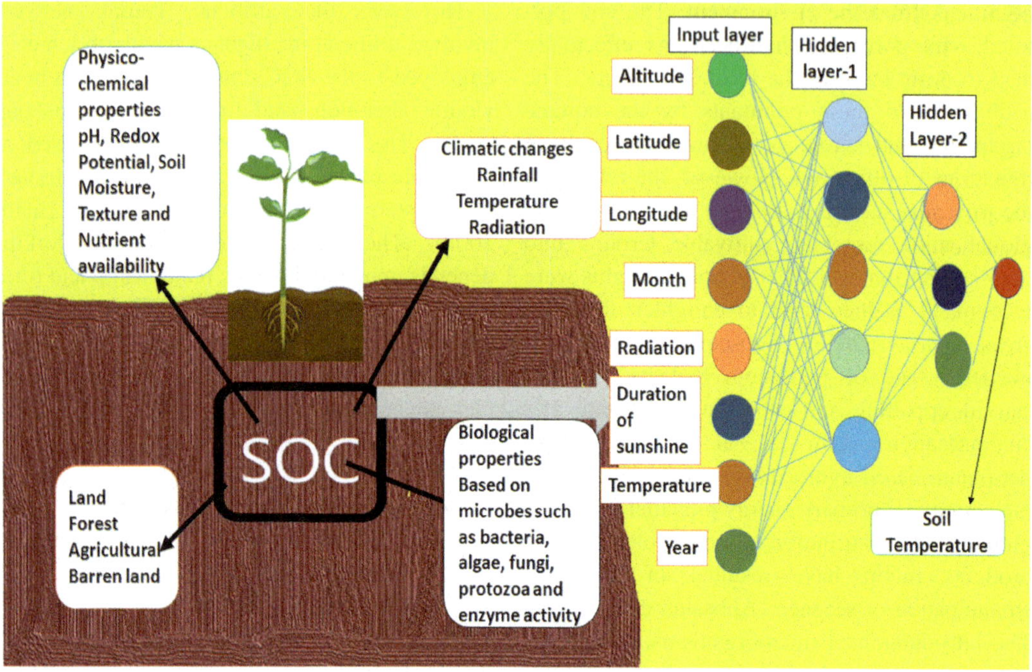

Fig. 26.6 The SOC properties estimated with ANN for predicting the soil temperature

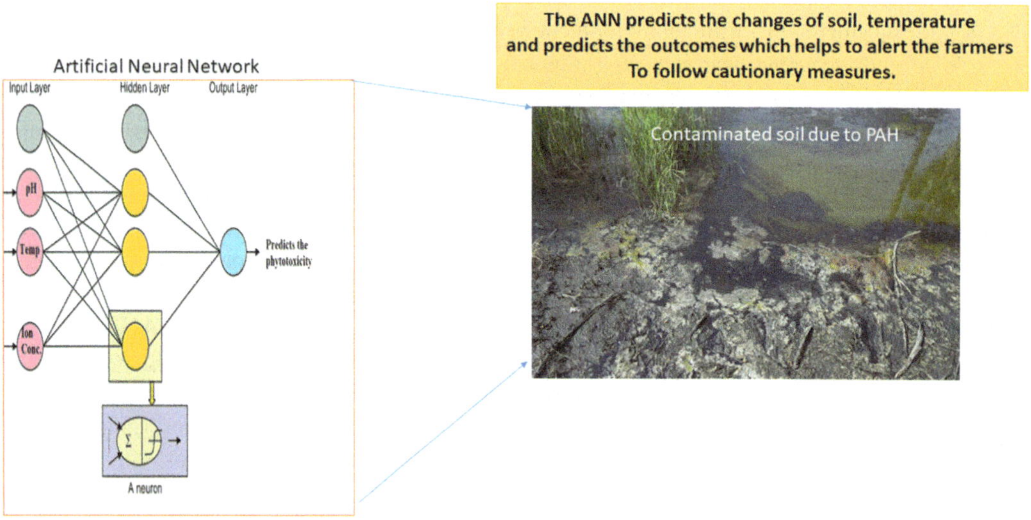

Fig. 26.7 ANN used for estimation of soil parameters

neurons (Jenkins 1997). The input and output patterns involve the training sets to provide known target outputs (Rafiq et al. 2001; Hambli 2009; Hambli et al. 2006) (Fig. 26.7). The ANN models can predict the soil parameters relationship changes quickly and with much ease in comparison with the statistical data. Hence, today, this technology is in demand in all fields of science due to its precision to predict.

26.8 ANN Model Structure for Predicting Environmental Soil Properties

The ANN model used for estimating soil properties are backpropagation (BP) with non-linearity data sets (Li 1998). In the BP technique, every node is connected to the adjacent node. The input data are fed with the parameters to check the toxicity, and the hidden layer is the major layer in determining the complexity of the modelling. However, finally, the output provides fine-tuned value which predicts the toxicity. The ANN set was developed to assess the six layered structure depicted in the Fig. 26.8, where the input layer possesses six nodes which entangle each other for prediction like neurons in brain to assess the hidden layer, based on the properties. Nevertheless, the final output layer gives the prediction about the water content in soil along with other soil parameters. Thus, the important factor explained by the employment of ANN model structure is that it gives precise result about the soil properties and in turn predicts future outcomes too.

26.9 Conclusion

Although, it is obvious that conventional methods have been employed for treating the contaminated soil with petroleum hydrocarbons but the major drawback is that these are most expensive methodologies which require very costly equipments for treatment of contaminated soil. Therefore, with the intense research in science, this led to introduction of cheapest and safer method of treating contaminated terrain with pollutants. Nevertheless, the previous literature reports have emphasized the bioremediation by in situ and exsitu methods as the effective tools to detect and decontaminate the pollutants from soil, but today with the advancement of artificial intelligence, the application of ANN model for the rapid and precise prediction of heavy metal concentration in soil is much reliable. However, the input data for ANN are based on soil parameters so as to predict the final target output based on soil factors. Nevertheless, ANN model can be used in combination with machine like models at the research level as a process to define the chief parameters and further research has to be undertaken in direction to reach phenomenal accuracy.

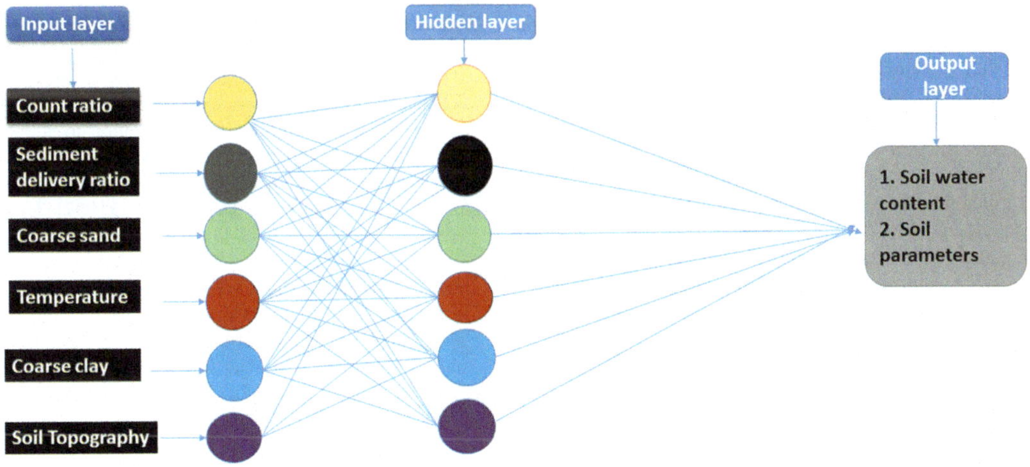

Fig. 26.8 The structure of ANN depicts the prediction of high tenacity of soil

References

Abbasian F, Lockington R, Mallavarapu M, Naidu R (2015) A comprehensive review of aliphatic hydrocarbon biodegradation by bacteria. Appl Biochem Biotechnol 176:670–699

Andreolli M, Lampis S, Brignoli P, Vallini G (2015) Bioaugmentation and biostimulation as strategies for the bioremediation of a burned woodland soil contaminated by toxic hydrocarbons: a comparative study. J Environ Econ Manag 153:121–131

Bação F, Lobo V, Painho M (2005) Self-organizing maps as substitutes for k-means clustering. International conference on computational science. Springer, Berlin, Heidelberg, pp 476–483

Chivenge PP, Murwira HK, Giller KE, Mapfumo P, Six J (2007) Long-term impact of reduced tillage and residue management on soil carbon stabilization: implications for conservation agriculture on contrasting soils. Soil Tillage Research 94:328–337

Degroot M (1986) Probability and statistics, 2nd edn. Addison-Wesley Press, Reading, MA

Domingo JL (1994) Metal-induced developmental toxicity in mammals: a review. J Toxicol Environ Health 42:123–141

Ehsani AH, Quiel F (2008) Application of self organizing map and SRTM data to characterize yardangs in the Lut desert. Iran Remote Sens Environ 112(7):3284–3294

Ferentinou MD, Sakellariou MG (2007) Computational intelligence tools for the prediction of slope performance. Comput Geotech 34(5):362–384

Floch C, Chevremont AC, Joanico K, Capowiez K, Criquet S (2011) Indicators of pesticide contamination: Soil enzyme compared to functional diversity of bacterial communities via Biolog® Ecoplates. Eup J Soil Biology 47(4):256–263

Florinsky IV, Eilers RG, Manning GR, Fuller LG (2002) Prediction of soil properties by digital terrain modeling. Environ Model Softw 17:295

Florinsky IV, McMahon S, Burton DL (2004) Topographic control of soil microbial activity: a case study of denitrifiers. Geoderma 119:33

Fraser S (2006) Data mining geoscientific data sets using self organizing maps: mastering the data explosion in the earth and environmental sciences. Extended Abstracts 5–7

Gurney K (1997) An introduction to neural networks. Taylor & Francis Inc Bristol, PA, USA

Hambli R (2009) Statistical damage analysis of extrusion processes using finite element method and neural networks simulation. Finite Elements Anal Des 45 (10):640–649

Hambli R, Chamekh A, Bel Hadj Salah H (2006) Real-time deformation of structure using finite element and neural networks in virtual reality applications. Finite Elem Anal Des 42(11):985–991

Hatzichristos T (2004) Delineation of demographic regions with GIS and computational intelligence. Environ Plann B Plann Des 31(1):39–49

Haykin S (1994) Neural networks: a comprehensive foundation. IEEE Press, New York

Hosokawa M, Hoshi T (2001) Landform classification method using self-organizing map and its application to earthquake damage evaluation. In: IGARSS 2001. Scanning the present and resolving the future. Proceedings IEEE 2001 international geoscience and remote sensing symposium, vol. 4, pp 1684–1686

Huang Y (2009) Advances in artificial neural networks—methodological development and application. Algorithms 2:973–1007

Ijah UJ, Antai SP (2003) The potential use of chicken-drop microorganisms for oil spill remediation. Environmentalist 23:89–95

Jenkins WM (1997) An introduction to neural computing for the structural engineer. J Struct Eng 75(3):38–41

Ji CY (2000) Land-use classification of remotely sensed data using Kohonen self organizing feature map neural networks. Photogramm Eng Remote Sens 66 (12):1451–1460

Kaski S, Kohonen T (1996) Exploratory data analysis by the self-organizing map: structures of welfare and poverty in the world. In: Apostolos-Paul N, Refenes Y, Abu-Mostafa Y, Moody J, Weigend A (eds) Neural Networks in Financial Engineering. World Scientific, Singapore, pp 498–507

Kaski S, Nikkilä J and Kohonen T (1998) Methods for interpreting a self-organized map in data analysis. In: Proceedings of ESANN'98, 6th european symposium on artificial neural networks, D-Facto, Bruges, Belgium

Kim MY, Cho H (2004) Three-dimensional map building for mobile robot navigation environments using a self-organizing neural network. J Robot Syst 21:323–343

Kohonen T (2001) Self-organizing maps. Springer-Verlag, Berlin Heidelberg, Germany, p 502

Koua E, Kraak MJ (2004) Geo-visualization to support the exploration of large health and demographic survey data. Int J Health Geo-Graphics 3(1):12

Krauss G, Kindangen JI, Depecker P (1997) Using artificial neural networks to predict interior velocity coefficients. Build Environ 32:295–303

Kumar V, Shahi SK, Singh S (2018) Bioremediation: an eco-sustainable approach for restoration of contaminated sites. In: Microbial bioprospecting for sustainable development, Springer Singapore, pp 115–136

Kuppusamy S, Maddela NR, Megharaj M, Venkateswarlu K (2020) Ecological impacts of total petroleum hydrocarbons. In: Total petroleum hydrocarbons, Springer International Publishing, vol 1, Cham, Switzerland, pp 95–138

Lee S, Lathrop RG (2006) Subpixel analysis of landsat ETM+ using self- organizing map (SOM) neural networks for urban land cover characterization. IEEE Trans Geosci Remote Sens 44(6):1642–1654

Leeper GW, Uren NC (1993) Soil science: an introduction, 5th edn. Melbourne University Press, Melbourne
Li X, Liu L, Wang Y, Luo G, Chen X, Yang X (2013) Heavy metal contamination of urban soil in an old industrial city (Shenyang) in Northeast China. Geoderma 192:50–58
Li Z, Eastman JR (2006) Commitment and typicality measurements for the self organizing map. UCGIS 2006 Summer Assembly, Vancouver, Washington
Li ZY (1998) Supervised classification of multispectral remote sensing image using B-P neural network. J Infrared Millimeter Waves 17:153–156
Lim MW, Von Lau E, Poh PE (2016) A comprehensive guide of remediation technologies for oil contaminated soil-present works and future directions. Mar Pollut Bull 109:14–45
Mele PM, Crowley DE (2008) Application of self-organizing maps for assessing soil biological quality. Agr Ecosyst Environ 126(3–4):139–152
Melesse AM, Hanley RS (2005) Artificial neural network application for multiecosystem carbon flux simulation. Ecol Model 189:305–314
Minasny B, McBratney AB (2002) The neuro-m methods for fitting neural network parametric pedotransfer functions. Soil Sci Soc Am J 66:352–361
Moreno JL, Bastida F, Ros M, Hernández T, García C (2009) Soil organic carbon buffers heavy metal contamination on semiarid soils: effects of different metal threshold levels on soil microbial activity. Eur J Soil Biol 45(3):220–228
Mubiru J, Banda EJ (2008) Estimation of monthly average daily global solar irradiation using artificial neural networks. Solar Energy 82:181187
Nie M, Zhang X, Wang J, Jiang L, Yang J, Quan Z (2009) Rhizosphere effects on soil bacterial abundance and diversity in the yellow river deltaic ecosystem as influenced by petroleum contamination and soil salinization. Soil Biol Biochem 41(12):2535–2542
Okolo JC, Amadi EN, Odu CTI (2005) Effects of soil treatments containing poultry manure on crude oil degradation in a sandy loam soil. Appl Ecol Environ Res 3:47–53
Oyana TJ, Boppidi D Yan J, Lwebuga-Mukasa JS (2005) Exploration of geographic information systems (GIS)-based medical databases with self- organizing maps (SOM): A case study of adult asthma. In: Proceedings of the 8th International conference on geo computation, university of michigan, Ann Arbor, USA
Penn BS (2005) Using self-organizing maps to visualize high-dimensional data. Comput Geosci 31(5):531–544
Quintella CM, Mata AMT, Lima LCP (2019) Overview of bioremediation with technology assessment and emphasis on fungal bioremediation of oil contaminated soils. J Environ Manag 241:156–166
Rafiq MY, Bugmann G, Easterbrook DJ (2001) Neural network design for engineering applications. Comput Struct 79(17):1541–1552
Rimmer DL, Vizard CG, Pless-Mulloli T, Singleton I, Air VS, Keatinge ZAF (2006) Metal contamination of urban soils in the vicinity of a municipal waste incinerator: one source among many. Sci Total Environ 356(1–3):207–216
Ritter L, Solomon KR, Forget J, Stemeroff M, O'Leary C (2007) Persistent organic pollutants. United Nations Environmental Programme
Rumelhart DE, McClelland JL (1986) Parallel recognition in modern computers. processing: explorations in the microstructure of cognition, MIT Press/ Bradford Books, Cambridge, MA
Scelza R, Rao MA, Gianfreda L (2008) Response of an agricultural soil to pentachlorophenol (PCP) contamination and the addition of compost or dissolved organic matter. Soil Biol Biochem 40(9):2162–2169
Seiffert U, Michaelis B (1995) Three-dimensional self-organizing maps for classification of image properties. In: 2nd New Zealand two-stream international conference on artificial neural networks and expert systems (ANNES'95), IEEE Computer Society, Dunedin, New Zealand
Silipo R (1999) Neural Networks. In: Berthold M, Hand DJ (eds) Intelligent data analysis: an introduction. Springer-Verlag, New York, pp 432–447
Sinanoglu C (2004) The analysis of the effects of surface roughness of shafts on journal bearings using recurrent hybrid neural network. Ind Lubrication Tribology 56:32433
Soleimani M (2014) Comparison of biological and thermal remediation methods in decontamination of oil polluted soils. J Bioremed Biodegrad 5
Varjani SJ (2017) Microbial degradation of petroleum hydrocarbons. Biores Technol 223:277–286
Wan W, Fraser D (1993) M2dSOMAP: clustering and classification of remotely sensed imagery by combining multiple Kohonen self-organizing maps and associative memory. In: Proceedings of 1993 international joint conference on neural networks, supermolecular science division electrotechni, IJCNN '93 Nagoya, Japan
Wan W, Fraser D (1994) A self-organizing map model for spatial and temporal contextual classification. In: IGARSS'94, surface and atmospheric remote sensing: technologies, data, analysis and interpretation, IEEE Pasadena, California, USA
Yakubu MB (2007) Biodegradation of Lagoma crude oil using pig dung. Afr J Biotechnol 6(24):2821–2822
Yun L, Uchimura K (2007) Using self-organizing map for road network extraction from ikonos Imagery. Int J Innov Comput Inf Control 3(3):641–656
Zhang C, McGrath D (2004) Geostatistical and GIS analyses on soil organic carbon concentrations in grassland of southeastern Ireland from two different periods. Geoderma 119:261–275
Zhao Z, Chow TL, Rees HW, Yang Q, Xing Z, Meng FR (2009) Predict soil texture distributions using an artificial neural network model. Comput Electron Agric 65:36–48

Will Climate Change Alter the Efficiency of Bioremediation?

27

Anandkumar Naorem

Abstract

The twenty-first century is marked with challenges such as increasing industrialization, clearing up of natural ecosystems, environmental pollution and rise of population in every second. These problems are threatening the delivery of ecosystem services and global food security. Human health is always at risk due to the increasing new diseases related to the exposure to extreme levels of pollutants. Bioremediation utilizes the inherent capacity of the microorganisms to degrade and decontaminate/detoxify the pollutants in the presence of optimum environmental conditions. Although very limited scientific evidences have quantified and reported the direct or indirect effects of climate change on bioremediation, it is highly important to discuss the climate change-related environmental parameters and its associated effects on soil microorganisms involved in bioremediation process. As a change in soil moisture or temperature markedly affects the crucial soil processes such as decomposition of soil organic matter and nutrient cycling, it will definitely affect the soil microbial activities. Therefore, with this background, this chapter discusses an overview of bioremediation, environmental factors that affect bioremediation and the possible effects of climate change on bioremediation.

Keyword

Biostimulation · Chemotaxis · Genetically engineered microorganisms · Soil moisture · Soil pH · Temperature

27.1 Bioremediation: An Eco-Friendly Tool for a Sustainable Ecosystem

During the last few decades, there has been rapid increasing of industrialization and deforestation coupled with significant land-use changes. These have resulted in environmental pollution that has degraded the air, water and soil quality. The accumulation of significant amounts of heavy metals in plants through the soil system is a major concern in human health. In addition to this, organic and inorganic pollutants are continuously causing major threats to both the soil and human health (Samant et al. 2018). The modern conventional agriculture utilizes huge quantities of chemicals in the form of pesticides and fertilizers, which further affects the soil health and agricultural productivity. Misuse of these chemicals has led to enormous forms of environmental pollution. Few pollutants are

A. Naorem (✉)
Indian Council of Agricultural Research-Central Arid Zone Research Institute, Regional Research Station, Bhuj, Gujarat 370105, India
e-mail: anandkumar.naorem@icar.gov.in

J. A. Malik (ed.), *Advances in Bioremediation and Phytoremediation for Sustainable Soil Management*,
https://doi.org/10.1007/978-3-030-89984-4_27

easily degraded and detoxified with the help of soil microbes while others remain for longer period of time and enter the food chain and further biomagnified (Pandey et al. 2019). Conventional remediation techniques such as excavation, land-filling and pyrolysis are not fully efficient and could lead to secondary contamination, if not implemented properly. Therefore, it has been advocated that the soil beneficial microorganisms must be explored for its bioremediating capacity and employ in cleaning up of these pollutants. The use of soil microorganisms for pollutant detoxification/decontamination is not only eco-friendly but also cost-effective and more efficient than the conventional counterparts. This process of utilizing soil microorganisms in remediation of the toxic pollutants is referred as "Bioremediation". In other words, bioremediation can be described as an environmental tool used to restore the polluted environment and prevent further pollution. The soil microorganisms utilize the pollutants as their energy or nutrient sources and further transforms the pollutants into a less toxic or non-toxic forms. However, there are few environmental pollutants that are resistant to microbial decomposition, and therefore, the in-depth exploration of specific soil microorganisms capable of detoxifying these resistant pollutants is continuing (Gangola et al. 2019). The bioremediation efficiency of the soil microbes depends on several factors including the chemical nature and amount of the pollutants, soil and environmental conditions.

Based on the elimination of toxic pollutants, bioremediation can be divided into two types: in situ and *ex situ*. The *in situ* technique allows to detoxify the pollutants at their respective places with less disturbances. As it does not require transportation of the pollutants, there is lesser risk of secondary contamination. The minimal disturbance to the soil in *in situ* bioremediation makes it a safer mode for pollutant remediation. On the other hand, *ex situ* bioremediation involves excavation of the pollutants or polluted sites and transportation to a treatment site. It operates under artificial environment so that the resistant pollutants can be degraded or detoxified under controlled conditions.

27.2 Approaches to Enhance Bioremediation

If there is a possibility of reduced bioremediation with increasing climate change, there are approaches that can also enhance bioremediation. These approaches include chemotaxis, use of biosurfactants, genetically engineered microorganisms and biostimulation.

27.2.1 Chemotaxis

Chemotaxis is the movement of soil microorganisms in a direction where there is an increasing or decreasing chemical gradient in the soil. Here, it refers to the movement of the soil microorganisms towards the pollutants. Chemotaxis is another form of in situ bioremediation where the soil microorganisms act on the pollutants at their respective sites. It increases the bioavailability of the pollutants and increases the bioremediation rate. For example, Law and Aitken (2003) reported a chemotactic bacterial strain could degrade naphthalene more rapidly than the non-chemotactic mutant. Bacteria are easily accessed to hydrophobic organic pollutants and develop biofilm around the pollutants. The pollutants are absorbed as nutrient sources by the bacteria through biofilm formation and attachment to the hydrophobic pollutants. In order to enhance the chemotaxis of the bacteria for more effective bioremediation process, it is important to further understand the underlying molecular mechanisms of chemotaxis-mediated bioremediation (Olson et al. 2004).

27.2.2 Biofilm or Biosurfactants

Some hydrophobic organic pollutants are not easily available to the soil microorganisms due to which the bioremediation rate is reduced. Biosurfactants can overcome this problem by bridging the gap between the microbes and the pollutants. Odukkathil and Vasudevan (2013) reported the elevated bioremediation (30–45% increase) of chlorinated endosulfan by *Bacillus subtilis* MTCC1427 with the help of biosurfactants. Therefore, biosurfactants have been found as good enhancer in detoxifying n-alkanes and PAHs (Garcia-Junco et al. 2003). Furthermore, soil microorganisms can also produce a slimy layer around the pollutants which is called biofilm. The biofilm increases the survival rate of the soil microbes even in the presence of the toxic compounds in the environment and maintains the bioremediation ability of the soil microbes.

27.2.3 Biostimulation

Bioremediation can be carried out in natural or with human intervention. The degradation of the pollutants with the help of native soil microorganisms without any direct intervention is known as natural attenuation (Leal et al. 2017). The pollutants are detoxified to less toxic compounds with the help of native soil microbes through several processes (Megharaj et al. 2011). When such natural attenuation is carefully monitored, it is referred to monitored natural attenuation. Another form of bioremediation with human intervention is biostimulation, in which the environmental conditions are monitored and adjusted, the nutrients and electron acceptors for the microbes are supplied externally to enhance the biodegradation process. Biostimulation does not inoculate additional microorganisms to the soil. It is simply the manipulation of the environmental factors so that the native soil microorganisms could grow and perform their biodegradation process more effectively. The sites polluted with hydrocarbons, when acted by bioremediating microbes, deplete the major essential nutrients such as nitrogen and phosphorus. Therefore, biostimulation also involves application of such nutrients to the contaminated sites (Sarkar et al. 2005).

27.2.4 Bioaugmentation

Bioaugmentation refers to the artificial inoculation of the pollutant degrading microbes in the soil to increase the bioremediation process (Thompson et al. 2005). Bioaugmentation is carried out when the native soil microorganisms could not degrade the pollutants or when the population of the specific biodegrading microbes is low in the soil (Dzionek et al. 2016). In addition to these, there are certain criteria that need to be fulfilled before inoculating the soil microbe in the polluted site such as (Mohammed et al. 2007):

(1) must be able to survive and reproduce in the polluted site,
(2) must be compatible with the native soil microbes,
(3) must be easily accessible to the pollutants,
(4) must have high *in situ* biodegradation capacity.

The major problem with bioaugmentation is that the microbial strains might show high biodegrading capacity in laboratory conditions but could not perform well in the field conditions. This is attributed to the complex nature of the soil system, predation or competition with the native microbial species. Therefore, it is essential to conduct the experiments in different location trials with different environmental conditions. But still the soil and environmental conditions are so complex that it might be difficult to predict the efficiency of the inoculated strains. In this regard, microbial consortium comprising of several compatible soil biodegrading microbes could be inoculated so that at least one of them survives and performs the biodegradation process. Another advantage of the microbial consortium is that one microbial species can utlilize or detoxify the toxic intermediate products released by one strain (Heinaru et al. 2005). Bioremediation can be limited due to poor accessibility of the pollutants to the microbes. Movement of soil organisms such as earthworms can help in transportation and distribution of the inoculated bacteria or native soil microbes to the site of pollution.

27.2.5 Genetically Engineered Microorganisms

Soil bacteria are modified through the manipulation of their genetic constituents from other organisms to increase its degradation capability. The modified soil microbe is referred as "genetically engineered microorganisms (GEM)". It includes the improvement of bioavailability of the pollutants to the soil microbe, development of new pathways or enhancement of the established process, increase in catabolic activities, etc. GEMs used in *in situ* treatment of the pollutants are reported to be not so impactful because *in situ* treatment with GEM could increase the risk of horizontal gene transfer that causes migration of GEM in the environment (Naik and Duraphe 2012).

27.3 Effects of Climate Change on Bioremediation Efficiency

Climate change mostly refers to the change in abiotic factors such as higher rainfall, extreme droughts, increased temperature, humidity, etc. Bioremediation employs soil microbes to detoxify the pollutants, and therefore, it requires a constricted range of physicochemical environment for efficient degradation of the pollutants. The variations that could be brought by climate change will certainly affect the bioremediation efficiency of the soil microorganisms. Change in temperature, soil water content, aeration, etc. can significantly alter the growth and development of the soil microorganisms. These changes can influence both the structure and composition of the soil microbes and certainly modify the activities of microbes involved in biodegradation of pollutants. The microbial cells are stressed during these abiotic changes and unable to generate energy to perform bioremediation. In some cases, when the microbes are subjected to stressed conditions, their metabolic activity is reduced accordingly and develop the stress tolerance. This transition froms an active to a dormant state reduces the bioremediation potential of the microorganism. Therefore, the possible effects of climate change on the soil processes have invited several theoretical and experimental trials on the effect of climate change on bioremediation (Fig. 27.1).

27.3.1 Temperature

The enzyme-catalyzed reactions in bioremediation process are temperature sensitive, and therefore, a significant increase or decrease of temperature will affect the biodegradation activity and rate. Temperature also regulates the decomposition of soil organic matter and nutrient cycling that further affects the soil microbial activity. Lower soil temperature generally restricts the bioremediation rate. On the other hand, there is an increase in biodegradation rate with the increase in soil temperature up to 65^0C due to higher microbial metabolic activity. Warming can stimulate the functional genes involved in carbon degradation (Bardgett et al. 2013). It might be temporary and then the microbes adapts to the warming. However, higher temperature might prove harmful to the growth and development of certain microorganisms. Backman et al. (2004) tested the impact of temperature on the bioremediation potential of *Arthrobacter chlorophenolicus* A6, a bacterium capable of degrading 4-chlorophenol. Two levels of temperature, *i.e.* 5 and 28 °C, were applied to the test bacterium. At 28 °C, the cell integrity of the bacterium was lost accompanied by the reduction of the metabolic activity, thereby showing significant reduction in biodegradation of 4-chlorophenol. On the other hand, the incubation of the test bacterium at 5 °C could not significantly affect the biodegradation rate as most of the cells remained intact at this temperature. The survival of these kind of bacteria in cold climate without affecting the biodegradation efficiency is partly attributed to the production of cold-induced proteins that overcomes the effects of the stressed conditions (Givskov et al. 1994). However, it depends on the type and species of the soil microorganisms, whether it can survive in harsh conditions.

Increase in temperature can also increase soil respiration, microbial biomass and soil organic matter decomposition, but this effect is generally of short duration. The increase in soil microbial activity with the increase in temperature may lead to decline in labile carbon pool as it is easily

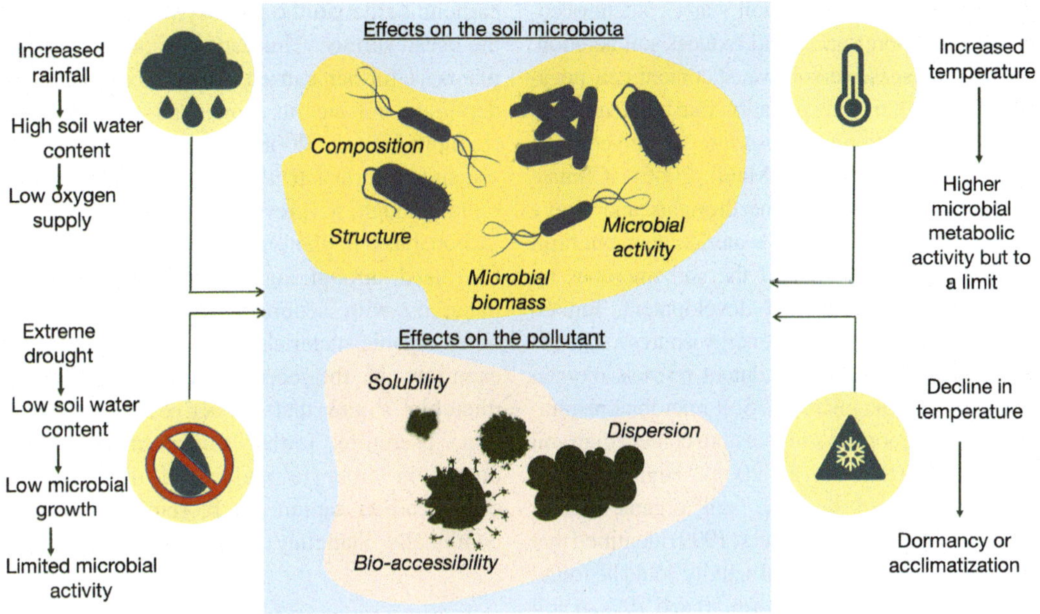

Fig. 27.1 Possible effects of climate change on the soil microbiota (involved in bioremediation) and pollutants. The direct effects of climate change are depicted in italics

available to the soil microbes. It may be followed by acclimatization, change in microbial composition or alteration of microbial biomass to adapt to the changing environmental conditions (Classen et al. 2015). The results from long-term and short-term experiments generated contradictory values regarding the effects of temperature on bioremediation ability of the microbes. It indicates that the interaction of the soil microorganisms with the change in temperature is complex and must be further studied.

The effects of increasing temperature on soil microbial biodegradation rate can be rather minimal, if we analyse the predicted change in global temperature. Raftery et al. (2017) predicted that by the end of twenty-first century, there will be an increase of global temperature by 1–3 °C, which may not significantly affect the soil microbial activity. It can be dangerous for humanity but not for the bioremediation. However, these effects cannot be ruled out and must be further analysed. This is because the toxicity of some pollutants can increase with higher temperature (Noyes et al. 2009).

27.3.2 Soil pH

Soil pH affects the growth and development of microorganisms. However, a pH of 6.5–8.5 is a safe range where the bioremediation capacity is generally not affected. As soil pH regulates the nutrient availability and solubility in the soil, it will certainly affect the biodegradation capacity of the soil microorganisms. Even the bioavailability of major essential nutrients such as phosphorus are pH-sensitive. Due to the unfavourable pH of the surrounding soil in the polluted sites, nitrogen and phosphorus are generally found deficient that may further limits the biodegradation of the pollutants. Therefore, it is advisable to add nitrogen and phosphorus in usable form in adequate amounts (Malik 2006).

27.3.3 Soil Water

Soil water regulates the transport of nutrients across the microbial system. It also helps in ejection of microbial waste of the cell during the detoxification process. However, higher soil water content occupies the soil pore spaces and reduces soil aeration. Therefore, excessive soil water content can negatively affects the bioremediation rate by creating an anaerobic condition, unless the bioremediation needs such conditions (Malik 2006). Climate change can lead to extreme droughts and heavy rainfall or floods. Low soil water content can limit the biodegrading ability of the soil microbes by restricting its growth and development, limited diffusion of nutrients and energy sources. On the other hand, higher water content restricts oxygen supply to aerobic microbes. Soil microbial respiration depends more on soil moisture content than soil temperature (Silva et al. 2008). High moisture content can also decrease soil organic matter decomposition. Franzluebbers (1999) identified that maximum aerobic microbial activity could be found at soil moisture levels between 50 and 70% of soil water holding capacity. Most of the bioremediation process functions well between the soil moisture levels of 50–75% of water holding capacity (Ajlan 2016). In case of remediating oil-polluted soils, 30–90% soil moisture level is required based on the type of soil and the chemical nature of the pollutants. Therefore, the soil moisture levels must be adjusted in order to enhance the bioremediation process.

Increase or decline in annual rainfall can significantly affect the soil microbial activity through alternate wetting and drying cycles. The wetting–drying cycle can alter the bioavailability of the pollutants (Shelton and Parkin 1991). Soil moisture content not only influence soil aeration but also the solubility, transportation and bioaccessibility of the pollutants to the soil microbes. Heavy precipitation and floods often leads to soil erosion rate and runoff, thus dispersing the pollutants from the original site to other uncontaminated locations.

27.4 Effects of Ocean Carbon Sequestration on Bioremediation

The presence of effective microbial strains also allows carbon sequestration in oceans that capture the anthropogenic atmospheric carbondioxide via the photosynthetic pathway in the ocean surface. This carbon captured in ocean plants is further carried down to deeper ocean depth and enhance its sequestration in the ocean (Lampitt et al. 2008). This sinking of the organic materials to the deeper ocean must be rapid enough to prevent the loss of captured carbon back to the atmosphere. The sinking is facilitated through aggregation of the organic materials with sediments or association with other organic materials. The increase of carbon capturing in the ocean must also affect the microbial strains that are active in bioremediation. Therefore, further studies must be conducted in order to understand the effects of ocean carbon capture on bioremediation under continually changing climate.

27.5 Conclusion

The increasing climate change could lead to significant variations in abiotic components of the environment. The soil microorganisms require optimum environmental conditions for their growth, development and metabolic activities. As climate change will alter the abiotic factors, the soil microbial activities are predicted to be affected significantly, thereby affecting the bioremediation process. In order to understand the potential adverse effects of climate change on bioremediation, it is important to identify and study how quickly the climate change affects the soil processes and its ecosystem services. As soil-microbe-plant interaction is one of the complex relationships in the natural ecosystem, it is crucial to identify the response of each components due to climate change effects. The microbial adaptation to increased stressed condition must be focused to unravel the underlying mechanisms of climate change-induced bioremediation. As climate change is closely related to abiotic changes, it must also influence the structure and community of the soil microbiota. The abundance and decline of specific microbial species with the increasing climate change effects must be explored.

References

Ajlan A (2016) A review on bioremediation. Microbiology 1:103

Backman A, Maraha N, Jansson JK (2004) Impact of temperature on the physiological status of a potential bioremediation inoculant, *Arthrobacter chlorophenolicus* A6. Appl Environ Microbiol 70(5):2952–2958. https://doi.org/10.1128/AEM.70.5.2952

Bardgett RD, Manning P, Morriën E, De Vries FT (2013) Hierarchical responses of plant-soil interactions to climate change: consequences for the global carbon cycle. J Ecol 101(2):334–343

Classen A, Sundqvist M, Henning JA, Newman GS, Moore JA, Cregger MA, Patterson CM (2015) Direct and indirect effects of climate change on soil microbial and soil microbial-plant interactions: what lies ahead? Ecosphere 6(8):1–21

Dzionek A, Wojcieszynska D, Guzik U (2016) Natural carriers in bioremediation: a review. Electron J Biotechn 23:28–36

Franzluebbers AJ (1999) Microbial activity in response to water-filled pore space of variably eroded southern Piedmont soils. Appl Soil Ecol 11:91–101

Gangola S, Joshi S, Kumar S, Pandey SC (2019) Comparative analysis of fungal and bacterial enzymes in biodegradation of xenobiotic compounds. In: Bhatt P (ed) Smart bioremediation technologies: microbial enzymes. Academic Press, Cambridge, MA, pp 169–189

Garcia-Junco M, Gomez-Lahoz C, Niqui-Arroyo JL et al (2003) Biosurfactant- and biodegradation-enhanced partitioning of polycyclic aromatic hydrocarbons from nonaqueous-phase liquids. Environ Sci Technol 37:2988–2996

Givskov M, Eberl L, Møller S, Poulsen LK, Molin S (1994) Responses to nutrient starvation in *Pseudomonas putida* KT2442: analysis of general cross-protection, cell shape, and macromolecular content. J Bacteriol 176:7–14

Heinaru E, Merimaa M, Viggor S, Lehiste M, Leito I, Truu J, Heinaru A (2005) Biodegradation efficiency of functionally important populations selected for bioaugmentation in phenol- and oil-polluted area. Fems Microbiol Ecol 51(3):363–373

Lampitt RS, Achterberg EP, Anderson TR, Hughes JA, Iglesias-Rodriguez MD, Kelly-Gerreyn BA, Yool A (2008) Ocean fertilization: a potential means of geoengineering? Phil Trans R Soc A 366 (1882):3939–3945

Law AM, Aitken MD (2003) Bacterial chemotaxis to naphthalene desorbing from a nonaqueous liquid. Appl Environ Microbiol 69:5968–5973

Leal AJ, Rodrigues EM, Leal PL, Júlio ADL, Fernandes RDCR, Borges AC, Tótola MR (2017) Changes in the microbial community during bioremediation of gasolina-contaminated soil. Braz J Microbiol 48 (2):342–351

Malik A (2006) Bioremediation. In: Environmental microbiology. National Science Digital Library, (xth five year plan network project of NISCAIR (CSIR), UGC, MHRD, New Delhi, India

Megharaj M, Ramakrishnan B, Venkateswarlu K, Sethunathan N, Naidu R (2011) Bioremediation approaches for organic pollutants: a critical perspective. Environ Int 37(8):1362–1375

Mohammed D, Ramsubhag A, Beckles D (2007) An assessment of the biodegradation of petroleum hydrocarbons in contaminated soil using non-indigenous, commercial microbes. Water Air Soil Poll 182:349–356

Naik M, Duraphe M (2012) Review paper on–parameters affecting bioremediation. Int J Life Sci Pharma Res 2: L77–L80

Noyes PD, McElwee MK, Miller HD, Clark BW, Van Tiem LA, Walcott KC, Levin ED (2009) The toxicology of climate change: environmental contaminants in a warming world. Environ Int 35(6):971–986

Odukkathil G, Vasudevan N (2013) Enhanced biodegradation of endosulfan and its major metabolite endosulfate by a biosurfactant producing bacterium. J Environ Sci Health Part B 48:462–469

Olson MS, Ford RM, Smith JA, Fernandez EJ (2004) Quantification of bacterial chemotaxis in porous media using magnetic resonance imaging. Environ Sci Technol 38(14):3864–3870

Pandey SC, Pandey A, Joshi T, Pande V, Sati D, Samant M (2019) Microbiological monitoring in the biodegradation of food waste. In: Gunjal AB, Waghmode MS, Patil NN, Bhatt P (eds) Global initiatives for waste reduction and cutting food loss. IGI Global, USA, pp 116–140

Raftery AE, Zimmer A, Frierson DM, Startz R, Liu P (2017) Less than 2 °C warming by 2100 unlikely. Nat Clim Change 7(9):637–641

Samant M, Pandey SC, Pandey A (2018) Impact of hazardous waste material on environment and their management strategies. In: Bhatt P, Sharma A (eds) Microbial biotechnology in environmental monitoring and cleanup. IGI Global, USA, pp 175–192

Sarkar D, Ferguson M, Datta R, Birnbaum S (2005) Bioremediation of petroleum hydrocarbons in contaminated soils: comparison of biosolids addition, carbon supplementation, and monitored natural attenuation. Environ Poll 136(1):187–195

Shelton DR, Parkin TB (1991) Effect of moisture on sorption and biodegradation of carbofuran in soil. J Agr Food Chem 39:2063–2068

Silva CC, Guido ML, Ceballos JM, Marsch R, Dendooven L (2008) Production of carbon dioxide and nitrous oxide in alkaline saline soil of Texcoco at different water contents amended with urea: a laboratory study. Soil Biol Biochem 40(7):1813–1822

Thompson IP, Van der Gast CJ, Ciric L, Singer AC (2005) Bioaugmentation for bioremediation: the challenge of strain selection. Environ Microbiol 7(7):909–915

Correction to: Production of Safer Vegetables from Heavy Metals Contaminated Soils: The Current Situation, Concerns Associated with Human Health and Novel Management Strategies

Hafiz Muhammad Tauqeer, Veysel Turan, and Muhammad Iqbal

Correction to:
Chapter 19 in: J. A. Malik (ed.), *Advances in Bioremediation and Phytoremediation for Sustainable Soil Management*, https://doi.org/10.1007/978-3-030-89984-4_19

The original version of the book was inadvertently published with incorrect affiliations in chapter 19 for the authors Veysel Turan and Muhammad Iqbal. Corrections to the Previously Published Version have been updated.

The updated version of this chapter can be found at https://doi.org/10.1007/978-3-030-89984-4_19

J. A. Malik (ed.), *Advances in Bioremediation and Phytoremediation for Sustainable Soil Management*,
https://doi.org/10.1007/978-3-030-89984-4_28

Index

J. A. Malik (ed.), *Advances in Bioremediation and Phytoremediation for Sustainable Soil Management*, https://doi.org/10.1007/978-3-030-89984-4